Statistics and Computing

Series Editors:

J. Chambers

D. Hand

W. Härdle

For other titles published in this series, go to
http://www.springer.com/series/3022

James E. Gentle

Computational Satistics

J.E. Gentle
Department of Computational & Data Sciences
George Mason University
4400, University Drive
Fairfax, VA 22030-4444
USA
jgentle@gmu.edu

ISBN 978-1-4614-2929-6 e-ISBN 978-0-387-98144-4
DOI 10.1007/978-0-387-98144-4
Springer Dordrecht Heidelberg London New York

Printed on acid-free paper

Springer is part of Springer Science+Business Media (www.springer.com)

To María

Preface

This book began as a revision of *Elements of Computational Statistics*, published by Springer in 2002. That book covered computationally-intensive statistical methods from the perspective of statistical applications, rather than from the standpoint of statistical computing.

Most of the students in my courses in computational statistics were in a program that required multiple graduate courses in numerical analysis, and so in my course in computational statistics, I rarely covered topics in numerical linear algebra or numerical optimization, for example. Over the years, however, I included more discussion of numerical analysis in my computational statistics courses. Also over the years I have taught numerical methods courses with no or very little statistical content. I have also accumulated a number of corrections and small additions to the elements of computational statistics. The present book includes most of the topics from *Elements* and also incorporates this additional material. The emphasis is still on computationally-intensive *statistical* methods, but there is a substantial portion on the numerical methods supporting the statistical applications.

I have attempted to provide a broad coverage of the field of computational statistics. This obviously comes at the price of depth.

Part I, consisting of one rather long chapter, presents some of the most important concepts and facts over a wide range of topics in intermediate-level mathematics, probability and statistics, so that when I refer to these concepts in later parts of the book, the reader has a frame of reference.

Part I attempts to convey the attitude that *computational inference*, together with exact inference and asymptotic inference, is an important component of statistical methods.

Many statements in Part I are made without any supporting argument, but references and notes are given at the end of the chapter. Most readers and students in courses in statistical computing or computational statistics will be familiar with a substantial proportion of the material in Part I, but I do not recommend skipping the chapter. If readers are already familiar with the material, they should just read faster. The perspective in this chapter is that of the "big picture". As is often apparent in oral exams, many otherwise good students lack a basic understanding of what it is all about.

A danger is that the student or the instructor using the book as a text will too quickly gloss over Chapter 1 and miss some subtle points.

Part II addresses statistical computing, a topic dear to my heart. There are many details of the computations that can be ignored by most statisticians, but no matter at what level of detail a statistician needs to be familiar with the computational topics of Part II, there are two simple, higher-level facts that all statisticians should be aware of and which I state often in this book:

> *Computer numbers are not the same as real numbers, and the arithmetic operations on computer numbers are not exactly the same as those of ordinary arithmetic.*

and

> *The form of a mathematical expression and the way the expression should be evaluated in actual practice may be quite different.*

Regarding the first statement, some of the differences in real numbers and computer numbers are summarized in Table 2.1 on page 98.

A prime example of the second statement is the use of the normal equations in linear regression, $X^{\mathrm{T}}Xb = X^{\mathrm{T}}y$. It is quite appropriate to write and discuss these equations. We might consider the elements of $X^{\mathrm{T}}X$, and we might even write the least squares estimate of β as $b = (X^{\mathrm{T}}X)^{-1}X^{\mathrm{T}}y$. That does not mean that we ever actually compute $X^{\mathrm{T}}X$ or $(X^{\mathrm{T}}X)^{-1}$, although we may compute functions of those matrices or even certain elements of them.

The most important areas of statistical computing (to me) are

- computer number systems
- algorithms and programming
- function approximation and numerical quadrature
- numerical linear algebra
- solution of nonlinear equations and optimization
- generation of random numbers.

These topics are the subjects of the individual chapters of Part II.

Part III in six relatively short chapters addresses methods and techniques of computational statistics. I think that any exploration of data should begin with graphics, and the first chapter in Part III, Chapter 8, presents a brief overview of some graphical methods, especially those concerned with multidimensional data. The more complicated the structure of the data and the higher the dimension, the more ingenuity is required for visualization of the data; it is, however, in just those situations that graphics is most important. The orientation of the discussion on graphics is that of computational statistics; the emphasis is on discovery, and the important issues that should be considered in making presentation graphics are not addressed.

Chapter 9 discusses methods of projecting higher-dimensional data into lower dimensions. The tools discussed in Chapter 9 will also be used in Part IV for clustering and classification, and, in general, for exploring structure in data. Chapter 10 covers some of the general issues in function *estimation*, building on the material in Chapter 4 on function *approximation*.

Chapter 11 is about Monte Carlo simulation and some of its uses in computational inference, including Monte Carlo tests, in which artificial data are generated according to a hypothesis. Chapters 12 and 13 discuss computational inference using resampling and partitioning of a given dataset. In these methods, a given dataset is used, but Monte Carlo sampling is employed repeatedly on the data. These methods include randomization tests, jackknife techniques, and bootstrap methods, in which data are generated from the empirical distribution of a given sample, that is, the sample is resampled.

Identification of interesting features, or "structure", in data is an important activity in computational statistics. In Part IV, I consider the problem of identification of structure and the general problem of estimation of probability densities. In simple cases, or as approximations in more realistic situations, structure may be described in terms of functional relationships among the variables in a dataset.

The most useful and complete description of a random data-generating process is the associated probability density, if it exists. Estimation of this special type of function is the topic of Chapters 14 and 15, building on general methods discussed in earlier chapters, especially Chapter 10. If the data follow a parametric distribution, or rather, if we are willing to assume that the data follow a parametric distribution, identification of the probability density is accomplished by estimation of the parameters. Nonparametric density estimation is considered in Chapter 15.

Features of interest in data include clusters of observations and relationships among variables that allow a reduction in the dimension of the data. I discuss methods for statistical learning in Chapter 16, building on some of the basic measures introduced in Chapter 9.

Higher-dimensional data have some surprising and counterintuitive properties, and I discuss some of the interesting characteristics of higher dimensions.

In Chapter 17, I discuss asymmetric relationships among variables. For such problems, the objective often is to estimate or predict a response for a given set of explanatory or predictive variables, or to identify the class to which an observation belongs. The approach is to use a given dataset to develop a model or a set of rules that can be applied to new data. Statistical modeling may be computationally intensive because of the number of possible forms considered or because of the recursive partitioning of the data used in selecting a model. In computational statistics, the emphasis is on *building* a model rather than just estimating the parameters in the model. Parametric estimation, of course, plays an important role in building models.

Many of the topics addressed in this book could easily be (and are) subjects for full-length books. My intent is to describe these methods in a general manner and to emphasize commonalities among them. Decompositions of matrices and of functions are examples of basic tools that are used in a variety of settings in computational statistics. Decompositions of matrices, which are introduced on page 28 of Chapter 1, play a major role in many computations in linear algebra and in statistical analysis of linear models. The decompositional approach to matrix computations has been chosen as one of the Top 10 developments in algorithms in the twentieth century. (See page 138.) The PDF decomposition of a function so that the function has a probability density as a factor, introduced on page 37 of Chapter 1, plays an important role in many statistical methods. We encounter this technique in Monte Carlo methods (pages 192 and 418), in function estimation (Chapters 10 and 15), and in projection pursuit (Chapter 16).

My goal has been to introduce a number of topics and devote some suitable proportion of pages to each. I have given a number of references for more in-depth study of most of the topics. The references are not necessarily chosen because they are the "best"; they're just the ones I'm most familiar with. A general reference for a slightly more detailed coverage of most of the topics in this book is the handbook edited by Wolfgang Härdle, Yuichi Mori, and me (Gentle, Härdle, and Mori, 2004).

The material in Chapters 2, 5, and 9 relies heavily on my book on *Matrix Algebra* (Gentle, 2007), and some of the material in Chapters 7 and 11 is based on parts of my book on *Random Number Generation* (Gentle, 2003).

Each chapter has a section called "notes and further reading". The content of these is somewhat eclectic. In some cases, I had fun writing the section, so I went on at some length; in other cases, my interest level was not adequate for generation of any substantial content.

A Little History

While I have generally tried to keep up with developments in computing, and I do not suffer gladly old folks who like to say "well, the way we used to do it was ...", occasionally while writing this book, I looked in *Statistical Computing* to see what Bill Kennedy and I said thirty years ago about the things I discuss in Part II. The biggest change in computing of course has resulted from the personal computer. "Computing" now means a lot more than it did thirty years ago, and there are a lot more people doing it. Advances in display devices has been a natural concurrence with the development of the PC, and this has changed statistical graphics in a quantum way.

While I think that the PC *sui generis* is the Big Thing, the overall advance in computational power is also important. There have been many evolutionary advances, basically on track with Moore's law (so long as we adjust the number of months appropriately). The net result of the evolutionary advance in speed has been enormous. Some people have suggested that statistical methods/approaches should undergo fundamental changes every time there is an increase of one order of magnitude in computational speed and/or storage. Since 1980, and roughly in accordance with Moore's law, there have been 4 such increases. I leave to others an explicit interpretation of the relevance of this fact to the field of statistics, but it is clear that the general increase in the speed of computations has allowed the development of useful computationally-intensive methods. These methods together constitute the field of computational statistics. Computational inference as an approximation is now generally as accepted as asymptotic inference (more readily by many people).

At a more technical level, standardization of hardware and software has yielded meaningful advances. In the 1970's over 75% of the computer market was dominated by the IBM 360/370 systems. Bill and I described the arithmetic implemented in this computer. It was in base 16 and did not do rounding. The double precision exponent had the same range as that of single precision. The IBM Fortran compilers (G and H) more-or-less conformed to the Fortran 66 standard (and they chose the one-trip for null DO-loops). Pointers and dynamic storage allocation were certainly not part of the standard. PL/I was a better language/compiler and IBM put almost as many 1970s dollars in pushing it as US DoD in 1990s dollars pushing Ada. And of course, there was JCL!

The first eight of the Top 10 algorithms were in place thirty years ago, and we described statistical applications of at least five of them. The two that were not in place in 1980 do not have much relevance to statistical applications. (OK, I know somebody will tell me soon how important these two algorithms are, and how they have just been used to solve some outstanding statistical problem.)

One of the Top 10 algorithms, dating to the 1940s, is the basis for MCMC methods, which began receiving attention by statisticians around 1990, and in the past twenty years has been one of the hottest areas in statistics.

I could go on, but I tire of telling "the way we used to do it". Let's learn what we need to do it the best way now.

Data

I do not make significant use of any "real world" datasets in the book. I often use "toy" datasets because I think that is the quickest way to get the essential characteristics of a method. I sometimes refer to the datasets that are available in R or S-Plus, and in some exercises, I point to websites for various real world datasets.

Many exercises require the student to generate artificial data. While such datasets may lack any apparent intrinsic interest, I believe that they are often the best for learning how a statistical method works. One of my firm beliefs is

If I understand something, I can simulate it.

Learning to simulate data with given characteristics means that one understands those characteristics. Applying statistical methods to simulated data may lack some of the perceived satisfaction of dealing with "real data", but it helps us better to understand those methods and the principles underlying them.

A Word About Notation

I try to be very consistent in notation. Most of the notation is "standard". Appendix C contains a list of notation, but a general summary here may be useful. Terms that represent mathematical objects, such as variables, functions, and parameters, are generally printed in an italic font. The exceptions are the standard names of functions, operators, and mathematical constants, such as sin, log, Γ (the gamma function), Φ (the normal CDF), E (the expectation operator), d (the differential operator), e (the base of the natural logarithm), and so on.

I tend to use Greek letters for parameters and English letters for almost everything else, but in some cases, I am not consistent in this distinction.

I do not distinguish vectors and scalars in the notation; thus, "x" may represent either a scalar or a vector, and x_i may represent either the ith element of an array or the ith vector in a set of vectors. I use uppercase letters for matrices and the corresponding lowercase letters with subscripts for elements of the matrices. I do not use boldface except for emphasis or for headings.

I generally use uppercase letters for random variables and the corresponding lowercase letters for realizations of the random variables. Sometimes I am not completely consistent in this usage, especially in the case of random samples and statistics.

I describe a number of methods or algorithms in this book. The descriptions are in a variety of formats. Occasionally they are just in the form of text; the algorithm is described in (clear?!) English text. Often they are presented in the form of pseudocode in the form of equations with simple for-loops, such as for the accumulation of a sum of corrected squares on page 116, or in pseudocode that looks more like Fortran or C. (Even if C-like statements are used, I almost always begin the indexing at the 1st element; that is, at the *first* element, not the zeroth element. The exceptions are for cases in which the index also represents a power, such as in a polynomial; in such cases, the 0th element is the first element. I call this "0 equals first" indexing.) Other times the algorithms are called "Algorithm x.x" and listed as a series of steps, as on page 218. There is a variation of the "Algorithm x.x" form. In one form the algorithm is given a name and its input is listed as input arguments, for example MergeSort, on page 122. This form is useful for recursive algorithms because it allows for an easy specification of the recursion. Pedagogic considerations (if not laziness!) led me to use a variety of formats for presentation of algorithms; the reader will likely encounter a variety of formats in literature in statistics and computing, and some previous exposure should help to make the format irrelevant.

Use in the Classroom

Most statistics students will only take one or two courses in the broad field of computational statistics. I have tried at least to introduce the major areas of the field, but this means, of course, that depth of coverage of most areas has been sacrificed.

The chapters and sections in the book vary considerably in their lengths, and this sometimes presents a problem for an instructor to allocate the coverage over the term. The number of pages is a better, but still not very accurate, measure of the time required to cover the material.

There are several different types of courses for which this book could be used, either as the primary text or as a supplement.

Statistical Computing Courses

Most programs in statistics in universities in the United States include a course called "statistical computing". There are two kinds of courses called "statistical computing". One kind is "packages and languages for data analysis". This book would not be of much use in such a course.

The other kind of course in statistical computing is "numerical methods in statistics". Part II of this book is designed for such a course in statistical computing. Selected chapters in Parts III and IV could also be used to illustrate and motivate the topics of those six chapters. Chapter 1 could be covered as necessary in a course in statistical computing, but that chapter should not be skipped over too lightly.

One of the best ways to learn and understand a computational method is to implement the method in a computer program. Executing the program provides immediate feedback on the correctness. Many of the exercises in Part II require the student to "write a program in Fortran or C". In some cases, the purpose is to identify design issues and how to handle special datasets, but in most cases the purpose is to ensure that the method is understood; hence, in most cases, instead of Fortran or C, a different language could be used, even a higher-level one such as R. Those exercises also help the student to develop a facility in programming. Programming is the best way to learn programming. (Read that again; yes, that's what I mean. It's like learning to type.)

Computational Statistics Courses

Another course often included in statistics programs is one on "computationally intensive statistical methods", that is, what I call "computational statistics". This type of course, which is often taught as a "special topics" course, varies widely. The intent generally is to give special attention to such statistical methods as the bootstrap or to such statistical applications as density estimation. These topics often come up in other courses in statistical theory and methods, but because of the emphasis in these courses, there is no systematic development of the computationally-intensive methods. Parts III and IV of this book are designed for courses in computational statistics. I have taught such a course for a number of years, and I find that the basic material of Chapter 1 bears repeating (although it is prerequisite for the course that I teach). Some smattering of Part II, especially random number generation in Chapter 7, may also be useful in such a course, depending on whether or not the students have a background in statistical computing (meaning "numerical methods in statistics").

Modern Applied Statistics Courses
The book, especially Parts III and IV, could also be used as a text in a course on "modern applied statistics". The emphasis would be on modeling and statistical learning; the approach would be exploration of data.

Exercises

The book contains a number of exercises that reinforce the concepts discussed in the text or, in some cases, extend those concepts. Appendix D provides solutions or comments for several of the exercises.

Some of the exercises are rather open-ended, asking the student to "explore". Some of the "explorations" are research questions.

One weakness of students (and lots of other people!) is the ability to write clearly. Writing is improved by practice and by criticism. Several of the exercises, especially the "exploration" ones, end with the statement: "Summarize your findings in a clearly-written report." Grading such exercises, including criticism of the writing, usually requires more time—so a good trick is to let students "grade" each others' work. Writing and editing are major activities in the work of statisticians (not just the academic ones!), and so what better time to learn and improve these activities than during the student years.

In most classes I teach in computational statistics, I give Exercise A.3 in Appendix A (page 656) as a term project. It is to replicate and extend a Monte Carlo study reported in some recent journal article. Each student picks an article to use. The statistical methods studied in the article must be ones that the student understands, but that is the only requirement as to the area of statistics addressed in the article. I have varied the way in which the project is carried out, but it usually involves more than one student working together. A simple way is for each student to referee another student's first version (due midway through the term) and to provide a report for the student author to use in a revision. Each student is both an author and a referee. In another variation, I have students be coauthors.

Prerequisites

It is not (reasonably) possible to itemize the background knowledge required for study of this book. I could claim that the book is self-contained, in the sense that it has brief introductions to almost all of the concepts, but that would not be fair. An intermediate-level background in mathematics and statistics is assumed.

The book also assumes some level of computer literacy, and the ability to "program" in some language such as R or Matlab. "Real" programming ability is highly desirable, and many of the exercises in Part II require real programming.

Acknowledgements

I thank John Kimmel of Springer for his encouragement and advice on this book and other books he has worked with me on. I also thank the many readers who have informed me of errors in other books and who have otherwise provided comments or suggestions for improving my exposition.

I thank my wife María, to whom this book is dedicated, for everything.

I used LaTeX 2_ε to write the book, and I used R to generate the graphics. I did all of the typing, programming, etc., myself, so all mistakes are mine. I would appreciate receiving notice of errors as well as suggestions for improvement.

Notes on this book, including errata, are available at
`http://mason.gmu.edu/~jgentle/cmstatbk/`

Fairfax County, Virginia

James E. Gentle
April 24, 2009

Contents

Appendices

Part I

Preliminaries

Introduction to Part I

The material in Part I is basic to the field of computational statistics. It includes some intermediate-level mathematics, and probability and statistics. Many readers will be familiar with at least some of this material.

Although the presentation is rather terse, the single chapter in this part is somewhat longer than most of the chapters in this book. That is due both to the diversity of the topics and to their importance in the subsequent chapters.

We discuss the general objectives in statistical analyses and, in particular, those objectives for which computationally intensive methods are appropriate.

After the introductory discussion of exploratory data analysis, we begin with some definitions and general discussions of useful measures in vector spaces and some of the operations on vectors, functions, and matrices.

When data are organized into a matrix, the mathematical properties of the matrix can reveal a lot about the structure of the data. We therefore briefly describe some of the important properties of matrices. We will encounter various aspects of properties of matrices later. In Chapter 5 of Part II we discuss computational methods, in Chapter 9 of Part III we describe various transformations of data using matrix algebra, and in Chapters 16 and 17 of Part IV we discuss methods of matrix algebra for understanding statistical relationships among variables or observations.

We then describe briefly some of the methods of statistical inference that are applicable generally whether in computational statistics or not. Although much of this discussion may appear rather elementary, it does presuppose some general familiarity with statistical theory and methods; otherwise the material would be insufficient for the subsequent developments in the book.

We emphasize that *computational inference* is often a useful alternative to the asymptotic inference used in many of the standard statistical methods.

The empirical cumulative distribution function (ECDF) plays a very basic role in statistical inference, especially in computational inference in such methods as the bootstrap. Despite the fundamental nature of the ECDF, it is not often given its due in textbooks on statistical inference.

Many methods in statistical analysis can be couched as solutions to optimization problems. In Section 1.8 we emphasize this perspective, and briefly discuss some of the implications for the statistical properties of the methods.

Many statements made in Part I lack supporting arguments. The purpose of this part is to state the highlights that are assumed in later parts of the book. References to more complete presentations and other notes are given at the end of the chapter.

1

Mathematical and Statistical Preliminaries

The purpose of an exploration of data may be rather limited, and it may be ad hoc, or the purpose may be more general, perhaps to gain understanding of some natural phenomenon.

The questions addressed in the data exploration may be somewhat open-ended. The process of understanding often begins with general questions about the structure of the data. At any stage of the analysis, our understanding is facilitated by means of a *model.*

A model is a description that embodies our current understanding of a phenomenon. In an operational sense, we can formulate a model either as a description of a *data-generating process*, or as a prescription for processing data.

The model is often expressed as a set of equations that relate data elements to each other. It may include probability distributions for the data elements.

If any of the data elements are considered to be realizations of random variables, the model is a *stochastic model.*

A model should not limit our analysis; rather, the model should be able to evolve. The process of understanding involves successive refinements of the model. The refinements proceed from vague models to more specific ones. An exploratory data analysis may begin by mining the data to identify interesting properties. These properties generally raise questions that are to be explored further.

A class of models may have a common form within which the members of the class are distinguished by values of *parameters*. For example, the class of normal probability distributions has a single form of a probability density function that has two parameters. Within this family of probability distributions, these two parameters completely characterize the distributional properties. If this form of model is chosen to represent the properties of a dataset, we may seek confidence intervals for values of the two parameters or perform statistical tests of hypothesized values of these two parameters.

In models that are not as mathematically tractable as the normal probability model — and many realistic models are not — we may need to use compu-

J.E. Gentle, *Computational Statistics*, Statistics and Computing,
DOI: 10.1007/978-0-387-98144-4_1,

tationally intensive methods involving simulations, resamplings, and multiple views to make inferences about the parameters of a model. These methods are part of the field of *computational statistics.*

1.1 Discovering Structure in Data

The components of statistical datasets are "observations" and "variables" or "features". In general, "data structures" are ways of organizing data to take advantage of the relationships among the variables constituting the dataset. Data structures may express hierarchical relationships, crossed relationships (as in "relational" databases), or more complicated aspects of the data (as in "object-oriented" databases). Data structures, or more generally, database management, is a relatively mature area of computer science.

In data analysis, "structure in the data" is of interest. Structure in the data includes such nonparametric features as modes, gaps, or clusters in the data, the symmetry of the data, and other general aspects of the shape of the data. Because many classical techniques of statistical analysis rely on an assumption of normality of the data, the most interesting structure in the data may be those aspects of the data that deviate most from normality.

In identifying and studying structure, we use both numerical measures and graphical views.

Multiple Analyses and Multiple Views

There are many properties that are more apparent from graphical displays of the data.

Although it may be possible to express the structure in the data in terms of mathematical models, prior to attempting to do so, graphical displays may be used to discover qualitative structure in the data. Patterns observed in the data may suggest explicit statements of the structure or of relationships among the variables in the dataset. The process of building models of relationships is an iterative one, and graphical displays are useful throughout the process. Graphs comparing data and the fitted models are used to refine the models.

Effective use of graphics often requires multiple views. For multivariate data, plots of individual variables or combinations of variables can be produced quickly and used to get a general idea of the properties of the data. The data should be inspected from various perspectives. Instead of a single histogram to depict the general shape of univariate data, for example, multiple histograms with different bin widths and different bin locations may provide more insight. (See Figure 8.4 on page 347.)

Sometimes, a few data points in a display can completely obscure interesting structure in the other data points. This is the case when the Euclidean distances between various pairs of data points differ greatly. A zooming window to restrict the scope of the display and simultaneously restore the scale to an appropriate viewing size can reveal structure. A zooming window can be used with any graphics software whether the software supports it or not; zooming can be accomplished by deletion of the points in the dataset outside of the window.

Scaling the axes can also be used effectively to reveal structure. The relative scale is called the "aspect ratio". In Figure 1.1, which is a plot of a bivariate dataset, we form a zooming window that deletes a single observation. The greater magnification and the changed aspect ratio clearly show a relationship between X and Y in a region close to the origin that may not hold for the full range of data. A simple statement of this relationship would not extrapolate outside the window to the outlying point.

The use of a zooming window is not "deletion of outliers"; it is focusing in on a subset of the data and is done independently of whatever is believed about the data outside of the window.

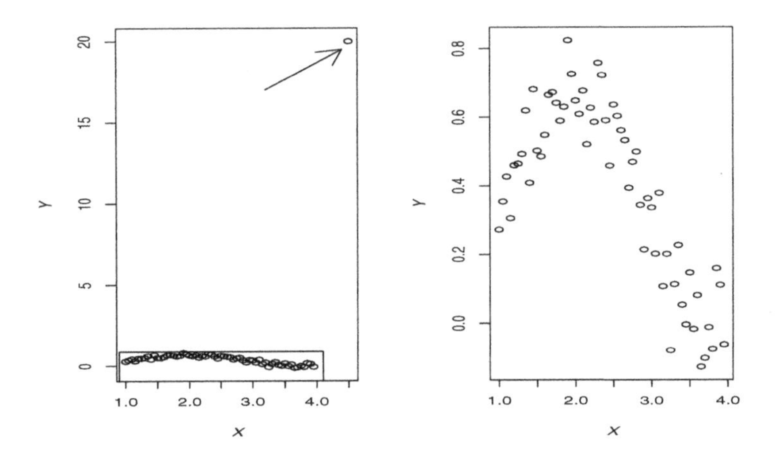

Fig. 1.1. Scales Matter

Although the zooming window in Figure 1.1 reveals structure by changing the aspect ratio as it focused on a subset of the data, views with different aspect ratios may reveal meaningless structure just by differentially changing the scales of measurement of the variables. Such structural characteristics of data are sometimes just *artificial structure.* Artificial structure is structure that can be changed meaningfully by univariately rescaling the data. Many multivariate statistical analyses reveal artificial structure. Clustering (see Section 16.1) and principal component analysis (see Section 16.3), for example, are sensitive to artificial structure. This, of course, presents a real challenge to the data analyst. The meaning of "artificial" is somewhat subjective; what is artificial to one person or in one application may be meaningful in another application. *Data analysis cannot be conducted in the abstract.*

One type of structure that may go undetected is that arising from the order in which the data were collected. For data that are recognized as a time series by the analyst, this is obviously not a problem, but often there is a time dependency in the data that is not recognized immediately. "Time" or "location" may not be an explicit variable on the dataset, even though it may be an important variable. The index of the observation within the dataset may be a surrogate variable for time, and characteristics of the data may vary as the index varies. Often it is useful to make plots in which one axis is the index number of the observations. For univariate data $x_1, x_2, \ldots$, quick insights can be obtained by a "4-plot" (Filliben, 1982) that consists of the following four plots, as in Figure 1.2:

- plot of x_i versus i to see if there is any trend in the way the data are ordered, which is likely to be the order in which the data were collected;
- plot of x_{i+1} versus x_i to see if there are systematic lags (again, this is done because of possible effects of the order in which the data were collected);
- histogram;
- normal probability plot of the data.

The DATAPLOT program distributed freely by NIST implements 4-plots; see
`http://www.itl.nist.gov/div898/software.htm`

The patterns of the data seen in Figure 1.2 are interesting. The shape in the upper left plot may lead us to expect the data-generating process is periodic, however, the nature of the periodicity is not very clear. The two lower plots seem to indicate that the process is more-or-less normal. The upper right plot shows the strongest structure, and so we should pursue what it may suggest, namely to look at first-order differences. Following up on this, in this particular dataset, we would see that the first-order differences seem to be uncorrelated and to follow a normal distribution. (The dataset is, in fact, an artificially-generated stationary martingale with normal marginals. Such a data-generating process can yield some unexpected patterns, and in the more interesting cases of nonstationarity and nonnormality the data can be very difficulty to analyze. Martingales often can be used effectively in modeling the behavior of stock prices.)

More subtle time dependencies are those in which the values of the variables are not directly related to time, but relationships among variables are changing over time. The identification of such time dependencies is much more difficult, and often requires fitting a model and plotting residuals. Another strictly graphical way of observing changes in relationships over time is by using a sequence of graphical displays. The DATAPLOT program includes a "6-plot", which helps in exploring relationships between two variables that may be changing over time, and whether the stochastic component of the relationship follows a normal distribution.

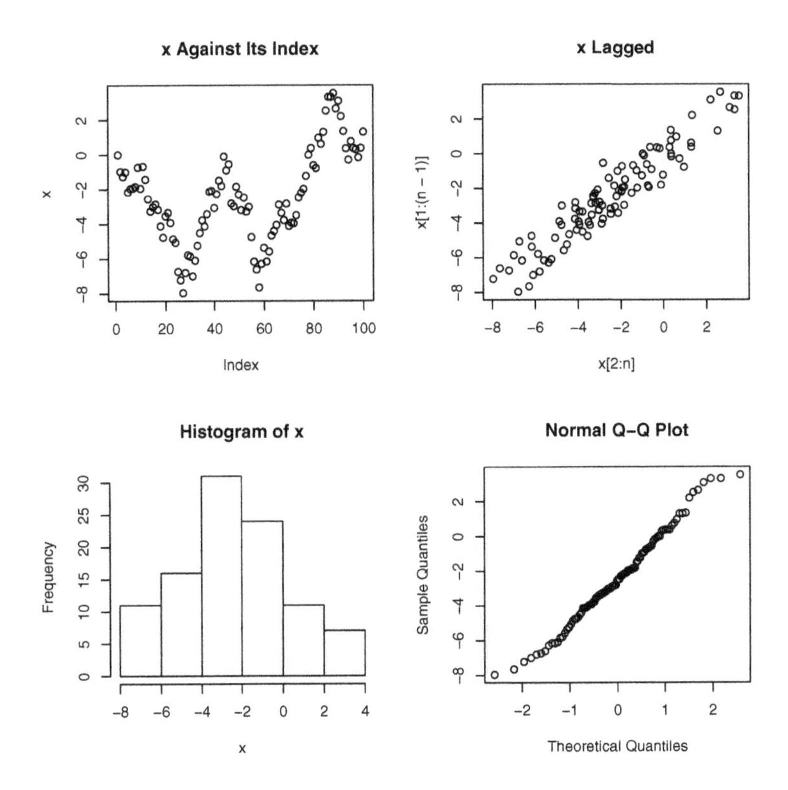

Fig. 1.2. 4-Plot

Simple Plots May Reveal the Unexpected

Although in "computational statistics", the emphasis is generally on interactive graphical displays or on displays of high-dimensional data, as we have seen, very simple plots are also useful. A simple plot of the data will often reveal structure or other characteristics of the data that numerical summaries do not.

An important property of data that is often easily seen in a graph is the unit of measurement. Data on continuous variables are often rounded or measured on a coarse grid. This may indicate other problems in the collection of the data. The horizontal lines in Figure 1.3 indicate that the data do not come from a continuous distribution, even if the data analyst thinks they did, and is using a model based on the assumption that they did. Whether we can use methods of data analysis that assume continuity depends on the coarseness of the grid or measurement; that is, on the extent to which the data are discrete or the extent to which they have been discretized.

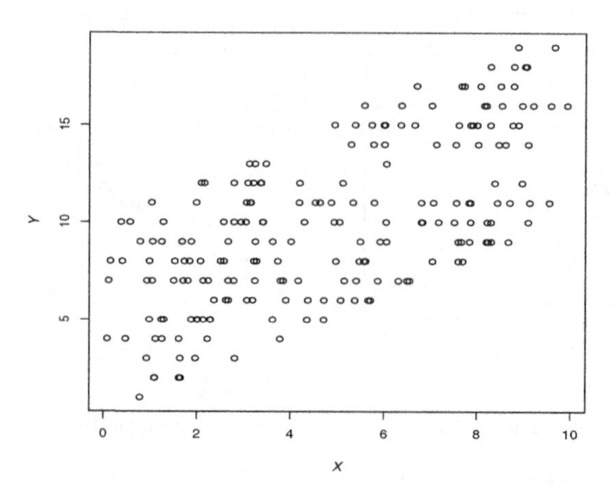

Fig. 1.3. Discrete Data, Rounded Data, or Data Measured Imprecisely

We discuss graphical methods further in Chapter 8. The emphasis is on the use of graphics for discovery. The field of statistical graphics is much broader, of course, and includes many issues of design of graphical displays for conveying (rather than discovering) information.

1.2 Mathematical Tools for Identifying Structure in Data

While the features of interest in a dataset may be such things as colors or other nominal properties, in order to carry out a meaningful statistical analysis, the data must be numeric or else must have been mapped into the real number system.

We think of a set of n observations on a single variable or feature as a vector in the n-dimensional vector space of real numbers, which we denote by IR^n. Likewise, we think of a set of m variables or features that are associated with a single observational unit as a vector in the m-dimensional vector space of real numbers, IR^m. The matrix whose elements are the n observations on the m variables is in the space that we denote by $\mathrm{IR}^{n\times m}$. I do not use different notation to distinguish a vector from a scalar (that is, from a single real number); hence, "x" may represent either a scalar or a vector, and "0" may represent the scalar zero or a vector of zeros. I usually use an upper-case letter to represent a matrix (but upper-case is also used for other things, such as random variables).

Useful Measures in Vector Spaces

Three basic mathematical objects that we will work with are vectors, matrices, and functions. These are special classes of objects. They are members of *vector spaces*, which are mathematical structures consisting of a set and two operations,

- an operation of addition for any two elements in the class that yields an element in the class
- an operation of multiplication of any element in the class by a real scalar that yields an element in the class,

and a special member called the additive identity, which when added to any element in the vector space yields that same element.

In the following, we will generally refer to the members of a vector space as "elements", although occasionally we will call them "vectors", although they may not be vectors in the usual sense.

For any class of objects with these two operations and an additive identity, that is, for vector spaces, we can define three useful types of functions. These are inner products, norms, and metrics.

Inner Products

The *inner product* of two elements x and y, denoted by $\langle x, y\rangle$, is any function into $\mathrm{I\!R}$ that satisfies the following conditions:

- nonnegativity for (x, x):

$$\text{for all } x \neq 0, \quad \langle x, x\rangle > 0; \tag{1.1}$$

- mapping of $(x, 0)$ and $(0, x)$:

$$\text{for all } x, \quad \langle 0, x\rangle = \langle x, 0\rangle = \langle 0, 0\rangle = 0; \tag{1.2}$$

- commutativity:

$$\text{for all } x, y, \quad \langle x, y\rangle = \langle y, x\rangle; \tag{1.3}$$

- factoring of scalar multiplication in inner products:

$$\text{for all } x, y, \text{ and for all } a \in \mathrm{I\!R}, \quad \langle ax, y\rangle = a\langle x, y\rangle; \tag{1.4}$$

- relation of addition in the vector space to addition of inner products:

$$\text{for all } x, y, z, \quad \langle x + y, z\rangle = \langle x, z\rangle + \langle y, z\rangle. \tag{1.5}$$

(In the above statements, "for all" means for all elements in the set for which the function is defined.)

An important property of inner products is the Cauchy-Schwarz inequality:

$$\langle x, y\rangle \leq \langle x, x\rangle^{1/2}\langle y, y\rangle^{1/2}. \tag{1.6}$$

This is easy to see by first observing that for every real number t,

$$\begin{aligned} 0 &\leq (\langle(tx+y), (tx+y)\rangle) \\ &= \langle x, x\rangle t^2 + 2\langle x, y\rangle t + \langle y, y\rangle \\ &= at^2 + bt + c, \end{aligned}$$

where the constants a, b, and c correspond to the inner products in the preceding equation. This nonnegative quadratic in t cannot have two distinct real roots, hence the discriminant, $b^2 - 4ac$, must be less than or equal to zero; that is,

$$\left(\frac{1}{2}b\right)^2 \leq ac.$$

By substituting and taking square roots, we get the Cauchy-Schwarz inequality. It is also clear from this proof that equality holds only if $x = 0$ or if $y = rx$ for some scalar r.

The inner product or *dot product* of two vectors x and y in IR^n, also denoted by $x^{\mathrm{T}}y$, is defined by

$$\langle x, y\rangle = x^{\mathrm{T}}y = \sum_{i=1}^{n} x_i y_i. \tag{1.7}$$

Notice that $\langle x, y\rangle = \langle y, x\rangle$ and $x^{\mathrm{T}}y = (x^{\mathrm{T}}y)^{\mathrm{T}} = y^{\mathrm{T}}x$.

The inner product or dot product of the real functions f and g over the domain D, denoted by $\langle f, g\rangle_D$ or usually just by $\langle f, g\rangle$, is defined as

$$\langle f, g\rangle_D = \int_D f(x)g(x)\,\mathrm{d}x \tag{1.8}$$

if the (Lebesgue) integral exists. By checking the defining properties of an inner product, it is easy to see that the functions defined in equations (1.7) and (1.8) are norms (exercise).

Dot products of functions (as well as of vectors and matrices) over the complex number field are defined in terms of integrals (or sums) of complex conjugates,

$$\langle f, g\rangle_D = \int_D f(x)\bar{g}(x)\,\mathrm{d}x,$$

if the integral exists. The notation $\bar{g}(\cdot)$ denotes the complex conjugate of the function $g(\cdot)$. Often, even if the vectors and matrices in data analysis have real elements, many functions of interest are complex. In this book, however, we will generally assume the functions are real-valued, and so we do not write inner product using the complex conjugate.

To avoid questions about integrability, we generally restrict attention to functions whose dot products with themselves exist; that is, to functions that are square Lebesgue integrable over the region of interest. The set of such square integrable functions is denoted $L^2(D)$. In many cases, the range of integration is the real line, and we may use the notation $L^2(\mathbb{R})$, or often just L^2, to denote that set of functions and the associated inner product.

We can also define an inner product for matrices, but the matrix inner product is not so useful for our purposes, so we will not define it here.

Norms

The "size" of an object with multiple elements can be quantified by a real-valued function called a *norm*, often denoted by $\|\cdot\|$.

A norm is any function into $\mathbb{R}$ that satisfies the following conditions:

- nonnegativity:
$$\text{for all } x \neq 0, \quad \|x\| > 0; \tag{1.9}$$
- mapping of the identity:
$$\text{for } x = 0, \quad \|x\| = 0; \tag{1.10}$$
- relation of scalar multiplication to real multiplication:
$$\text{for all } a \in \mathbb{R} \text{ and for all } x, \quad \|ax\| = |a|\|x\|; \tag{1.11}$$
- triangle inequality:
$$\text{for all } x, y, \quad \|x+y\| \leq \|x\| + \|y\|. \tag{1.12}$$

(Again, "for all" means for all elements in the set for which the function is defined.)

For matrix norms we usually also require an additional condition:

- consistency property:
$$\text{for all conformable matrices } A, B, \quad \|AB\| \leq \|A\|\|B\|. \tag{1.13}$$

There are various kinds of norms. One of the most useful norms is that induced by an inner product:

$$\|x\| = \sqrt{\langle x, x\rangle}. \tag{1.14}$$

(In Exercise 1.3a you are asked to show that the function defined from an inner product in this way is indeed a norm.)

A useful class of norms, called L_p norms, for $p \geq 1$, are defined as

$$\|x\|_p = \left(\left\langle |x|^{p/2}, |x|^{p/2}\right\rangle\right)^{1/p}. \tag{1.15}$$

(In the expression above, $|x|^{p/2}$ means the object of the same type as x whose elements are the absolute values of the corresponding elements of x raised to the $p/2$ power. For the n-vector x, $|x|^{p/2} = (|x_1|^{p/2}, \ldots, |x_n|^{p/2})$.)

As above, we often use a subscript to designate a particular norm, for example, $\|x\|_p$ or $\|A\|_F$. For vectors, the norm arising from the inner product (1.14) is denoted by $\|x\|_2$, although it is so commonly used that the simple notation $\|x\|$ usually refers to this vector norm. It is sometimes called the Euclidean norm or the L_2 norm.

For matrices, the norm arising from the inner product (1.14) is called the *Frobenius norm*, denoted by $\|A\|_F$, and is equal to $(\sum_{i,j} a_{ij}^2)^{1/2}$. Usually a matrix norm denoted without a subscript is the Frobenius matrix norm, but this is not as universally used as the notation without subscript to denote the Euclidean vector norm. (The Frobenius matrix norm is also sometimes called the "Euclidean norm", but in the case of matrices that term is more widely used to refer to the L_2 matrix norm defined below.)

Matrix norms can also be defined in terms of vector norms, and the L_2 vector norm results in a matrix norm that is different from the Frobenius norm. For clarity, we will denote a vector norm as $\|\cdot\|_v$ and a matrix norm as $\|\cdot\|_M$. (This notation is meant to be generic; that is, $\|\cdot\|_v$ represents any vector norm.) The matrix norm $\|\cdot\|_M$ *induced* by $\|\cdot\|_v$ is defined by

$$\|A\|_M = \max_{x \neq 0} \frac{\|Ax\|_v}{\|x\|_v}. \tag{1.16}$$

It is easy to see that an induced norm is indeed a matrix norm. The first three properties of a norm are immediate, and the consistency property can be verified by applying the definition (1.16) to AB and replacing Bx with y; that is, using Ay.

We usually drop the v or M subscript, and the notation $\|\cdot\|$ is overloaded to mean either a vector or matrix norm. (Overloading of symbols occurs in many contexts, and we usually do not even recognize that the meaning is context-dependent. In computer language design, overloading must be recognized explicitly because the language specifications must be explicit.)

From equation (1.16) we have the L_2 matrix norm:

$$\|A\|_2 = \max_{\|x\|_2=1} \|Ax\|_2.$$

The induced norm of A given in equation (1.16) is sometimes called the *maximum magnification* by A. The expression looks very similar to the maximum eigenvalue, and indeed it is in some cases.

For any vector norm and its induced matrix norm, we see from equation (1.16) that

$$\|Ax\| \leq \|A\| \, \|x\| \tag{1.17}$$

because $\|x\| \geq 0$.

Metrics

The *distance* between two elements of a vector space can be quantified by a *metric*, which is a function Δ from pairs of elements in a vector space into $\mathbb{R}$ satisfying the properties

- nonnegativity:

$$\text{for all } x, y \text{ with } x \neq y, \quad \Delta(x, y) > 0; \tag{1.18}$$

- mapping of the identity:

$$\text{for all } x, \quad \Delta(x, x) = 0; \tag{1.19}$$

- commutativity:

$$\text{for all } x, y, \quad \Delta(x, y) = \Delta(y, x); \tag{1.20}$$

- triangle inequality:

$$\text{for all } x, y, z, \quad \Delta(x, z) \leq \Delta(x, y) + \Delta(y, z). \tag{1.21}$$

(Again, "for all" means for all elements in the set for which the function is defined.)

There are various kinds of metrics, and we often use a subscript to designate a particular metric, for example, $\Delta_1(x, y)$ or $\Delta_p(x, y)$. Many useful metrics are induced by norms. One of the most useful induced metrics for vectors is the one induced by the Euclidean norm:

$$\Delta(x, y) = \|x - y\|_2. \tag{1.22}$$

Any norm applied in this way defines a metric (Exercise 1.3b).

For vectors, the metric (1.22) is called the Euclidean metric (or "Euclidean distance") or the L_2 metric. Notice that this metric is the square root of the inner product $(x - y)^{\mathrm{T}}(x - y)$. This suggests a generalization to

$$\sqrt{(x - y)^{\mathrm{T}} A (x - y)} \tag{1.23}$$

for a given positive definite matrix A. This is the *elliptic metric.* If A is the inverse of a variance-covariance matrix, the quantity in (1.23) is sometimes called the Mahalanobis distance between x and y. The expression without the square root is often used; it may also be called the Mahalanobis distance, but more properly is called the Mahalanobis squared distance (see also page 392). Notice that if $A = I$, the Mahalanobis squared distance is the square of the Euclidean distance. The Mahalanobis squared distance is not a metric. (Because of the square, it does not satisfy the triangle inequality.)

Norms and metrics play an important role in identifying structure in data.

Linear Combinations and Linear Independence

A very common operation in working with vectors is the addition of a scalar multiple of one vector to another vector,

$$z = ax + y, \tag{1.24}$$

where a is a scalar and x and y are vectors conformable for addition. Viewed as a single operation with three operands, this is called an "axpy" for obvious reasons. (Because the Fortran programs to perform this operation were called `saxpy` and `daxpy`, the operation is also sometimes called "saxpy" or "daxpy".) The axpy operation is called a *linear combination*. Such linear combinations of vectors are the basic operations in most areas of linear algebra. The composition of axpy operations is also an axpy; that is, one linear combination followed by another linear combination is a linear combination. In general, for the vectors $v_1, \ldots, v_k$ and scalars $a_1 = \cdots = a_k = 0$, we form

$$z = a_1 v_1 + \cdots + a_k v_k. \tag{1.25}$$

Any linear combination such as (1.25) can be decomposed into a sequence of axpy operations.

If a given vector can be formed by a linear combination of one or more vectors, the set of vectors (including the given one) is said to be linearly dependent; conversely, if in a set of vectors no one vector can be represented as a linear combination of any of the others, the set of vectors is said to be *linearly independent*.

Linear independence is one of the most important concepts both in linear algebra and in statistics.

We can see that the definition of a linearly independent set of vectors $\{v_1, \ldots, v_k\}$ is equivalent to stating that if

$$a_1 v_1 + \cdots a_k v_k = 0, \tag{1.26}$$

then $a_1 = \cdots = a_k = 0$.

If the set of vectors $\{v_1, \ldots, v_k\}$ is not linearly independent, then it is possible to select a *maximal linearly independent subset*; that is, a subset of $\{v_1, \ldots, v_k\}$ that is linearly independent and has maximum cardinality. We do this by selecting an arbitrary vector, v_{i_1}, and then seeking a vector that is independent of v_{i_1}. If there are none in the set that is linearly independent of v_{i_1}, then a maximum linearly independent subset is just the singleton, because all of the vectors must be a linear combination of just one vector (that is, a scalar multiple of that one vector). If there is a vector that is linearly independent of v_{i_1}, say v_{i_2}, we next seek a vector in the remaining set that is independent of v_{i_1} and v_{i_2}. If one does not exist, then $\{v_{i_1}, v_{i_2}\}$ is a maximal subset because any other vector can be represented in terms of these two and hence, within any subset of three vectors, one can be represented in terms of the two others. Thus, we see how to form a maximal linearly independent subset, and we see there is a unique maximum cardinality of any subset of linearly independent vectors.

Basis Sets

If each vector in the vector space $\mathcal{V}$ can be expressed as a linear combination of the vectors in some set G, then G is said to be a *generating set* or *spanning set* of $\mathcal{V}$. If, in addition, all linear combinations of the elements of G are in $\mathcal{V}$, the vector space is the *space generated by* G.

A set of linearly independent vectors that generate or span a space is said to be a *basis* for the space. The cardinality of a basis set for a vector space consisting of n-vectors is n. The cardinality of a basis set may be countably infinite, such as in the case of a vector space (or "linear space") of functions.

Normalized Vectors

The Euclidean norm of a vector corresponds to the length of the vector x in a natural way; that is, it agrees with our intuition regarding "length". Although, as we have seen, this is just one of many vector norms, in most applications it is the most useful one. (I must warn you, however, that occasionally I will carelessly but naturally use "length" to refer to the order of a vector; that is, the number of elements. This usage is common in computer software packages such as R and SAS IML, and software necessarily shapes our vocabulary.)

Dividing a given vector by its length *normalizes* the vector, and the resulting vector with length 1 is said to be *normalized*; thus

$$\tilde{x} = \frac{1}{\|x\|}x \tag{1.27}$$

is a normalized vector. Normalized vectors are sometimes referred to as "unit vectors", although we will generally reserve this term for a special kind of normalized vector that has 0s in all positions except one and has a 1 in that position. A normalized vector is also sometimes referred to as a "normal vector". I use "normalized vector" for a vector such as $\tilde{x}$ in equation (1.27) and use the "normal vector" to denote a vector that is orthogonal to a subspace.

Orthogonal Vectors and Orthogonal Vector Spaces

Two vectors v_1 and v_2 such that

$$\langle v_1, v_2 \rangle = 0 \tag{1.28}$$

are said to be *orthogonal*, and this condition is denoted by $v_1 \perp v_2$. (Sometimes we exclude the zero vector from this definition, but it is not important to do so.) Normalized vectors that are all orthogonal to each other are called *orthonormal* vectors. (If the elements of the vectors are from the field of complex numbers, orthogonality and normality are defined in terms of the dot products of a vector with a complex conjugate of a vector.)

A set of nonzero vectors that are mutually orthogonal are necessarily linearly independent. To see this, we show it for any two orthogonal vectors and then indicate the pattern that extends to three or more vectors. Suppose v_1

and v_2 are nonzero and are orthogonal; that is, $\langle v_1, v_2 \rangle = 0$. We see immediately that if there is a scalar a such that $v_1 = av_2$, then a must be nonzero and we have a contradiction because $\langle v_1, v_2 \rangle = a\langle v_1, v_1 \rangle \neq 0$. For three mutually orthogonal vectors, v_1, v_2, and v_3, we consider $v_1 = av_2 + bv_3$ for a or b nonzero, and arrive at the same contradiction.

Orthogonalization Transformations

Given m nonnull, linearly independent vectors, $x_1, \ldots, x_m$, it is easy to form m orthonormal vectors, $\tilde{x}_1, \ldots, \tilde{x}_m$, that span the same space. A simple way to do this is sequentially. First normalize x_1 and call this $\tilde{x}_1$. Now, suppose that x_2 is represented in an orthogonal coordinate system in which one axis is $\tilde{x}_1$, and determine the coordinate of x_2 for that axis. This means the point of intersection of an orthogonal line from x_2 to $\tilde{x}_1$. This is an orthogonal projection, so next, orthogonally project x_2 onto $\tilde{x}_1$ and subtract this projection from x_2. The result is orthogonal to $\tilde{x}_1$; hence, normalize this and call it $\tilde{x}_2$. These first two steps are

$$
\begin{aligned}
\tilde{x}_1 &= \frac{1}{\|x_1\|} x_1, \\
\tilde{x}_2 &= \frac{1}{\|x_2 - \langle \tilde{x}_1, x_2 \rangle \tilde{x}_1\|} (x_2 - \langle \tilde{x}_1, x_2 \rangle \tilde{x}_1).
\end{aligned}
\tag{1.29}
$$

These are called *Gram-Schmidt transformations*. We discuss them further beginning on page 219.

Series Expansions in Basis Sets

Basis sets are useful because we can represent any element in the vector space uniquely as a linear combination of the elements in the basis set. If $\{v_1, v_2, \ldots\}$ is a given basis set for a vector space containing the element x, then there are unique constants $c_1, c_2, \ldots$ such that

$$
x = \sum_k c_k v_k. \tag{1.30}
$$

(In this expression and many of those that follow, I do not give the limits of the summation. The index k goes over the full basis set, which we assume to be countable, but not necessarily finite.)

In the case of finite-dimensional vector spaces, the set $\{v_1, v_2, \ldots\}$ is finite; its cardinality is the dimension of the vector space.

The reason that basis sets and expansions in a basis set are important is for the use of expansions in approximating and estimating vectors, matrices, and functions. Approximations and estimation are major topics in later chapters of this book.

A basis set whose elements are normalized and mutually orthogonal is usually the best to work with because they have nice properties that facilitate computations, and there is a large body of theory about their properties.

If the basis set is orthonormal, we can easily determine the coefficients c_k in the expansion (1.30):

$$c_k = \langle x, v_k \rangle. \tag{1.31}$$

The coefficients $\{c_k\}$ are called the *Fourier coefficients* of x with respect to the orthonormal basis $\{v_k\}$.

If x has the expansion above, the square of the L_2 norm of the function is the sum of squares of the Fourier coefficients:

$$\begin{aligned} \langle x, x \rangle &= \left\langle \sum_k c_k v_k, \sum_k c_k v_k \right\rangle \\ &= \sum_k |c_k|^2. \end{aligned} \tag{1.32}$$

In applications, we *approximate* an element of a vector space using a truncated orthogonal series. In this case, we are interested in the residual,

$$x - \sum_{k=1}^{j} c_k v_k, \tag{1.33}$$

where j is less than the upper bound on the index in equation (1.30).

Series Expansions of Functions

The basic objects such as inner products, norms, and metrics can be defined for functions in terms of integrals, just as these objects are defined for vectors in terms of sums. With functions, of course, we need to consider the existence of the integrals, and possibly make some slight changes in definitions.

We will change the notation slightly for functions, and then use it consistently in later chapters such as Chapters 4 and 10 where we approximate or estimate functions. We will start the index at 0 instead of 1, and we will use q_k to denote the kth function in a generic basis set. We call this "0 equals first" indexing.

After these basic objects are in place, we can define concepts such as linear independence and orthogonality just as we have done above. For a given class of functions, we may also be able to identify a basis set. If the class is very restricted, such as say, the class of all real polynomials of degree k or less over a finite interval $[a, b]$, then the basis set may be finite and rather simple. For more interesting classes of functions, however, the basis set must be infinite. (See Section 4.2 for some basis sets for functions.) For approximating functions using an infinite basis set, it is obviously necessary to use a truncated series.

Because, in practice we deal with truncated series, the error due to that truncation is of interest. For the function f, the error due to finite truncation at j terms of the infinite series is the residual function $f - \sum_{k=0}^{j} c_k q_k$.

The *mean squared error* over the domain D is the scaled, squared L_2 norm of the residual,

$$\frac{1}{d}\left\| f - \sum_{k=0}^{j} c_k q_k \right\|^2, \tag{1.34}$$

where d is some measure of the domain D. (If the domain is the interval $[a, b]$, for example, one choice is $d = b - a$.)

A very important property of Fourier coefficients is that they yield the minimum mean squared error for an expansion in any given subset of a basis set of functions $\{q_i\}$; that is, for any other constants, $\{a_i\}$, and any j,

$$\left\| f - \sum_{k=0}^{j} c_k q_k \right\|^2 \leq \left\| f - \sum_{k=0}^{j} a_k q_k \right\|^2 \tag{1.35}$$

(see Exercise 1.4).

Another important property of the residuals, analogous to that of the linear least squares estimator is that the residual or error, $f - \sum_{k=0}^{j} c_k q_k$, is orthogonal to the approximation, that is,

$$\left\langle \sum_{k=0}^{j} c_k q_k,\ f - \sum_{k=0}^{j} c_k q_k \right\rangle = 0. \tag{1.36}$$

Partial sums of squares of Fourier coefficients, $\sum_{k=0}^{j} c_k^2$, for any j are bounded by $\|f\|^2$, that is,

$$\sum_{k=0}^{j} |c_k|^2 \leq \|f\|^2. \tag{1.37}$$

This is called *Bessel's inequality*, and, it follows from equation (1.32) or from

$$\begin{aligned} 0 &\leq \left\| f - \sum_{k=0}^{j} c_k q_k \right\|^2 \\ &= \|f\|^2 - \sum_{k=0}^{j} |c_k|^2. \end{aligned}$$

The optimality of Fourier coefficients, the orthogonality of the residual, and Bessel's inequality, that is, equations (1.35), (1.36), and (1.37), apply for orthogonal series expansions in any vector space.

There are some additional special considerations for expansions of functions, which we will address in Chapter 4.

Properties of Functions and Operations on Functions

There are many important properties of functions, such as continuity and differentiability, that we will assume from time to time.

Another important property of some functions is convexity. A function f is a *convex function* if for any two points x and y in the domain of f and $w \in (0, 1)$, then

$$f(wx + (1 - w)y) \leq wf(x) + (1 - w)f(y); \tag{1.38}$$

that is, by the definition of convexity, f is convex if its value at the weighted average of two points does not exceed the weighted average of the function at those two points. If the inequality (1.38) is strict, then the function is said to be *strictly convex.*

If f is a convex function, then $-f$ is said to be a *concave function.* Many interesting functions in statistical applications are concave.

If f is convex over D then there is a b such that for any x and t in D,

$$b(x - t) + f(t) \leq f(x). \tag{1.39}$$

Notice that for a given b, $L(x) = b(x - t) + f(t)$ is a straight line through the point $(t, f(t))$, with slope b.

For functions over the same domain, the axpy operation, such as in the expansions in basis sets, is one of the most common operations on functions.

If the domain of the function f is a subset of the range of the function g, then the *composition* of f and g, denoted $f \circ g$, is defined as

$$f \circ g(x) = f(g(x)). \tag{1.40}$$

The *convolution* of two functions f and g is the function, which we denote as $f * g$, defined as

$$f * g(x) = \int f(x - t)g(t) \, \mathrm{d}t, \tag{1.41}$$

if the integral exists. Note that the range of integration (in the appropriate dimension) must be such that the integrand is defined over it. (Either of the functions may be defined to be zero over subdomains, of course.)

We often refer to a function with an argument that is a function as a *functional.* Function transforms, such as Fourier transforms and probability characteristic functions, are functionals. In probability theory, many parameters are defined as functionals; see examples on page 31.

Kernel Functions

A function specifying operations involving two variables is often called a *kernel function.* The function f in the integrand in the definition of the convolution above is a simple type of kernel. In that case, the two variables are combined into a single variable, but to emphasize the role of the two arguments, we could write $K(x,t) = f(x-t)$.

Properties of a particular process involving some kernel function $K(x, y)$ can often be studied by identifying particular properties of the kernel.

If a kernel K is defined over the same region for each of its arguments and if it has the property $K(x, y) = K(y, x)$ for all x and y for which it is defined, then K is said to be *symmetric*.

We will encounter kernel functions in various applications in later chapters. Kernels are often called "filters", especially in applications in function approximation or estimation. The conditional PDF $p_{Y|Z}$ in equation (1.69) below can be thought of as a kernel. A kernel of that form is often called a "transition kernel". The properties of transition kernels are important in Markov chain Monte Carlo methods, which we will discuss in later chapters.

Properties of Matrices and Operations on Matrices

A common data structure for statistical analysis is a rectangular array; rows represent individual observational units, or just "observations", and columns represent the variables or features that are observed for each unit. If the values of the variables are elements of a field, for example if they are real numbers, the rectangular array is a matrix, and the mathematics of matrices can be useful in the statistical analysis. (If the values of the variables are other things, such as "red" or "green", or "high" or "low", those values can be mapped to real numbers, and again, we have a matrix, but the algebra of matrices may or may not be useful in the analysis.) We will concentrate on situations in which numeric values appropriately represent the observational data.

If the elements of a matrix X represent numeric observations on variables in the structure of a rectangular array as indicated above, the mathematical properties of X carry useful information about the observations and about the variables themselves. In addition, mathematical operations on the matrix may be useful in discovering structure in the data. These operations include various transformations and factorizations that we discuss in Chapters 5 and 9. We also address some of the computational details of operations on matrices in Chapter 5.

Symmetric Matrices

A matrix A with elements a_{ij} is said to be symmetric if each element a_{ji} has the same value as a_{ij}. Symmetric matrices have useful properties that we will mention from time to time.

Symmetric matrices provide a generalization of the inner product. If A is symmetric and x and y are conformable vectors, then the *bilinear form* $x^{\mathrm{T}}Ay$ has the property that $x^{\mathrm{T}}Ay = y^{\mathrm{T}}Ax$, and hence this operation on x and y is commutative, which is one of the properties of an inner product.

More generally, a bilinear form is a kernel function of the two vectors, and a symmetric matrix corresponds to a symmetric kernel.

An important type of bilinear form is $x^{\mathrm{T}}Ax$, which is called a *quadratic form.*

Nonnegative Definite and Positive Definite Matrices

A real symmetric matrix A such that for any real conformable vector x the quadratic form $x^{\mathrm{T}}Ax$ is nonnegative, that is, such that

$$x^{\mathrm{T}}Ax \geq 0, \tag{1.42}$$

is called a *nonnegative definite matrix.* We denote the fact that A is nonnegative definite by

$$A \succeq 0.$$

(Note that we consider the zero matrix, $0_{n\times n}$, to be nonnegative definite.)

A symmetric matrix A such that for any (conformable) vector $x \neq 0$ the quadratic form

$$x^{\mathrm{T}}Ax > 0 \tag{1.43}$$

is called a *positive definite matrix.* We denote the fact that A is positive definite by

$$A \succ 0.$$

(Recall that $A \geq 0$ and $A > 0$ mean, respectively, that all elements of A are nonnegative and positive.) When A and B are symmetric matrices of the same order, we write $A \succeq B$ to mean $A - B \succeq 0$ and $A \succ B$ to mean $A - B \succ 0$. Nonnegative and positive definite matrices are very important in applications. We will encounter them often in this book.

A kernel function K defined as

$$K(x, y) = x^{\mathrm{T}}Ay, \tag{1.44}$$

is said to be nonnegative or positive definite if A has the corresponding property. (More general types of kernel may also be described as nonnegative or positive definite, but the meaning is similar to the meaning in the bilinear form of equation (1.44).)

In this book we use the terms "nonnegative definite" and "positive definite" only for symmetric matrices or kernels. In other literature, these terms may be used more generally; that is, for any (square) matrix that satisfies (1.42) or (1.43).

Systems of Linear Equations

One of the most common uses of matrices is to represent a system of linear equations

$$Ax = b. \tag{1.45}$$

Whether or not the system (1.45) has a solution (that is, whether or not for a given A and b there is an x such that $Ax = b$) depends on the number of linearly independent rows in A (that is, considering each row of A as being a vector). The number of linearly independent rows of a matrix, which is also the number of linearly independent columns of the matrix, is called the *rank* of the matrix. A matrix is said to be of *full rank* if its rank is equal to either its number of rows or its number of columns. A square full rank matrix is called a *nonsingular* matrix. We call a matrix that is square but not full rank *singular.*

The system (1.45) has a solution if and only if

$$\text{rank}(A|b) \leq \text{rank}(A), \tag{1.46}$$

where $A|b$ is the matrix formed from A by adjoining b as an additional column. (This and other facts cited here are proved in standard texts on linear algebra.) If a solution exists, the system is said to be *consistent.* (The common regression equations, which we will encounter in many places throughout this book, do not satisfy the condition (1.46).)

We now briefly discuss the solutions to a consistent system of the form of (1.45).

Matrix Inverses

If the system $Ax = b$ is consistent then

$$x = A^{-}b \tag{1.47}$$

is a solution, where A^{-} is any matrix such that

$$AA^{-}A = A, \tag{1.48}$$

as we can see by substituting $A^{-}b$ into $AA^{-}Ax = Ax$.

Given a matrix A, a matrix A^{-} such that $AA^{-}A = A$ is called a *generalized inverse* of A, and we denote it as indicated. If A is square and of full rank, the generalized inverse, which is unique, is called the *inverse* and is denoted by A^{-1}. It has a stronger property than (1.48): $AA^{-1} = A^{-1}A = I$, where I is the identity matrix.

To the general requirement $AA^{-}A = A$, we successively add three requirements that define special generalized inverses, sometimes called respectively g_2, g_3, and g_4 inverses. The "general" generalized inverse is sometimes called a g_1 inverse. The g_4 inverse is called the Moore-Penrose inverse.

For a matrix A, a *Moore-Penrose inverse*, denoted by A^{+}, is a matrix that has the following four properties.

1. $AA^+A = A$. Any matrix that satisfies this condition is called a generalized inverse, and as we have seen above is denoted by A^-. For many applications, this is the only condition necessary. Such a matrix is also called a g_1 *inverse*, an *inner pseudoinverse*, or a *conditional inverse.*
2. $A^+AA^+ = A^+$. A matrix A^+ that satisfies this condition is called an *outer pseudoinverse.* A g_1 inverse that also satisfies this condition is called a g_2 *inverse* or *reflexive generalized inverse*, and is denoted by A^*.
3. A^+A is symmetric.
4. AA^+ is symmetric.

The Moore-Penrose inverse is also called the *pseudoinverse*, the *p-inverse*, and the *normalized generalized inverse.*

We can see by construction that the Moore-Penrose inverse for any matrix A exists and is unique. (See, for example, Gentle, 2007, page 102.)

The Matrix $X^{\mathrm{T}}X$

When numerical data are stored in the usual way in a matrix X, the matrix $X^{\mathrm{T}}X$ often plays an important role in statistical analysis. A matrix of this form is called a *Gramian* matrix, and it has some interesting properties.

First of all, we note that $X^{\mathrm{T}}X$ is symmetric; that is, the $(ij)^{\mathrm{th}}$ element, $\sum_k x_{k,i}x_{k,j}$ is the same as the $(ji)^{\mathrm{th}}$ element. Secondly, because for any y, $(Xy)^{\mathrm{T}}Xy \geq 0$, $X^{\mathrm{T}}X$ is nonnegative definite.

Next we note that

$$X^{\mathrm{T}}X = 0 \quad \Longleftrightarrow \quad X = 0. \tag{1.49}$$

The implication from right to left is obvious. We see the left to right implication by noting that if $X^{\mathrm{T}}X = 0$, then the ith diagonal element of $X^{\mathrm{T}}X$ is zero. The ith diagonal element is $\sum_j x_{ji}^2$, so we have that x_{ji} for all j and i; hence $X = 0$.

Another interesting property of a Gramian matrix is that, for any matrices B and C (that are conformable for the operations indicated),

$$BX^{\mathrm{T}}X = CX^{\mathrm{T}}X \quad \Longleftrightarrow \quad BX^{\mathrm{T}} = CX^{\mathrm{T}}. \tag{1.50}$$

The implication from right to left is obvious, and we can see the left to right implication by writing

$$(BX^{\mathrm{T}}X - CX^{\mathrm{T}}X)(B^{\mathrm{T}} - C^{\mathrm{T}}) = (BX^{\mathrm{T}} - CX^{\mathrm{T}})(BX^{\mathrm{T}} - CX^{\mathrm{T}})^{\mathrm{T}},$$

and then observing that if the left-hand side is null, then so is the right-hand side, and if the right-hand side is null, then $BX^{\mathrm{T}} - CX^{\mathrm{T}} = 0$ because $X^{\mathrm{T}}X = 0 \quad \Longrightarrow \quad X = 0$, as above. Similarly, we have

$$X^{\mathrm{T}}XB = X^{\mathrm{T}}XC \quad \Longleftrightarrow \quad X^{\mathrm{T}}B = X^{\mathrm{T}}C. \tag{1.51}$$

The generalized inverses of $X^{\mathrm{T}}X$ have useful properties. First, we see from the definition, for any generalized inverse $(X^{\mathrm{T}}X)^{-}$, that $((X^{\mathrm{T}}X)^{-})^{\mathrm{T}}$ is also a generalized inverse of $X^{\mathrm{T}}X$. (Note that $(X^{\mathrm{T}}X)^{-}$ is not necessarily symmetric.) Also, we have, from equation (1.50),

$$X(X^{\mathrm{T}}X)^{-}X^{\mathrm{T}}X = X. \tag{1.52}$$

This means that $(X^{\mathrm{T}}X)^{-}X^{\mathrm{T}}$ is a generalized inverse of X.

The Moore-Penrose inverse of X has an interesting relationship with a generalized inverse of $X^{\mathrm{T}}X$:

$$XX^{+} = X(X^{\mathrm{T}}X)^{-}X^{\mathrm{T}}. \tag{1.53}$$

This can be established directly from the definition of the Moore-Penrose inverse.

An important property of $X(X^{\mathrm{T}}X)^{-}X^{\mathrm{T}}$ is its invariance to the choice of the generalized inverse of $X^{\mathrm{T}}X$. Suppose G is any generalized inverse of $X^{\mathrm{T}}X$. Then, from equation (1.52), we have $X(X^{\mathrm{T}}X)^{-}X^{\mathrm{T}}X = XGX^{\mathrm{T}}X$, and from the implication (1.50), we have

$$XGX^{\mathrm{T}} = X(X^{\mathrm{T}}X)^{-}X^{\mathrm{T}}; \tag{1.54}$$

that is, $X(X^{\mathrm{T}}X)^{-}X^{\mathrm{T}}$ is invariant to the choice of generalized inverse (which of course, it must be for the Moore-Penrose inverse to be unique, as we stated above).

The matrix $X(X^{\mathrm{T}}X)^{-}X^{\mathrm{T}}$ has a number of interesting properties in addition to those mentioned above. We note

$$\begin{aligned}\left(X(X^{\mathrm{T}}X)^{-}X^{\mathrm{T}}\right)\left(X(X^{\mathrm{T}}X)^{-}X^{\mathrm{T}}\right) &= X(X^{\mathrm{T}}X)^{-}(X^{\mathrm{T}}X)(X^{\mathrm{T}}X)^{-}X^{\mathrm{T}} \\ &= X(X^{\mathrm{T}}X)^{-}X^{\mathrm{T}}, \end{aligned} \tag{1.55}$$

that is, $X(X^{\mathrm{T}}X)^{-}X^{\mathrm{T}}$ is *idempotent.* (A matrix A is idempotent if $AA = A$. It is clear that the only idempotent matrix that is of full rank is the identity I.) Any real symmetric idempotent matrix is a *projection matrix.*

The most familiar application of the matrix $X(X^{\mathrm{T}}X)^{-}X^{\mathrm{T}}$ is in the analysis of the linear regression model $y = X\beta+\epsilon$. This matrix projects the observed vector y onto a lower-dimensional subspace that represents the fitted model:

$$\hat{y} = X(X^{\mathrm{T}}X)^{-}X^{\mathrm{T}}y. \tag{1.56}$$

Projection matrices, as the name implies, generally transform or project a vector onto a lower-dimensional subspace. We will encounter projection matrices again in Chapter 9.

Eigenvalues and Eigenvectors

Multiplication of a given vector by a square matrix may result in a scalar multiple of the vector. If A is an $n \times n$ matrix, v is a vector not equal to 0, and c is a scalar such that

$$Av = cv, \tag{1.57}$$

we say v is an *eigenvector* of A and c is an *eigenvalue* of A.

We should note how remarkable the relationship $Av = cv$ is: The effect of a matrix multiplication of an eigenvector is the same as a scalar multiplication of the eigenvector. The eigenvector is an *invariant* of the transformation in the sense that its direction does not change under the matrix multiplication transformation. This would seem to indicate that the eigenvector and eigenvalue depend on some kind of deep properties of the matrix, and indeed, this is the case.

We immediately see that if an eigenvalue of a matrix A is 0, then A must be singular.

We also note that if v is an eigenvector of A, and t is any nonzero scalar, tv is also an eigenvector of A. Hence, we can normalize eigenvectors, and we often do.

If A is symmetric there are several useful facts about its eigenvalues and eigenvectors. The eigenvalues and eigenvector of a (real) symmetric matrix are all real. The eigenvectors of a symmetric matrix are (or can be chosen to be) mutually orthogonal. We can therefore represent a symmetric matrix A as

$$A = VCV^{\mathrm{T}}, \tag{1.58}$$

where V is an orthogonal matrix whose columns are the eigenvectors of A and C is a diagonal matrix whose $(ii)^{\mathrm{th}}$ element is the eigenvalue corresponding to the eigenvector in the ith column of V. This is called the *diagonal factorization* of A.

If A is a nonnegative (positive) definite matrix, and c is an eigenvalue with corresponding eigenvector v, if we multiply both sides of the equation $Av = cv$, we have $v^{\mathrm{T}}Av = cv^{\mathrm{T}}v \geq 0 (> 0)$, and since $v^{\mathrm{T}}v > 0$, we have $c \geq 0 (> 0)$. That is to say, the eigenvalues of a nonnegative definite matrix are nonnegative, and the eigenvalues of a positive definite matrix are positive.

The maximum modulus of any eigenvalue in a given matrix is of interest. This value is called the *spectral radius*, and for the matrix A, is denoted by $\rho(A)$:

$$\rho(A) = \max |c_i|, \tag{1.59}$$

where the c_i's are the eigenvalues of A.

The spectral radius is very important in many applications, from both computational and statistical standpoints. The convergence of some iterative algorithms, for example, depend on bounds on the spectral radius. The spectral radius of certain matrices determines important characteristics of stochastic processes.

Two interesting properties of the spectral radius of the matrix $A = (a_{ij})$ are

$$\rho(A) \leq \max_j \sum_i |a_{ij}|, \tag{1.60}$$

and

$$\rho(A) \leq \max_i \sum_j |a_{ij}|. \tag{1.61}$$

The spectral radius of the square matrix A is related to the L_2 norm of A by

$$\|A\|_2 = \sqrt{\rho(A^{\mathrm{T}}A)}. \tag{1.62}$$

We refer the reader to general texts on matrix algebra for proofs of the facts we have stated without proof, and for many other interesting and important properties of eigenvalues, which we will not present here.

Singular Values and the Singular Value Decomposition

Computations with matrices are often facilitated by first decomposing the matrix into multiplicative factors that are easier to work with computationally, or else reveal some important characteristics of the matrix. Some decompositions exist only for special types of matrices, such as symmetric matrices or positive definite matrices. One of most useful decompositions, and one that applies to all types of matrices, is the singular value decomposition. We discuss it here, and in Section 5.3 we will discuss other decompositions.

An $n \times m$ matrix A can be factored as

$$A = UDV^{\mathrm{T}}, \tag{1.63}$$

where U is an $n \times n$ orthogonal matrix, V is an $m \times m$ orthogonal matrix, and D is an $n \times m$ diagonal matrix with nonnegative entries. (An $n \times m$ diagonal matrix has $\min(n,m)$ elements on the diagonal, and all other entries are zero.)

The number of positive entries in D is the same as the rank of A. The factorization (1.63) is called the *singular value decomposition* (SVD) or the *canonical singular value factorization* of A. The elements on the diagonal of D, d_i, are called the *singular values* of A. We can rearrange the entries in D so that $d_1 \geq d_2 \geq \cdots$, and by rearranging the columns of U correspondingly, nothing is changed.

If the rank of the matrix is r, we have $d_1 \geq \cdots \geq d_r > 0$, and if $r < \min(n,m)$, then $d_{r+1} = \cdots = d_{\min(n,m)} = 0$. In this case

$$D = \begin{bmatrix} D_r & 0 \\ 0 & 0 \end{bmatrix},$$

where $D_r = \mathrm{diag}(d_1, \ldots, d_r)$.

From the factorization (1.63) defining the singular values, we see that the singular values of A^{T} are the same as those of A.

For a matrix with more rows than columns, in an alternate definition of the singular value decomposition, the matrix U is $n \times m$ with orthogonal columns, and D is an $m \times m$ diagonal matrix with nonnegative entries. Likewise, for a matrix with more columns than rows, the singular value decomposition can be defined as above but with the matrix V being $m \times n$ with orthogonal columns and D being $m \times m$ and diagonal with nonnegative entries.

If A is symmetric its singular values are the absolute values of its eigenvalues.

The Moore-Penrose inverse of a matrix has a simple relationship to its SVD. If the SVD of A is given by UDV^{T}, then its Moore-Penrose inverse is

$$A^{+} = VD^{+}U^{\mathrm{T}}, \tag{1.64}$$

as is easy to verify. The Moore-Penrose inverse of D is just the matrix D^{+} formed by inverting all of the positive entries of D and leaving the other entries unchanged.

Square Root Factorization of a Nonnegative Definite Matrix

If A is a nonnegative definite matrix (which, in this book, means that it is symmetric), its eigenvalues are nonnegative, so we can write $S = C^{\frac{1}{2}}$, where S is a diagonal matrix whose elements are the square roots of the elements in the C matrix in the diagonal factorization of A in equation (1.58). Now we observe that $(VSV^{\mathrm{T}})^2 = VCV^{\mathrm{T}} = A$; hence, we write

$$A^{\frac{1}{2}} = VSV^{\mathrm{T}}, \tag{1.65}$$

and we have $(A^{\frac{1}{2}})^2 = A$.

1.3 Data-Generating Processes; Probability Distributions

The model for a data-generating process often includes a specification of a random component that follows some probability distribution. Important descriptors or properties of a data-generating process or probability distribution include the cumulative distribution function (CDF), the probability density function (PDF), and the expected value of the random variable or of certain functions of the random variable. It is assumed that the reader is familiar with the basics of probability distributions at an advanced calculus level, but in the following we will give some definitions and state some important facts concerning CDFs and PDFs.

For a random variable Y, the CDF, which we often denote with the same symbols as we use to denote the distribution itself, is a function whose argument y is a real quantity of the same order as Y and whose value is the probability that Y is less than or equal to y; that is,

$$P_\theta(y) = \Pr(Y \leq y \mid \theta). \tag{1.66}$$

We also sometimes use the symbol that represents the random variable as a subscript on the symbol for the CDF, for example, $P_Y(y)$.

A CDF is defined over the entire real line IR, or over the entire space IR^d; it is nondecreasing in y; and it is bounded between 0 and 1.

The notation for a CDF is usually an upper-case letter, and the notation for the corresponding PDF is usually the same letter in lower case. We also may use a subscript on the letter representing the PDF, for example, $p_Y(y)$. The PDF is the derivative of the CDF (with respect to the appropriate measure), and so

$$P_Y(y) = \int_{t \leq y} p_Y(t)\, \mathrm{d}t. \tag{1.67}$$

The CDF or PDF of a joint distribution is denoted in an obvious fashion, for example, $p_{YZ}(y, z)$.

If $Y = (Y_1, Y_2)$ with CDF $P_Y(y)$, where Y_1 is a d_1-vector and Y_2 is a d_2-vector, then $P_{Y_1}(y_1) = \Pr(Y_1 \leq y_1)$ is called the *marginal* CDF of Y_1, and is given by

$$P_{Y_1}(y_1) = \int_{\mathrm{IR}^{d_2}} P_Y(y_1, y_2)\, \mathrm{d}y_2. \tag{1.68}$$

The CDF or PDF of a conditional distribution is denoted in an obvious fashion, for example, $p_{Y|Z}(y|z)$. Conditional functions are kernels, and a conditional PDF may be interpreted as a transition density from the conditioning random variable to the other random variable. We have the familiar relationship

$$p_{YZ}(y, z) = p_{Y|Z}(y|z) p_Z(z). \tag{1.69}$$

The region in which the PDF is positive is called the *support of the distribution.*

The expected value of any (reasonable) function T of the random variable Y is denoted by $\mathrm{E}(T(Y))$, and if $p(y)$ is the PDF of the random variable, is defined as

$$\mathrm{E}(T(Y)) = \int_{\mathrm{IR}^d} T(y) p(y) \mathrm{d}y, \tag{1.70}$$

for a d-dimensional random variable. In the simplest form of this expression, T is the identity, and $\mathrm{E}(Y)$ is the *mean* of the random variable Y.

Transforms of the CDF

There are three transforms of the CDF that are useful in a wide range of statistical applications: the *moment generating function*, the *cumulant generating function*, and the *characteristic function.* They are all expected values of a function of the variable of the transform.

The characteristic function, which is similar to a Fourier transform exists for all CDFs. It is

$$\varphi(t) = \int_{\mathrm{IR}^d} \exp(\mathrm{i}t^{\mathrm{T}} y)\, \mathrm{d}P(y). \tag{1.71}$$

A related function is the moment generating function,

$$M(t) = \int_{\mathrm{IR}^d} \exp(t^{\mathrm{T}}y) \, \mathrm{d}P(y), \tag{1.72}$$

if the integral exists for t in an open set that includes 0. In this case, note that $M(t) = \varphi(t)$. The moment generating function is also called the *Laplace transform*, although the usual definition of the Laplace transform is for an integral over the positive reals, and the argument of the Laplace transform is the negative of the argument t in the moment generating function.

One of the useful properties of the characteristic function and the moment generating function is the simplicity of those functions for linear combinations of a random variable. For example, if we know the moment generating function of a random variable Y to be $M_Y(t)$ and we have the mean $\widehat{Y}$ of a random sample of Y of size n, then the moment generating function of $\widehat{Y}$ is just $M_{\widehat{Y}}(t) = (M_Y(t/n))^n$.

Finally, the cumulant generating function is

$$K(t) = \log(M(t)), \tag{1.73}$$

if $M(t)$ exists.

These functions are useful as generators of raw moments or cumulants. For example, (assuming $M(t)$ and $\mathrm{E}(Y)$ exist)

$$\varphi'(0) = M'(0) = \mathrm{E}(Y).$$

These functions are also useful in series expansions and approximations of the PDF or of other functions.

Statistical Functions of the CDF

In many models of interest, a parameter can be expressed as a functional of the probability density function or of the cumulative distribution function of a random variable in the model. The mean of a distribution, for example, can be expressed as a functional M of the CDF P:

$$M(P) = \int_{\mathrm{IR}^d} y \, \mathrm{d}P(y). \tag{1.74}$$

(Notice, following convention, we are using the same symbol M for the mean functional that we use for the moment generating function. Using the same symbol for in two ways, we have $M(P) = M'(0)$.)

A functional that defines a parameter is called a *statistical function.*

For random variables in IR, the raw moment functionals

$$M_r(P) = \int_{\mathrm{IR}} y^r \, \mathrm{d}P(y), \tag{1.75}$$

and the quantile functionals

$$\Xi_\pi(P) = P^{-1}(\pi), \tag{1.76}$$

are useful. (For a discrete distribution, the inverse of the CDF must be defined because the function does not have an inverse. There are various ways of doing this; see equation (1.142) for one way.)

An expected value, as in equation (1.70) is a functional of the CDF. Other expectations are useful descriptors of a probability distribution; for example,

$$\begin{aligned}\Sigma(P) &= \int_{\mathrm{IR}^d} \left(y - \int_{\mathrm{IR}^d} t\,\mathrm{d}P(t)\right)\left(y - \int_{\mathrm{IR}^d} t\,\mathrm{d}P(t)\right)^{\mathrm{T}} \mathrm{d}P(y) \\ &= (\mathrm{E}(Y - \mathrm{E}(Y)))\,(\mathrm{E}(Y - \mathrm{E}(Y)))^{\mathrm{T}} \\ &= \mathrm{E}(YY^{\mathrm{T}}) - (\mathrm{E}(Y))\,(\mathrm{E}(Y))^{\mathrm{T}},\end{aligned} \tag{1.77}$$

which is the variance-covariance of Y, or just the variance of Y. The off-diagonal elements in the matrix $\Sigma(P)$ are the pairwise covariances of the elements of Y.

The variance-covariance is the second-order central moment. For univariate random variables, higher-order central moments similar to equation (1.75) are useful. For vector-valued random variables, moments higher than the second are not very useful, except by considering the elements one at a time. (What would be the third central moment analogous to the second central moment defined in equation (1.77)?)

The covariances themselves may not be so useful for random variables that do not have a normal distribution. Another approach to measuring the relationship between pairs of random variables is by use of copulas. A copula is a function that relates a multivariate CDF to lower dimensional marginal CDFs. The most common applications involve bivariate CDFs and their univariate marginals, and that is the only one that we will use here. A two-dimensional copula is a function C that maps $[0,1]^2$ onto $[0,1]$ with the following properties for every $u \in [0,1]$ and every $(u_1,u_2),(v_1,v_2) \in [0,1]^2$ with $u_1 \le v_1$ and $u_2 \le v_2$:

$$C(0,u) = C(u,0) = 0, \tag{1.78}$$

$$C(1,u) = C(u,1) = u, \tag{1.79}$$

and

$$C(u_1,u_2) - C(u_1,v_2) - C(v_1,u_2) + C(v_1,v_2) \ge 0. \tag{1.80}$$

The arguments to a copula C are often taken to be CDFs, which of course take values in $[0,1]$. The usefulness of copulas derive from *Sklar's theorem*:

> Let P_{YZ} be a bivariate CDF with marginal CDFs P_Y and P_Z. Then there exists a copula C such that for every $y,z \in \mathrm{IR}$,
>
> $$P_{YZ}(y,z) = C(P_Y(y), P_Z(z)). \tag{1.81}$$

If P_Y and P_Z are continuous everywhere, then C is unique; otherwise C is unique over the support of the distributions defined by P_Y and P_Z.
Conversely, if C is a copula and P_Y and P_Z are CDFs, then the function $P_{YZ}(y, z)$ defined by equation (1.81) is a CDF with marginal CDFs $P_Y(y)$ and $P_Z(z)$.

Thus, a copula is a joint CDF of random variables with U(0, 1) marginals. The proof of this theorem is given in Nelsen (2007), among other places.

For many bivariate distributions the copula is the most useful way to relate the joint distribution to the marginals, because it provides a separate description of the individual distributions and their association with each other.

One of the most important uses of copulas is to combine two marginal distributions to form a joint distribution with known bivariate characteristics. We can build the joint distribution from a marginal and a conditional.

We begin with two U(0, 1) random variables U and V. For a given association between U and V specified by the copula $C(u, v)$, from Sklar's theorem, we can see that

$$P_{U|V}(u|v) = \frac{\partial}{\partial v} C(u, v)|_v. \tag{1.82}$$

We denote $\frac{\partial}{\partial v} C(u, v)|_v$ by $C_v(u)$.

Certain standard copulas have been used in specific applications. The copula that corresponds to a bivariate normal distribution with correlation coefficient ρ is

$$C_{N\rho}(u, v) = \int_{-\infty}^{\Phi^{-1}(u)} \int_{-\infty}^{\Phi^{-1}(v)} \phi_\rho(t_1, t_2)\, \mathrm{d}t_2 \mathrm{d}t_1, \tag{1.83}$$

where $\Phi(\cdot)$ is the standard normal CDF, and $\phi_\rho(\cdot, \cdot)$ is the bivariate normal PDF with means 0, variances 1, and correlation coefficient ρ. This copula is usually called the Gaussian copula and has been wisely used in financial applications.

A general class of copulas is called *extreme value* copulas. They have the scaling property for all $t > 0$,

$$C(u^t, v^t) = (C(u, v))^t. \tag{1.84}$$

An extreme value copula can also be written as

$$C(u, v) = \exp\left(\log(uv) A\left(\frac{\log(u)}{\log(uv)}\right)\right), \tag{1.85}$$

for some convex function $A(t)$, called the dependence function, from $[0, 1]$ to $[1/2, 1]$ with the property that $\max(t, 1-t) < A(t) < 1$ for all $t \in [0, 1]$.

A specific extreme value copula that is widely used (also in financial applications, for example) is the Gumbel copula:

$$C_{\rm G}(u,v) = \exp\left(- \left((-\log(u))^\theta + (-\log(v))^\theta \right)^{1/\theta} \right), \qquad (1.86)$$

where $\theta \geq 1$.

Another general class of copulas is called *Archimedean* copulas. These are the copulas that can be written in the form

$$C(u,v) = f^{-1}(f(u) + f(v)), \qquad (1.87)$$

where f, called the Archimedean generator, is a continuous, strictly decreasing, convex function from the unit interval $[0,1]$ to the positive reals, $\mathbb{R}_+$, such that $f(1) = 0$.

One of the widely-used Archimedean copulas is the Joe copula:

$$C_{\rm J}(u,v) = 1 - \left((1-u)^\theta + (1-v)^\theta - (1-u)^\theta(1-v)^\theta \right)^{1/\theta}, \qquad (1.88)$$

where $\theta \geq 1$.

The association determined by a copula is not the same as that determined by a correlation; that is, two pairs of random variables may have the same copula but different correlations. Kendall's τ (a correlation based on ranks; see Lehmann, 1975, for example) is fixed for a given copula. (There are minor differences in Kendall's τ based on how ties are handled; but for continuous distributions, the population value of Kendall's τ is related to a copula by $\tau = 4{\rm E}(C(U,V)) - 1$.

We will see examples of the use of copulas in random number generation in Chapter 7 (see Exercise 7.5). Copulas are often used in the Monte Carlo methods of computational finance, especially in the estimation of value at risk; see, for example, Rank and Siegl (2002).

Families of Probability Distributions

It is useful to group similar probability distributions into families.

A family of distributions with probability measures P_θ for $\theta \in \Theta$ is called a *parametric family* if $\Theta \subset \mathbb{R}^d$ for some fixed positive integer d and θ fully determines the measure. In that case, we call θ the *parameter* and we call Θ the parameter space. A parameter can often easily be defined as a functional of the CDF.

A family that cannot be indexed in this way is called a nonparametric family. In nonparametric methods, our analysis usually results in some general description of the distribution, rather than in a specification of the distribution.

The type of a family of distributions depends on the parameters that characterize the distribution. A "parameter" is a real number that can take on more than one value within a parameter space. If the parameter space contains only one point, the corresponding quantity characterizing the distribution is not a parameter.

Many common families are multi-parameter, and specialized subfamilies are defined by special values of one or more parameters. For example, in a very widely-used notation, the family of gamma distributions is characterized by three parameters, γ, called the "location"; β, called the "scale"; and α, called the "shape". Its PDF is $(\Gamma(\alpha))^{-1}\beta^{-\alpha}(x-\gamma)^{\alpha-1}\mathrm{e}^{-(x-\gamma)/\beta}\mathrm{I}_{[\gamma,\infty)}(x)$. This is sometimes called the "three-parameter gamma", because often γ is taken to be a fixed value, usually 0.

Specific values of the parameters determine special subfamilies of distributions. For example, in the three-parameter gamma, if α is fixed at 1, the resulting distribution is the two-parameter exponential, and if, additionally, γ is fixed at 0, the resulting distribution is an exponential distribution.

Several of the standard parametric families are shown in Tables B.1 and B.2 beginning on page 660. The most important of these families is the normal or Gaussian family. We often denote its CDF by Φ and its PDF by ϕ. The form of the arguments indicates various members of the family; for example, $\phi(x|\mu,\sigma^2)$ is the PDF of a univariate normal random variable with mean μ and variance σ^2, and $\phi(x)$ is the PDF of a standard univariate normal random variable with mean 0 and variance 1.

An important family of distributions is the *exponential class*, which includes a number of parametric families. The salient characteristic of a family of distributions in the *exponential class* is the way in which the parameter and the value of the random variable can be separated in the density function. Another important characteristic of the exponential family is that the support of a distribution in this family does not depend on any "unknown" parameter.

A member of a family of distributions in the exponential class is one with density that can be written in the form

$$p_\theta(y) = \exp\left((\eta(\theta))^{\mathrm{T}}T(y) - \xi(\theta)\right) h(y), \tag{1.89}$$

where $\theta \in \Theta$.

The exponential class is also called the "exponential family", but do not confuse an "exponential class" in the sense above with the "exponential family", which are distributions with density $\lambda\mathrm{e}^{-\lambda x}\,\mathrm{I}_{(0,\infty)}(y)$.

Notice that all members of a family of distributions in the exponential class have the same support. Any restrictions on the range may depend on y through $h(y)$, but they cannot depend on the parameter.

The form of the expression depends on the *parametrization*; that is, the particular choice of the form of the parameters.

As noted above, if a parameter is assigned a fixed value, then it ceases to be a parameter. This is important, because what are considered to be parameters determine the class of a particular family. For example, the three-parameter gamma is not a member of the exponential class; however, the standard two-parameter gamma, with γ fixed at 0, is a member of the exponential class.

In addition to the standard parametric families shown in Tables B.1 and B.2, there are some general families of probability distributions that are very useful in computational statistics. These families, which include the Pearson, the Johnson, Tukey's generalized lambda, and the Burr, cover wide ranges of shapes and have a variety of interesting properties that are controlled by a few parameters. Some are designed to be particularly simple to simulate. We discuss these families of distributions in Section 14.2.

Mixture Distributions

In applications it is often the case that a single distribution does not model the observed data adequately. In many such cases, however, a mixture of two or more standard distributions from the same or different parametric families does provide a good model.

If we have m distributions with PDFs p_j, we can form a new PDF as

$$p_M(y) = \sum_{j=1}^{m} \omega_j p_j(y \mid \theta_j), \tag{1.90}$$

where $\omega_j \geq 0$ and $\sum_{j=1}^{m} \omega_j = 1$. If all of the PDFs are from the same parametric family the individual densities would be $p(y \mid \theta_j)$.

If all of the densities have the same form, we can easily extend the idea of a mixture distribution to allow the parameter to change continuously, so instead of $\omega_j p(y \mid \theta_j)$, we begin with $\omega(\theta) p(y \mid \theta)$. If we have, analogously to the properties above, $\omega(\theta) \geq 0$ over some relevant range of θ, say Θ, and $\int_\Theta \omega(\theta)\,\mathrm{d}\theta = 1$, then $\omega(\theta) p(y \mid \theta)$ is the joint PDF of two random variables, and the expression analogous to (1.90),

$$p_M(y) = \int_\Theta \omega(\theta) p(y \mid \theta)\,\mathrm{d}\theta, \tag{1.91}$$

is a marginal PDF. This type of mixture distribution is central to Bayesian analysis, as we see in equations (1.104) and (1.105).

A linear combination such as equation (1.90) provides great flexibility, even if the individual densities $p_j(y \mid \theta_j)$ are from a restricted class. For example, even if the individual densities are all normals, which are all symmetric, a skewed distribution can be formed by a proper choice of the ω_j and $\theta_j = (\mu_j, \sigma_j^2)$.

We must be clear that this is a mixture of distributions of random variables, $Y_1, \ldots, Y_m$, not a linear combination of the random variables themselves. Some linear properties carry over the same for mixtures as for linear combinations. For example, if $Y_j \sim p_j$ and $Y \sim p$ in equation (1.90), and

$$Z = \sum_{j=1}^{m} \omega_j Y_j,$$

then $\mathrm{E}(Z) = \mathrm{E}(Y) = \sum_{j=1}^{m} \omega_j \mathrm{E}(Y_j)$, assuming the expectations exist, but the distribution of Z is not the same as that of Y.

The important linear transforms defined above, the moment generating function and the characteristic function, carry over as simple linear combinations. The cumulant generating function can then be evaluated from the moment generating function using its definition (1.73). This is one of the reasons that these transforms are so useful.

The PDF Decomposition

Probability distributions have useful applications even in situations where there is no obvious data-generating process.

If f is a function such that $\int_D f(x)\,\mathrm{d}x < \infty$, then for some function $g(x)$, we can write

$$f(x) = g(x)p_X(x) \tag{1.92}$$

where p_X is the probability density function of a random variable X with support over the relevant domain of f. The decomposition of the function f in this way is called *probability density function decomposition* or *PDF decomposition*.

The PDF decomposition allows us to use methods of statistical estimation for approximations involving f. We will see examples of the PDF decomposition in various applications, such as Monte Carlo methods (pages 192 and 418), function estimation (Chapters 10 and 15), and projection pursuit (Chapter 16).

1.4 Statistical Inference

For statistical inference, we generally assume that we have a *sample* of observations $Y_1, \ldots, Y_n$ on a random variable Y. A *random sample*, which we will usually just call a "sample", is a set of independent and identically distributed (i.i.d.) random variables. We will often use Y to denote a random sample on the random variable Y. (This may sound confusing, but it is always clear from the context.) A *statistic* is any function of Y that does not involve any unobservable values. We may denote the actual observed values as $y_1, \ldots, y_n$ since they are not random variables.

We assume that the sample arose from some data-generating process or, equivalently, as a random sample from a probability distribution. Our objective is to use the sample to make inferences about the process. We may assume that the specific process, call it P_θ, is a member of some family of probability distributions $\mathcal{P}$. For statistical inference, we fully specify the family $\mathcal{P}$ (it can be a very large family), but we assume some aspects of P_θ are unknown. (If the distribution P_θ that yielded the sample is fully known, while there may be some interesting questions about probability, there are no interesting statistical questions.) Our objective in statistical inference is to determine a specific $P_\theta \in \mathcal{P}$, or some subfamily $\mathcal{P}_\theta \subset \mathcal{P}$, that could likely have generated the sample.

The distribution may also depend on other observable variables. In general, we assume we have observations $Y_1, \ldots, Y_n$ on Y, together with associated observations on any related variable X or x. We denote the observed values as $(y_1, x_1), \ldots, (y_n, x_n)$, or just as $y_1, \ldots, y_n$. In this context, a *statistic* is any function that does not involve any unobserved values.

In statistical inference, we distinguish observable random variables and "parameters", but we are not always careful in referring to parameters. We think of two kinds of parameters: "known" and "unknown". A statistic is a function of observable random variables that does not involve any unknown parameters. The "θ" in the expression P_θ above may be a parameter, perhaps a vector, or it may just be some index that identifies the distribution within a set of possible distributions.

Types of Statistical Inference

There are three different types of inference related to the problem of determining the specific $P_\theta \in \mathcal{P}$: *point estimation*, *hypothesis tests*, and *confidence sets*. In point estimation, the *estimand* is often some function of the basic parameter θ. We often denote the estimand in general as $g(\theta)$. Hypothesis tests and confidence sets are associated with probability statements that depend on P_θ. We will briefly discuss them in Section 1.5.

In parametric settings, each type of inference concerns a *parameter*, θ, that is assumed to be in some *parameter space*, $\Theta \subset \mathrm{IR}^k$. If Θ is not a closed set, it is more convenient to consider the closure of Θ, denoted by $\overline{\Theta}$, because sometimes a good estimator may actually be on the boundary of the open set Θ. (If Θ is closed, $\overline{\Theta}$ is the same set, so we can always just consider $\overline{\Theta}$.)

A related problem in estimation is *prediction*, in which the objective is to estimate the expected value of a random variable, given some information about past realizations of the random variable and possibly, covariates associated with those realizations.

Performance of Statistical Methods for Inference

There are many properties of a method of statistical inference that may be relevant. In the case of point estimation a function of of the parameter θ, for example, we may use an estimator $T(Y)$ based on the sample Y. Relevant properties of $T(Y)$ include its bias, its variance, and its mean squared error. The *bias* of $T(Y)$ for $g(\theta)$ is

$$\mathrm{Bias}(T, g(\theta)) = \mathrm{E}(T(Y)) - g(\theta). \tag{1.93}$$

When it is clear what is being estimated, we often use the simpler notation $\mathrm{Bias}(T)$.

If this quantity is 0, then $T(Y)$ is said to be *unbiased* for $g(\theta)$. The *mean squared error (MSE)* of $T(Y)$ is

$$\mathrm{MSE}(T, g(\theta)) = \mathrm{E}((T(Y) - g(\theta))^2). \tag{1.94}$$

Again, if it is clear what is being estimated, we often use the simpler notation MSE(T). Note that

$$\begin{aligned}
\mathrm{E}((T(Y) - g(\theta))^2) &= \mathrm{E}((T(Y) - \mathrm{E}(T(Y)) + \mathrm{E}(T(Y)) - g(\theta))^2) \\
&= (\mathrm{E}(T(Y) - g(\theta)))^2 + \mathrm{E}((T(Y) - \mathrm{E}(T(Y))^2) \\
&= (\mathrm{Bias}(T))^2 + \mathrm{V}(T); \qquad (1.95)
\end{aligned}$$

that is, the MSE is the square of the bias plus the variance.

We may also be interested in other distributional properties of an estimator, for example, its median. When $T(Y)$ is used as an estimator for $g(\theta)$, if

$$\mathrm{Med}(T(Y)) = g(\theta), \tag{1.96}$$

where Med(X) is the median of the random variable X, we say that $T(Y)$ is *median-unbiased* for $g(\theta)$. A useful fact about median-unbiasedness is that if T is median-unbiased for θ, and h is a monotone increasing function, then $h(T)$ is median-unbiased for $h(\theta)$. This, of course, does not hold in general for bias defined in terms of expected values. (If the CDF of the random variable is not strictly increasing, or if in the last statement h is not strictly increasing, we may have to be more precise about the definition of the median; see equation (1.142) on page 62.)

Important research questions in statistics often involve identifying which statistical methods perform the "best", in the sense of being unbiased and having the smallest variance, of having the smallest MSE, or of having other heuristically appealing properties. Often the underlying probability distributions are complicated, and statistical methods are difficult to compare mathematically. In such cases, Monte Carlo methods such as discussed in Appendix A may be used. In various exercises in this book, such as Exercise 1.18, you are asked to use Monte Carlo simulation to compare statistical procedures.

There are several important properties of statistics that determine the usefulness of those statistics in statistical inference. One of the most useful properties of a statistic is sufficiency. (The previous sentences use the term "statistic" to denote an observation or a function of observations that does not involve an unknown quantity. Unfortunately, the plural "statistics" can mean different things.)

Large Sample Properties

We are often interested in the performance of statistical procedures as the sample size becomes unboundedly large. When we consider the large sample properties, we often denote the statistics with a subscript representing the sample size, for example, T_n.

If T_n is a statistic from a sample of size n, and if $\lim_{n\to\infty} \mathrm{E}(T_n) = \theta$, then T_n is said to be unbiased in the limit for θ. If $\lim_{n\to\infty} \mathrm{E}((T_n - \theta)^2) = 0$, then T_n is said to be *consistent in mean-squared error*. There are other kinds of statistical consistency, but consistency in mean-squared error is the most commonly used.

If T_n is a statistic from a sample of size n, and if $\mathrm{E}(T_n) = \theta + \mathrm{O}(n^{-1/2})$, then T_n is said to be *first-order accurate* for θ; if $\mathrm{E}(T_n) = \theta + \mathrm{O}(n^{-1})$, it is *second-order accurate*. (See page 670 for the definition of $\mathrm{O}(\cdot)$. Convergence of T_n or of $\mathrm{E}(T_n)$ can also be expressed as a stochastic convergence of T_n, in which case we use the notation $\mathrm{O_P}(\cdot)$.)

The order of the mean squared error is an important characteristic of an estimator. For good estimators of location, the order of the mean squared error is typically $\mathrm{O}(n^{-1})$. Good estimators of probability densities, however, typically have mean squared errors of at least order $\mathrm{O}(n^{-4/5})$ (see Chapter 15).

Sufficiency

Let Y be a sample from a population $P \in \mathcal{P}$. A statistic $T(Y)$ is *sufficient* for $P \in \mathcal{P}$ if and only if the conditional distribution of Y given T does not depend on P. In similar fashion, we define sufficiency for a parameter or for an element in a vector of parameters. Sufficiency depends on $\mathcal{P}$, the family of distributions. If a statistic is sufficient for $\mathcal{P}$, it may not be sufficient for a larger family, $\mathcal{P}_1$, where $\mathcal{P} \subset \mathcal{P}_1$.

In general terms, sufficiency implies the conditional independence from the parameter of the distribution of any other function of the random variable, given the sufficient statistic.

The reason that sufficiency is such an important property is that it may allow reduction of data without sacrifice of information.

Another, more specific property of sufficiency is that the statistical properties of a given method of inference that is based on a statistic that is not sufficient can often be improved by conditioning the statistic on a sufficient statistic, if one is available. A well-known instance of this fact is stated in the Rao-Blackwell theorem, one version of which states:

> Let Y be a random sample from a distribution $P_\theta \in \mathcal{P}$, and let $S(Y)$ be sufficient for $\mathcal{P}$ and have finite variance. Let $T(Y)$ be an unbiased estimator for $g(\theta)$ with finite variance. Let
>
> $$\widetilde{T} = \mathrm{E}(T(Y)|S(Y)). \tag{1.97}$$
>
> Then $\widetilde{T}$ is unbiased for $g(\theta)$ and
>
> $$\mathrm{V}(\widetilde{T}) \leq \mathrm{V}(T). \tag{1.98}$$

There are several other ways of stating essentially equivalent results about a statistic that is conditioned on a sufficient statistic. A more general statement in a decision-theoretic framework for estimation is that if the loss function is convex, a statistic conditioned on a sufficient statistic has no greater risk than the statistic unconditioned. See Lehmann and Casella (1998) for a statement in terms of convex loss functions and a proof.

Five Approaches to Statistical Inference

If we assume that we have a random sample of observations $Y_1, \ldots, Y_n$ on a random variable Y from some distribution P_θ, which is a member of some family of probability distributions $\mathcal{P}$, our objective in statistical inference is to determine a specific $P_\theta \in \mathcal{P}$, or some subfamily $\mathcal{P}_\theta \subset \mathcal{P}$, that could likely have generated the sample.

Five approaches to statistical inference are

- use of the empirical cumulative distribution function (ECDF)
- definition and use of a loss function.
- use of a likelihood function
- fitting expected values
- fitting a probability distribution

These approaches are not mutually exclusive.

The computational issues in statistical inference are varied. In most approaches an optimization problem is central, and many of the optimization problems cannot be solved analytically. Some approaches, such as those using the ECDF, lead to computationally intensive methods that use simulated datasets. The use of a loss function may yield very complicated integrals representing average loss. These cannot be evaluated analytically, and so are often evaluated using Monte Carlo methods.

We will discuss use of the ECDF more fully in Section 1.7, and in the rest of this section, we will briefly discuss the other approaches listed above. In Exercise 1.21 you are asked to obtain various estimates based on these approaches. You should pay particular attention to the specific model or assumptions that underlie each approach.

A Decision-Theoretic Approach; Loss and Risk

In the decision-theoretic approach to statistical inference, we call the inference a *decision* or an *action*, and we identify a *cost* or *loss* that depends on the decision and the true (but unknown) state of nature modeled by $P \in \mathcal{P}$.

Obviously, we try to take an action that minimizes the expected loss.

We call the set of allowable actions or decisions the *action space* or decision space, and we denote it as $\mathcal{A}$. We base the inference on the random variable X; hence, the decision is a mapping from $\mathcal{X}$, the range of X, to $\mathcal{A}$.

If we observe data X, we take the action $T(X) = a \in \mathcal{A}$. The statistical procedure that leads to $T(\cdot)$ is the decision rule.

Loss Function

A *loss function*, L, is a mapping from $\mathcal{P} \times \mathcal{A}$ to $[0, \infty)$. The value of the function at a given distribution $P \in \mathcal{P}$ for the action a is $L(P, a)$.

If $\mathcal{P}$ is indexed by θ, we can write the value of the function at a given value θ for the action a as $L(\theta, a)$.

Depending on the parameter space Θ, the action space $\mathcal{A}$, and our objectives, the loss function often is a function only of $g(\theta) - a$; that is, we may have $L(\theta, a) = L(g(\theta) - a)$.

The loss function generally should be nondecreasing in $|g(\theta) - a|$. A loss function that is convex has nice mathematical properties. A particularly nice loss function, which is strictly convex, is the "squared-error loss":

$$L_2(\theta, a) = (g(\theta) - a)^2. \tag{1.99}$$

Any strictly convex loss function over an unbounded interval is unbounded. It is not always realistic to use an unbounded loss function. A common bounded loss function is the 0-1 loss function, which may be

$$\begin{aligned} L_{0-1}(\theta, a) &= 0 \quad \text{if } |g(\theta) - a| \le \alpha(n) \\ L_{0-1}(\theta, a) &= 1 \quad \text{otherwise.} \end{aligned}$$

Risk Function

To choose an action rule T so as to minimize the loss function is not a well-defined problem. We can make the problem somewhat more precise by considering the expected loss based on the action $T(X)$, which we define to be the *risk*:

$$R(P, T) = \mathrm{E}\big(L(P, T(X))\big). \tag{1.100}$$

The problem may still not be well defined. For example, to estimate $g(\theta)$ so as to minimize the risk function is still not a well-defined problem. We can make the problem precise either by imposing additional restrictions on the estimator or by specifying in what manner we want to minimize the risk.

Optimal Decision Rules

We compare decision rules based on their risk with respect to a given loss function and a given family of distributions. If a decision rule T_* has the property

$$R(P, T_*) \le R(P, T) \quad \forall\, P \in \mathcal{P}, \tag{1.101}$$

for all T, then T_* is called an *optimal* decision rule.

Approaches to Minimizing the Risk

We use the principle of minimum risk in the following restricted ways. In all cases, the approaches depend, among other things, on a given loss function.

- We may first place a restriction on the estimator and then minimize risk subject to that restriction. For example, we may

  - require unbiasedness
  - require equivariance.
- We may minimize some global property of the risk ("global" over the values of θ). For example, we may
  - minimize maximum risk
  - minimize average risk.

These various ways of minimizing the risk lead to some familiar classical procedures of statistical inference, such as UMVUE (uniformly minimum variance unbiased estimation), UMPT (uniformly most powerful test), and minimax rules.

To minimize an average risk over the parameter space requires some definition of an averaging function. If we choose the averaging function as $\Lambda(\theta)$, with $\int_\Theta \mathrm{d}\Lambda(\theta) = 1$, then the average risk is $\int_\Theta R(\theta, T)\mathrm{d}\Lambda(\theta)$.

The decision that minimizes the average risk with respect to $\Lambda(\theta)$ is called the Bayes rule, and the minimum risk, $\int_\Theta R(\theta, T_\Lambda)\,\mathrm{d}\Lambda(\theta)$, is called the Bayes risk.

Bayesian Inference

The averaging function $\Lambda(\theta)$ allows various interpretations, and it allows the flexibility of incorporating prior knowledge or beliefs. The regions over which $\Lambda(\theta)$ is large will be given more weight; therefore the estimator will be pulled toward those regions.

In formal Bayes procedures, we call the averaging function the prior probability density for the parameter, which we consider to be a random variable in its own right. Thus, we think of the probability distribution of the *observable* random variable Y as a *conditional* distribution, given the *unobservable* random parameter variable, $\Theta = \theta$. We then form the joint distribution of θ and Y, and then the conditional distribution of θ given Y, called the posterior distribution. We can summarize the approach in a Bayesian statistical analysis as beginning with these steps:

1. identify the conditional distribution of the observable random variable; assuming the density exists, call it
$$p_{Y|\Theta}(y|\theta) \tag{1.102}$$
2. identify the prior (marginal) distribution of the parameter; assuming the density exists, call it
$$p_\Theta(\theta) \tag{1.103}$$
3. identify the joint distribution; if densities exist, it is
$$p_{Y,\Theta}(y, \theta) = p_{Y|\Theta}(y|\theta)p_\Theta(\theta) \tag{1.104}$$

4. determine the marginal distribution of the observable; if densities exist, it is

$$p_Y(y) = \int_\Theta p_{Y,\Theta}(y, \theta)\mathrm{d}\theta \tag{1.105}$$

5. determine the posterior conditional distribution of the parameter given the observable random variable; this is the posterior; if densities exist, it is

$$p_{\Theta|y}(\theta|x) = p_{Y,\Theta}(y, \theta)/p_Y(y). \tag{1.106}$$

The posterior conditional distribution is then the basis for whatever decisions are to be made.

The Bayes rule is determined by minimizing the risk, where the expectation is taken with respect to the posterior distribution. This expectation is often a rather complicated integral, and Monte Carlo methods, specifically, Markov chain Monte Carlo (MCMC) techniques, are generally used to evaluate the rule or to study the posterior distribution. We will discuss these techniques in Chapter 11 and their applications in Chapter 17.

Likelihood

Given a sample $Y_1, \ldots, Y_n$ from distributions with probability densities $p_i(y)$, where all PDFs are defined with respect to a common σ-finite measure, the *likelihood function* is

$$L_n(p_i\,;\,Y) = \prod_{i=1}^n p_i(Y_i). \tag{1.107}$$

(Any nonnegative function proportional to $L_n(p_i\,;\,Y)$ is a likelihood function, but it is common to speak of $L_n(p_i\,;\,Y)$ as "the" likelihood function.) We can view the sample either as a set of random variables or as a set of constants, the realized values of the random variables. Thinking of the likelihood as a function of realized values, we usually use lower-case letters.

The *log-likelihood function* is the log of the likelihood:

$$l_{L_n}(p_i\,;\,y) = \log L_n(p_i \,|\, y_i), \tag{1.108}$$

It is a sum rather than a product.

The n subscript serves to remind us of the sample size, and this is often very important in use of the likelihood or log-likelihood function particularly because of their asymptotic properties. We often drop the n subscript, however. Also, we often drop the L subscript on the log-likelihood. (I should also mention that some authors use the upper and lower cases in the opposite way from that given above.)

In many cases of interest, the sample is from a single parametric family. If the PDF is $p(y\,;\,\theta)$ then the likelihood and log-likelihood functions are written as

$$L(\theta\,;\,y) = \prod_{i=1}^{n} p(y_i\,;\,\theta), \tag{1.109}$$

and

$$l(\theta\,;\,y) = \log L(\theta\,;\,y). \tag{1.110}$$

We sometimes write the expression for the likelihood without the observations: $L(\theta)$ or $l(\theta)$.

The Parameter Is the Variable

Note that the likelihood is a function of θ for a given y, while the PDF is a function of y for a given θ.

While if we think of θ as a fixed, but unknown, value, it does not make sense to think of a function of that particular value, and if we have an expression in terms of that value, it does not make sense to perform operations such as differentiation with respect to that quantity. We should think of the likelihood as a function of some dummy variable t, and write $L(t\,;\,y)$ or $l(t\,;\,y)$.

The likelihood function arises from a probability density, but it is not a probability density function. It does not in any way relate to a "probability" associated with the parameters or the model.

Although non-statisticians will often refer to the "likelihood of an observation", in statistics, we use the term "likelihood" to refer to a model or a distribution *given observations*.

In a multiparameter case, we may be interested in only some of the parameters. There are two ways of approaching this, use of a profile likelihood or of a conditional likelihood.

Let $\theta = (\theta_1, \theta_2)$. If θ_2 is fixed, the likelihood $L(\theta_1\,;\,\theta_2, y)$ is called a *profile likelihood* or *concentrated likelihood* of θ_1 for given θ_2 and y.

If the PDFs can be factored so that one factor includes θ_2 and some function of the sample, $S(y)$, and the other factor, given $S(y)$, is free of θ_2, then this factorization can be carried into the likelihood. Such a likelihood is called a *conditional likelihood* of θ_1 given $S(y)$.

Maximum Likelihood Estimation

The *maximum likelihood estimate* (MLE) of θ, $\widehat{\theta}$, is defined as

$$\widehat{\theta} = \arg\max_{t \in \overline{\Theta}} L(t\,;\,y), \tag{1.111}$$

where $\overline{\Theta}$ is the closure of the parameter space.

The MLE in general is not unbiased for its estimand. A simple example is the MLE of the variance in a normal distribution with unknown mean. If $Y_1, \ldots, Y_n \sim i.i.d.\mathrm{N}(\mu, \sigma^2)$, it is easy to see from the definition (1.111) that the MLE of σ^2, that is, of the variance of Y is

$$\widehat{\mathrm{V}(Y)} = \frac{1}{n}\sum_{i=1}^{n}(Y_i - \overline{Y})^2. \tag{1.112}$$

Thus the MLE is $(n-1)S^2/n$, where S^2 is the usual sample variance:

$$S^2 = \frac{1}{n-1}\sum_{i=1}^{n}(Y_i - \overline{Y})^2. \tag{1.113}$$

Notice that the MLE of the variance depends on the distribution. (See Exercise 1.16d.)

The MLE may have smaller MSE than an unbiased estimator, and, in fact, that is the case for the MLE of σ^2 in the case of a normal distribution with unknown mean compared with the estimator S^2 of σ^2.

We will discuss statistical properties of maximum likelihood estimation beginning on page 70, and some of the computational issues of MLE in Chapter 6.

Score Function

In statistical inference, we often use the information about how the likelihood or log-likelihood would vary if θ were to change. (As we have indicated, "θ" sometimes plays multiple roles. I like to think of it as a fixed but unknown value and use "t" or some other symbol for variables that can take on different values. Statisticians, however, often use the same symbol to represent something that might change.) For a likelihood function (and hence, a log-likelihood function) that is differentiable with respect to the parameter, a function that represents this change and plays an important role in statistical inference is the *score function*:

$$s_n(\theta\,;\,y) = \frac{\partial l(\theta\,;\,y)}{\partial\theta}. \tag{1.114}$$

Likelihood Equation

In statistical estimation, if there is a point at which the likelihood attains its maximum (which is, of course, the same point at which the log-likelihood attains its maximum) that point obviously is of interest; it is the MLE in equation (1.111).

If the likelihood is differentiable with respect to the parameter, the roots of the score function are of interest whether or not they correspond to MLEs. The score function equated to zero,

$$\frac{\partial l(\theta\,;\,y)}{\partial\theta} = 0, \tag{1.115}$$

is called the *likelihood equation*. The derivative of the likelihood equated to zero, $\partial L(\theta\,;\,y)/\partial\theta = 0$, is also called the likelihood equation.

Equation (1.115) is an *estimating equation*; that is, its solution, if it exists, is an estimator. *Note that it is not necessarily an MLE; it is a root of the likelihood equation, or RLE.*

It is often useful to define an estimator as the solution of some estimating equation. We will see other examples of estimating equations in subsequent sections.

Likelihood Ratio

When we consider two different possible distributions for a sample y, we have two different likelihoods, say L_0 and L_1. (Note the potential problems in interpreting the subscripts; here the subscripts refer to the two different distributions. For example L_0 may refer to $L(\theta_0 \mid y)$ in a notation consistent with that used above.) In this case, it may be of interest to compare the two likelihoods in order to make an inference about the two possible distributions. A simple comparison, of course, is the ratio, and indeed

$$\frac{L(\theta_0\,;\,y)}{L(\theta_1\,;\,y)}, \tag{1.116}$$

or L_0/L_1 in the simpler notation, is called the *likelihood ratio* with respect to the two possible distributions. Although in most contexts we consider the likelihood to be a function of the parameter for given, fixed values of the observations, it may also be useful to consider the likelihood ratio to be a function of y.

The most important use of the likelihood ratio is as the basis for statistical tests that are constructed following the Neyman-Pearson lemma for a simple null hypothesis versus a simple alternative hypothesis (see page 53). If the likelihood is monotone in θ_1, we can extend the simple hypotheses of the Neyman-Pearson lemma to certain composite hypotheses. Thus, a *monotone likelihood ratio* is an important property of a distribution.

The likelihood ratio, or the log of the likelihood ratio, plays an important role in statistical inference. Given the data y, the log of the likelihood ratio is called the *support of the hypothesis* that the data came from the population that would yield the likelihood L_0 versus the hypothesis that the data came from the population that would yield the likelihood L_1. The support of the hypothesis clearly depends on both L_0 and L_1, and it ranges over $\mathbb{R}$. The support is also called the *experimental support.*

Likelihood Principle

The *likelihood principle* in statistical inference asserts that all of the information which the data provide concerning the relative merits of two hypotheses (two possible distributions that give rise to the data) is contained in the likelihood ratio of those hypotheses and the data. An alternative statement of the likelihood principle is that, if for x and y,

$$\frac{L(\theta\,;\,x)}{L(\theta\,;\,y)} = c(x, y) \quad \forall\, \theta,$$

where $c(x, y)$ is constant for given x and y, then any inference about θ based on x should be in agreement with any inference about θ based on y.

Fitting Expected Values

Given a random sample $Y_1, \ldots, Y_n$ from distributions with probability densities $p_{Y_i}(y_i; \theta)$, where all PDFs are defined with respect to a common σ-finite measure, if we have that $\mathrm{E}(Y_i) = g_i(\theta)$, then a reasonable approach to estimation of θ may be to choose a value $\widehat{\theta}$ that makes the differences $\mathrm{E}(Y_i) - g_i(\theta)$ close to zero.

We must define the sense in which the differences are close to zero. A simple way to do this is to define a nonnegative scalar-valued function of scalars, $\rho(u, v)$, that is increasing in the absolute difference of its arguments. We then define

$$S(\theta, y) = \sum_{i=1}^{n} \rho(y_i, \theta), \tag{1.117}$$

and a reasonable estimator is

$$\widehat{\theta} = \arg\min_{t \in \overline{\Theta}} S(t, y). \tag{1.118}$$

One simple choice for the function is $\rho(u, v) = (u - v)^2$. In this case, the estimator is called the *least squares* estimator. Another choice, which is more difficult mathematically is $\rho(u, v) = |u - v|$. In this case, the estimator is called the *least absolute values* estimator.

Compare the minimum residual estimator in equation (1.118) with the maximum likelihood estimate of θ, defined in equation (1.111).

If the Y_i are i.i.d., then all $g_i(\theta)$ are the same, say $g(\theta)$.

In common applications, we have *covariates*, $Z_1, \ldots, Z_n$, and the $\mathrm{E}(Y_i)$ have a constant form that depends on the covariate: $\mathrm{E}(Y_i) = g(Z_i, \theta)$.

As with solving the maximization of the likelihood, the solution to the minimization problem (1.118) may be obtained by solving

$$\frac{\partial S(\theta\,;\,y)}{\partial \theta} = 0. \tag{1.119}$$

Like equation (1.115), equation (1.119) is an *estimating equation*; that is, its solution, if it exists, is an estimator. There may be various complications, of course; for example, there may be multiple roots of (1.119).

Fitting Probability Distributions

In an approach to statistical inference based on information theory, the true but unknown distribution is compared with information in the sample using a *divergence measure* between the population distribution and the sample distribution. The divergence measure may also be used to compare two hypothesized distributions. A general type of divergence measure is called *ϕ-divergence* and for the PDFs p and q of random variables with a common support D is defined as

$$\int_D \phi\left(\frac{p(y)}{q(y)}\right) \mathrm{d}y, \tag{1.120}$$

if the integral exists. The ϕ-divergence is also called the f-divergence.

The ϕ-divergence is in general not a metric because it is not symmetric. One function is taken as the base from which the other function is measured.

A specific instance of ϕ-divergence that is widely used is the *Kullback-Leibler measure,*

$$\int_{\mathrm{IR}} p(y) \log\left(\frac{p(y)}{q(y)}\right) \mathrm{d}y. \tag{1.121}$$

Functions of Parameters and Functions of Estimators

Suppose that instead of estimating the parameter θ, we wish to estimate $g(\theta)$, where $g(\cdot)$ is some function. If the function $g(\cdot)$ is monotonic or has certain other regularity properties, it may be the case that the estimator that results from the minimum residuals principle or from the maximum likelihood principle is invariant; that is, the estimator of $g(\theta)$ is merely the function $g(\cdot)$ evaluated at the solution to the optimization problem for estimating θ. The statistical properties of a T for estimating θ, however, do not necessarily carry over to $g(T)$ for estimating $g(\theta)$.

As an example of why a function of an unbiased estimator may not be unbiased, consider a simple case in which T and $g(T)$ are scalars, and the function g is convex (see page 21).

Now consider $\mathrm{E}(g(T))$ and $g(\mathrm{E}(T))$. If g is a convex function, then *Jensen's inequality* states that

$$\mathrm{E}(g(T)) \leq g(\mathrm{E}(T)), \tag{1.122}$$

This is easy to see by using the definition of convexity and, in particular, equation (1.39). We have for some b and any x and t,

$$b(x-t) + g(t) \leq g(x).$$

Now, given this, let $x = \mathrm{E}(T)$ and take expectations of both sides of the inequality.

The implication of this is that even though T is unbiased for θ, $g(T)$ may not be unbiased for $g(\theta)$. Jensen's inequality is illustrated in Figure 1.4.

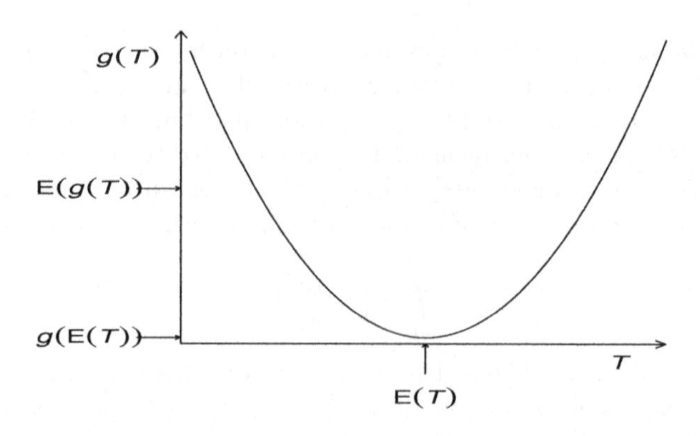

Fig. 1.4. Jensen's Inequality

If the function is strictly convex, then Jensen's inequality (1.122) is also strict, and so if T is unbiased for θ then $g(T)$ is biased for $g(\theta)$.

An opposite inequality obviously also applies to a concave function, in which case the bias is positive.

It is often possible to adjust $g(T)$ to be unbiased for $g(\theta)$; and properties of T, such as sufficiency for θ, may carry over to the adjusted $g(T)$. Some of the applications of the jackknife and the bootstrap that we discuss later are in making adjustments to estimators of $g(\theta)$ that are based on estimators of θ.

The variance of $g(T)$ can often be approximated in terms of the variance of T. We will consider this for the more general case in which T and θ are m-vectors, and T is mean-square consistent for θ. Assume $g(T)$ is a k-vector. In a simple but common case, we may know that T in a sample of size n has an approximate normal distribution with mean θ and some variance-covariance matrix, say $\mathrm{V}(T)$, of order n^{-1}, and g is a smooth function (that is, it can be approximated by a truncated Taylor series about θ):

$$g(T) \approx g(\theta) + \mathrm{J}_g(\theta)(T - \theta) + \frac{1}{2}(T - \theta)^{\mathrm{T}}\mathrm{H}_g(\theta)(T - \theta),$$

where J_g and H_g are respectively the Jacobian and the Hessian of g. Because the variance of T is $\mathrm{O}(n^{-1})$, the remaining terms in the expansion go to zero in probability at the rate of at least n^{-1}.

This yields the approximations

$$\mathrm{E}(g(T)) \approx g(\theta) \tag{1.123}$$

and

$$\mathrm{V}(g(T)) \approx \mathrm{J}_g(\theta)\,\mathrm{V}(T)\left(\mathrm{J}_g(\theta)\right)^{\mathrm{T}}. \tag{1.124}$$

This method of approximation of the variance is called the *delta method.*

A common form of a simple estimator that may be difficult to analyze and may have unexpected properties is a ratio of two statistics,

$$R = \frac{T}{S},$$

where S is a scalar. An example is a studentized statistic, in which T is a sample mean and S is a function of squared deviations. If the underlying distribution is normal, a statistic of this form may have a well-known and tractable distribution. In particular, if T is a mean and S is a function of an independent chi-squared random variable, the distribution is that of a Student's t. If the underlying distribution has heavy tails, however, the distribution of R may have unexpectedly light tails. An asymmetric underlying distribution may also cause the distribution of R to be very different from a Student's t distribution. If the underlying distribution is positively skewed, the distribution of R may be negatively skewed (see Exercise 1.14).

Types of Statistical Inference

We began this section with an outline of the types of statistical inference that include point estimation, confidence sets, and tests of hypotheses. (Notice this section has the same title as the section beginning on page 38.) There is another interesting categorization of statistical inference.

When the exact distribution of a statistic is known (based, of course, on an assumption of a given underlying distribution of a random sample), use of the statistic for inferences about the underlying distribution is called exact inference.

Often the exact distribution of a statistic is not known, or is too complicated for practical use. In that case, we may resort to approximate inference. It is important to note how the terms "exact" and "approximate" are used here. The terms are used in the context of *assumptions.* We do not address reality.

There are basically three types of approximate inference.

One type occurs when a simple distribution is very similar to another distribution. For example, the simpler Kumaraswamy distribution, with PDF $\alpha\beta x^{\alpha-1}(1-x)^{\beta-1}$ over $[0,1]$, may be used as an approximation to the beta distribution because it does not involve the more complicated beta functions.

Asymptotic inference is a commonly used type of approximate inference. In asymptotic approximate inference we are interested in the properties of a sequence of statistics $T_n(Y)$ as the sample size n increases. We focus our attention on the sequence $\{T_n\}$ for $n = 1, 2, \ldots$, and, in particular, consider the properties of $\{T_n\}$ as $n \to \infty$. Because asymptotic properties are often easy to work out, those properties are often used to identify a promising statistical method. How well the method performs in the real world of finite, and possibly small samples is a common topic in statistical research.

Another type of approximate inference, called *computational inference*, is used when an unknown distribution can be simulated by resampling of the given observations. Computational inference is a major topic of the present book.

1.5 Probability Statements in Statistical Inference

There are two important instances in statistical inference in which statements about probability are associated with the decisions of the inferential methods. In hypothesis testing, under assumptions about the distributions, we base our inferential methods on probabilities of two types of errors. In confidence intervals the decisions are associated with probability statements about coverage of the parameters. For both cases the probability statements are based on the distribution of a random sample, $Y_1, \ldots, Y_n$.

In computational inference, probabilities associated with hypothesis tests or confidence intervals are estimated by simulation of a hypothesized data-generating process or by resampling of an observed sample.

Tests of Hypotheses

Often statistical inference involves testing a "null" hypothesis, H_0, about the parameter. In a simple case, for example, we may test the hypothesis

$$H_0: \quad \theta = \theta_0$$

versus an alternative hypothesis

$$H_1: \quad \theta = \theta_1.$$

We do not know which hypothesis is true, but we want a statistical test that has a very small probability of rejecting the null hypothesis when it is true and a high probability of rejecting it when it is false. There is a tradeoff between these two decisions, so we will put an upper bound on the probability of rejecting the null hypothesis when it is true (called a "Type I error"), and under that constraint, seek a procedure that minimizes the probability of the other type of error ("Type II"). To be able to bound either of these probabilities, we must know (or make assumptions about) the true underlying data-generating process.

Thinking of the hypotheses in terms of a parameter θ that indexes these two densities by θ_0 and θ_1, for a sample $X = x$, we have the likelihoods associated with the two hypotheses as $L(\theta_0; x)$ and $L(\theta_1; x)$. We may be able to define an α-level critical region for nonrandomized tests in terms of the ratio of these likelihoods: Let us assume that a positive number k exists such that there is a subset of the sample space C with complement with respect to the sample space $\overline{C}$, such that

$$\begin{aligned} \frac{L(\theta_1; x)}{L(\theta_0; x)} &\geq k \quad \forall x \in C \\ \frac{L(\theta_1; x)}{L(\theta_0; x)} &\leq k \quad \forall x \in \overline{C} \end{aligned} \tag{1.125}$$

and

$$\alpha = \Pr(X \in C \mid H_0).$$

(Notice that such a k and C may not exist.)

The Neyman-Pearson Fundamental Lemma tells us that this test based on the *likelihood ratio* is the most powerful nonrandomized test of the simple null H_0 that specifies the density p_0 for X versus the simple alternative H_1 that specifies the density p_1. Let's consider the form of the Lemma that does not involve a randomized test; that is, in the case that an exact α-level nonrandomized test exists, as assumed above. Let k and C be as above. Then the Neyman-Pearson Fundamental Lemma states that C is the best critical region of size α for testing H_0 versus H_1.

Although it applies to a simple alternative (and hence "uniform" properties do not make much sense), the Neyman-Pearson Lemma gives us a way of determining whether a *uniformly most powerful* (UMP) test exists, and if so how to find one. We are often interested in testing hypotheses in which either or both of Θ_0 and Θ_1 are continuous regions of $\mathbb{R}$ (or $\mathbb{R}^k$).

We must look at the likelihood ratio as a function both of θ and x. The question is whether, for given θ_0 and any $\theta_1 > \theta_0$ (or equivalently any $\theta_1 < \theta_0$), the likelihood is monotone in some function of x; that is, whether the family of distributions of interest is parameterized by a scalar in such a way that it has a *monotone likelihood ratio* (see page 47). In that case, it is clear that we can extend the test in (1.125) to test to be uniformly most powerful for testing $H_0 : \theta = \theta_0$ against an alternative $H_1 : \theta > \theta_0$ (or $\theta_1 < \theta_0$).

The straightforward way of performing the test involves use of a test statistic, T, computed from a random sample of data, $Y_1, \ldots, Y_n$. Associated with T is a rejection region C such that if the null hypothesis is true, $\Pr(T \in C)$ is some preassigned (small) value, α, and $\Pr(T \in C)$ is greater than α if the null hypothesis is not true. Thus, C is a region of more "extreme" values of the test statistic if the null hypothesis is true. If $T \in C$, the null hypothesis is rejected. It is desirable that the test have a high probability of rejecting the null hypothesis if indeed the null hypothesis is not true. The probability of rejection of the null hypothesis is called the power of the test.

A procedure for testing that is mechanically equivalent to this is to compute the test statistic t and then to determine the probability that T is more extreme than t. In this approach, the realized value of the test statistic determines a region C_t of more extreme values. The probability that the test statistic is in C_t if the null hypothesis is true, $\Pr(T \in C_t | \mathrm{H}_0)$, is called the "p-value" or "significance level" of the realized test statistic.

If the distribution of T under the null hypothesis is known, the critical region or the p-value can be determined. If the distribution of T is not known, some other approach must be used. A common method is to use some approximation to the distribution. The objective is to approximate a quantile of T under the null hypothesis. The approximation is often based on an asymptotic distribution of the test statistic. In Monte Carlo tests, discussed in Section 11.2, the quantile of T is estimated by simulation of the distribution of the underlying data.

Confidence Intervals

Our usual notion of a confidence interval relies on a frequency approach to probability, and it leads to the definition of a $1 - \alpha$ confidence interval for the (scalar) parameter θ as the random interval (T_L, T_U) that has the property

$$\Pr(T_L \leq \theta \leq T_U) = 1 - \alpha. \tag{1.126}$$

This is also called a $(1 - \alpha)100\%$ confidence interval. The endpoints of the interval, T_L and T_U, are functions of a sample, $Y_1, \ldots, Y_n$. The interval (T_L, T_U) is not uniquely determined.

The concept extends easily to vector-valued parameters. Rather than taking vectors T_L and T_U, however, we generally define an ellipsoidal region, whose shape is determined by the covariances of the estimators.

A realization of the random interval, say (t_L, t_U), is also called a confidence interval. Although it may seem natural to state that the "probability that θ is in (t_L, t_U) is $1 - \alpha$", this statement can be misleading unless a certain underlying probability structure is assumed.

In practice, the interval is usually specified with respect to an estimator of θ, say T. If we know the sampling distribution of $T - \theta$, we may determine c_1 and c_2 such that

$$\Pr(c_1 \leq T - \theta \leq c_2) = 1 - \alpha; \tag{1.127}$$

and hence

$$\Pr(T - c_2 \leq \theta \leq T - c_1) = 1 - \alpha.$$

If exactly one of T_L or T_U in equation (1.126) is chosen to be infinite or to be a boundary point on the parameter space, the confidence interval is one-sided. (In either of those cases, the T_L or T_U would be a degenerate random variable. Furthermore, the values must respect the relation $T_L < T_U$.) For two-sided confidence intervals, we may seek to make the probability on either side of T to be equal, to make $c_1 = -c_2$, and/or to minimize $|c_1|$ or $|c_2|$. This is similar in spirit to seeking an estimator with small variance.

For forming confidence intervals, we generally use a function of the sample that also involves the parameter of interest, $f(T, \theta)$. The confidence interval is then formed by separating the parameter from the sample values.

Whenever the distribution depends on parameters other than the one of interest, we may be able to form only conditional confidence intervals that depend on the value of the other parameters. A class of functions that are particularly useful for forming confidence intervals in the presence of such nuisance parameters are called *pivotal* values, or pivotal functions. A function $f(T, \theta)$ is said to be a pivotal function if its distribution does not depend on any unknown parameters. This allows exact confidence intervals to be formed for the parameter θ. We first form

$$\Pr\Big(f_{\alpha/2} \;\leq\; f(T,\theta) \;\leq\; f_{1-\alpha/2}\Big) \;=\; 1-\alpha, \tag{1.128}$$

where $f_{\alpha/2}$ and $f_{1-\alpha/2}$ are quantiles of the distribution of $f(T, \theta)$; that is,

$$\Pr(f(T,\theta) \leq f_\pi) = \pi.$$

If, as in the case considered above, $f(T, \theta) = T - \theta$, the resulting confidence interval has the form

$$\Pr\Big(T - f_{1-\alpha/2} \;\leq\; \theta \;\leq\; T - f_{\alpha/2}\Big) \;=\; 1-\alpha.$$

For example, suppose that $Y_1, \ldots, Y_n$ is a random sample from a $\mathrm{N}(\mu, \sigma^2)$ distribution, and $\overline{Y}$ is the sample mean. The quantity

$$f(\overline{Y},\mu) = \frac{\sqrt{n(n-1)}\,(\overline{Y}-\mu)}{\sqrt{\sum\;(Y_i-\overline{Y})^2}} \tag{1.129}$$

has a Student's t distribution with $n - 1$ degrees of freedom, no matter what is the value of σ^2. This is one of the most commonly used pivotal values.

The pivotal value in equation (1.129) can be used to form a confidence value for θ by first writing

$$\Pr\left(t_{\alpha/2} \;\leq\; f(\overline{Y},\mu) \;\leq\; t_{1-\alpha/2}\right) \;=\; 1-\alpha,$$

where t_π is a percentile from the Student's t distribution. Then, after making substitutions for $f(\overline{Y}, \mu)$, we form the familiar confidence interval for μ:

$$\Big(\overline{Y} - t_{1-\alpha/2}\, s/\sqrt{n},\quad \overline{Y} - t_{\alpha/2}\, s/\sqrt{n}\Big), \tag{1.130}$$

where s^2 is the usual sample variance, $\sum(Y_i - \overline{Y})^2/(n-1)$.

Other similar pivotal values have F distributions. For example, consider the usual linear regression model in which the n-vector random variable Y has a $\mathrm{N}_n(X\beta, \sigma^2 I)$ distribution, where X is an $n \times m$ known matrix, and the

m-vector β and the scalar σ^2 are unknown. A pivotal value useful in making inferences about β is

$$g(\widehat{\beta}, \beta) = \frac{\left(X(\widehat{\beta} - \beta)\right)^{\mathrm{T}} X(\widehat{\beta} - \beta)/m}{(Y - X\widehat{\beta})^{\mathrm{T}}(Y - X\widehat{\beta})/(n - m)}, \tag{1.131}$$

where

$$\widehat{\beta} = (X^{\mathrm{T}}X)^{+}X^{\mathrm{T}}Y.$$

The random variable $g(\widehat{\beta}, \beta)$ for any finite value of σ^2 has an F distribution with m and $n - m$ degrees of freedom.

For a given parameter and family of distributions, there may be multiple pivotal values. For purposes of statistical inference, such considerations as unbiasedness and minimum variance may guide the choice of a pivotal value to use. Alternatively, it may not be possible to identify a pivotal quantity for a particular parameter. In that case, we may seek an approximate pivot. A function is asymptotically pivotal if a sequence of linear transformations of the function is pivotal in the limit as $n \to \infty$.

If the distribution of T is known, c_1 and c_2 in equation (1.127) can be determined. If the distribution of T is not known, some other approach must be used. A method for computational inference, discussed in Section 13.3, is to use "bootstrap" samples from the ECDF.

1.6 Modeling and Computational Inference

The process of building models involves successive refinements. The evolution of the models proceeds from vague, tentative models to more complete ones, and our understanding of the process being modeled grows in this process.

A given model usually contains parameters that are chosen to fit a given set of data. Other models of different forms or with different parameters may also be fit. The models are compared on the basis of some criterion that indicates their goodness-of-fit to the available data. The process of fitting and evaluating is often done on separate partitions of the data. It is a general rule that the more parameters a model of a given form has, the better the model will appear to fit any given set of data. The model building process must use criteria that avoid the natural tendency to overfit. We discuss this type of problem further in Section 12.2.

The usual statements about statistical methods regarding bias, variance, and so on are made in the context of a model. It is not possible to measure bias or variance of a procedure to *select* a model, except in the relatively simple case of selection from some well-defined and simple set of possible models. Only within the context of rigid assumptions (a "metamodel") can we do a precise statistical analysis of model selection. Even the simple cases of selection of variables in linear regression analysis under the usual assumptions about the distribution of residuals (and this is a highly idealized situation) present more problems to the analyst than are generally recognized.

Descriptive Statistics, Inferential Statistics, and Model Building

We can distinguish statistical activities that involve:

- data collection;
- descriptions of a given dataset;
- inference within the context of a model or family of models; and
- model selection.

In any given application, it is likely that all of these activities will come into play. Sometimes (and often, ideally!), a statistician can specify how data are to be collected, either in surveys or in experiments. We will not be concerned with this aspect of the process in this text.

Once data are available, either from a survey or designed experiment, or just observational data, a statistical analysis begins by considering general descriptions of the dataset. These descriptions include ensemble characteristics, such as averages and spreads, and identification of extreme points. The descriptions are in the form of various summary statistics and graphical displays. The descriptive analyses may be computationally intensive for large datasets, especially if there are a large number of variables. The computationally intensive approach also involves multiple views of the data, including consideration of a large number of transformations of the data. We discuss these methods in various chapters of Part III.

A stochastic model is often expressed as a PDF or as a CDF of a random variable. In a simple linear regression model with normal errors,

$$Y = \beta_0 + \beta_1 x + E, \tag{1.132}$$

for example, the model may be expressed by use of the probability density function for the random variable E. (Notice that Y and E are written in uppercase because they represent random variables.)

If E in equation (1.132) has a normal distribution with variance σ^2, which we would denote by

$$E \sim \mathrm{N}(0, \sigma^2),$$

then the probability density function for Y is

$$p(y) = \frac{1}{\sqrt{2\pi}\sigma} e^{-(y-\beta_0-\beta_1 x)^2/(2\sigma^2)}. \tag{1.133}$$

In this model, x is an observable covariate; σ, β_0, and β_1 are unobservable (and, generally, unknown) parameters; and 2 and π are constants. Statistical inference about parameters includes estimation or tests of their values or statements about their probability distributions based on observations of the elements of the model.

The elements of a stochastic model include observable random variables, observable covariates, unobservable parameters, and constants. Some random variables in the model may be considered to be "responses". The covariates may be considered to affect the response; they may or may not be random variables. The parameters are variable within a class of models, but for a specific data model the parameters are constants. The parameters may be considered to be unobservable random variables, and in that sense, a specific data model is defined by a realization of the parameter random variable. In the model, written as

$$Y = f(x; \beta) + E, \tag{1.134}$$

we identify a "systematic component", $f(x; \beta)$, and a "random component", E.

The selection of an appropriate model may be very difficult, and almost always involves not only questions of how well the model corresponds to the observed data, but also the tractability of the model. The methods of computational statistics allow a much wider range of tractability than can be contemplated in mathematical statistics.

Statistical analyses generally are undertaken with the purpose of making a decision about a dataset, about a population from which a sample dataset is available, or in making a prediction about a future event. Much of the theory of statistics developed during the middle third of the twentieth century was concerned with formal inference; that is, use of a sample to make decisions about stochastic models based on probabilities that would result if a given model was indeed the data-generating process. The heuristic paradigm calls for rejection of a model if the probability is small that data arising from the model would be similar to the observed sample. This process can be quite tedious because of the wide range of models that should be explored and because some of the models may not yield mathematically tractable estimators or test statistics. Computationally intensive methods include exploration of a range of models, many of which may be mathematically intractable.

In a different approach employing the same paradigm, the statistical methods may involve direct simulation of the hypothesized data-generating process rather than formal computations of probabilities that would result under a given model of the data-generating process. We refer to this approach as *computational inference*. We discuss methods of computational inference in Chapters 11, 12, and 13. In a variation of computational inference, we may not even attempt to develop a model of the data-generating process; rather, we build decision rules directly from the data. This is often the approach in clustering and classification, which we discuss in Chapter 16. Computational inference is rooted in classical statistical inference, which was briefly summarized in Section1.4, but which must be considered as a prerequisite for the present book. In subsequent sections of the current chapter, we discuss general techniques used in statistical inference.

1.7 The Role of the Empirical Cumulative Distribution Function

Methods of statistical inference are based on an assumption (often implicit) that a discrete uniform distribution with mass points at the observed values of a random sample is asymptotically the same as the distribution governing the data-generating process. Thus, the distribution function of this discrete uniform distribution is a model of the distribution function of the data-generating process.

For a given set of univariate data, $y_1, \ldots, y_n$, the *empirical cumulative distribution function*, or ECDF, is

$$P_n(y) = \frac{\#\{y_i, \text{ s.t. } y_i \leq y\}}{n}.$$

The ECDF is a CDF in its own right. It is the CDF of the discrete distribution with n or fewer mass points, one at each sample value, and with a probability mass at each point corresponding to the number of sample values at that point times n^{-1}. If all of the sample points are unique, it is the CDF of the discrete uniform distribution. The ECDF is the basic function used in many methods of computational inference. It contains all of the information in the sample.

Although the ECDF has similar definitions for univariate and multivariate random variables, it is most useful in the univariate case.

An equivalent expression for univariate random variables, in terms of intervals on the real line, is

$$P_n(y) = \frac{1}{n} \sum_{i=1}^{n} \mathrm{I}_{(-\infty,y]}(y_i), \tag{1.135}$$

where I is the indicator function. (See page 669 for the definition and some of the properties of the indicator function. The measure $\mathrm{dI}_{(-\infty,a]}(x)$, which we use in equation (1.139) below, is particularly interesting.)

It is easy to see that the ECDF is pointwise unbiased for the CDF; that is, if the y_i are independent realizations of random variables Y_i, each with CDF $P(\cdot)$, for a given y,

$$\begin{aligned}
\mathrm{E}\big(P_n(y)\big) &= \mathrm{E}\left(\frac{1}{n} \sum_{i=1}^{n} \mathrm{I}_{(-\infty,y]}(Y_i)\right) \\
&= \frac{1}{n} \sum_{i=1}^{n} \mathrm{E}\left(\mathrm{I}_{(-\infty,y]}(Y_i)\right) \\
&= \Pr\left(Y \leq y\right) \\
&= P(y).
\end{aligned} \tag{1.136}$$

Similarly, we find

$$\mathrm{V}(P_n(y)) = P(y)(1 - P(y))/n; \tag{1.137}$$

indeed, at a fixed point y, $nP_n(y)$ is a binomial random variable with parameters n and $\pi = P(y)$. Because P_n is a function of the order statistics, which form a complete sufficient statistic for P, there is no unbiased estimator of $P(y)$ with smaller variance.

We also define the *empirical probability density function* (EPDF) as the derivative of the ECDF:

$$p_n(y) = \frac{1}{n}\sum_{i=1}^{n}\delta(y - y_i), \tag{1.138}$$

where δ is the Dirac delta function. The EPDF is just a series of spikes at points corresponding to the observed values. It is not as useful as the ECDF. It is, however, unbiased at any point for the probability density function at that point.

The ECDF and the EPDF can be used as estimators of the corresponding population functions, but there are better estimators (see Chapter 15).

Estimation Using the ECDF

As we have seen, there are many ways to construct an estimator and to make inferences about the population. If we are interested in a measure of the population that is expressed as a statistical function (see page 31), we may use data to make inferences about that measure by applying the statistical function to the ECDF. An estimator of a parameter that is defined in this way is called a *plug-in estimator*. A plug-in estimator for a given parameter is the same functional of the ECDF as the parameter is of the CDF.

For the mean of the model, for example, we use the estimate that is the same functional of the ECDF as the population mean in equation (1.74),

$$\begin{aligned} M(P_n) &= \int_{-\infty}^{\infty} y\,\mathrm{d}P_n(y) \\ &= \int_{-\infty}^{\infty} y\,\mathrm{d}\frac{1}{n}\sum_{i=1}^{n}\mathrm{I}_{(-\infty,y]}(y_i) \\ &= \frac{1}{n}\sum_{i=1}^{n}\int_{-\infty}^{\infty} y\,\mathrm{d}\mathrm{I}_{(-\infty,y]}(y_i) \\ &= \frac{1}{n}\sum_{i=1}^{n} y_i \\ &= \bar{y}. \end{aligned} \tag{1.139}$$

The sample mean is thus a plug-in estimator of the population mean. Such an estimator is called a *method of moments estimator*. This is an important type of plug-in estimator. For a univariate random variable Y, the method of moments results in estimates of the parameters $\mathrm{E}(Y^r)$ that are the corresponding sample moments.

Statistical properties of plug-in estimators are generally relatively easy to determine, and often the statistical properties are optimal in some sense.

In addition to point estimation based on the ECDF, other methods of computational statistics make use of the ECDF. In some cases, such as in bootstrap methods, the ECDF is a surrogate for the CDF. In other cases, such as Monte Carlo methods, an ECDF for an estimator is constructed by repeated sampling, and that ECDF is used to make inferences using the observed value of the estimator from the given sample.

A functional, Θ, denotes a specific functional form of a CDF or ECDF. Any functional of the ECDF is a function of the data, so we may also use the notation $\Theta(Y_1, \ldots, Y_n)$. Often, however, the notation is cleaner if we use another letter to denote the function of the data; for example, $T(Y_1, \ldots, Y_n)$, even if it might be the case that

$$T(Y_1, \ldots, Y_n) = \Theta(P_n).$$

We will also often simplify the notation further by using the same letter that denotes the functional of the sample to represent the random variable computed from a random sample; that is, we may write

$$T = T(Y_1, \ldots, Y_n).$$

As usual, we will use t to denote a realization of the random variable T.

Use of the ECDF in statistical inference does not require many assumptions about the distribution. Other methods, such as MLE and others discussed in Section 1.4, are based on specific information or assumptions about the data-generating process.

Linear Functionals and Estimators

A functional Θ is *linear* if, for any two functions f and g in the domain of Θ and any real number a,

$$\Theta(af + g) = a\Theta(f) + \Theta(g). \tag{1.140}$$

A statistic is linear if it is a linear functional of the ECDF. A linear statistic can be computed from a sample using an online algorithm, and linear statistics from two samples can be combined by addition. Strictly speaking, this definition excludes statistics such as means, but such statistics are *essentially linear* in the sense that they can be combined by a linear combination if the sample sizes are known.

Quantiles

A useful distributional measure for describing a univariate distribution with CDF P is is a quantity y_π, such that

$$\Pr(Y \leq y_\pi) \geq \pi, \text{ and } \Pr(Y \geq y_\pi) \geq 1 - \pi, \tag{1.141}$$

for $\pi \in (0, 1)$. This quantity is called a π quantile.

For an absolutely continuous distribution with CDF P,

$$y_\pi = P^{-1}(\pi).$$

If P is not absolutely continuous, or in the case of a multivariate random variable, y_π may not be unique.

For a univariate distribution with CDF P, we define the π quantile as a unique value by letting

$$y_{\pi+} = \min_y \{y, \text{ s.t. } P(y) \geq \pi\}$$

and

$$y_{\pi-} = \min_y \{y, \text{ s.t. } P(y) \leq \pi \text{ and } P(y) > P(x) \text{ for } y > x\},$$

and then

$$y_\pi = y_{\pi-} + \frac{\pi - P(y_{\pi-})}{P(y_{\pi+}) - P(y_{\pi-})}(y_{\pi+} - y_{\pi-}). \tag{1.142}$$

For discrete distributions, the π quantile may be a quantity that is not in the support of the distribution.

It is clear that y_π is a functional of the CDF, say $\Xi_\pi(P)$. For an absolutely continuous distribution, the functional is very simple:

$$\Xi_\pi(P) = P^{-1}(\pi). \tag{1.143}$$

For a univariate random variable, the π quantile is a single point. For a d-variate random variable, a similar definition leads to a $(d-1)$-dimensional object that is generally nonunique. Quantiles are not so useful in the case of multivariate distributions.

Empirical Quantiles

For a given sample, the order statistics constitute an obvious set of empirical quantiles. The probabilities from the ECDF that are associated with the order statistic $y_{(i)}$ is i/n, which leads to a probability of 1 for the largest sample value, $y_{(n)}$, and a probability of $1/n$ for the smallest sample value, $y_{(1)}$. (The notation $y_{(i)}$ denotes the ith order statistic. We also sometimes incorporate the sample size in the notation: $y_{(i:n)}$ to indicates the ith order statistic in a sample of size n.

If $Y_{(1)}, \ldots, Y_{(n)}$ are the order statistics in a random sample of size n from a distribution with PDF $p_Y(\cdot)$ and CDF $P_Y(\cdot)$, then the PDF of the ith order statistic is

$$p_{Y_{(i)}}(y_{(i)}) = \binom{n}{i} \left(P_Y(y_{(i)})\right)^{i-1} p_Y(y_{(i)}) \left(1 - P_Y(y_{(i)})\right)^{n-i}. \tag{1.144}$$

This expression is easy to derive by considering the ways the ith element can be chosen from a set of n, the probability that $i - 1$ elements are less than or equal to this element, the density of the element itself, and finally the probability that $n - i$ elements are greater than this element.

Order statistics are not independent. The joint density of all order statistics is

$$n! \prod p(y_{(i)}) \mathrm{I}_{y_{(1)} \leq \cdots \leq y_{(n)}}(y_{(1)}, \ldots, y_{(n)}). \tag{1.145}$$

Interestingly, the order statistics from a U$(0,1)$ distribution have beta distributions. As we see from equation (1.144), the ith order statistic in a sample of size n from a U$(0,1)$ distribution has a beta distribution with parameters i and $n - i + 1$. Because of this simple distribution, it is easy to determine properties of the order statistics from a uniform distribution. For example the expected value of the ith order statistic in a sample of size n from U$(0,1)$ is

$$\mathrm{E}(U_{(i:n)}) = \frac{i}{n+1}, \tag{1.146}$$

and the variance is

$$\mathrm{V}(U_{(i:n)}) = \frac{i(n-i+1)}{(n+1)^2(n+2)}. \tag{1.147}$$

Order statistics have interesting, and perhaps unexpected properties. Consider a sample of size 25 from a standard normal distribution. Some simple facts about the maximum order statistic $Y_{(25)}$ are worth noting. First of all, the distribution of $Y_{(25)}$ is not symmetric. Secondly, the expected value of the standard normal CDF, Φ, evaluated at $Y_{(25)}$ is not 0.960 (24/25) or 0.962 (25/26), and of course, it is certainly not 1, as is the value of the ECDF at $Y_{(25)}$. Notice that if $\mathrm{E}(\Phi(Y_{(25)})) = \mathrm{E}(U_{(25)})$, where $U_{(25)}$ is the maximum order statistic in a sample of size 25, the value would be 25/26, but the expected value does not carry over through nonlinear functions. Because Φ is a strictly increasing function, however, we do have

$$\mathrm{Med}(\Phi(Y_{(25)})) = \mathrm{Med}(U_{(25)}), \tag{1.148}$$

where Med(X) is the median of the random variable X. (This comes from the fact that median-unbiasedness carries over to monotone increasing functions.) Filliben (1975) suggested fitting quantiles by equation (1.148). For the median of the ith order statistic in a sample of size n from a U$(0,1)$ distribution, he suggested an expression of the form

$$\frac{i - \gamma}{n - 2\gamma + 1}. \tag{1.149}$$

Filliben then fit various approximations to the order statistics and came up with the fit

$$\mathrm{Med}(U_{(i:n)})) = \begin{cases} 1 - 2^{-1/n} & i = 1 \\ \frac{i-0.3175}{n+0.365} & i = 2, \ldots, n-1 \\ 2^{-1/n} & i = n. \end{cases} \tag{1.150}$$

By Filliben's rule for using the median of the uniform order statistics and fitting them as above, we have $\Phi(Y_{(25)}) \approx 0.973$; other reasonable empirical adjustments may yield values as large as 0.982.

The foregoing raises the question as to what probability should correspond to the ith order statistic, $y_{(i)}$, in a sample of size n. The probability is often approximated as some adjustment of i/n as in equation (1.149), but clearly it depends on the underlying distribution.

We use empirical quantiles in Monte Carlo inference, in nonparametric inference, and in graphical displays for comparing a sample with a standard distribution or with another sample. The most common of the graphs is the q-q plot discussed beginning on page 348.

Estimation of Quantiles

Empirical quantiles can be used as estimators of the population quantiles, but there are generally other estimators that are better, as we can deduce from basic properties of statistical inference. The first thing that we note is that the extreme order statistics have very large variances if the support of the underlying distribution is infinite. We would therefore not expect them alone to be the best estimator of an extreme quantile unless the support is finite.

A fundamental principle of statistical inference is that a sufficient statistic should be used, if one is available. No order statistic alone is sufficient, except for the minimum or maximum order statistic in the case of a distribution with finite support. The set of all order statistics, however, is always sufficient. Because of the Rao-Blackwell theorem (see page 40), this would lead us to expect that some combination of order statistics would be a better estimator of any population quantile than a single estimator.

The Kaigh-Lachenbruch estimator (see Kaigh and Lachenbruch, 1982) and the Harrell-Davis estimator (see Harrell and Davis, 1982), use weighted combinations of order statistics. The Kaigh-Lachenbruch estimator uses weights from a hypergeometric distribution, and the Harrell-Davis estimator uses weights from a beta distribution. The Kaigh-Lachenbruch weights arise intuitively from combinatorics, and the Harrell-Davis come from the fact that for any continuous CDF P if Y is a random variable from the distribution with

CDF P, then $U = P(Y)$ has a U(0,1) distribution, and the order statistics from a uniform have beta distributions.

The Harrell-Davis estimator for the π quantile uses the beta distribution with parameters $\pi(n+1)$ and $(1-\pi)(n+1)$. Let $P_{\beta_\pi}(\cdot)$ be the CDF of the beta distribution with those parameters. The Harrell-Davis estimator for the π quantile is

$$\widehat{y_\pi} = \sum_{i=1}^{n} w_i y_{(i)}, \tag{1.151}$$

where

$$w_i = P_{\beta_\pi}(i/n) - P_{\beta_\pi}((i-1)/n). \tag{1.152}$$

Estimators of the form of linear combinations of order statistics, such as the Harrell-Davis or Kaigh-Lachenbruch quantile estimators, are called "L statistics". In Exercise 1.18 you are asked to study the relative performance of the sample median and the Harrell-Davis estimator as estimators of the population median.

1.8 The Role of Optimization in Inference

Important classes of estimators are defined as points at which some function that involves the parameter and the random variable achieves an optimum with respect to a variable in the role of the parameter in the function. There are, of course, many functions that involve the parameter and the random variable. One example of such a function is the probability density function itself, and as we have seen optimization of this function is the idea behind maximum likelihood estimation.

In the use of function optimization in inference, once the objective function is chosen, observations on the random variable are taken and are then considered to be fixed; the parameter in the function is considered to be a variable (the "decision variable", in the parlance often used in the literature on optimization). The function is then optimized with respect to the parameter variable. The nature of the function determines the meaning of "optimized"; if the function is the probability density, for example, "optimized" would logically mean "maximized", which leads to maximum likelihood estimation.

In discussing the use of optimization in statistical estimation, we must be careful to distinguish between a symbol that represents a fixed parameter and a symbol that represents a "variable" parameter. When we denote a probability density function as $p(y \mid \theta)$, we generally expect "θ" to represent a fixed, but possibly unknown, parameter. In an estimation method that involves optimizing this function, however, θ is a variable placeholder. In the following discussion, we will generally consider a variable t in place of θ. We also use t_0, t_1, and so on to represent specific fixed values of the variable. In an iterative algorithm, we use $t^{(k)}$ to represent a fixed value in the kth iteration. We do not always do this, however, and sometimes, as other authors do, we will use

θ to represent the true value of the parameter on which the random variable observed is conditioned — but we consider it changeable. We may also use θ_0, θ_1, and so on, to represent specific fixed values of the variable, or in an iterative algorithm, $\theta^{(k)}$ to represent a fixed value in the kth iteration.

Some Comments on Optimization

The solution to an optimization problem is in some sense "best" for that problem and its objective functions; this may mean it is considerably less good for some other optimization problem. It is often the case, therefore, that an optimal solution is not robust to assumptions about the phenomenon being studied. Use of optimization methods is likely to magnify the effects of the assumptions.

In the following pages we discuss two of the general approaches to statistical inference that we mentioned on page 41 in which optimization is used. One is to minimize deviations of observed values from what a model would predict. This is an intuitive procedure which may be chosen without regard to the nature of the data-generating process. The justification for a particular form of the objective function, however, may arise from assumptions about a probability distribution underlying the data-generating process.

Another common way in which optimization is used in statistical inference is in maximizing the likelihood. The correct likelihood function depends on the probability distribution underlying the data-generating process, which, of course, is not known and can only be assumed. How good or how poor the maximum likelihood estimator is depends on both the true distribution and the assumed distribution.

In the discussion below, we briefly describe particular optimization techniques that assume that the objective function is a continuous function of the decision variables, or the parameters. We also assume that there are no a priori constraints on the values of the parameters. Techniques appropriate for other situations, such as for discrete optimization and constrained optimization, are available in the general literature on optimization.

We must also realize that mathematical expressions below do not necessarily imply computational methods. *This is a repeating theme of this book.* There are many additional considerations for the numerical computations. A standard example of this point is in the solution of the linear full-rank system of n equations in n unknowns: $Ax = b$. While we may write the solution as $x = A^{-1}b$, we would almost never compute the solution by forming the inverse and then multiplying b by it.

Estimation by Minimizing Residuals

In many applications, we can express the expected value of a random variable as a function of a parameter (which might be a vector, of course):

$$\mathrm{E}(Y) = g(\theta). \tag{1.153}$$

The expectation may also involve covariates, so in general we may write $g(x, \theta)$. The standard linear regression model is an example: $\mathrm{E}(Y) = x^{\mathrm{T}}\beta$. If the covariates are observable, they can be subsumed into $g(\theta)$.

The more difficult and interesting problems, of course, involve the determination of the form of the function $g(\theta)$. Here, however, we concentrate on the simpler problem of determining an appropriate value of θ, assuming that the form of the function g is known.

If we can obtain observations $y_1, \ldots, y_n$ on Y (and observations on the covariates if there are any), a reasonable estimator of θ is a value $\widehat{\theta}$ that minimizes the residuals,

$$r_i(t) = y_i - g(t), \tag{1.154}$$

over all possible choices of t, where t is a variable placeholder. This approach makes sense because we expect the observed y's to be close to $g(\theta)$.

There are, of course, several ways we could reasonably "minimize the residuals". In general, we seek a value of t to minimize some norm of $r(t)$, the n-vector of residuals. The optimization problem is

$$\min_t \|r(t)\|. \tag{1.155}$$

We often choose the norm as the L_p norm, so we minimize a function of an L_p norm of the residuals,

$$s_p(t) = \sum_{i=1}^{n} |y_i - g(t)|^p, \tag{1.156}$$

for some $p \geq 1$, to obtain an L_p *estimator*. Simple choices are the sum of the absolute values and the sum of the squares. The latter choice yields the *least squares estimator*. More generally, we could minimize

$$s_\rho(t) = \sum_{i=1}^{n} \rho(y_i - g(t))$$

for some nonnegative function $\rho(\cdot)$ to obtain an *"M estimator"*. (The name comes from the similarity of this objective function to the objective function for some maximum likelihood estimators.)

Standard techniques for optimization can be used to determine estimates that minimize various functions of the residuals, that is, for some appropriate function of the residuals $s(\cdot)$, to solve

$$\min_t s(t). \tag{1.157}$$

Except for special forms of the objective function, the algorithms to solve expression (1.157) are iterative, such as Newton's method, which we discuss on page 266.

The function $s(\cdot)$ is usually chosen to be differentiable, at least piecewise.

Statistical Properties of Minimum-Residual Estimators

There are, of course, two considerations. One involves the actual computations. We discuss those in Chapter 6. The other involves the statistical properties of the estimators.

It is generally difficult to determine the variance or other high-order statistical properties of an estimator defined as above (that is, defined as the minimizer of some function of the residuals). In many cases, all that is possible is to approximate the variance of the estimator in terms of some relationship that holds for a normal distribution. (In robust statistical methods, for example, it is common to use a "scale estimate" expressed in terms of some mysterious constant times a function of some transformation of the residuals.)

There are two issues that affect both the computational method and the statistical properties of the estimator defined as the solution to the optimization problem. One issue has to do with the acceptable values of the parameter θ. In order for the model to make sense, it may be necessary that the parameter be in some restricted range. In some models, a parameter must be positive, for example. In these cases, the optimization problem has constraints. Such a problem is more difficult to solve than an unconstrained problem. Statistical properties of the solution are also more difficult to determine. More extreme cases of restrictions on the parameter may require the parameter to take values in a countable set. Obviously, in such cases, Newton's method cannot be used because the derivatives cannot be defined. In those cases, a combinatorial optimization algorithm must be used instead. Other situations in which the function is not differentiable also present problems for the optimization algorithm. In such cases, if the domain is continuous, a descending sequence of simplexes can be used.

The second issue involves the question of a unique global solution to the optimization problem (1.157). It may turn out that the optimization problem has local minima. This depends on the nature of the function $f(\cdot)$ in equation (1.153). Local minima present problems for the computation of the solution because the algorithm may get stuck in a local optimum. Local minima also present conceptual problems concerning the appropriateness of the estimation criterion itself. As long as there is a unique global optimum, it seems reasonable to seek it and to ignore local optima. It is not so clear what to do if there are multiple points at which the global optimum is attained. That is not a question specifically for methods of computational statistics; it is fundamental to the heuristic of minimizing residuals.

Least Squares Estimation

Least squares estimators are generally more tractable than estimators based on other functions of the residuals. They are more tractable both in terms of solving the optimization problem to obtain the estimate, and in approximating statistical properties of the estimators, such as their variances.

Assume in equation (1.153) that t (and hence, θ) is an m-vector and that $f(\cdot)$ is a smooth function. Letting y be the n-vector of observations, we can write the *least squares objective function* corresponding to equation (1.156) as

$$s(t) = \big(r(t)\big)^{\mathrm{T}} r(t). \tag{1.158}$$

Often in applications, the residuals in equation (1.154) are not given equal weight for estimating θ. This may be because the reliability or precision of the observations may be different. For *weighted least squares*, instead of equation (1.158) we have the objective function

$$s_w(t) = \sum_{i=1}^{n} w_i \big(r_i(t)\big)^2. \tag{1.159}$$

Variance of Least Squares Estimators

If the distribution of Y has finite moments, the sample mean $\overline{Y}$ is a consistent estimator of $g(\theta)$. Furthermore, the minimum residual norm $\big(r(\widehat{\theta})\big)^{\mathrm{T}} r(\widehat{\theta})$ divided by $(n-m)$ is a consistent estimator of the variance of Y, say σ^2; that is, of

$$\sigma^2 = \mathrm{E}(Y - g(\theta))^2.$$

A consistent estimator of σ^2 is

$$\widehat{\sigma^2} = \big(r(\widehat{\theta})\big)^{\mathrm{T}} r(\widehat{\theta})/(n-m).$$

This estimator, strictly speaking, is not a least squares estimator of σ^2. It is based on least squares estimators of another parameter. (In the linear case, the consistency of $\widehat{\sigma^2}$, in fact, its unbiasedness, is straightforward. In other cases, it is not so obvious. The proof can be found in texts on nonlinear regression or on generalized estimating equations.)

The variance-covariance of the least squares estimator $\widehat{\theta}$, however, is not easy to work out, except in special cases. It obviously involves σ^2. In the simplest case, g is linear and Y has a normal distribution, and we have the familiar linear regression estimates of θ and σ^2 and of the variance of the estimator of θ.

Without the linearity property, however, even with the assumption of normality, it may not be possible to write a simple expression for the variance-covariance matrix of an estimator that is defined as the solution to the least squares optimization problem. Using a linear approximation, however, we may estimate an approximate variance-covariance matrix for $\widehat{\theta}$ as

$$\Big(\big(\mathrm{J}_r(\widehat{\theta})\big)^{\mathrm{T}} \mathrm{J}_r(\widehat{\theta})\Big)^{-1} \widehat{\sigma^2}. \tag{1.160}$$

Compare this linear approximation to the expression for the estimated variance-covariance matrix of the least squares estimator $\widehat{\beta}$ in the linear regression model $\mathrm{E}(Y) = X\beta$, in which $\mathrm{J}_r(\widehat{\beta})$ is just X.

Taking derivatives of $\nabla s(t)$, we express the Hessian of s in terms of the Jacobian of r as

$$\mathrm{H}_s(t) = \left(\mathrm{J}_r(t)\right)^{\mathrm{T}} \mathrm{J}_r(t) + \sum_{i=1}^{n} r_i(t) \mathrm{H}_{r_i}(t).$$

If the residuals are small, the Hessian is approximately equal to the cross-product of the Jacobian, so an alternate expression for the estimated variance-covariance matrix is

$$\left(\mathrm{H}_s(\widehat{\theta})\right)^{-1} \widehat{\sigma^2}. \tag{1.161}$$

See Exercises 1.19 and 1.20 for comparisons of these two expressions.

Although there may be some differences in the performance of these two variance estimators, they usually depend on aspects of the model that are probably not well understood. Which expression for the variance estimator is used often depends on the computational method used. The expression (1.161) is more straightforward than (1.160) if Newton's method (equation (6.29) on page 266) or a quasi-Newton method is used instead of the Gauss-Newton method (equation (6.61) on page 292) for the solution of the least squares problem because in these methods the Hessian or an approximate Hessian is used in the computations.

Estimation by Maximum Likelihood

One of the most commonly used approaches to statistical estimation is *maximum likelihood.* The concept has an intuitive appeal, and the estimators based on this approach have a number of desirable mathematical properties, at least for broad classes of distributions.

Given a sample $y_1, \ldots, y_n$ from a distribution with probability density or probability mass function $p(y \mid \theta)$, a reasonable estimate of θ is the value that maximizes the joint density or joint probability with variable t at the observed sample value: $\prod_i p(y_i \mid t)$. We define the *likelihood function* as a function of a variable in place of the parameter:

$$L_n(t\,;\, y) = \prod_{i=1}^{n} p(y_i \mid t). \tag{1.162}$$

Note the reversal in roles of variables and parameters. The likelihood function appears to represent a "posterior probability", but, as emphasized by R. A. Fisher who made major contributions to the use of the likelihood function in inference, that is not an appropriate interpretation.

Just as in the case of estimation by minimizing residuals, the more difficult and interesting problems involve the determination of the form of the function $p(y_i \mid \theta)$. In these sections, as above, however, we concentrate on the simpler problem of determining an appropriate value of θ, assuming that the form of p is known.

The value of t for which L attains its maximum value is the *maximum likelihood estimate* (MLE) of θ for the given data, y. The data — that is, the realizations of the variables in the density function — are considered as fixed, and the parameters are considered as variables of the optimization problem,

$$\max_t L_n(t\,;\,y). \tag{1.163}$$

This optimization problem can be much more difficult than the optimization problem (1.155) that results from an estimation approach based on minimization of some norm of a residual vector. As we discussed in that case, there can be both computational and statistical problems associated either with restrictions on the set of possible parameter values or with the existence of local optima of the objective function. These problems also occur in maximum likelihood estimation.

Applying constraints in the optimization problem to force the solution to be within the set of possible parameter values is called *restricted maximum likelihood estimation*, or REML estimation. In addition to problems due to constraints or due to local optima, other problems may arise if the likelihood function is bounded. The conceptual difficulties resulting from an unbounded likelihood are much deeper. In practice, for computing estimates in the unbounded case, the general likelihood principle may be retained, and the optimization problem redefined to include a penalty that keeps the function bounded. Adding a penalty to form a bounded objective function in the optimization problem, or to dampen the solution is called *penalized maximum likelihood estimation.*

For a broad class of distributions, the maximum likelihood criterion yields estimators with good statistical properties. The conditions that guarantee certain optimality properties are called the "regular case".

Although in practice, the functions of residuals that are minimized are almost always differentiable, and the optimum occurs at a stationary point, this is often not the case in maximum likelihood estimation. A standard example in which the MLE does not occur at a stationary point is a distribution in which the range depends on the parameter, and the simplest such distribution is the uniform $\mathrm{U}(0,\theta)$. In this case, the MLE is the max order statistic.

Maximum likelihood estimation is particularly straightforward for distributions in the exponential class, that is, those with PDFs of the form in equation (1.89) on page 35. Whenever $\mathcal{Y}$ does not depend on θ, and $\eta(\cdot)$ and $\xi(\cdot)$ are sufficiently smooth, the MLE has certain optimal statistical properties. This family of probability distributions includes many of the familiar distributions, such as the normal, the binomial, the Poisson, the gamma, the Pareto, and the negative binomial.

The *log-likelihood function*,

$$l_{L_n}(\theta\,;\,y) = \log L_n(\theta\,;\,y), \tag{1.164}$$

is a sum rather than a product. The form of the log-likelihood in the exponential family is particularly simple:

$$l_{L_n}(\theta\,;\,y) = \sum_{i=1}^{n} \theta^{\mathrm{T}} g(y_i) - n\,a(\theta) + c,$$

where c depends on the y_i but is constant with respect to the variable of interest.

The logarithm is monotone, so the optimization problem (1.163) can be solved by solving the maximization problem with the log-likelihood function:

$$\max_{t} l_{L_n}(t\,;\,y). \tag{1.165}$$

We usually drop the subscript n in the notation for the likelihood and the log-likelihood, and we often work with the likelihood and log-likelihood as if there is only one observation. (A general definition of a likelihood function is any nonnegative function that is proportional to the density or the probability mass function; that is, it is the same as the density or the probability mass function except that the arguments are switched, and its integral or sum over the domain of the random variable need not be 1.)

The log-likelihood function relates directly to useful concepts in statistical inference. If it exists, the derivative of the log-likelihood is the relative rate of change, with respect to the parameter placeholder t, of the probability density function at a fixed observation. If θ is a scalar, some positive function of the derivative such as its square or its absolute value is obviously a measure of the effect of change in the parameter or in the estimate of the parameter. More generally, an outer product of the derivative with itself is a useful measure of the changes in the components of the parameter at any given point in the parameter space:

$$\nabla l_L(\theta\,;\,y)\ \left(\nabla l_L(\theta\,;\,y)\right)^{\mathrm{T}}.$$

The average of this quantity with respect to the probability density of the random variable Y,

$$I(\theta\,|\,Y) = \mathrm{E}_\theta\left(\nabla l_L(\theta\,|\,Y)\ \left(\nabla l_L(\theta\,|\,Y)\right)^{\mathrm{T}}\right), \tag{1.166}$$

is called the *information matrix*, or the Fisher information matrix, that an observation on Y contains about the parameter θ. (As we mentioned when discussing the score function, "θ" sometimes plays multiple roles. I like to think of it as a fixed but unknown value and use "t" or some other symbol for variables that can take on different values. Statisticians, however, often use the same symbol to represent something that might change.)

The expected value of the square of the first derivative is the expected value of the negative of the second derivative:

$$\mathrm{E}\left(\nabla l_L(\theta\,;\,y)\ \left(\nabla l_L(\theta\,;\,y)\right)^{\mathrm{T}}\right) = -\,\mathrm{E}\left(\mathrm{H}_{l_L}(\theta\,;\,y)\right). \tag{1.167}$$

This is interesting because the expected value of the second derivative, or an approximation of it, can be used in a Newton-like method to solve the maximization problem. We will discuss this in Chapter 6.

In some cases a covariate x_i may be associated with the observed y_i, and the distribution of Y with given covariate x_i has a parameter μ that is a function of x_i and θ. (The linear regression model is an example, with $\mu_i = x_i^{\mathrm{T}}\theta$.) We may in general write $\mu = x_i(\theta)$.

Sometimes we may be interested in the MLE of θ_i given a fixed value of θ_j. Separating the arguments of the likelihood or log-likelihood function in this manner leads to what is called *profile likelihood*, or *concentrated likelihood.*

Statistical Properties of MLE

As with estimation by minimizing residuals, there are two considerations in maximum likelihood estimation. One involves the actual computations, which we discuss in Chapter 6. The other involves the statistical properties of the estimators.

Under suitable regularity conditions we referred to earlier, maximum likelihood estimators have a number of desirable properties. For most distributions used as models in practical applications, the MLEs are consistent. Furthermore, in those cases, the MLE $\widehat{\theta}$ is asymptotically normal (with mean θ) with variance-covariance matrix

$$\left(\mathrm{E}_\theta\left(-\mathrm{H}_{l_L}(\theta \mid Y)\right)\right)^{-1}, \tag{1.168}$$

which is the inverse of the Fisher information matrix. A consistent estimator of the variance-covariance matrix is the inverse of the Hessian at $\widehat{\theta}$. (Note that there are two kinds of asymptotic properties and convergence issues. Some involve the iterative algorithm, and the others are the usual statistical asymptotics in terms of the sample size.)

An issue that goes to the statistical theory, but is also related to the computations, is that of multiple maxima. Here, we recall the last paragraph of the discussion of the statistical properties of minimum residual estimators, and the following is from that paragraph with the appropriate word changes. It may turn out that the optimization problem (1.165) has local maxima. This depends on the nature of the function $f(\cdot)$ in equation (1.164). Local maxima present problems for the computation of the solution because the algorithm may get stuck in a local optimum. Local maxima also present conceptual problems concerning the appropriateness of the estimation criterion itself. As long as there is a unique global optimum, it seems reasonable to seek it and to ignore local optima. It is not so clear what to do if there are multiple points at which the global optimum is attained. That is not a question specifically for methods of computational statistics; it is fundamental to the heuristic of maximizing a likelihood.

Notes and Further Reading

The introductory material on vectors and matrices in Section 1.2 will evolve in later chapters. In Chapter 5 we will discuss computational issues regarding vectors and matrices, and in Chapter 9 we will describe some linear transformations that are useful in statistical analysis. A full-fledged course in "matrices for statisticians" would be useful for someone working in computational statistics.

The material on data-generating processes and statistical inference in Sections 1.3, 1.4 and 1.5 is generally considered to be prerequisite for the present book. A few terms are defined in those sections, but many terms are mentioned without definition, and theorems are referenced without proof. The reader should become familiar with all of those terms and facts because they may be used later in the book. This material is covered in detail in Bickel and Doksum (2001), Casella and Berger (2002), and Hogg et al. (2004), or at a slightly higher level by Lehmann and Casella (1998) and Lehmann and Romano (2005). Statistical theory is based on probability theory. There are many good books on probability theory. The one I use most often is Billingsley (1995).

In addition to the general references on mathematical statistics and statistical inference, the reader should have texts on applied statistics and nonparametric statistics available. A useful text on applied statistics is Kutner, Nachtsheim, and Neter (2004), and one on nonparametric methods based on ranks is Lehmann (1975, reprinted 2006).

There are many subtle properties of likelihood that I do not even allude to in Section 1.4 or Section 1.8. Maximum likelihood estimation is particularly simple in certain "regular" cases (see Lehmann and Casella, 1998, page 485, for example). Various nonregular cases are discussed by Cheng and Traylor (1995).

The information-theoretic approach based on divergence measures mentioned on page 49 is described in some detail in the book by Pardo (2005).

For issues relating to building regression models, as discussed in Section 1.6, see Kennedy and Bancroft (1971), Speed and Yu (1993), and Harrell (2001). A Bayesian perspective is presented in Chapter 6 of Gelman et al. (2004).

For a more thorough coverage of the properties of order statistics alluded to in Section 1.7, see David and Nagaraja (2004).

Dielman, Lowry, and Pfaffenberger (1994) provide extensive comparisons of various quantile estimators, including the simple order statistics. Their results were rather inconclusive, because of the dependence of the performance of the quantile estimators on the shape of the underlying distribution. This is to be expected, of course. If a covariate is available, it may be possible to use it to improve the quantile estimate. This is often the case in simulation studies. See Hesterberg and Nelson (1998) for a discussion of this technique.

Section 1.8 shows that most statistical procedures can be set up as an optimization problem. This is explored more fully in Chapter 1 of Gentle (2009). We discuss some issues in numerical optimization in Chapter 6.

Exercises

1.1. a) How would you describe, in nontechnical terms, the structure of the dataset displayed in Figure 1.1, page 7?
 b) How would you describe the structure of the dataset in more precise mathematical terms? (Obviously, without having the actual data, your equations must contain unknown quantities. The question is meant to make you think about *how* you would do this—that is, what would be the components of your model.)

1.2. Show that the functions defined in equations (1.7) and (1.8) are norms, by showing that they satisfy the defining properties of an inner product given on page 11.

1.3. Inner products, norms, and metrics.
 a) Prove that if $\langle x, y\rangle$ is an inner product, then $\sqrt{\langle x, x\rangle}$ is a norm; that is, it satisfies the properties of a norm listed on page 13 for $x \in \mathrm{IR}^n$.
 b) Prove the if $\|x\|$ satisfies the properties of a norm listed on page 13 for $x \in \mathrm{IR}^n$, then $d(x, y) = \|x - y\|$ satisfies the properties of a metric listed on page 14 for $x, y \in \mathrm{IR}^n$.

1.4. Prove that the Fourier coefficients form the finite expansion in basis elements with the minimum mean squared error (that is, prove inequality (1.35) on page 20). *Hint:* Write $\|x - a_1 v_1\|^2$ as a function of a_1, $\langle x, x\rangle - 2a_0\langle x, v_1\rangle + a_0^2\langle v_1, v_1\rangle$, differentiate, set to zero for the minimum, and determine $a_1 = c_1$ (equation (1.31)). Continue this approach for $a_2, a_3, \ldots, a_k$, or else induction can be used from a_2 on.

1.5. Matrix norms.
 Consider the system of linear equations, $Ax = b$:

$$\begin{aligned} 1.000x_1 + 0.500x_2 &= 1.500, \\ 0.667x_1 + 0.333x_2 &= 1.000. \end{aligned} \tag{1.169}$$

 What are the norms $\|A\|_1$, $\|A\|_2$, and $\|A\|_\infty$?
 We will consider this example further in Exercise 5.1.

1.6. Work out the moment generating function for the mean of a random sample of size n from a $\mathrm{N}(\mu, \sigma^2)$ distribution.

1.7. Let Y and Z have marginal distributions as exponential random variables with parameters α and β respectively. Consider a joint distribution of Y and Z difined by a Gaussian copula (equation (1.83), page 33). What is the correlation between Y and Z? (See also Exercise 7.5c on page 330.)

1.8. Assume a random sample $Y_1, \ldots, Y_n$ from a normal distribution with mean μ and variance σ^2. Determine an unbiased estimator of σ based on the sample variance, S^2, given in equation (1.113). (Note that S^2 is sufficient and unbiased for σ^2.)

1.9. Both the binomial and normal families of distributions are in the exponential class. Show this by writing their PDFs in the form of equation (1.89) on page 35. (The PDFs of these and other distributions are given in Appendix B.)

1.10. For the random variable Y with a distribution in the exponential class and whose density is expressed in the form of equation (1.89), and assuming that the first two moments of $T(Y)$ exist and that $\xi(\cdot)$ is twice differentiable, show that

$$\mathrm{E}(T(Y)) = \nabla\xi(\theta)$$

and

$$\mathrm{V}(T(Y)) = \mathrm{H}_\xi(\theta).$$

Hint: First, assume that we can interchange differentiation with respect to θ and integration with respect to y, and show that

$$\mathrm{E}(\nabla \log(p(Y \mid \theta))) = 0,$$

where the differentiation is with respect to θ. (To show this, write out the derivative of the logarithm, cancel the PDF in the integrand, interchange the integral and the derivative, and differentiate the resulting constant to get 0.)

1.11. Derive equation (1.112) on page 46.

1.12. Discuss (compare and contrast) pivotal and sufficient functions. (Start with the basics: Are they statistics? In what way do they both depend on some universe of discourse, that is, on some family of distributions?)

1.13. Use the pivotal value $g(\widehat{\beta}, \beta)$ in equation (1.131) on page 56 to form a $(1-\alpha)100\%$ confidence region for β in the usual linear regression model.

1.14. Assume that $\{X_1, X_2\}$ is a random sample of size 2 from an exponential distribution with parameter θ. Consider the random variable formed as a Student's t,

$$T = \frac{\overline{X} - \theta}{\sqrt{S^2/2}},$$

where $\overline{X}$ is the sample mean and S^2 is the sample variance,

$$\frac{1}{n-1}\sum(X_i - \overline{X})^2.$$

(Note that $n = 2$.)

a) Show that the distribution of T is negatively skewed (although the distribution of X is positively skewed).
b) Give a heuristic explanation of the negative skewness of T.

The properties illustrated in the exercise relate to the robustness of statistical procedures that use Student's t. While those procedures may be robust to some departures from normality, they are often not robust to skewness. These properties also have relevance to the use of statistics like a Student's t in the bootstrap.

1.15. Show that the variance of the ECDF at a point y is the expression in equation (1.137) on page 60. *Hint:* Use the definition of the variance in terms of expected values, and represent $\mathrm{E}\Big(\big(P_n(y)\big)^2\Big)$ in a manner similar to how $\mathrm{E}\big(P_n(y)\big)$ was represented in equations (1.136).

1.16. The variance functional.
   a) Express the variance of a random variable as a functional of its CDF as was done in equation (1.74) for the mean.
   b) What is the same functional of the ECDF?
   c) What is the plug-in estimate of the variance?
   d) Is the plug-in estimate of the variance an MLE? (The answer is *no*, in general. Why not? For example, what is the MLE of the variance in a gamma(α, β), given a random sample from that distribution? See Appendix B for the PDF and the mean and variance of a gamma distribution.)
   e) What are the statistical properties of the plug-in estimator of the variance? (Is it unbiased? Is it consistent? etc.)

1.17. Give examples of
   a) a parameter that is defined by a linear functional of the distribution function (see equation (1.140)), and
   b) a parameter that is not a linear functional of the distribution function.
   c) Is the variance a linear functional of the distribution function?

1.18. Comparison of estimators of the population median.
Conduct a small Monte Carlo study to compare the MSE of the sample median with the MSE of the Harrell-Davis estimator (equation (1.151)) of the sample median. First, write a function to compute this estimator for any given sample size and given probability. For example, in R:

```
hd <- function(y,p){
      n <- length(y)
      a <- p*(n+1)
      b <- (1-p)*(n+1)
      q <-sum(sort(y)*(pbeta((1:n)/n,a,b)-
                         pbeta((0:(n-1))/n,a,b)))

      q
}
```

Use samples of size 25, and use 1000 Monte Carlo replicates. In each case, for each replicate, generate a pseudorandom sample of size 25, compute the two estimators of the median and obtain the squared error, using the known population value of the median. The average of the squared errors over the 1000 replicates is your Monte Carlo estimate of the MSE. (See Section 7.6 for information on software for generating random deviates.)

a) Use a normal distribution with mean 0 and variance 1. The median is 0.
b) Use a Cauchy distribution with center 0 and scale 1. The median is 0.
c) Use a gamma distribution with shape parameter 2 and scale parameter 3. There is no closed-form expression for the median, but
```
qgamma(.5, 3, 7)
```
yields 0.382.

Summarize your findings in a clearly-written report. What are the differences in relative performance of the sample median and the Harrell-Davis quantile estimator as estimators of the population median? What characteristics of the population seem to have an effect on the relative performance?

1.19. Consider the least squares estimator of β in the usual linear regression model, $\mathrm{E}(Y) = X\beta$.
a) Use expression (1.160) on page 69 to derive the variance-covariance matrix for the estimator.
b) Use expression (1.161) to derive the variance-covariance matrix for the estimator.

1.20. Assume a random sample $y_1, \ldots, y_n$ from a gamma distribution with parameters α and β.
a) What are the least squares estimates of α and β? (Recall $\mathrm{E}(Y) = \alpha\beta$ and $\mathrm{V}(Y) = \alpha\beta^2$.)
b) What is an approximation value of the variance-covariance matrix? Use both expression (1.160) and expression (1.161).
c) Formulate the optimization problem for determining the MLE of α and β. Does this problem have a closed-form solution?
d) What is an approximation of the variance-covariance matrix? (Use expression (1.168), page 73.)

1.21. Summary of types of estimators.
a) Assume a random sample $Y_1, \ldots, Y_n$ from a normal distribution with mean μ and variance σ^2.
i. What is the MLE of μ?
ii. What is the plug-in estimate of μ, when μ is defined by the functional M in equation (1.74)?
iii. What is the plug-in estimate of μ, when μ is defined by the functional $\Xi_{.5}$ in equation (1.76)?
iv. What is the least squares estimate of μ?
v. What is the least absolute values estimate of μ?
vi. What is the Bayes estimate of μ under the assumed prior PDF

$$p_M(\mu) = \frac{1}{\sqrt{2\pi}\sigma_\mathrm{p}} \mathrm{e}^{-(\mu-\mu_\mathrm{p})^2/2\sigma_\mathrm{p}^2}$$

and with a squared-error loss?

vii. *Only if you know UMVUE theory*: What is the UMVUE of μ?

b) Assume a random sample $Y_1, \ldots, Y_n$ from a double exponential distribution with mean μ and variance $2/\lambda^2$.
   i. What is the MLE of μ?
   ii. What is the plug-in estimate of μ, when μ is defined by the functional M in equation (1.74)?
   iii. What is the plug-in estimate of μ, when μ is defined by the functional $\Xi_{.5}$ in equation (1.76)?
   iv. What is the least squares estimate of μ?
   v. What is the least absolute values estimate of μ?

   Note the similarities and the differences in your answers.

Part II

Statistical Computing

Introduction to Part II

The terms "computational statistics" and "statistical computing" are sometimes used interchangeably. The latter term, however, is often used more specifically to refer to the actual computations, both numerical and nonnumerical. The emphasis of Part II is on the computations themselves.

Statistical computing includes relevant areas of numerical analysis, the most important of which are computer number systems, algorithms and programming, function approximation and numerical quadrature, numerical linear algebra, solution of nonlinear equations and optimization, and generation of random numbers. These topics are the subjects of the individual chapters of Part II.

No matter at what level of detail a statistician needs to be familiar with the computational topics of this part, there are two simple, higher-level facts all statisticians should be aware of:

> *Computer numbers are not the same as real numbers, and the arithmetic operations on computer numbers are not exactly the same as those of ordinary arithmetic.*

and

> *The form of a mathematical expression and the way the expression should be evaluated in actual practice may be quite different.*

These statements appear word for word in several places in this book. The material in Chapter 2 illustrates the first statement in some detail. As for the second statement, the difference between an expression and a computing method can easily be illustrated by the problem of obtaining the solution to the linear system of equations $Ax = b$. Assuming A is square and of full rank, the solution can be written as $A^{-1}b$. This is a simple expression, and it is certainly appropriate to use it to denote the solution. This expression may imply that to solve the linear system, we first determine A^{-1} and then multiply it by b on the right to obtain $A^{-1}b$. *This is not the way to obtain the solution.*

In Chapter 5 we will describe how the solution should be obtained. *It does not involve inverting the matrix.* This is just one example that a convenient form of a mathematical expression and the way the expression should be evaluated may be different.

Statistical computations, while motivated by computations in the field of real numbers, $\mathbb{R}$, do not in actual practice conform to the rules of arithmetic in a field. (A *field* is a mathematical structure consisting of a set of two closed, associative, and commutative operations, usually called "addition" that has an additive identity for which each element has an additive inverse, and "multiplication" that has a multiplicative identity for which each element except the additive inverse has a multiplicative inverse, and such that multiplication distributes over addition.) In both the field $\mathbb{R}$ and the computer number arithmetic system, which we will denote as $\mathbb{F}$, there are two basic operations, and the arithmetic operations in the computer *simulate* those in $\mathbb{R}$. A computer engineer may identify a different set of "basic" operations, but those differences are not relevant for our purposes here; the essential facts are that addition on the computer *simulates* addition in the numbers of interest, $\mathbb{R}$, and multiplication on the computer *simulates* multiplication in $\mathbb{R}$. The result of an arithmetic operation in the computer may not yield the same value as the operation that it simulates. Furthermore, the two important properties of arithmetic in $\mathbb{R}$, which are common to all fields, that is, associativity of both operations and distributivity of multiplication over addition, do not hold in computer operations. These facts are very significant for statistical computing.

The mathematical properties of the two structures $\mathbb{R}$ and $\mathbb{F}$ are important, and they are essential to the elements within each structure. In Chapter 2 we describe standards that computer arithmetic must follow. In these standards there are six basic operations, and the standard requires that each of these operations be correct to within rounding. (Note that the exceptions mentioned above involve more than one operation.)

How much a computer user needs to know about the way the computer works depends on the complexity of the use and the extent to which the necessary operations of the computer have been encapsulated in software that is oriented toward the specific application. Although some of the details we discuss will not be important for the computational scientist or for someone doing routine statistical computing, the consequences of those details are important, and the serious computer user must be at least vaguely aware of the consequences. The fact that multiplying two positive integers on the computer can yield a negative number should cause anyone who programs a computer to take care.

We next address, in Chapter 3, some basic issues related to computations, such as algorithm/data interaction, programming principles and so on.

After these two general chapters, the next four chapters address the numerical analysis for the four main classes of problems alluded to above.

2

Computer Storage and Arithmetic

Data represent information at various levels. The form of data, whether numbers, characters, or picture elements, provide different perspectives. Data of whatever form are represented by groups of 0s and 1s, called *bits* from the words "binary" and "digits". (The word was coined by John Tukey.) For representing simple text (that is, strings of characters with no special representation), the bits are usually taken in groups of eight, called *bytes*, or in groups of sixteen, and associated with a specific character according to a fixed coding rule. Because of the common association of a byte with a character, those two words are often used synonymously.

For representing characters in bytes, "ASCII" (pronounced "askey", from American Standard Code for Information Interchange), was the first standard code widely used. At first only English letters, Arabic numerals, and a few marks of punctuation had codes. Gradually over time more and more symbols were given codified representations. Also, because the common character sets differ from one language to another (both natural languages and computer languages), there are several modifications of the basic ASCII code set. When there is a need for more different characters than can be represented in a byte (2^8), codes to associate characters with larger groups of bits are necessary. For compatibility with the commonly used ASCII codes using groups of 8 bits, these codes usually are for groups of 16 bits. These codes for "16-bit characters" are useful for representing characters in some Oriental languages, for example. The Unicode Consortium has developed a 16-bit standard, called Unicode, that is widely used for representing characters from a variety of languages. For any ASCII character, the Unicode representation uses eight leading 0s and then the same eight bits as the ASCII representation.

An important consideration in the choice of a method to represent data is the way data are communicated within a computer and between the computer and peripheral components such as data storage units. Data are usually treated as a fixed-length sequence of bits. The basic grouping of bits in a computer is sometimes called a "word" or a "storage unit". The lengths of words or storage units commonly used in computers are 32 or 64 bits.

J.E. Gentle, *Computational Statistics*, Statistics and Computing,
DOI: 10.1007/978-0-387-98144-4_2,

Like the ASCII standard for representation of characters, there are also some standards for representation of, and operations on, numeric data. The Institute of Electrical and Electronics Engineers (IEEE) and, subsequently, the International Electrotechnical Commission (IEC) have been active in promulgating these standards, and the standards themselves are designated by an IEEE number and/or an IEC number.

The two mathematical models that are often used for numeric data are the ring of integers, $\mathbb{Z}$, and the field of reals, $\mathbb{R}$. We use two computer models, $\mathbb{I}$ and $\mathbb{F}$, to simulate these mathematical entities. Neither $\mathbb{I}$ nor $\mathbb{F}$ is a simple mathematical construct, such as a ring or field.

2.1 The Fixed-Point Number System

Because an important set of numbers is a finite set of reasonably sized integers, efficient schemes for representing these special numbers are available in most computing systems. The scheme is usually some form of a base 2 representation and may use one computer storage unit (this is most common), two storage units, or one half of a storage unit. For example, if a storage unit consists of 32 bits and one storage unit is used to represent an integer, the integer 5 may be represented in binary notation using the low-order bits, as shown in Figure 2.1.

0	0	0	0	0	0	0	0	0	0	0	0	0	0	0	0	0	0	0	0	0	0	0	0	0	0	0	0	0	1	0	1

Fig. 2.1. The Value 5 in a Binary Representation

The sequence of bits in Figure 2.1 represents the value 5, using one storage unit. The character "5" is represented in the ASCII code shown previously, 00110101.

If the set of integers includes the negative numbers also, some way of indicating the sign must be available. The first bit in the bit sequence (usually one storage unit) representing an integer is usually used to indicate the sign; if it is 0, a nonnegative number is represented; if it is 1, a negative number is represented.

Special representations for numeric data are usually chosen so as to facilitate manipulation of data. A common method for representing negative integers, called "twos-complement representation". The twos-complement representation makes arithmetic operations particularly simple. In twos-complement representation, the sign bit is set to 1 and the remaining bits are set to their opposite values (0 for 1; 1 for 0), and then 1 is added to the result. If the bits for 5 are ...00101, the bits for -5 would be ...11010 + 1, or ...11011. If there are k bits in a storage unit (and one storage unit is used to represent a single integer), the integers from 0 through $2^{k-1} - 1$ would be represented in ordinary binary notation using $k - 1$ bits. An integer i in the interval $[-2^{k-1}, -1]$ would be represented by the same bit pattern by which the nonnegative integer $2^{k-1} - i$ is represented, except the sign bit would be 1.

The sequence of bits in Figure 2.2 represents the value -5 using twos-complement notation in 32 bits, with the leftmost bit being the sign bit and the rightmost bit being the least significant bit; that is, the 1 position. The ASCII code for "-5" consists of the codes for "$-$" and "5"; that is, 00101101 00110101.

1	1	1	1	1	1	1	1	1	1	1	1	1	1	1	1	1	1	1	1	1	1	1	1	1	1	1	1	1	0	1	1

Fig. 2.2. The Value -5 in a Twos-Complement Representation

It is easy to see that the largest integer that can be represented in the twos-complement form is $2^{k-1} - 1$ and that the smallest integer is -2^{k-1}.

A representation scheme such as that described above is called *fixed-point* representation or *integer* representation, and the set of such numbers is denoted by $\mathbb{I}$. The notation $\mathbb{I}$ is also used to denote the system built on this set. This system is similar in some ways to a mathematical system called a ring, which is what the integers $\mathbb{Z}$ are. (A ring is similar to a field, except there is no requirement for multiplicative inverses, and the requirement that multiplication be commutative is usually dropped.)

There are several variations of the fixed-point representation. The number of bits used and the method of representing negative numbers are two aspects that often vary from one computer to another. Even within a single computer system, the number of bits used in fixed-point representation may vary; it is typically one storage unit or half of a storage unit.

Fixed-Point Operations

The operations of addition, subtraction, and multiplication for fixed-point numbers are performed in an obvious way that corresponds to the similar operations on the ring of integers. Subtraction is addition of the additive inverse. (In the usual twos-complement representation we described earlier, all fixed-point numbers have additive inverses except -2^{k-1}.) Because there is no multiplicative inverse, however, division is not multiplication by the inverse. The result of division with fixed-point numbers is the result of division with the corresponding real numbers rounded toward zero. This is not considered an arithmetic exception.

As we indicated above, the set of fixed-point numbers together with addition and multiplication is not the same as the ring of integers, if for no other reason than that the set is finite. Under the ordinary definitions of addition and multiplication, the set is not closed under either operation. The computer operations of addition and multiplication, however, are defined so that the set is closed. These operations occur as if there were additional higher-order bits and the sign bit were interpreted as a regular numeric bit. The result is then whatever would be in the standard number of lower-order bits. If the lost higher-order bits are necessary, the operation is said to *overflow*. The result depends on the specific computer architecture. Aside from the interpretation of the sign bit, the result is essentially the same as would result from a modular reduction. In many systems the sign bit is interpreted as an ordinary sign even if it is mathematically inconsistent with the correct result of the operation. (For example, addition of two large positive integers could result in a negative integer because of overflow into the sign bit.) There are some special-purpose algorithms that actually use this modified modular reduction, although such algorithms would not be portable across different computer systems.

2.2 The Floating-Point Number System

In a fixed-point representation, all bits represent values greater than or equal to 1; the *base point* or *radix point* is at the far right, before the first bit. In a fixed-point representation scheme using k bits, the range of representable numbers is of the order of 2^k, usually from approximately -2^{k-1} to 2^{k-1}. Numbers outside of this range cannot be represented directly in the fixed-point scheme. Likewise, nonintegral numbers cannot be represented directly. Large numbers and fractional numbers are generally represented in a scheme similar to what is sometimes called "scientific notation" or in a type of logarithmic notation. Because within a fixed number of digits the radix point is not fixed, this scheme is called *floating-point* representation, and the set of such numbers is denoted by $\mathbb{F}$. The notation $\mathbb{F}$ is also used to denote the system built on this set. (The "system" includes operations in addition to the set itself.)

A floating-point number is also sometimes called "real". Both computer "integers", $\mathbb{I}$, and "reals", $\mathbb{F}$, represent useful subsets of the corresponding mathematical entities, $\mathbb{Z}$ and $\mathbb{R}$, but while the computer numbers called "integers" do constitute a fairly simple subset of the integers, the computer numbers called "real" do not correspond to the real numbers in a natural way. In particular, the floating-point numbers do not occur uniformly over the real number line.

Within the allowable range, a mathematical integer is exactly represented by a computer fixed-point number, but a given real number, even a rational number, of any size may or may not have an exact representation by a floating-point number. This is the familiar situation where fractions such as $\frac{1}{3}$ have no finite representation in base 10. The simple rule, of course, is that the number must be a rational number whose denominator in reduced form factors into only primes that appear in the factorization of the base. In base 10, for example, only rational numbers whose factored denominators contain only

2s and 5s have an exact, finite representation; and in base 2, only rational numbers whose factored denominators contain only 2s have an exact, finite representation.

For a given real number x, we will occasionally use the notation

$$[x]_{\mathrm{c}} \tag{2.1}$$

to indicate the floating-point number that is "closest" to x, and we will refer to the exact value of a floating-point number as a *computer number*. That is, $[x]_{\mathrm{c}}$ is a computer number, but x is a computer number if and only if $x = [x]_{\mathrm{c}}$. We will also use the phrase "computer number" to refer to the value of a computer fixed-point number. While the definition of $[x]_{\mathrm{c}}$ requires that $|[x]_{\mathrm{c}} - x| \leq |y - x|$ for any $y \in \mathbb{F}$, standard-conforming computers have four different rounding modes, as we describe on page 94. What we have defined here for $[x]_{\mathrm{c}}$ is "round to nearest".

It is important to understand that computer numbers $\mathbb{I}$ and $\mathbb{F}$ are finite. The set of fixed-point numbers $\mathbb{I}$ is a proper subset of $\mathbb{Z}$. The set of floating-point numbers is almost a proper subset of $\mathbb{R}$, but it is not a subset because it contains some numbers not in $\mathbb{R}$; see the special floating-point numbers discussed on page 94. There are many concepts in $\mathbb{R}$, such as irrationality, that do not exist in $\mathbb{F}$. (There are no irrational numbers in $\mathbb{F}$.)

Our main purpose in using computers, of course, is not to evaluate functions of the set of computer floating-point numbers or the set of computer integers; the main immediate purpose usually is to perform operations in the field of real (or complex) numbers or occasionally in the ring of integers. (And, in the famous dictum of Richard Hamming, "the purpose of computing is insight, not numbers".) Doing computations on the computer, then, involves using the sets of computer numbers to simulate the sets of reals or integers.

The Parameters of the Floating-Point Representation

The parameters necessary to define a floating-point representation are the *base* or *radix*, the range of the *mantissa* or *significand*, and the range of the *exponent*. Because the number is to be represented in a fixed number of bits, such as one storage unit or word, the ranges of the significand and exponent must be chosen judiciously so as to fit within the number of bits available. If the radix is b and the integer digits d_i are such that $0 \leq d_i < b$, and there are enough bits in the significand to represent no more than p digits, then a real number is approximated by

$$\pm 0.d_1 d_2 \cdots d_p \times b^e, \tag{2.2}$$

where e is an integer. This is the standard model for the floating-point representation. (The d_i are called "digits" from the common use of base 10.)

The number of bits allocated to the exponent e must be sufficient to represent numbers within a reasonable range of magnitudes; that is, so that the smallest number in magnitude that may be of interest is approximately $b^{e_{\min}}$ and the largest number of interest is approximately $b^{e_{\max}}$, where $e_{\min}$ and $e_{\max}$ are, respectively, the smallest and the largest allowable values of the exponent. Because $e_{\min}$ is likely negative and $e_{\max}$ is positive, the exponent requires a sign. In practice, most computer systems handle the sign of the exponent by defining a *bias* and then subtracting the bias from the value of the exponent evaluated without regard to a sign.

In order to ensure a unique representation for all numbers (except 0), most floating-point systems require that the leading digit in the significand be nonzero unless the magnitude is less than $b^{e_{\min}}$. A number with a nonzero leading digit in the significand is said to be *normalized.*

The most common value of the base b is 2, although 16 and even 10 are sometimes used. If the base is 2, in a normalized representation, the first digit in the significand is always 1; therefore, it is not necessary to fill that bit position, and so we effectively have an extra bit in the significand. The leading bit, which is not represented, is called a "hidden bit". This requires a special representation for the number 0, however.

In a typical computer using a base of 2 and 64 bits to represent one floating-point number, 1 bit may be designated as the sign bit, 52 bits may be allocated to the significand, and 11 bits allocated to the exponent. The arrangement of these bits is somewhat arbitrary, and of course the physical arrangement on some kind of storage medium would be different from the "logical" arrangement. A common logical arrangement assigns the first bit as the sign bit, the next 11 bits as the exponent, and the last 52 bits as the significand. (Computer engineers sometimes label these bits as $0, 1, \ldots$, and then get confused as to which is the ith bit. When we say "first", we mean "first", whether an engineer calls it the "0th" or the "1st".) The range of exponents for the base of 2 in this typical computer would be 2,048. If this range is split evenly between positive and negative values, the range of orders of magnitude of representable numbers would be from -308 to 308. The bits allocated to the significand would provide roughly 16 decimal places of precision.

Figure 2.3 shows the bit pattern to represent the number 5, using $b = 2$, $p = 24$, $e_{\min} = -126$, and a bias of 127, in a word of 32 bits. The first bit on the left is the sign bit, the next 8 bits represent the exponent, 129, in ordinary base 2 with a bias, and the remaining 23 bits represent the significand beyond the leading bit, known to be 1. (The binary point is to the right of the leading bit that is not represented.) The value is therefore $+1.01 \times 2^2$ in binary notation.

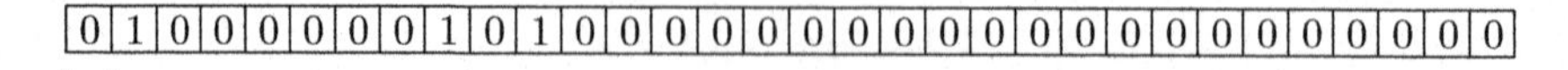

Fig. 2.3. The Value 5 in a Floating-Point Representation

As mentioned above, the set of floating-point numbers is not uniformly distributed over the ordered set of the reals. (Exercise 2.9a and its partial solution on page 677 may help you to see how the spacing varies.) There are the same number of floating-point numbers in the interval $[b^i,\, b^{i+1}]$ as in the interval $[b^{i+1},\, b^{i+2}]$ for any integer $e_{\min} \leq i \leq e_{\max} - 2$, even though the second interval is b times as long as the first. Figures 2.4 through 2.6 illustrate this. The fixed-point numbers, on the other hand, are uniformly distributed over their range, as illustrated in Figure 2.7.

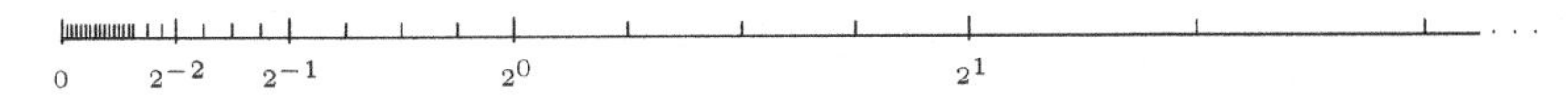

Fig. 2.4. The Floating-Point Number Line, Nonnegative Half

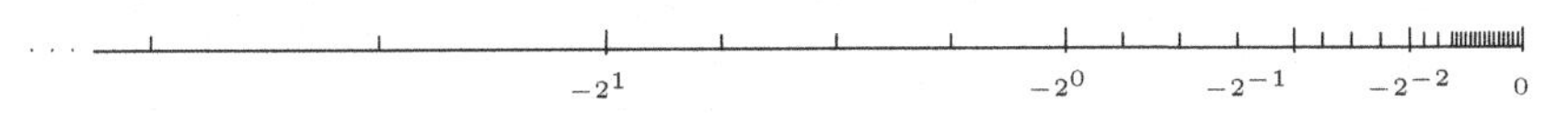

Fig. 2.5. The Floating-Point Number Line, Nonpositive Half

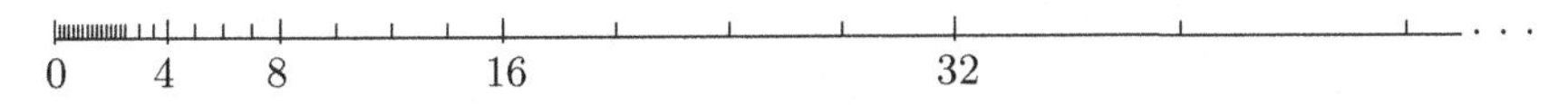

Fig. 2.6. The Floating-Point Number Line, Nonnegative Half; Another View

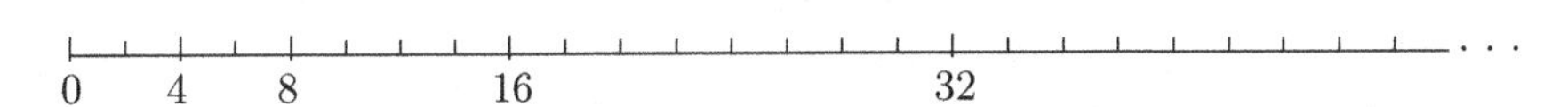

Fig. 2.7. The Fixed-Point Number Line, Nonnegative Half

The density of the floating-point numbers is generally greater closer to zero. Notice that if floating-point numbers are all normalized, the spacing between 0 and $b^{e_{\min}}$ is $b^{e_{\min}}$ (that is, there is no floating-point number in that open interval), whereas the spacing between $b^{e_{\min}}$ and $b^{e_{\min}+1}$ is $b^{e_{\min}-p+1}$. Most systems do not require floating-point numbers less than $b^{e_{\min}}$ in magnitude to be normalized. This means that the spacing between 0 and $b^{e_{\min}}$ can be $b^{e_{\min}-p}$, which is more consistent with the spacing just above $b^{e_{\min}}$. When these nonnormalized numbers are the result of arithmetic operations, the result is called "graceful" or "gradual" underflow.

The spacing between floating-point numbers has some interesting (and, for the novice computer user, surprising!) consequences. For example, if 1 is repeatedly added to x, by the recursion

$$x^{(k+1)} = x^{(k)} + 1,$$

the resulting quantity does not continue to get larger. Obviously, it could not increase without bound because of the finite representation. It does not eventually become Inf (see page 94). It does not even approach the largest number representable! (This is assuming that the parameters of the floating-point representation are reasonable ones.) In fact, if x is initially smaller in absolute value than $b^{e_{\max}-p}$ (approximately), the recursion

$$x^{(k+1)} = x^{(k)} + c$$

will converge to a stationary point for any value of c smaller in absolute value than $b^{e_{\max}-p}$.

The way the arithmetic is performed would determine these values precisely; as we shall see below, arithmetic operations may utilize more bits than are used in the representation of the individual operands.

The spacings of numbers just smaller than 1 and just larger than 1 are particularly interesting. This is because we can determine the *relative spacing* at any point by knowing the spacing around 1. These spacings at 1 are sometimes called the "machine epsilons", denoted $\epsilon_{\min}$ and $\epsilon_{\max}$ (not to be confused with $e_{\min}$ and $e_{\max}$ defined earlier). It is easy to see from the model for floating-point numbers on page 89 that

$$\epsilon_{\min} = b^{-p} \tag{2.3}$$

and

$$\epsilon_{\max} = b^{1-p}; \tag{2.4}$$

see Figure 2.8. The more conservative value, $\epsilon_{\max}$, sometimes called "the machine epsilon", ϵ or ϵ_{mach}, provides an upper bound on the rounding that occurs when a floating-point number is chosen to represent a real number. A floating-point number near 1 can be chosen within $\epsilon_{\max}/2$ of a real number that is near 1. This bound, $\frac{1}{2}b^{1-p}$, is called the *unit roundoff*.

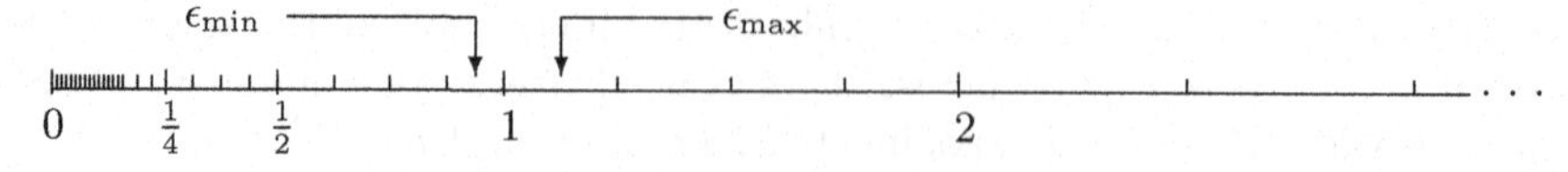

Fig. 2.8. Relative Spacings at 1: "Machine Epsilons"

These machine epsilons are also called the "smallest relative spacing" and the "largest relative spacing" because they can be used to determine the relative spacing at the point x (see Figure 2.8).

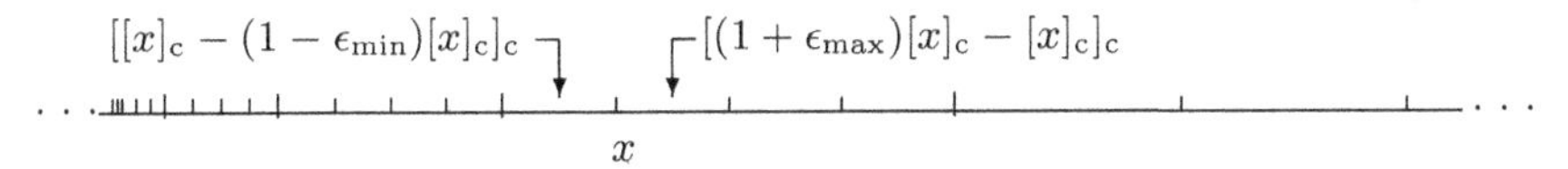

Fig. 2.9. Relative Spacings

If x is not zero, the relative spacing at x is approximately

$$\frac{x - (1 - \epsilon_{min})x}{x} \tag{2.5}$$

or

$$\frac{(1 + \epsilon_{max})x - x}{x}. \tag{2.6}$$

Notice that we say "approximately". First of all, we do not even know that x is representable. Although $(1 - \epsilon_{min})$ and $(1 + \epsilon_{max})$ are members of the set of floating-point numbers by definition, that does not guarantee that the product of either of these numbers and $[x]_c$ is also a member of the set of floating-point numbers. However, the quantities $[(1 - \epsilon_{min})[x]_c]_c$ and $[(1 + \epsilon_{max})[x]_c]_c$ are representable (by the definition of $[\cdot]_c$ as a floating point number approximating the quantity within the brackets); and, in fact, they are respectively the next smallest number than $[x]_c$ (if $[x]_c$ is positive, or the next largest number otherwise) and the next largest number than $[x]_c$ (if $[x]_c$ is positive). The spacings at $[x]_c$ therefore are

$$[x]_c - [(1 - \epsilon_{min})[x]_c]_c \tag{2.7}$$

and

$$[(1 + \epsilon_{max})[x]_c - [x]_c]_c. \tag{2.8}$$

As an aside, note that this implies it is probable that

$$[(1 - \epsilon_{min})[x]_c]_c = [(1 + \epsilon_{min})[x]_c]_c.$$

In practice, to compare two numbers x and y, we do not ask if

$$x == y. \tag{2.9}$$

We must compare $[x]_c$ and $[y]_c$. We consider x and y different if

$$[|y|]_c < [|x|]_c - [\epsilon_{min}[|x|]_c]_c \tag{2.10}$$

or if

$$[|y|]_c > [|x|]_c + [\epsilon_{max}[|x|]_c]_c. \tag{2.11}$$

The relative spacing at any point obviously depends on the value represented by the least significant digit in the significand. This digit (or bit) is called the "unit in the last place", or "ulp". The magnitude of an ulp depends of course on the magnitude of the number being represented. Any real number within the range allowed by the exponent can be approximated within $\frac{1}{2}$ ulp by a floating-point number.

Standardization of Floating-Point Representation

Although different computers represent numeric data in different ways, there has been some attempt to provide standards for the range and precision of floating-point quantities. The IEEE Standard 754-1985 (IEEE, 1985) is a binary standard that specifies the exact layout of the bits for two different precisions, "single" and "double". In both cases, the standard requires that the radix be 2. For single precision, p must be 24, $e_{\max}$ must be 127, and $e_{\min}$ must be -126. For double precision, p must be 53, $e_{\max}$ must be 1023, and $e_{\min}$ must be -1022.

The IEEE Standard 754 also defines two additional precisions, "single extended" and "double extended". For each of the extended precisions, the standard sets bounds on the precision and exponent ranges rather than specifying them exactly. The extended precisions have larger exponent ranges and greater precision than the corresponding precision that is not "extended".

The standard also defines four rounding modes: *round down*, *round up*, *round toward zero*, and *round to nearest*, each with the obvious meaning. The standard requires that round to nearest be the default rounding mode. (In the case of a tie, round to nearest chooses a 0 in the least significant position.) The standard requires that the result of add, subtract, multiply, divide, remainder, and square root be correct to the specified rounding mode.

The IEEE Standard 754-1985 has been revised and is now IEEE Standard 754-2008. This standard now also allows a radix of 10; that is, it provides a standard for decimal storage and arithmetic operations.

Special Floating-Point Numbers

It is convenient to be able to represent certain special numeric entities, such as infinity or "indeterminate" (0/0), which do not have ordinary representations in any base-digit system. Although 8 bits are available for the exponent in the single-precision IEEE binary standard, $e_{\max} = 127$ and $e_{\min} = -126$. This means there are two unused possible values for the exponent; likewise, for the double-precision standard, there are two unused possible values for the exponent. These extra possible values for the exponent allow us to represent certain special floating-point numbers.

An exponent of $e_{\max} + 1$ allows us to represent $\pm\infty$ or the indeterminate value. A floating-point number with this exponent and a significand of 0 represents $\pm\infty$ (the sign bit determines the sign, as usual). This value is called Inf or -Inf.

Numerical operations with Inf or $-$Inf yield values consistent with those in the extended real number system; that is, if $\mathtt{x} \in \mathbb{F}$ and $0<\mathtt{x}<$Inf, then x$*$Inf=Inf and $-$x$*$Inf=$-$Inf. We also have Inf+Inf=Inf and Inf$*$Inf=Inf, but 0$*$Inf, Inf$-$Inf, and Inf/Inf are indeterminate.

A floating-point number with the exponent $e_{\max}+1$ and a nonzero significand represents an indeterminate numerical value, such as $\frac{0}{0}$, or else a missing value. A missing value is an element of whatever type that has not been assigned a value. An indeterminate numerical value is called "not-a-number", or "NaN", and a missing value is called "not-available", or "NA".

Any numerical operation involving a NaN and valid numerical values results in a NaN. Any operation involving a NA results in a NA.

Working with NaNs or NAs requires software to identify these values. Ordinary computer language components, such as for determining whether two variables have equal values, cannot directly determine whether or not a variable has a value of NaN or NA. If a variable `x` has a value of NaN, it is not true that `x = NaN`. (The value is indeterminate, so it is not true that it equals itself.) Special functions should be used to determine if a value is NaN or NA. Although it is not part of the standard definitions of the languages, most Fortran and C compilers include a function `isnan` to test for a NaN. Many C compilers include an additional function `isinf` to test for $\pm\infty$. Neither of these languages normally includes a function to test for unassigned values. R includes a function `is.nan` to test for a NaN (it is false for a NA) and a function `is.na` to test for a NA or a NaN (it is true for either a NA or a NaN).

Determining the Numerical Characteristics of a Particular Computer

Computer designers have a great deal of latitude in how they choose to represent data. The ASCII standards of ANSI and ISO have provided a common representation for individual characters. The IEEE Standards 754-1985 and 754-2008 referred to previously (IEEE, 1985) brought some standardization to the representation of floating-point data, but do not specify how the available bits are to be allocated among the sign, exponent, and significand.

The environmental inquiry program MACHAR can be used to determine the characteristics of a specific computer's floating-point representation and its arithmetic. The program, which is available in CALGO from `netlib` (see page 692 in the Bibliography), was written in Fortran 77 and has been translated into C and R. In R, the results on a given system are stored in the variable `.Machine`. Other R objects that provide information on a computer's characteristics are the variable `.Platform` and the function `capabilities`.

Computer Operations on Numeric Data

As we have emphasized above, the numerical quantities represented in the computer are used to simulate or approximate more interesting quantities, namely the real numbers or perhaps the integers. Obviously, because the sets (that is, of computer numbers and real numbers) are not the same, we could not define operations on the computer numbers that would yield the same field as the familiar field of the reals. In fact, because of the nonuniform spacing of floating-point numbers, we would suspect that some of the fundamental properties of a field may not hold. Depending on the magnitudes of the quantities involved, it is possible, for example, that if we compute ab and ac and then $ab+ac$, we may not get the same thing as if we compute $(b+c)$ and then $a(b+c)$. Just as we use the computer quantities to simulate real quantities, we define operations on the computer quantities to simulate the familiar operations on real quantities. Designers of computers attempt to define computer operations so as to correspond closely to operations on real numbers, but we must not lose sight of the fact that the computer uses a different arithmetic system.

The basic operational objective in numerical computing, of course, is that a computer operation, when applied to computer numbers, yields computer numbers that approximate the number that would be yielded by a certain mathematical operation applied to the numbers approximated by the original computer numbers. Just as we introduced the notation

$$[x]_{\mathrm{c}}$$

on page 89 to denote the computer floating-point number approximation to the real number x, we occasionally use the notation

$$[\circ]_{\mathrm{c}} \tag{2.12}$$

to refer to a computer operation that simulates the mathematical operation $\circ$. Thus,

$$[+]_{\mathrm{c}}$$

represents an operation similar to addition but that yields a result in a set of computer numbers. (We use this notation only where necessary for emphasis, however, because it is somewhat awkward to use it consistently.) The failure of the familiar laws of the field of the reals, such as the distributive law cited above, can be anticipated by noting that

$$[[a]_{\mathrm{c}}\ [+]_{\mathrm{c}}\ [b]_{\mathrm{c}}]_{\mathrm{c}} \neq [a+b]_{\mathrm{c}}, \tag{2.13}$$

or by considering the simple example in which all numbers are rounded to one decimal and so $\frac{1}{3}+\frac{1}{3} \neq \frac{2}{3}$ (that is, $.3+.3 \neq .7$).

The three familiar laws of the field of the reals (commutativity of addition and multiplication, associativity of addition and multiplication, and distribution of multiplication over addition) result in the independence of the order in which operations are performed; the failure of these laws implies that the order of the operations may make a difference. When computer operations are performed sequentially, we can usually define and control the sequence fairly easily. If the computer performs operations in parallel, the resulting differences in the orders in which some operations may be performed can occasionally yield unexpected results.

Because the operations are not closed, special notice may need to be taken when the operation would yield a number not in the set. Adding two numbers, for example, may yield a number too large to be represented well by a computer number, either fixed-point or floating-point. When an operation yields such an anomalous result, an *exception* is said to exist.

Floating-Point Operations

As we have seen, real numbers within the allowable range may or may not have an exact floating-point operation, and the computer operations on the computer numbers may or may not yield numbers that represent exactly the real number that would result from mathematical operations on the numbers. If the true result is r, the best we could hope for would be $[r]_{\mathrm{c}}$. As we have mentioned, however, the computer operation may not be exactly the same as the mathematical operation being simulated, and furthermore, there may be several operations involved in arriving at the result. Hence, we expect some error in the result.

Summary: Comparison of Reals and Floating-Point Numbers

For most applications, the system of floating-point numbers simulates the field of the reals very well. It is important, however, to be aware of some of the differences in the two systems. There is a very obvious useful measure for the reals, namely the Lebesgue measure, μ, based on lengths of open intervals. An approximation of this measure is appropriate for floating-point numbers, even though the set is finite. The finiteness of the set of floating-point numbers means that there is a difference in the cardinality of an open interval and a closed interval with the same endpoints. The uneven distribution of floating-point values relative to the reals (Figures 2.4 and 2.5) means that the cardinalities of two interval-bounded sets with the same interval length may be different. On the other hand, a counting measure does not work well at all.

Some general differences in the two systems are exhibited in Table 2.1. The last four properties in Table 2.1 are properties of a field. The important facts are that $\mathbb{R}$ is an uncountable field and that $\mathbb{F}$ is a more complicated finite mathematical structure.

2.3 Errors

If the computed value is $\tilde{r}$ (for the true scalar value r), we speak of the *absolute error*,

$$|\tilde{r} - r|, \tag{2.14}$$

and the *relative error*,

$$\frac{|\tilde{r} - r|}{|r|} \tag{2.15}$$

(so long as $r \neq 0$). An important objective in numerical computation obviously is to ensure that the error in the result is small.

Table 2.1. Differences in Real Numbers and Floating-Point Numbers

	$\mathbb{R}$	$\mathbb{F}$
cardinality:	uncountable	finite
measure:	$\mu((x,y)) = \|x-y\|$ $\mu((x,y)) = \mu([x,y])$	$\nu((\mathtt{x},\mathtt{y})) = \nu([\mathtt{x},\mathtt{y}]) = \|\mathtt{x}-\mathtt{y}\|$ $\exists\, \mathtt{x},\mathtt{y},\mathtt{z},\mathtt{w} \ni \|\mathtt{x}-\mathtt{y}\| = \|\mathtt{z}-\mathtt{w}\|$, but $\#(\mathtt{x},\mathtt{y}) \neq \#(\mathtt{z},\mathtt{w})$
continuity:	if $x < y$, $\exists z \ni x < z < y$ and $\mu([x,y]) = \mu((x,y))$	$\mathtt{x} < \mathtt{y}$, but no $\mathtt{z} \ni \mathtt{x} < \mathtt{z} < \mathtt{y}$ and $\#[\mathtt{x},\mathtt{y}] > \#(\mathtt{x},\mathtt{y})$
convergence	$\sum_{x=1}^{\infty} x$ diverges	$\sum_{\mathtt{x}=1}^{\infty} \mathtt{x}$ converges, if interpreted as $(\cdots((1+2)+3)\cdots)$
closure:	$x, y \in \mathbb{R} \Rightarrow x + y \in \mathbb{R}$ $x, y \in \mathbb{R} \Rightarrow xy \in \mathbb{R}$	not closed wrt addition not closed wrt multiplication (exclusive of infinities)
operations with an identity, a or $\mathtt{a}$:	$a = 0$, unique $a + x = x$, for any x $x - x = a$, for any x	$\mathtt{a}+\mathtt{x} = \mathtt{b}+\mathtt{x}$, but $\mathtt{b} \neq \mathtt{a}$ $\mathtt{a}+\mathtt{x} = \mathtt{x}$, but $\mathtt{a}+\mathtt{y} \neq \mathtt{y}$ $\mathtt{a}+\mathtt{x} = \mathtt{x}$, but $\mathtt{x}-\mathtt{x} \neq \mathtt{a}$
associativity:	$x, y, z \in \mathbb{R} \Rightarrow$ $(x+y)+z = x+(y+z)$ $(xy)z = x(yz)$	 not associative not associative
distributivity:	$x, y, z \in \mathbb{R} \Rightarrow$ $x(y+z) = xy + xz$	 not distributive

As mentioned above, if r is the result of one of the six basic operations (addition, subtraction, multiplication, division, remaindering, and extraction of a square root), then in conformance to IEEE Standard 754, $\tilde{r}$ must be $[r]_c$ (under default rounding). In this case, the absolute error is

$$|[r]_c - r|,$$

which is as good as can be desired. Obviously, if more that one operation is involved in obtaining the result, it may be the case that $\tilde{r} \neq [r]_c$, and there is no general way to ensure that $\tilde{r} = [r]_c$.

In Section 3.1, we expand on the discussion of computational errors, and in that and the subsequent section, we mention some ways of reducing errors.

Addition of Several Numbers

When several numbers x_i are to be summed, it is likely that as the operations proceed serially, the magnitudes of the partial sum and the next summand will be quite different. In such a case, the full precision of the next summand is lost. This is especially true if the numbers are of the same sign. As we mentioned earlier, a computer program to implement serially the algorithm implied by $\sum_{i=1}^{\infty} i$ will converge to some number much smaller than the largest floating-point number.

If the numbers to be summed are not all the same constant (and if they are constant, just use multiplication!), the accuracy of the summation can be increased by first sorting the numbers and summing them in order of increasing magnitude. If the numbers are all of the same sign and have roughly the same magnitude, a pairwise "fan-in" method may yield good accuracy. In the fan-in method, the n numbers to be summed are added two at a time to yield $\lceil n/2 \rceil$ partial sums. The partial sums are then added two at a time, and so on, until all sums are completed. It is likely that the numbers to be added will be of roughly the same magnitude at each stage. Remember we are assuming they have the same sign initially; this would be the case, for example, if the summands are squares.

Another way that is even better is due to W. Kahan:

$$\begin{aligned}
&s = x_1 \\
&a = 0 \\
&\text{for } i = 2, \ldots, n \\
&\{ \\
&\quad y = x_i - a \\
&\quad t = s + y \\
&\quad a = (t - s) - y \\
&\quad s = t \\
&\}.
\end{aligned} \tag{2.16}$$

Often the nature of the addends is such that the sum just cannot be computed. (See Exercise 2.2a, for which this statement is a tautology!) In other cases, it may be helpful to change the problem. For example, consider the evaluation of e^x using the Taylor series

$$e^x = 1 + x + x^2/2! + x^3/3! + \cdots \tag{2.17}$$

This example is used only for illustration; *this is not the way to evaluate* e^x.

Suppose $x = 20$. The R code shown in Figure 2.10 yields 485165100, which is correct to 7 digits. (Actually, it yields 485165195., which is correct to 9 digits.)

```
stop <- 100
ex <- 1
xi <- 1
ifac <- 1
for (i in 1:stop) {
  xi <- x*xi
  ifac <- i*ifac
  ex <- ex+xi/ifac
  }
ex
```

Fig. 2.10. Code to Compute e^x Using a Taylor Series Approximation

Now, let $x = -20$. Using the code above, we get 4.992639e-09. The answer to 7 digits is 2.061154e-09. The solution is correct to only 1 digit; the relative error is over 100%. Notice, however, if we compute $1/e^{20}$, we get 1/485165195, or 2.061154e-09, which is again correct to 7 digits.

The problem in the evaluation of the series (2.17) arises not just from the varying magnitude of the terms, but also in the signs of the terms, resulting in *cancellation*. Cancellation can have even more dramatic effects. We call it catastrophic cancellation.

Catastrophic Cancellation

Another type of error that results from the finite precision of floating-point numbers is *catastrophic cancellation*. This can occur when two rounded values of approximately equal magnitude and opposite signs are added. If the values are exact, cancellation can also occur, but it is *benign*. After catastrophic cancellation occurs, the digits left are just the digits that represented the rounding.

Suppose $x \approx y$ and that $[x]_c = [y]_c$. The computed result will be zero, whereas the correct (rounded) result is $[x - y]_c$. The relative error is 100%. This error is caused by rounding, but it is different from the "rounding error" discussed above. Although the loss of information arising from the rounding error is the culprit, the rounding would be of little consequence were it not for the cancellation.

Additions of quantities of approximately equal magnitude and opposite signs may arise often in numerical computations. To avoid catastrophic cancellation, we must carefully consider all additions, and when catastrophic cancellation may occur, rearrange the computations if possible. For a simple example, consider the problem of computing the roots of a quadratic polynomial,

$$ax^2 + bx + c.$$

In the quadratic formula

$$x = \frac{-b \pm \sqrt{b^2 - 4ac}}{2a}, \tag{2.18}$$

the square root of the discriminant, $(b^2 - 4ac)$, may be approximately equal to b in magnitude, meaning that one of the roots is close to zero and, in fact, may be computed as zero. The solution is to compute only one of the roots, x_1, by the formula (the "$-$" root if b is positive and the "$+$" root if b is negative) and then compute the other root, x_2 by the relationship $x_1 x_2 = c/a$.

Catastrophic cancellation results from only a few operations, however, the effects of smaller cancellations may accumulate. In the example above for evaluating e^{-20} using the Taylor series, we have

$$1 - 20 + 200 - 1333.\ldots + 6666.\ldots - 26666.\ldots + \cdots .$$

The partial sums are

$$1, -19, 181, -1152.\ldots, +5514.\ldots, -21152.\ldots, +67736.\ldots, -186231.\ldots.$$

The practical question of interest is how can we tell that there are problems. In the simple example above, if we monitor the computations, we find that on step 99 the computed approximation is 6.138260e-09. Getting 4.992639e-09 tells us clearly that the process is not working.

Notes and Further Reading

Some of the material in this chapter is based on Chapter 10 in Gentle (2007). In particular, Table 2.1 and some of the illustrations of bit-level representations are from that book.

The details of representation and computation with computer numbers discussed in this chapter may be important only to a subset of people working in computational statistics. Every statistician, however, should understand the conclusion of the section:

> *Computer numbers are not the same as real numbers, and the arithmetic operations on computer numbers are not exactly the same as those of ordinary arithmetic.*

Standardization of numerical representation and operations has made the work of applied numerical analysts much easier. There are not many computers produced nowadays that do not implement the IEEE floating point standard.

The work of Jim Cody and Velvel Kahan was instrumental in getting computer manufacturers to improve the numerical operations. Kahan played the key role in formulating the IEEE Standard 754. Once this standard was adopted, computer manufacturers quickly began ensuring that their representation of numeric quantities and the numeric operations in their CPUs conformed to the standards.

More detail on computer arithmetic is covered by Kulisch (2008) and more specifically for the IEEE Standard 754 by Overton (2001).

Topics not addressed in this chapter include higher precision computations, including "exact" computations for rational numbers, and interval arithmetic, in which each of the computer numbers used throughout the data input and computations are interval bounds for the exact real number. Walster (2005) and Moore et al. (2009) give general descriptions of interval data types and discussions of ways they can be used in computations. The book edited by Hu at al. (2008) contains chapters on the use of interval arithmetic in various problems in numerical analysis, and the journal *Reliable Computing* is devoted to research in this area. The book edited by Einarsson (2005) contains chapters on several different issues in the accuracy of computations resulting from the use of higher precision, of interval arithmetic, and of software features meant to ensure a higher degree of reliability.

Exercises

2.1. In the IEEE Standard 754 single precision format, what is the value of 0.1 in the default rounding mode? What is the value in the round down mode? What is the value in the round up mode? Answering these questions requires representing the solution in base 2. Note an interesting fact: $1/10$ is a rational number; therefore its representation in any integer base is a "repeating fraction", just as $1/3$ is a repeating decimal fraction. ($1/10$ is a repeating fraction in base 10, of course. The repeating decimal sequence happens to be "0".)

2.2. An important attitude in the computational sciences is that the computer is to be used as a tool of exploration and discovery. The computer should be used to check out "hunches" or conjectures, which then later should be subjected to analysis in the traditional manner. There are limits to this approach, however. An example is in limiting processes. Because the computer deals with finite quantities, the results of a computation may be misleading. Explore each of the situations below, using C or Fortran. A few minutes or even seconds of computing should be enough to give you a feel for the nature of the computations.

In these exercises, you may write computer programs in which you perform tests for equality. A word of warning is in order about such tests, however. If a test involving a quantity `x` is executed soon after the computation of `x`, the test may be invalid within the set of floating-point numbers with which the computer nominally works. This is because the test may be performed using the extended precision of the computational registers.

a) Consider the question of the convergence of the series

$$\sum_{i=1}^{\infty} i.$$

Obviously, this series does not converge in $\mathbb{R}$. Suppose, however, that we begin summing this series using floating-point numbers. Will the computations overflow? If so, at what value of i (approximately)? Or will the series converge in $\mathbb{F}$? If so, to what value, and at what value of i (approximately)? In either case, state your answer in terms of the standard parameters of the floating-point model, b, p, $e_{\min}$, and $e_{\max}$ (page 89).

b) Consider the question of the convergence of the series

$$\sum_{i=1}^{\infty} 2^{-2i}.$$

Same questions as above.

c) Consider the question of the convergence of the series

$$\sum_{i=1}^{\infty} \frac{1}{i}.$$

Same questions.

d) Consider the question of the convergence of the series

$$\sum_{i=1}^{\infty} \frac{1}{i^p},$$

for $p \geq 1$. Same questions, except address the effect of the value of the variable p.

2.3. We know, of course, that the harmonic series in Exercise 2.2c does not converge (although the naive program to compute it does). It is, in fact, true that

$$\begin{aligned} H_n &= \sum_{i=1}^{n} \frac{1}{i} \\ &= f(n) + \gamma + \mathrm{o}(1), \end{aligned}$$

where f is an increasing function and γ is Euler's constant. For various n, compute H_n. Determine a function f that provides a good fit and obtain an approximation of Euler's constant.

2.4. Machine characteristics.
   a) Write a program to determine the smallest and largest relative spacings. Use it to determine them on the machine you are using.
   b) Write a program to determine whether your computer system implements gradual underflow.
   c) Write a program to determine the bit patterns of $+\infty$, $-\infty$, and NaN on a computer that implements the IEEE binary standard. (This may be more difficult than it seems.)

2.5. What is the numerical value of the rounding unit ($\frac{1}{2}$ ulp) in the IEEE Standard 754 double precision?

2.6. Consider the standard model (2.2) for the floating-point representation:

$$\pm 0.d_1 d_2 \cdots d_p \times b^e,$$

with $e_{\min} \leq e \leq e_{\max}$. Your answers may depend on an additional assumption or two. Either choice of (standard) assumptions is acceptable.
   a) How many floating-point numbers are there?
   b) What is the smallest positive number?
   c) What is the smallest number larger than 1?
   d) What is the smallest number X, such that $X + 1 = X$?
   e) Suppose $p = 4$ and $b = 2$ (and $e_{\min}$ is very small and $e_{\max}$ is very large). What is the next number after 20 in this number system?

2.7. a) Define parameters of a floating-point model so that the number of numbers in the system is less than the largest number in the system.
   b) Define parameters of a floating-point model so that the number of numbers in the system is greater than the largest number in the system.

2.8. Suppose that a certain computer represents floating point numbers in base 10, using five decimal places for the mantissa, one decimal places for the exponent, one decimal place for the sign of exponent, and one decimal place for the sign of the number.
   a) What is the "smallest relative spacing" and the "largest relative spacing"? (Your answer may depend on certain additional assumptions about the representation; state any assumptions.)
   b) What is the largest number g, such that $417 + g = 417$?
   c) Discuss the associativity of addition using numbers represented in this system. Give an example of three numbers, a, b, and c, such that using this representation, $(a + b) + c \neq a + (b + c)$, unless the operations are chained. Then show how chaining could make associativity hold for some more numbers, but still not hold for others.
   d) Compare the maximum rounding error in the computation $x+x+x+x$ with that in $4 * x$. (Again, you may wish to mention the possibilities of chaining operations.)

2.9. Consider the same floating-point system of Exercise 2.8.

a) Let X be a U(0, 1) random variable. Develop a probability model for the representation $[X]_c$. (This is a discrete random variable. How many mass points does it have?)

b) Let X and Y be random variables uniformly distributed over the same interval as above. Develop a probability model for the representation $[X+Y]_c$. (How many mass points does it have?)

c) Develop a probability model for $[X]_c$ $[+]_c$ $[Y]_c$. (How many mass points does it have?)

2.10. Give an example to show that the sum of three floating-point numbers can have a very large relative error.

2.11. a) Write a single program in Fortran or C to compute

i.
$$\sum_{i=0}^{5} \binom{10}{i} 0.25^i 0.75^{10-i}$$

ii.
$$\sum_{i=0}^{10} \binom{20}{i} 0.25^i 0.75^{20-i}$$

iii.
$$\sum_{i=0}^{50} \binom{100}{i} 0.25^i 0.75^{100-i}.$$

iv.
$$\sum_{i=75}^{100} \binom{100}{i} 0.25^i 0.75^{100-i}.$$

b) Generalize this problem to write a single program in Fortran or C that will compute $\Pr(X \leq x|n, \pi)$ or $1 - \Pr(X \leq x|n, \pi)$ where X is a binomial random variable with parameters n and π.

2.12. We can think of the algorithm given in the R code in Figure 2.10 as an iterative algorithm in `i`. At each value of `i`, there is a difference in the value of `ex` and the true value e^x. (The exact value of this difference is the *truncation error.*) Modify the code (or use different code) to determine the relative error in `ex` for each value of `i`. For $x = 20$, make a plot of the relative error and the number of iterations (that is, of `i`) for 1 to 100 iterations. Now, repeat this for $x = -20$. Notice that the rounding error completely overwhelms the truncation error in this case.

3

Algorithms and Programming

We will use the term "algorithm" rather loosely but always in the general sense of a *method* or a *set of instructions* for doing something. Formally, an "algorithm" must terminate. Sometimes we may describe an algorithm that may not terminate simply following steps in our description. Whether we expressly say so or not, there should always be a check on the number of steps, and the algorithm should terminate after some large number of steps no matter what. Algorithms are sometimes distinguished as "numerical", "seminumerical", and "nonnumerical", depending on the extent to which operations on real numbers are simulated.

Algorithms and Programs

Algorithms are expressed by means of a flowchart, a series of steps, or in a computer language or pseudolanguage. The expression in a computer language is a source program or module; hence, we sometimes use the words "algorithm" and "program" synonymously.

The program is the set of computer instructions that implement the algorithm. A poor implementation can render a good algorithm useless. A good implementation will preserve the algorithm's accuracy and efficiency, and will detect data that are inappropriate for the algorithm. A *robust algorithm* is applicable over a wide rand of data to which it is applied. A *robust program*, which is more important, is one that will detect input data that are inappropriate either for the algorithm or for the implementation in the given program.

The exact way an algorithm is implemented in a program depends of course on the programming language, but it also may depend on the computer and associated system software. A program that will run on most systems without modification is said to be *portable*, and this is an important property because most useful programs will be run on a variety of platforms.

The two most important aspects of a computer algorithm are its accuracy and its efficiency. Although each of these concepts appears rather simple on the surface, each is actually fairly complicated, as we shall see.

J.E. Gentle, *Computational Statistics*, Statistics and Computing,
DOI: 10.1007/978-0-387-98144-4_3,

Data Structures

The efficiency of a program can be greatly affected by the type of structures used to store and operate on the data. As the size and complexity of the problem increase, the importance of appropriate database structures increases. In some data-intensive problems, the data structure can be the single most important determinant of the overall efficiency of a program. The data structure is not just a method of organization of data; it also can identify appropriate algorithms for addressing the problem.

In many applications the data are organized naturally as a list, which is a simple linearly ordered structure. The two main types of lists are *stacks*, in which data elements are added and removed from the same end (last in first out, LIFO), and *queues*, in which new data elements are added to one end and elements already in the list are removed from the other end (first in first out, FIFO). In many cases the space allocated to a given list is limited a priori, so as the list grows, another region of storage must be used. This results in a *linked list*, in which each list except one must contain, in addition to its data elements, a link or pointer to the next list. In the extreme case of this structure, each sublist contains only one piece of data and the link to the next sublist.

The next most basic data structure is a *tree*, which is a finite set whose elements (called "nodes") consist of a special element called a "root" and, if there is more than one element, a partition of the remaining elements such that each member of the partition is a tree. If there are no remaining elements, that is, if the tree contains only one element, which by definition is a root, that element is also called a "leaf". Many problems in modeling, classification, and clustering require a tree data structure (see Chapter 16).

A generalization of the tree is a *graph*, in which the nodes are usually called "vertices", and there is no fixed method of partitioning. The other type of component of this structure consists of connections between pairs of vertices, called "edges". Two vertices may be connected symmetrically, connected asymmetrically, or not connected. Graphs are useful in statistical applications primarily for identification of an appropriate method for addressing a given problem.

There are many variations of these basic structures, including special types of trees and graphs (binary trees, heaps, directed graphs, and so on).

An important problem in statistical computing may be another aspect of the data organization, one that relates to the hardware resources. There are various types of storage in the computer, and how fast the data can be accessed in the various types may affect the efficiency of a program. We will not consider these issues in any detail in this book.

3.1 Error in Numerical Computations

An "accurate" algorithm is one that gets the "right" answer. Knowing that the right answer may not be representable and that rounding within a set of operations may result in variations in the answer, we often must settle for an answer that is "close". As we discuss in Section 2.3 for scalar quantities, we measure error, or closeness, as either the absolute error or the relative error of a computation.

Another way of considering the concept of "closeness" is by looking backward from the computed answer and asking what perturbation of the original problem would yield the computed answer exactly. This approach is called *backward error analysis.* The backward analysis is followed by an assessment of the effect of the perturbation on the solution. Although backward error analysis may not seem as natural as "forward" analysis (in which we assess the difference between the computed and true solutions), it is easier to perform because all operations in the backward analysis are performed in $\mathbb{F}$ instead of in $\mathbb{R}$. Each step in the backward analysis involves numbers in the set $\mathbb{F}$, that is, numbers that could actually have participated in the computations that were performed. Because the properties of the arithmetic operations in $\mathbb{R}$ do not hold and, at any step in the sequence of computations, the result in $\mathbb{R}$ may not exist in $\mathbb{F}$, it is very difficult to carry out a forward error analysis.

There are other complications in assessing errors. Suppose the answer is a vector, such as a solution to a linear system. How do we modify the definitions of absolute and relative errors on page 97? The obvious answer is to use a vector norm, but what norm do we use to compare the closeness of vectors? Another, more complicated situation for which assessing correctness may be difficult is random number generation. It would be difficult to assign a meaning to "accuracy" for such a problem.

The basic source of error in numerical computations is the inability to work with the reals. The field of reals is simulated with a finite set. This has several consequences. A real number is rounded to a floating-point number; the result of an operation on two floating-point numbers is rounded to another floating-point number; and passage to the limit, which is a fundamental concept in the field of reals, is not possible in the computer.

Rounding errors that occur just because the result of an operation is not representable in the computer's set of floating-point numbers are usually not too bad. Of course, if they accumulate through the course of many operations, the final result may have an unacceptably large accumulated rounding error.

Another, more pernicious, effect of rounding can occur in a single operation, resulting in catastrophic cancellation (see page 100).

Measures of Error and Bounds for Errors

If the result of computer operations that should yield the real number r instead yield $\tilde{r}$, we define absolute error, $|\tilde{r} - r|$, and relative error, $|\tilde{r} - r|/|r|$ (so long as $r \neq 0$). The result, however, may not be a simple real number; it may consist of several real numbers. For example, in statistical data analysis, the numerical result, $\tilde{r}$, may consist of estimates of several regression coefficients, various sums of squares and their ratio, and several other quantities. We may then be interested in some more general measure of the difference of $\tilde{r}$ and r,

$$\Delta(\tilde{r}, r),$$

where $\Delta(\cdot, \cdot)$ is a nonnegative, real-valued function. This is the absolute error, and the relative error is the ratio of the absolute error to $\Delta(r, r_0)$, where r_0 is a baseline value, such as 0.

If r is a vector, the measure may be based on some norm, and in that case, $\Delta(\tilde{r}, r)$ may be $\|(\tilde{r} - r)\|$. A norm tends to become larger as the number of elements increases, so instead of using a raw norm, it may be appropriate to scale the norm to reflect the number of elements being computed.

However the error is measured, for a given algorithm, we would like to have some knowledge of the amount of error to expect or at least some bound on the error. Unfortunately, almost any measure contains terms that depend on the quantity being evaluated. Given this limitation, however, often we can develop an upper bound on the error. In other cases, we can develop an estimate of an "average error" based on some assumed probability distribution of the data comprising the problem.

In Monte Carlo methods that we will discuss in later chapters, we estimate the solution based on a "random" sample, so just as in ordinary statistical estimation, we are concerned about the variance of the estimate. We can usually derive expressions for the variance of the estimator in terms of the quantity being evaluated, and of course we can estimate the variance of the estimator using the realized random sample. The standard deviation of the estimator provides an indication of the distance around the computed quantity within which we may have some confidence that the true value lies. The standard deviation is sometimes called a "probabilistic error bound".

Order of Error

It is often useful to identify the "order of the error" whether we are concerned about error bounds, average expected error, or the standard deviation of an estimator. In general, we speak of the order of one function in terms of another function as a common argument of the functions approaches a given value. A function $f(t)$ is said to be of order $g(t)$ at t_0, written $\mathrm{O}(g(t))$ ("big O of $g(t)$"), if there exists a positive constant M such that

$$|f(t)| \leq M|g(t)| \quad \text{as } t \to t_0.$$

This is the *order of convergence* of one function to another function at a given point. Notice that this is pointwise convergence; we compare the functions near the point t_0.

If our objective is to compute $f(t)$ and we use an approximation $\tilde{f}(t)$, the order of the *error due to the approximation* is the order of the convergence. In this case, the argument of the order of the error may be some variable that defines the approximation. For example, if $\tilde{f}(t)$ is a finite series approximation to $f(t)$ using, say, k terms, we may express the error as $\mathrm{O}(h(k))$ for some function $h(k)$. Typical orders of errors due to the approximation may be $\mathrm{O}(1/k)$, $\mathrm{O}(1/k^2)$, or $\mathrm{O}(1/k!)$. An approximation with order of error $\mathrm{O}(1/k!)$ is to be preferred over one order of error $\mathrm{O}(1/k)$, for example, because the error is decreasing more rapidly. The order of error due to the approximation is only one aspect to consider; roundoff error in the representation of any intermediate quantities must also be considered.

The special case of convergence to the constant zero is often of interest. A function $f(t)$ is said to be "little o of $g(t)$" at t_0, written $\mathrm{o}(g(t))$, if

$$f(t)/g(t) \to 0 \quad \text{as } t \to t_0.$$

If the function $f(t)$ approaches 0 at t_0, $g(t)$ can be taken as a constant and $f(t)$ is said to be $\mathrm{o}(1)$.

Big O and little o convergences are defined in terms of dominating functions. In the analysis of algorithms, it is often useful to consider analogous types of convergence in which the function of interest dominates another function. This type of relationship is similar to a lower bound. A function $f(t)$ is said to be $\Omega(g(t))$ ("big omega of $g(t)$") if there exists a positive constant m such that

$$|f(t)| \geq m|g(t)| \quad \text{as } t \to t_0.$$

Likewise, a function $f(t)$ is said to be "little omega of $g(t)$" at t_0, written $\omega(g(t))$, if

$$g(t)/f(t) \to 0 \quad \text{as } t \to t_0.$$

Usually the limit on t, that is, t_0, in order expressions is either 0 or ∞, and because it is obvious from the context, mention of it is omitted. The order of the error in numerical computations usually provides a measure in terms of something that can be controlled in the algorithm, such as the point at which an infinite series is truncated in the computations. The measure of the error usually also contains expressions that depend on the quantity being evaluated, however.

Error of Approximation

Some algorithms are exact, such as an algorithm to multiply two matrices that just uses the definition of matrix multiplication. Other algorithms are approximate because the result to be computed does not have a finite closed-form expression. An example is the evaluation of the normal cumulative distribution function. One way of evaluating this is by using a rational polynomial approximation to the distribution function. Such an expression may be evaluated with very little rounding error, but the expression has an *error of approximation*.

When solving a differential equation on the computer, the differential equation is often approximated by a difference equation. Even though the differences used may not be constant, they are finite and the passage to the limit can never be effected. This kind of approximation leads to a *discretization error*. The amount of the discretization error has nothing to do with rounding error. If the last differences used in the algorithm are δt, then the error is usually of order $O(\delta t)$, even if the computations are performed exactly.

Another type of error of approximation occurs when the algorithm uses a series expansion. The series may be exact, and in principle the evaluation of all terms would yield an exact result. The algorithm uses only a smaller number of terms, and the resulting error is *truncation error*. (See Exercise 2.12.)

Often the exact expansion is an infinite series, and we approximate it with a finite series. When a truncated Taylor series is used to evaluate a function at a given point x_0, the order of the truncation error is the derivative of the function that would appear in the first unused term of the series, evaluated at x_0.

We need to have some knowledge of the magnitude of the error. For algorithms that use approximations, it is often useful to express the order of the error in terms of some quantity used in the algorithm or in terms of some aspect of the problem itself. We must be aware, however, of the limitations of such measures of the errors or error bounds. For an oscillating function, for example, the truncation error may never approach zero over any nonzero interval.

Consistency Checks for Identifying Numerical Errors

Even though the correct solution to a problem is not known, we would like to have some way of assessing the accuracy of our computations. Sometimes a convenient way to do this in a given problem is to perform *internal consistency checks*.

When the computations result in more than one value, say a vector, an internal consistency test may be an assessment of the agreement of various parts of the output. Relationships among the output are exploited to ensure that the individually computed quantities satisfy these relationships. Other internal consistency tests may be performed by comparing the results of the solutions of two problems with a known relationship.

Another simple internal consistency test that is applicable to many problems is the use of two different levels of precision in the computations. In using this approach, one must be careful to make sure that the input data are the same. Rounding of the input data in the lower precision may cause incorrect output to result, but that is not the fault of the computational algorithm.

Internal consistency tests cannot confirm that the results are correct; they can only give an indication that the results are incorrect.

3.2 Algorithms and Data

The performance of an algorithm may depend on the data. We have seen that even the simple problem of computing the roots of a quadratic polynomial, $ax^2 + bx + c$, using the quadratic formula, equation (2.18), can lead to catastrophic cancellation. For many values of a, b, and c, the quadratic formula works perfectly well. Data that are likely to cause computational problems are referred to as ill-conditioned data, and, more generally, we speak of the "condition" of data. The concept of condition is understood in the context of a particular set of operations. Heuristically, data for a given problem are ill-conditioned if small changes in the data may yield large changes in the solution.

Consider the problem of finding the roots of a high-degree polynomial, for example. Wilkinson (1959) gave an example of a polynomial that is very simple on the surface yet whose solution is sensitive to small changes of the values of the coefficients:

$$\begin{aligned} f(x) &= (x-1)(x-2)\cdots(x-20) \\ &= x^{20} - 210x^{19} + \cdots + 20!. \end{aligned} \tag{3.1}$$

While the solution is easy to see from the factored form, the solution is very sensitive to perturbations of the coefficients. For example, changing the coefficient 210 to $210 + 2^{-23}$ changes the roots drastically; in fact, ten of them are now complex. Of course, the extreme variation in the magnitudes of the coefficients should give us some indication that the problem may be ill-conditioned.

Condition of Data

We attempt to quantify the condition of a set of data for a particular set of operations by means of a *condition number*. Condition numbers are defined to be positive and in such a way that large values of the numbers mean that the data or problems are ill-conditioned. A useful condition number for the problem of finding roots of a function can be defined to be increasing as the reciprocal of the absolute value of the derivative of the function in the vicinity of a root. We will discuss this kind of condition number on pages 257 and 260.

In the solution of a linear system of equations, the coefficient matrix determines the condition of this problem. The most commonly used condition number is the number associated with a matrix with respect to the problem of solving a linear system of equations. We will discuss this kind of condition number beginning on page 207.

Condition numbers are only indicators of possible numerical difficulties for a given problem. They must be used with some care. For example, according to the condition number for finding roots based on the derivative (see page 257), Wilkinson's polynomial, equation (3.1), is well-conditioned.

Robustness of Algorithms

A very poor algorithm may be able to give accurate solutions in the presence of well-conditioned data. Our interest is in identifying algorithms that give accurate solutions when the data are ill-conditioned.

The ability of an algorithm to handle a wide range of data and either to solve the problem as requested or else to determine that the condition of the data does not allow the algorithm to be used is called the *robustness* of the algorithm.

A robust algorithm does not give a "wrong" answer; if it cannot give a "right" answer, it says so and stops.

Stability of Algorithms

Another desirable property of algorithms is *stability*. Stability is quite different from robustness. An algorithm is said to be *stable* if it always yields a solution that is an *exact* solution to a perturbed problem; that is, for the problem of computing $f(x)$ using the input data x, an algorithm is stable if the result it yields, $\widetilde{f}(x)$, is such that

$$\widetilde{f}(x) = f(x + \delta x)$$

for some small perturbation δx of x. Stated another way, an algorithm is stable if small perturbations in the input or in intermediate computations do not result in large differences in the results.

The concept of stability for an algorithm should be contrasted with the concept of condition for a problem or a dataset. If a problem is ill-conditioned, even a stable algorithm (a "good algorithm") will produce results with large differences for small differences in the specification of the problem. This is because the exact results have large differences. An algorithm that is not stable, however, may produce large differences for small differences in the computer description of the problem, which may involve rounding, in the input even in well-conditioned problems. Perturbations to the input data may occur because of truncation or discretization.

The concept of stability arises from backward error analysis. The stability of an algorithm may depend on how continuous quantities are discretized, such as when a range is gridded for solving a differential equation.

Reducing the Error in Numerical Computations

An objective in designing an algorithm to evaluate some quantity is to avoid accumulated rounding error and to avoid catastrophic cancellation. In the discussion of floating-point operations above, we have seen two examples of how an algorithm can be constructed to mitigate the effect of accumulated rounding error (using equations (2.16) on page 99 for computing a sum) and to avoid possible catastrophic cancellation in the evaluation of the expression (2.18) for the roots of a quadratic equation.

Another example familiar to statisticians is the computation of the sample sum of squares:

$$\sum_{i=1}^{n}(x_i - \bar{x})^2 = \sum_{i=1}^{n} x_i^2 - n\bar{x}^2. \tag{3.2}$$

This quantity is $(n-1)s^2$, where s^2 is the sample variance.

Either expression in equation (3.2) can be thought of as describing an algorithm. The expression on the left-hand side implies the "two-pass" algorithm:

$$
\begin{array}{l}
a = x_1 \\
\text{for } i = 2, \ldots, n \\
\{ \\
\quad a = x_i + a \\
\} \\
a = a/n \\
b = (x_1 - a)^2 \\
\text{for } i = 2, \ldots, n \\
\{ \\
\quad b = (x_i - a)^2 + b \\
\}.
\end{array}
\tag{3.3}
$$

This algorithm yields $\bar{x} = a$ and then $(n-1)s^2 = b$. Each of the sums computed in this algorithm may be improved by using equations (2.16). A major problem with this algorithm, however, is the fact that it requires two passes through the data. Because the quantities in the second summation are squares of residuals, they are likely to be of relatively equal magnitude. They are of the same sign, so there will be no catastrophic cancellation in the early stages when the terms being accumulated are close in size to the current value of b. There will be some accuracy loss as the sum b grows, but the addends $(x_i - a)^2$ remain roughly the same size. The accumulated rounding error, however, may not be too bad.

The expression on the right-hand side of equation (3.2) implies the "one-pass" algorithm:

$$
\begin{array}{l}
a = x_1 \\
b = x_1^2 \\
\text{for } i = 2, \ldots, n \\
\{ \\
\quad a = x_i + a \\
\quad b = x_i^2 + b \\
\} \\
a = a/n \\
b = b - na^2.
\end{array}
\tag{3.4}
$$

This algorithm requires only one pass through the data, but if the x_i's have magnitudes larger than 1, the algorithm has built up two relatively large quantities, b and na^2. These quantities may be of roughly equal magnitudes; subtracting one from the other may lead to catastrophic cancellation.

Another algorithm is shown in equations (3.5) below. It requires just one pass through the data, and the individual terms are generally accumulated fairly accurately.

$$
\begin{aligned}
& a = x_1 \\
& b = 0 \\
& \text{for } i = 2, \ldots, n \\
& \{ \\
& \quad d = (x_i - a)/i \\
& \quad a = d + a \\
& \quad b = i(i-1)d^2 + b \\
& \}.
\end{aligned}
\tag{3.5}
$$

A condition number that quantifies the sensitivity in s, the sample standard deviation, to the data, the x_i's, is

$$
\kappa = \frac{\sum_{i=1}^n x_i^2}{\sqrt{n-1}s}, \tag{3.6}
$$

where s^2 is the sample variance, as above. This is a measure of the "stiffness" of the data. It is clear that if the mean is large relative to the variance, this condition number will be large, and a dataset with a large mean relative to the variance is said to be *stiff*. (Recall that we define condition numbers so that large values imply ill-conditioning. Also recall that condition numbers must be interpreted with some care.) Notice that the condition number κ achieves its minimum value of approximately s for data with zero mean. Hence, for data $y_1, \ldots, y_n$, if we form $x_i = y_i - \bar{y}$ and the computations for $\bar{y}$ and $y_i - \bar{y}$ are exact, then the data in the last part of the algorithm in equations (3.3) would be well-conditioned.

Often when a finite series is to be evaluated, it is necessary to accumulate a subset of terms of the series that have similar magnitudes, and then combine it with similar partial sums. It may also be necessary to scale the individual terms by some very large or very small multiplicative constant while the terms are being accumulated and then remove the scale after some computations have been performed.

3.3 Efficiency

The *efficiency* of an algorithm refers to its usage of computer resources. The two most important resources are the processing units and the memory ("storage"). The amount of time the processing units are in use and the amount of memory required are the key measures of efficiency. In the following, we will generally emphasize the time the processing units are in use, rather than the amount of storage used.

A processing unit does more than just arithmetic; it also must perform "fetch" and "store" from and to memory. A common operation that takes time but does not involve arithmetic is "exchange a and b" (as in sorting methods, described on page 122; of course, in the sorting application there was also a comparison that had to executed prior to the exchange). The exchange saves storage space, but the simple operation of exchange involves fetching a, storing it somewhere, fetching b and putting it where a was, and then putting a where b was. By contrast, the operation "add a and b and store the result as c" involves fetching a, fetching b, adding them, and storing the result as c. Some computers are designed to perform these operations with a minimum of fetches and stores; nevertheless, the operations must be considered part of the overall operation.

A limiting factor for the time the processing units are in use is the number and type of operations required. Some operations take longer than others; for example, the operation of adding floating-point numbers may take more time than the operation of adding fixed-point numbers. This, of course, depends on the computer system and on what kinds of floating-point or fixed-point numbers we are dealing with. If we have a measure of the size of the problem, we can characterize the performance of a given algorithm by specifying the number of operations of each type or just the number of operations of the slowest type.

In numerical computations, the most important types of computation are usually the floating-point operations. The actual number of floating-point operations divided by the number of seconds required to perform the operations is called the *flops* (floating-point operations per second) rate.

Measuring Efficiency

Often, instead of the exact number of operations, we use the *order* of the number of operations in terms of the measure of problem size. If n is some measure of the size of the problem, an algorithm has order $\mathrm{O}(f(n))$ if, as $n \to \infty$, the number of computations $\to cf(n)$, where c is some constant that does not depend on n. For example, to multiply two $n \times n$ matrices in the obvious way requires $\mathrm{O}(n^3)$ multiplications and additions; to multiply an $n \times m$ matrix and an $m \times p$ matrix requires $\mathrm{O}(nmp)$ multiplications and additions. In the latter case, n, m, and p are all measures of the size of the problem.

Notice that in the definition of order there is a constant c. The order of an algorithm is a measure of how well the algorithm "scales"; that is, the extent to which the algorithm can deal with truly large problems. Two algorithms that have the same order may have different constants and in that case are said to "differ only in the constant".

In addition to the constant c there may be some overhead work to set up the problem. If we let $h(n)$ represent overhead work and $g(n)$ represent the remainder of work for a problem of size n, then the total amount of work is $g(n) + h(n)$. If the overhead work does not grow very fast as the problem grows, that is, if $h(n) = \mathrm{o}g(n)$, then we may have

$$g(n) + h(n) \to g(n) \to cf(n),$$

and in this case the algorithm is $\mathrm{O}(f(n))$. The constant c is relevant for evaluating the large-scale properties of the algorithm, but $h(n)$ may also be relevant for evaluating the speed of the algorithm in any real application.

Let n be a measure of the problem size, and let b and q be constants. An algorithm of order $\mathrm{O}(b^n)$ has *exponential order*, one of order $\mathrm{O}(n^q)$ has *polynomial order*, and one of order $\mathrm{O}(\log n)$ has *log order*. Notice that for log order it does not matter what the base is. Also, notice that $\mathrm{O}(\log n^q) = \mathrm{O}(\log n)$. For a given task with an obvious algorithm that has polynomial order, it is often possible to modify the algorithm to address parts of the problem so that in the order of the resulting algorithm one n factor is replaced by a factor of $\log n$. This often happens in a divide and conquer strategy, as we discuss below.

Although it is often relatively easy to determine the order of an algorithm, an interesting question in algorithm design involves the *order of the problem*; that is, the order of the most efficient algorithm possible. A problem of polynomial order is usually considered tractable, whereas one of exponential order may require a prohibitively excessive amount of time for its solution. An interesting class of problems are those for which a solution can be verified in polynomial time yet for which no polynomial algorithm is known to exist. Such a problem is called a *nondeterministic polynomial*, or NP, problem. "Nondeterministic" does not imply any randomness; it refers to the fact that no polynomial algorithm for determining the solution is known. Most interesting NP problems can be shown to be equivalent to each other in order by reductions that require polynomial time. Any problem in this subclass of NP problems is equivalent in some sense to all other problems in the subclass and so such a problem is said to be *NP-complete*. Some common types of problems that are NP-complete are combinatorial optimization (Section 6.3), data partitioning (Chapter 12), and tessellations, spanning trees, and other methods discussed in Chapter 16.

For many problems it is useful to measure the size of a *problem* in some standard way and then to identify the order of an *algorithm* for the problem with separate components. A common measure of the size of a problem is L, the length of the stream of data elements. An $n \times n$ matrix would have length proportional to $L = n^2$, for example. To multiply two $n \times n$ matrices in the obvious way requires $\mathrm{O}(L^{3/2})$ multiplications and additions, as we mentioned above.

The order of an algorithm (or, more precisely, the "order of *operations* of an algorithm") is an asymptotic measure of the operation count as the size of the problem goes to infinity. The order of an algorithm is important, but in practice the actual count of the operations is also important. In practice, an algorithm whose operation count is approximately n^2 may be more useful than one whose count is $1000(n \log n + n)$, although the latter would have order $O(n \log n)$, which is much better than that of the former, $O(n^2)$. When an algorithm is given a fixed-size task many times, the finite efficiency of the algorithm becomes very important.

The number of computations required to perform some tasks depends not only on the size of the problem but also on the data. For example, for most sorting algorithms, it takes fewer computations (comparisons) to sort data that are already almost sorted than it does to sort data that are completely unsorted. We sometimes speak of the *average* time and the *worst-case* time of an algorithm. (It is not always easy to define "average".) For some algorithms, these may be very different, whereas for other algorithms or for some problems these two may be essentially the same.

Our main interest is usually not in how many computations occur but rather in how long it takes to perform the computations. Because some computations can take place simultaneously, even if all kinds of computations required the same amount of time, the *order of time* could be different from the order of the number of computations.

In addition to the actual processing, the data may need to be copied from one storage position to another. Data movement slows the algorithm and may cause it not to use the processing units to their fullest capacity. When groups of data are being used together, blocks of data may be moved from ordinary storage locations to an area from which they can be accessed more rapidly (called "caching"). The efficiency of a program is enhanced if all operations that are to be performed on a given block of data are performed one right after the other. Sometimes a higher-level language prevents this from happening.

Although there have been orders of magnitude improvements in the speed of computers because the hardware is better, the order of time required to solve a problem is almost entirely dependent on the algorithm. The improvements in efficiency resulting from hardware improvements are generally differences only in the constant. The practical meaning of the order of the time must be considered, however, and so the constant may be important.

In addition to the efficiency of an algorithm, an important issue is how fast the program implementing the algorithm runs. Because of data movement and other reasons, a program that implements a fast algorithm may be slow. We address this issue further on page 136.

Recursion

In addition to techniques to improve the efficiency and the accuracy of computations, there are also special methods that relate to the way we build programs or store data. Before proceeding to consider ways of improving efficiency, we consider *recursion* of algorithms, which is often useful in organizing algorithms and programs.

The algorithm for computing the mean and the sum of squares (3.5) on page 116 can be derived as a recursion. Suppose we have the mean a_k and the sum of squares s_k for k elements $x_1, x_2, \ldots, x_k$, and we have a new value x_{k+1} and wish to compute a_{k+1} and s_{k+1}. The obvious solution is

$$a_{k+1} = a_k + \frac{x_{k+1} - a_k}{k+1}$$

and

$$s_{k+1} = s_k + \frac{k(x_{k+1} - a_k)^2}{k+1}.$$

These are the same computations as in equations (3.5).

Another example of how viewing the problem as an update problem can result in an efficient algorithm is in the evaluation of a polynomial of degree d,

$$p_d(x) = c_d x^d + c_{d-1} x^{d-1} + \cdots + c_1 x + c_0.$$

Doing this in a naive way would require $d-1$ multiplications to get the powers of x, d additional multiplications for the coefficients, and d additions. If we write the polynomial as

$$p_d(x) = x(c_d x^{d-1} + c_{d-1} x^{d-2} + \cdots + c_1) + c_0,$$

we see a polynomial of degree $d-1$ from which our polynomial of degree d can be obtained with but one multiplication and one addition; that is, the number of multiplications is equal to the increase in the degree — not two times the increase in the degree. Generalizing, we have

$$p_d(x) = x(\cdots x(x(c_d x + c_{d-1}) + \cdots) + c_1) + c_0, \tag{3.7}$$

which has a total of d multiplications and d additions. The method for evaluating polynomials in equation (3.7) is called *Horner's method.* (See page 169 for more on evaluation of polynomials, and equation (4.41) on that page for a slightly different form of equation (3.7).)

A computer subprogram that implements recursion invokes itself. Not only must the programmer be careful in writing the recursive subprogram, but the programming system must maintain call tables and other data properly to allow for recursion. Once a programmer begins to understand recursion, there may be a tendency to overuse it. To compute a factorial, for example, the inexperienced C programmer may write the code in Figure 3.1.

The problem is that this C program is implemented by storing a stack of statements. Because n may be relatively large, the stack may become quite large and inefficient. It is just as easy to write the function as a simple loop, and it would be a much better piece of code.

Both C and Fortran allow for recursion. Many versions of Fortran have supported recursion for years, but it was not part of the Fortran standards before Fortran 90.

```
float Factorial(int n)
  {
   if(n==0)
     return 1;
   else
     return n*Factorial(n-1);
  }
```

Fig. 3.1. Recursive Code

Improving Efficiency

There are many ways to attempt to improve the efficiency of an algorithm. Often the best way is just to look at the task from a higher level of detail and attempt to construct a new algorithm. Many obvious algorithms are serial methods that would be used for hand computations, and so are not the best for use on the computer.

Divide and Conquer

An effective general method of developing an efficient algorithm is called *divide and conquer.* In this method, the problem is broken into subproblems, each of which is solved, and then the subproblem solutions are combined into a solution for the original problem. In some cases, this can result in a net savings either in the number of computations, resulting in an improved order of computations, or in the number of computations that must be performed serially, resulting in an improved order of time.

Let the time required to solve a problem of size n be $t(n)$, and consider the recurrence relation

$$t(n) = pt(n/p) + cn \tag{3.8}$$

for p positive and c nonnegative. Then $t(n) = \mathrm{O}(n \log n)$.

The basic fact is that recursively dividing a problem in half is an $\mathrm{O}(\log n)$ operation. Divide and conquer strategies can sometimes be used together with a simple method that would be $\mathrm{O}(n^2)$ if applied directly to the full problem to reduce the order to $\mathrm{O}(n \log n)$.

One of the simplest examples of a divide and conquer approach is in sorting. The simple sorting problem is given an n-vector x, determine a vector s such each element of x is an element of s, but $s_1 \leq s_2 \leq \cdots \leq s_n$. One obvious method works on pairs, starting with x_1 and x_2 and puts x_2 in the first position if $x_2 < x_1$, then proceeds to consider the value in the ith position, call it $x_i^{(ij)}$, starting with $j = 1$ and comparing it with $x_j^{(ij)}$, exchanging the values if $x_i^{(ij)} < x_j^{(ij)}$, incrementing j and continuing this process until $j = i - 1$, then incrementing i, resetting j to 1, and continuing the process (until $i = n$). This is called a "bubble sort". It is an $\mathrm{O}(n^2)$ algorithm. You are asked to analyze this method in Exercise 3.9. A divide and conquer algorithm for sorting, however, successively divides the data to be sorted into two smaller datasets and thereby becomes $O(n \log n)$.

There are several divide and conquer sorting algorithms that follow the same general approach but differ in their details. One such algorithm is generically called mergesort. It follows two distinct steps, one of which is divide and conquer and one of which is a simple merge of sorted lists. Mergesort provides a good example of how a problem that at first glance appears to be of $O(n^2)$ can be solved in $O(n \log n)$ steps. In sorting methods, the question of auxiliary storage is an issue. In the bubble sort method described above there is no extra storage required, but in our description of MergeSort below we will use about $n/2$ units of extra storage.

The divide and conquer idea of MergeSort depends on the fact that two sorted lists each of length $n/2$ can be merged into a sorted list in n operations. One way of doing this is given in Algorithm 3.1, where auxiliary storage of size $n/2$ is used. In Algorithms 3.1 and 3.2, we pass three indexes, `lo`, `hi`, and `mid`, to a single vector `list` that is to be sorted.

Algorithm 3.1 Merge(list,lo,mid,hi).

```
work[1:(mid-lo+1)] = list[lo:mid]
i=1; j=lo; k=mid
while (j<k && k<=hi) {
   if (work[i]<=list[k]) {
      list[j] = work[i]
      i=i+1; j=j+1
      }
   else {
      list[j]=list[k]
      j=j+1; k=k+1
   }
}
while (j<k) {
   list[j] = work[i]
   j=j+1; i=i+1
}
```

∎

Once we can merge two lists of roughly the same length, as in Algorithm 3.1, we merely need a way of recursively dividing the problem into problems of sorting shorter lists, and Algorithm 3.2 does this.

Algorithm 3.2 MergeSort(list,lo,hi).

```
if (lo<hi) {
   mid = (lo+hi)/2
   MergeSort(list,lo,mid)
```

```
        MergeSort(list,(mid+1),hi)
        Merge(list, lo, mid, hi)
    }
```

■

Note what is done by these algorithms. The basic idea of divide and conquer should emerge clearly from a study of the MergeSort algorithm. At the beginning the vector is split into two; the subvector on the left is spit into two; this continues until the subvectors contain only one element each (and of course they are each sorted); these are merged into a sorted vector with two elements; then the operations move to the right to work on the subvectors containing only one element each (the fourth and fifth elements from the left in the original vector); these are merged into a sorted vector with two elements; then the two two-element sorted vectors are merged into a sorted vector with four elements; and so on.

If the number of elements is not a power of 2, at some point in the MergeSort process, the subvectors will not be of equal length. (That is what the last loop in Merge takes care of.) The overall efficiency is not affected very much, and of course, in any event it is still $\mathrm{O}(n \log n)$.

There are several other sorting methods that are $\mathrm{O}(n \log n)$. They vary in the constant, in the overhead, in relative worst-case to average-case performance, and also in the amount of auxiliary storage required. Probably the most widely used sorting method is *Quicksort*. To sort a list Quicksort does the following:

1. select an element in the list as a "pivot"
2. rearrange the elements in the list into a left sublist and a right sublist such that no element in the left sublist is larger than the pivot, and no element in the right sublist is smaller than the pivot
3. recursively sort (by going back to step 1) the left and the right sublists.

Quicksort has poor worst-case performance, but its average-case performance (under most reasonable "average-case" models) is the best of any known sorting method.

Another example of a divide and conquer algorithm is Strassen's algorithm for matrix multiplication (see Gentle, 2007, page 437). It is an $\mathrm{O}(n^{\log_2 7})$ algorithm for a problem in which the standard algorithm is $\mathrm{O}(n^3)$. While the savings may not seem like a lot, for large n the difference between n^3 and $n^{2.81}$ can be significant.

The "fan-in algorithm" (see page 99) is an example of a divide and conquer strategy that allows $\mathrm{O}(n)$ operations to be performed in $\mathrm{O}(\log n)$ time if the operations can be performed simultaneously. The number of operations does not change materially; the improvement is in the time.

Divide and conquer algorithms are particularly appropriate for implementation in parallel computing environments. The number of processors may be an important consideration. In the fan-in algorithm, for example, the improvement in order is dependent on the unrealistic assumption that as the problem size increases without bound, the number of processors also increases without bound. Divide and conquer strategies do not require multiple processors for their implementation, of course.

Convolutions

We discussed the convolution $f * g$ of two functions on page 21. If f and g are PDFs of stochastically independent random variables U and V, then $f * g$ is the PDF of $U + V$. (This is true whether the random variables are continuous or discrete. See a text on mathematical statistics if this fact is not familiar.) In many cases, transforms, either moment generating functions or characteristic functions, can be used to work out the distribution of $U + V$ more easily. This follows from the fact that if $\varphi_1(t)$ is the characteristic function of U and $\varphi_2(t)$ is that of V, the characteristic function of $U + V$ is just $\varphi_1(t)\varphi_2(t)$. The PDF can be obtained by inverting the characteristic function. In some cases, this is easy, possibly because its form is recognized as some standard function.

Similar to the convolution of two functions, the discrete convolution of two vectors x and y is the vector, which we denote as $x * y$, whose m^{th} element is

$$x * y_m = \sum_j x_{m-j} y_j. \tag{3.9}$$

Note that the limits of the summation must be such that the vectors are defined over the range of their indices. The convolution is an inner product.

Discrete Transforms

Some computational methods can be performed more efficiently by first making a *transform* on the operands, performing a related operation on the transformed data, and then inverting the transform. These problems arise in various areas, such as time series, density estimation, and signal processing. The canonical problem for which a transform is useful is the evaluation of a convolution of two objects that are indexed over the similar domains.

A discrete transform produces the elements of an n-vector $\tilde{x}$ from a given n-vector x by an inner product with another vector. (In the context of discrete transforms, the argument of the "functions" is the index of the vectors.) Because they have a wide range of applications, it is important to be able to compute transforms with high efficiency.

The index of summation of a discrete transform often occurs in the exponent of one of the factors in the summands. For that reason, it is convenient to use "0 equals first" indexing, as we did with series expansion of functions on page 20.

One of the most important and most commonly used discrete transforms is the discrete Fourier transform (DFT), which, for the n-vector x is

$$\tilde{x}_m = \sum_{j=0}^{n-1} x_j \mathrm{e}^{\frac{2\pi \mathrm{i}}{n} jm}, \quad \text{for } m = 0, \ldots, n-1. \tag{3.10}$$

The inverse, that is, x_j for $j = 0, \ldots, n-1$ in terms of the $\tilde{x}$'s, is

$$x_j = \frac{1}{n} \sum_{m=0}^{n-1} \tilde{x}_m \mathrm{e}^{-\frac{2\pi \mathrm{i}}{n} jm}.$$

Equation (3.10) is readily seen to be equivalent to the linear transformation of x by the matrix A, that is, $\tilde{x}_m = Ax$, where

$$A_{mj} = \mathrm{e}^{\frac{2\pi \mathrm{i}}{n} jm}.$$

Multiplying an n-vector by an $n \times n$ matrix is essentially an $\mathrm{O}(n^2)$ task. If the matrix has special form, however, it may be possible to reduce the number of computations. This is the idea of the *fast Fourier Transform (FFT)*.

Another way of looking at the transform is to think of $x_0, \ldots, x_{n-1}$ as the coefficients of a polynomial in z:

$$x(z) = x_{n-1} z^{n-1} + \cdots + x_1 z + x_0. \tag{3.11}$$

The Fourier element $\tilde{x}_m$ is the value of $x(z)$ at $z = \left(\mathrm{e}^{\frac{2\pi \mathrm{i}}{n}}\right)^m$. For simplicity, let $w = \mathrm{e}^{\frac{2\pi \mathrm{i}}{n}}$.

Now, by Horner's method (equation (3.7)) we know we need no more than n operations to evaluate an nth degree polynomial. We can do better than this, however, when we consider the very special form of the points at which we need to evaluate the polynomial. These points are called *primitive roots of unity*. A number w is a primitive n^{th} root of unity if $w \neq 1$, $w^n = 1$, and $\sum_{p=0}^{n-1} w^{jp} = 0$ for $j = 1, \ldots, n-1$. As examples, -1 is a primitive n^{th} root of unity for $n = 2$, i is a primitive n^{th} root of unity for $n = 4$. In general, $\mathrm{e}^{\frac{2\pi \mathrm{i}}{n}}$ is a primitive n^{th} root of unity, as we can see by checking the conditions.

The relevance of this for the FFT are two facts about primitive n^{th} roots of unity when n is an even integer. If $n = 2m$ and w is a primitive n^{th} root of unity, then $-w^j = w^{j+m}$, and, secondly, w^2 is a primitive m^{th} root of unity. Both of these are easily seen from the definition.

Let us *assume n is even* and let $m = n/2$. Break the polynomial (3.11) into two parts:

$$\begin{aligned} x(z) = x_{n-1} z^{n-1} &+ x_{n-3} z^{n-3} + \cdots + x_1 z \\ &x_{n-2} z^{n-2} + \cdots + x_2 z^2 + x_0. \end{aligned}$$

Letting $y = z^2$, we can write $x(z)$ as the sum of two polynomials,

$$c(y) = z(x_{n-1} y^{n-1} + x_{n-3} y^{n-2} + \cdots + x_1)$$

and

$$b(y) = (x_{n-2}y^{n-1} + x_{n-4}y^{n-2} + \cdots + x_0).$$

Our interest is in computing the Fourier coefficients, $x(w^j)$ for $j = 0, \ldots, n-1$, We can write the first half, for $j = 0, \ldots, k-1$, as

$$x(w^j) = c(w^{2j})w^j + b(w^{2j}),$$

and the second half as

$$x(w^{j+k}) = -c(w^{2j})w^j + b(w^{2j}),$$

again for $j = 0, \ldots, k-1$.

We have now divided the problem in half. The FFT continues this process of dividing what was an $\mathrm{O}(n^2)$ operation into an $\mathrm{O}(n \log n)$ operation, just as we did with the sorting operation above.

As with many divide and conquer methods, the FFT can be conveniently expressed recursively. We do this Algorithm 3.3. Following a notational convention for Fourier transforms, we denote the $\tilde{x}$ as X. In the algorithm, we assume $n = 2^p$, w is a primitive n^{th} root of unity, X is a complex array of length n with values w^j for $j = 0, \ldots, n-1$, and B, C, and ws are complex arrays.

Algorithm 3.3 FFT(n,x,w,X)

```
if n=1  X=x[0]
else {
   k = n/2
   b = (x[n-2],...,x[2],x[0])
   c = (x[n-1],...,x[3],x[1])
   FFT(k,b,w^2,B)
   FFT(k,c,w^2,C)
   ws[0] = 1/w
   for j=0 to n-1 {
      ws[j+1] = w*ws[j]
      X[j] = B[j]+wp[j+1]*C[j]
      X[j+k] = B[j]-wp[j+1]*C[j]
   }
}
```

∎

Greedy Methods

Some algorithms are designed so that each step is as efficient as possible, without regard to what future steps may be part of the algorithm. An algorithm that follows this principle is called a *greedy algorithm*. A greedy algorithm is often useful in the early stages of computation for a problem or when a problem lacks an understandable structure. An example of a greedy algorithm is a steepest descent method (see page 265).

Bottlenecks and Limits

There is a maximum flops rate possible for a given computer system. This rate depends on how fast the individual processing units are, how many processing units there are, and how fast data can be moved around in the system. The more efficient an algorithm is, the closer its achieved flops rate is to the maximum flops rate.

For a given computer system, there is also a maximum flops rate possible for a given problem. This has to do with the nature of the tasks within the given problem. Some kinds of tasks can utilize various system resources more easily than other tasks. If a problem can be broken into two tasks, T_1 and T_2, such that T_1 must be brought to completion before T_2 can be performed, the total time required for the problem depends more on the task that takes longer. This tautology has important implications for the limits of efficiency of algorithms.

The efficiency of an algorithm may depend on the organization of the computer, the implementation of the algorithm in a programming language, and the way the program is compiled.

High-Performance Computing

In "high-performance" computing, major emphasis is placed on computational efficiency. The architecture of the computer becomes very important, and the programs are often designed to take advantage of the particular characteristics of the computer on which they are to run.

The three main architectural elements are memory, processing units, and communication paths. A controlling unit oversees how these elements work together.

There are various ways memory can be organized. There is usually a hierarchy of types of memory with different speeds of access. The controlling unit will attempt to retain data in a high-speed memory area, often called a cache, if it is anticipated that the data will be used in subsequent computations. The various levels of memory can also be organized into banks with separate communication links to the processing units.

There are various types of processing units. Three general types are called central processing units (CPU), vector processors (VP), and graphics processing units (GPU). A CPU is the standard type, but the term covers a wide range of designs. A CPU usually has separate areas for floating-point and fixed-point operations. The number of operations it directly implements varies. A RISC ("reduced instruction set computer") can perform only a relatively small number of operations, so a complex operation may require multiple operations. The tradeoff is in the speed of execution of any single operation. Vector processors are particularly suited for computations on all of the elements in a linear array or on the elements in two linear arrays. Finally, GPUs generally provide only a limited number of different operations, but they are designed for a large number of simultaneous operations. They are said to be ideally suited for "super computing" on the "desktop"; that is, they are inexpensive and environmentally robust. When they are used for general-purpose numerical computations, the name "graphics processing unit" is merely a legacy referring to the applications for which this type of unit was originally developed. Any processing unit may consist of multiple processors within the same unit, often called a core.

If more than one processing unit is available, it may be possible to perform operations simultaneously. In this case, the amount of time required may be drastically smaller for an efficient parallel algorithm than it would for the most efficient serial algorithm that utilizes only one processor at a time. An analysis of the efficiency must take into consideration how many processors are available, how many computations can be performed in parallel, and how often they can be performed in parallel.

The most effective way of decreasing the time required for solving a computational problem is to perform the computations in parallel if possible. There are some computations that are essentially serial, but in almost any problem there are subtasks that are independent of each other and can be performed in any order. Parallel computing remains an important research area.

3.4 Iterations and Convergence

We use the word "iteration" in a nontechnical sense to mean a step in a sequence of computations. The iteration may be one single arithmetic operation, but usually it refers to a group of operations.

We may speak of iterations in any algorithm, but there are certain types of algorithms that are "iterative"; that is, methods in which groups of computations form successive approximations to the desired solution. This usually means a loop through a common set of instructions in which each pass through the loop changes the initial values of operands in the instructions. When we refer to an iterative algorithm or iterative method, we mean this type of algorithm.

In an algorithm that is not iterative, there are a fixed, finite number of operations or iterations. The algorithm terminates when that number is reached. In an iterative algorithm, the number of iterations may not be known in advance; the algorithm terminates when it appears that the problem has been solved, or when it is decided that the method used is not going to solve the problem.

We use the word "converge", and the various derivatives of this root word, to refer to a condition in the progress of an algorithm in which the values no longer change.

The steps in simple summations, in which we have a sequence of partial sums, are iterations. A finite sequence converges to the exact sum (it is hoped).

An infinite sequence may or may not converge. Our interest, clearly, is in evaluating sequences that do converge. The computations themselves may not be a reliable indicator of whether or not the sequence converges; see Exercise 2.2.

In working with iterative methods, we will generally use the notation $x^{(k)}$ to refer to the computed value of x at the kth iteration.

Testing for Convergence

The term "algorithm" refers to a method that terminates in a finite number of steps. We should never implement a method on the computer unless we know that this will occur, no matter what kind of data is input to the method.

In iterative numerical algorithms, the most important issues are convergence of the algorithm and convergence of the algorithm to the "correct" solution to the problem. If the algorithm converges, presumably it terminates, but if it does not converge, we can always make it terminate by placing a limit on the number of iterations.

In the actual computations, an iterative algorithm terminates when some *convergence criterion* or *stopping criterion* is satisfied.

An example is to declare that an algorithm has converged when

$$\Delta\left(x^{(k)}, x^{(k-1)}\right) \leq \epsilon, \tag{3.12}$$

where $\Delta(x^{(k)}, x^{(k-1)})$ is some measure of the difference of $x^{(k)}$ and $x^{(k-1)}$ and ϵ is a small positive number. Because x may not be a single number, we must consider general measures, usually metrics, of the difference of $x^{(k)}$ and $x^{(k-1)}$. It may be the case, however, that even if $x^{(k)}$ and $x^{(k-1)}$ happen to be close, if the computations were allowed to continue, $x^{(k+1)}$ would be very different from $x^{(k)}$.

Even though the computed values may stop changing, that is, *algorithmic convergence* has occurred, the question of convergence to the correct value remains. (Recall the "convergent" series of Exercise 2.2.) The assessment of convergence to the "correct" value is often an ad hoc process that depends on some analysis of the problem. Perhaps a perturbation of the problem can be used to assess correctness, as suggested on page 112. Also, if something about the behavior of the function at the point of convergence is known, in addition to first order comparisons as in equation (3.12), higher order differences may be useful.

There are basically three reasons to terminate an algorithm.

- It is known that the problem is solved.
- The iterations converge, that is, the computed values quit changing.
- The number of iterations is too large, or the problem appears to diverge.

The first two of these indicate algorithmic convergence. The first one has limited applicability. For most problems we do not know when it is solved.

An example of when we do know is solving an equation; specifically, finding the roots of an equation. If the problem is to solve for x in $f(x) = 0$, and if

$$\left|f(x^{(k)})\right| \leq \epsilon, \tag{3.13}$$

for some reasonable ϵ, then we can take $x^{(k)}$ as the best solution we are likely to get. Is $x^{(k)}$ the correct solution however? Possibly. Consider $f(x) = \mathrm{e}^{-x}$, however. Clearly, $f(x) = 0$ has no solution, but $x^{(k)} = -\log(\epsilon)$ will satisfy the convergence criterion.

The second reason can be used in a wide range of cases, but we must be careful because the iterates may not change monotonically, as we indicated above.

If termination occurs because of the third reason, we say that the algorithm did not converge.

On closer consideration of the criteria in (3.12) and (3.13) an important point becomes obvious. We should choose ϵ with some care. Consideration of this leads us to the realization that it is problem dependent. Without even knowing the objectives of the problem, which, of course, would determine how we really should make the decisions, we realize that often the data themselves determine the important magnitudes. In either case, therefore, instead of the absolute comparison with ϵ, perhaps we should use a relative comparison using some other value, say ϵ_r:

$$\Delta\left(x^{(k)}, x^{(k-1)}\right) \leq \epsilon_r \left|x^{(k-1)}\right|, \tag{3.14}$$

for example.

The point of this discussion is although assessment of convergence is very important, it is difficult and, unfortunately, may be somewhat ad hoc.

A computer program implementing an iterative algorithm should allow the user to set convergence and termination criteria.

An iterative algorithm usually should have more than one stopping criterion. Often a maximum number of iterations is set so that the algorithm will be sure to terminate whether it converges or not. In any event, it is always a good idea, in addition to stopping criteria based on convergence of the solution, to have a stopping criterion that is independent of convergence and that limits the number of operations.

Rate of Convergence

In addition to the question of how to decide when an algorithm has actually converged and we should stop the iterations, we may be interested in how fast the iterations are converging.

The *convergence ratio* of the sequence $x^{(k)}$ to a constant x_0 is

$$\lim_{k\to\infty} \frac{\Delta(x^{(k+1)}, x_0)}{\Delta(x^{(k)}, x_0)} \tag{3.15}$$

if this limit exists. If the convergence ratio is greater than 0 and less than 1, the sequence is said to converge *linearly*. If the convergence ratio is 0, the sequence is said to converge *superlinearly*.

The convergence rate is often a function of k, say $g(k)$. The convergence is then expressed as an order in k, $\mathrm{O}(g(k))$.

We can often determine the order of convergence experimentally, merely by fitting a curve to the fraction in expression (3.15) for $k = 1, 2, \ldots$.

Other measures of the rate of convergence are based on

$$\lim_{k\to\infty} \frac{\Delta(x^{(k+1)}, x_0)}{(\Delta(x^{(k)}, x_0))^r} = c \tag{3.16}$$

(again, assuming the limit exists; i.e., $c < \infty$). In equation (3.16), the exponent r is called the *rate of convergence*, and the limit c is called the *rate constant*. If $r = 2$ (and c is finite), the sequence is said to converge *quadratically*. It is clear that for any $r > 1$ (and finite c), the convergence is superlinear.

Convergence defined in terms of equation (3.16) is sometimes referred to as "Q-convergence" because the criterion is a quotient. Types of convergence may then be referred to as "Q-linear", "Q-quadratic", and so on.

Speeding Up Convergence by Extrapolation

An iterative algorithm involves a certain amount of work at each step. For a given application, the amount of this work is relatively constant for a given approach.

We can speed up an iterative algorithm either by changing the computations in a given step so as to reduce the amount of work at each step, or else by reducing the number of steps until convergence.

Sometimes the amount of work in each step can be reduced by using some approximations. This is the idea, for example, in some of the quasi-Newton methods discussed beginning on page 269. When approximations are used, the number of steps may increase but the overall work may decrease.

We now look at some ways of reducing the number of steps by perhaps doing a small amount of extra work in each step.

We must be aware, however, that generally there are no modifications of an algorithm that are guaranteed to speed up the convergence. Even worse, an algorithm with relatively reliable convergence properties may lose these properties after modifications; that is, we may begin with an iterative algorithm that is relatively robust, and after modifications, will fail to converge for a wider range of problems.

We will now consider two general methods. The first is a simple approach called *Aitken's Δ^2-extrapolation*, or *Aitken acceleration*.

We begin with a convergent sequence $\{x_k\}$, and consider the forward difference

$$\Delta x_k = x_{k+1} - x_k.$$

Now we apply the difference operator a second time to get

$$\begin{aligned}\Delta^2 x_n &= \Delta(\Delta x_n) \\ &= (x_{n+2} - x_{n-1}) - (x_{n+1} - x_n) \\ &= x_{n+2} - 2x_{n-1} + x_n.\end{aligned}$$

Now, assume $\{x_k\}$ converges linearly to x, and that $x_k \neq x$ for all $k \geq 0$. If there exists r with $|r| < 1$ such that

$$\lim_{k\to\infty} \frac{x - x_{k+1}}{x - x_k} = r,$$

we can improve the speed of convergence by using the sequence $\{\tilde{x}_k\}$ defined by

$$\begin{aligned}\tilde{x}_k &= x_k - \frac{(\Delta x_k)^2}{\Delta^2 x_k} \\ &= x_k - \frac{(x_{k+1} - x_k)^2}{x_{k+2} - 2x_{k-1} + x_k}. \qquad (3.17)\end{aligned}$$

We will consider an example of this on page 246.

Use of the sequence $\{\tilde{x}_k\}$ in place of $\{x_k\}$ is called Aitken acceleration or Aitken's Δ^2 process.

The only guarantee is that the Aitken sequence will converge faster. If the original sequence $\{x_k\}$ is linearly convergent, in most cases $\{\tilde{x}_k\}$ will only be linearly convergent also, but its asymptotic error constant will be smaller than that of the original sequence.

When Aitken's acceleration is combined with a fixed-point method, the resulting process is called *Steffensen acceleration* (see page 246).

Another acceleration method arises in the context of a discrete grid over a continuous domain. It is based on the idea of decreasing the grid size. Numerical computations are performed on a discrete set that approximates the reals or $\mathbb{R}^d$. This may result in *discretization errors.* By "discretization error", we do not mean a rounding error resulting from the computer's finite representation of numbers. The discrete set used in computing some quantity such as an integral is often a grid. If h is the interval width of the grid, the computations may have errors that can be expressed as a function of h. For example, if the true value is x and, because of the discretization, the *exact value* that would be computed is x_h, then we can write

$$x = x_h + e(h).$$

For a given algorithm, suppose the error $e(h)$ is proportional to some power of h, say h^n, and so we can write

$$x = x_h + ch^n \qquad (3.18)$$

for some constant c.

Now, suppose we use a different discretization, with interval length rh where $0 < r < 1$. We have

$$x = x_{rh} + c(rh)^n \tag{3.19}$$

and, after subtracting from equation (3.18), we have

$$0 = x_h - x_{rh} + c(h^n - (rh)^n)$$

or

$$ch^n = \frac{(x_h - x_{rh})}{r^n - 1}. \tag{3.20}$$

This analysis relies on the assumption that the error in the discrete algorithm is proportional to h^n. Under this assumption, ch^n in equation (3.20) is the discretization error in computing x, using exact computations, and is an estimate of the error due to discretization in actual computations. A more realistic regularity assumption is that the error is $\mathrm{O}(h^n)$ as $h \to 0$; that is, instead of (3.18), we have

$$x = x_h + ch^n + \mathrm{O}(h^{n+\alpha}) \tag{3.21}$$

for $\alpha > 0$.

Whenever this regularity assumption is satisfied, equation (3.20) provides us with an inexpensive improved estimate of x:

$$x_R = \frac{x_{rh} - r^n x_h}{1 - r^n}. \tag{3.22}$$

It is easy to see that $|x - x_R|$ is less than the absolute error using an interval size of either h or rh.

The process described above is called *Richardson extrapolation*, and the value in equation (3.22) is called the Richardson extrapolation estimate. Richardson extrapolation is also called "Richardson's deferred approach to the limit". It has general applications in numerical analysis, but is most widely used in numerical quadrature. We will encounter it on page 188, where we use it to develop Romberg integration by accelerating simpler quadrature rules.

Extrapolation can be extended beyond just one step as in the presentation above.

Reducing the computational burden by using extrapolation is very important in higher dimensions. In many cases, for example in direct extensions of quadrature rules, the computational burden grows exponentially with the number of dimensions. This is sometimes called "the curse of dimensionality" and can render a fairly straightforward problem in one or two dimensions unsolvable in higher dimensions.

A direct extension of Richardson extrapolation in higher dimensions would involve extrapolation in each direction, with an exponential increase in the amount of computation. An approach that is particularly appealing in higher dimensions is called *splitting extrapolation*, which avoids independent extrapolations in all directions.

3.5 Programming

Although I like to think of programming as a science, there are many elements of programming that resemble art. Programming is only learned by programming. (*Read that again.*)

The advancement of science depends on high-quality software. Most scientific software is not developed ab initio by professional programmers. Most evolves from rudimentary, ad hoc programs. This type of development presents a serious risk; the range of applicability (robustness) may be quite limited. The writer of the program may or may not be aware of this. If the writer continues to use the program this limitation may become apparent very quickly. The real danger, however, usually comes from the usage by the writer's colleagues. After a few years of usage on easy problems or on problems very similar to the one for which the program was originally written, the program has stood the "test of time" and may be generally accepted as a solid piece of software. There is a significant amount of anecdotal evidence that much of the code incorporated in R packages evolved in this way, and it is not robust.

Of lesser concern is the extent to which the computer program utilizes coding methods to speed up its execution. Although an algorithm may state, for example, "for (i in 1 to n) do ..." the competent programmer may write a do-loop that includes a number, say k, of successive steps within the do-loop, and then has a small bit of code at the end of the loop to handle the n mod k remaining cases. This is called *unrolling the loop*, and, depending on the computer hardware, can result in significant speedup.

No matter how a code is written, the compiler may cause the order of execution to be different from what the programmer expected. A compiler attempts to maximize the speed of execution. Some compilers work harder than others at optimizing the code. These are called optimizing compilers. Some optimizing compilers will unroll do-loops, for example. As one might guess, it is not easy for a program (the compiler) to decide how the statements in another program should best be executed. The first practical optimizing compiler was selected by *Computing in Science & Engineering* as one of the Top 10 algorithms of the twentieth century; see page 138.

Translating Mathematics into Computer Programs

Although one of the important mantras of statistical computing is that a mathematical expression does not necessarily imply a reasonable computational method, it is often the case that the mathematical expression is at the appropriate level of abstraction. An expression such as $A^{\mathrm{T}}B$, for example, may prompt a Fortran or C programmer to envision writing code to implement loops to compute

$$\sum_i \sum_j \sum_k a_{ki} b_{kj}. \tag{3.23}$$

If the computer language supports a construct directly similar to $A^{\mathrm{T}}B$, as Fortran (but not C) and R do, it is likely that use of that construct will result in a much more efficient program than use of the nested loops.

The point is that while the mathematical expression does not specify the computations, we should begin with code (or pseudocode) that is similar to the mathematical expression, and then refine the code for accuracy, stability, and efficiency.

Computations without Storing Data

For computations involving large sets of data, it is desirable to have algorithms that sequentially use a single data record, update some cumulative data, and then discard the data record. Such an algorithm is called a *real-time* algorithm, and operation of such an algorithm is called *online* processing. An algorithm that has all of the data available throughout the computations is called a *batch* algorithm.

An algorithm that generally processes data sequentially in a similar manner as a real-time algorithm but may have subsequent access to the same data is called an *online* algorithm or an *"out-of-core"* algorithm. (This latter name derives from the erstwhile use of "core" to refer to computer memory.) Any real-time algorithm is an online or out-of-core algorithm, but an online or out-of-core algorithm may make more than one pass through the data. (Some people restrict "online" to mean "real-time" as we have defined it above.)

If the quantity t is to be computed from the data $x_1, x_2, \ldots, x_n$, a real-time algorithm begins with a quantity $t^{(0)}$ and from $t^{(0)}$ and x_1 computes $t^{(1)}$. The algorithm proceeds to compute $t^{(k+1)}$ using x_{k+1} and so on, never retaining more than just the current value, $t^{(k)}$. The quantities $t^{(k)}$ may of course consist of multiple elements. The point is that the number of elements in $t^{(k)}$ is independent of n.

Many summary statistics can be computed in online processes. For example, the algorithms discussed beginning on page 115 for computing the sample sum of squares are real-time algorithms. The algorithm in equations (3.3) requires two passes through the data, so it is not a real-time algorithm, although it is out-of-core. There are stable online algorithms for other similar statistics, such as the sample variance-covariance matrix. The least squares linear regression estimates can also be computed by a stable one-pass algorithm that, incidentally, does not involve computation of the variance-covariance matrix (or the sums of squares and cross products matrix). There is no real-time algorithm for finding the median. The number of data records that must be retained and reexamined depends on n.

In addition to the reduced storage burden, a real-time algorithm allows a statistic computed from one sample to be updated using data from a new sample. A real-time algorithm is necessarily $\mathrm{O}(n)$.

Measuring the Speed of a Program

Beginning on page 117 we discussed the efficiency of algorithms. For them to be useful, algorithms must be translated into computer programs, and it is ultimately the speed of these programs that matter. The arrangement of loops, the movement of data, and other factors can degrade the performance of an algorithm. One of the useful tools in programming is a function to measure the time the computer spends on a given task — this is not the elapsed time, because the computer may be working on other tasks.

It is important that the programmer know which parts of the program are more computationally intensive. Of course, this may vary for different datasets or problems. Any program intended for frequent use should be *profiled* over different problems; that is, for a range of problems, the proportional average execution time for each module should be empirically measured. Even the proportional times may be different on different types of computers, so for important, widely-used programs, profiling should be performed on a range of computers.

There is an intrinsic function in Fortran 95, `cpu_time(time)` that returns the current processor time in seconds. (If the processor is unable to provide a timing, a negative value is returned instead.) The exact nature of the timing is implementation dependent; a parallel processor might return an array of times corresponding to the various processors. In C, `clock()` in the `time.h` library returns the number of clock steps, and an associated constant `CLOCKS_PER_SEC` can be used to convert clock steps to seconds. In R, `proc.time()` returns the "user" time, the "system time", and the elapsed time in seconds. The distinction between "user" time and the "system time" depends on the operating system and the hardware platform. There is another useful timing function in R, `system.time`, which gives the same three times for the evaluation of an expression. (The "expression" is a program module to be timed. It must be specified as an argument to `system.time`.)

Computer functions to provide timing information are notoriously unreliable. They should always be used over multiple runs and average times taken. The resolution of all of these functions is system dependent. In most cases, the time for a certain computation is obtained by subtracting two different calls to the timing routine, although doing this can only yield an elapsed system time.

Code Development

An important aspect of statistical computing is the formulation of both the data and the computational methods in a way that can be used by the computer. This can be done by using some application such as a spreadsheet program, or it can be done by writing a program in a programming language. We will not address use of higher level application programs for computational statistics in this book.

In general we distinguish two types of programming languages: the languages in which programs are compiled before execution, such as C and Fortran, and the languages that issue immediate commands to the computer, such as Octave and R. We sometimes refer to interactive systems such as Octave and R as "higher-level" languages. Systems such as Octave and R also allow a sequence of commands to be issued together, so we can think of "programs" or "scripts" in these higher-level systems. For large-scale computations we should use a compiled language, because the execution is much faster. There are, of course, many general issues and many more details to consider. We will not address them here, but computational and programming aspects will be a theme throughout this book.

3.6 Computational Feasibility

Data must be stored, transported, sorted, searched, and otherwise rearranged, and computations must be performed on it. The size of the dataset largely determines whether these actions are feasible. Huber (1994, 1996) proposed a classification of datasets by the number of bytes required to store them (see also Wegman, 1995). Huber described as "tiny" those requiring on the order of 10^2 bytes; as "small" those requiring on the order of 10^4 bytes; as "medium" those requiring on the order of 10^6 bytes (one megabyte); as "large", 10^8 bytes; "huge", 10^{10} bytes (10 gigabytes); and as "massive", 10^{12} bytes (one terabyte). ("Tera" in Greek means "monster".) This log scale of two orders of magnitude is useful to give a perspective on what can be done with data. Online or out-of-core algorithms are generally necessary for processing massive datasets.

For processing massive datasets, the order of computations is a key measure of feasibility. We can quickly determine that a process whose computations are $\mathrm{O}(n^2)$ cannot be reasonably contemplated for massive (10^{12} bytes) datasets. If computations can be performed at a rate of 10^{12} per second (teraflop), it would take over three years to complete the computations. (A rough order of magnitude for quick "year" computations is $\pi \times 10^7$ seconds equals approximately one year.) A process whose computations are $\mathrm{O}(n \log n)$ could be completed in 230 milliseconds for a massive dataset. This remarkable difference in time required for $\mathrm{O}(n^2)$ and $\mathrm{O}(n \log n)$ processes is the reason that the fast Fourier transform (FFT) algorithm was such an important advance.

Exponential orders can make operations even on tiny (10^2 bytes) datasets infeasible. A process whose computations require time of $\mathrm{O}(2^n)$ may not be completed in four centuries.

Sometimes, it is appropriate to reduce the size of the dataset by forming groups of data. "Bins" can be defined, usually as nonoverlapping intervals covering IR^d, and the number of observations falling into each bin can be determined. This process is linear in the number of observations. The amount of information loss, of course, depends on the sizes of the bins. Binning of data has long been used for reducing the size of a dataset, and earlier books on statistical analysis usually had major sections dealing with "grouped data".

Another way of reducing the size of a dataset is by sampling. This must be done with some care, and often, in fact, sampling is not a good idea. Sampling is likely to miss the unusual observations just because they are relatively rare, but it is precisely these outlying observations that are most likely to yield new information.

Advances in computer hardware continue to expand what is computationally feasible. It is interesting to note, however, that the order of computations is determined by the problem to be solved and by the algorithm to be used, not by the hardware. Advances in algorithm design have reduced the order of computations for many standard problems, while advances in hardware have not changed the order of the computations. Hardware advances change the constant in the order of time.

Notes and Further Reading

Algorithms

A good general and comprehensive coverage of computer algorithms is in the very large book by Cormen et al. (2001). Garey and Johnson (1979) did an early study of the class of NP-complete problems, and their book contains an extensive list of problems that are known to be NP-hard.

The January/February, 2000, issue of *Computing in Science & Engineering* was devoted to the Top 10 Algorithms of the twentieth century. Guest editors Jack Dongarra and Francis Sullivan discuss the role that efficient computational algorithms have played in the advancement of science. As we have pointed out, the replacement of an $\mathrm{O}(n^2)$ algorithm with an $\mathrm{O}(n \log n)$ algorithm, as is often the case in divide and conquer strategies, has allowed problems to be solved that could not be solved before. The special issue contains brief articles on all of the Top 10 algorithms. Some of the Top 10 are rather simple methods that can be described in just a few steps and others require many steps. Some are collections of several methods, and instead of being algorithms, they are general approaches. Some can be implemented in a computer program reliably by an amateur, while others involve many numerical subtleties.

The Top 10, in the chronological order of their development are

- Metropolis algorithm for Monte Carlo
- simplex method for linear programming
- Krylov subspace iteration methods
- the decompositional approach to matrix computations
- the Fortran optimizing compiler
- QR algorithm for computing eigenvalues
- quicksort algorithm

- fast Fourier transform
- integer relation detection
- fast multipole method

We have discussed two of these algorithms and mentioned another in this chapter and we will encounter others in later chapters of Part II.

Developing algorithms can be an enjoyable exercise in problem solving. The book by Bentley (2000) provides very readable descriptions of a number of problems, and then gives elegant algorithms for them. (A teaser is given in Exercise 3.9; a solution for which is provided on page 679.)

Extrapolation

Richardson's ideas for extrapolation appeared in 1910, but it took fifty years before the technique was widely used. Today most numerical computations that use discrete approximations to evaluate a quantity that is defined in terms of a continuous function, such as an integral, use some form of extrapolation.

The use of extrapolation in higher dimensions is even more important, but it should not be done as independent extrapolations in each of the dimensions. The splitting extrapolation method is described in some detail by Liem, Lü, and Shih (1995).

Programming and Software

Statisticians depend on high-quality software, and many statisticians are amateur software developers for their own computing needs. Much of the software used by any statistician, even one working in statistical computing, however, is written by someone else, possibly a team at a commercial or semi-commercial software company.

Reviews of software, whether published formally or distributed through blogs, are useful, but the target is moving and evaluations frequently go awry. McCullough (1999) discusses some of the methods of reviewing statistical software, and the complexities of this activity.

Often a statistician needs to write new code to address a specific problem or to implement a new statistical method. In this case a programming language and possibly a programming system must be used. The common programming languages for statistical applications are R, S-Plus, and Matlab, or Fortran, C, and C++. Obviously, there are many other languages to choose from, but the user should consider the length of time the program may be used. (I personally have written thousands of lines of PL/I that several years later I had to have high-paid programmers translate into Fortran.)

Any serious programming effort needs a programming system for the chosen language. A programming system includes a language-aware editor, a good debugger, and a version control system. A programming system is built around some specific application program interface (API), and understanding the details of the API is necessary for effective programming.

Thinking of the various components of code as representing objects, and thinking of the objects systematically as members of classes with common characteristics is a good way to approach code design and development. Virius (2004) provides a general description of the object-oriented approach, "object oriented programming", (OOP). There is a dogma associated with OOP (that reminds me of the dogmas of BCLSs and structured programming described in Kennedy and Gentle, 1980), but like earlier movements, the attitudes engendered are more important than strict adherence to the religion.

The issue of ensuring reliable and correct software during the development of the software is important. It should not be necessary to correct unreliable software after the fact. Chapter 8 of the book edited by Einarsson (2005) specifically addresses facilities in various languages such as Fortran or C for implementing reliable and correct software. Chapter 1 of that book also contains a number of interesting examples where unreliable or incorrect software has endangered lives and cost a lot of money to correct.

A good book on programming is Lemmon and Schafer (2005), who provide many guidelines for developing software in Fortran 95. They also describe system integration, primarily in the context of Microsoft Visual Basic NET system.

Chambers (2008) discusses principles of programming, with particular emphasis on the R system. Much of the material in both Lemmon and Schafer (2005) and Chambers (2008) applies to programming in any language. Chambers also covers system integration with less dependence on a particular platform.

Software development is not just about writing programs. There are many issues including interface design, algorithm selection and implementation, documentation, and maintainability that the amateur software developer is not likely to consider. Klinke (2004) discusses many issues of the design of the interface, and the book by McConnell (2004) provides a good coverage of this as well as other issues, such as code design and documentation.

R and S-Plus

The software system called S was developed at Bell Laboratories in the mid-1970s by John Chambers and colleagues. S was both a data-analysis system and an object-oriented programming language.

S-Plus is an enhancement of S, including graphical interfaces, more statistical analysis functionality, and support.

The freely available package called R provides generally the same functionality in the same language as S (see Gentleman and Ihaka, 1997). The R system includes a useful feature for incorporation of new "packages" into the program. See
`http://www.r-project.org/`

for a current description of R and a listing of packages available over the web. A large number of research workers have contributed packages to the R system. Many of these packages have overlapping capabilities, and, unfortunately, some are not of high-quality. The most serious flaw is usually lack of robustness.

There are a number of useful books on R or on various statistical methods using R, such a Gentleman (2009), Murrell (2006), and Rizzo (2007). The Springer series Use R! includes a number of books such as Albert (2007), Sarkar (2008), and Spector (2008) that address specific topics relating to R.

Computations in Parallel

I have said very little about parallel computations in this book. It is, however, clearly the most important approach for increasing computational power, and it is absolutely necessary for addressing very computationally intensive problems. Parallel computers are becoming more widely available. Even personal computers generally have multiple processing units.

Someone doing research and development in statistical computing must be fluent in parallel computations. Nakano (2004) provides a good summary of the techniques.

The first issue for parallel computations is how to divide up the work (the program) and the data. The most common model is SPMD, "single-program-multiple-data". In SPMD the most important issue is passing of data, and for that a standard API, called the Message Passing Interface (MPI) has been developed. There are various Fortran and C libraries implementing the MPI.

There are evolving standards for compiler support of parallel computations. In Fortran, the most important concept is co-arrays. Various Fortran 95 compilers have supported this for years, and they are now codified in the Fortran 2008 standard. This allows for direct memory copies, achieving the same result as message passing, but at a significantly faster rate. (As of this writing, this standard has not been published.)

There is current work on a C standard called Uniform Parallel C (UPC) to promote portability of parallel codes in C.

The main challenge is to be able to develop long-lasting software that takes advantage of multiple processors, and the development of standards should help to preserve the value of parallel code.

Data Structure

Another important topic that I have not discussed in much detail is database structure. The brevity of my discussion of data structures is not indicative of its importance. Detailed consideration of data structures is outside of the scope of this book. For further information on this topic, the reader is referred to general books on algorithms, such as Cormen et al. (2001), which address data structures that are central to the construction of many algorithms. Some of the issues of database structure that are most relevant in statistical applications are discussed by Boyens, Günther, and Lenz (2004).

Exercises

3.1. In standard mathematical libraries there are functions for $\log(x)$ and $\exp(x)$, called `log` and `exp` respectively. There is a function in the IMSL Libraries to evaluate $\log(1+x)$ and one to evaluate $(\exp(x)-1)/x$. (The names in Fortran, for single precision, are `alnrel` and `exprl`.)
   a) Explain why the designers of the libraries included those functions, even though `log` and `exp` are available.
   b) Give an example in which the standard log loses precision. Evaluate it using `log` in the standard math library of Fortran or C. Now evaluate it using a Taylor series expansion of $\log(1+x)$.

3.2. Suppose you have a program to compute the cumulative distribution function for the chi-squared distribution. The input for the program is x and ν, the degrees of freedom, and the output is $\Pr(X \leq x)$. Suppose you are interested in probabilities in the extreme upper range and high accuracy is very important. What is wrong with the design of the program for this problem?

3.3. Errors in computations.
   a) Explain the difference in truncation and cancellation.
   b) Why is cancellation not a problem in multiplication?

3.4. Assume we have a computer system that can maintain 7 digits of precision (base 10). Evaluate the sum of squares for the data set $\{9000, 9001, 9002\}$.
   a) Use the algorithm in (3.4), page 115.
   b) Use the algorithm in (3.5).
   c) Now assume there is one guard digit. Would the answers change?

3.5. Develop algorithms similar to (3.5) to evaluate the following.
   a) The weighted sum of squares:

$$\sum_{i=1}^{n} w_i(x_i - \bar{x})^2$$

   b) The third central moment:

$$\sum_{i=1}^{n} (x_i - \bar{x})^3$$

   c) The sum of cross products:

$$\sum_{i=1}^{n} (x_i - \bar{x})(y_i - \bar{y})$$

Hint: Look at the difference in partial sums,

$$\sum_{i=1}^{j}(\cdot) - \sum_{i=1}^{j-1}(\cdot)$$

3.6. Given the recurrence relation

$$t(n) = pt(n/p) + cn,$$

for p positive and c nonnegative. Show that $t(n)$ is $\mathrm{O}(n \log n)$. *Hint*: First assume n is a power of p.

3.7. Suppose all of the n addends in a summation are positive. Why is the computation of the sum by a fan-in algorithm likely to have less roundoff error than computing the sum by a standard serial algorithm? (This does not have anything to do with the parallelism, and the reason does not involve catastrophic cancellation.)

3.8. Consider the problem of computing `w = x - y + z`, where each of `x`, `y`, and `z` is nonnegative. Write a robust expression for this computation.

3.9. Searching.
Given an array of length n containing real numbers, find a subarray with maximum sum. (If all of the contents are nonnegative, obviously the full array is a subarray that satisfies the requirement.) The subarray may not be unique. The best algorithm is $\mathrm{O}(n)$.

3.10. Sorting.
Write pseudo code similar to for the bubble sort method described on page 122 for a single vector. Show that its operation count is $\mathrm{O}(n^2)$.

3.11. Sorting.
Write either a C function or a Fortran subroutine that accepts a vector and sorts it using Quicksort (page 123). Test your program to ensure that it is working correctly.

3.12. Sorting and merging.

a) Consider a very simple sorting method for an array a, consisting of n elements, which is to be sorted in place. *(This algorithm is not a good one; it is just used for illustration. It is similar to the bubble sort.)*

0. Set $i = 1$.
1. Set $r = i$.
2. For $j = i + 1$ to n, if $a(i) > a(j)$, then $r = j$.
3. If $r > i$, then
    interchange $a(i)$ and $a(r)$ and go to step 1;
  otherwise
    if $i < n$, then
      set $i = i + 1$ and go to step 1,
    otherwise
      end.

Describe carefully how you might implement this algorithm on k processors with a shared memory. What is the order of this algorithm? Discuss the algorithm critically. Consider the case in which $k \approx n$.

b) Now consider the problem of merging two sorted lists. The arrays b and c are each sorted and we wish to merge them into a new list d that is sorted. Describe carefully how you might do this using k processors with a shared memory.

3.13. Consider again the series

$$\sum_{i=1}^{\infty} \frac{1}{i^p},$$

for $p \geq 1$ from Exercise 2.2d. For $p = 1.5, 2.0, 2.5, \ldots 4.0$, experimentally determine the order of convergence. Now express the order of convergence in terms of n and p.

3.14. While a series of computations may converge, the question of the convergence of the error remains. In Exercise 2.12, using the algorithm in Figure 2.10, we saw that for $x = 20$, the Taylor series approximation yielded a fairly good value for e^x. We know the order of errors of a truncated Taylor series, in terms of derivatives. Now, experimentally determine the order of the error for approximating e^x with the Taylor series for $x = 20$ in terms of the number of iterations (that is, the number of terms in the Taylor series).

3.15. Design and write either a C function or a Fortran subroutine that accepts two addends in vectors and a required precision level, and that returns the sum in a vector to the precision required. (Think of the elements of the vectors as being digits in some base.) Write a user-oriented description of your module.

3.16. Develop a set of test programs that will probe the accuracy of a given module to compute a sample sum of squares. You should consider various ways that the given module could go wrong and the various types of data that could cause it to have problems, and provide tests for all of them.

3.17. Assume the problem P has solution s (unknown, of course). An algorithm/program F is available to solve P. Ideally, of course, $F(P) = s$. Discuss the issues and the methods you would employ to determine that $F(P)$ is an adequate approximation to s.

3.18. Write either a C function or a Fortran subroutine that accepts the three coefficients of a quadratic polynomial and evaluates one of the roots by means of the quadratic formula, equation (2.18), and computes the other root in an appropriate manner. (See the discussion on page 101.) Write your function or subroutine as a standard software part. Write the part specification very carefully, but succinctly, as comments in the programming language. (What do you do if $b^2 < 4ac$? You do not have to provide a solution in this case, i.e., complex roots, but your software part must *handle* that case.)

3.19. Design and write either a C function or a Fortran subroutine that uses a real-time algorithm for the method you developed in Exercise 3.5c to compute the mean vector and variance-covariance matrix for multivariate data in the standard $n \times m$ rectangular layout (that is, as a matrix). Your program module should accept as input

- the number of variables
- the number of observations input in the current invocation (this is a number between 0 and n) – call this number n_i (a 0 value only makes sense if the current invocation is the final one for the given problem, and only wrap-up operations are to be performed)
- a subset of the rows of the overall data matrix (this is an $n_i \times m$ matrix)
- an indicator of whether this invocation is the first one for the given problem, an intermediate one, or the final one
- the total number of observations that have been processed before the current invocation – and output as the updated total, to include the observations in the current invocation
- the vector of means of the observations that have been processed before the current invocation – and output as the updated means, to include the observations in the current invocation
- the matrix of sums of squares and cross-products of the observations that have been processed before the current invocation – and output as the updated sums of squares and cross-products to include the observations in the current invocation. On the final invocation, the sums of squares and cross-products should be scaled by the appropriate divisor to form variances and covariances.

3.20. Discuss design issues for your program module of Exercise 3.19 if the data may contain missing values.

3.21. In statistical data analysis, it is common to have some missing data. This may be because of nonresponse in a survey questionnaire or because an experimental or observational unit dies or discontinues participation in the study. When the data are recorded, some form of missing-data indicator must be used. Discuss the use of NaN as a missing-value indicator. What are some advantages and disadvantages?

3.22. Timing.
Write a program to time the C function or Fortran subroutine (or you can use R) that you wrote in Exercise 3.11 to sort a vector using Quicksort. Experiment with the performance of your code using input vectors with various orderings. (You must use large vectors to be able to tell any difference, but you must also use different sizes of vectors to assess the performance.) Examine the timing as a function of the size of the problem, on average.
What is the worst case problem for Quicksort? Examine the timing as a function of the size of the problem for the worst cases.
Summarize your findings in a clearly-written report.

4

Approximation of Functions and Numerical Quadrature

Often in applied mathematics, we encounter a functional model that we express as

$$y = f(x),$$

but yet we do not know f fully, or else we cannot evaluate it in closed form.

We may only have some general knowledge about the shape of f and know its value at certain points, $f(x_1), f(x_2), \ldots$. In that case, our objective may be just to determine some "nice" function, say a polynomial, that goes through those known values. We say it "interpolates" those points.

Even if we know the values at certain points, we may take a slightly different approach in which we determine a nice function that goes "close" to the points, but does not necessarily interpolate them. We say it "smoothes" the points. In this case there is an adjustment term at each of the points that is not interpolated.

Alternatively the function may be parameterized, and we have full knowledge of the function except for the values of the parameters; that is, we form the model as

$$y = f(x; \theta),$$

and we know $f(x; \theta)$ all except for θ. If there is a value of θ so that the function interpolates all known values, that value provides us full knowledge of f (conditional on the known values). It may be the case that there is no value of θ such that $f(x; \theta)$ fits all of the known values (x_i, y_i). In this case we must recognize that $y = f(x; \theta)$ is not really correct; it needs some kind of adjustment term.

If our objective is to interpolate the known points, there are several ways we can do that, so we set some reasonable criteria about the form of the interpolating function, and then proceed. We will discuss ways of interpolation in this chapter.

If our objective is to smooth the data or if the functional model cannot interpolate all of the known points, then we must deal with an adjustment term. There are two possible approaches. One is set some reasonable criteria

J.E. Gentle, *Computational Statistics*, Statistics and Computing,
DOI: 10.1007/978-0-387-98144-4_4,

about the form of the smoothing function and about what kinds of adjustment terms to allow, and then proceed. That approach is a topic of this chapter; the objective is to *approximate* the function. In another approach, we assume that the adjustment terms are random variables, and so we can use statistical techniques to *estimate* the function; that is the topic of Chapter 10.

There are many reasons for approximating a function. One reason for doing this, which we will address in later sections of this chapter, is to evaluate an integral involving the function. Another reason, which we address in Chapter 8, is to draw lines in a graphical display. An additional reason for approximating a function is to put the function in a form that is amenable for estimation; that is, we *approximate* a function and then *estimate* the approximating function.

Before proceeding to the main topic of this chapter, that is, methods of approximation of functions, we review some general topics of linear spaces developed in Section 1.2 for the special case of function spaces. This discussion leads to the notions of basis functions in Section 4.2. In Sections 4.3, 4.4, and 4.5, we discuss methods of approximation, first using a basis set of orthogonal polynomials and truncated series in those basis functions, then using finite sets of spline basis functions, and then using a kernel filter. In Sections 4.6 and 4.7 we discuss numerical quadrature. Some methods of quadrature are based on the function approximations.

Inner Products, Norms, and Metrics

The inner products, norms, and metrics that we defined beginning on page 11 are useful in working with functions, but there are a few differences that result from the fact that the "dimension" is essentially uncountable.

The inner product of functions is naturally defined in terms of integrals of the products of the functions, analogously to the definition of the dot product of vectors in equation (1.7). Just as the inner product of vectors is limited to vectors with the same length, the inner product of functions is defined for functions over some fixed range of integration (possibly infinite).

Also, just as we sometimes define vector inner products and vector norms in terms of a weight vector or matrix, we likewise define inner products for scalar-valued functions with respect to a weight function, $w(x)$, or with respect to the measure μ, where $\mathrm{d}\mu = w(x)\mathrm{d}x$,

$$\langle f, g\rangle_{(\mu;D)} = \int_D f(x)\bar{g}(x)w(x)\,\mathrm{d}x, \tag{4.1}$$

if the integral exists. Often, both the weight and the range are assumed to be fixed, and the simpler notation $\langle f, g\rangle$ is used. We remind the reader that in this book, we will generally assume the functions are real-valued, and so we usually do not write inner products using the complex conjugate. The inner product in equation (4.1) is also called the dot product, just as in the analogous case with vectors.

We can define norms of functions in the same general way we have defined any norm. There are two differences we must be aware of, however. First of all is the question of integrability of the integrand that defines the norm. While for finite-dimensional vectors, this was not a problem, we must identify the class of functions for which we are defining the norm so as to ensure integrability. We will use the notation $L^2(D)$ to represent the space of functions that take values in the same vector space and that are square-integrable over D; that is, $f \in L^2(D)$ if and only if

$$\int_D (f(x))^2 w(x) \mathrm{d}x < \infty. \tag{4.2}$$

In the following, we will often assume that the function is scalar-valued. If the function is not scalar-varlued, then the integrands in equations (4.1) and (4.2) would be replaced by norms of vectors.

Secondly, the nonnegativity property of a norm is somewhat awkward in working with functions. This is because the inner product defined as an integral would not induce a norm that satisfies the nonnegativity property. It is clearly possible that $f(x_0) > 0$ for some value x_0, yet $\int_D (f(x))^2 w(x)\,\mathrm{d}x = 0$.

We note, however, that if D_1 is a set of x for which $f(x) > 0$ and $\int_{D_1}(f(x))^2 w(x)\,\mathrm{d}x = 0$, then $\int_{D_1} w(x)\,\mathrm{d}x = 0$; that is, D_1 is a set with measure 0 with respect to the weighting in the definition of the inner product.

With this in mind, we often define a more useful function, a *pseudonorm*, in which the nonnegativity requirement is relaxed to allow a zero value of the pseudonorm to imply only that the function is 0 almost everywhere, where *almost everywhere* means "everywhere except on a set of measure 0" (with respect to some measure). Any norm is obviously a pseudonorm, but not vice versa. Unless the distinction is important, however, we often refer to a pseudonorm just as a "norm". We may use the term "pseudonorm" for technical correctness, but in computations, the distinction is generally meaningless. The whole space $\mathbb{F}$ has Lebesgue measure 0. (See Section 2.2.)

A common L_p function pseudonorm is the L_2 pseudonorm, which is denoted by $\|f\|_2$. Because this pseudonorm is so commonly used, we often denote it simply by $\|f\|$. This pseudonorm is related to the inner product:

$$\|f\|_2 = \langle f, f \rangle^{1/2}. \tag{4.3}$$

The space consisting of the set of functions whose L_2 pseudonorms over $\mathbb{R}$ exist together with this pseudonorm itself is denoted L^2. (To be more precise, the measure μ from equation (4.1) is the Lebesgue measure. Notice that the space L^2 consists of both a set of functions S and the special function, the pseudonorm, whose domain is S.)

Another common L_p function pseudonorm is the L_∞ pseudonorm, especially as a measure of the difference between two functions. This pseudonorm, which is called the *Chebyshev norm* or the *uniform norm*, is the limit of equation (1.15) on page 13 as $p \to \infty$ (with the inner product in equation (4.1)). In most cases, this pseudonorm has the simpler relationship

$$\|f\|_\infty = \sup_D |f(x)w(x)|. \tag{4.4}$$

(Notice that the expression on the right side of equation (4.4) is actually a norm instead of just a pseudonorm. We may not want to define $\|f\|_\infty$ in this manner, however, because f could be very large on some discrete points, and so the sup would not capture the relevant size of f. On the other hand, we could define the class of relevant functions in such a way that this is not an issue.)

To emphasize the measure of the weighting function, the notation $\|f\|_\mu$ is sometimes used. (The ambiguity of the possible subscripts on $\|\cdot\|$, whether they refer to a type of norm or to the measure, is usually resolved by the context.) For functions over finite domains, the weighting function is most often the identity.

A *normalized function* is one whose norm is 1. Although this term can be used with respect to any norm, it is generally reserved for the L_2 norm (that is, the norm arising from the inner product). A normalized function is also sometimes called a "normal function", but we usually use that latter term to refer to a function whose integral (over a relevant range, usually $\mathbb{R}$) is 1. Density functions and weight functions are often normalized (that is, scaled to be normal).

Sequences; Complete Spaces

For approximation methods, it may be important to know that a sequence of functions (or vectors) within a given space converges to a function (or vector) in that space.

A sequence $\{f_i\}$ in an inner product space is said to converge to f in a given norm $\|\cdot\|$ if, given $\epsilon > 0$, there exists an integer M such that $\|f_i - f\| \leq \epsilon$ for all $i \geq M$. This convergence of the norm is uniform convergence; that is, at all points. We also often consider pointwise convergence of a sequence of functions, which depends on the argument of each function in the sequence.

A sequence is said to be a Cauchy sequence if, given $\epsilon > 0$, there exists an integer M such that $\|f_i - f_j\| \leq \epsilon$ for all $i, j \geq M$.

A space in which every Cauchy sequence converges to a member of the space is said to be *complete.*

A complete space together with a norm defined on the space is called a *Banach space.* A closed Banach space in which the norm arises from an inner product, as in equation (4.3), is called a *Hilbert space.*

The finite-dimensional vector space $\mathbb{R}^d$ and the space of square-integrable functions L^2 are both Hilbert spaces. (See a text on real analysis, such as Hewitt and Stromberg, 1965.) They are, by far, the two most important Hilbert spaces for our purposes. The convergence properties of the iterative methods we often employ in smoothing and in optimization methods generally derive from the fact that we limit our domain of discourse to Hilbert spaces.

Roughness

We use the terms "roughness" and "variation" in referring to functions in a nontechnical sense, and the terms are more-or-less synonymous. We often, however, refer to a specific measure of roughness or variation. (Notice that we try to avoid use of the term "variance" in a nontechnical sense.)

A reasonable measure of the variation of a scalar function is

$$\mathcal{V}(f) = \int_D \left(f(x) - \int_D f(t)\mathrm{d}t \right)^2 \mathrm{d}x. \tag{4.5}$$

This quantity is the variance of $f(Y)$, where Y is a random variable with a uniform distribution over D (see expression (4.86) on page 194).

If the integral $\int_D f(x)\mathrm{d}x$ is constrained to be some constant (such as 1 in the case that $f(x)$ is a probability density), then the variation can be measured by the square of the L_2 norm,

$$\mathcal{S}(f) = \int_D \|f(x)\|_2^2 \,\mathrm{d}x. \tag{4.6}$$

Another intuitive measure of the roughness of a twice-differentiable and integrable univariate function f is the integral of the square of the second derivative:

$$\mathcal{R}(f) = \int_D \|f''(x)\|_2^2 \,\mathrm{d}x. \tag{4.7}$$

A function constructed so as to approximate a given function often is very rough. We sometimes constrain the approximating function in some way so that its roughness is small or else that it is similar in magnitude to the roughness of the given function if we know it or if we have an approximation for it.

Linear Operators

Approximations of functions are often formed by use of a *functional* or an *operator*.

A functional is a mapping of a function space into a vector space; for our purposes we will consider a functional to be a mapping of a function into the finite-dimensional vector space $\mathbb{R}^d$. An operator is a mapping of a function space into a function space.

We are interested in properties of functionals and operators because those properties relate to the magnitude of the error in function approximation.

We will denote functions using an upper case letter, and operators using a calligraphic font, for example, we may write $v = L(f)$ and $g = \mathcal{L}(f)$, where f is a function, v is a real vector, and g is a function. Depending on the emphasis or the need for clarity, we also may write the functions with formal arguments, $g(t) = \mathcal{L}(f(x))$; or with actual arguments, $g(t_0) = \mathcal{L}(f(x_0))$; and we may write the operator without parentheses, $g = \mathcal{L}f$ or $g(t) = \mathcal{L}f(x)$.

For a given function f, the measure of the goodness of the function g to approximate f is the functional $L(g) = \|f - g\|$.

The most commonly used functionals and operators are linear. The functional L is a *linear functional* if for any two functions f_1 and f_2 within the domain of L and for any constant c,

$$L(cf_1 + f_2) = cL(f_1) + L(f_2). \tag{4.8}$$

Likewise, the operator $\mathcal{L}$ is a *linear operator* if for any two functions f_1 and f_2 within the domain of $\mathcal{L}$ and for any constant c,

$$\mathcal{L}(cf_1 + f_2) = c\mathcal{L}(f_1) + \mathcal{L}(f_2). \tag{4.9}$$

An example of a linear operator on functions with domain $[a, b]$ is the one that results in a straight line through the function values at a and b. A similar, more general linear operator is

$$\mathcal{L}(f) = \sum_{i=1}^{n} c_i f(x_i). \tag{4.10}$$

The Lagrange interpolating polynomial, which we consider below in equation (4.16), is an example of this linear operator.

Another example is a finite Taylor series approximation of a differentiable function:

$$\mathcal{L}(f) = f(x_0) + (t - x_0)f'(x_0). \tag{4.11}$$

Convolutions and covariances are also important linear operators.

Norms of Operators

Norms of functionals and operators measure their variation. Because of the way we use functionals and operators in approximation, the most useful norms are often those that capture maximal deviations. Hence, we define the norms as Chebyshev norms. We define the norm of functionals and operators in terms of their use on normalized functions; that is, on functions whose norm is 1. The norm $\|\cdot\|_f$ used to define the normalized function is not necessarily a Chebyshev norm. The *norm of functional*, $\|L\|$, in terms of a normalized function, is the vector norm:

$$\|L\| = \max_{\|f\|_f=1} |L(f)|. \tag{4.12}$$

The *norm of an operator*, $\|\mathcal{L}\|$, in terms of a normalized function, is the function norm:

$$\|\mathcal{L}\| = \sup_{\|f\|_f=1} \|\mathcal{L}(f)\|. \tag{4.13}$$

It is clear that these norms satisfy the properties that define a norm (see page 13).

For example, the norm of the linear interpolant operator mentioned above that is the straight line between the points $(a, f(a))$ and $(b, f(b))$ is easily seen to be

$$\max(|f(a)|, |f(b)|). \tag{4.14}$$

The norm of the interpolant operator serves to measure the sup error of the approximation.

4.1 Function Approximation and Smoothing

We often need to approximate a given function f by another function $\widetilde{f}$. This may be because we know f only at some specific points; it may be because the approximation $\widetilde{f}$ may be easier to work with than f; or it may be because we have good ways of estimating $\widetilde{f}$, but do not have a direct way of estimating f.

By "to approximate f" we may mean to approximate an integral of f, to approximate a derivative of f, or just to approximate some values of f that may be unknown or may be difficult to evaluate directly.

There are a number of ways we can approximate the function f:

- globally, as a truncated series of other, basic functions
- globally, as a ratio of other, basic functions
- piecewise, using different forms of another, basic function in different regions of the domain
- globally, with a weighting function that is centered at different places within the domain.

In Sections 4.2 and 4.3 we will consider the first approach. This is reminiscent of the discussion beginning on page 18 for representing vectors as linear combinations of the elements in a basis set.

An approximation that sometimes provides a better fit is a ratio of two truncated power series or of two polynomials:

$$f(x) \approx \frac{p(x)}{q(x)}.$$

This type of approximation, which is an instance of a rational approximation, is called a *Padé approximation.* Although Padé approximation may be useful, particularly when the function to be approximated contains poles, we will not cover this approach.

The most common piecewise approach to approximating a function is to use splines, which are functions with different forms in different regions of the domain but which join smoothly at the juncture of two adjacent regions. We will discuss this method in Section 4.4.

A global approach that allows different treatment at all points within the domain is to use a moving weight function. This is the idea behind kernel methods, which we will discuss in Section 4.5.

In Section 4.6 we consider the use of function approximation for evaluation of definite integrals. We also consider in that section other approaches to numerical quadrature.

The approximating function $\widetilde{f}$ should be easy to evaluate, easy to differentiate, and easy to integrate. It should also of course be easy to determine, given a function f that we wish to approximate.

How well one function approximates another function is usually measured by a norm of the difference in the functions over the relevant range. If $\widetilde{f}$ is used to approximate f, the Chebyshev norm, $\|\widetilde{f} - f\|_\infty$, is likely to be the norm of interest. This is the norm most often used in numerical analysis when the objective is interpolation or quadrature and when we make assumptions about continuity of the functions or their derivatives. *Chebyshev approximation* is approximation in which this norm is minimized over a set of approximating functions $\widetilde{f}$.

In problems with noisy data, or when $\widetilde{f}$ may be very different from f, $\|\widetilde{f} - f\|_2$ may be the more appropriate norm. This is the norm most often used in estimating probability density functions (see Chapter 15) or in projection pursuit (see Section 16.5), for example.

To use $\widetilde{f}$ as an approximation for a function f, given a known set of values $\{(y_i, x_i)\}$ such that $y_i = f(x_i)$, we may require that $\widetilde{f}(x_i) = f(x_i)$ at each of the known points. This is interpolation. Alternatively, we may require that $\widetilde{f}(x_i) \approx f(x_i)$ and that $\widetilde{f}$ not be very rough, by some measure of roughness. This is smoothing.

Models for Interpolation

The model $y = f(x)$ can be viewed as expressing an exact relationship for a fixed set of values. Given a set $\{(x_i, y_i)\}$ of values for which $y_i = f(x_i)$, it is reasonable to require that the approximating function $\widetilde{f}$ be such that $\widetilde{f}(x_i) = f(x_i)$. That exact fit would provide an approximation at other values that x may assume. This approach of fitting a function to the given y and x is called *interpolation*.

There are several possibilities for choosing a continuous function to interpolate the data. An *interpolant* is likely not to be very smooth or else it may exhibit wide variation. The requirement to fit all data values exactly may also mean that the relationship is not a (single-valued) function. It is unlikely that a single easily-defined function, other than a polynomial, could interpolate the data. A polynomial of degree $n - 1$ can, of course, interpolate n data points (assuming no two points have the same ordinate values), but such a polynomial may have wild swings; that is, it may be very rough.

If a functional form for $\widetilde{f}$ is chosen, say $\widetilde{f}(x; \theta)$, where θ is a parameter, it may be difficult or impossible to determine a value of θ that would interpolate a given dataset; that is, a value of θ that would yield an equality at each data point. For interpolation, there must be considerable freedom to choose the function $\widetilde{f}$. Rather than a single function, however, we may choose a piecewise set of functions, each of which interpolates a subsequence of the given points that have adjacent abscissas. Two piecewise interpolating polynomials are shown in Figure 4.1.

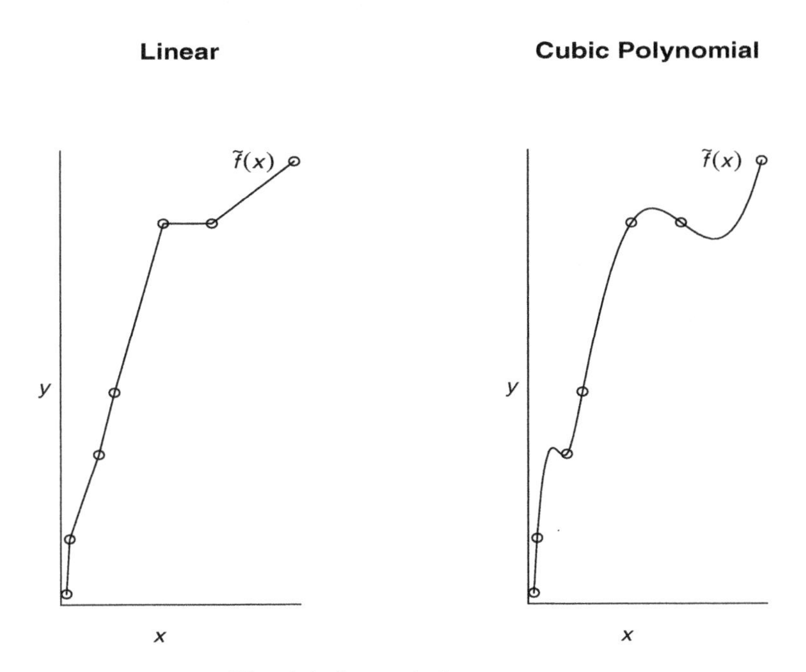

Fig. 4.1. Interpolation

A general form of an interpolating function for two-dimensional data can be built from *Lagrange polynomials.* Given $x_1, x_2, \ldots, x_n$, we define $n-1$ Lagrange polynomials each of degree $n-1$, so that the jth Lagrange polynomial is

$$l_j(x) = \prod_{i \neq j} \frac{x - x_i}{x_j - x_i}. \tag{4.15}$$

It is clear that

$$l_j(x_i) = \begin{cases} 1 \text{ if } i = j \\ 0 \text{ otherwise;} \end{cases}$$

therefore, if the function values at the given points are $y_1, y_2, \ldots, y_n$, the function

$$h(x) = \sum_{i=1}^{n} l_i(x) y_i \tag{4.16}$$

is an interpolating polynomial of degree $n - 1$.

A single Lagrange polynomial of degree $n - 1$ could be used over the full range, but it would be very rough. Another way is to use different piecewise Lagrange polynomials over subsets of the points. For a set of k adjacent points, a polynomial of degree $k - 1$ could interpolate them. The piecewise linear function in Figure 4.1 is a piecewise Lagrange polynomial of degree 1 each piece of which is defined on two points, and the interpolating curve shown on the right in Figure 4.1 is a piecewise cubic polynomial interpolating three adjacent points.

The sharp increase at the first two points in Figure 4.1 causes large values for the derivatives. Instead of joining the polynomials at data points, we could force them to be joined between points. A polynomial of degree $k - 1$ can interpolate k points whether or not two of the points are endpoints of the polynomial.

The breakpoints are called knots, and this kind of interpolation, in which polynomials of degree $k - 1$ can interpolate k points and such that the polynomials join smoothly at the knots, is called spline interpolation, and we will consider it in more detail in Section 4.4.

Error in Function Approximation

From equation (4.10) on page 152, the norm of the Lagrange interpolating polynomial (4.16) is the sum of the norms of the individual Lagrange polynomials; that is, if

$$\mathcal{L}(f)(x) = \sum_{i=1}^{n} f(x_i) l_i(x),$$

then

$$\|\mathcal{L}\| = \sum_{i=1}^{n} \|l_i(x)\|. \tag{4.17}$$

This is easily seen by observing

$$\begin{aligned}
\|\mathcal{L}\| &= \left\| \sum_{i=1}^{n} f(x_i) l_i(x) \right\| \\
&\leq \sum_{i=1}^{n} |f(x_i)| \, \|l_i(x)\| \\
&\leq \max |f(x_i)| \sum_{i=1}^{n} \|l_i(x)\| \\
&\leq \|f\| \sum_{i=1}^{n} \|l_i(x)\|.
\end{aligned}$$

Take f as a function such that $\|f\| = 1$, and the result follows.

With this, we have the sup norm on the error of approximation by a Lagrange interpolating polynomial for any function with finite Chebyshev norm.

Models for Smoothing Data

A function that interpolates the data can be very wiggly. A continuous function with continuous low-order derivatives may be more useful, even if it does not interpolate the points exactly. The process of selecting a relatively simple model that provides a good approximation to the data is called "smoothing".

One way to smooth data is to choose some simple functional form $\widetilde{f}$ with a parameter θ that can chosen so that the observed values y_i are close to the smoothed values $\widetilde{f}(x_i;\widehat{\theta})$, for some $\widehat{\theta}$. In Figure 4.2, we see two different smoothing models for the same data. In the plot on the left, we have a simple straight line that approximates the data. This functional form has a parameter θ that is a 2-vector (the slope and the intercept). No matter how the parameter is chosen, the straight line does not fit the observations very well. A different functional form is used in the plot on the right in Figure 4.2. This approximation seems to fit the data better, and it captures an important apparent structure in the data. (Although it is not important for our purposes here, the function on the right is $y = (\Gamma(\alpha)\beta^\alpha)^{-1}x^{\alpha-1}\mathrm{e}^{-x/\beta}$, and the values of α and β are chosen by least squares.)

For approximation, just as for interpolation, we could also use different functional forms over different regions of the data.

Multivariate Approximation

Conceptually, most of the discussion applies immediately to multivariate functions defined over subspaces of IR^d. Practical difficulties, however, prevent direct methods of multivariate approximation from being very useful.

In multivariate function approximation, the function values are usually known at points on a grid in IR^d. The most common way of approximating a multivariate function is by successive univariate approximations. Sometimes successive univariate approximations are easy to construct. For example, if the function to be approximated is the bivariate function $f(x, y)$, and the values of the function are known at a rectangular grid of points $(x_1, \ldots, x_n) \times (y_1, \ldots, y_m)$, suppose we wish to approximate $f(x, y)$ at the point (x^*, y^*). In successive univariate approximation, we first approximate $f(x_i, y)$ at each given point x_i by a function $g_{x_i}(y)$. (This is univariate approximation in the variable y.) Now, with the values $g_{x_i}(y^*)$ for each i, we approximate $f(x^*, y^*)$ using univariate approximation in the variable x. Often, of course, the grid of known values is unstructured, and we could not use this grid-line approach of successive univariate approximations because for a given x_i there may be only one known value of y.

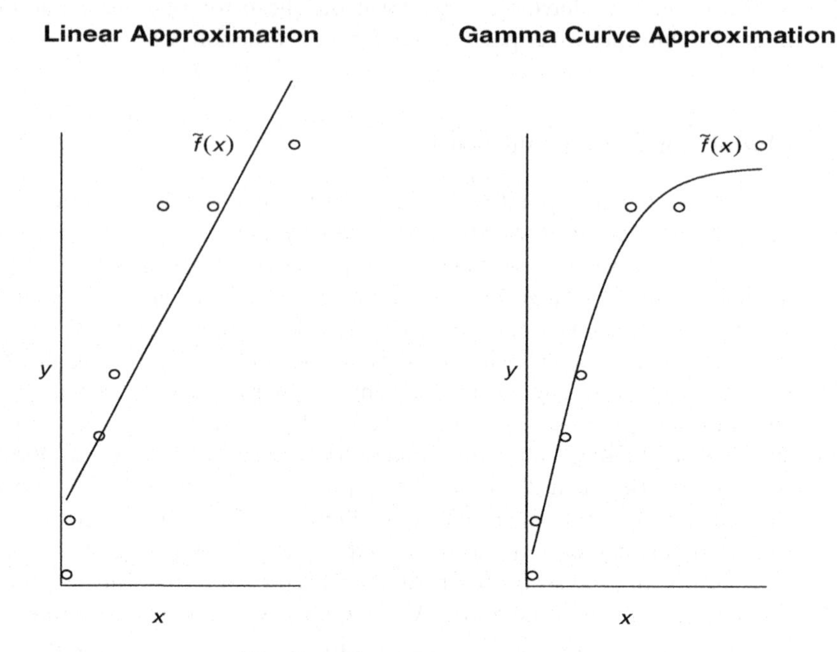

Fig. 4.2. Smoothing

The method of successive univariate approximation extends to higher dimensions, but the number of computations obviously grows exponentially.

Evaluation of Special Functions

Some of the most important and most common numerical computations in scientific applications are the evaluation of the so-called "special functions". These functions, such as the exponential and logarithmic functions, the trigonometric functions, and functions that arise in the solutions of differential equations, are rarely evaluated for their own sake; rather, their evaluations are generally performed as part of some larger computational problem. The functions are often evaluated repeatedly in the solution of the larger problem. It is therefore important not only that they be evaluated accurately, but that the computations be very efficient.

Abramowitz and Stegun (1964) describe computational methods for evaluating many special functions. The basic algorithms they discussed are in many cases still the best ways of evaluating the special functions. Abramowitz and Stegun classified special function into various groups that remain the general organization of numerical libraries for the functions. The Guide to Available Mathematical Software (GAMS) (see the bibliography), follows the general structure of Abramowitz and Stegun (1964). There is also another grouping of special functions in GAMS for probability distributions.

We will not consider the particular algorithms for the special functions; rather, we discuss general techniques that are applicable to various types of functions. For a given function, of course, some methods are better than others, and it is important that very good methods be used on special functions that arise frequently in numerical computations.

Evaluation of Distribution Functions and Their Inverses

Application of most methods of statistical inference involve computation of probabilities or of rejection regions corresponding to an estimator or a test statistic. Two simple examples are the computation of the probability that a standard normal random variable Z is larger in absolute value than some given value z_0, and the computation of the value t_α such that a Student's t random variable with given degrees of freedom would have a given probability of being greater than t_α. For the more common distributions, tables with three or four decimals of precision for these probabilities and critical values have been available for many years. For setting confidence intervals and doing significance tests, this level of precision for the distribution function of the random variable being used is quite adequate.

It is not always the case, however, that three or four decimals of precision is adequate for a distribution function. Perhaps the simplest example of the need for high precision is in the evaluation of the distribution function of an order statistic. The distribution function of the kth order statistic in a sample of size n from a population with distribution function $P(\cdot)$ is

$$\Pr(X_{(k)} \leq x_0) = \sum_{j=k}^{n} \binom{n}{j} (P(x_0))^j (1 - P(x_0))^{n-j}.$$

Obviously, if this relationship is used, even if only three or four decimals of precision is required for the distribution function of the order statistic, it is necessary to have much greater precision in the evaluation of $P(\cdot)$.

In addition to the computational problems resulting from the need for higher precision in computing a standard, relatively simple distribution function, distribution functions such as for the doubly noncentral F random variable are exceedingly complicated and require different algorithms for evaluation at different points in the argument/parameter space.

We will not consider the particular algorithms for the particular distributions; rather, we discuss general techniques that are applicable to various types of distribution functions. For a given distribution, of course, some methods are better than others, and it is important that very good methods be available for distributions that arise frequently in statistics.

4.2 Basis Sets in Function Spaces

If each function in a linear space can be expressed as a linear combination of the functions in a set G, then G is said to be a *generating set* or a *spanning set* for the linear space.

Linear independence for functions is defined similarly to linear independence of vectors, as on page 16. A difference is the phrase "almost everywhere". A set of functions $f_1, f_2, \ldots$ is *linearly independent* if

$$a_1 f_1 + a_2 f_2 + \cdots = 0 \text{ almost everywhere}$$

implies that $a_1 = a_2 = \cdots = 0$. A simple set of linearly independent functions over any real interval is the set of monomials, $1, x, x^2, \ldots$.

If the functions in the generating set are linearly independent then the set is a *basis set.* The basis sets for finite-dimensional vector spaces are finite; for most function spaces of interest, the basis sets are infinite. The set of monomials is a basis set for a large class of functions.

Expansions in Monomials

As in our discussion of vectors on page 18, our objective is to represent a function of interest as a linear combination of the functions in some basis set. A common example is the Taylor series expansion of a univariate function (whose Taylor expansion exists) about some given point in its domain. This is an expansion in the basis set of monomials:

$$f(x) = f(x_0) + \sum_{k=1}^{\infty} \frac{f^{(k)}(x_0)}{k!}(x - x_0)^k, \tag{4.18}$$

where $f^{(k)}(x_0)$ is the kth derivative of f evaluated at x_0.

The *Taylor series approximation* of $f(x)$ is

$$\widetilde{f}(x) = f(x_0) + \sum_{k=1}^{j} \frac{f^{(k)}(x_0)}{k!}(x - x_0)^k. \tag{4.19}$$

For $j = 2$, this approximation has an immediate extension to a multivariate function.

If f is nonnegative, letting $g(x) = \log(f(x))$, we can express f as

$$f(x) = \mathrm{e}^{g(x)}. \tag{4.20}$$

Many interesting functions in statistical applications are concave. If f is concave and differentiable, and x_0 is a point at which f is maximum, then $f'(x_0) = g'(x_0) = 0$; hence, in this case, we have another approximation of f in terms of g at x_0:

$$\widetilde{f}(x) = \exp\left(g'(x_0) + \frac{(x-x_0)^2}{2} g''(x_0)\right), \tag{4.21}$$

which we recognize as $\exp(g'(x_0))$ times the normal PDF with mean x_0 and variance $-1/g''(x_0)$:

$$\widetilde{f}(x) = \exp(g'(x_0))\, \phi(x \mid x_0, -1/g''(x_0)). \tag{4.22}$$

If in addition to its being nonnegative, we also assume that it has a finite integral over $(-\infty, \infty)$, that is, f is essentially a PDF, then this type of expansion can be extended so as to lead to more general expansions of PDFs as in equation (4.31) on page 165.

Because of the factor of the form $\exp(t)$, the approximation $\widetilde{f}(x)$ is an *exponentially tilted* measure. Exponential tilting is often a useful transformation of PDFs, so as to form more tractable functions.

The approximation (4.22) also leads us immediately to the *Laplace approximation* for the integral of a nonnegative concave function f, which is

$$\int f(x)\mathrm{d}x \approx \mathrm{e}^{g(x_0)} \left(-\frac{2\pi}{g''}(x_0)\right)^{1/2}. \tag{4.23}$$

This approximation is often useful in numerical quadrature, a topic which we consider in more detail in Section 4.6.

Series Expansions in Orthogonal Basis Functions

A set of functions $\{q_k\}$ is *orthogonal over the domain D with respect to the nonnegative weight function $w(x)$* if the inner product with respect to $w(x)$ of q_k and q_l, $\langle q_k, q_l \rangle$, is 0 if $k \neq l$; that is,

$$\int_D q_k(x)\bar{q}_l(x)w(x)\mathrm{d}x = 0 \quad k \neq l. \tag{4.24}$$

If, in addition,

$$\int_D q_k(x)\bar{q}_k(x)w(x)\mathrm{d}x = 1,$$

the functions are called *orthonormal.*

In the following, we will be concerned with real functions of real arguments, so we can take $\bar{q}_k(x) = q_k(x)$.

The weight function can also be incorporated into the individual functions to form a different set,

$$\widetilde{q}_k(x) = q_k(x)w^{1/2}(x).$$

This set of functions also spans the same function space and is orthogonal over D with respect to a constant weight function.

Basis sets consisting of orthonormal functions are generally easier to work with and can be formed from any basis set using the Gram-Schmidt function transformations (see pages 18 and 219).

As in the case of basis sets for vectors, it is often desirable to use a basis set of orthonormal functions. We represent a function of interest, $f(x)$, over some domain D, as a linear combination of orthonormal functions, $q_0(x), q_1(x), \ldots$:

$$f(x) = \sum_{k=0}^{\infty} c_k q_k(x). \tag{4.25}$$

There are various ways of constructing the q_k functions. We choose a set $\{q_k\}$ that spans some class of functions over the given domain D. A set of orthogonal basis functions is often the best choice because they have nice properties that facilitate computations and also there is a large body of theory about their properties.

If the function $f(x)$ is continuous and integrable over a domain D, the orthonormality property allows us to determine the coefficients c_k in the expansion (4.25), just as in equation (1.31):

$$c_k = \langle f, q_k \rangle, \tag{4.26}$$

and the coefficients $\{c_k\}$ are called the *Fourier coefficients* of f with respect to the orthonormal functions $\{q_k\}$.

We then approximate a function using a truncated orthogonal series:

$$f(x) \approx \sum_{k=0}^{j} c_k q_k(x). \tag{4.27}$$

The important quantity that provides a measure of the goodness of the approximation is proportional to the *mean squared error*,

$$\left\| f - \sum_{k=0}^{j} c_k q_k \right\|^2, \tag{4.28}$$

and an important property of Fourier coefficients is that they yield the minimum mean squared error for a given subset of basis functions $\{q_i\}$. (See equation (1.35) and Exercise 1.4.)

Estimation

In applications of statistical data analysis, after forming the approximation, we may then *estimate* the coefficients from equation (4.26) after doing a PDF decompisition (see page 404.) Note the difference in "approximation" and "estimation". Function estimation is the topic of Chapter 10 and, for special types of functions, in Chapter 15. Expected values can be estimated using observed or simulated values of the random variable and the approximation of the probability density function.

Complete Series

By Bessel's inequality (1.37) on page 20, we see that the monotone sequence $\{\sum_{k=0}^{n} |c_k|^2\}$ is bounded by $\|f\|^2$. This implies that $\{\sum_{k=0}^{n} c_k q_k\}$ is a Cauchy sequence in $L^2(\mathrm{IR})$ (or even in $L^2(\mathrm{IR}^d)$), and since $L^2(\mathrm{IR})$ is a complete space the sequence must converge in the L_2 sense to a member of $L^2(\mathrm{IR})$. We of course want it to converge to f, the function we are trying to approximate.

In order to insure that, we need to require that the orthogonal system $\{q_k\}$ have another property. We say that an orthogonal system $\{q_k\}$ is *complete* in $L^2(D)$ if no nontrivial $q \in L^2(D)$ is orthogonal to all the q_k's; that is, if $\langle q, q_k \rangle = 0$, for $k = 0, 1, 2, \ldots$ for $q \in L^2(D)$, then $q = 0$ almost everywhere. A system that is complete in a given function space is a generating set for that function space.

For any function $f \in L^2(D)$ with $D \subset \mathrm{IR}^d$, a linear combination formed by Fourier coefficients with functions from a complete orthonormal system converges to f; that is, if $\{q_k\}$ is a complete orthonormal system in $L^2(D)$, and $c_k = \langle f, q_k \rangle$, then

$$\left\| f - \sum_{k=1}^{n} c_k q_k \right\| \to 0 \quad \text{as } n \to \infty. \tag{4.29}$$

To see this, we first note that $\{\sum_{k=0}^{\infty} c_k q_k\}$ converges to some member, say g, of $L^2(D)$, and hence,

$$f - \sum_{k=0}^{\infty} c_k q_k \to f - g.$$

(This is because $\{\sum_{k=0}^{n} c_k q_k\}$ is a Cauchy sequence and $L^2(D)$ is a complete space.) Now the Fourier coefficients of g, as in equation (4.26), are given by

$$\begin{aligned} \langle g, q_k \rangle &= \lim_{n \to \infty} \left\langle \sum_{k=0}^{n} c_k q_k, q_k \right\rangle \\ &= c_k. \end{aligned}$$

These are the same as the Fourier coefficients of f, and hence the coefficients of $f - g$ are all zero. By the completeness of the system $\{q_k\}$, $f = g$ almost everywhere. Because $\{\sum_{k=0}^{\infty} c_k q_k\}$ converges to g, it must also converge to f.

The theorem expressed by (4.29) has a converse: If all of the coefficients are zero, then f must be zero almost everywhere, and hence the system is complete.

In addition to the requirement of completeness, the basis functions are generally chosen to be easy to use in computations. As mentioned before, the monomials form a set of basis functions for a large class of functions, and a common type of expansion of many functions in that basis set is a Taylor series expansion. The monomials of course are not orthogonal. Common examples of orthogonal basis sets include the Fourier trigonometric functions $\sin(kt)$ and

$\cos(kt)$ for $k = 1, 2, \ldots$, orthogonal polynomials such as Legendre or Hermite, and wavelets. We discuss Fourier series below and orthogonal polynomials in Section 4.3.

Another approach to function approximation is to partition the domain of the function into regions within which good approximations can be achieved by functions that are easy to work with, such as low-degree polynomials. This approach leads to the use of splines. We discuss splines in Section 4.4.

Fourier Series

While any expansion in an orthonormal basis set such as equation (4.25) may be called a Fourier series, this term is often used explicitly to refer to an expansion in sines and cosines. Because, for $j, k = 0, 1, 2, \ldots$,

$$\int_{-\pi}^{\pi} \sin(jx)\sin(kx)\mathrm{d}x = \pi\delta_{jk},$$

$$\int_{-\pi}^{\pi} \cos(jx)\cos(kx)\mathrm{d}x = \pi\delta_{jk},$$

$$\int_{-\pi}^{\pi} \sin(jx)\cos(kx)\mathrm{d}x = 0,$$

$$\int_{-\pi}^{\pi} \sin(jx)\mathrm{d}x = 0,$$

and

$$\int_{-\pi}^{\pi} \cos(jx)\mathrm{d}x = 0,$$

where δ_{jk} is the Kronecker delta, and for any $(x_0, y_0) \in [-\pi, \pi] \times [-1, 1]$, there is a j and k such that $y_0 = \sin(jx_0)$ and $y_0 = \cos(kx_0)$, the set

$$\{\sin(jx), \cos(kx)\ ;\ j, k = 0, 1, 2, \ldots\} \tag{4.30}$$

is a complete orthogonal system in $[-\pi, \pi]$, and consequently in any finite interval $[a, b]$. This is called the Fourier basis set or the trigonometric basis set.

Because of the periodic nature of the trigonometric functions, a Fourier series is often used to approximate periodic functions, although the series can also be used to approximate other functions. The Fourier series may be used in the estimation of probability density functions, as we mention in Section 15.5. It is also of course related to the Fourier transform, referred to in Chapter 3, and also to the characteristic function of a probability distribution.

Expansion of Probability Density Functions

While we may expand a probability density function in terms of an orthogonal system (as we will do in Section 15.5), it is often useful to represent a complicated or unknown probability density function as a series of functions related to a known and well-understood distribution. Alternatively, the characteristic function of the complicated or unknown distribution may be represented as a series of moments or cumulants of a known distribution. The most common distribution, of course, is the standard normal. There are three related series expansions based on a normal distribution. These are the Gram-Charlier series, the Edgeworth series, and the Cornish-Fisher expansion. The general form of these expansions is

$$p(x) = \phi(x) + \frac{1}{6}\gamma_1\phi'''(x) + \frac{1}{24}\gamma_2\phi''''(x) + \cdots, \tag{4.31}$$

where $\phi(x)$ is the PDF of the standard normal distribution. (Compare this with equation (4.22) on page 161.)

The normal distribution pervades statistical theory and methods for two reasons. The first is that this distribution serves so well to model natural phenomena. Even if finite samples follow some other distribution, it is likely that following a suitable transformation, the normal distribution is a good asymptotic approximation. The second reason follows from the first. A wealth of methods have been developed that are directly applicable to the normal distribution. If another distribution can be related to the normal, the vast array of statistical methods for the normal distribution may become available for that other distribution.

Saddlepoint Approximations

A series such as Edgeworth or Gram-Charlier may be expressed in terms of derivatives of the characteristic function or of the cumulant generating function. (Recall that the derivatives of the characteristic function, or of the moment generating function if it exists, evaluated at zero yield the raw moments of a distribution. The derivatives of the cumulant generating function, if it exists, evaluated at zero yield the cumulants of a distribution, and cumulants are uniquely determined by the moments.) Given an expansion in the moments, or given the characteristic function, it may be of interest to determine (or approximate) the PDF.

An important theorem that relates a univariate PDF to the associated characteristic function provides the inverse of equation (1.71) on page 30. We have

$$p(x) = \frac{1}{2\pi}\int_{-\infty}^{\infty} \mathrm{e}^{-\mathrm{i}tx}\varphi(t)\mathrm{d}t. \tag{4.32}$$

This is called the *inversion formula*, and is similar to the inversion formula for the Fourier transform. See Billingsley (1995) for a proof.

If the cumulant function $K(t)$ (see equation (1.73)) exists, we have

$$p(x) = \frac{1}{2\pi} \int_{-\infty}^{\infty} \mathrm{e}^{K(\mathrm{i}t)-\mathrm{i}tx} \mathrm{d}t. \tag{4.33}$$

We can express the integral in equation (4.33) in terms of a real integrand in a neighborhood of 0 by a change of variable, $r = it$. The imaginary unit goes into the limits of integration and into the Jacobian:

$$p(x) = \frac{1}{2\pi\mathrm{i}} \int_{\epsilon-\mathrm{i}\infty}^{\epsilon+\mathrm{i}\infty} \mathrm{e}^{K(r)-rx} \mathrm{d}r. \tag{4.34}$$

Next, we seek a point $r_0(x)$ such that

$$K'(r_0(x)) = x, \tag{4.35}$$

and expanding $K(r) - rx$ about this r_0 in a truncated Taylor series, we have the approximation

$$K(r) - rx \approx K(r_0) - r_0 x + \frac{K''(r_0)}{2}(r - r_0)^2.$$

Equation (4.35) is called the *saddlepoint equation.* This yields

$$p(x) \approx \frac{1}{\sqrt{2\pi K''(r_0)}} \mathrm{e}^{K(r_0)-r_0 x}. \tag{4.36}$$

There are some technical details that have been ignored here; see Daniels (1954) for a more precise development.

Notice that r_0 is a function in x, as we originally wrote it. The point $r_0(x)$ in the complex plane is a saddlepoint; hence the approximation (4.36) is called a *saddlepoint approximation* to the PDF.

The saddlepoint approximation can often be improved by *renormalizing* it so that its integral is 1, as that of a PDF should be. In practice, the integration to determine this normalizing constant must be performed by numerical quadrature, using methods similar to those discussed in Section 4.6 or 4.7.

The saddlepoint approximation is useful in the approximation of densities of various useful statistics such as the mean, a maximum likelihood estimator, a likelihood ratio statistic, and a score statistic. It can also be used in approximating tail probabilities for various distributions.

An example of the use of a saddlepoint approximation is to approximate the PDF of the mean from a mixture distribution. As we mentioned in the general description of mixtures on page 36, the moment generating function for a mixture can be formed easily from the moment generating functions of the individual distributions, and then the cumulant generating function can be determined from the moment generating functions (assuming all exist). In Exercise 4.5 you are led through a step by step derivation of the saddlepoint approximation of the PDF of the expected value of a mixture of two normal distributions.

4.3 Orthogonal Polynomials

The most useful type of basis function depends on the nature of the function being approximated or estimated. Orthogonal polynomials are useful for a very wide range of functions.

Various systems of orthonormal functions can be constructed as polynomials. Because a system of nontrivial polynomials of degrees $0, 1, 2, \ldots$ is independent and complete in any finite nonnull interval of L^2 (and hence, can be orthonormalized), by the theorem in expression (4.29), we know that orthogonal polynomials can be used to approximate any function in L^2 to any degree of accuracy. The familiar *Weierstrass approximation theorem* is explicit for polynomials:

> Let f be a continuous real function defined on $[a, b]$ and let $\epsilon > 0$ be given. Then there exists a polynomial p with real coefficients such that $|f(x) - p(x)| < \epsilon$ for all x in $[a, b]$.

This theorem is proved in many texts on real analysis, such as Hewitt and Stromberg (1965).

In the following, we use the notation $p_i(x)$ or $q_i(x)$ to denote a general polynomial of nonnegative integral degree i; hence, the first item in the sequence has an index of 0. (In the previous sections, we have used $q_i(x)$ to denote any member of an orthogonal basis set, and we began the index at 1 instead of 0. Later in this section, we will discuss specific types of polynomials, and we will use different letters to represent them.)

We often work with unnormalized orthogonal polynomials; hence, the reader must be careful to note whether we are using an orthonormal sequence or one that is possibly not normalized. The reason we often use unnormalized orthogonal polynomials is partly historical, but also because of the simplicity of the coefficients in some standard systems. Any scalar multiple of any member of an orthogonal system leaves the system orthogonal (but not orthonormal). The *normalizing factor* is the scalar that normalizes a given polynomial.

Systems of orthogonal polynomials can be developed from series solutions to differential equations, or by orthogonalizing a set of independent polynomials. We will use the latter approach.

Construction of Orthogonal Polynomials

The simplest set of linearly independent polynomials, that is, the monomials,

$$1,\ x,\ x^2,\ \ldots \tag{4.37}$$

can be orthogonalized and normalized by Gram-Schmidt transformations,

$$
\begin{aligned}
q_0(x) &= 1 \\
q_1(x) &= \frac{x - \langle 1, x \rangle}{\|x - \langle 1, x \rangle\|} \\
q_2(x) &= \frac{x^2 - \langle 1, x^2 \rangle - \langle q_1(x), x^2 \rangle q_1(x)}{\|x^2 - \langle 1, x^2 \rangle - \langle q_1(x), x^2 \rangle q_1(x)\|},
\end{aligned} \tag{4.38}
$$

and so on (see page 18). Sometimes the polynomials are not normalized, but it is usually better to work with orthonormal polynomials.

The specific inner products in the Gram-Schmidt transformations determine the specific form of the system of polynomials. The inner product depends on the domain and on the weight function. Orthogonal polynomials of real variables are their own complex conjugates, so the inner products involve just the polynomials themselves.

In some applications it is important that the orthogonal polynomials have full sets of distinct real roots. Also, applications are often simpler if the coefficient of the term of largest degree is 1. As mentioned above, we sometimes work with unnormalized polynomials. Within a given system, it is generally not possible both to normalize the polynomials and to scale them so that the coefficient of the term of largest degree is 1.

Orthogonal vectors can be formed by evaluating orthogonal polynomials over a grid. These orthogonal vectors are discrete versions of the corresponding polynomials. Orthogonal vectors are useful in forming independent linear hypotheses in analysis of variance.

Relations among the Members of an Orthogonal System

It is clear that for the kth polynomial in the orthogonal sequence, we can choose a constant r_k that does not involve x, such that

$$
q_k(x) - r_k x q_{k-1}(x)
$$

is a polynomial of degree $k-1$. Now, because any polynomial of degree $k-1$ can be represented by a linear combination of the first k members of any sequence of orthogonal polynomials (which necessarily includes a polynomial of degree at least $k-1$), we can write

$$
q_k(x) - r_k x q_{k-1}(x) = \sum_{i=0}^{k-1} c_i q_i(x).
$$

Because of orthogonality, however, all c_i for $i \leq k-3$ must be 0. Therefore, collecting terms, we have, for some constants r_k, s_k, and t_k, the three-term recursion that applies to any sequence of orthogonal polynomials:

$$
q_k(x) = (r_k x + s_k) q_{k-1}(x) - t_k q_{k-2}(x), \quad \text{for } k = 2, 3, \ldots. \tag{4.39}
$$

The coefficients r_k, s_k, and t_k in this recursion formula depend on the specific sequence of orthogonal polynomials, of course.

The three-term recursion formula applies for an unnormalized orthogonal sequence, and so it also applies to a orthonormal sequence. The coefficients would be different, of course.

This three-term recursion formula can also be used to develop a formula for the sum of products of orthogonal polynomials. For $q_i(x)$ and $q_i(y)$ in an orthonormal sequence, with a_j the coefficient of the jth power in the polynomial, we derive

$$\sum_{i=0}^{k} q_i(x)q_i(y) = \frac{a_k}{a_{k+1}} \frac{q_{k+1}(x)q_k(y) - q_k(x)q_{k+1}(y)}{x - y}. \tag{4.40}$$

This expression, which is called the Christoffel-Darboux formula, is useful in evaluating the product of arbitrary functions that have been approximated by finite series of orthogonal polynomials. Notice that the equation (4.40) applies to normalized orthogonal polynomials. For unnormalized orthogonal polynomials, it can easily be modified to include the normalizing factors.

Computations Involving Polynomials

Horner's method, which we mentioned on page 120 as an example of a recursive algorithm, is the most efficient way to evaluate a general polynomial. There may be other issues such as accuracy, however. In a polynomial of moderate degree, it is quite possible that the magnitude of the individual terms will vary considerably, and that could result in accumulated rounding error or even catastrophic cancellation.

The polynomial $p_d(x) = c_d x^d + \cdots + c_1 x + c_0$ together with constants $a_1, \ldots, a_d$ can be written as

$$p_d(x) = (x - a_1)(\cdots(x - a_{d-2})((x - a_{d-1})(c_d(x - a_d) + c_{d-1}) + \cdots) + c_1) + c_0. \tag{4.41}$$

This is called the *nested Newton form.* It has the computational efficiency of Horner's method, and also, for careful choice of the "centers" a_i, it has good numerical stability. The centers are chosen so as to keep the magnitude of the product similar to that of c_0.

If a function f is approximated by a truncated expansion,

$$f(x) \approx p_j(x) = \sum_{k=0}^{j} c_k q_k(x), \tag{4.42}$$

it is necessary to evaluate all $j + 1$ polynomials and their sum. If we have the coefficients r_k, s_k, and t_k in the three-term recurrence formula (4.39), we can use the nested Newton form (without centering) to evaluate $p_j(x)$ in equation (4.42). We show the steps in Algorithm 4.1 for $j \geq 2$.

Algorithm 4.1 Evaluation of a Truncated Expansion in Orthogonal Polynomials at x

1. Let $f_j = c_j$.
2. Let $f_{j-1} = c_{j-1} + q_j(r_{j-1}x - s_{j-1})$.
3. For $k = j-2, j-3, \ldots, 0$,
   let $f_k = c_k + q_{k+1}(r_k x - s_k) - q_{k+2}t_{k+1}$.
4. Set $p_k(x) = f_0$. ∎

Standard Systems of Univariate Orthogonal Polynomials

A system of orthogonal polynomials is defined by the weight function and the domain. The main thing that determines which system to use is the domain, although the shape of the weight function may be important in achieving better finite series approximations.

There are several widely-used complete systems of univariate orthogonal polynomials. The different systems are characterized by the one-dimensional intervals over which they are defined and by their weight functions. The Legendre, Chebyshev, and Jacobi polynomials are defined over $[-1, 1]$, and hence can be scaled into any finite interval. The weight function of the Jacobi polynomials is more general, so a finite sequence of them may fit a given function better, but the Legendre and Chebyshev polynomials are simpler and so are often used. The Laguerre polynomials are defined over the half line $[0, \infty)$ and the Hermite polynomials are defined over the reals, $(-\infty, \infty)$. Table 4.1 summarizes the ranges and weight functions. The weight functions correspond to common PDFs. Note that any finite range $[a, b]$ can be shifted and scaled into $[-1,\ 1]$, and any half finite range $[a,\ \infty)$ or $[-\infty,\ b)$ can be shifted and, possibly, scaled (by -1) into $[0,\ \infty)$.

Table 4.1. Orthogonal Polynomials

Polynomial Series	Range	Weight Function
Legendre	$[-1,\ 1]$	1 (uniform)
Chebyshev	$[-1,\ 1]$	$(1-x^2)^{1/2}$ (finite Chebyshev)
Jacobi	$[-1,\ 1]$	$(1-x)^\alpha(1+x)^\beta$ (beta)
Laguerre	$[0,\ \infty)$	$x^{\alpha-1}\mathrm{e}^{-x}$ (gamma)
Hermite	$(-\infty,\ \infty)$	e^{-x^2} (normal)

Most of these systems have particularly simple expressions for the coefficients in the recurrence relation (4.39), so they are relatively simple to compute. The kth-degree polynomial in each system has k distinct real roots.

The usefulness of the standard orthogonal polynomials derives from their use in approximations and also from their use as solutions to standard classes of differential equations.

These systems of orthogonal polynomials are described below. For some systems, different forms of the weight functions are used in the literature. The properties of the orthogonal polynomials are essentially the same for the differing forms of the weight functions, but the coefficients of the polynomials are different. In some cases, also, the polynomials may be normalized.

The first system we consider is the Jacobi system. There are two special and simpler cases of the Jacobi system, with which we will begin.

Legendre Polynomials

The *Legendre polynomials* have a constant weight function and are defined over the interval $[-1, 1]$. Building the Legendre polynomials from the monomials (4.37), it is easy to see that the first few unnormalized Legendre polynomials are

$$\begin{array}{ll} P_0(x) = 1 & P_1(x) = x \\ P_2(x) = (3x^2 - 1)/2 & P_3(x) = (5x^3 - 3x)/2 \\ P_4(x) = (35x^4 - 30x^2 + 3)/8 & P_5(x) = (63x^5 - 70x^3 + 15x)/8. \end{array} \tag{4.43}$$

Graphs of these polynomials are shown in Figure 4.3.

The normalizing constant for the kth Legendre polynomial is determined by noting that

$$\int_{-1}^{1} (P_k(t))^2 \mathrm{d}t = \frac{2}{2k+1}, \tag{4.44}$$

and hence it is $(P_k(1))^{1/2}$.

The recurrence formula (4.39) for the Legendre polynomials, for $k \geq 2$, is

$$P_k(x) - \frac{2k-1}{k} x P_{k-1}(x) + \frac{k-1}{k} P_{k-2}(x) = 0. \tag{4.45}$$

Notice that for the recursion formula (4.39),

$$r_k = (2k-1)k, \quad s_k = 0, \quad t_k = (k-1)/k. \tag{4.46}$$

These are the quantities to use in Algorithm 4.1.

Notice that if the Legendre polynomials (or, in general, the Jacobi polynomials) are to be used over the finite interval $[a, b]$, it is necessary to make a change of variable:

$$y = (b-a)x/2 + (b+a)/2. \tag{4.47}$$

This transformation would change the normalizing constant (by the Jacobian of the transformation) and also the coefficient r_k in the recursion formula (4.39).

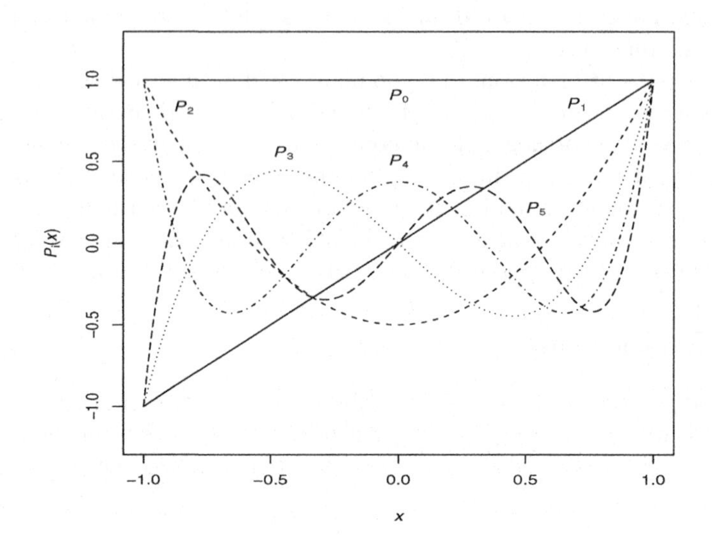

Fig. 4.3. Legendre Polynomials

As mentioned above, orthogonal polynomials can also be developed as solutions to differential equations. The standard series of orthogonal polynomials arise from differential equations that are important in applied mathematics. The Legendre polynomials are the solutions to *Legendre's equation*,

$$(1 - x^2)u'' - 2xu' + k(k + 1)u = 0. \tag{4.48}$$

This equation describes an inverse r^2 potential.

Also as mentioned above, vectors whose elements are orthogonal polynomials evaluated over a grid are discrete versions of the orthogonal polynomials. The discrete Legendre polynomials can be formed easily by setting x_i to the values of a grid over $[-1, 1]$, forming a Vandermonde matrix evaluated at those grid points, and then forming the QR decomposition of the matrix. (The first column of the Vandermonde matrix is 1, the second column is x, the third is x^2, and so on.) The discrete Legendre polynomials are often used in statistical analysis of linear models. (They form contrasts.)

Chebyshev Polynomials

The *Chebyshev polynomials* have a weight function proportional to the Chebyshev density, $w(x) = (1 - x^2)^{-1/2}$. They are defined over the interval $[-1, 1]$. The first few Chebyshev polynomials are

$$\begin{array}{ll} T_0(x) = 1 & T_1(x) = x \\ T_2(x) = 2x^2 - 1 & T_3(x) = 4x^3 - 3x \\ T_4(x) = 8x^4 - 8x^2 + 1 & T_5(x) = 16x^5 - 20x^3 + 5x. \end{array} \tag{4.49}$$

The normalizing constant for the kth Chebyshev polynomial is $T_k(1)$, similar to that for the Legendre polynomial.

The recurrence formula for the Chebyshev polynomials is

$$T_k(x) - 2xT_{k-1}(x) + T_{k-2}(x) = 0. \tag{4.50}$$

Notice that for the recursion formula (4.39), $r_k = 2$, $s_k = 0$, and $t_k = 1$. This means that Algorithm 4.1 is especially efficient for the Chebyshev polynomials. The sequence of Chebyshev polynomials is the only sequence for which the coefficients in formula (4.39) are the same for all k.

These polynomials are sometimes called Chebyshev polynomials of the first kind; similar polynomials with weight function $w(x) = (1 - x^2)^{1/2}$ are called Chebyshev polynomials of the second kind.

Jacobi Polynomials

The *Jacobi polynomials* are defined over the interval $[-1, 1]$, with a beta weight function, $w(x) = (1 - x)^{\alpha}(1 + x)^{\beta}$, for $\alpha, \beta > -1$. The Legendre polynomials are Jacobi polynomials with $\alpha = \beta = 0$, and the Chebyshev polynomials are Jacobi with $\alpha = \beta = -1/2$.

For various values of α and β the weight function can assume a wide variety of shapes. If α is large and β is small, for example, the weight function is large near -1 and small near 1. This may be an appropriate weight to use to construct orthogonal polynomials for approximating a function that varies more near -1 than it does near 1.

Laguerre Polynomials

The *Laguerre polynomials* are defined over $[0, \infty)$, with a gamma weight function, $w(x) = x^{\alpha-1}e^{-x}$. The jth Laguerre polynomial is often denoted by $L_j^{(\alpha-1)}(x)$. The parameter α provides some flexibility for fitting functions of different shapes. The most commonly used series of Laguerre polynomials have $\alpha = 1$, however, and in this case, the notation $L_j(x)$ is used. The first few Laguerre polynomials with $\alpha = 1$ are

$$\begin{aligned} L_0(x) &= 1 \\ L_1(x) &= -x + 1 \\ L_2(x) &= (x^2 - 4x + 2)/2 \\ L_3(x) &= (-x^3 + 9x^2 - 18x + 6)/6 \\ L_4(x) &= (x^4 - 16x^3 + 72x^2 - 96x + 24)/24 \\ L_5(x) &= (-x^5 + 25x^4 - 200x^3 + 600x^2 - 600x + 120)/120. \end{aligned} \tag{4.51}$$

The recurrence formula (4.39) for these Laguerre polynomials is

$$L_k(x) - \left(\frac{2k-1}{k} + \frac{x}{k}\right)L_{k-1}(x) + \frac{k-1}{k}L_{k-2}(x) = 0.$$

Hermite Polynomials

The *Hermite polynomials* are defined over $(-\infty, \infty)$ with a weight function proportional to the error function density, $w(x) = \mathrm{e}^{-x^2}$. Building the Hermite polynomials from the monomials, it is easy to see that the first few Hermite polynomials are

$$\begin{array}{ll} H_0(x) = 1 & H_1(x) = 2x \\ H_2(x) = 4x^2 - 2 & H_3(x) = 8x^3 - 12x \\ H_4(x) = 16x^4 - 48x^2 + 12 & H_5(x) = 32x^5 - 160x^3 + 120x. \end{array} \quad (4.52)$$

In an alternative definition of Hermite polynomials, the normal weight function, $w(x) = \mathrm{e}^{-x^2/2}$, is used. This form is more widely used in statistical applications than the form defined above. These polynomials also called Chebyshev-Hermite polynomials. (This distinction does not seem to be common, and most statisticians refer to the alternate polynomials just as "Hermite". I will follow this usage, although usually I will remind the reader when I use the alternate form.) The alternate Hermite polynomials are sometimes denoted in the same way as the ones defined above, $H_0, H_1, \ldots$, although they are often denoted by $He_0, He_1, \ldots$. We will use the notation $H_0^{\mathrm{e}}, H_1^{\mathrm{e}}, \ldots$. The polynomials are related by

$$H_k^{\mathrm{e}}(x) = 2^{-k/2}H_k(x/\sqrt{2}).$$

The alternate Hermite polynomials are often used in statistical applications because the weight function is proportional to the normal density. Also, because the coefficient of the term of largest degree is 1, some applications involve simpler expressions. The first few alternate Hermite polynomials are

$$\begin{array}{ll} H_0^{\mathrm{e}}(x) = 1 & H_1^{\mathrm{e}}(x) = x \\ H_2^{\mathrm{e}}(x) = x^2 - 1 & H_3^{\mathrm{e}}(x) = x^3 - 3x \\ H_4^{\mathrm{e}}(x) = x^4 - 6x^2 + 3 & H_5^{\mathrm{e}}(x) = x^5 - 10x^3 + 15x. \end{array} \quad (4.53)$$

The recurrence formula (4.39) for these alternate Hermite polynomials is particularly simple:

$$H_k^{\mathrm{e}}(x) - xH_{k-1}^{\mathrm{e}}(x) + (k-1)H_{k-2}^{\mathrm{e}}(x) = 0. \quad (4.54)$$

One use of the Hermite polynomials in in Gram-Charlier or Edgeworth expansions of PDFs of the form of equation (4.31). The Edgeworth expansion of a PDF $p(x)$ of a distribution with finite moments $\mu_2, \mu_3, \ldots$, and $\mu_1 = 0$, is

$$p(x) = \frac{1}{\sqrt{2\pi}}\left(1 + \frac{1}{2}(\mu_2 - 1)H_2^{\mathrm{e}}(x) + \frac{1}{6}\mu_3 H_3^{\mathrm{e}}(x) + \frac{1}{24}(\mu_4 - 6\mu_4 + 3)H_4^{\mathrm{e}}(x) + \cdots\right). \tag{4.55}$$

This is derived by expanding the characteristic function of the distribution. Another use of Hermite polynomials we encounter in Chapter 16 is in providing an index of the difference in the distribution of a given random variable from a standard normal distribution for use in projection pursuit.

Orthogonal Functions Related to the Orthogonal Polynomials

Sometimes applications are simpler when the weight function is incorporated into the orthogonal functions. The resulting functions are orthogonal with respect to a constant weight.

In the case of the Hermite polynomials, for example, this results in the *exponentially tilted* polynomials,

$$H_k^{\mathrm{f}}(x) = H_k(x)\mathrm{e}^{-x^2/2}. \tag{4.56}$$

The set $H_0^{\mathrm{f}}, H_1^{\mathrm{f}}, H_2^{\mathrm{f}}, \ldots$ is an orthogonal set with respect to a constant weight:

$$\int_{-\infty}^{\infty} H_i^{\mathrm{f}}(x)H_j^{\mathrm{f}}(x)\mathrm{d}x = 0 \quad \text{for } i \neq j.$$

These orthogonal functions are called *Hermite functions.*

It is more convenient to have a constant weight function, of course, but the exponential tilting may also make the computations simpler for some functions that are being approximated (see also equation (4.22)).

Even and Odd functions

An *even function* is a function f such that $f(-x) = f(x)$, and an *odd function* is one such that $f(-x) = -f(x)$. The Legendre, Chebyshev, and Hermite polynomials are either even or odd functions, depending on their degree. In each case, a polynomial of even degree is an even function and one of odd degree is an odd function; for example, the Chebyshev polynomials satisfy the relation

$$T_i(-x) = (-1)^i T_i(x).$$

Expansion of Functions in Orthogonal Polynomials

Complicated functions or functions that are intractable for certain operations can often be approximated with a finite sum of orthogonal polynomials. An important application of this type of approximation is in evaluation of integrals by expansion of the integrand using orthogonal polynomials. This method of numerical integration is called Gaussian quadrature, and is discussed in Section 4.6, page 190.

Another reason that we study the expansion of functions in orthogonal systems is for use in function estimation, which we discuss in Chapter 10 for general functions and in Chapter 15 for probability density functions.

To represent a given function in a standard series of orthogonal polynomials, the first consideration is the domain of the function. The three types of domain and the possible polynomials to use are

- a finite interval, $[a, b]$, Jacobi polynomials (or the special cases, Legendre and Chebyshev);
- a half-infinite interval, $[a, \infty]$ or $[-\infty, b]$, Laguerre polynomials;
- an infinite interval, $[-\infty, \infty]$, Hermite polynomials.

An Example

As an example of the use of orthogonal polynomials to approximate a given function, consider the expansion of $f(x) = \mathrm{e}^{-x}$ over the interval $[-1, 1]$. Because the range is finite, we use the Jacobi system of polynomials, and in this case, we do not need to make a transformation to the basic interval of the definition of the Jacobi polynomials.

In this example we use the Legendre polynomials.

The Fourier coefficients are determined by equation (4.26). In using this formula, we can either normalize the polynomials (using equation (4.44)) or we can include the normalizing factor in the Fourier coefficients. Using the normalized polynomials, we have $c_0 = (\mathrm{e}^1 - \mathrm{e}^{-1})/2$ and we obtain the remaining c_i's by integration by parts; $c_1 = -2\mathrm{e}^{-1}\sqrt{3/2}$, and so on. After we have the Fourier coefficients, we identify the recurrence coefficients (equation (4.46)), and finally we use Algorithm 4.1 to compute the approximation at the point x.

Forming a grid in x over $[-1, 1]$, and evaluating the approximation at each grid point, we can construct graphs of the function and the approximation. Figure 4.4 shows the exact function f, and the different truncated series approximations using up to six terms ($j = 0, 1, \ldots, 5$),

$$\widetilde{f}(x) = \sum_{k=0}^{j} c_k P_k(x). \tag{4.57}$$

Each truncated series is the best linear combination of the Legendre polynomials (in terms of the L_2 norm) of the function using no more than $j + 1$ terms. Notice that the convergence is very slow after $j = 1$.

Smoothing Data with Orthogonal Polynomials

In the previous section we considered the approximation of a function with known form by a series of orthogonal polynomials. In many applications, we

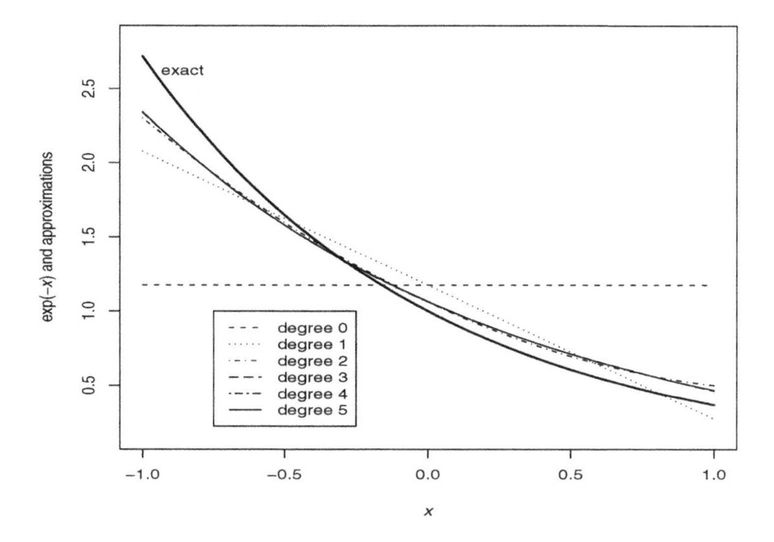

Fig. 4.4. Approximations with Legendre Polynomials

do not have a function with a closed form; we may have a discrete function composed of observations with two components corresponding to an argument, x_i, and a function value, y_i. If we assume the data represent exact values, we may interpolate the data to form a continuous function. If, however, the data are assumed to arise from a process with noise, we may build a smooth approximation of the function as a finite series of orthogonal polynomials,

$$\widetilde{f}(x) = \sum_{k=1}^{j} c_k q_k(x).$$

Because we do not know the form of $f(x)$, we choose the c_k so as to minimize the differences

$$y_i - \sum_{k=1}^{j} c_k q_k(x_i). \tag{4.58}$$

Instead of a function norm, as in equation (4.28), we consider a norm of the vector:

$$\left\| y_i - \sum_{k=1}^{j} c_k q_k(x_i) \right\|. \tag{4.59}$$

The norm is most often chosen as the L_2 norm, and the resulting approximation is the least squares fit. Other appropriate norms include the L_1 norm, resulting in an approximation with the least absolute deviations, and the L_∞

norm, resulting in an approximation with the minimum maximum deviation. The latter type of fit is called a minimax or Chebyshev approximation.

Multivariate Orthogonal Polynomials

Multivariate orthogonal polynomials can be formed easily as tensor products of univariate orthogonal polynomials. The tensor product of the functions $f(x)$ over D_x and $g(y)$ over D_y is a function of the arguments x and y over $D_x \times D_y$:

$$h(x, y) = f(x)g(y).$$

If $\{q_{1,k}(x_1)\}$ and $\{q_{2,l}(x_2)\}$ are sequences of univariate orthogonal polynomials, a sequence of bivariate orthogonal polynomials can be formed as

$$q_{kl}(x_1, x_2) = q_{1,k}(x_1)q_{2,l}(x_2). \tag{4.60}$$

These polynomials are orthogonal in the same sense as in equation (4.24), where the integration is over the two-dimensional domain. Similarly as in equation (4.25), a bivariate function can be expressed as

$$f(x_1, x_2) = \sum_{k=0}^{\infty}\sum_{l=0}^{\infty} c_{kl}q_{kl}(x_1, x_2), \tag{4.61}$$

with the coefficients being determined by integrating over both dimensions.

Although obviously such product polynomials, or radial polynomials, would emphasize features along coordinate axes, they can nevertheless be useful for representing general multivariate functions. Often, it is useful to apply a rotation of the coordinate axes, as we discuss in Section 9.1, beginning on page 373.

The weight functions, such as those for the Jacobi polynomials, that have various shapes controlled by parameters can also often be used in a mixture model of the function of interest. This is the way the Bernstein polynomials (8.4) are used in Bézier curves, as discussed on page 342, except in that case the coefficients are determined to approximate a fixed set of points, subject to some smoothness conditions. The weight function for the Hermite polynomials can be generalized by a linear transformation (resulting in a normal weight with mean μ and variance σ^2), and the function of interest may be represented as a mixture of general normals.

4.4 Splines

The approach to function approximation that we pursued in the previous section makes use of a finite subset of an infinite basis set consisting of polynomials of degrees $p = 0, 1, \ldots$. This approach yields a smooth approximation

$\widetilde{f}(x)$. ("Smooth" means an approximation that is continuous and has continuous derivatives. These are useful properties of the approximation.) The polynomials in $\widetilde{f}(x)$, however, cause oscillations that may be undesirable. The approximation oscillates a number of times one less than the highest degree of the polynomial used. Also, if the function being approximated has quite different shapes in different regions of its domain, the global approach of using the same polynomials over the full domain may not be very effective.

Another approach is to subdivide the interval over which the function is to be approximated and then on each subinterval use polynomials with low degree. The approximation at any point is a sum of one or more piecewise polynomials. Even with polynomials of very low degree, if we use a large number of subintervals, we can obtain a good approximation to the function. Zero-degree polynomials, for example, would yield a piecewise constant function that could be very close to a given function if enough subintervals are used. Using more and more subintervals, of course, is not a very practical approach. Not only is the approximation a rather complicated function, but it may be discontinuous at the interval boundaries. We can achieve smoothness of the approximation by imposing continuity restrictions on the piecewise polynomials and their derivatives. This is the approach in *spline* approximation and smoothing.

The polynomials are of degree no greater than some specified number, often just 3. This means, of course, that the class of functions for which these piecewise polynomials form a basis is the set of polynomials of degree no greater than the degree of polynomial in the basis; hence, we do not begin with an exact representation as in equation (4.25).

In spline approximation, the basis functions are polynomials over given intervals and zero outside of those intervals. The polynomials have specified contact at the endpoints of the intervals; that is, their derivatives of a specified order are continuous at the endpoints. The endpoints are called "knots". The finite approximation therefore can be smooth and, with the proper choice of knots, is close to the function being approximated at any point. The approximation, $\widetilde{f}(x)$, formed as a sum of such piecewise polynomials $b_k(x)$ is called a "spline":

$$\widetilde{f}(x) = \sum_{k=1}^{j} c_k b_k(x). \tag{4.62}$$

The "order" of a spline is the number of free parameters in each interval. (For polynomial splines, the order is the degree plus 1.)

There are three types of spline basis functions commonly used:

- *truncated power functions* (or just power functions). For k knots and degree p, there are $k + p + 1$ of these:

  $$1, x, ..., x^p, ((x - z_1)_+)^p, ..., ((x - z_k)_+)^p,$$

  where $(t)_+$ means t if t is positive and 0 otherwise. Sometimes, the constant is not used, so there are only $k + p$ functions. These are nice when we are adding or deleting knots. Deletion of the ith knot, z_i, is equivalent to removal of the basis function $((x - z_i)_+)^p$.

- *B-splines.* B-splines are probably the most widely used set of splines, and they are available in many software packages. The IMSL Library, for example, contains three routines for univariate approximations using B-splines, with options for variable knots or constraints, and routines for two- and three-dimensional approximations using tensor product B-splines. The influence of any particular B-spline coefficient extends over only a few intervals, so B-splines can provide good fits to functions that are not smooth. The B-spline functions also tend to be better conditioned than the power functions. The mathematical development of B-splines is more complicated than the power functions.
- *"natural" polynomial splines.* These basis functions are such that the second derivative of the spline expansion is 0 for all x beyond the boundary knots. This condition can be imposed in various ways. An easy way is just to start with any set of basis functions and replace the degrees of freedom from two of them with the condition that every basis function have zero second derivative for all x beyond the boundary knots. For natural cubic splines with k knots, there are k basis functions. There is nothing "natural" about the natural polynomial splines. A way of handling the end conditions that is usually better is to remove the second and the penultimate knots and to replace them with the requirement that the basis functions have contact one order higher. (For cubics, this means that the third derivatives match.)

Some basis functions for various types of splines over the interval $[-1, 1]$ are shown in Figure 4.5.

Interpolating Splines

Splines can be used for interpolation, approximation, and estimation. An interpolating spline fit matches each of a given set of points. Each point is usually taken as a knot, and the continuity conditions are imposed at each point. It makes sense to interpolate points that are known to be exact.

The reason to use an interpolating spline is usually to approximate a function at points other than those given (maybe for quadrature), so applied mathematicians may refer to the results of the interpolating spline as an "approximation". An interpolating spline is used when a set of points are assumed to be known exactly (more or less).

Consider again the example of approximating the function $f(x) = \mathrm{e}^{-x}$ over the interval $[-1, 1]$ using natural cubic spline interpolation with 2, 3, and 4 knots. Graphs of the function and the approximations are shown in Figure 4.6. Notice that the approximation is very good with 4 knots. The approximations were computed using the R function `spline`.

Compare the use of splines with the use of orthogonal polynomials on page 177 to approximate this same function. We do not need to compute

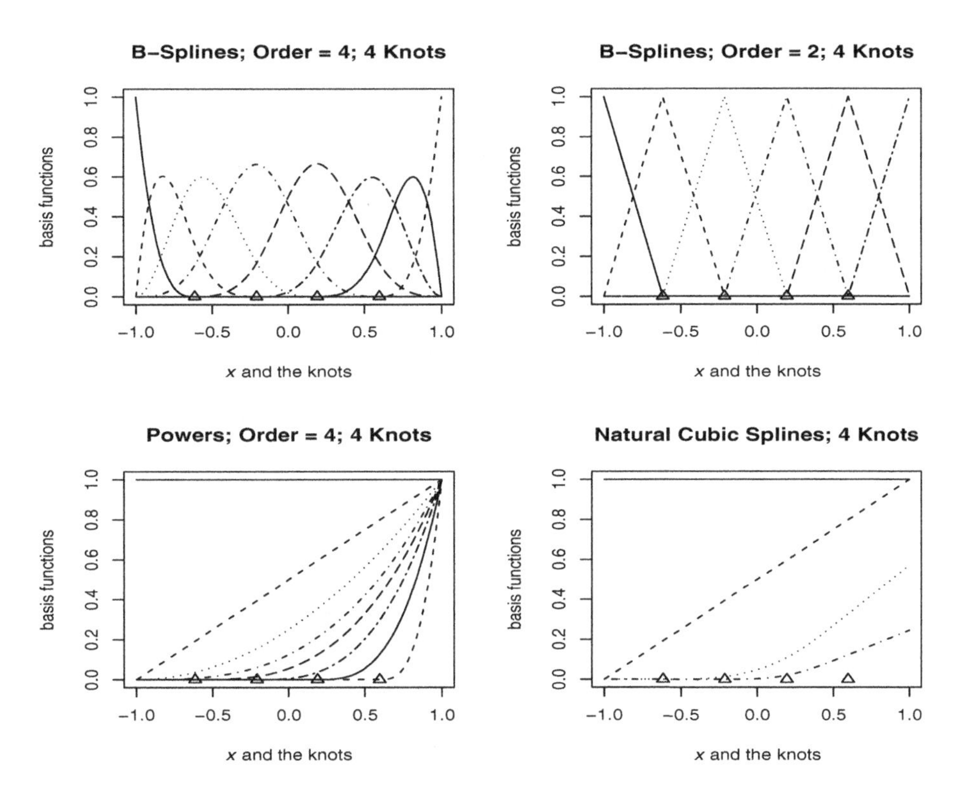

Fig. 4.5. Spline Basis Functions

any coefficients that depend on integrals. Furthermore, we get a much better approximation, even though we use only a small number of known function values.

Smoothing Splines

The other way of using splines is for approximation or smoothing. The individual points may be subject to error, so the spline may not go through any of the given points. In this usage, the splines are evaluated at each abscissa point, and the ordinates are fitted by some criterion (such as least squares) to the spline.

Choice of Knots in Smoothing Splines

The choice of knots is a difficult problem when the points are measured subject to error. One approach is to include the knots as decision variables in the fitting optimization problem. This approach may be ill-posed. A common

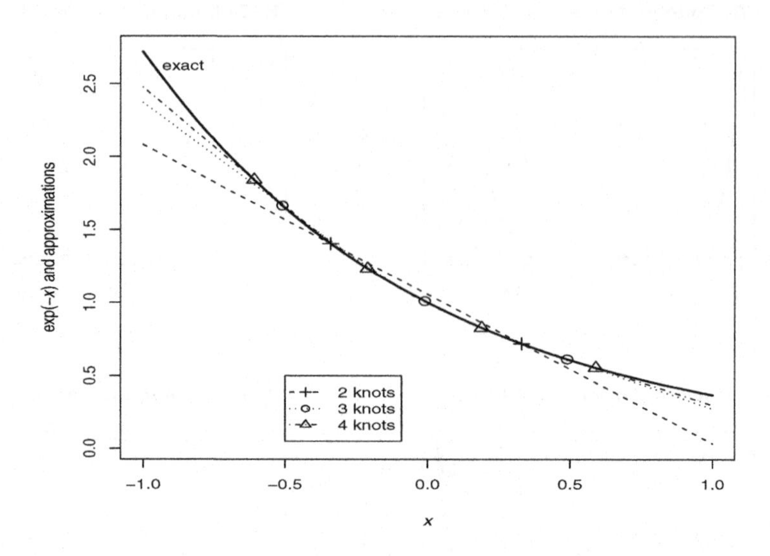

Fig. 4.6. Approximations with Natural Cubic Splines with Different Numbers of Knots

approach is to add (pre-chosen) knots in a stepwise manner. Another approach is to use a regularization method (addition of a component to the fitting optimization objective function that increases for roughness or for some other undesirable characteristic of the fit).

Multivariate Splines

Multivariate splines are easily formed as tensor products of univariate splines in the same way as the bivariate orthogonal polynomials were formed from univariate polynomials in equation (4.60). Although conceptually, this is straightforward, there are a number of practical difficulties in applications.

4.5 Kernel Methods

Another approach to function approximation and estimation is to use a *filter* or *kernel* function to provide local weighting of given points. The basic method of this approach is to convolve the given function, $f(x)$, with a filter or the kernel, $K(t)$, (see page 21):

$$K * f(x) = \int_D f(y)K(x-y)\,\mathrm{d}y, \tag{4.63}$$

if the integral exists. If $K(t)$ a unimodal function that decreases rapidly away from a central point, then $K * f(x)$ is "close" to $f(x)$.

In practice, we use a given set of points, $x_1, \dots, x_j$, in a discrete version of equation (4.63):

$$\widetilde{f}(x) = \sum_{k=1}^{j} f(x_k) K(x - x_k). \tag{4.64}$$

The kernel approximation does not interpolate this set of points, but each of those points exerts the strongest influence on the approximation at nearby points x.

Kernel Functions

Some examples of univariate kernel functions are shown in equations (4.65) through (4.67).

$$\text{uniform}: \; K_{\mathrm{u}}(t) = \frac{1}{2\lambda} \mathrm{I}_{[-\lambda,\lambda]}(t) \tag{4.65}$$

$$\text{quadratic}: \; K_{\mathrm{q}}(t) = \frac{3}{\lambda^2(6-2\lambda)} (\lambda - t^2) \mathrm{I}_{[-\lambda,\lambda]}(t) \tag{4.66}$$

$$\text{normal}: \; K_{\mathrm{n}}(t) = \frac{1}{\sqrt{2\pi}} \mathrm{e}^{-(t/\lambda)^2/2} \tag{4.67}$$

Notice that all of these kernels are nonnegative and integrate to 1, hence, they are PDFs. Often, multivariate kernels are formed as products of these or other univariate kernels.

As in the discussion on page 21, a kernel is actually a function of two arugments, $K(x, y)$, but often the two arguments are combined into a single argument as in the kernels above. A bilinear form (see page 23) is one of the most common types of kernel.

Kernel Windows

In kernel methods, the locality of influence is controlled by a *smoothing parameter*. In equations (4.65) through (4.67), the λ is the smoothing parameter. We sometimes also refer to the *window* or the *window width* around the point of interest. In equations (4.65) and (4.66), the window is a finite interval. In equation (4.67) the window is the real line, but we nevertheless sometimes speak of the "window" in a vague way as a synonym for the smoothing parameter. The choice of the size of the window is the most important issue in the use of kernel methods. The window width must be great enough to allow multiple known points or observations to enter in the sum of equation (4.64). In practice, this generally means that use of kernels for approximation is limited to situations in which there are a large number of known values or observations. Probability density function estimation usually is only done in such situations, and kernel methods are very useful in that case, as we see in Section 15.3.

For a given choice of the size of the window, the argument of the kernel function is transformed to reflect the size. The transformation is accomplished using a positive definite matrix, V, whose determinant measures the volume (size) of the window.

In Exercise 4.13 you are asked to use kernels to approximate the function $f(x) = \mathrm{e}^{-x}$ over the interval $[-1, 1]$, as we have done with orthogonal polynomials and with splines.

Multivariate Kernels

Kernel methods extend immediately to higher dimensions. The kernel is often chosen as a product kernel of a univariate kernel:

$$K_d(t_1, \ldots, t_d) = \prod_{j=1}^{d} K(t_j). \tag{4.68}$$

4.6 Numerical Quadrature

One of the most common mathematical operations in scientific computing is quadrature, the evaluation of a definite integral. It is used to determine volume, mass, or total charge, for example. In the evaluation of probabilities, of expectations, and of marginal or conditional densities, integration is the basic operation.

Most of the integrals and differential equations of interest in real-world applications do not have closed-form solutions; hence, their solutions must be approximated numerically.

There are two ways of approximating an integral. One type of approximation is based on direct approximation of the Riemann sum, which we take as the basis for the definition of the integral. The other type of approximation is based on an approximation of the function using one of the methods discussed above.

We begin with approximations that are based on Riemann sums. We also generally limit the discussion to univariate integrals.

Evaluation of a Single Integral

Although some of the more interesting problems are multivariate and the region of integration is not rectangular, we begin with the simple integral,

$$I = \int_a^b f(x)\,\mathrm{d}x. \tag{4.69}$$

There are various definitions of the integral (4.69), each of which makes certain assumptions about the integrand $f(x)$ that are required for the existence of the integral. The Riemann integral is defined as the limit of the *Riemann sums*:

$$\frac{1}{n}\sum_{i=1}^{n}(x_i - x_{i-1})f(\widetilde{x}_i), \tag{4.70}$$

where $a = x_0 < x_1 < \cdots < x_n = b$ and $\widetilde{x}_i \in [x_{i-1}, x_i]$. Where $\widetilde{x}_i$ is within the interval would make no difference in the limit if the function is well-behaved. When the location of $\widetilde{x}_i$ makes a difference, the Riemann integral may not be undefined, but in most applications of numerical quadrature, the Riemann integral does exist.

This definition extends to multiple integrals in a natural way.

One way of approaching the problem of evaluating (4.69) is to approximate it directly by a sum of areas under the curve. The Riemannian definition of the integral leads to a set of rectangles, the sum of whose areas approximates the integral. More generally an approximation of the integral results from a piecewise approximation of $f(x)$ using simpler functions that can be integrated in closed form. If the piecewise approximants are step functions, the approximation is similar to a Riemann sum.

The Trapezoid Rule

Instead of a simple step function, the function $f(x)$ may be approximated as shown in Figure 4.7 by a piecewise linear function $p_1(x)$ that agrees with f at each of the points $a = x_0 < x_1 < x_2 < \ldots < x_n = b$. In that case, the integral (4.69) can be approximated by a sum of integrals,

$$\int_a^b f(x)\,\mathrm{d}x \approx \sum_{i=0}^{n}\int_{x_i}^{x_{i+1}} p_1(x)\,\mathrm{d}x,$$

each of which is particularly easy to evaluate. Because p_1 is linear over each interval, the integral in the ith interval is just the area of the trapezoid, that is,

$$h\big(f(x_i) + f(x_{i+1})\big)/2.$$

The integral (4.69) is therefore approximated by

$$T(f) = h\big(f(a) + 2f(x_1) + 2f(x_2) + \cdots + 2f(x_{n-1}) + f(b)\big)/2. \tag{4.71}$$

The expression (4.71) is called the *trapezoid rule*.

Figure 4.7 shows how the areas in the trapezoids may be used to approximate areas under the curve.

A simple choice for the points is to make them equally spaced, that is,

$$(x_{i+1} - x_i) = (b - a)/n.$$

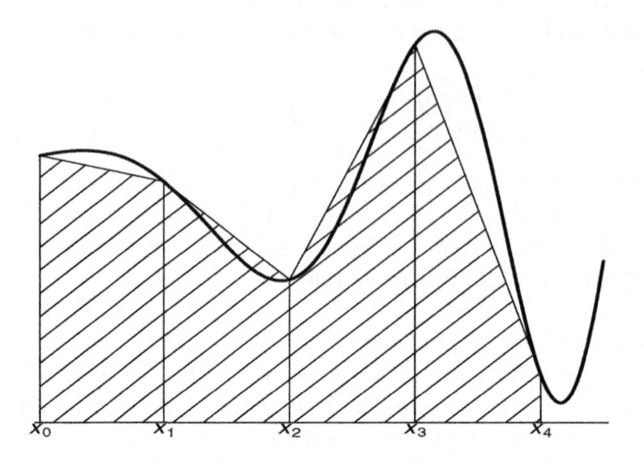

Fig. 4.7. The Trapezoid Rule

The number of points n, or the width of the interval, say h,

$$h = (b - a)/n,$$

is a tuning parameter of the quadrature algorithm. Subject to the rounding induced by working with floating point numbers instead of real numbers, the larger is n, or the smaller is h, the better the approximation of the finite sum to the integral. Because of the rounding, however, after a certain level of refinement, no further gains can be achieved by simply making the intervals smaller.

Many other quadrature rules can be built using this same idea of an approximating function that agrees with f at each of some set of points $a = x_0 < x_1 < x_2 < \ldots < x_n = b$. Quadrature formulas that result from this kind of approach are called *Newton-Cotes formulas.* (Roger Cotes was an eighteenth century English mathematician who worked closely with Newton.)

Simpson's Rules

Rather than the linear functions of the trapezoid rule, a more accurate approximation would probably result from use of quadratic functions that agree with f at each of three successive points. If $p_2(x)$ is a quadratic that agrees with $f(x)$ at the equally-spaced points x_i, $x_i + h$, $x_i + 2h$, then the piece of the integral (4.69) from x_i to x_{i+2} can be approximated by

$$\int_{x_i}^{x_{i+2}} p_2(x)\,\mathrm{d}x = \frac{1}{3}h\Big(f(x_i) + 4f(x_i+h) + f(x_i+2h)\Big). \tag{4.72}$$

If an even number of intervals is chosen, and the integrals like (4.72) are summed, the integral (4.69) is approximated by

$$\frac{1}{3}h\Big(f(a) + 4f(x_1) + 2f(x_2) + 4f(x_3) + \cdots + 4f(x_{n-1}) + f(b)\Big). \tag{4.73}$$

The formula (4.73) is called Simpson's $\frac{1}{3}$ rule.

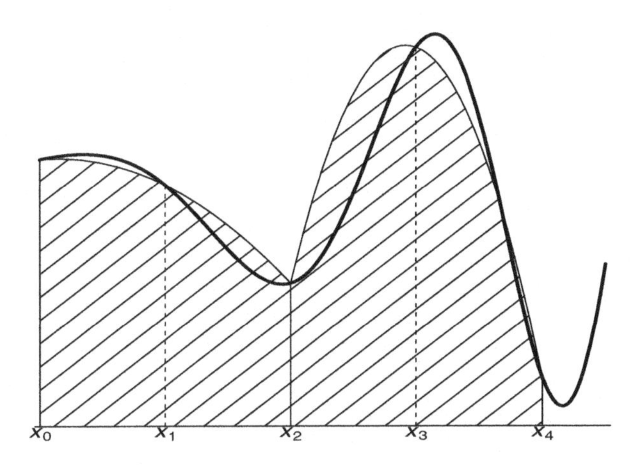

Fig. 4.8. Simpson's 1/3 Rule

Figure 4.8 shows how the areas under the quadratics may be used to approximate areas under the curve. Comparison with Figure 4.7 indicates that approximation by a higher degree polynomial (quadratic instead of linear) does not always guarantee an improvement.

A better approximation can be constructed using the same idea, except instead of using three points at a time and fitting a quadratic, we use four points at a time, and fit a cubic function. For equal-spaced points, the analog to (4.72) is

$$\int_{x_i}^{x_{i+3}} p_3(x)\,\mathrm{d}x = \frac{3}{8}h\Big(f(x_i) + 3f(x_i+h) + 3f(x_i+2h) + f(x_i+3h)\Big),$$

and over the entire range of integration that has been divided into a multiple of three equal-length intervals, we have, analogously to (4.73),

$$\frac{3}{8}h\Big(f(a)+3f(x_1)+3f(x_2)+2f(x_3)+3f(x_4)+\cdots+3f(x_{n-1})+f(b)\Big). \quad (4.74)$$

The formula (4.74) is called Simpson's $\frac{3}{8}$ rule.

Error in Newton-Cotes Quadrature

It is important to analyze the error in numerical computations. As we discussed in Section 3.1, there are generally multiple sources of error. The error in approximations must be considered separately from the error in rounding, although at some level of discretization, the rounding error may prevent any decrease in approximation error, even though the approximation is really the source of the error.

Quadrature operators are linear, and the results of linear operators and functions on page 151 can be useful in analyzing errors in numerical quadrature.

For algorithms that use a discrete number system to approximate quantities that depend on a continuous number system, such as integrals, our objective is generally to express the error in terms of the order of some function of a discretizing unit. In Newton-Cotes quadrature, this means an expression of the form $\mathrm{O}(g(h))$.

In the examples shown in Figures 4.7 and 4.8, the actual error is the total area of the regions between the polygonal line and the curve defined by f. Of course, if we could evaluate this, we would not have an error. To approximate the error as an integral in that region, we consider a polynomial of degree n over that region, because we could have a single such polynomial that corresponds to f at each of the break points. We omit the details here, which involve an expansion of the polynomial in finite differences similar to a Taylor series, evaluation of the error in one interval, and addition of the errors. (See Kennedy and Gentle, 1980, pages 86 through 89.) The error in use of the trapezoid rule can then be expressed as

$$-\frac{1}{12}(b-a)h^2 f''(x^*)$$

for some $x^* \in [a, b]$. This is not very useful in practice. It is important, however to note that the error is $\mathrm{O}(h^2)$.

Using similar approaches we can determine that the error for both Simpson's $\frac{1}{3}$ rule and $\frac{3}{8}$ rule is $\mathrm{O}(h^4)$.

Extrapolation in Quadrature Rules

We can use Richardson extrapolation (see page 133) to improve the approximation in Newton-Cotes formulas. In the trapezoid rule, for example, we consider various numbers of intervals. Let T_{0k} represent the value of the expression in equation (4.71) when $n = 2^k$; that is, when there are n intervals, and we examine the formula for $1, 2, 4, \ldots$ intervals, that is, $T_{00}, T_{01}, T_{02}, \ldots$. Now, we use Richardson extrapolation to form

$$T_{1,k} = (4T_{0,k+1} - T_{0,k})/3. \tag{4.75}$$

Generalizing this, we define

$$T_{m,k} = (4^m T_{m-1,k} - T_{m-1,k-1})/(4^m - 1). \tag{4.76}$$

Notice in $T_{m,k}$, m represents the extent of extrapolation, and k determines the number of intervals. (This kind of scheme is used often in numerical analysis. It can be represented as a triangular table in which the ith row consists of the i terms $T_{0,i-1}, \ldots, T_{i-1,0}$.)

This application of Richardson extrapolation in quadrature with the trapezoid rule is called *Romberg quadrature.*

Each $T_{m,k}$ is an approximation of the integral. The approximation is exact for a function that is a piecewise polynomial of degree $2m+2$ on the subintervals of length $(b-a)/2^k$. Not only are the $T_{m,k}$ good approximations of the integral, their relative values give some idea of the convergence of the approximation (assuming in a general way, that the integrand is ever better approximated by polynomials of higher degree). As to be expected with higher degree polynomials, however, at some point the wild fluctuations of the polynomials result in significant rounding error. The extent of the rounding is assessed by the relative values of $T_{m,k}$ for successive values of m. If the values change significantly, it may be due to rounding error. On the other hand, it could be due to a large fluctuation in the integrand, so three successive values should be inspected before deciding to terminate the extrapolation process.

Adaptive Quadrature Rules

It is never obvious how to choose the interval width in Newton-Cotes formulas. Obviously if the interval width is too large, finer structure in the integrand will be missed. On the other hand, if the interval width is too small, in addition to increased cost of evaluation of the integrand, rounding error can become significant. There are various ways of trying to achieve a balance between accuracy and number of function evaluations. In most cases these involve approximation of the integral over different subintervals with different widths used in each of the subintervals. Initially, this may identify subintervals of the domain of integration that require smaller widths in the Newton-Cotes formulas (that is, regions in which the integrand is rougher). Evaluations at different widths over the different subintervals may lead to a good choice of both subintervals and widths within the different subintervals. This kind of approach is called *adaptive quadrature.*

From the description, it should be obvious that it is not easy to do adaptive quadrature well. Software that does a good job is very complicated, and amateurs need not attempt to develop it. A very good routine for adaptive quadrature is `dcadre` in the IMSL Library.

Quadrature Following Expansions of the Integrand

The approximation of a given function is formed by some linear combination of other functions.

With a basis set $\{q_k\}$, the function $f(x)$ in the space spanned by the basis set can be represented exactly as

$$f(x) = \sum_{k=1}^{\infty} c_k q_k(x), \tag{4.77}$$

and so a finite series expansion such as

$$f(x) \approx \sum_{k=1}^{j} c_k q_k(x) \tag{4.78}$$

may be used. When the c_k in equation (4.78) are the Fourier coefficients, that is, the coefficients from equation (4.77), the approximation in equation (4.78) is the least squares approximation of the given form; that is, using the same basis functions (see page 162).

One of the simplest examples of quadrature following an expansion of the function is Laplace approximation that uses the Taylor series of a function exponential. We considered a special case of this approximation for a nonnegative concave function on page 161. Jensen (1995) covers Laplace approximations in more general settings.

Gaussian Quadrature

We now briefly discuss another approach to the evaluation of the integral (4.69) called *Gaussian quadrature.* Gaussian quadrature uses the idea of expansion of the integrand. Like the Newton-Cotes approaches, Gaussian quadrature arises from the Riemann sum (4.70), except here we interpret the interval widths as weights:

$$\sum_{i=0}^{n} w(x_i) f(x_i). \tag{4.79}$$

In Newton-Cotes rules, we generally choose the intervals and the points within the intervals where we evaluate the function to be equally spaced. In Gaussian quadrature, we put more emphasis on choosing the points, and by so doing, we need a smaller number of points. Whether or not this is a good idea of course depends on how we choose the points and how we define $w(x_i)$.

If f is a polynomial of degree $2n-1$, it is possible to represent the integral $\int_a^b f(x)\mathrm{d}x$ exactly in the form (4.79). We can illustrate this easily for $n = 2$. Before proceeding, however, let us first map f over $[a, b]$ onto $[-1, 1]$. (This clearly can be done by a simple change of variables. We do it just for simplification of the problem.) Now consider the integration of $f(x) = x^3 + x^2 + x + 1$. We want to determine w_1, w_2, x_1, x_2. We have

$$\begin{aligned} \int_{-1}^{1} \mathrm{d}x &= 2 = w_1 + w_2 \\ \int_{-1}^{1} x\mathrm{d}x &= 0 = w_1 x_1 + w_2 x_2 \\ \int_{-1}^{1} x^2\mathrm{d}x &= 2/3 = w_1 x_1^2 + w_2 x_2^2 \\ \int_{-1}^{1} x^3\mathrm{d}x &= 0 = w_1 x_1^3 + w_2 x_2^3. \end{aligned} \tag{4.80}$$

Solving this system of equations yields

$$\begin{aligned} w_1 &= 1 \\ w_2 &= 1 \\ x_1 &= -1/\sqrt{3} \\ x_2 &= 1/\sqrt{3}. \end{aligned} \tag{4.81}$$

This simple example illustrates that the idea is feasible, but indicates that there may be some difficulty if we need a large number of points. Gaussian quadrature is based on the formula (4.79) in which the x_i's and w_i's are chosen so that the approximation is correct when f is a polynomial, and it can provide a approximation in many cases with a relatively small n. Often only 5 or 6 points provide a good approximation.

To make this a useful method for any given (reasonable) integrand over a finite range, the obvious approach is to represent the function as a series in a standard sequence of orthogonal polynomials as described in Section 4.3. (That is, the q_k in equation (4.77) are polynomials.) This results in the approximation

$$f(x) \approx g(x), \tag{4.82}$$

where g is a polynomial.

If $q_0, q_1, \ldots$ is a sequence of polynomials orthogonal to the weight $w(w)$ over (a, b), with full sets of distinct real roots in (a, b), and if we choose the x_i in (4.79) as the distinct roots of $q_n(x)$, then the weights are given by

$$w_i = -\frac{c_{n+1}}{c_n} \frac{1}{q_{n+1}(x_i) q_n'(x_i)},$$

where c_j is the coefficient of the term of degree j in q_j. These weights are all positive. The derivation of this requires the kind of integration and algebra used in deriving the system of equations (4.80) and getting the solution in equations (4.81).

Error in Gaussian Quadrature

The error in Gaussian quadrature is

$$\int_a^b f(x)\,\mathrm{d}x - \sum_{i=0}^n w_i g(x_i),$$

which we can write as

$$\int_a^b g(x)w(x)\,\mathrm{d}x - \sum_{i=0}^n w_i g(x_i) = \frac{g^{(2n)}(x^*)}{(2n)!c_n^2}, \tag{4.83}$$

for some point x^* in (a, b).

As with many expressions for errors in numerical computations, we can feel good when we have the expression, but its usefulness in applications may be very limited.

A problem with Gaussian quadrature is that it is not easy to use the results for n to compute results for $\tilde{n}$, and hence the kinds of extrapolation and adaptation we discussed above for Newton-Cotes quadrature are not very useful for Gaussian quadrature.

4.7 Monte Carlo Methods for Quadrature

In the Monte Carlo method of quadrature we first formulate the integral to be evaluated as an expectation of a function of a random variable, then simulate realizations of the random variable, and take the average of the function evaluated at those realizations. This is analogous to a standard method of statistical estimation, in which we use a sample mean to estimate a parameter of the distribution of a random variable. In applications, the realizations of the random variable are pseudorandom numbers; nevertheless, our analysis relies on statistical estimation theory.

The deterministic methods of quadrature, such as Newton-Cotes and Gaussian, yield *approximations*; Monte Carlo quadrature yields *estimates*. Use of Monte Carlo methods for quadrature is sometimes called stochastic integration.

An advantage of Monte Carlo quadrature is that the nature of the domain of integration is not as critical as in the other quadrature methods we have discussed above. We consider an integral similar to (4.69), except with a more general domain of integration, D. To estimate the integral using Monte Carlo, we first formulate the integral as

$$\begin{aligned} I &= \int_D f(x)\,\mathrm{d}x \\ &= \int_D h(x)p_X(x)\,\mathrm{d}x, \end{aligned} \tag{4.84}$$

where p_X is the probability density function of a random variable X with support on D. We encountered this decomposition of the function f in Chapter 1. It is called probability density function decomposition or PDF decomposition.

This step may benefit from some familiarity with probability density functions. For one-dimensional interval domains, appropriate probability density functions are similar to the weight functions for orthogonal polynomials shown in Table 4.1, page 170, depending on whether neither, one, or both limits of integration are infinite.

If D is the interval (a, b), as in (4.69), a and b are finite, the trivial uniform density may always be used. The uniform density over $[a, b]$ is the constant $1/(b-a)$, so one possible formulation (4.84) is

$$I = (b-a)\int_a^b f(x)\frac{1}{b-a}\,\mathrm{d}x.$$

As we will see below, however, this may not be a good choice because the performance of the Monte Carlo method degrades for $h(x)$ with large variation, so the Monte Carlo estimates will be better (in a sense to be defined below) if $h(x)$ is nearly constant. Also, if a or b is infinite, the uniform density cannot be used because of the $(b-a)$ factor.

The decomposition in (4.84) results in

$$\begin{aligned} I &= \int_D h(x)p_X(x)\,\mathrm{d}x \\ &= \mathrm{E}(h(X)), \end{aligned}$$

where $\mathrm{E}(h(X))$ is the expectation (or "average" over the full distribution of X) of the function $h(X)$.

If $x_1, x_2, \ldots, x_m$ is a random sample (or pseudorandom sample) of the random variable X, the sample average,

$$\overline{h(x_i)} = \frac{1}{m}\sum_{i=1}^m h(x_i),$$

is an estimate of the integral, $\mathrm{E}(h(X))$, or I. We often denote an estimate of I as $\widehat{I}$, so in this case

$$\widehat{I} = \frac{1}{m}\sum_{i=1}^m h(x_i). \tag{4.85}$$

If we formulate the estimator $\widehat{I}$ as a sum of functions of independent random variables, each with density p_X, instead of a sum of realizations of random variables, the estimator itself is a random variable. (Note the distinction in "estimate" and "estimator".) An obviously desirable property of this random variable is that its expectation be equal to the quantity being estimated. Assuming the expectations exist, this is easily seen to be the case:

$$\begin{aligned}
\mathrm{E}(\widehat{I}) &= \mathrm{E}\Big(\frac{1}{m}\sum_{i=1}^{m} h(X_i)\Big) \\
&= \frac{1}{m}\sum_{i=1}^{m} \mathrm{E}(h(X_i)) \\
&= \frac{1}{m}\sum_{i=1}^{m} I \\
&= I.
\end{aligned}$$

We therefore say the estimator is *unbiased.*

Variance of Monte Carlo Estimators

Monte Carlo methods are sampling methods; therefore the estimates that result from Monte Carlo procedures have associated *sampling errors*. The fact that the estimate is not equal to its expected value (assuming the estimator is unbiased) is not an "error" or a "mistake"; it is just a result of the variance of the random (or pseudorandom) data. The sampling errors mean that we get different estimates of the integral if we evaluate it on different occasions.

In the case of scalar functions, the variance of the estimator $\widehat{I}$ is a rather complicated function involving the original integral (assuming the integrals exist):

$$\begin{aligned}
\mathrm{V}\left(\widehat{I}\right) &= \frac{1}{m}\mathrm{E}\left(\left(h(X) - \mathrm{E}(h(X))\right)^2\right) \\
&= \frac{1}{m}\int_D \left(h(x) - \int_D h(y)p_X(y)\,\mathrm{d}y\right)^2 p_X(x)\,\mathrm{d}x. \qquad (4.86)
\end{aligned}$$

If $p_X(x)$ is constant, that is, if the sampling is uniform over D, then the expression (4.86) is merely the variance of the mean of the roughness defined in equation (4.5) on page 151. Loosely speaking, this variance is a measure of how variable the Monte Carlo estimates would be if we were to evaluate the integral on different occasions.

We see that the magnitude of the variance depends on the variation in

$$h(x) - \int_D h(y)p_X(y)\,\mathrm{d}y,$$

which depends in turn on the variation in $h(x)$. If $h(x)$ is constant, the variance of $\widehat{I}$ is 0. Of course, in this case, we do not need to do the Monte Carlo estimation; we have the solution $I = h(\cdot)$.

While the variance in (4.86) is complicated, we have a very simple estimate of the variance; it is the sample variance of the elements composing the estimate of the integral, divided by the sample size m:

$$\widehat{\mathrm{V}}(\widehat{I}) = \frac{1}{m}\left(\frac{1}{m-1}\sum_{i=1}^{m}\left(h(x_i) - \overline{h(x_i)}\right)^2\right). \tag{4.87}$$

The second factor in this expression is the sample variance of the observations $h(x_i)$.

An important fact to be observed in equation (4.86) is that a similar expression would hold if the integrand was multivariate. Therefore, the variance of the Monte Carlo estimate is independent of the dimensionality. This is one of the most important properties of Monte Carlo quadrature.

Reducing the Variance

As we see from equation (4.86) the variance of the Monte Carlo estimator is linear in m^{-1}; hence, the variance is reduced by increasing the Monte Carlo sample size. More effective methods of variance reduction include use of antithetic variates, importance sampling, and stratified sampling, as discussed in Section 11.5, beginning on page 425.

Combining Monte Carlo Estimators

The Monte Carlo estimator (4.85) is linear in $h(x_i)$. This implies that the estimator can be evaluated as separate partial sums, either computed in parallel or computed at different times. Separate computations yield separate estimators, $\widehat{I}_1, \widehat{I}_2, \ldots, \widehat{I}_k$, which can be combined to yield

$$\widehat{I} = \sum_{i=1}^{k} a_i \widehat{I}_i, \tag{4.88}$$

where the a_i are constants. If each of the $\widehat{I}_i$ is unbiased, this estimator is unbiased so long as $\sum_{i=1}^{k} a_i = 1$:

$$\begin{aligned}
\mathrm{E}(\widehat{I}) &= \mathrm{E}\Big(\sum_{i=1}^{k} a_i \widehat{I}_i\Big) \\
&= \sum_{i=1}^{k} a_i \mathrm{E}(\widehat{I}_i) \\
&= \sum_{i=1}^{k} a_i I \\
&= I.
\end{aligned}$$

If all of the individual estimators are uncorrelated, the variance of the combined estimator is

$$\sum_{i=1}^{k} a_i^2 \mathrm{V}(\widehat{I}_i).$$

To minimize the variance of the linear combination (4.88), the a_i's are chosen inversely proportional to the variances of the component random variables, that is,

$$a_i = \frac{c}{\mathrm{V}(\widehat{I}_i)},$$

for some $c > 0$.

Error in Monte Carlo Quadrature

As we have emphasized, Monte Carlo quadrature differs from quadrature methods such as Newton-Cotes methods and Gaussian quadrature in a fundamental way; Monte Carlo methods involve random (or pseudorandom) sampling. The expressions in the Mont Carlo quadrature formulas do not involve any approximations, so questions of bounds of the error of approximation do not arise. Instead of error bounds or order of the error as some function of the integrand as we discuss for the deterministic methods on pages 188 and 191, we use the variance of the random estimator to indicate the extent of the uncertainty in the solution.

The square root of the variance, that is, the standard deviation of the estimator, is a good measure of the range within which different estimators of the integral may fall. Under certain assumptions, using the standard deviation of the estimator, we can define statistical "confidence intervals" for the true value of the integral I. Loosely speaking, a confidence interval is an interval about an estimator $\widehat{I}_i$ that in repeated sampling would include the true value I a specified portion of the time. (The specified portion is the "level" of the confidence interval, and is often chosen to be 90% or 95%. Obviously, all other things being equal, the higher the level of confidence the wider must be the interval.)

Because of the dependence of the confidence interval on the standard deviation the standard deviation is sometimes called a "probabilistic error bound". The word "bound" is misused here, of course, but in any event, the standard deviation does provide some measure of a sampling "error".

The important thing to note from equation (4.86) is the order of error in the Monte Carlo sample size; it is $\mathrm{O}(m^{-\frac{1}{2}})$. This results in the usual diminished returns of ordinary statistical estimators; to halve the error, the sample size must be quadrupled.

We should be aware of a very important aspect of this discussion of error bounds for the Monte Carlo estimator. It applies to random numbers. The pseudorandom numbers we actually use only simulate the random numbers, so "unbiasedness" and "variance" must be interpreted carefully.

Variations of Monte Carlo Quadrature

The method of estimating an integral described above is sometimes called "crude Monte Carlo". Another method, which may be more familiar, called "hit-or-miss" Monte Carlo is not to be recommended (see Gentle, 2003, Exercise 7.2, page 271).

Another Monte Carlo method can be developed as suggested in Exercise 4.15, page 202. To estimate the integral

$$I = \int_a^b f(x)\,\mathrm{d}x$$

first generate a random sample of uniform order statistics $x_{(1)}, x_{(2)}, \ldots, x_{(n)}$ on the interval (a, b), and define $x_{(0)} = a$ and $x_{(n+1)} = b$. Then estimate I as

$$\widehat{I} = \frac{1}{2}\left(\sum_{i=1}^{n}(x_{(i+1)} - x_{(i-1)})f(x_{(i)}) + (x_{(2)} - a)f(x_{(1)}) + (b - x_{(n-1)})f(x_{(n)})\right). \tag{4.89}$$

This method is similar to approximation of the integral by Riemann sums, except in this case the intervals are random.

Higher Dimensions

The Monte Carlo quadrature methods extend directly to multivariate integrals, although, obviously, it takes larger samples to fill the space. It is, in fact, only for multivariate integrals that Monte Carlo quadrature should ordinarily be used. The preference for Monte Carlo in multivariate quadrature results from the independence of the pseudoprobabilistic error bounds and the dimensionality mentioned above.

An important property of the standard deviation of a Monte Carlo estimate of a definite integral is that the order in terms of the number of function evaluations is independent of the dimensionality of the integral so the order of the error remains $\mathrm{O}(m^{-\frac{1}{2}})$. On the other hand, the usual error bounds for numerical quadrature are $\mathrm{O}((g(n))^{-\frac{1}{d}})$, where d is the dimensionality, and $g(n)$ is the order for one-dimensional quadrature.

Notes and Further Reading

I have made frequent reference to Hewitt and Stromberg (1965). This is just because that is where I first learned real analysis. Many newer and more readily accessible texts would serve just as well.

Function Approximation and Computations Involving Polynomials

Extensive discussions of function approximation are available in texts on numerical methods, such as Rice (1993).

Horner's method is so called because William George Horner described it in 1819. The method, however, was known to Isaac Newton many years earlier. Newton also was aware of the need to shift the values in a polynomial prior to raising them to a power, and the form described by Newton utilizes the nesting of Horner's method.

Function Expansions

A hundred years ago, expansion of functions, especially of probability density functions, or of general functions following a PDF decomposition, were widely studied and used. The Gram-Charlier series, the Edgeworth series, and the Cornish-Fisher expansion were very important topics in mathematical statistics. These expansions, of course, remain useful, but their use seems to wax and wane, and, at best, remain among the techniques in the background memory of most statisticians and applied mathematicians.

The paper by Barndorff-Nielsen and Cox (1979) revived interest and application of expansions, and brought the saddlepoint approximation method of Daniels (1954) to the wider attention of statisticians. Although approximations similar to the saddlepoint approximation had been used in various applications previously, Daniels derived it in its currently-used form and illustrated its usefulness for the density of a sample mean. The book by Jensen (1995) and the article by Goutisand and Casella (1999) provide good introductions to the saddlepoint method.

Special Functions

GAMS is a good source of information about software for evaluating the special functions. Programs for evaluation of special functions are available in the IMSL Libraries (a function is available for each entry in the list above), in the Maple and Mathematica packages, and in CALGO (see page 692), as well as in more specialized collections, such as Cody (1993) or Cody and Coonen (1993).

Spanier and Oldham (1987) and Thompson (1997) provide general descriptions of many special functions. Both books discuss relationships among the special functions and describe methods for evaluating the special functions. They also contain many graphs of the functions. Abramowitz and Stegun (1964) provide tables of the values of special functions for many arguments. Note that an update of this book is currently under production. The new version is called the Digital Library of Mathematical Functions (DLMF). See

`http://dlmf.nist.gov/`

The extent and the form in which DLMF will exist in hardcopy is not clear. A portion of DLMF, supplement with discussions of the computational methods, is available in Gil, Segura, and Temme (2007).

Orthogonal Systems

The standard treatment of orthogonal polynomials is Szegö (1958), in which several other systems are described and more properties of orthogonal polynomials are discussed. A general reference on multivariate orthogonal polynomials is Dunkl and Yu (2001).

A type of orthogonal system that I mentioned, but did not discuss, are wavelets. For this I refer the reader to Walter and Ghorai (1992) or to Vidakovic (2004).

Splines

De Boor (2002) provides a comprehensive development of splines and an extensive discussions of their properties. The emphasis is on B-splines and he gives several Fortran routines for using B-splines and other splines.

A good introduction to multivariate splines is given by Chui (1988).

Numerical Quadrature

Evans and Schwartz (2000) provide a good summary of methods for numerical quadrature, including both the standard deterministic methods of numerical analysis and Monte Carlo methods.

The most significant difficulties in numerical quadrature occur in multiple integration. The papers in the book edited by Flournoy and Tsutakawa (1991) provide good surveys of specific methods, especially ones with important applications in statistics.

Monte Carlo Quadrature

Monte Carlo quadrature, of course, requires a source of random numbers. Section 7.6 describes software for generation of pseudorandom numbers. In higher dimensional quadrature, rather than the usual pseudorandom numbers, it may be better to use quasirandom numbers. Software for quasirandom number generation is not as widely available, but a reference is given on page 322.

Exercises

4.1. For any function f with finite, nonzero norm, show that the L_1 norm of $f_a(ax)$, for any given $a \neq 0$, is the same as the L_1 norm of $f(x)$, and show that the L_2 norm of $f_a(ax)$ is not the same as the L_2 norm of $f(x)$.

4.2. Let

$$p(x) = \frac{1}{\sqrt{2\pi}\sigma} \mathrm{e}^{-x^2/(2\sigma^2)}$$

(the normal density with mean equal to 0).

a) Compute $\mathcal{R}(p)$ (from equation (4.6) on page 151).

b) Compute $\mathcal{S}(p')$.

c) Compute $\mathcal{S}(p'') = \mathcal{R}(p)$.

4.3. Develop an extension of the roughness definition given in equation (4.7) for functions of more than one variable. (You obviously use the Hessian. How do you map it to $\mathrm{I\!R}$?)

4.4. Derive equation (4.26) from equation (4.25).

4.5. Consider a mixture of two normal distributions $\mathrm{N}(\mu_1, \sigma_1^2)$ and $\mathrm{N}(\mu_2, \sigma_2^2)$ with mixing parameter ω (that is, ω are from the first distribution and $1 - \omega$ are from the second distribution).

a) Determine the moment generating function for each component in the mixture. (This is a standard result; it is $M(t) = \exp(\mu_i t + \sigma_i^2 t/2)$.)

b) Determine the moment generating function for the mean of each component in the mixture.

c) Determine the moment generating function for the mean of the mixture distribution.

d) Determine the cumulant generating function for the mean of the mixture distribution.

e) For a fixed value of the mean, say $\bar{x}$, determine r_0 that solves the saddlepoint equation (4.35).

f) Determine the saddlepoint approximation for the PDF of the mean of the mixture distribution.

g) Use Monte Carlo with a normal PDF to estimate the normalizing constant of your approximation. (See Section 4.7.)

4.6. Let $\{q_k : k = 1, \ldots, m\}$ be a set of orthogonal functions. Show that

$$\left\| \sum_{k=1}^{m} q_k \right\|^2 = \sum_{k=1}^{m} \|q_k\|^2,$$

where $\|\cdot\|$ represents an L_2 norm. What is the common value of the expressions above if the q_k are orthonormal?

Would a similar equation hold for a general L_p norm?

4.7. Suppose that the Legendre polynomials are to be used to approximate a function over the interval $[0, 10]$.

a) What are the normalizing factor?

b) What is the recurrence formula?

4.8. Using the recurrence equation (4.50) and beginning with $T_0(t) = 1$ and $T_1(t) = t$, derive the first four Chebyshev polynomials, $T_0(t)$, $T_1(t)$, $T_2(t)$, and $T_3(t)$, which are given in (4.49).

4.9. Show that the normalizing constant for the kth Chebyshev polynomial is $T_k(1)$.

4.10. Approximate $f(t) = \mathrm{e}^t$ over $[-1, 1]$ as

$$\sum_{k=0}^{5} c_k T_k(t),$$

where the $T_k(t)$ are the Chebyshev polynomials. (Compare this with the example on page 176 that uses Legendre polynomials to approximate e^{-t}.)

a) Write a program to use Algorithm 4.1 to compute the approximation at a given point t.
b) Graph the function and your approximation.
c) Determine the error at $t = 0$.
d) Determine the integrated squared error.
e) Would some more general sequence of Jacobi polynomials form a better approximation? Why do you think so? What values of α and β might be more appropriate?
f) For reasonable values of α and β from the previous question, derive $J_0^{(\alpha,\beta)}(t)$, $J_1^{(\alpha,\beta)}(t)$, $J_2^{(\alpha,\beta)}(t)$, and $J_3^{(\alpha,\beta)}(t)$. Now, approximate $f(t) = \mathrm{e}^t$ over $[-1, 1]$ with your polynomials and determine the error at $t = 0$ and the integrated squared error.

4.11. Compute roughness measures of your Chebyshev-polynomial approximation in Exercise 4.10. Compute $\mathcal{S}(\widehat{f})$ in equation (4.6), $\mathcal{V}(\widehat{f})$ in equation (4.5), and $\mathcal{R}(\widehat{f})$ in equation (4.7).

4.12. Now, assume that we have 19 data points, (x_i, y_i), with $x_1 = -.9$, $x_2 = -.8 \cdots x_{19} = .9$, and $y_i = \mathrm{e}^{-x_i}$. We want to fit a function $y = f(x)$, but we do not know the form of f; all we know are the 19 data points. Use expression (4.58) in (4.59) to obtain the Fourier coefficients for an expansion in the first 6 Chebyshev polynomials that yields a least squares fit to the given data.

4.13. Use kernel method to approximate the function $f(x) = \mathrm{e}^{-x}$ over the interval $[-1, 1]$, as we did in the text with orthogonal polynomials and with splines. Use both $K_{\mathrm{u}}(t)$ and $K_{\mathrm{n}}(t)$ with λ chosen as 1/2, 1/4, and 1/8. Evaluate $f(x)$ at enough points to have at least two known points within one window at any point in the interval.

Notice that this (and our other examples with this function) are artificial, in the sense that we would rarely in applications encounter a problem just like this — if we can evaluate $f(x)$, we likely would not be interested in approximating it (although, if our approximations were simpler to compute, we might want to do this). The point of the examples and of the exercise, however, is to assess the performance of the approximation method in more realistic situations, in which we do not know the function everywhere, we only know it at a few select points. This, of course, is the type of situation we face in statistical estimation.

4.14. Show that equation (4.75) results from Richardson extrapolation (equation (3.22)) of the T_{0k} representing the values of the expression in equation (4.71) with successively smaller intervals.

4.15. Consider the following Monte Carlo method to evaluate the integral:

$$I = \int_a^b f(x)\,\mathrm{d}x.$$

Generate a random sample of uniform order statistics $x_{(1)}, x_{(2)}, \ldots, x_{(n)}$ on the interval (a, b), and define $x_{(0)} = a$ and $x_{(n+1)} = b$. Estimate I by equation (4.89) on page 197. This method is similar to approximation of the integral by Riemann sums except that in this case the intervals are random. Determine the variance of $\widehat{I}$. What is the order of the variance in terms of the sample size? How would this method compare in efficiency with the crude Monte Carlo method?

5

Numerical Linear Algebra

Many scientific computational problems involve vectors and matrices. It is necessary to work with either the elements of vectors and matrices individually or with the arrays themselves. Programming languages such as C provide the capabilities for working with the individual elements but not directly with the arrays. Fortran and higher-level languages such as Octave or Matlab and R allow direct manipulation with vectors and matrices.

The distinction between the set of real numbers, $\mathbb{R}$, and the set of floating-point numbers, $\mathbb{F}$, that we use in the computer has important implications for numerical computations. An element x of a vector or matrix is approximated by $[x]_{\mathrm{c}}$, and a mathematical operation $\circ$ is simulated by a computer operation $[\circ]_{\mathrm{c}}$. As we emphasized in Section 2.2, the familiar laws of algebra for the field of the reals do not hold in $\mathbb{F}$.

These distinctions, of course, carry over to arrays of floating-point numbers that represent real numbers, and the mathematical properties of vectors and matrices may not hold for their computer counterparts. For example, the dot product of a nonzero vector with itself is positive, but $\langle x_{\mathrm{c}}, x_{\mathrm{c}} \rangle_{\mathrm{c}} = 0$ does not imply $x_{\mathrm{c}} = 0$. (This is reminiscent of the considerations that led us to discuss pseudonorms on page 149, but the issues here are entirely different.)

The elements of vectors and matrices are represented as ordinary numeric data in either fixed-point or floating-point representation. In the following, we will consider the floating-point representation and the computations in $\mathbb{F}$.

Storage Modes

The elements of an array are generally stored in a logically contiguous area of the computer's memory. What is logically contiguous may not be physically contiguous, however.

Because accessing data from memory in a single pipeline may take more computer time than the computations themselves, computer memory may be organized into separate modules, or *banks*, with separate paths to the central processing unit. Logical memory is *interleaved* through the banks; that is,

J.E. Gentle, *Computational Statistics*, Statistics and Computing,
DOI: 10.1007/978-0-387-98144-4_5,

two consecutive logical memory locations are in separate banks. In order to take maximum advantage of the computing power, it may be necessary to be aware of how many interleaved banks the computer system has, but we will not consider such details here.

There are no convenient mappings of computer memory that would allow matrices to be stored in a logical rectangular grid, so matrices are usually stored either as columns strung end-to-end (a "column-major" storage) or as rows strung end-to-end (a "row-major" storage). Sometimes it is necessary to know which way the matrix is stored in the computer's logical address space; that is, whether $a_{i,j}$ is stored logically next to $a_{i+1,j}$ or to $a_{i,j+1}$. (Physically, in the hardware, it may be next to neither of these.)

For some software to deal with matrices of varying sizes, the user must specify the length of one dimension of the array containing the matrix. (In general, the user must specify the lengths of all dimensions of the array except one.) In Fortran subroutines, it is common to have an argument specifying the leading dimension (number of rows), and in C functions it is common to have an argument specifying the column dimension. In an object-oriented system, this information is bundled in the object, and it is the object itself (the matrix, rather than a computer memory address) that is passed from one program module to another.

Notation

It is assumed that the reader is generally familiar with the basics of linear algebra, at least to the level covered in the relevant parts of Section 1.2.

An n-vector is an ordered structure with n real elements. We denote the sapce of n-vectors along with the axpy and inner product operators as $\mathbb{R}^n$. We identify the elements of a vector x by a display of the form

$$x = (x_1, \ldots, x_n).$$

There is no need to call this display a "transpose". How x is displayed has no relevance for how matrix-vector operations are interpreted. For matrix-vector operations, we interpret a vector as an $n \times 1$ matrix and then use the usual matrix-matrix rules for operations. Stating this another way, we interpret vectors as "column vectors", although we display them horizontally.

An $n \times m$ matrix A is an element of $\mathbb{R}^{n \times m}$. It is often denoted as $A = (a_{ij})$. Its transpose, denoted by A^{T}, is (a_{ji}). The Moore-Penrose generalized inverse is denoted by A^+. If the inverse exists, that is, if A is square and of full rank, it is denoted by A^{-1}.

The ith row of the matrix $A = (a_{ij})$ is denoted by a_{i*}, and the jth column is denoted by a_{*j}. Both a_{i*} and a_{*j} are treated as ordinary vectors. They are both "column" vectors, as are all vectors in this book.

Sparsity

If a matrix has many elements that are zeros, and if the positions of those zeros are easily identified, many operations on the matrix can be speeded up.

Matrices with many zero elements are called *sparse matrices*. They occur often in certain types of problems; for example, in the solution of differential equations, and in statistical designs of experiments.

The first consideration is how to represent the matrix and to store the matrix and the location information. Different software systems may use different schemes to store sparse matrices. An important consideration is how to preserve the sparsity during intermediate computations. We mention one way this may be done in an iterative algorithm on page 226, however, most of the computational issues for dealing with sparse matrices are beyond the scope of this book,

5.1 General Computational Considerations for Vectors and Matrices

Because many of the computations in linear algebra are sums of elements in a list, the discussion of such computations beginning on page 99 must be borne in mind. Catastrophic cancellation is of special concern.

One common situation that gives rise to numerical errors in computer operations is when a quantity x is transformed to $t(x)$ but the value computed is unchanged:

$$[t(x)]_c = [x]_c; \tag{5.1}$$

that is, the operation actually accomplishes nothing. A simple type of transformation that has this problem is just the addition

$$t(x) = x + \epsilon, \tag{5.2}$$

where $|\epsilon|$ is much smaller than $|x|$. If all we wish to compute is $x + \epsilon$, the fact that we get x is probably not important. Usually, however, this simple computation is part of some larger set of computations in which ϵ was computed. This, therefore, is the situation we want to anticipate and avoid.

Another instance of this problem is the addition to x of a computed quantity y that overwhelms x in magnitude. In this case, we may have

$$[x + y]_c = [y]_c. \tag{5.3}$$

Again, this is a situation we want to anticipate and avoid.

In later sections in this chapter we will consider various types of computations in numerical linear algebra. We distinguish these methods as being either *direct*, meaning that the number of computations is set a priori, or *iterative*, meaning that the results of the computations in each step determine whether to perform additional computations. We discuss iterative methods in Section 5.4. (Recall from our discussion in Chapter 3 that the steps in a direct method may also be called "iterations".)

In the remainder of this section, we discuss the very important issue of identifying the level of accuracy we can expect in computations involving matrices. This depends on the *condition* of the data.

Condition

A measure of the worst-case numerical error in numerical computation involving a given mathematical entity is the "condition" of that entity for the particular computations. The condition, quantified in some way, provides a bound on the relative norms of a "correct" solution to a linear system and a solution to a nearby problem. Hence, the condition of data depends on the particular computations to be performed. For example, the "stiffness" measure in equation (3.6) is an appropriate condition measure of the extent of the numerical error to be expected in computing variances.

Many computations in linear algebra are related to the basic problem of solving a system of equations:

$$Ax = b. \tag{5.4}$$

This is the canonical problem to which much of this chapter is devoted.

If A is square and nonsingular, the solution is $x = A^{-1}\tilde{b}$. Actual computations, however, yield the solution $\tilde{x}$, which we might identify as the solution to a nearby problem: Solve

$$A\tilde{x} = \tilde{b}, \tag{5.5}$$

where $\tilde{x} = x + \delta x$ and $\tilde{b} = b + \delta b$. Here we are using the symbol δ, not as a multiplier, but as a perturbation operation; that is, δx is a perturbation about x. If the original problem is well behaved, we would expect that if δb is small, then δx is small and the solution $\tilde{x}$ is "close to" x.

We quantify the condition of the matrix by a *condition number*. To develop this quantification for the problem of solving linear equations, consider a linear system $Ax = b$, with A nonsingular and $b \neq 0$, as above. Now perturb the system slightly by adding a small amount, δb, to b, and let $\tilde{b} = b + \delta b$. The system has a solution $\tilde{x} = \delta x + x = A^{-1}\tilde{b}$. (Notice that δb and δx do not necessarily represent scalar multiples of the respective vectors.) If the system is well-conditioned, for any reasonable norm, if $\|\delta b\|/\|b\|$ is small, then $\|\delta x\|/\|x\|$ is likewise small.

From $\delta x = A^{-1}\delta b$ and the inequality (1.17) (on page 14), for an induced norm on A, we have

$$\|\delta x\| \leq \|A^{-1}\| \, \|\delta b\|. \tag{5.6}$$

Likewise, because $b = Ax$, we have

$$\frac{1}{\|x\|} \leq \|A\| \frac{1}{\|b\|}, \tag{5.7}$$

and equations (5.6) and (5.7) together imply

$$\frac{\|\delta x\|}{\|x\|} \leq \|A\| \, \|A^{-1}\| \frac{\|\delta b\|}{\|b\|}. \tag{5.8}$$

This provides a bound on the change in the solution $\|\delta x\|/\|x\|$ in terms of the perturbation $\|\delta b\|/\|b\|$.

The bound in equation (5.8) motivates us to define the *condition number with respect to inversion* denoted by $\kappa(\cdot)$ as

$$\kappa(A) = \|A\| \, \|A^{-1}\| \tag{5.9}$$

for nonsingular A. The specific condition number therefore depends on the specific norm.

In the context of linear algebra, the condition number with respect to inversion is so dominant in importance that we generally just refer to it as the "condition number". A condition number is a useful measure of the condition of A for the problem of solving a linear system of equations. There are other condition numbers useful in numerical analysis, however, such as the condition number for computing the sample variance or a condition number for a root of a function.

We can write equation (5.8) as

$$\frac{\|\delta x\|}{\|x\|} \leq \kappa(A) \frac{\|\delta b\|}{\|b\|}, \tag{5.10}$$

or analogously as

$$\frac{\|\delta b\|}{\|b\|} \leq \kappa(A) \frac{\|\delta x\|}{\|x\|}. \tag{5.11}$$

These inequalities are sharp, as we can see by letting $A = I$.

Because the condition number is an upper bound on a quantity that we would not want to be large, a large condition number is "bad".

Notice that our definition of the condition number does not specify the norm; it only requires that the norm be an induced norm. (An equivalent definition does not rely on the norm being an induced norm.) We sometimes specify a condition number with regard to a particular norm, and just as we sometimes denote a specific norm by a special symbol, we may use a special symbol to denote a specific condition number. For example, $\kappa_p(A)$ may denote the condition number of A in terms of an L_p norm. Most of the properties of condition numbers (but not their actual values) are independent of the norm used.

An interesting relationship for the L_2 condition number is

$$\kappa_2(A) = \frac{\max_{x \neq 0} \frac{\|Ax\|}{\|x\|}}{\min_{x \neq 0} \frac{\|Ax\|}{\|x\|}}, \tag{5.12}$$

which can be shown directly from the definition (5.9) of condition number and of the L_2 norm and from properties of eigenvalues. The numerator and denominator in equation (5.12) look somewhat like the maximum and minimum eigenvalues. Indeed, the L_2 condition number of a nonsingular square matrix is just the ratio of the largest eigenvalue in absolute value to the smallest (see Gentle, 2007, page 131).

Some useful facts about condition numbers are:

$$\kappa(A) = \kappa(A^{-1}), \tag{5.13}$$

$$\kappa(cA) = \kappa(A), \quad \text{for } c \neq 0, \tag{5.14}$$

$$\kappa(A) \geq 1, \tag{5.15}$$

$$\kappa_1(A) = \kappa_\infty(A^{\mathrm{T}}), \tag{5.16}$$

$$\kappa_2(A^{\mathrm{T}}) = \kappa_2(A), \tag{5.17}$$

$$\begin{aligned} \kappa_2(A^{\mathrm{T}}A) &= \kappa_2^2(A) \\ &\geq \kappa_2(A). \end{aligned} \tag{5.18}$$

All of these facts follow immediately from the definitions or from properties of the matrix norms.

Equation (5.18) is of some interest, and there are similar results for other condition numbers. The point is that the condition number of $A^{\mathrm{T}}A$ is larger, possibly much larger, than the condition number of A. (Recall that a matrix that appears often in regression analysis is $X^{\mathrm{T}}X$.)

Even though the condition number provides a very useful indication of the condition of the problem of solving a linear system of equations, it can be misleading at times. Consider, for example, the coefficient matrix

$$A = \begin{bmatrix} 1 & 0 \\ 0 & \epsilon \end{bmatrix}, \tag{5.19}$$

where $\epsilon < 1$. The condition numbers are

$$\kappa_1(A) = \kappa_2(A) = \kappa_\infty(A) = \frac{1}{\epsilon},$$

and so if ϵ is small, the condition number is large. It is easy to see, however, that small changes to the elements of A or b in the system $Ax = b$ do not cause undue changes in the solution (which is our heuristic definition of ill-conditioning). In fact, the simple expedient of multiplying the second row of A by $1/\epsilon$ (that is, multiplying the second equation, $a_{21}x_1 + a_{22}x_2 = b_2$, by $1/\epsilon$) yields a linear system that is very well-conditioned.

This kind of apparent ill-conditioning is called *artificial ill-conditioning.* It is due to the different rows (or columns) of the matrix having a very different *scale*; the condition number can be changed just by scaling the rows or columns. This usually does not make a linear system any better or any worse conditioned, but this fact emphasizes the importance of scaling in data analysis. (Scaling has implications not only for numerical computations; it also affects the results of many multivariate analyses, even if the computations are exact. As we mentioned in Section 1.1 scaling induces artificial structure and it affects such analytic methods as clustering and principal component analysis.)

Condition of Singular or Nonsquare Matrices

We have discussed condition in the context of the solution of a full-rank, consistent linear system. The same kinds of issues of numerical accuracy arise in non-full-rank systems and in overdetermined systems. A general condition number for such matrices can be defined as an extension of the L_2 condition number in equation (5.12) based on singular values. The singular value condition number of a general matrix A is

$$\kappa_{\mathrm{sv}}(A) = \frac{\sigma_1}{\sigma_k}, \tag{5.20}$$

where σ_1 is the largest singular value of A and σ_k is the smallest positive singular value of A.

5.2 Gaussian Elimination and Elementary Operator Matrices

The most common direct method for the solution of linear systems is Gaussian elimination. The basic idea in this method is to form equivalent sets of equations, beginning with the system to be solved, $Ax = b$, and ending with a system $Ux = Tb$, where U is an upper triangular matrix, and T is some matrix that makes the system equivalent to the original one.

Consider the individual equations

$$\begin{aligned} a_{1*}^{\mathrm{T}}x &= b_1 \\ a_{2*}^{\mathrm{T}}x &= b_2 \\ \ldots &= \ldots \\ a_{n*}^{\mathrm{T}}x &= b_n, \end{aligned} \tag{5.21}$$

where a_{j*} is the jth row of A. (Recall that a_{j*} is a vector and all vectors are "column" vectors.) An equivalent set of equations can be formed by a sequence of *elementary operations* on the equations in the given set.

These elementary operations on equations are essentially the same as the elementary operations on the rows of matrices. The two most important kinds of elementary operations are an interchange of two equations,

$$\begin{aligned} a_{j*}^{\mathrm{T}}x = b_j &\leftarrow a_{k*}^{\mathrm{T}}x = b_k, \\ a_{k*}^{\mathrm{T}}x = b_k &\leftarrow a_{j*}^{\mathrm{T}}x = b_j, \end{aligned} \tag{5.22}$$

and a replacement of a single equation with a sum of it and a scalar multiple of another equation,

$$a_{j*}^{\mathrm{T}}x = b_j \quad \leftarrow \quad a_{j*}^{\mathrm{T}}x + c a_{k*}^{\mathrm{T}}x = b_j + cb_k. \tag{5.23}$$

(The operation (5.23) is an axpy with $a = c$, $x = b_k$, and $y = b_j$.)

These operations can be effected by premultiplication by *elementary operator matrices*, which are matrices formed by performing the indicated operation on the identity matrix. The elementary operator matrix that exchanges rows j and k, which we denote as E_{jk}, is the identity matrix with rows j and k interchanged. The elementary operator matrix that performs the operation (5.23), which we denote as $E_{jk}(c)$, is the identity matrix with the 0 in position (j, k) replaced by c.

The elementary operation on the equation

$$a_{2*}^{\mathrm{T}}x = b_2$$

in which the first equation is combined with it using $c = -a_{21}/a_{11}$ will yield an equation with a zero coefficient for x_1. The sequence of equivalent equations, beginning with $Ax = b$, is

$$\begin{aligned} E_{21}(c_2^{(1)})Ax &= E_{21}(c_2^{(1)})b \\ E_{31}(c_3^{(1)})E_{21}(c_2^{(1)})Ax &= E_{31}(c_3^{(1)})E_{21}(c_2^{(1)})b \\ &\vdots \\ E_{n1}(c_n^{(1)})\cdots E_{31}(c_3^{(1)})E_{21}(c_2^{(1)})Ax &= E_{n1}(c_n^{(1)})\cdots E_{31}(c_3^{(1)})E_{21}(c_2^{(1)})b, \end{aligned} \tag{5.24}$$

where

$$c_i^{(1)} = -a_{i1}/a_{11}.$$

At this stage, the equations (5.21) are

$$\begin{aligned} a_{1*}^{\mathrm{T}}x &= b_1 \\ \left(\tilde{a}_{2*}^{(1)}\right)^{\mathrm{T}} x &= b_2 \\ \ldots &= \ldots \\ \left(\tilde{a}_{n*}^{(1)}\right)^{\mathrm{T}} x &= b_n, \end{aligned} \tag{5.25}$$

where the vector $\tilde{a}_{j*}^{(1)}$ has a zero in its first position.

In Gaussian elimination we continue this process by using elementary operator matrices of the form $E_{i2}(c_i^{(2)})$, where $i \geq 3$ and $c_i^{(2)} = -\tilde{a}_{i2}^{(1)}/\tilde{a}_{22}^{(1)}$. After $n-2$ such operations, we have a system of equations similar to equations (5.25), in which now for $j \geq 3$, the vector $\tilde{a}_{j*}^{(2)}$ has zeros in its first two positions.

Continuing this process, we form systems of equations with more and more zeros as coefficients of x's. Finally, we have a completely triangular system, Ux, on the left side. This system is easy to solve because the coefficient matrix is upper triangular. The last equation in the system yields

$$x_n = \frac{\tilde{b}_n^{(n-1)}}{\tilde{a}_{nn}^{(n-1)}}.$$

By back substitution, we get

$$x_{n-1} = \frac{\tilde{b}_{n-1}^{(n-2)} - \tilde{a}_{n-1,n}^{(n-2)} x_n}{\tilde{a}_{n-1,n-1}^{(n-2)}},$$

and we obtain the rest of the x's in a similar manner.

Gaussian elimination consists of two steps: the forward reduction, which is of order $\mathrm{O}(n^3)$, and the back substitution, which is of order $\mathrm{O}(n^2)$.

While Gaussian elimination is mathematically equivalent to a sequence of matrix multiplications, the actual computations would not appear to be matrix multiplications. This reminds us that:

The form of a mathematical expression and the way the expression should be evaluated in actual practice may be quite different.

Furthermore, there are many details of the computations that must be performed carefully. It is clear that if, at some step in the process above, $\tilde{a}_{kk}^{(k-1)} = 0$, we would have to do something differently. If this happens, then we would have to interchange two rows before proceeding. (This is called pivoting; see below.) But before proceeding with this simple fix, we recall:

Computer numbers are not the same as real numbers, and the arithmetic operations on computer numbers are not exactly the same as those of ordinary arithmetic.

If $\tilde{a}_{kk}^{(k-1)} = 0$, it may be the case that the computations in the first $k-1$ steps did not yield an exact 0 in the (k, k) position.

In arithmetic with floating-point numbers, checking for an exact 0 rarely makes sense.

Other problems may arise. Suppose, for example, that $\tilde{a}_{kk}^{(k-1)}$ is very small in absolute value, and some $\tilde{a}_{ik}^{(k-1)}$ is very large. In that case, it is quite possible that

$$\begin{aligned}\left[c_i^{(k)}\right]_c &= \left[\tilde{a}_{ik}^{(k-1)}\right]_c / \left[\tilde{a}_{kk}^{(k-1)}\right]_c \\ &= \mathrm{Inf}. \end{aligned} \tag{5.26}$$

Another type of problem may arise if at some stage $c_i^{(k)} \tilde{a}_{kj}^{(k-1)} \approx -\tilde{a}_{ij}^{(k-1)}$. This is the standard setup for catastrophic cancellation (see page 100). The resulting value $\left[\tilde{a}_{ij}^{(k)}\right]_c$ may have only one or two units of precision.

Pivoting

The divisors $a_{kk}^{(k-1)}$s are called "pivot elements". The obvious problem with the method of Gaussian elimination mentioned above arises if some of the pivot elements are zero (or very small in magnitude).

Suppose, for example, we have the equations $Ax = b$, where

$$\begin{aligned} 0.0001x_1 + x_2 &= 1, \\ x_1 + x_2 &= 2. \end{aligned} \tag{5.27}$$

The solution is $x_1 = 1.0001$ and $x_2 = 0.9999$. Suppose we are working with three digits of precision (so our solution is $x_1 = 1.00$ and $x_2 = 1.00$). After the first step in Gaussian elimination, we have

$$\begin{aligned} 0.0001x_1 + \qquad\qquad x_2 &= \qquad\quad 1, \\ -10,000x_2 &= -10,000, \end{aligned}$$

and so the solution by back substitution is $x_2 = 1.00$ and $x_1 = 0.000$. The L_2 condition number of the coefficient matrix is 2.618, so even though the coefficients vary greatly in magnitude, we certainly would not expect any difficulty in solving these equations.

A simple solution to this potential problem is to interchange the equation having the small leading coefficient with an equation below it. Thus, in our example, we first form

$$\begin{aligned} x_1 + x_2 &= 2, \\ 0.0001x_1 + x_2 &= 1, \end{aligned}$$

so that after the first step we have

$$\begin{aligned} x_1 + x_2 &= 2, \\ x_2 &= 1, \end{aligned}$$

and the solution is $x_2 = 1.00$ and $x_1 = 1.00$, which is correct to three digits.

Another strategy would be to interchange the column having the zero or small leading coefficient with a column to its right. Both the row interchange and the column interchange strategies could be used simultaneously, of course. These processes, which obviously do not change the solution, are called *pivoting*. The equation or column to move into the active position may be chosen in such a way that the magnitude of the new diagonal element is the largest possible.

Performing only row interchanges, so that at the kth stage the equation with

$$\max_{i=k}^{n} |a_{ik}^{(k-1)}|$$

is moved into the kth row, is called *partial pivoting*. Performing both row interchanges and column interchanges, so that

$$\max_{i=k;j=k}^{n;n} |a_{ij}^{(k-1)}|$$

is moved into the kth diagonal position, is called *complete pivoting*.

An *elementary permutation matrix* can be used to interchange rows or columns in a matrix. An elementary permutation matrix that interchanges rows p and q in another matrix is the identity with the pth and q^{th} rows interchanged. It is denoted by E_{pq}. So E_{pq} is the identity, except the pth row is the qth unit vector e_q and the qth row is the pth unit vector e_p. Note that $E_{pq} = E_{qp}$. Thus, for example, if the given matrix is $4 \times m$, to interchange the second and third rows, we use

$$E_{23} = E_{32} = \begin{bmatrix} 1 & 0 & 0 & 0 \\ 0 & 0 & 1 & 0 \\ 0 & 1 & 0 & 0 \\ 0 & 0 & 0 & 1 \end{bmatrix}.$$

It is easy to see from the definition that an elementary permutation matrix is symmetric. Note that the notation E_{pq} does not indicate the size of the elementary permutation matrix; that must be specified in the context.

Premultiplying a matrix A by a (conformable) E_{pq} results in an interchange of the pth and q^{th} rows of A as we see above. Any permutation of rows of A can be accomplished by successive premultiplications by elementary permutation matrices. Note that the order of multiplication matters. Although a given permutation can be accomplished by different elementary permutations, the number of elementary permutations that effect a given permutation is always either even or odd; that is, if an odd number of elementary permutations results in a given permutation, any other sequence of elementary permutations to yield the given permutation is also odd in number. Any given permutation can be effected by successive interchanges of adjacent rows.

Postmultiplying a matrix A by a (conformable) E_{pq} results in an interchange of the pth and q^{th} columns of A:

$$\begin{bmatrix} a_{11} & a_{12} & a_{13} \\ a_{21} & a_{22} & a_{23} \\ a_{31} & a_{32} & a_{33} \\ a_{41} & a_{42} & a_{43} \end{bmatrix} \begin{bmatrix} 1 & 0 & 0 \\ 0 & 0 & 1 \\ 0 & 1 & 0 \end{bmatrix} = \begin{bmatrix} a_{11} & a_{13} & a_{12} \\ a_{21} & a_{23} & a_{22} \\ a_{31} & a_{33} & a_{32} \\ a_{41} & a_{43} & a_{42} \end{bmatrix}.$$

Note that

$$A = E_{pq}E_{pq}A = AE_{pq}E_{pq};$$

that is, as an operator, an elementary permutation matrix is its own inverse operator: $E_{pq}E_{pq} = I$.

Because all of the elements of a permutation matrix are 0 or 1, the trace of an $n \times n$ elementary permutation matrix is $n - 2$.

The product of elementary permutation matrices is also a *permutation matrix* in the sense that it permutes several rows or columns. For example, premultiplying A by the matrix $Q = E_{pq}E_{qr}$ will yield a matrix whose p^{th} row is the rth row of the original A, whose q^{th} row is the pth row of A, and whose r^{th} row is the qth row of A. We often use the notation E_{π} to denote a more general permutation matrix. This expression will usually be used generically, but sometimes we will specify the permutation, π.

A general permutation matrix (that is, a product of elementary permutation matrices) is not necessarily symmetric, but its transpose is also a permutation matrix. It is not necessarily its own inverse, but its permutations can be reversed by a permutation matrix formed by products of elementary permutation matrices in the opposite order; that is,

$$E_\pi^{\mathrm{T}} E_\pi = I.$$

In complete pivoting, we may permute both rows and columns, so we often have a representation such as

$$B = E_{\pi_1} A E_{\pi_2},$$

where E_{π_1} is a permutation matrix to permute the rows and E_{π_2} is a permutation matrix to permute the columns.

The pivoting in the simple example of equation (5.27) would be accomplished by multiplying both sides of the equation by the matrix

$$\begin{bmatrix} 0 & 1 \\ 1 & 0 \end{bmatrix}.$$

It is always important to distinguish descriptions of effects of actions from the actions that are actually carried out in the computer. Pivoting is interchanging rows or columns. In the computer, a row or a column is determined by the index identifying the row or column. All we do for pivoting is to keep track of the indices that we have permuted; we do not move data around in the computer's memory. This is another, trivial instance of the dictum:

The form of a mathematical expression and the way the expression should be evaluated in actual practice may be quite different.

There are many more computations required in order to perform complete pivoting than are required to perform partial pivoting. Gaussian elimination with complete pivoting can be shown to be stable; that is, the algorithm yields an exact solution to a slightly perturbed system, $(A + \delta A)x = b$. (We discuss stability on page 114.) For Gaussian elimination with partial pivoting, there are examples that show that it is not stable. These examples are somewhat contrived, however, and experience over many years has indicated that Gaussian elimination with partial pivoting is stable for most problems occurring in practice. For this reason, together with the computational savings, Gaussian elimination with partial pivoting is one of the most commonly used methods for solving linear systems.

Partial pivoting does not require as many computations as complete pivoting does, and there are modifications of partial pivoting that result in stable algorithms (see Gentle, 2007, page 210).

Nonfull Rank and Nonsquare Systems

The existence of an x that solves the linear system $Ax = b$ depends on that system being consistent; it does not depend on A being square or of full rank. The methods discussed above apply even if A is nonsquare or non-full rank.

In applications, it is often annoying that many software developers do not provide capabilities for handling nonfull-rank or nonsquare systems. Many of the standard programs for solving systems provide solutions only if A is square and of full rank. This is a poor software design decision.

5.3 Matrix Decompositions

In Chapter 1 we described two types of matrix factorization or decomposition, the singular value decomposition (SVD) in equation (1.63), and the square root factorization of positive definite matrices in equation (1.65). The use of decompositions in matrix computations has been listed as one of the top 10 algorithms of the twentieth century (see page 138).

The term "decomposition" could refer to an additive decomposition or a multiplicative decomposition, which we also call a factorization. The most important decompositions are factorizations, and when we use the term "decomposition" in regard to matrices, we will almost always mean "factorization".

We will now discuss some other types of factorization or decomposition of matrices, and then in Table 5.1 summarize important matrix factorizations.

Gaussian Elimination and the *LU* Decomposition

Generalizing the computations in equations (5.24), we perform elementary operations on the second through the nth equations to yield a set of equivalent equations in which all but the first have zero coefficients for x_1.

Next, we perform elementary operations using the second equation with the third through the nth equations, so that the new third through the nth equations have zero coefficients for x_2.

Let U denote the upper triangular matrix $E_{n,n-1}(c_n) \cdots E_{32}(c_2)E_{21}(c_1)A$, and L denote the inverse of the matrix $E_{n,n-1}(c_n) \cdots E_{32}(c_2)E_{21}(c_1)$, then we can write the last system as $Ux = L^{-1}b$.

This back substitution is equivalent to forming

$$x = U^{-1}L^{-1}b, \tag{5.28}$$

or $x = A^{-1}b$ with $A = LU$. The expression of A as LU is called the LU *decomposition* or the LU *factorization* of A. An LU factorization exists and is unique for nonnegative definite matrices. For more general matrices, the factorization may not exist, and the conditions for the existence are not so easy to state. (Golub and Van Loan (1996), for example, describe the conditions.)

QR Factorization

A very useful factorization is

$$A = QR, \tag{5.29}$$

where Q is orthogonal and R is upper triangular or trapezoidal. This is called the QR factorization.

If A is square and of full rank, R has the form

$$\begin{bmatrix} \mathtt{X} & \mathtt{X} & \mathtt{X} \\ 0 & \mathtt{X} & \mathtt{X} \\ 0 & 0 & \mathtt{X} \end{bmatrix}.$$

If A is nonsquare, R is nonsquare, with an upper triangular submatrix. If A has more columns than rows, R is trapezoidal and can be written as $[R_1 \,|\, R_2]$, where R_1 is upper triangular.

If A is $n \times m$ with more rows than columns, which is the case in common applications of QR factorization, then

$$R = \begin{bmatrix} R_1 \\ 0 \end{bmatrix}, \tag{5.30}$$

where R_1 is $m \times m$ upper triangular.

When A has more rows than columns, we can likewise partition Q as $[Q_1 \,|\, Q_2]$, and we can use a version of Q that contains only relevant rows or columns,

$$A = Q_1 R_1, \tag{5.31}$$

where Q_1 is an $n \times m$ matrix whose columns are orthonormal. This form is called a "skinny" QR. It is more commonly used than a full QR decomposition with a square Q.

It is interesting to note that the Moore-Penrose inverse of A with full column rank is immediately available from the QR factorization:

$$A^+ = \left[R_1^{-1} \; 0 \right] Q^{\mathrm{T}}. \tag{5.32}$$

Nonfull Rank Matrices

If A is square but not of full rank, R has the form

$$\begin{bmatrix} \mathtt{X} & \mathtt{X} & \mathtt{X} \\ 0 & \mathtt{X} & \mathtt{X} \\ 0 & 0 & 0 \end{bmatrix}. \tag{5.33}$$

In the common case in statistical applications in which A has more rows than columns, if A is not of full (column) rank, R_1 in equation (5.30) will have the form shown in matrix (5.33).

If A is not of full rank, we apply permutations to the columns of A by multiplying on the right by a permutation matrix. The permutations can be taken out by a second multiplication on the right. If A is of rank r ($\leq m$), the resulting decomposition consists of three matrices: an orthogonal Q, a T with an $r \times r$ upper triangular submatrix, and a permutation matrix E_π^{T},

$$A = QTE_\pi^{\mathrm{T}}. \tag{5.34}$$

The matrix T has the form

$$T = \begin{bmatrix} T_1 & T_2 \\ 0 & 0 \end{bmatrix}, \tag{5.35}$$

where T_1 is upper triangular and is $r \times r$. The decomposition in equation (5.34) is not unique because of the permutation matrix. The choice of the permutation matrix is the same as the pivoting that we discussed in connection with Gaussian elimination. A generalized inverse of A is immediately available from equation (5.34):

$$A^{-} = P \begin{bmatrix} T_1^{-1} & 0 \\ 0 & 0 \end{bmatrix} Q^{\mathrm{T}}, \tag{5.36}$$

where P is the permutation matrix E_π.

Additional orthogonal transformations can be applied from the right-hand side of the $n \times m$ matrix A in the form of equation (5.34) to yield

$$A = QRU^{\mathrm{T}}, \tag{5.37}$$

where R has the form

$$R = \begin{bmatrix} R_1 & 0 \\ 0 & 0 \end{bmatrix}, \tag{5.38}$$

where R_1 is $r \times r$ upper triangular, Q is $n \times n$ and as in equation (5.34), and U^{T} is $n \times m$ and orthogonal. (The permutation matrix in equation (5.34) is also orthogonal, of course.) The decomposition (5.37) is unique, and it provides the unique Moore-Penrose generalized inverse of A:

$$A^{+} = U \begin{bmatrix} R_1^{-1} & 0 \\ 0 & 0 \end{bmatrix} Q^{\mathrm{T}}. \tag{5.39}$$

(Compare equation (1.64) on page 29 relating the SVD to the Moore-Penrose inverse.)

It is often of interest to know the rank of a matrix. Given a decomposition of the form of equation (5.34), the rank is obvious, and in practice, this QR decomposition with pivoting is a good way to determine the rank of a matrix. The QR decomposition is said to be "rank-revealing". The computations are quite sensitive to rounding, however, and the pivoting must be done with some care.

The QR factorization is particularly useful in computations for overdetermined systems, and in other computations involving nonsquare matrices.

There are three good methods for obtaining the QR factorization: Householder transformations or reflections; Givens transformations or rotations; and the (modified) Gram-Schmidt procedure, all of which we discuss in Chapter 9. Different situations may make one of these procedures better than the two others. The Householder transformations described in the next section are probably the most commonly used. If the data are available only one row at a time, the Givens transformations are very convenient. Whichever method is used to compute the QR decomposition, at least $2n^3/3$ multiplications and additions are required. The operation count is therefore about twice as great as that for an LU decomposition.

Cholesky Factorization

If the matrix A is symmetric and *positive definite* (that is, if $x^{\mathrm{T}}Ax > 0$ for all $x \neq 0$), another important factorization is the *Cholesky* decomposition. In this factorization,

$$A = T^{\mathrm{T}}T, \tag{5.40}$$

where T is an upper triangular matrix with positive diagonal elements. We occasionally denote the Cholesky factor of A (that is, T in the expression above) as A_{C}.

The factor T in the Cholesky decomposition is sometimes called the *square root*, but we have defined a different matrix as the square root, $A^{\frac{1}{2}}$, on page 29. The Cholesky factor is more useful in practice, but the square root has more applications in the development of the theory.

A factor of the form of T in equation (5.32) is unique up to the sign, just as a square root is. To make the Cholesky factor unique, we require that the diagonal elements be positive. The elements along the diagonal of T will be square roots. Notice, for example, that t_{11} is $\sqrt{a_{11}}$.

Algorithm 5.1 is a method for constructing the Cholesky factorization.

Algorithm 5.1 Cholesky Factorization

1. Let $t_{11} = \sqrt{a_{11}}$.
2. For $j = 2, \ldots, n$, let $t_{1j} = a_{1j}/t_{11}$.
3. For $i = 2, \ldots, n$,

   {

   let $t_{ii} = \sqrt{a_{ii} - \sum_{k=1}^{i-1} t_{ki}^2}$, and
   for $j = i+1, \ldots, n$,

   {

   let $t_{ij} = (a_{ij} - \sum_{k=1}^{i-1} t_{ki}t_{kj})/t_{ii}$

   }

   }. ∎

It can be shown that the elements $a_{ii} - \sum_{k=1}^{i-1} t_{ki}^2$ in this algorithm are nonnegative if A is nonnegative definite. (See Gentle, 2007, page 194.)

There are other algorithms for computing the Cholesky decomposition. The method given in Algorithm 5.1 is sometimes called the inner product formulation because the sums in step 3 are inner products. The algorithms for computing the Cholesky decomposition are numerically stable. Although the order of the number of computations is the same, there are only about half as many computations in the Cholesky factorization as in the LU factorization. Another advantage of the Cholesky factorization is that there are only $n(n+1)/2$ unique elements as opposed to n^2+n in the LU decomposition.

The Cholesky decomposition can also be formed as $\widetilde{T}^{\mathrm{T}} D\widetilde{T}$, where D is a diagonal matrix that allows the diagonal elements of $\widetilde{T}$ to be computed without taking square roots.

The Cholesky decomposition also exists for a nonnegative definite matrix that is not of full rank. This is accomplished by a simple modification in Algorithm 5.1. For any t_{ii} that is zero, we merely fill the corresponding row of the matrix T with zeros and proceed.

Table 5.1. Matrix Factorizations

Factorization	Restrictions	Properties of Factors
SVD, page 28	none	U orthogonal
$A_{nm} = U_{nn}D_{nm}V_{mm}^{\mathrm{T}}$		V orthogonal
		D nonnegative diagonal
variations:	for symmetric A, $A = VCV^{\mathrm{T}}$	
LU, page 215	A square, (others)	L full-rank lower triangular
$A_{nn} = L_{nn}U_{nn}$		U upper triangular
variations:	with partial pivoting, $A = LUP$	
	with full pivoting, $P_1AP_2 = LU$	
	$A = LDU$, with D diagonal and $u_{ii} = 1$	
QR, page 216	none	Q orthogonal
$A_{nm} = Q_{nn}R_{nm}$		R upper triangular
variations:	skinny QR for $n > m$, $A = Q_1R_1$	
Cholesky, page 216	A nonnegative definite	L full-rank lower triangular
$A_{nn} = L_{nn}U_{nn}$		U upper triangular
diagonal, page 27	A symmetric	V orthogonal
$A_{nn} = V_{nn}C_{nn}V_{nn}^{\mathrm{T}}$		C diagonal
square root, page 29	A nonnegative definite	$A_{nn}^{\frac{1}{2}}$ nonnegative definite
$A_{nn} = (A_{nn}^{\frac{1}{2}})^2$		

"Modified" and "Classical" Gram-Schmidt Transformations

Pivoting, discussed on page 212, is a method for avoiding a situation like that in equation (5.3). In Gaussian elimination, for example, we do an addition,

$x+y$, where the y is the result of having divided some element of the matrix by some other element and x is some other element in the matrix. If the divisor is very small in magnitude, y is large and may overwhelm x as in equation (5.3).

Another example of how to avoid a situation similar to that in equation (5.1) is the use of the correct form of the Gram-Schmidt transformations, which we give in Algorithm 5.2 on page 220.

Given two nonnull, linearly independent vectors, x_1 and x_2, it is easy to form two orthonormal vectors, $\tilde{x}_1$ and $\tilde{x}_2$, that span the same space:

$$\begin{aligned}\tilde{x}_1 &= \frac{x_1}{\|x_1\|_2}, \\ \tilde{x}_2 &= \frac{(x_2 - \tilde{x}_1^{\mathrm{T}} x_2 \tilde{x}_1)}{\|x_2 - \tilde{x}_1^{\mathrm{T}} x_2 \tilde{x}_1\|_2}.\end{aligned} \tag{5.41}$$

These are called *Gram-Schmidt transformations.* It is easy to confirm by multiplication that $\tilde{x}_1$ and $\tilde{x}_2$ are orthonormal. Further, because they are orthogonal and neither is 0, they must be independent; hence, they span the same space as x_1 and x_2. We can see that they are independent also by observing that

$$[\tilde{x}_1 \tilde{x}_2] = A\,[x_1 x_2]\,,$$

where A is an upper triangular (that is, full rank) matrix.

The Gram-Schmidt transformations can be continued with all of the vectors in the linearly independent set. There are two straightforward ways equations (5.41) can be extended. One method generalizes the second equation in an obvious way:

for $k = 2, 3 \ldots,$

$$\tilde{x}_k = \left(x_k - \sum_{i=1}^{k-1} \langle \tilde{x}_i, x_k \rangle \tilde{x}_i \right) \Bigg/ \left\| x_k - \sum_{i=1}^{k-1} \langle \tilde{x}_i, x_k \rangle \tilde{x}_i \right\|. \tag{5.42}$$

In this method, at the kth step, we orthogonalize the kth vector by computing its residual with respect to the plane formed by all the previous $k - 1$ orthonormal vectors.

Another way of extending the transformations of equations (5.41) is, at the kth step, to compute the residuals of all remaining vectors with respect just to the kth normalized vector. We describe this method explicitly in Algorithm 5.2.

Algorithm 5.2 Gram-Schmidt Orthonormalization of a Set of Linearly Independent Vectors, $x_1, \ldots, x_m$

0. For $k = 1, \ldots, m$,

   {

   set $\tilde{x}_i = x_i$.

   }

1. Ensure that $\tilde{x}_1 \neq 0$;
   set $\tilde{x}_1 = \tilde{x}_1/\|\tilde{x}_1\|$.
2. If $m > 1$, for $k = 2, \ldots, m$,
   {
     for $j = k, \ldots, m$,
     {
       set $\tilde{x}_j = \tilde{x}_j - \langle \tilde{x}_{k-1}, \tilde{x}_j \rangle \tilde{x}_{k-1}$.
     }
     ensure that $\tilde{x}_k \neq 0$;
     set $\tilde{x}_k = \tilde{x}_k/\|\tilde{x}_k\|$.
   }

Although the method indicated in equation (5.42) is mathematically equivalent to this method, the use of Algorithm 5.2 is to be preferred for computations because it is less subject to rounding errors. (This may not be immediately obvious, although a simple numerical example can illustrate the fact — see Exercise 5.3c. We will not digress here to consider this further, but the difference in the two methods has to do with the relative magnitudes of the quantities in the subtraction. The method of Algorithm 5.2 is sometimes called the "modified" Gram-Schmidt method. We will discuss this method again on page 219.) This is an instance of an important principle:

The form of a mathematical expression and the way the expression should be evaluated in actual practice may be quite different.

5.4 Iterative Methods

As we mentioned earlier, we distinguish computational methods for matrices as being either *direct*, meaning that the number of computations is fixed a priori, or *iterative*, meaning that the results of the computations in each step determine whether to perform additional computations. The methods we have discussed so far in this chapter are direct. Iterative methods are especially useful in very large linear systems. They are also usually the favored methods for sparse systems.

Iterative methods are based on a sequence of approximations that (it is hoped) converge to the correct solution. The most important considerations in an iterative method involve its convergence, in terms of both speed and accuracy.

A fundamental trade-off in iterative methods is between the amount of work expended in getting a good approximation at each step and the number of steps required for convergence.

The Gauss-Seidel Method with Successive Overrelaxation

One of the simplest iterative procedures for solving a system of linear equations is the *Gauss-Seidel method.* In this method, we begin with an initial approximation to the solution, $x^{(0)}$. We then compute an update for the first element of x:

$$x_1^{(1)} = \frac{1}{a_{11}} \left(b_1 - \sum_{j=2}^{n} a_{1j} x_j^{(0)} \right).$$

Continuing in this way for the other elements of x for $i = 2, \ldots, n$, we get the next approximation to the solution, $x^{(1)}$. After getting the approximation $x^{(1)}$, we then continue this same kind of iteration for $x^{(2)}, x^{(3)}, \ldots$, in which we compute the ith element as

$$x_i^{(k)} = \frac{1}{a_{ii}} \left(b_i - \sum_{j=1}^{i-1} a_{ij} x_j^{(k)} - \sum_{j=i+1}^{n} a_{ij} x_j^{(k-1)} \right), \tag{5.43}$$

where no sums are performed if the upper limit is smaller than the lower limit.

We continue the iterations until a convergence criterion is satisfied. This criterion may be of the form

$$\Delta \left(x^{(k)}, x^{(k-1)} \right) \le \epsilon,$$

where $\Delta \left(x^{(k)}, x^{(k-1)} \right)$ is a measure of the difference of $x^{(k)}$ and $x^{(k-1)}$, such as $\|x^{(k)} - x^{(k-1)}\|$. We may also base the convergence criterion on $\|r^{(k)} - r^{(k-1)}\|$, where $r^{(k)} = b - Ax^{(k)}$.

The Gauss-Seidel iterations can be thought of as beginning with a re-arrangement of the original system of equations as

$$\begin{array}{lllllll}
a_{11}x_1 & & & & = & b_1 & - a_{12}x_2 \cdots - a_{1n}x_n \\
a_{21}x_1 & + & a_{22}x_2 & & = & b_2 & \cdots - a_{2n}x_n \\
\vdots & + & \vdots & \vdots & & \vdots & \\
a_{(n-1)1}x_1 & + & a_{(n-1)2}x_2 & + \cdots & = & b_{n-1} & - a_{nn}x_n \\
a_{n1}x_1 & + & a_{n2}x_2 & + \cdots + a_{nn}x_n & = & b_n. &
\end{array}$$

In this form, we identify three matrices: a diagonal matrix D, a lower triangular L with 0s on the diagonal, and an upper triangular U with 0s on the diagonal:

$$(D + L)x = b - Ux.$$

We can write this entire sequence of Gauss-Seidel iterations in terms of these three fixed matrices:

$$x^{(k+1)} = (D + L)^{-1}(-Ux^{(k)} + b). \tag{5.44}$$

This method will converge for any arbitrary starting value $x^{(0)}$ if and only if the spectral radius of $(D + L)^{-1}U$ is less than 1. (See Golub and Van Loan, 1996, for a proof of this.) Moreover, the rate of convergence increases with decreasing spectral radius.

Successive Overrelaxation

The Gauss-Seidel method may be unacceptably slow, so it may be modified so that the update is a weighted average of the regular Gauss-Seidel update and the previous value. This kind of modification is called *successive overrelaxation*, or *SOR*. Instead of equation (5.44), the update is given by

$$\frac{1}{\omega}(D+L)\,x^{(k+1)} = \frac{1}{\omega}\big((1-\omega)D - \omega U\big)x^{(k)} + b, \tag{5.45}$$

where the relaxation parameter ω is usually chosen to be between 0 and 1. For $\omega = 1$ the method is the ordinary Gauss-Seidel method; see Exercises 5.2c, 5.2e, and 5.2f.

Conjugate Gradient Methods for Symmetric Positive Definite Systems

In the Gauss-Seidel methods the convergence criterion is based on successive differences in the solutions $x^{(k)}$ and $x^{(k-1)}$ or in the residuals $r^{(k)}$ and $r^{(k-1)}$. Other iterative methods focus directly on the magnitude of the residual

$$r^{(k)} = b - Ax^{(k)}. \tag{5.46}$$

We seek a value $x^{(k)}$ such that the residual is small (in some sense). Methods that minimize $\|r^{(k)}\|_2$ are called minimal residual (MINRES) methods or generalized minimal residual (GMRES) methods.

For a system with a symmetric positive definite coefficient matrix A, it turns out that the best iterative method is based on minimizing the "conjugate" L_2 norm

$$\|r^{(k)\mathrm{T}}A^{-1}r^{(k)}\|_2.$$

A method based on this minimization problem is called a *conjugate gradient method.*

The problem of solving the linear system $Ax = b$ is equivalent to finding the minimum of the function

$$f(x) = \frac{1}{2}x^{\mathrm{T}}Ax - x^{\mathrm{T}}b. \tag{5.47}$$

By setting the derivative of f to 0, we see that a stationary point of f occurs at the point x where $Ax = b$.

If A is positive definite, the (unique) minimum of f is at $x = A^{-1}b$, and the value of f at the minimum is $-\frac{1}{2}b^{\mathrm{T}}Ab$. The minimum point can be approached iteratively by starting at a point $x^{(0)}$, moving to a point $x^{(1)}$ that yields a smaller value of the function, and continuing to move to points yielding smaller values of the function. The kth point is $x^{(k-1)} + \alpha^{(k-1)}p^{(k-1)}$, where $\alpha^{(k-1)}$ is a scalar and $p^{(k-1)}$ is a vector giving the direction of the movement. Hence, for the kth point, we have the linear combination

$$x^{(k)} = x^{(0)} + \alpha^{(1)}p^{(1)} + \cdots + \alpha^{(k-1)}p^{(k-1)}.$$

At the point $x^{(k)}$, the function f decreases most rapidly in the direction of the negative gradient, $-\nabla f(x^{(k)})$, which is just the residual,

$$-\nabla f(x^{(k)}) = r^{(k)}.$$

If this residual is 0, no movement is indicated because we are at the solution.

Moving in the direction of steepest descent may cause a very slow convergence to the minimum. (The curve that leads to the minimum on the quadratic surface is obviously not a straight line. The direction of steepest descent changes as we move to a new point $x^{(k+1)}$.) A good choice for the sequence of directions $p^{(1)}, p^{(2)}, \ldots$ is such that

$$(p^{(k)})^{\mathrm{T}} A p^{(i)} = 0, \quad \text{for } i = 1, \ldots, k-1. \tag{5.48}$$

Such a vector $p^{(k)}$ is said to be A-conjugate to $p^{(1)}, p^{(2)}, \ldots p^{(k-1)}$. Given a current point $x^{(k)}$ and a direction to move $p^{(k)}$ to the next point, we must also choose a distance $\alpha^{(k)}\|p^{(k)}\|$ to move in that direction. We then have the next point,

$$x^{(k+1)} = x^{(k)} + \alpha^{(k)}p^{(k)}. \tag{5.49}$$

(Notice that here, as often in describing algorithms in linear algebra, we use Greek letters, such as α, to denote scalar quantities.)

A conjugate gradient method for solving the linear system is shown in Algorithm 5.3. The paths defined by the directions $p^{(1)}, p^{(2)}, \ldots$ in equation (5.48) are called the conjugate gradients.

Algorithm 5.3 The Conjugate Gradient Method for Solving the Symmetric Positive Definite System $Ax = b$, Starting with $x^{(0)}$

0. Input stopping criteria, ϵ and k_{max}.
   Set $k = 0$; $r^{(k)} = b - Ax^{(k)}$; $s^{(k)} = Ar^{(k)}$; $p^{(k)} = s^{(k)}$; and $\gamma^{(k)} = \|s^{(k)}\|^2$.
1. If $\gamma^{(k)} \leq \epsilon$, set $x = x^{(k)}$ and terminate.
2. Set $q^{(k)} = Ap^{(k)}$.
3. Set $\alpha^{(k)} = \frac{\gamma^{(k)}}{\|q^{(k)}\|^2}$.
4. Set $x^{(k+1)} = x^{(k)} + \alpha^{(k)}p^{(k)}$.
5. Set $r^{(k+1)} = r^{(k)} - \alpha^{(k)}q^{(k)}$.
6. Set $s^{(k+1)} = Ar^{(k+1)}$.
7. Set $\gamma^{(k+1)} = \|s^{(k+1)}\|^2$.
8. Set $p^{(k+1)} = s^{(k+1)} + \frac{\gamma^{(k+1)}}{\gamma^{(k)}}p^{(k)}$.
9. If $k < k_{\mathrm{max}}$,
   set $k = k + 1$ and go to step 1;
   otherwise
   issue message that
   "algorithm did not converge in k_{max} iterations". ∎

This algorithm is a simple example of Newton's method (see page 249), which specifies the directions of the steps as $Ar^{(k)}$, where $r^{(k)}$ is the residual, $b - Ax^{(k)}$, at the kth step.

There are various ways in which the computations in Algorithm 5.3 could be arranged. Although any vector norm could be used in Algorithm 5.3, the L_2 norm is the most common one.

This method, like other iterative methods, is more appropriate for large systems. ("Large" in this context means bigger than 1000×1000.)

In exact arithmetic, the conjugate gradient method should converge in n steps for an $n \times n$ system. In practice, however, its convergence rate varies widely, even for systems of the same size. Its convergence rate generally decreases with increasing L_2 condition number (which is a function of the maximum and minimum nonzero eigenvalues), but that is not at all the complete story. The rate depends in a complicated way on all of the eigenvalues. The more spread out the eigenvalues are, the slower the rate. For different systems with roughly the same condition number, the convergence is faster if all eigenvalues are in two clusters around the maximum and minimum values.

Preconditioning

In order to achieve acceptable rates of convergence for iterative algorithms, it is often necessary to precondition the system; that is, to replace the system $Ax = b$ by the system

$$M^{-1}Ax = M^{-1}b$$

for some suitable matrix M. The choice of M involves some art, and we will not consider the issues further here.

Restarting and Rescaling

In many iterative methods, not all components of the computations are updated in each iteration. As we mentioned in Chapter 3, there is sometimes a tradeoff between the number of iterations required for convergence and the amount of work done in each iteration.

An approximation to a given matrix or vector may be adequate during some sequence of computations without change, but then at some point the approximation is no longer close enough, and a new approximation must be computed. An example of this is in the use of quasi-Newton methods in optimization in which an approximate Hessian is updated (see Chapter 6). We may, for example, just compute an approximation to the Hessian every few iterations, perhaps using second differences, and then use that approximate matrix for a few subsequent iterations.

Preservation of Sparsity

In computations involving large sparse systems, we may want to preserve the sparsity, even if that requires using approximations. Fill-in (when a zero position in a sparse matrix becomes nonzero) would cause loss of the computational and storage efficiencies of software for sparse matrices.

In forming a preconditioner for a sparse matrix A, for example, we may choose a matrix $M = \widetilde{L}\widetilde{U}$, where $\widetilde{L}$ and $\widetilde{U}$ are approximations to the matrices in an LU decomposition of A. These matrices are constructed so as to have zeros everywhere A has, and $A \approx \widetilde{L}\widetilde{U}$. This is called incomplete factorization, and often, instead of an exact factorization, an approximate factorization may be more useful because of computational efficiency.

Iterative Refinement

Once an approximate solution $x^{(0)}$ to the linear system $Ax = b$ is available, iterative refinement can yield a solution that is closer to the true solution. The residual

$$r = b - Ax^{(0)}$$

is used for iterative refinement. Clearly, if $h = A^{+}r$, then $x^{(0)} + h$ is a solution to the original system.

The problem considered here is not just an iterative solution to the linear system discussed above. Here, we assume $x^{(0)}$ was computed accurately given the finite precision of the computer. In this case, it is likely that r cannot be computed accurately enough to be of any help. If, however, r can be computed using a higher precision, then a useful value of h can be computed. This process can then be iterated as shown in Algorithm 5.4.

Algorithm 5.4 Iterative Refinement of the Solution to $Ax = b$, Starting with $x^{(0)}$

0. Input stopping criteria, ϵ and $k_{\max}$.
   Set $k = 0$.
1. Compute $r^{(k)} = b - Ax^{(k)}$ in higher precision.
2. Compute $h^{(k)} = A^{+}r^{(k)}$.
3. Set $x^{(k+1)} = x^{(k)} + h^{(k)}$.
4. If $\|h^{(k)}\| \leq \epsilon \|x^{(k+1)}\|$, then
   set $x = x^{(k+1)}$ and terminate; otherwise,
   if $k < k_{\max}$,
      set $k = k + 1$ and go to step 1;
   otherwise,
      issue message that
      "algorithm did not converge in $k_{\max}$ iterations". ∎

In step 2, if A is of full rank then A^{+} is A^{-1}. Also, as we have emphasized already, the fact that we write *an expression such as* $A^{+}r$ *does not mean that we compute* A^+. The norm in step 4 is usually chosen to be the ∞ norm. The algorithm may not converge, so it is necessary to have an alternative exit criterion, such as a maximum number of iterations.

The use of iterative refinement as a general-purpose method is severely limited by the need for higher precision in step 1. On the other hand, if computations in higher precision can be performed, they can be applied to step 2—or just in the original computations for $x^{(0)}$. In terms of both accuracy and computational efficiency, using higher precision throughout is usually better.

5.5 Updating a Solution to a Consistent System

In applications of linear systems, it is often the case that after the system $Ax = b$ has been solved, the right-hand side is changed and the system $Ax = c$ must be solved. If the linear system $Ax = b$ has been solved by a direct method using one of the factorizations discussed above, the factors of A can be used to solve the new system $Ax = c$. If the right-hand side is a small perturbation of b, say $c = b + \delta b$, an iterative method can be used to solve the new system quickly, starting from the solution to the original problem.

If the coefficient matrix in a linear system $Ax = b$ is perturbed to result in the system $(A + \delta A)x = b$, it may be possible to use the solution x_0 to the original system efficiently to arrive at the solution to the perturbed system. One way, of course, is to use x_0 as the starting point in an iterative procedure. Often, in applications, the perturbations are of a special type, such as

$$\widetilde{A} = A - uv^{\mathrm{T}},$$

where u and v are vectors. (This is a "rank-one" perturbation of A, and when the perturbed matrix is used as a transformation, it is called a "rank-one" update. As we have seen, a Householder reflection is a special rank-one update.) Assuming A is an $n \times n$ matrix of full rank, it is easy to write $\widetilde{A}^{-1}$ in terms of A^{-1}:

$$\widetilde{A}^{-1} = A^{-1} + \alpha(A^{-1}u)(v^{\mathrm{T}}A^{-1}) \tag{5.50}$$

with

$$\alpha = \frac{1}{1 - v^{\mathrm{T}}A^{-1}u}.$$

These are called the Sherman-Morrison formulas. $\widetilde{A}^{-1}$ exists so long as $v^{\mathrm{T}}A^{-1}u \neq 1$. Because $x_0 = A^{-1}b$, the solution to the perturbed system is

$$\tilde{x}_0 = x_0 + \frac{(A^{-1}u)(v^{\mathrm{T}}x_0)}{(1 - v^{\mathrm{T}}A^{-1}u)}.$$

If the perturbation is more than rank one (that is, if the perturbation is

$$\widetilde{A} = A - UV^{\mathrm{T}}, \tag{5.51}$$

where U and V are $n \times m$ matrices with $n \geq m$), a generalization of the Sherman-Morrison formula, sometimes called the Woodbury formula, is

$$\widetilde{A}^{-1} = A^{-1} + A^{-1}U(I_m - V^{\mathrm{T}}A^{-1}U)^{-1}V^{\mathrm{T}}A^{-1}. \tag{5.52}$$

The solution to the perturbed system is easily seen to be

$$\tilde{x}_0 = x_0 + A^{-1}U(I_m - V^{\mathrm{T}}A^{-1}U)^{-1}V^{\mathrm{T}}x_0.$$

As we have emphasized many times, we rarely compute the inverse of a matrix, and so the Sherman-Morrison-Woodbury formulas are not used directly. Having already solved $Ax = b$, it should be easy to solve another system, say $Ay = u_i$, where u_i is a column of U. If m is relatively small, as it is in most applications of this kind of update, there are not many systems $Ay = u_i$ to solve. Solving these systems, of course, yields $A^{-1}U$, the most formidable component of the Sherman-Morrison-Woodbury formula. The system to solve is of order m also.

Occasionally the updating matrices in equation (5.51) may be used with a weighting matrix, so we have $\widetilde{A} = A - UWV^{\mathrm{T}}$. An extension of the Sherman-Morrison-Woodbury formula is

$$(A - UWV^{\mathrm{T}})^{-1} = A^{-1} + A^{-1}U(W^{-1} - V^{\mathrm{T}}A^{-1}U)^{-1}V^{\mathrm{T}}A^{-1}. \tag{5.53}$$

This is sometimes called the Hemes formula.

Another situation that requires an update of a solution occurs when the system is augmented with additional equations and more variables:

$$\begin{bmatrix} A & A_{12} \\ A_{21} & A_{22} \end{bmatrix} \begin{bmatrix} x \\ x_+ \end{bmatrix} = \begin{bmatrix} b \\ b_+ \end{bmatrix}.$$

A simple way of obtaining the solution to the augmented system is to use the solution x_0 to the original system in an iterative method. The starting point for a method based on Gauss-Seidel or a conjugate gradient method can be taken as $(x_0, 0)$, or as $(x_0, x_+^{(0)})$ if a better value of $x_+^{(0)}$ is known.

In many statistical applications, the systems are overdetermined, with A being $n \times m$ and $n > m$. In the next section, we consider the general problem of solving overdetermined systems by using least squares, and then we discuss updating a least squares solution to an overdetermined system.

5.6 Overdetermined Systems; Least Squares

Linear models are often used to express a relationship between one observable variable, a "response", and another group of observable variables, "predictor variables". Consider a simple linear model in an equation of the form $y = b_0 + b^{\mathrm{T}}x$. The model is unlikely to fit exactly any set of observed values of responses and predictor variables. This may be due to effects of other predictor variables that are not included in the model, measurement error, the relationship among the variables being nonlinear, or some inherent randomness in the system.

In such applications, we generally take a larger number of observations than there are variables in the system; thus, with each set of observations on the response and associated predictors making up one equation, we have a system with more equations than variables. This results in an overdetermined system of linear equations, so instead of having the canonical problem of equation (5.4), we have a situation that cannot fit a simple linear equations relating the y and x.

An overdetermined system may be written as

$$Xb \approx y, \tag{5.54}$$

where X is $n \times m$ and $\text{rank}(X|y) > m$; that is, the system is not consistent. We have changed the notation slightly from the consistent system (5.4) $Ax = b$ that we have been using because now we have in mind statistical applications, and in those the notation $y \approx X\beta$ is more common. The problem is to determine a value of b that makes the approximation close in some sense. In applications of linear systems, we refer to this as "fitting" the system, which is referred to as a "model".

Overdetermined systems arise frequently in fitting equations to data. The usual linear regression model is an overdetermined system and we discuss statistical regression problems further in Chapter 17.

We should not confuse *statistical inference* with *fitting equations to data*, although the latter task is a component of the former activity, but in this section, we consider some of the more mechanical and computational aspects of the problem.

Least Squares Solution of an Overdetermined System

Although there may be no b that will make the system in (5.54) an equation, the system can be written as the equation

$$Xb = y - r, \tag{5.55}$$

where r is an n-vector of possibly arbitrary residuals or "errors".

A *least squares* solution $\widehat{b}$ to the system in (5.54) is one such that the Euclidean norm of the vector of residuals is minimized; that is, the solution to the problem

$$\min_b \|y - Xb\|_2. \tag{5.56}$$

The least squares solution is also called the "ordinary least squares" (OLS) fit.

By rewriting the square of this norm as

$$(y - Xb)^{\mathrm{T}}(y - Xb), \tag{5.57}$$

differentiating, and setting it equal to 0, we see that the minimum (of both the norm and its square) occurs at the $\widehat{b}$ that satisfies the square system

$$X^{\mathrm{T}}X\widehat{b} = X^{\mathrm{T}}y. \tag{5.58}$$

The system (5.58) is called the *normal equations.* As we mentioned on page 208, the condition number of $X^{\mathrm{T}}X$ is the square of the condition number of X. Because of this, it may be better to work directly on X in (5.54) rather than to use the normal equations. The normal equations are useful expressions, however, whether or not they are used in the computations. This is another case where *a formula does not define an algorithm.* We should note, of course, that any information about the stability of the problem that the Gramian may provide can be obtained from X directly.

Special Properties of Least Squares Solutions

The least squares fit to the overdetermined system has a very useful property with two important consequences. The least squares fit partitions the space into two interpretable orthogonal spaces. As we see from equation (5.58), the residual vector $y - X\widehat{b}$ is orthogonal to each column in X:

$$X^{\mathrm{T}}(y - X\widehat{b}) = 0. \tag{5.59}$$

A consequence of this fact for models that include an intercept is that the sum of the residuals is 0. (The residual vector is orthogonal to the 1 vector.) Another consequence for models that include an intercept is that the least squares solution provides an exact fit to the mean.

These properties are so familiar to statisticians that some think that these facts are essential characteristics of any regression modeling; they are not. We will see in later sections that they do not hold for other approaches to fitting the basic model $y \approx Xb$. The least squares solution, however, has some desirable statistical properties under fairly common distributional assumptions. We discuss statistical aspects of least squares solutions in Chapter 17, beginning on page 604.

Weighted Least Squares

One of the simplest variations on fitting the linear model $Xb \approx y$ is to allow different weights on the observations; that is, instead of each row of X and corresponding element of y contributing equally to the fit, the elements of X and y are possibly weighted differently.

The relative weights can be put into an n-vector w and the squared norm in equation (5.57) replaced by a quadratic form in $\mathrm{diag}(w)$. More generally, we form the quadratic form as

$$(y - Xb)^{\mathrm{T}} W (y - Xb), \tag{5.60}$$

where W is a positive definite matrix. Because the weights apply to both y and Xb, there is no essential difference in the weighted or unweighted versions of the problem.

The use of the QR factorization for the overdetermined system in which the weighted norm (5.60) is to be minimized is similar to the development above. It is exactly what we get if we replace $y - Xb$ in equation (5.61) by $W_{\mathrm{C}}(y - Xb)$, where W_{C} is the Cholesky factor of W.

There are other variations on ordinary least squares for fitting the linear model, and we will discuss some of them in Section 17.3.

We now continue to address some of the computational issues of least squares.

Least Squares with a Full Rank Coefficient Matrix

If the $n \times m$ matrix X is of full column rank, the least squares solution, from equation (5.58), is $\widehat{b} = (X^{\mathrm{T}}X)^{-1}X^{\mathrm{T}}y$ and is obviously unique. A good way to compute this is to form the QR factorization of X.

First we write $X = QR$, as in equation (5.29) on page 216, where R is as in equation (5.30),

$$R = \begin{bmatrix} R_1 \\ 0 \end{bmatrix},$$

with R_1 an $m \times m$ upper triangular matrix. The residual norm (5.57) can be written as

$$\begin{aligned}(y - Xb)^{\mathrm{T}}(y - Xb) &= (y - QRb)^{\mathrm{T}}(y - QRb) \\ &= (Q^{\mathrm{T}}y - Rb)^{\mathrm{T}}(Q^{\mathrm{T}}y - Rb) \\ &= (c_1 - R_1 b)^{\mathrm{T}}(c_1 - R_1 b) + c_2^{\mathrm{T}} c_2,\end{aligned} \tag{5.61}$$

where c_1 is a vector with m elements and c_2 is a vector with $n - m$ elements, such that

$$Q^{\mathrm{T}}y = \begin{pmatrix} c_1 \\ c_2 \end{pmatrix}. \tag{5.62}$$

Because quadratic forms are nonnegative, the minimum of the residual norm in equation (5.61) occurs when $(c_1 - R_1 b)^{\mathrm{T}}(c_1 - R_1 b) = 0$; that is, when $(c_1 - R_1 b) = 0$, or

$$R_1 b = c_1. \tag{5.63}$$

We could also use the same technique of differentiation to find the minimum of equation (5.61) that we did to find the minimum of equation (5.57).

Because R_1 is triangular, the system is easy to solve: $\widehat{b} = R_1^{-1}c_1$. From equation (5.32), we have

$$X^{+} = \left[\, R_1^{-1} \; 0 \,\right],$$

and so we have

$$\widehat{b} = X^{+}y. \tag{5.64}$$

We also see from equation (5.61) that the minimum of the residual norm is $c_2^{\mathrm{T}} c_2$. This is called the *residual sum of squares* in the least squares fit.

Least Squares with a Coefficient Matrix Not of Full Rank

If X is not of full rank (that is, if X has rank $r < m$), the least squares solution is not unique, and in fact a solution is any vector $\widehat{b} = (X^{\mathrm{T}}X)^{-}X^{\mathrm{T}}y$, where $(X^{\mathrm{T}}X)^{-}$ is any generalized inverse. This is a solution to the normal equations (5.58). The residual corresponding to this solution is

$$y - X(X^{\mathrm{T}}X)^{-}X^{\mathrm{T}}y = (I - X(X^{\mathrm{T}}X)^{-}X^{\mathrm{T}})y.$$

The residual vector is invariant to the choice of generalized inverse, as we see from equation (1.54) on page 26.

An Optimal Property of the Solution Using the Moore-Penrose Inverse

The solution corresponding to the Moore-Penrose inverse is unique because, as we have seen, that generalized inverse is unique. That solution is interesting for another reason, however: the b from the Moore-Penrose inverse has the minimum L_2-norm of all solutions.

To see that this solution has minimum norm, first factor X, as in equation (5.37) on page 217,

$$X = QRU^{\mathrm{T}},$$

and form the Moore-Penrose inverse as in equation (5.39):

$$X^{+} = U \begin{bmatrix} R_1^{-1} & 0 \\ 0 & 0 \end{bmatrix} Q^{\mathrm{T}}.$$

Then

$$\widehat{b} = X^{+}y \tag{5.65}$$

is a least squares solution, just as in the full rank case. Now, let

$$Q^{\mathrm{T}}y = \begin{pmatrix} c_1 \\ c_2 \end{pmatrix},$$

as in equation (5.62), except ensure that c_1 has exactly r elements and c_2 has $n - r$ elements, and let

$$U^{\mathrm{T}}b = \begin{pmatrix} z_1 \\ z_2 \end{pmatrix},$$

where z_1 has r elements. We proceed as in the equations (5.61). We seek to minimize $\|y - Xb\|_2$ (which is the square root of the expression in equations (5.61)); and because multiplication by an orthogonal matrix does not change the norm, we have

$$\begin{aligned} \|y - Xb\|_2 &= \|Q^{\mathrm{T}}(y - XUU^{\mathrm{T}}b)\|_2 \\ &= \left\| \begin{pmatrix} c_1 \\ c_2 \end{pmatrix} - \begin{bmatrix} R_1 & 0 \\ 0 & 0 \end{bmatrix} \begin{pmatrix} z_1 \\ z_2 \end{pmatrix} \right\|_2 \\ &= \left\| \begin{pmatrix} c_1 - R_1 z_1 \\ c_2 \end{pmatrix} \right\|_2 . \end{aligned} \tag{5.66}$$

The residual norm is minimized for $z_1 = R_1^{-1}c_1$ and z_2 arbitrary. However, if $z_2 = 0$, then $\|z\|_2$ is also minimized. Because $U^{\mathrm{T}}b = z$ and U is orthogonal, $\|\widehat{b}\|_2 = \|z\|_2$, and so $\|\widehat{b}\|_2$ is the minimum among all least squares solutions.

Updating a Least Squares Solution of an Overdetermined System

In regression applications, after fitting the linear model, we may obtain additional observations. Alternatively, we may decide to include more predictor variables in the model. The original overdetermined system is modified by adding either some rows or some columns to the coefficient matrix X. This corresponds to including additional equations in the system,

$$\begin{bmatrix} X \\ X_+ \end{bmatrix} b \approx \begin{bmatrix} y \\ y_+ \end{bmatrix},$$

or to adding variables,

$$\begin{bmatrix} X & X_+ \end{bmatrix} \begin{bmatrix} b \\ b_+ \end{bmatrix} \approx y.$$

In either case, if the QR decomposition of X is available, the decomposition of the augmented system can be computed readily. Consider, for example, the addition of k equations to the original system $Xb \approx y$, which has n approximate equations. With the QR decomposition, for the original full rank system, putting $Q^{\mathrm{T}}X$ and $Q^{\mathrm{T}}y$ as partitions in a matrix, we have

$$\begin{bmatrix} R_1 & c_1 \\ 0 & c_2 \end{bmatrix} = Q^{\mathrm{T}} \begin{bmatrix} X & y \end{bmatrix}.$$

Augmenting this with the additional rows yields

$$\begin{bmatrix} R & c_1 \\ 0 & c_2 \\ X_+ & y_+ \end{bmatrix} = \begin{bmatrix} Q^{\mathrm{T}} & 0 \\ 0 & I \end{bmatrix} \begin{bmatrix} X & y \\ X_+ & y_+ \end{bmatrix}. \tag{5.67}$$

All that is required now is to apply orthogonal transformations, such as Givens rotations, to the system (5.67) to produce

$$\begin{bmatrix} R_* & c_{1*} \\ 0 & c_{2*} \end{bmatrix},$$

where R_* is an $m \times m$ upper triangular matrix and c_{1*} is an m-vector as before but c_{2*} is an $(n - m + k)$-vector.

The updating is accomplished by applying m rotations to system (5.67) so as to zero out the $(n + q)^{\text{th}}$ row for $q = 1, 2, \ldots, k$. These operations go through an outer loop with $p = 1, 2, \ldots, n$ and an inner loop with $q = 1, 2, \ldots, k$. The operations rotate R through a sequence $R^{(p,q)}$ into R_*, and they rotate X_+ through a sequence $X_+^{(p,q)}$ into 0. We consider these rotations further on page 377. As we see there, at the p, q step, the rotation matrix Q_{pq} corresponding to equation (9.4) has

$$\cos\theta = \frac{R_{pp}^{(p,q)}}{r}$$

and

$$\sin\theta = \frac{\left(X_+^{(p,q)}\right)_{qp}}{r},$$

where

$$r = \sqrt{\left(R_{pp}^{(p,q)}\right)^2 + \left(\left(X_+^{(p,q)}\right)_{qp}\right)^2}.$$

Other Solutions of Overdetermined Systems

A solution to an inconsistent, overdetermined system

$$Xb \approx y,$$

where X is $n \times m$ and $\text{rank}(X|y) > m$, is some value b that makes $y - Xb$ close to zero. We define "close to zero" in terms of a norm on $y - Xb$. The most common norm, of course, is the L_2 norm as in expression (5.56), and the minimization of this norm is straightforward, as we have seen. In addition to the simple analytic properties of the L_2 norm, the least squares solution has some desirable statistical properties under fairly common distributional assumptions, as we have seen.

There are various norms that may provide a reasonable fit. In addition to the use of the L_2 norm, that is, an ordinary least squares (OLS) fit, there are various other ways of approaching the problem. We will return to this topic, and consider variations on least squares as well as use of other norms in fitting a linear model in Section 17.3.

As we have stated before, *we should not confuse statistical inference with fitting equations to data, although the latter task is a component of the former activity.* In statistical applications, we need to make statements (that is, assumptions) about relevant probability distributions. These probability distributions, together with the methods used to collect the data, may indicate specific methods for fitting the equations to the given data.

5.7 Other Computations with Matrices

There are several other kinds of computational problems that we will not address in this book. One important example is the extraction of eigenvalues. The most common method for this problem is the QR method, which was selected as one of the Top 10 algorithms of the twentieth century (see page 138). The QR method for computing eigenvalues is described in Gentle (2007), Section 7.4.

In the next section we briefly discuss the problem of determining a reduced-rank matrix that approximates a given matrix. We close with a section on the use of consistency checks, as discussed on page 112, for the specific problem of solving a linear system.

Matrix Approximation

We may wish to approximate the matrix A with a matrix A_r of rank $r \leq \text{rank}(A)$. The singular value decomposition provides an easy way to do this,

$$A_r = UD_rV^{\mathrm{T}},$$

where D_r is the same as D, except with zeros replacing all but the r largest singular values. It can be shown that A_r is the rank r matrix closest to A as measured by the Frobenius norm,

$$\|A - A_r\|_{\mathrm{F}},$$

(see Gentle, 2007). This kind of matrix approximation is the basis for dimension reduction by principal components. We discuss principal components in Chapter 16.

Consistency Checks for Identifying Numerical Errors

In real-life applications, the correct solution is not known, but we would still like to have some way of assessing the accuracy using the data themselves. Sometimes a convenient way to do this in a given problem is to perform internal consistency tests. An internal consistency test may be an assessment of the agreement of various parts of the output. Relationships among the output are exploited to ensure that the individually computed quantities satisfy these relationships. Other internal consistency tests may be performed by comparing the results of the solutions of two problems with a known relationship.

The solution to the linear system $Ax = b$ has a simple relationship to the solution to the linear system $Ax = b + ca_j$, where a_j is the jth column of A and c is a constant. A useful check on the accuracy of a computed solution to $Ax = b$ is to compare it with a computed solution to the modified system. Of course, if the expected relationship does not hold, we do not know which solution is incorrect, but it is probably not a good idea to trust either. If the expected relationships do not obtain, the analyst has strong reason to doubt the accuracy of the computations.

Another simple modification of the problem of solving a linear system with a known exact effect is the permutation of the rows or columns. Although this perturbation of the problem does not change the solution, it does sometimes result in a change in the computations, and hence it may result in a different computed solution. This obviously would alert the user to problems in the computations.

Another simple internal consistency test that is applicable to many problems is to use two levels of precision in the computations. In using this test, one must be careful to make sure that the input data are the same. Rounding of the input data may cause incorrect output to result, but that is not the fault of the computational algorithm.

Internal consistency tests cannot confirm that the results are correct; they can only give an indication that the results are incorrect.

Notes and Further Reading

More complete coverage of the computational issues in linear algebra are covered in Čížková and Čížek (2004), in Part III of Gentle (2007), and in Golub and Van Loan (1996).

Computational methods for sparse matrices are discussed in some detail in Saad (2003).

Software for Numerical Linear Algebra

Mathematical software for linear algebra has traditionally been some of the best software, from the earlier days when libraries in various programming languages were widely distributed to the interpretive systems that allowed direct manipulation of vectors and matrices. Currently, there are several relatively mature interactive systems, including Matlab and Octave from an applied mathematics heritage, and S-Plus and R that emphasize statistical applications. There continues to be a need for specialized software for very large linear systems or for rather specialized applications. There are many libraries for these needs in Fortran or C/C++ that are freely available. A list of such software is maintained at

`http://www.netlib.org/utk/people/JackDongarra/la-sw.html`

The R system is widely used by statisticians. This system provides a wide range of operations for vectors and matrices. It treats vectors as special kind of list and matrices as special kinds of rectangular arrays. An impact of this is in writing functions that accept matrices; a vector will not be accepted just as a matrix with one dimension being 1. A vector has a `length` attribute but not a `dim` attribute.

Certain aspects of the result of operations that involve both vectors and matrices do not correspond with what the user might expect. For example, if `A` is a matrix and `x` is a vector, `A%*%x` is an array instead of a vector; `A%*%t(x)` is not an allowable operation but `x%*%A` is allowable and is the same as `t(x)%*%A` and they are both arrays instead of vectors; and `x%*%x` is the same as `t(x)%*%x` and they are both arrays instead of scalars. The functions `as.vector` and `as.matrix` can be used to convert the results to the expected class.

Exercises

5.1. Matrix norms and condition numbers.
In the system of linear equations (1.169) in Exercise 1.5 on page 75 (see also solution on page 677), the solution is easily seen to be $x_1 = 1.000$ and $x_2 = 1.000$. Figure 5.1 illustrates the original system and this modified one.
Now consider a small change in the right-hand side:

$$\begin{aligned} 1.000x_1 + 0.500x_2 &= 1.500, \\ 0.667x_1 + 0.333x_2 &= 0.999. \end{aligned} \tag{5.68}$$

This system has solution $x_1 = 0.000$ and $x_2 = 3.000$.
Alternatively, consider a small change in one of the elements of the coefficient matrix:

$$\begin{aligned} 1.000x_1 + 0.500x_2 &= 1.500, \\ 0.667x_1 + 0.334x_2 &= 1.000. \end{aligned} \tag{5.69}$$

The solution now is $x_1 = 2.000$ and $x_2 = -1.000$.
In both cases, small changes of the order of 10^{-3} in the input (the elements of the coefficient matrix or the right-hand side) result in relatively large changes (of the order of 1) in the output (the solution). Solving the system (either one of them) is an ill-conditioned problem.
The nature of the data that cause ill-conditioning depends on the type of problem. In this case, the problem is that the lines represented by the equations are almost parallel, as seen in Figure 5.1, and so their point of

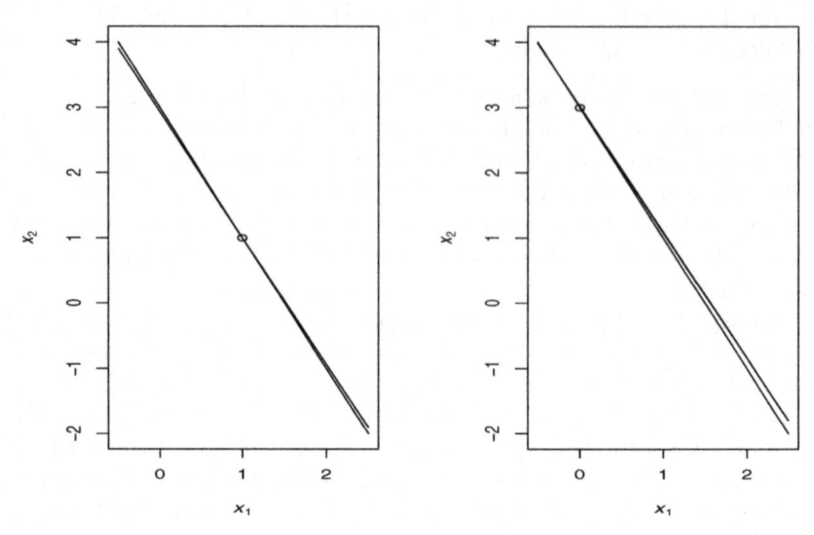

Fig. 5.1. Almost Parallel Lines: Ill-Conditioned Coefficient Matrices, Equations (1.169) and (5.68)

intersection is very sensitive to slight changes in the coefficients defining the lines.

The problem can also be described in terms of the angle between the lines. When the angle is small, but not necessarily 0, we refer to the condition as "collinearity".

We would expect that these properties of the system of equations would be reflected in the condition number of the coefficient matrix A.

Evaluate $\kappa_1(A)$, $\kappa_2(A)$, and $\kappa_\infty(A)$. (Notice that the condition numbers are not exactly the same, but they are close. Notice also that the condition numbers are of the order of magnitude of the ratio of the output perturbation to the input perturbation in those equations. These numbers are "large" only in a relative sense to 10^{-3}. They are not nearly large enough to cause any problems on the computer.)

5.2. Consider the system of linear equations

$$\begin{aligned} x_1 + 4x_2 + x_3 &= 12, \\ 2x_1 + 5x_2 + 3x_3 &= 19, \\ x_1 + 2x_2 + 2x_3 &= 9. \end{aligned}$$

a) Solve the system using Gaussian elimination with partial pivoting.
b) Solve the system using Gaussian elimination with complete pivoting.
c) Determine the D, L, and U matrices of the Gauss-Seidel method (equation (5.44), page 222) and determine the spectral radius of

$$(D + L)^{-1}U.$$

d) We stated that the convergence rate of the Gauss-Seidel method increases as the spectral radius ρ of $(D+L)^{-1}U$ decreases. We also know that $\rho(cA) = c\rho(A)$. Why don't we just scale the problem to increase the convergence rate?

e) Do two steps of the Gauss-Seidel method starting with $x^{(0)} = (1,1,1)$, and evaluate the L_2 norm of the difference of two successive approximate solutions.

f) Do two steps of the Gauss-Seidel method with successive overrelaxation using $\omega = 0.1$, starting with $x^{(0)} = (1,1,1)$, and evaluate the L_2 norm of the difference of two successive approximate solutions.

g) Do two steps of the conjugate gradient method starting with $x^{(0)} = (1,1,1)$, and evaluate the L_2 norm of the difference of two successive approximate solutions.

5.3. Gram-Schmidt orthonormalization.

a) Write a program module (in Fortran, C, R or S-Plus, Octave or Matlab, or whatever language you choose) to implement Gram-Schmidt orthonormalization using Algorithm 5.2. Your program should be for an arbitrary order and for an arbitrary set of linearly independent vectors.

b) Write a program module to implement Gram-Schmidt orthonormalization using equations (5.41) and (5.42).

c) Experiment with your programs. Do they usually give the same results? Try them on a linearly independent set of vectors all of which point "almost" in the same direction. Do you see any difference in the accuracy? Think of some systematic way of forming a set of vectors that point in almost the same direction. One way of doing this would be, for a given x, to form $x + \epsilon e_i$ for $i = 1, \ldots, n-1$, where e_i is the ith unit vector, that is, the vector with 0s in all positions except the ith position, which is 1, and ϵ is a small positive number. The difference can even be seen in hand computations for $n = 3$. Take $x_1 = (1, 10^{-6}, 10^{-6})$, $x_2 = (1, 10^{-6}, 0)$, and $x_3 = (1, 0, 10^{-6})$.

5.4. Generalized inverses.

a) Confirm that $A^+ = \left[R_1^{-1} \; 0 \right] Q^{\mathrm{T}}$ (equation (5.32)).

b) With A decomposed as in equation (5.34), confirm that

$$A^- = E_\pi \begin{bmatrix} T_1^{-1} & 0 \\ 0 & 0 \end{bmatrix} Q^{\mathrm{T}}$$

is a generalized inverse of A

5.5. Solving an overdetermined system $Xb = y$, where X is $n \times m$.

a) Count how many multiplications and additions are required to form $X^{\mathrm{T}}X$. (A multiplication or addition such as this is performed in floating point on a computer, so the operation is called a "flop". Sometimes a flop is considered a combined operation of multiplication and addition; at other times, each is considered a separate flop. The distinction is not important here; just count the total number.)

b) Count how many flops are required to form $X^{\mathrm{T}}y$.
c) Count how many flops are required to solve $X^{\mathrm{T}}Xb = X^{\mathrm{T}}y$ using a Cholesky decomposition.
d) Count how many flops are required to form a QR decomposition of X using reflectors.
e) Count how many flops are required to form a $Q^{\mathrm{T}}y$.
f) Count how many flops are required to solve $R_1 b = c_1$ (equation (5.63), page 231).
g) If n is large relative to m, what is the ratio of the total number of flops required to form and solve the normal equations using the Cholesky method to the total number required to solve the system using a QR decomposition? Why is the QR method generally preferred?

6

Solution of Nonlinear Equations and Optimization

As we discussed in Section 1.8, most problems in statistical inference can be posed as optimization problems.

An optimization problem is to solve

$$\underset{x \in D}{\arg\min}\, f(x) \tag{6.1}$$

for x. The scalar-valued function f is called the *objective function*. The variable x, which is usually a vector, is called the decision variable, and its elements are called *decision variables*. The domain D of the decision variables is called the *feasible set*.

In this chapter, after some preliminary general issues, we will consider different methods, which depend on the nature of D. First we consider problems in which D is continuous. The methods for continuous D often involve solving nonlinear equations, so we discuss techniques for solving equations in Section 6.1, before going on to the topic of optimization over continuous domains in Section 6.2. In Section 6.3 we consider the problem of optimization over a discrete domain. We mention a variety of methods, but consider only one of these, simulated annealing, in any detail.

In Section 6.2 we assume that $D = \mathrm{I\!R}^m$, that is, we assume that there are no constraints on the decision variables, and in Section 6.3 we likewise generally ignore any constraints imposed by D. In Section 6.4, we consider the necessary changes to accommodate constraints in D.

In the final sections we consider some specific types of optimization problems that arise in statistical applications.

Categorizations of Optimization Problems

This basic problem (6.1) has many variations. A simple variation is the maximization problem, which is addressed by using $-f(x)$. The general methods do not depend on whether we are minimizing or maximizing. We use the term

J.E. Gentle, *Computational Statistics*, Statistics and Computing,
DOI: 10.1007/978-0-387-98144-4_6,

"optimum" generally to mean either a minimum or maximum, depending on how the problem is stated.

If D is essentially irrelevant (that is, if D includes all x for which $f(x)$ is defined), the problem is called an *unconstrained* optimization problem. Otherwise, it is a *constrained* optimization problem.

Two distinct classes of problems can be associated with the cardinality of D. In one class, D is continuous, dense, and uncountable. In the other class, D is discrete and countable. The techniques for solving problems in these two classes are quite different, although techniques for one type can be used as approximations in the other type.

In this chapter, we will address all of these types of problems, but not with equal time. We will give most attention to the case of continuous D. The nature of f determines how the problem must be addressed, and this is especially true in the case of continuous D. If f is linear, the unconstrained problem is either trivial or ill-posed. If f is linear and the problem is constrained, the nature of D determines what kinds of algorithms are appropriate. If D is a convex polytope, for example, the problem is a *linear program.* The *simplex method*, which is a discrete stepping algorithm, is the most common way to solve linear programming problems. (This method was chosen as one of the Top 10 algorithms of the twentieth century; see page 138.) Linear programs arise in a limited number of statistical applications, for example, in linear regression fitting by minimizing either the L_1 or the L_∞ norm of the residuals. We will only briefly discuss linear programming in this chapter.

In the more general case of f over a continuous domain, the continuity of f is probably the most important issue. If f is arbitrarily discontinuous, there is very little that can be done in a systematic fashion. For general methods in continuous domains, we will limit our consideration to continuous objective functions. After continuity, the next obvious issue is the differentiability of f, and for some methods, we make assumptions about its differentiability. These assumptions are of two types: about *existence*, which affects theoretical properties of optimization methods, and, assuming they exist, about the *cost of computation* of derivatives, which affects the choice of method for solving the problem.

The other broad class of optimization problems are those in which D is discrete and countable. These are essentially problems in combinatorics. They must generally be attacked in very different ways from the approaches used in continuous domains.

Optimization problems over either dense or countable domains may have more than one solution; that is, there may be more than one point at which the minimum is attained. In the case of dense domains, the set of optimal points may even be uncountable. A more common case is one in which within certain regions of the domain, there is a local optimum, and within another region, there is another local optimum. We use the terms "global optimum" to refer to an optimum over the domain D. We use the term "local optimum" to refer to an optimum within some region of D. ("Region" is not precisely defined, and of course, in any event, it must depend on the nature of D.

It is generally difficult for an optimization method to detect the existence of local optima. In fact, in the case of optimization problems over dense domains, after a given problem is solved, we could always look back over the iterations and change the given objective function so that it retains its original continuity and differentiability characteristics and the iterations would proceed exactly as they did for the given problem, yet the new objective function is arbitrarily smaller at some point other than the computed solution. We will not go through the details to prove this, but it should be obvious because we can stretch any function over the "holes" in a grid and still retain continuity and differentiability. This fact means that we could never have a computer algorithm that is guaranteed to determine a global optimum.

Two other issues involve the evaluation of the objective function. The objective function may be "noisy"; that is, for given x, we may compute or observe $f(x) + \epsilon$. The noise may be due either to computational approximations or to the underlying model of the problem we are attempting to solve. From the standpoint of the mechanical techniques of optimization, the source of the noise is not relevant. The existence of the noise, however, may have implications for the organization of the computations in the algorithm. Another issue is whether the evaluation of the objective function is easy and cheap, or difficult and expensive. The cost of evaluating the objective function may affect the way we arrange the computations within an iteration of an optimization algorithm.

Another categorization relevant to optimization concerns the nature of the algorithm, rather than of the problem. An algorithm may involve deterministic steps, or it may follow random paths. We refer to algorithms of the latter type as "stochastic algorithms". They are particularly useful in problems that have multiple optima, and in combinatorial problems, in which they are used to sample points in the domain.

Testing for Convergence

On page 129 in Chapter 3, we discussed some of the issues in testing for convergence in iterative algorithms. As we indicated there, the problem is difficult and the methods are often necessarily ad hoc.

Convergence tests usually involve comparisons such as (3.12) or (3.14):

$$\Delta\left(x^{(k)}, x^{(k-1)}\right) \leq \epsilon,$$

or

$$\Delta\left(x^{(k)}, x^{(k-1)}\right) \leq \epsilon_r \left|x^{(k-1)}\right|.$$

Other tests may involve change in the function; that is,

$$\Delta\left(f\left(x^{(k)}\right), f\left(x^{(k-1)}\right)\right) \leq \epsilon,$$

or

$$\Delta\left(f\left(x^{(k)}, x^{(k-1)}\right)\right) \leq \epsilon_r \left|f\left(x^{(k-1)}\right)\right|.$$

These tests are only for algorithmic convergence, of course.

As we discuss algorithms in the following sections, for simplicity, we will not include the convergence tests in the algorithms themselves. Instead, we will refer to a logical value of a function "converged$(\cdot)$" that has an argument or multiple arguments on which convergence tests are based:

$$\text{converged}(d_1, d_2, \ldots). \tag{6.2}$$

This logical function may have several built-in tolerance factors and may employ various tests. For example, if $d_1 = x^{(k)}$ and $d_2 = x^{(k-1)}$, one test in converged$(d_1, d_2, \ldots)$ might be based on the comparison

$$\left|x^{(k)} - x^{(k-1)}\right| \leq \epsilon.$$

In the descriptions of algorithms, we will use the term "converged" to indicate that some convergence criterion has been satisfied. For example, a step in an algorithm may contain the phrase

"if converged$(d_1, d_2, \ldots)$..."

which would mean that the convergence test is based on the current values of $d_1, d_2, \ldots$. It does not specify the convergence test, however.

An algorithm should always be defined in such a way that it will terminate. (The formal definition of *algorithm* requires that it terminate in a finite number of steps.) This means that an iterative algorithm should always have a limit on the number of iterations. In the iterative algorithms we describe in this chapter, we often have a step "Set $k = k + 1$", where k is an iteration counter. At the point at which k is incremented, any program implementing the algorithm should include a test for the limit on the number of iterations. Because this test is not included explicitly in the algorithms described in this chapter, technically whether or not they are "algorithms" depends on the problems to which they are applied.

When an algorithm terminates because of reaching the limit on the number of iterations, we say that it did not converge.

6.1 Finding Roots of Equations

We will describe several general methods for solving a system of nonlinear equations.

For convergence tests in the case of finding roots of an equation, in addition to measures of changes within the domain of the function, we also often can use a comparison such as (3.13):

$$\left|f(x^{(k)})\right| \leq \epsilon,$$

so in statements of algorithms, we may use the phrase

"if converged$\big(d, f(x^{(k)})\big)$..."

A program implementing any of these algorithms should be set to terminate if the number of iterations exceeds some fixed number before algorithmic convergence has occurred. In this case, the program should inform the user that the algorithm has not converged.

Basic Methods for a Single Equation

We first consider methods for a single equation in a scalar variable. For the problem of a single equation, there are several methods:

- simple fixed-point method,
- bisection method,
- Newton's method,
- secant method,
- regula falsi method.

We also introduce and briefly discuss stochastic approximation in the context of a single equation.

Also for a single equation, we consider the condition of the problem, and define a condition number for finding a root of an equation.

Each of the methods discussed may be the best for some given problem, and it is important to understand how these methods work. There are some specialized methods, such as for finding the roots of a polynomial, but we will not discuss them.

Consider a scalar-valued function f of a scalar variable x. Our objective is to find a value of x for which

$$f(x) = 0. \tag{6.3}$$

If there is no closed form for the inverse $f^{-1}(\cdot)$, and if f is continuous, then the solution is effected by an iterative process. As mentioned above, the iterative process must have a *convergence criterion* to decide when the solution is "close enough", and the number of iterations allowed must also be bounded.

In the following discussion, we assume f is a continuous function, and that a solution x_0 exists. We will sometimes use x without the subscript. There may be multiple solutions, of course, and in some cases we may wish to know if there are multiple solutions, and if so, to find all of them. If f can be factored, we may reduce the function and continue to find additional roots. That is rarely the case, so the more common approach is to use different starting points and hope that the iterative algorithm will converge to a different solution.

We will illustrate various methods using the function

$$f(x) = x^3 - 4x^2 + 18x - 115, \tag{6.4}$$

which has a single root at $x = 5$. (There are special algorithms for roots of polynomials, but we will not discuss them here.)

Fixed-Point Method

A general type of iteration for problems such as (6.3) is called a *fixed-point method*, or *fixed-point iteration.* In this problem the fixed-point method uses the fact that at the solution

$$x_0 = f(x_0) + x_0.$$

The fixed-point iteration is then

$$x_0^{(k+1)} = f\left(x_0^{(k)}\right) + x_0^{(k)}, \tag{6.5}$$

after starting with any value $x_0^{(0)}$.

This iterative process can be speeded up by use of Aitken's Δ^2-extrapolation (see page 132), as shown in Algorithm 6.1. Aitken's extrapolation in this setting is also called Steffensen's method.

Algorithm 6.1 Steffensen's Fixed-Point Method to Find a Root of an Equation

0. Set $k = 0$, and determine an approximation $x^{(0)}$.
1. Set $f_1 = f(x^{(k)}) + x^{(k)}$.
2. Set $f_2 = f(f_1) + f_1$.
3. Set $d = f_2 - f_1$.
4. If converged$(f_2, f_1, f(x^{(k)}))$,
   terminate and return the solution as $x^{(k)}$.
5. Set $s = (s_1 - x^{(k)})/d$.
6. Set $x^{(k+1)} = f_2 + d/(s - 1)$.
7. Set $k = k + 1$ and go to step 1. ∎

Notice that the basic iteration formula in Algorithm 6.1 is

$$x^{(k+1)} = x^{(k)} - \frac{f(x^{(k)})}{f(x^{(k)}) + f(x^{(k)}) - f(x^{(k)})}. \tag{6.6}$$

Bisection Method

One of the simplest iterative methods for solving $f(x) = 0$ is the *bisection method.* The method begins with two values that bracket the solution, and then tightens the interval by halves. We assume that there are values x_l, x_u, and x_0, with $x_l < x_u$ and $x_l \leq x_0 \leq x_u$, such that

$$f(x_l) \leq 0,$$

$$f(x_u) \geq 0,$$

and

$$f(x_0) = 0.$$

If $f(x_l) \geq 0$ and $f(x_u) \leq 0$, we merely relabel the points. The method is shown in Algorithm 6.2.

Algorithm 6.2 Bisection to Find a Root of an Equation

0. Set $k = 0$, and
   find an interval $[x_l,\ x_u]$ in which a solution lies.
1. Set $k = k + 1$ and set $x^{(k)} = (x_u + x_l)/2$.
2. If $\text{sign}\left(f\left(x^{(k)}\right)\right) = \text{sign}(f(x_l))$, then
   2.a. set $x_l = x^{(k)}$;
   otherwise
   2.b. set $x_u = x^{(k)}$.
3. If $\text{converged}\big(x_u, x_l, f(x^{(k)})\big)$,
   terminate and return the solution as $x^{(k)}$.
   otherwise,
   set $k = k + 1$ and go to step 1. ∎

As an example, we will now use the bisection method to find a root of equation (6.4). The steps are shown in Figure 6.1. The interval is successively halved, first by moving the upper bound down, then moving the lower bound up, then moving the lower bound up again, and so on. In each step the approximation to the solution $x_0^{(k)}$ is the midpoint of an interval, and then becomes an endpoint of the interval in the next step.

The bisection method is very easy to understand and to implement. The solution always remains within a known interval. After k steps, the length of that interval is 2^{-k} times its initial length, so the error of the approximation is of order 2^{-k}. Each iteration gains one more bit of accuracy. Because the ratios of the lengths of successive intervals is constant, the bisection method converges linearly. The iterations beginning with those shown in Figure 6.1, and continuing until 11 digits of accuracy are shown in Table 6.1. The length of the interval is 7 initially. After 35 steps, it is approximately $7 \cdot 2^{-35}$.

The stopping rule in Algorithm 6.2 is based on the length of the interval, among other things. If one of the stopping rules is $x_u - x_l \leq \epsilon$, it is clear that, beginning with x_l and x_u, the algorithm terminates after exactly

$$\lceil \log_2(x_u - x_l)/\epsilon) \rceil$$

steps.

The bisection method requires that the function be continuous within the initial interval. The function need not be differentiable, however.

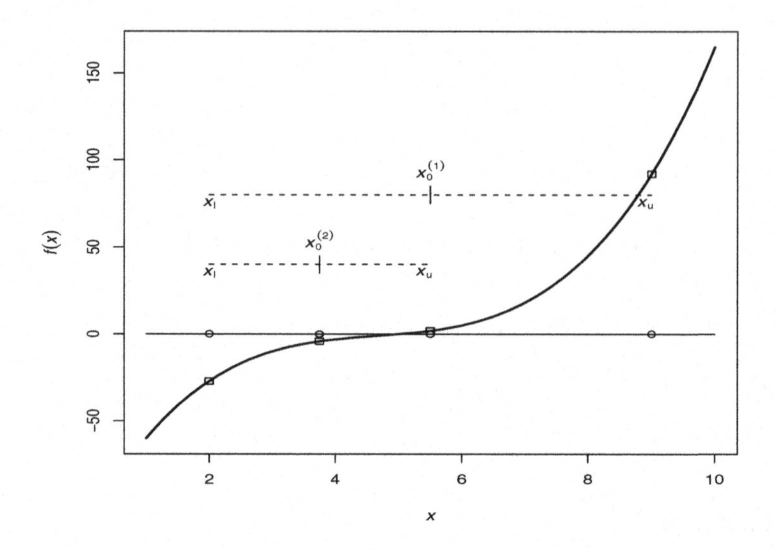

Fig. 6.1. Bisection to Find x_0, so that $f(x_0) = 0$

Table 6.1. Bisection Iterations

k	x_l	x_u	k	x_l	x_u
0	2.00000000000000	9.00000000000000	18	4.99998855590820	5.00001525878906
1	2.00000000000000	5.50000000000000	19	4.99998855590820	5.00000190734863
2	3.75000000000000	5.50000000000000	20	4.99999523162842	5.00000190734863
3	4.62500000000000	5.50000000000000	21	4.99999856948853	5.00000190734863
4	4.62500000000000	5.06250000000000	22	4.99999856948853	5.00000023841858
5	4.84375000000000	5.06250000000000	23	4.99999940395355	5.00000023841858
6	4.95312500000000	5.06250000000000	24	4.99999982118607	5.00000023841858
7	4.95312500000000	5.00781250000000	25	4.99999982118607	5.00000002980232
8	4.98046875000000	5.00781250000000	26	4.99999992549419	5.00000002980232
9	4.99414062500000	5.00781250000000	27	4.99999997764826	5.00000002980232
10	4.99414062500000	5.00097656250000	28	4.99999997764826	5.00000000372529
11	4.99755859375000	5.00097656250000	29	4.99999999068677	5.00000000372529
12	4.99926757812500	5.00097656250000	30	4.99999999720603	5.00000000372529
13	4.99926757812500	5.00012207031250	31	4.99999999720603	5.00000000046566
14	4.99969482421875	5.00012207031250	32	4.99999999883585	5.00000000046566
15	4.99990844726563	5.00012207031250	33	4.99999999965075	5.00000000046566
16	4.99990844726563	5.00001525878906	34	4.99999999965075	5.00000000005821
17	4.99996185302734	5.00001525878906	35	4.99999999985448	5.00000000005821

Newton's Method

Newton's method for a differential function is based on the first-order Taylor series of the function about a point near the solution:

$$f(x) \approx f\left(x_0^{(k)}\right) + \left(x - x_0^{(k)}\right) f'\left(x_0^{(k)}\right). \tag{6.7}$$

As before, the solution is approached through the iterates, $x_0^{(k)}, x_0^{(k+1)}, \ldots$. The update is obtained by solving the Taylor series approximation

$$f\left(x_0^{(k+1)}\right) \approx f\left(x_0^{(k)}\right) + \left(x_0^{(k+1)} - x_0^{(k)}\right) f'\left(x_0^{(k)}\right),$$

in which we assume that

$$f\left(x_0^{(k+1)}\right) = 0.$$

If $f'(x_0^{(k)}) \neq 0$, this approximation yields

$$x_0^{(k+1)} = x_0^{(k)} - \frac{f\left(x_0^{(k)}\right)}{f'\left(x_0^{(k)}\right)}. \tag{6.8}$$

Newton's method uses the slope of the function at one point to choose the next point, which is the direction of a smaller value of the function, indicated by the slope. The method is given in Algorithm 6.3.

Algorithm 6.3 Newton's Method to Find a Root of an Equation

0. Set $k = 0$, and determine an approximation $x^{(0)}$.
1. Solve for $x^{(k+1)}$ in

$$f'\left(x^{(k)}\right)\left(x^{(k+1)} - x^{(k)}\right) = -f\left(x^{(k)}\right)$$

that is, set

$$x^{(k+1)} = x^{(k)} - \frac{f\left(x^{(k)}\right)}{f'\left(x^{(k)}\right)},$$

if $f'\left(x^{(k)}\right)^{-1}$ exists.

2. If converged$\left(x^{(k+1)}, x^{(k)}, f(x^{(k+1)})\right)$,
   terminate and return the solution as $x^{(k+1)}$.
   otherwise,
   set $k = k + 1$ and go to step 1. ∎

The stopping rule in Algorithm 6.3 is based on the interval between two successive approximations, just as the stopping rule of the bisection method is based on the length of the interval. As in the other cases of root finding, $\left|f\left(x_0^{(k)}\right)\right|$ could also be used as a stopping criterion.

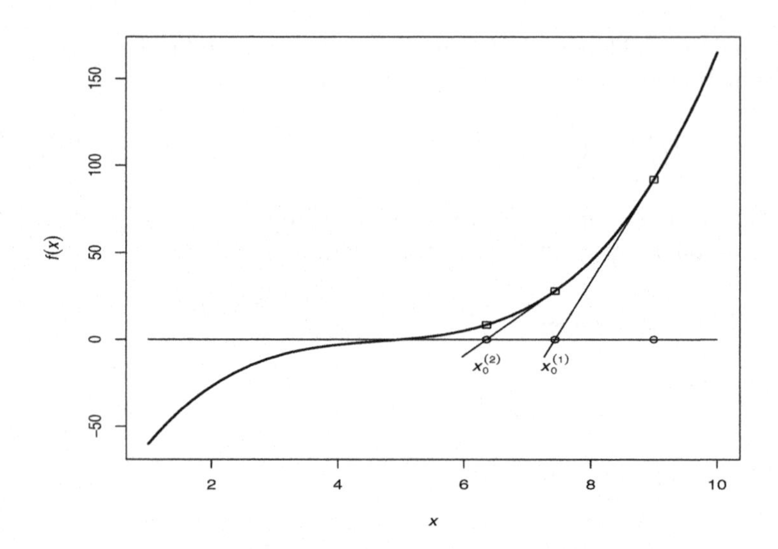

Fig. 6.2. Newton's Method to Find x_0, so that $f(x_0) = 0$

Newton's method is easy to understand and to implement if the derivative is available. In Figure 6.2, we show Newton's method applied to the same function we used bisection on in Figure 6.1. In the example in Figure 6.2, Newton's method proceeds in an orderly fashion toward the zero of the function.

Notice in Figure 6.2 that the derivatives (the slopes) are decreasing, as the solution is approached from the right side. This could cause some problems with the method, because the denominator in step 1 of Algorithm 6.3 becomes small. In our example problem (6.4), the derivative,

$$f'(x) = 3x^2 - 8x + 18,$$

is not zero at the solution. (See Exercise 6.6, page 301. The derivative of the function in Exercise 6.6b is zero at the solution.)

A modification of Newton's method is to use a numerical approximation to the derivative:

$$f'\left(x_0^{(k)}\right) \approx \frac{f\left(x_0^{(k)} + h\right) - f\left(x_0^{(k)}\right)}{h}. \tag{6.9}$$

This is sometimes called the "discrete Newton's method". It is also essentially the same as the secant method discussed below.

Newton's method can also be speeded up by use of Aitken's Δ^2-extrapolation. In this case, starting with x_0, two steps of Newton's method are use to compute

$$x_1 = x_0 - \frac{f(x_0)}{f'(x_0)}$$

and

$$x_2 = x_1 - \frac{f(x_1)}{f'(x_1)},$$

and then Aitken's Δ^2 process is used to compute

$$\tilde{x}_0 = x_0 - \frac{(\Delta x_0)^2}{\Delta^2 x_0},$$

which is continued by setting $x_0 = \tilde{x}_0$ and repeating the previous steps. Aitken's extrapolation in this setting is also called Steffensen's method.

Convergence or Failure of Newton's Method

To investigate the convergence of Newton's method, consider the first-order Taylor series with remainder, expanded about a point near the solution, $x_0^{(k)}$, and evaluated at the solution x_0:

$$\begin{aligned} f(x_0) &= f\left(x_0^{(k)}\right) + \left(x_0 - x_0^{(k)}\right) f'\left(x_0^{(k)}\right) + \frac{1}{2}\left(x_0 - x_0^{(k)}\right)^2 f''(\xi) \\ &= 0. \end{aligned}$$

Using equation (6.8), we have

$$\frac{\left(x_0 - x_0^{(k+1)}\right)}{\left(x_0 - x_0^{(k)}\right)^2} = \frac{1}{2}\frac{f''(\xi)}{f'\left(x_0^{(k)}\right)}.$$

So, if the limit, as $k \to \infty$, of the ratio on the right exists, the convergence is quadratic (see page 131). It is clear that if $f'\left(x_0^{(k)}\right) = 0$ at any point, the method may fail.

Even if the derivatives are not zero, however, Newton's method may diverge unless the starting point is sufficiently close to the solution. Two ways in which Newton's method can go wrong are illustrated in Figures 6.3 and 6.4.

In both of these examples, the failure of Newton's method occurs because the starting point is too far away from the zero. The possibility of this occurring makes the choice of starting value very important. In the bisection method, we do not have to be concerned about this, so long as we can find values that bracket the solution.

Secant Method

The *secant method* is similar to Newton's method in using the slope to determine successive points in the iteration. Newton's method uses the derivative

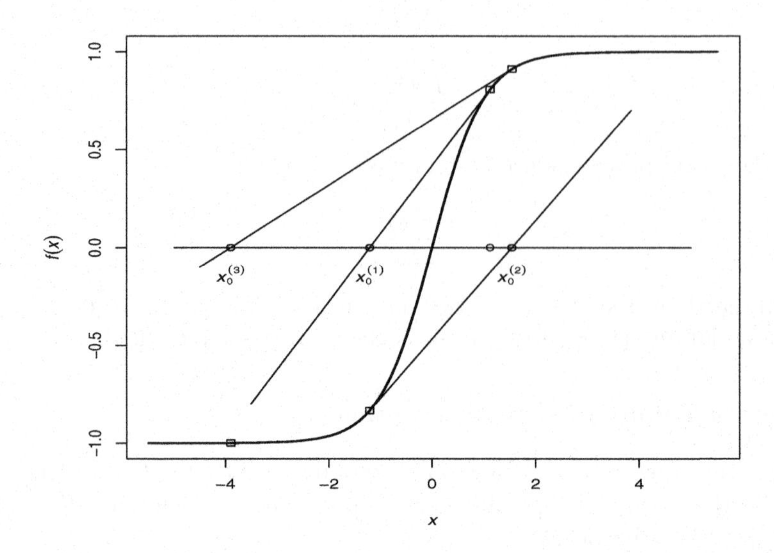

Fig. 6.3. Failure of Newton's Method

or the tangent at a given point, and the secant method uses the slope of the

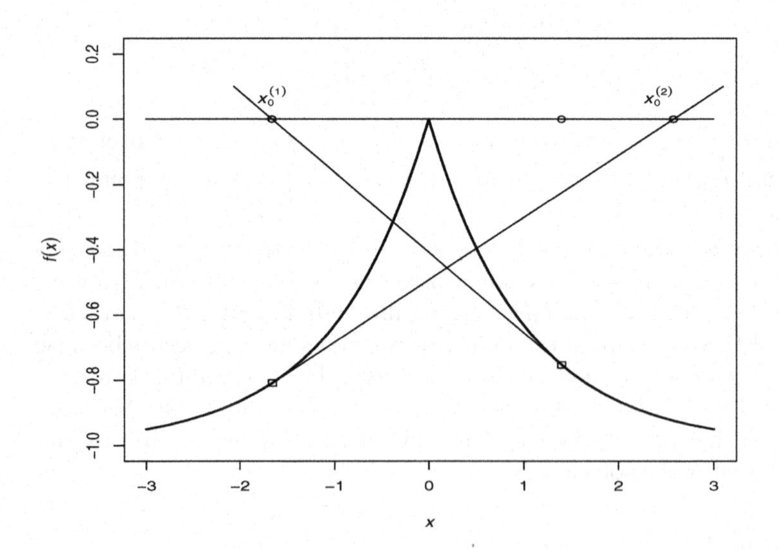

Fig. 6.4. Failure of Newton's Method; Another Example

function between two given points to choose the next point. The method is given in Algorithm 6.4.

Algorithm 6.4 Secant Method to Find a Root of an Equation

0. Set $k = 1$, and determine approximations $x^{(0)}$ and $x^{(1)}$.
1. Set $x^{(k+1)} = x^{(k)} - \dfrac{f\left(x^{(k)}\right)\left(x^{(k)} - x^{(k-1)}\right)}{f\left(x^{(k)}\right) - f\left(x^{(k-1)}\right)}$.
2. If converged$\left(x^{(k+1)}, x^{(k)}, f\left(x^{(k+1)}\right)\right)$,
   terminate and return the solution as $x^{(k+1)}$.
   otherwise,
   set $k = k + 1$ and go to step 1. ∎

The intersection of the line between the two points on the function and the x-axis is taken as the next point at which to evaluate the function, as we see in Figure 6.5. The choice of $x_0^{(0)}$ and $x_0^{(1)}$ is arbitrary, although just as in Newton's method, if they are too far away from the solution, the method may not converge.

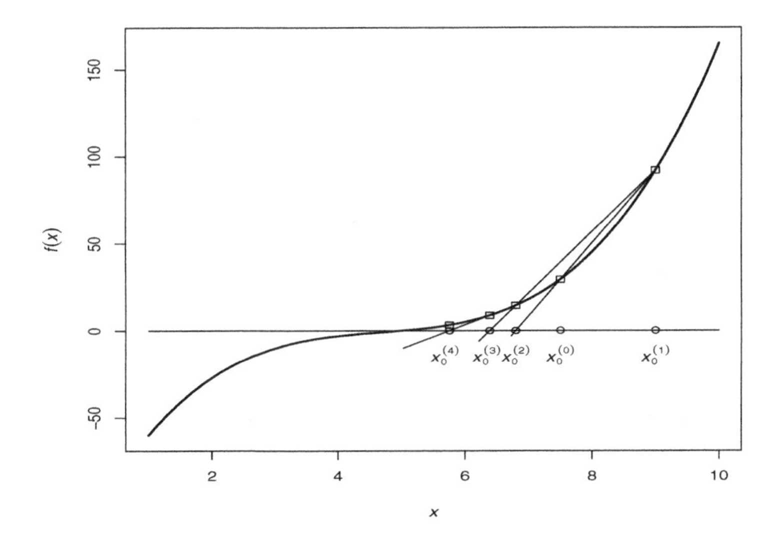

Fig. 6.5. Secant Method to Find x_0, so that $f(x_0) = 0$

The two points in the secant method may or may not bracket a root.

Regula Falsi Method

The *regula falsi* or *false position* method is similar to the secant method, except that the two starting points are chosen so as to bracket a solution, and as in the bisection method, each successive point is chosen so that it, together with one of the two previous points, brackets a solution. The method given in Algorithm 6.5 is a slight modification of the ordinary regula falsi method, and is sometimes called the "modified" regula falsi method. Because the "unmodified" regula falsi method (which omits steps 2.a.ii and 2.b.ii in Algorithm 6.5) should not even be used, we just refer to the method given here as regula falsi. Algorithm 6.5 is also sometimes called the "Illinois method".

Algorithm 6.5 Regula Falsi to Find a Root of an Equation

0. Set $k = 0$;
   find an interval $[x_l,\ x_u]$ in which a solution lies;
   set $f_l = f(x_l)$;
   set $f_u = f(x_u)$; and
   set $x^{(0)} = x_l$.
1. Set $x^{(k+1)} = \frac{x_l f_u - x_u f_l}{f_u - f_l}$.
2. If $f_l f\left(x^{(k+1)}\right) < 0$, then
   2.a.i. set $x_u = x^{(k+1)}$ and $f_u = f\left(x^{(k+1)}\right)$.
   2.a.ii. if $f\left(x^{(k)}\right) f\left(x^{(k+1)}\right) > 0$, then set $f_l = f_l/2$.
   Otherwise,
   2.b.i. set $x_l = x^{(k+1)}$ and $f_l = f\left(x^{(k+1)}\right)$.
   2.b.ii. if $f\left(x^{(k)}\right) f\left(x^{(k+1)}\right) > 0$, then set $f_u = f_u/2$.
3. If converged$\left(x_u, x_l, f\left(x^{(k+1)}\right)\right)$,
   terminate and return the solution as $x^{(k+1)}$.
   otherwise,
   set $k = k + 1$ and go to step 1. ∎

The regula falsi method generally converges more slowly than the secant method, but it is more reliable, because the solution remains bracketed. Figure 6.6 illustrates two iterations of the method.

Stochastic Approximation

In practical applications we often cannot evaluate $f(x)$ precisely. Instead, we make observations that are contaminated with random errors or noise. At $x_0^{(k)}$, instead of $f\left(x_0^{(k)}\right)$, we observe

$$y_0^{(k)} = f\left(x_0^{(k)}\right) + \epsilon_k.$$

A fixed-point iteration of the form

$$x_0^{(k+1)} = x_0^{(k)} + \widehat{f}\left(x_0^{(k)}\right) \tag{6.10}$$

could be used, where $\widehat{f}\left(x_0^{(k)}\right)$ is an estimate of the value of f at $x_0^{(k)}$, based on some observations of $y_0^{(k)}$.

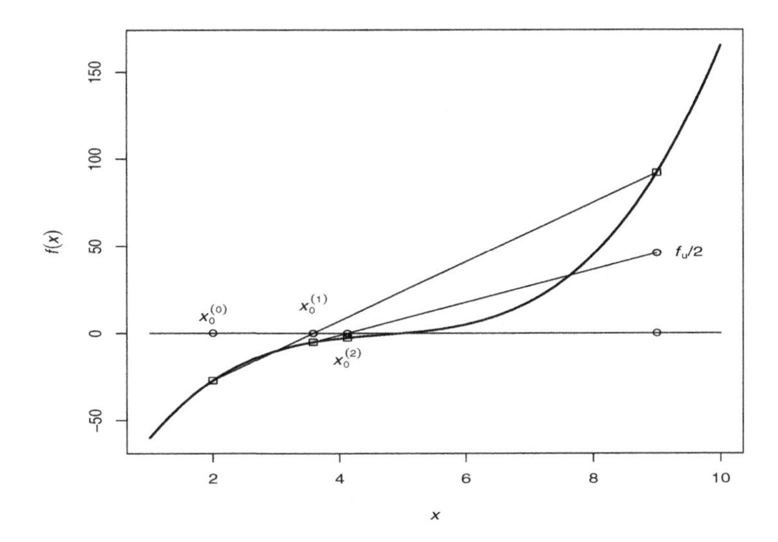

Fig. 6.6. Regula Falsi to Find x_0, so that $f(x_0) = 0$

Alternatively, the model of interest may be a random process, and we may be interested in some function of the random process, $f(x)$. For example, we may model an observable process by a random variable Y with probability density function $p_Y(y, x)$, where x is a parameter of the distribution. We may be interested in the mean of Y as a function of the parameter x,

$$f(x) = \int y \, p_Y(y, x) \, \mathrm{d}y.$$

If we know $p_Y(y, x)$ and can perform the integration, the problem of finding a zero of $f(x)$ (or, more generally, finding x such that the mean, $f(x)$, is some specified level curve) is similar to the other problems we have discussed. Often in practice we do not know $p_Y(y, x)$, but we are able to take observations on Y. These observations could be used to obtain $\widehat{f}\left(x_0^{(k)}\right)$, and the recursion (6.10) used to find x.

Each observation on Y is an estimate of $f(x)$, so the recursion (6.10) can be rather simple. For a sequence of observations on Y,

$$y_1, y_2, \ldots,$$

we use the recursion

$$x_0^{(k+1)} = x_0^{(k)} + \alpha^{(k)} y_k, \tag{6.11}$$

where $\alpha^{(k)}$ is a decreasing sequence of positive numbers similar to $1/f'(x_0^{(k)})$ in Newton's method (6.8), page 249, when the approach is from the left. Use of this recursion is called the Robbins-Monro procedure. Convergence in the Robbins-Monro procedure is stochastic, rather than deterministic, because of the random variables.

Multiple Roots

It is possible that the function has more than one root, and we may want to find them all. A common way of addressing this problem is to use different starting points in the iterative solution process. Plots of the points evaluated in the iterations may also be useful. In general, if the number of different roots is unknown, there is no way of finding all of them with any assurance.

Accuracy of the Solution

As with most problems in numerical computations, the accuracy we can expect in finding the roots of a function varies from problem to problem; some problems are better conditioned than others. A measure of the condition of the problem of finding the root x_0 can be developed by considering the error in evaluating $f(x)$ in the vicinity of x_0. Suppose a bound on this error is ϵ, so

$$\left|\widehat{f}(x_0) - f(x_0)\right| \leq \epsilon,$$

or

$$\left|\widehat{f}(x_0)\right| \leq \epsilon,$$

where $\widehat{f}(x_0)$ is the computed value approximating $f(x_0)$. Let $[x_l, x_u]$ be the largest interval about x_0 such that

$$|f(x)| \leq \epsilon, \quad \text{if } x \in [x_l, x_u]. \tag{6.12}$$

Within this interval, the computed value $\widehat{f}(x)$ can be either positive or negative just due to error in computing the value. A stable algorithm for finding the root of the function yields a value in the interval, but no higher accuracy can be expected. If $f(x)$ can be expanded in a Taylor series about x_0, we have

$$f(x) \approx f(x_0) + f'(x_0)(x - x_0),$$

or

$$f(x) \approx f'(x_0)(x - x_0).$$

Now applying the bound in (6.12) to the approximation, we have that the interval is approximately

$$x_0 \pm \frac{1}{f'(x_0)}\epsilon,$$

if the derivative exists and is nonzero. Therefore, if the derivative exists and is nonzero, a quantitative measure of the condition of the problem is

$$\frac{1}{f'(x_0)}. \tag{6.13}$$

This quantity is a *condition number of the function f with respect to finding the root x_0*. Of course, to know the condition number usually means to know the solution. Its usefulness in practice is limited to situations where it can be approximated. In Figure 6.7, we can see the sensitivity of a root-finding algorithm to the condition number.

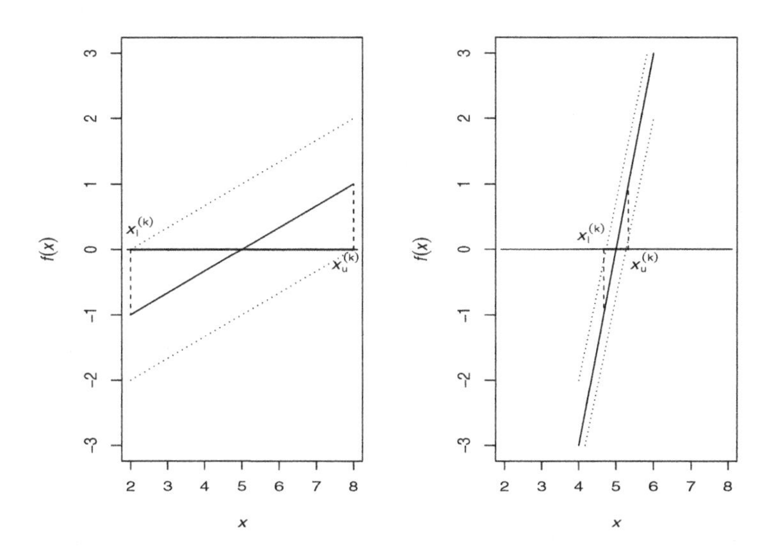

Fig. 6.7. Condition of the Root of $f(x) = 0$: Two Possibilities

Wilkinson (1959) considered the polynomial

$$f(x) = (x-1)(x-2)\cdots(x-20)$$

for studying rounding error in determining roots (see page 113). Very small perturbations in the coefficients of the polynomial lead to very large changes in the roots; hence, we referred to the problem as ill-conditioned. The derivative of that function in the vicinity of the roots is very large, so the condition number defined above in equation (6.13) would not indicate any conditioning problem. As we pointed out, however, the Wilkinson polynomial is ill-conditioned for the problem of finding its roots because of the extreme variation in the magnitude of the coefficients. This kind of situation is common in numerical analysis. Condition numbers do not always tell an accurate story; they should be viewed only as indicators, not as true measures of the condition.

Systems of Equations

If the argument of the function is an m-vector and the function value is an n-vector, equation (6.3),

$$f(x) = 0,$$

represents a system of equations:

$$\begin{aligned} f_1(x_1, x_2, \ldots, x_m) &= 0 \\ f_2(x_1, x_2, \ldots, x_m) &= 0 \\ \vdots \qquad\quad & \;\vdots \; \vdots \\ f_n(x_1, x_2, \ldots, x_m) &= 0. \end{aligned} \tag{6.14}$$

Each of the functions f_i is a scalar-valued function of the vector x. Solution of systems of nonlinear equations can be a significantly more computationally intensive problem than solution of a single equation.

Whether or not the system of equations (6.14) has a solution is not easy to determine. A nonlinear system that has a solution is said to be *consistent*, just as a consistent linear system. Unfortunately, we cannot write a simple necessary and sufficient condition as we did for the linear system in equation (1.46). If $n > m$, the system may be overdetermined, and it is very likely that no solution exists. In this case, a criterion, such as least squares, for a good approximate solution must be chosen. Even if $n = m$, we do not have easy ways of determining whether a solution exists, as we have for the linear system.

There are not as many different methods for solving a system of equations as those that we discussed for solving a single equation. We will only consider one approach, Newton's method, which is similar to the Newton's method we have described for a single equation. There are several variations of Newton's method. Rather than considering them here, we will defer that discussion to the applications of Newton's method to the main problem of interest, that is, optimization.

Newton's method requires the derivatives, so we will assume in the following that the functions are differentiable.

Newton's Method to Solve a System of Equations

As we have seen in the previous sections, the solution of nonlinear equations proceeds iteratively to points ever closer to zero. The derivative or an approximation to the derivative is used to decide which way to move from a given point. For a *scalar-valued* function of several variables, say $f_1(x)$, we must consider the slopes in various directions, that is, the gradient $\nabla f_1(x)$.

In a system of equations such as (6.14), we must consider all of the gradients; that is, the slopes in various directions of all of the scalar-valued functions. The matrix whose rows are the transposes of the gradients is called the

Jacobian. We denote the Jacobian of the function f by J_f. The transpose of the Jacobian, that is, the matrix whose columns are the gradients, is denoted by ∇f for the vector-valued function f. (Note that the symbol ∇ can denote either a vector or a matrix, depending on whether the function to which it is applied is scalar- or vector-valued.) Thus, the Jacobian for the system above is

$$\begin{aligned}\mathrm{J}_f &= \begin{bmatrix} \frac{\partial f_1}{\partial x_1} & \frac{\partial f_1}{\partial x_2} & \cdots & \frac{\partial f_1}{\partial x_m} \\ \frac{\partial f_2}{\partial x_1} & \frac{\partial f_2}{\partial x_2} & \cdots & \frac{\partial f_2}{\partial x_m} \\ & & \vdots & \\ \frac{\partial f_n}{\partial x_1} & \frac{\partial f_n}{\partial x_2} & \cdots & \frac{\partial f_n}{\partial x_m} \end{bmatrix} \\ &= (\nabla f)^{\mathrm{T}}. \end{aligned} \tag{6.15}$$

Notice that the Jacobian is a function, so we often specify the point at which it is evaluated, using the ordinary function notation, $\mathrm{J}_f(x)$. Newton's method described above for a single equation in one variable can be used to determine a vector x_0 that solves this system, if a solution exists, or to determine that the system does not have a solution.

For the vector-valued function in the system of equations (6.14), the first-order Taylor series about a point $x_0^{(k)}$ is

$$f(x) \approx f\left(x_0^{(k)}\right) + \mathrm{J}_f\left(x_0^{(k)}\right)\left(x - x_0^{(k)}\right).$$

This first-order Taylor series is the basis for Newton's method, shown in Algorithm 6.6.

Algorithm 6.6 Newton's Method to Solve a System of Equations
(Compare with Algorithm 6.3, page 249.)

0. Set $k = 0$, and determine an approximation $x^{(k)}$.
1. Solve for $x^{(k+1)}$ in

$$\mathrm{J}_f\left(x^{(k)}\right)\left(x^{(k+1)} - x^{(k)}\right) = f\left(x^{(k)}\right).$$

2. If converged$\left(x^{(k+1)} - x^{(k)}, f(x^{(k+1)})\right)$,
   terminate and return the solution as $x^{(k+1)}$.
   otherwise,
   set $k = k + 1$ and go to step 1. ∎

Note the similarity of this method to Algorithm 6.3, Newton's method to find the root for a single equation. In Algorithm 6.6, however, the convergence criterion would be based on $\|x^{(k+1)} - x^{(k)}\|$, for some appropriate norm.

Notice in general that m and n are not equal, and the system in step 1 is n equations in m unknowns. If, however, $m = n$, and the Jacobian is nonsingular, the solution in step 1 is

$$x_0^{(k+1)} = x_0^{(k)} - \left(\mathrm{J}_f\left(x_0^{(k)}\right)\right)^{-1} f\left(x_0^{(k)}\right). \tag{6.16}$$

It is important to remember that *this expression does not imply that the Jacobian matrix should be inverted.* Linear systems are not solved that way (see Sections 5.2 through 5.4.) Expressions involving the inverse of a matrix provide a compact representation, and so we often write equations such as (6.16).

Sometimes, just as the approximate derivative in equation (6.9) may be used for the single equation, the Jacobian is replaced by a finite-difference approximation,

$$\left(\frac{\partial f_i}{\partial x_j}\right) \approx \left(\frac{f_i(x_1, x_2, \ldots, x_j + h, \ldots x_m) - f_i(x_1, x_2, \ldots, x_j, \ldots x_m)}{h}\right), \tag{6.17}$$

for $h > 0$. Use of this approximation in place of the Jacobian is called the "discrete Newton's method". This, of course, doubles the number of function evaluations per iteration, but it does avoid the computation of the derivatives.

The number of computations in Newton's method may be reduced by assuming that the Jacobian (or the discrete approximation) does not change much from one iteration to the next. A value of the Jacobian may be used in a few subsequent iterations.

The number of computations can also be reduced if the Jacobian has a special structure, as is often the case in important applications, such as in solving systems of differential equations. It may be sparse or banded. In these cases, use of algorithms that take advantage of the special structure will reduce the computations significantly.

Other ways of reducing the computations in Newton's method use an estimate of the derivative that is updated within each iteration. This kind of method is called quasi-Newton. We will discuss quasi-Newton methods for optimization problems beginning on page 269. The ideas are the same.

The generalization of the condition number in equation (6.13) for a single equation is

$$\kappa\left(\mathrm{J}_f\left(x_0\right)\right), \tag{6.18}$$

for a matrix condition number κ for solving a linear system, as discussed beginning on page 207. The quantity in equation (6.18) is a condition number of the function f with respect to finding the root x_0.

If the ranges of the variables in a nonlinear system are quite different, the solution may not be very accurate. This is similar to the artificial ill-conditioning discussed on page 208. The accuracy can often be improved considerably by scaling the variables and the function values so that they all have approximately the same range. Scaling of a variable x_i is just a multiplicative transformation: $y_i = \sigma x_i$. Of course, the ranges of the values of the variables may not be known in advance, so it may be necessary to do some preliminary computations in order to do any kind of useful scaling.

6.2 Unconstrained Descent Methods in Dense Domains

We now return to the optimization problem (6.1). We denote a solution as x_*; that is,

$$x_* = \underset{x \in D}{\arg\min} f(x), \tag{6.19}$$

where x is an m-vector and f is a continuous real scalar-valued function.

In this section, we assume $f(x)$ is defined over IR^m and $D = \mathrm{IR}^m$; that is, it is a continuous, unconstrained optimization problem.

In this section we also generally *assume the function is differentiable in all variables*, and we often assume it is twice differentiable in all variables.

For a convex function f of a scalar variable, if its first derivative exists, the derivative is nondecreasing. If its second derivative f'' exists, then

$$f''(x) \geq 0 \quad \text{for all } x. \tag{6.20}$$

Strict convexity implies that the second derivative is positive. Likewise, the second derivative of a concave function is nonpositive, and it is negative if the function is strictly concave.

If f is convex, $-f$ is concave. A concave function is sometimes said to be "concave down", and a convex function is said to be "concave up".

For a differentiable function of a vector argument, the vector of partial derivatives provides information about the local shape of the function. This vector of derivatives is called the *gradient*, and is denoted by $\nabla f(x)$:

$$\nabla f(x) = \left(\frac{\partial f(x)}{\partial x_1}, \frac{\partial f(x)}{\partial x_2}, \cdots, \frac{\partial f(x)}{\partial x_m} \right). \tag{6.21}$$

(We often write a vector in the horizontal notation as in the equation above, but whenever we perform multiplication operations on vectors or subsetting operations on matrices, we consider a vector to be a column vector; that is, it behaves in many ways as a matrix with one column.)

For a convex function f of a vector variable, if its gradient exists, it is nondecreasing in each of its elements.

As in the scalar case, if a function f of a vector argument is twice-differentiable, more information about a stationary point can be obtained from the second derivatives, which are organized into a matrix, called the *Hessian*, denoted by H_f, and defined as

$$\begin{aligned} \mathrm{H}_f &= \nabla\big(\nabla f(x)\big) \\ &= \nabla^2 f(x) \\ &= \left(\frac{\partial^2 f(x)}{\partial x_i \partial x_j} \right) \\ &= \frac{\partial^2 f(x)}{\partial x \partial x^{\mathrm{T}}}. \end{aligned} \tag{6.22}$$

Notice that the Hessian is a function, so we often specify the point at which it is evaluated in the ordinary function notation, $\mathrm{H}_f(x)$. The symbol $\nabla^2 f(x)$ is also sometimes used to denote the Hessian, but because $\nabla^2 f(x)$ is often used to denote the Laplacian (which yields the diagonal of the Hessian), we will use $\mathrm{H}_f(x)$ to denote the Hessian.

For a convex function of a vector variable, if the Hessian exists, it is positive semidefinite, and the converse holds. Strict convexity implies that the Hessian is positive definite. This is analogous to the condition in inequality (6.20). These conditions are also sufficient.

Sometimes, rather than using the exact derivatives it is more efficient to use approximations such as finite differences. If the function is not differentiable, but is "well-behaved", the methods based on finite differences often also allow us to determine the optimum.

For the time being we will consider the problem of unconstrained optimization. The methods we describe are the basic ones whether constraints are present or not.

Solution of an optimization problem is usually an iterative process, moving from one point on the function to another. The basic things to determine are

- direction or path, p, in which to step and
- how far to step. (The step length is $\alpha\|p\|$, for the scalar α.)

Direction of Search

For a differentiable function, from any given point, an obvious direction to move is the negative gradient, or a direction that has an acute angle with the negative gradient. We call a vector p such that

$$p^{\mathrm{T}}\nabla f(x) < 0$$

a *descent direction* at the point x. For a function of a single variable, this direction of course is just the sign of the derivative of f at x.

If $\nabla f(x) \neq 0$, we can express p as

$$Rp = -\nabla f(x), \tag{6.23}$$

for some positive definite matrix R. A particular choice of R determines the direction. A method that determines the direction in this manner is called a "gradient method".

Numerical computations for quantities such as $p^{\mathrm{T}}\nabla f(x)$ that may be close to zero must be performed with some care. We sometimes impose the requirement

$$p^{\mathrm{T}}\nabla f(x) < -\epsilon,$$

for some positive number ϵ, so as to avoid possible numerical problems for quantities too close to zero.

Once a direction is chosen, the best step is the longest one for which the function continues to decrease.

These heuristic principles of choosing a “good” direction and a “long” step guide our algorithms, but we must be careful in applying the principles.

Line Searches

Although the first thing we must do is to choose a descent direction, in this section we consider the problem of choosing the length of a step in a direction that has already been chosen. In subsequent sections we return to the problem of choosing the direction.

We assume the direction chosen is a descent direction. The problem of finding a minimum in a given direction is similar to, but more complicated than, the problem of finding a zero of a function that we discussed in Section 6.1. In finding a root of a continuous function of a single scalar variable, two values can define an interval in which a root must lie. Three values are necessary to identify an interval containing a local minimum. Nearby points in a descent direction form a decreasing sequence, and any point with a larger value defines an interval containing a local minimum.

After a direction of movement $p^{(k)}$ from a point $x^{(k)}$ is determined, a new point, $x^{(k+1)}$, is chosen in that direction:

$$x^{(k+1)} = x^{(k)} + \alpha^{(k)} p^{(k)}, \tag{6.24}$$

where $\alpha^{(k)}$ is a positive scalar, called the *step length factor*. (The *step length* itself is $\|\alpha^{(k)} p^{(k)}\|$.)

Obviously, in order for the recursion (6.24) to converge, $\alpha^{(k)}$ must approach 0. A sequence of $\alpha^{(k)}$ that converges to 0, even in descent directions, clearly does not guarantee that the sequence $x^{(k)}$ will converge to x_*, however. This is easily seen in the case of the function of the scalar x,

$$f(x) = x^2,$$

starting with $x^{(0)} = 3$ and $\alpha^{(0)} = 1$, proceeding in the descent direction $-x$, and updating the step length factor as $\alpha^{(k+1)} = \frac{1}{2}\alpha^{(k)}$. The step lengths clearly converge to 0, and while the sequence $x^{(k)}$ goes in the correct direction, it converges to 1, not to the point of the minimum of f, $x_* = 0$.

Choice of the “best” $\alpha^{(k)}$ is an optimization problem in one variable:

$$\min_{\alpha^{(k)}} f\big(x^{(k)} + \alpha^{(k)} p^{(k)}\big), \tag{6.25}$$

for fixed $x^{(k)}$ and $p^{(k)}$. An issue in solving the original minimization problem for $f(x)$ is how to allocate the effort between determining a good $p^{(k)}$ and choosing a good $\alpha^{(k)}$. Rather than solving the minimization problem to find the best value of $\alpha^{(k)}$ for the kth direction, it may be better to get a reasonable approximation, and move on to choose another direction from the new point.

One approach to choosing a good value of $\alpha^{(k)}$ is to use a simple approximation to the one-dimensional function we are trying to minimize:

$$\rho(\alpha) = f\left(x^{(k)} + \alpha p^{(k)}\right).$$

A useful approximation is a second- or third-degree polynomial that interpolates $\rho(\alpha)$ at three or four nearby points. The minimum of the polynomial can be found easily, and the point of the minimum may be a good choice for $\alpha^{(k)}$.

A simpler approach, assuming $\rho(\alpha)$ is unimodal over some positive interval, say $[\alpha_l, \alpha_u]$, is just to perform a direct search along the path $p^{(k)}$. A bisection method or some other simple method for finding a zero of a function as we discussed in Section 6.1 could be modified and used.

Another approach for developing a direct search method is to choose two points α_1 and α_2 in $[\alpha_l, \alpha_u]$, with $\alpha_1 < \alpha_2$, and then, based on the function values of ρ, to replace the interval $I = [\alpha_l, \alpha_u]$ with either $I_l = [\alpha_l, \alpha_2]$ or $I_u = [\alpha_1, \alpha_u]$. In the absence of any additional information about ρ, we choose the points α_1 and α_2 symmetrically, in such a way that the lengths of both I_l and I_u are the same proportion, say τ, of the length of the original interval I. This leads to $\tau^2 = 1 - \tau$, the golden ratio. The search using this method of reduction is called the golden section search, and is given in Algorithm 6.7.

Algorithm 6.7 Golden Section Search

0. Set $\tau = \left(\sqrt{5} - 1\right)/2$ (the golden ratio).
   Set $\alpha_1 = \alpha_l + (1 - \tau)(\alpha_u - \alpha_l)$ and set $\alpha_2 = \alpha_l + \tau(\alpha_u - \alpha_l)$.
   Set $\rho_1 = \rho(\alpha_1)$ and $\rho_2 = \rho(\alpha_2)$.
1. If $\rho_1 > \rho_2$,
   1.a. set $\alpha_l = \alpha_1$,
   set $\alpha_1 = \alpha_2$,
   set $\alpha_2 = \alpha_l + \tau(\alpha_u - \alpha_l)$,
   set $\rho_1 = \rho_2$, and
   set $\rho_2 = \rho(\alpha_2)$;
   otherwise,
   1.b. set $\alpha_u = \alpha_2$,
   set $\alpha_2 = \alpha_1$,
   set $\alpha_1 = \alpha_l + (1 - \tau)(\alpha_u - \alpha_l)$,
   set $\rho_2 = \rho_1$, and
   set $\rho_1 = \rho(\alpha_1)$.
2. If converged(α_u, α_l),
   terminate and return the solution as α_1;
   otherwise,
   go to step 1. ∎

The golden section search is robust, but it is only linearly convergent, like the bisection method of Algorithm 6.2. (This statement about convergence applies just to this one-dimensional search, which is a subproblem in our optimization problem of interest.)

Another criterion for a direct search is to require

$$f\left(x^{(k)} + \alpha^{(k)} p^{(k)}\right) \leq f\left(x^{(k)}\right) + \tau \alpha^{(k)} \left(p^{(k)}\right)^{\mathrm{T}} \nabla f\left(x^{(k)}\right), \tag{6.26}$$

for some τ in $\left(0, \frac{1}{2}\right)$. This criterion is called the sufficient decrease condition, and the approach is called the Goldstein-Armijo method after two early investigators of the technique. After choosing τ, the usual procedure is to choose α as the largest value in $1, \frac{1}{2}, \frac{1}{4}, \frac{1}{8}, \ldots$ that satisfies the inequality.

If the step length is not too long, the descent at $x^{(k)}$ in the given direction will be greater than the descent in that direction at $x^{(k)} + \alpha^{(k)} p^{(k)}$. This leads to the so-called curvature condition:

$$\left| \left(p^{(k)}\right)^{\mathrm{T}} \nabla f\left(x^{(k)} + \alpha^{(k)} p^{(k)}\right) \right| \leq \eta \left| \left(p^{(k)}\right)^{\mathrm{T}} \nabla f\left(x^{(k)}\right) \right|, \tag{6.27}$$

for some η in $(0, 1)$.

Steepest Descent

We now turn to the problem of choosing a descent direction. Most methods we will consider are gradient methods, that is, they satisfy (6.23):

$$Rp = -\nabla f(x),$$

From a given point $x^{(k)}$, the function f decreases most rapidly in the direction of the negative gradient, $-\nabla f\left(x^{(k)}\right)$. A greedy algorithm uses this *steepest descent* direction; that is,

$$p^{(k)} = -\nabla f\left(x^{(k)}\right), \tag{6.28}$$

and so the update in equation (6.24) is

$$x^{(k+1)} = x^{(k)} - \alpha^{(k)} \nabla f\left(x^{(k)}\right).$$

The step length factor $\alpha^{(k)}$ is chosen by a line search method described beginning on page 263.

The steepest descent method is robust so long as the gradient is not zero. The method, however, is likely to change directions often, and the zigzag approach to the minimum may be quite slow (see Exercise 6.10a). For a function with circular contours, steepest descent proceeds quickly to the solution. For a function whose contours are ellipses, as the function in Exercise 6.10 (page 302), for example, the steepest descent steps will zigzag toward the solution. A matrix other than the identity may deform the elliptical contours so they are more circular. In Newton's method discussed next, we choose the Hessian as that matrix.

Newton's Method for Unconstrained Optimization

To find the minimum of the scalar-valued function $f(x)$, under the assumptions that f is convex and twice differentiable, we can seek the zero of $\nabla f(x)$ in the same way that we find a zero of a vector-valued function using the iteration in equation (6.16), page 260. We begin by forming a first-order Taylor series expansion of $\nabla f(x)$, which is the second-order expansion of $f(x)$. In place of a vector-valued function we have the gradient of the scalar-valued function, and in place of a Jacobian, we have the Hessian H_f, which is the Jacobian of the gradient.

This first-order Taylor series expansion of ∇f is equivalent to a second-order Taylor series expansion of f. Setting the gradient to zero, we obtain an iteration similar to equation (6.16):

$$x^{(k+1)} = x^{(k)} - \left(\mathrm{H}_f\left(x^{(k)}\right)\right)^{-1} \nabla f\left(x^{(k)}\right). \qquad (6.29)$$

Use of this recursive iteration is Newton's method. The method is also often called the Newton-Raphson method. (Joseph Raphson, was a late seventeenth century English mathematician, who developed this same iteration, unaware that Newton had used the same method several years earlier.)

In one dimension, the Newton recursion is just

$$\begin{aligned} x^{(k+1)} &= x^{(k)} - \frac{\nabla f\left(x^{(k)}\right)}{\nabla^2 f\left(x^{(k)}\right)} \\ &= x^{(k)} - \frac{f'\left(x^{(k)}\right)}{f''\left(x^{(k)}\right)}. \end{aligned}$$

The second-order Taylor series approximation to f about the point x_*,

$$f(x) \approx f(x_*) + (x - x_*)^{\mathrm{T}} \nabla f(x_*) + \frac{1}{2}(x - x_*)^{\mathrm{T}} \mathrm{H}_f(x_*)(x - x_*), \qquad (6.30)$$

is exact if f is a quadratic function. In that case, H_f is positive definite, and the terms in equation (6.29) exist and yield the solution immediately. When f is not quadratic, but is sufficiently regular, we can build a sequence of approximations by quadratic expansions of f about approximate solutions. This means, however, that the Hessian may not be positive definite and its inverse in (6.29) may not exist.

Once more, it is important to state that *we do not necessarily compute each term in an expression.*

> *The form of a mathematical expression and the way the expression should be evaluated in actual practice may be quite different.*

We choose mathematical expressions for their understandability; we choose computational method for their robustness, accuracy, and efficiency. Just as we commented on page 260 concerning inversion of the Jacobian, we comment here that we *do not* compute the Hessian and then compute its inverse, just because that appears in equation (6.29). We solve the linear systems

$$\mathrm{H}_f(x^{(k)})p^{(k)} = -\nabla f(x^{(k)}) \tag{6.31}$$

by more efficient methods such as Cholesky factorizations. Once we have the solution to equation (6.31), equation (6.29) becomes

$$x^{(k+1)} = x^{(k)} + p^{(k)}. \tag{6.32}$$

Newton's method, by scaling the path by the Hessian, is more likely to point the path in the direction of a local minimum, whereas the steepest descent method, in ignoring the second derivative, follows a path along the gradient that does not take into account the rate of change of the gradient. This is illustrated in Figure 6.8.

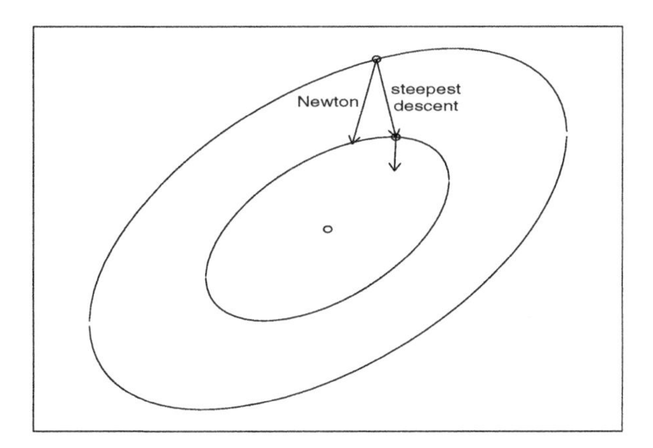

Fig. 6.8. Steepest Descent and Newton Steps

For functions that are close to a quadratic within a region close to the minimum, Newton's method can be very effective so long as the iterations begin close enough to the solution. In other cases Newton's method may be unreliable. The problems may be similar to those illustrated in Figures 6.3 and 6.4 (page 252) for finding a root.

One way of increasing the reliability of Newton's method is to use a damped version of the update (6.32),

$$x^{(k+1)} = x^{(k)} + \alpha^{(k)} p^{(k)},$$

for which a line search is used to determine an appropriate step length factor $\alpha^{(k)}$.

When the function is not quadratic, the Hessian may not be positive definite, and so a modified Cholesky factorization may be used. In this approach, positive quantities are added as necessary during the decomposition of the Hessian. This changes the linear system (6.31) to the system

$$\left(\mathrm{H}_f\left(x^{(k)}\right) + D^{(k)}\right) p^{(k)} = -\nabla f\left(x^{(k)}\right), \tag{6.33}$$

where $D^{(k)}$ is a diagonal matrix with nonnegative elements.

Another method of increasing the reliability of Newton's method is to restrict the movements to regions where the second-order Taylor expansion (6.30) is a good approximation. This region is called a "trust region". At the kth iteration, the second-order Taylor series approximation provides a scaled quadratic model $q^{(k)}$:

$$q^{(k)}(s) = f\left(x_*^{(k)}\right) + s^{\mathrm{T}} \nabla f\left(x_*^{(k)}\right) + \frac{1}{2} s^{\mathrm{T}} \mathrm{H}_f\left(x_*^{(k)}\right) s, \tag{6.34}$$

where $s = x - x_*^{(k)}$.

When the Hessian is indefinite, $q^{(k)}$ is unbounded below, so it is obviously not a good model of $f\left(x_*^{(k)} + s\right)$ if s is large. We therefore restrict $\|s\|$, or better, we restrict $\|D^{(k)} s\|$ for some scaling matrix $D^{(k)}$. For some $\tau^{(k)}$, we require

$$\|D^{(k)} s\| < \tau^{(k)}, \tag{6.35}$$

and we choose $s^{(k)}$ as the point where the quadratic $q^{(k)}$ achieves its minimum subject to this restriction. How much we should restrict s depends on how good the quadratic approximation is. If

$$\frac{f\left(x_*^{(k)}\right) - f\left(x_*^{(k)} + s^{(k)}\right)}{f\left(x_*^{(k)}\right) - q^{(k)}\left(s^{(k)}\right)}$$

is close to 1, that is, if the approximation is good, we increase $\tau^{(k)}$; if it is small or negative, we decrease $\tau^{(k)}$. Implementation of these methods requires some rather arbitrary choices of algorithm parameters.

Accuracy of Optimization Using Gradient Methods

The problem of finding a minimum of a function is somewhat more difficult than that of finding a zero of a function discussed in Section 6.1. Our intuition should tell us this is the case. In one dimension, a zero of a function can be determined by successively bracketing a zero with two points. An interval containing a minimum of a function requires three points to determine it.

Another way of comparing the accuracy of the solution of a nonlinear equation and the determination of the minimum of such an equation is to consider the Taylor expansion:

$$f(x) = f(\tilde{x}) + (x - \tilde{x})f'(\tilde{x}) + \frac{1}{2}(x - \tilde{x})^2 f''(\tilde{x}) + \cdots .$$

In the problem of finding a zero x_0, $f'(x_0)$ is generally nonzero, and for $\tilde{x}$ close to x_0, $(f(x) - f(\tilde{x}))$ is approximately proportional to $(x - \tilde{x})$, where the constant of proportionality is $f'(\tilde{x})$. A small value of the difference $(x - \tilde{x})$ results in a proportionate difference $(f(x) - f(\tilde{x}))$. On the other hand, in the problem of finding the minimum x_*, $f'(x_*)$ is zero, and for $\tilde{x}$ close to x_*, $(f(x) - f(\tilde{x}))$ is approximately proportional to $(x - \tilde{x})^2$, where the constant of proportionality is $f''(\tilde{x})$. A small value of the difference $(x - \tilde{x})$ results in a smaller difference $(f(x) - f(\tilde{x}))$. In finding roots of an equation we may set a convergence criterion proportional to the machine epsilon, ϵ_{mach}. In optimization problems, we often set a convergence criterion proportional to $\sqrt{\epsilon_{\text{mach}}}$.

Quasi-Newton Methods

All gradient descent methods determine the path of the step by the system of equations,

$$R^{(k)} p^{(k)} = -\nabla f\big(x^{(k)}\big). \tag{6.36}$$

The difference in the methods is the matrix $R^{(k)}$.

The steepest descent method chooses $R^{(k)}$ as the identity, I, in these equations. As we have seen, for functions with eccentric contours, the steepest descent method traverses a zigzag path to the minimum. Newton's method chooses $R^{(k)}$ as the Hessian, $\mathrm{H}_f\big(x_*^{(k)}\big)$, which results in a more direct path to the minimum. Aside from the issues of consistency of the resulting equation (6.33) and the general problems of reliability, a major disadvantage of Newton's method is the computational burden of computing the Hessian, which is $\mathrm{O}(m^2)$ function evaluations, and solving the system, which is $\mathrm{O}(m^3)$ arithmetic operations, at each iteration.

Instead of using the Hessian at each iteration, we may use an approximation, $B^{(k)}$. We may choose approximations that are simpler to update and/or that allow the equations for the step to be solved more easily. Methods using such approximations are called *quasi-Newton* methods or *variable metric* methods.

Because

$$\mathrm{H}_f\big(x^{(k+1)}\big)\big(x^{(k+1)} - x^{(k)}\big) \approx \nabla f\big(x^{(k+1)}\big) - \nabla f\big(x^{(k)}\big),$$

we choose $B^{(k+1)}$ so that

$$B^{(k+1)}\big(x^{(k+1)} - x^{(k)}\big) = \nabla f\big(x^{(k+1)}\big) - \nabla f\big(x^{(k)}\big). \tag{6.37}$$

This is called the secant condition. (Note the similarity to the secant method for finding a zero discussed in Section 6.1.)

We express the secant condition as

$$B^{(k+1)}s^{(k)} = y^{(k)}, \tag{6.38}$$

where

$$s^{(k)} = x^{(k+1)} - x^{(k)}$$

and

$$y^{(k)} = \nabla f(x^{(k+1)}) - \nabla f(x^{(k)}).$$

The system of equations in (6.38) does not fully determine $B^{(k)}$ of course. Because $B^{(k)}$ is approximating $\mathrm{H}_f(x^{(k)})$, we may want to require that it be symmetric and positive definite.

The most common approach in quasi-Newton methods is first to choose a reasonable starting matrix $B^{(0)}$ and then to choose subsequent matrices by additive updates,

$$B^{(k+1)} = B^{(k)} + B_a^{(k)},$$

subject to preservation of symmetry and positive definiteness.

The general steps in a quasi-Newton method are

0. Set $k = 0$ and choose $x^{(k)}$ and $B^{(k)}$.
1. Compute $s^{(k)}$ as $\alpha^{(k)}p^{(k)}$, where
   $B^{(k)}p^{(k)} = -\nabla f(x^{(k)})$.
2. Compute $x^{(k+1)}$ and $\nabla f(x^{(k+1)})$.
3. Check for convergence and stop if converged.
4. Compute $B^{(k+1)}$.
5. Set $k = k + 1$, and go to 1.

Within these general steps there are two kinds of choices to be made: the way to update the approximation $B^{(k)}$, and, as usual, the choice of the step length factor $\alpha^{(k)}$.

There are several choices for the update $B_a^{(k)}$ that preserve symmetry and positive definiteness (or at least nonnegative definiteness). One simple choice is the rank-one symmetric matrix

$$B_a^{(k)} = \frac{1}{(y^{(k)} - B^{(k)}s^{(k)})^{\mathrm{T}}s^{(k)}}\,(y^{(k)} - B^{(k)}s^{(k)})\,(y^{(k)} - B^{(k)}s^{(k)})^{\mathrm{T}}. \tag{6.39}$$

This update results in a symmetric matrix that satisfies the secant condition no matter what the previous matrix $B^{(k)}$ is. (You are asked to do the simple algebra to show this in Exercise 6.11.) If $B^{(k)}$ is positive definite, this update results in a positive definite matrix $B^{(k+1)}$ so long as $c^{(k)} \leq 0$, where $c^{(k)}$ is the denominator:

$$c^{(k)} = (y^{(k)} - B^{(k)}s^{(k)})^{\mathrm{T}}s^{(k)}.$$

Even if $c^{(k)} > 0$, positive definiteness can be preserved by shrinking $c^{(k)}$ to $\tilde{c}^{(k)}$ so that

$$\tilde{c}^{(k)} < \frac{1}{(y^{(k)} - B^{(k)}s^{(k)})^{\mathrm{T}} \, (B^{(k)})^{(-1)} \, (y^{(k)} - B^{(k)}s^{(k)})}.$$

Although this adjustment is not as difficult as it might appear, the computations to preserve positive definiteness and, in general, good condition of the $B^{(k)}$ account for a major part of the effort in quasi-Newton methods.

Other, more common choices for $B_a^{(k)}$ are the rank-two *Broyden updates* of the form

$$\begin{aligned} B_a^{(k)} &= -\frac{1}{(s^{(k)})^{\mathrm{T}}B^{(k)}s^{(k)}} \, B^{(k)}s^{(k)}(B^{(k)}s^{(k)})^{\mathrm{T}} \\ &\quad + \frac{1}{(y^{(k)})^{\mathrm{T}}s^{(k)}} \, y^{(k)}(y^{(k)})^{\mathrm{T}} \\ &\quad + \sigma^{(k)}\left((s^{(k)})^{\mathrm{T}}B^{(k)}s^{(k)}\right) v^{(k)}\left(v^{(k)}\right)^{\mathrm{T}}, \end{aligned} \tag{6.40}$$

where $\sigma^{(k)}$ is a scalar in $[0, 1]$, and

$$v^{(k)} = \frac{1}{(y^{(k)})^{\mathrm{T}}s^{(k)}} y^{(k)} - \frac{1}{(s^{(k)})^{\mathrm{T}}} B^{(k)}s^{(k)}B^{(k)}s^{(k)}.$$

Letting $\sigma^{(k)} = 0$ in (6.40) yields the Broyden-Fletcher-Goldfarb-Shanno (BFGS) update, which is one of the most widely used methods. If $\sigma^{(k)} = 1$, the method is called the Davidon-Fletcher-Powell (DFP) method.

The Broyden updates will preserve the positive definiteness of $B^{(k)}$ so long as

$$(y^{(k)})^{\mathrm{T}}s^{(k)} > 0.$$

This is the curvature condition (see (6.27) on page 265). If the curvature condition is not satisfied, $s^{(k)}$ could be scaled so as to satisfy this inequality. (Scaling $s^{(k)}$ of course changes $y^{(k)}$ also.) Alternatively, the update of $B^{(k)}$ can just be skipped, and the updated step is determined using the previous value, $B^{(k)}$. This method is obviously quicker, but it is not as reliable.

Inspection of either the rank-one updates (6.39) or the rank-two updates (6.40) reveals that the number of computations is $\mathrm{O}(m^2)$. If the updates are done to the inverses of the $B^{(k)}$'s or to their Cholesky factors, the computations required for the updated directions are just matrix-vector multiplications and hence can also be computed in $\mathrm{O}(m^2)$ computations.

It is easily seen that the updates can be done to the inverses of the $B^{(k)}$'s using the Sherman-Morrison formula (equation (5.50) on page 227) for rank-one updates, or the Woodbury formula (equation (5.52)) for more general updates. Using the Woodbury formula, the BFGS update, for example, results in the recursion,

$$\left(B^{(k+1)}\right)^{-1} =$$
$$\left(I - \frac{1}{(y^{(k)})^{\mathrm{T}} s^{(k)}} s^{(k)} (y^{(k)})^{\mathrm{T}}\right) \left(B^{(k)}\right)^{-1} \left(I - \frac{1}{(y^{(k)})^{\mathrm{T}} s^{(k)}} s^{(k)} (y^{(k)})^{\mathrm{T}}\right)$$
$$+ \frac{1}{(y^{(k)})^{\mathrm{T}} s^{(k)}} s^{(k)} (y^{(k)})^{\mathrm{T}}. \tag{6.41}$$

The best way of doing the inverse updates is to perform them on the Cholesky factors instead of on the inverses. The expression above for updating the inverse shows that this can be done.

Another important property of the quasi-Newton methods is that they can be performed without explicitly storing the $B^{(k)}$'s, which could be quite large in large-scale optimization problems. The storage required in addition to that for $B^{(k)}$ is for the vectors $s^{(k)}$ and $y^{(k)}$. If $B^{(k)}$ is a diagonal matrix, the total storage is $\mathrm{O}(m)$. In computing the update at the $(k+1)^{\text{th}}$ iteration, *limited-memory quasi-Newton* methods assume that $B^{(k-j)}$ is diagonal at some previous iteration. The update for the $(k+1)^{\text{th}}$ iteration can be computed by vector-vector operations beginning back at the $(k-j)^{\text{th}}$ iteration. In practice, diagonality is assumed at the fourth or fifth previous iteration; that is, j is taken as 4 or 5.

Quasi-Newton methods are available in most of the widely-used mathematical software packages. Broyden updates are the most commonly used in these packages, and of the Broyden updates, BFGS is probably the most popular. Some empirical results indicate, however, that the simple rank-one update (6.39) is often an adequate method.

Truncated Newton Methods

Another way of reducing the computational burden in Newton-type methods is to approximate the solution of the path direction

$$R^{(k)} p^{(k)} = -\nabla f\left(x^{(k)}\right),$$

where $R^{(k)}$ is either the Hessian, as in Newton's method, or an approximation, as in a quasi-Newton method. In a *truncated Newton method*, instead of solving for $p^{(k)}$, we get an approximate solution using only a few steps of an iterative linear equation solver, such as the conjugate gradient method. The conjugate gradient method is particularly suitable because it uses only matrix-vector products, so the matrix $R^{(k)}$ need not be stored. This can be very important in large-scale optimization problems that involve a large number of decision variables. How far to continue the iterations in the solution of the linear system is a major issue in tuning a truncated Newton method.

Nelder-Mead Simplex Method

The Nelder-Mead simplex method (Nelder and Mead, 1965) is a derivative-free, direct search method. The steps are chosen so as to ensure a local descent, but neither the gradient nor an approximation to it is used. In this method, to find the minimum of a function, f, of m variables, a set of $m+1$ extreme points (a *simplex*) is chosen to start with, and iterations proceed by replacing the point that has the largest value of the function with a point that has a smaller value. This yields a new simplex, and the procedure continues. The method is given in Algorithm 6.8 and illustrated for a bivariate function in Figure 6.9.

Algorithm 6.8 Nelder-Mead Simplex Method

0. Set tuning factors: reflection coefficient, $\alpha > 0$; expansion factor, $\gamma > 1$; contraction factor, $0 < \beta < 1$; and shrinkage factor, $0 < \delta < 1$.
   Choose an initial simplex, that is, $m+1$ extreme points (points on the vertices of a convex hull).
1. Evaluate f at each point in the current simplex, obtaining the values
   $$f_1 \le f_2 \le \cdots \le f_m \le f_{m+1}.$$
   Label the points correspondingly, that is, let x_{m+1} correspond to f_{m+1}, and so on.
2. Reflect the worst point: let $x_{\rm r} = (1+\alpha)x_{\rm a} - \alpha x_{m+1}$, where $x_{\rm a} = \sum_{i=1}^m x_i/m$, and let $f_{\rm r} = f(x_{\rm r})$.
3. If $f_1 \le f_{\rm r} \le f_m$, accept reflection: replace x_{m+1} by $x_{\rm r}$, and go to step 6.
4. If $f_{\rm r} < f_1$, compute expansion: $x_{\rm e} = \gamma x_{\rm r} + (1-\gamma)x_{\rm a}$.
   If $f(x_{\rm e}) < f_1$,
      4.a. accept expansion: replace x_{m+1} by $x_{\rm a}$;
   otherwise,
      4.b. replace x_{m+1} by $x_{\rm r}$.
   Go to step 6.
5. If $f_m < f_{\rm r} < f_{m+1}$, let $f_{\rm h} = f_{\rm r}$; otherwise, let $f_{\rm h} = f_{m+1}$. Let $x_{\rm h}$ be the corresponding point. Compute contraction: $x_{\rm c} = \beta x_{\rm h} + (1-\beta)x_{\rm a}$.
   If $f(x_{\rm c}) \le f(x_{\rm h})$,
      5.a. accept contraction: replace x_{m+1} by $x_{\rm c}$;
   otherwise,
      5.b. shrink simplex: for $i = 2, 3, \ldots, m+1$, replace x_i by
   $$\delta x_i + (1-\delta)x_1.$$
6. If convergence has not occurred (see below) or if a preset limit on the number of iterations has not been exceeded, go to step 1;
   otherwise, return the solution as x_1. ∎

There are three common ways of assessing convergence of the Nelder-Mead algorithm. All three, or variations of them, may be used together.

- The amount of variation in the function values at the simplex points. This is measured by the sample variance,

$$s_f^2 = \frac{1}{m+1} \sum (f_i - \bar{f})^2,$$

where $\bar{f}$ is the sample mean of $f_1, f_2, \ldots, f_{m+1}$. Convergence is declared if $s_f^2 < \epsilon$. This stopping criterion can lead to premature convergence, just because the simplex points happen to lie close to the same level curve of the function.

- The total of the norms of the differences in the points in the new simplex and those in the previous simplex. (In any iteration except shrinkage, there is only one point that is replaced.) This is one of several possible stopping criteria.
- The size of the simplex, as measured by

$$\frac{\max \|x_i - x_1\|}{\max(1, \|x_1\|)}.$$

The iterations are terminated when this measure is sufficiently small.

Figure 6.9 illustrates one iteration of the algorithm in a two-dimensional problem. In two dimensions, the iterations are those of a triangle tumbling downhill vertex over edge and deforming itself as it goes.

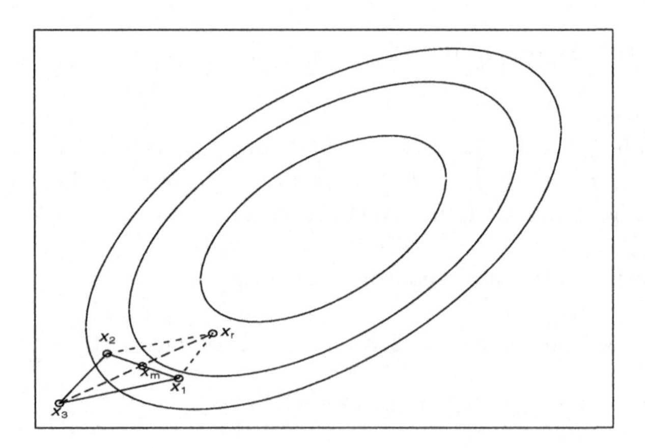

Fig. 6.9. One Nelder-Mead Iteration (In this step, x_2 becomes x_3; x_1 becomes x_2; and x_r becomes x_1.)

Although the Nelder-Mead algorithm may be slow to converge, it is a very useful method for several reasons. The computations in any iteration of the algorithm are not extensive. No derivatives are needed; in fact, not even the function values themselves are needed, only their relative values. The method is therefore well-suited to noisy functions; that is functions that cannot be evaluated exactly.

There have been many suggestions for improving the Nelder-Mead method. Most have concentrated on the stopping criteria or the tuning parameters. The various tuning parameters allow considerable flexibility, but there are no good general guidelines for their selection.

It is a simple matter to introduce randomness in the decisions made at various points in the Nelder-Mead algorithm, similar to what we do in stochastic methods, such as simulated annealing, which we discuss beginning on page 277. This may be useful for finding the global optimum of a function with many local optima. If some decisions are made randomly, however, the convergence criteria must be modified to reflect the fact that the iterations may no longer be strictly descending.

6.3 Unconstrained Combinatorial and Stochastic Optimization

If the objective function is differentiable and the derivatives are available, methods described in the previous section that make use of the gradient and Hessian or simple approximations to the gradient and Hessian are usually the most effective ones. Even if the derivatives are not available or do not exist everywhere for a continuous objective function, methods that use approximations to gradients are usually best.

If the objective function is not differentiable, or if it is very rough, some kind of direct search for the optimum may be necessary. In some cases the objective function is noisy, perhaps with an additive random error term that prevents exact evaluation. In these cases also it may not be effective to use gradient or approximate-gradient methods.

Another important type of optimization problem is one in which the decision variables are discrete. The solution may be a configuration of a finite set of points, that is, a graph. In the traveling salesperson problem, for example, we seek a configuration of cities that provides a path with minimal total length that visits each point in a set. In the vehicle routing problem, a fleet of vehicles stationed at a depot must make deliveries to a set of cities, and it is desired to route them so as to minimize the time required to make all the deliveries. In a resource scheduling problem, a set of machines or workers are to be assigned to a set of tasks, so as to minimize the time required to complete all the tasks, or so as to minimize idle time of the resources. These kinds of problems are examples of combinatorial optimization.

In combinatorial optimization problems it is often more natural to refer to the points in the domain as "states". The objective is to minimize a scalar-valued function of the states, $f(s)$. We will use "point" and "state" interchangeably in the following.

Although the state space is countable, it is often extremely large, and so we must use stochastic methods that do not consider every point in the space.

Search Methods

Direct search methods move from point to point using only the values of the function; they do not use derivative information, or approximations to derivatives. In some methods new points are chosen randomly, and then the decision to move to a new point is based on the relative values of the function at the old and new points. A tree or other graph of points may help to organize the points to visit in the search. There are several variations of direct searches. Some search methods use heuristics that mimic certain natural systems.

Sometimes, based on points that have already been evaluated, sets of other points can be ruled out. In tree-based search methods, such *fathoming* or *branch-and-bound* techniques may greatly enhance the overall efficiency of the search. "Tabu" methods keep lists of points that are not likely to lead to an optimum.

If the state space is relatively small, or if good fathoming or branch-and-bound techniques are available, it may be possible to do an exhaustive search; that is, a search in which every point in the state space is considered, either explicitly or implicitly. In other cases, a random selection scheme may be used that initially gives every point in the state space a positive probability of being considered. At the end of the iterations, however, not all points in the space may have been considered.

In all direct search methods the new points are accepted or rejected based on the objective function values. Some search methods allow iterations that do not monotonically decrease the objective function values. These methods are especially useful when there are local minima.

In these iterations, if the new point is better, then it is used for picking a point in the next iteration.

If the new point is not better, there are three possible actions:

- discard the point and find another one to consider
- accept the new point anyway
- declare the search to have converged

Random decisions may be made in two places in this general scheme. First, the new point may be chosen randomly. Of course, this does not necessarily mean with uniform probability. Any knowledge of the state space, or any information from previous iterations may be used to put a probability distribution on the state space.

Secondly, if the new candidate point is not better than the current point the decision to accept it may be made randomly. The probability distribution is Bernoulli and its parameter (probability of accepting) may depend on how much worse the candidate point is, and on the current count of iterations.

Convergence

In the methods for optimization of continuous functions over dense domains, the convergence criteria are based on interval lengths, which may be used in a norm, as indicated by our discussion of the "converged($d_1, d_2, \ldots$)" function on page 244. If termination of the algorithm occurs due to an excessive number of iterations, we declare that the algorithm had not converged.

In many cases in which the domain is discrete, there are no reasonable norms that can be used over the domain. We may, however, identify some ad hoc measure that indicates the amount of movement within the domain, and we can easily measure the distance that the objective function changes from one point to another. Nevertheless, often the main basis for terminating an algorithm for optimization over a discrete domain is the number of iterations. In that case, unless the algorithm is exhaustive, we cannot say that it has converged. This is a situation that is endemic to the problem. We must rely on indications that it is very likely that the algorithm converged to a correct solution.

Simulated Annealing

Simulated annealing is a method that simulates the thermodynamic process in which a metal is heated to its melting temperature and then is allowed to cool slowly so that its structure is frozen at the crystal configuration of lowest energy. In this process the atoms go through continuous rearrangements, moving toward a lower energy level as they gradually lose mobility due to the cooling. The rearrangements do not result in a monotonic decrease in energy, however. The density of energy levels at a given temperature ideally is exponential, the so-called Boltzmann distribution, with a mean proportional to the absolute temperature. (The constant of proportionality is called "Boltzmann's constant"). This is analogous to a sequence of optimization iterations that occasionally go uphill. If the function has local minima, going uphill occasionally is desirable.

Metropolis et al. (1953) developed a stochastic relaxation technique that simulates the behavior of a system of particles approaching thermal equilibrium. (This is the same paper that they described the Metropolis sampling algorithm, selected as one of the Top 10 algorithms of the twentieth century; see page 138.) The energy associated with a given configuration of particles is compared to the energy of a different configuration. If the energy of the new configuration is lower than that of the previous one, the new configuration is immediately accepted. If the new configuration has a larger energy, it is accepted with a nonzero probability. This probability is larger for small increases than for large increases in the energy level. One of the main advantages of simulated annealing is that the process is allowed to move away from a local optimum.

Although the technique is heuristically related to the cooling of a metal, it can be successfully applied to a broader range of problems. It can be used in many kinds of optimization problem, but it is particularly useful in problems that involve configurations of a discrete set, such as a set of particles whose configuration can continuously change, or a set of cities in which the interest is an ordering for shortest distance of traversal.

The Basic Algorithm

In simulated annealing, a "temperature" parameter controls the probability of moving uphill; when the temperature is high, the probability of acceptance of any given point is high, and the process corresponds to a pure random walk. When the temperature is low, however, the probability of accepting any given point is low; and in fact, only downhill points are accepted. The behavior at low temperatures corresponds to a gradient search.

As the iterations proceed and the points move lower on the surface (it is hoped), the temperature is successively lowered. An "annealing schedule" determines how the temperature is adjusted.

In the description of simulated annealing in Algorithm 6.9, recognizing the common applications in combinatorial optimization, we refer to the argument of the objective function as a "state", rather than as a "point". We also describe the convergence slightly differently from how we have done it in the deterministic algorithms. The steps in Algorithm 6.9 are generic. A particular step, such as "generate a new state ..." may mean quite different things in different problems. Following the general statement of the algorithm, we consider some specific methods.

Algorithm 6.9 Simulated Annealing

0. Set $k = 1$ and initialize state s.
1. Compute the temperature $T(k)$.
2. Set $i = 0$ and $j = 0$.
3. Generate a new state r and compute $\delta f = f(r) - f(s)$.
4. Based on δf, decide whether to move from state s to state r.
   If $\delta f \leq 0$,
      accept state r;
   otherwise,
      accept state r with a probability $P(\delta f, T(k))$.
   If state r is accepted, set $s = r$ and $i = i + 1$.
5. If i is equal to the limit for the number of successes at a given temperature, go to step 1.
6. Set $j = j + 1$. If j is less than the limit for the number of iterations at given temperature, go to step 3.

7. If $i = 0$,
   deliver s as the optimum; otherwise,
   if $k < k_{\max}$,
      set $k = k + 1$ and go to step 1;
   otherwise,
      issue message that
      'algorithm did not converge in $k_{\max}$ iterations'. ∎

For optimization of a continuous function over a region, the state is a point in that region. A new state or point may be selected by choosing a radius r and point on the d dimensional sphere of radius r centered at the previous point. For a continuous objective function, the movement in step 3 of Algorithm 6.9 may be a random direction to step in the domain of the objective function. In combinatorial optimization, the selection of a new state in step 3 may be a random rearrangement of a given configuration, as we mention below for the traveling salesperson problem.

Parameters of the Algorithm: The Probability Function

There are a number of tuning parameters that must be chosen in the simulated annealing algorithm. These include such relatively simple things as the number of repetitions or when to adjust the temperature. The probability of acceptance and the type of temperature adjustments present more complicated choices.

One approach is to assume that at a given temperature, T, the states have a known probability density (or set of probabilities, if the set of states is countable), $p_S(s, T)$, and then to define an acceptance probability to move from state s_k to s_{k+1} in terms of the relative change in the probability density from $p_S(s_k, T)$ to $p_S(s_{k+1}, T)$. In the original applications, the objective function was the energy of a given configuration, and the probability of an energy change of δf at temperature T is proportional to $\exp(-\delta f/T)$.

Even when there is no underlying probability model, the probability in step 4 of Algorithm 6.9 is often taken as

$$P(\delta f, T(k)) = e^{-\delta f/T(k)}, \tag{6.42}$$

although a completely different form could be used. The exponential distribution models energy changes in ensembles of molecules, but otherwise it has no intrinsic relationship to a given optimization problem.

The probability can be tuned in the early stages of the computations so that some reasonable proportion of uphill steps are taken. In some optimization problems, the value of the function at the optimum, f^*, is known, and the problem is only to determine the location of the optimum. In such cases a factor $(f - f^*)^g$ could be used in the probability of acceptance. If the value f^* is not known but a reasonable estimate is available, it could be used. The estimate could also be updated as the algorithm proceeds.

Parameters of the Algorithm: The Cooling Schedule

There are various ways the temperature can be updated in step 1.

The probability of the method converging to the global optimum depends on a slow decrease in the temperature. In practice, the temperature is generally decreased by some proportion of its current value:

$$T(k+1) = b(k)T(k), \tag{6.43}$$

for $0 < b(k) \leq 1$. We would like to decrease T as rapidly as possible, yet have a high probability of determining the global optimum. Geman and Geman (1984) showed that under the assumptions that the energy distribution is Gaussian and the acceptance probability is of the form (6.42), the probability of convergence goes to 1 if the temperature decreases as the inverse of the logarithm of the time, that is, if $b(k) = (\log(k))^{-1}$ in equation (6.43). Under the assumption that the energy distribution is Cauchy, a similar argument is based on $b(k) = k^{-1}$, and a uniform distribution over bounded regions is based on $b(k) = \exp(-c_k k^{1/d})$, where c_k is some constant, and d is the number of dimensions.

A constant temperature is often used in simulated annealing for optimization of continuous functions. A constant temperature may also be appropriate for optimization of noisy functions. The adjustments are usually taken as constants, rather than varying with k.

For functions of many continuous variables, it may be more efficient to use the basic simulated annealing approach on a sequence of lower-dimensional spaces. This approach can reduce the total number of computations, and would be particularly useful when the cost of evaluation of the function is very high.

In some cases it may desirable to exercise more control over the random walk that forms the basis of simulated annealing. For example, we may keep a list of "good" points, perhaps the m best points found so far. After some iterations, we may return to one or more of the good states and begin the walk anew.

We may use the number of times a point is visited to estimate the optimal solution.

Simulated annealing is often used in conjunction with other optimization methods, for example, to determine starting points for other optimization methods. Multiple starting points may allow the subsequent optimization method to identify more than one local optimum.

When gradient information is available, even in a limited form, simulated annealing is generally not as efficient as other methods that use that information. The main advantages of simulated annealing include its simplicity, its ability to move away from local optima, and the wide range of problems to which it can be applied.

It may be useful periodically to "re-anneal" by increasing the temperature. This might be done to get out of what might appear to be a local minimum.

In this case, the best value within that local area should be preserved in order to return to it in case no better points are found quickly.

Simulated annealing proceeds as a random walk through the domain of the objective function. There are many opportunities for parallelizing such a process. The most obvious is starting multiple walks on separate processors.

Applications

Simulated annealing has been successfully used in a range of optimization problems, including probability density smoothing classification, construction of minimum volume ellipsoids, and optimal experimental design.

The Canonical Example: The Traveling Salesperson Problem

The traveling salesperson problem can serve as a prototype of the problems in which the simulated annealing method has had good success. In this problem, a state is an ordered list of points ("cities"), and the objective function is the total distance between all the points in the order given (plus the return distance from the last point to the first point. One simple rearrangement of the list is the reversal of a sublist, that is, for example,

$$(1, \underline{2, 3, 4, 5, 6}, 7, 8, 9) \rightarrow (1, \underline{6, 5, 4, 3, 2}, 7, 8, 9).$$

Another simple rearrangement is the movement of a sublist to some other point in the list, for example,

$$(1, \underline{2, 3, 4, 5, 6}, 7, 8,_{\uparrow} 9) \rightarrow (1, 7, 8, \underline{2, 3, 4, 5, 6}, 9).$$

(Both of these rearrangements are called "2-changes", because in the graph defining the salesperson's circuit, exactly two edges are replaced by two others. The circuit is a Hamilton closed path.)

Evolutionary Algorithms

There are many variations of methods that use evolutionary strategies. These methods are inspired by biological evolution, and the descriptions of the methods often use terminology from biology. Genetic algorithms mimic the behavior of organisms in a competitive environment in which only the fittest and their offspring survive. Decision variables correspond to "genotypes" or "chromosomes"; a point or a state is represented by a string (usually a bit string); and new values of the decision variables are produced from existing points by "crossover" or "mutation". The set of points at any stage constitutes a "population". The points that survive from one stage to another are those yielding lower values of the objective function.

In most iterations it is likely that the new population includes a higher proportion of fit organisms (points yielding better values of the objective function) than the previous population, and that the best of the organisms is better than the best in the previous population.

Coding of Points

The first step in using a genetic algorithm is to define a coding of the points in the domain into strings that can be manipulated easily. One simple coding scheme is a binary representation of the index of each point. Of course, this must be preceded by an assignment of an index to each point. In some cases, if the points have an ordinal relationship, this indexing is natural. In other cases, such as the traveling salesperson problem, each point (or path) in the domain must be assigned a bit pattern following some heuristic scheme, or, lacking that, following arbitrary choices.

Evolution Method

Algorithm 6.10 provides an outline of a genetic algorithm. There are several decisions that must be made in order to apply the algorithm. The first, as mentioned above, is to decide how to represent the values of decision variables, that is, the states, in terms of chromosomes, and to decide how to evaluate the objective function in terms of a chromosome. Then, an initial population must be chosen.

Algorithm 6.10 Genetic Algorithm

0. Determine a representation of the problem, and define an initial population, $x_1^{(0)}, x_2^{(0)}, \ldots, x_n^{(0)}$. Set $k = 0$.
1. Compute the objective function (the "fitness") for each member of the population, $f(x_i^{(k)})$ and assign probabilities p_i to each item in the population, perhaps proportional to its fitness.
2. Choose (with replacement) a probability sample of size $m \leq n$. This is the reproducing population.
3. Randomly form a new population $x_1^{(k+1)}, x_2^{(k+1)}, \ldots, x_n^{(k+1)}$ from the reproducing population, using various mutation and recombination rules (see Table 6.2). This may be done using random selection of the rule for each individual of pair of individuals.
4. If convergence criteria are met, stop, and deliver $\arg\min_{x_i^{(k+1)}} f(x_i^{(k+1)})$ as the optimum; otherwise, set $k = k + 1$ and go to step 1. ∎

Mutation and Recombination Rules

There are several possibilities for producing a new generation of organisms from a given population. Some methods mimic sexual reproduction, that is, the combining of chromosomes from two organisms, and some methods are like asexual reproduction or mutation. A genetic algorithm may involve all of these methods, perhaps chosen randomly with fixed or varying probabilities.

Three simple methods are crossover, for combining two chromosomes, and inversion and mutation, for yielding a new chromosome from a single one. In crossover of two chromosomes each containing m bits, for a randomly selected j from 1 to l, the first j bits are taken from the chromosome of the first organism and the last $l-j$ bit are taken from the chromosome of the second organism. In inversion, for j and k randomly selected from 1 to l, the bits between positions j and k are reversed, while all others remain the same. In mutation, a small number of bits are selected randomly and are changed, from 0 to 1 or from 1 to 0. The number of bits to change may be chosen randomly, perhaps from a Poisson distribution, truncated at l. These operations are illustrated in Table 6.2.

Table 6.2. Reproduction Rules for a Genetic Algorithm

Generation k		Generation $k+1$
	Crossover	
$x_i^{(k)}$ 11001001		
	$\rightarrow$	$x_i^{(k+1)}$ 11011010
$x_j^{(k)}$ 00111010		
	Inversion	
$x_i^{(k)}$ 11101011	$\rightarrow$	$x_i^{(k+1)}$ 11010111
	Mutation	
$x_i^{(k)}$ 11101011	$\rightarrow$	$x_i^{(k+1)}$ 10111011
	Clone	
$x_i^{(k)}$ 11101011	$\rightarrow$	$x_i^{(k+1)}$ 11101011

In the example operations shown in Table 6.2, crossover occurs between the third and fourth bits; inversion occurs for the bits between (and including) the third and the sixth; and mutation occurs at the second and fourth bits.

As with simulated annealing, indeed, as with almost any optimization method, for a given problem, genetic algorithms may require a good deal of ad hoc tuning. In the case of genetic algorithms, there are various ways of encoding the problem, of adopting an overall strategy, and of combining organisms in a current population to yield the organisms in a subsequent population.

Genetic algorithms can be implemented in parallel rather directly.

Other Combinatorial Search Methods

There are a number of other methods of combinatorial optimization. One general type of method are guided direct search methods, in which at each stage there is an attempt to use the history of the search to choose new directions to explore.

Tabu search simulates the human memory process in maintaining a list of recent steps. The list is called a *tabu list.* The purpose of the list is to prevent the search from backtracking. Before a potential step is selected the search procedures checks the tabu list to determine if it is in the recent path to this point. The tabu list can be implemented by penalizing the objective function.

Artificial neural networks are another type of algorithm for decision making that is analogous to a biological process.

A number of other stochastic combinatorial search methods have been developed. Some of these methods derive from the stochastic approximations in the Robbins-Monro procedure (equation (6.11)).

6.4 Optimization under Constraints

The general optimization problem for a scalar-valued function in m variables with r constraints is

$$\begin{aligned} \min_x \quad & f(x) \\ \text{s.t. } & g(x) \leq b, \end{aligned} \tag{6.44}$$

where x is m-dimensional and $g(x) \leq b$ is a system of r inequalities. This formulation can include equality constraints by expressing an equality as two inequalities.

A point satisfying the constraints is called a *feasible point*, and the set of all such points is called the feasible region. For a given point x_j, a constraint g_i such that $g_i(x_j) = b_i$ is called an *active constraint.*

Any of the unconstrained optimization methods we have described can be modified to handle constraints by first insuring that the starting points satisfy the constraints and then explicitly incorporating checks at each iteration to insure that any new point also satisfies the constraints. If the new point does not satisfy the constraints, then some of the parameters of the algorithm may be adjusted and a new point generated (this is a possible approach in the Nelder-Mead simplex method, for example), or, in random methods such as simulated annealing, the new point is simply discarded and a new point chosen. Although this is a straightforward procedure, it is unlikely to be very efficient computationally.

Unconstrained methods can be used efficiently if a sequence of unconstrained problems that converges to a problem of interest can be defined. Although it may not be possible to evaluate the objective function in regions that are not feasible, this method can often be very effective.

Another approach to solving constrained problems is to incorporate the constraints into the objective function. One way in which this is done is by use of supplementary variables, as discussed below. Another way is to define transformations of the variables so that the objective function increases rapidly near constraint boundaries.

Constrained Optimization in Dense Domains

In a constrained optimization problem over a dense domain, the most important concerns are the shape of the feasible region and the smoothness of the objective function. The problem is much easier if the feasible region is convex, and fortunately many constrained real-world problems have convex feasible regions. The smoothness of the objective function is important, because if it is twice-differentiable, we may be able to use the known properties of derivatives at function optima to find those optima. For methods that incorporate the constraints into the objective function, the shape of the feasible region is important because the derivatives of the combined objective function depend on the functions defining the constraints.

Equality Constraints

We will first consider some simple problems. Equality constraints are generally much easier to handle than inequalities. This is a special case of problem (6.44) with a pair of inequalities, one a negative multiple of the other. The equality constraint problem is

$$\begin{aligned} \min_x \quad & f(x) \\ \text{s.t. } & g(x) = b. \end{aligned} \tag{6.45}$$

For any feasible point, all equality constraints are active constraints.

An optimization problem with equality constraints can often be transformed into an equivalent unconstrained optimization problem.

An important form of equality constraints are linear constraints, $Ax = b$, where A is an $r \times m$ (with $r \leq m$) matrix of rank s. With $g(x) = Ax$, we have

$$\begin{aligned} \min_x \quad & f(x) \\ \text{s.t. } & Ax = b. \end{aligned} \tag{6.46}$$

If the linear system is consistent (that is, $\text{rank}([A|b]) = s$), the feasible set is nonnull. The rank of A must be less than m, or else the constraints completely determine the solution to the problem. If the rank of A is less than r, however, some rows of A and some elements of b could be combined into a smaller number of constraints. We will therefore assume A is of full row rank; that is, $\text{rank}(A) = r$.

If x_c is any feasible point, that is, $Ax_c = b$, then any other feasible point can be represented as $x_c + p$, where p is any vector in the null space of A, $\mathcal{N}(A)$.

The dimension of $\mathcal{N}(A)$ is $m-r$, and its order is m. If B is an $m \times m-r$ matrix whose columns form a basis for $\mathcal{N}(A)$, all feasible points can be generated by $x_c + Bz$, where $z \in \mathrm{IR}^{m-r}$.

Hence, we need only consider the restricted variables

$$x = x_c + Bz, \tag{6.47}$$

and the function

$$h(z) = f(x_c + Bz). \tag{6.48}$$

The argument of this function is a vector with only $m - r$ elements, instead of m elements, as in the original function f. The unconstrained minimum of h, however, is the solution of the original constrained problem.

Now, if we assume differentiability, the gradient and Hessian of the reduced function can be expressed in terms of the the original function:

$$\begin{aligned} \nabla h(z) &= B^{\mathrm{T}} \nabla f(x_c + Bz) \\ &= B^{\mathrm{T}} \nabla f(x), \end{aligned} \tag{6.49}$$

and

$$\begin{aligned} \mathrm{H}_h(z) &= B^{\mathrm{T}} \mathrm{H}_f(x_c + Bz) B \\ &= B^{\mathrm{T}} \mathrm{H}_f(x) B. \end{aligned} \tag{6.50}$$

The relationship of the properties of stationary points to the derivatives are the conditions that determine a minimum of this reduced objective function; that is, x_* is a minimum if and only if

- $B^{\mathrm{T}} \nabla f(x_*) = 0$,
- $B^{\mathrm{T}} \mathrm{H}_f(x_*) B$ is positive definite, and
- $A x_* = b$.

These relationships then provide the basis for the solution of the optimization problem. Hence, the simple constrained optimization problem (6.45) can be solved using the same methods as discussed in Section 6.2.

Lagrange Multipliers

Consider again the equality-constrained problem (6.45) and the matrix B in equation (6.47). Because the $m \times m$ matrix $[B|A^{\mathrm{T}}]$ spans IR^m, we can represent the vector $\nabla f(x_*)$ as a linear combination of the columns of B and A^{T}, that is,

$$\nabla f(x_*) = (Bz_* | A^{\mathrm{T}} \lambda_*)^{\mathrm{T}},$$

where z_* is an $(m-r)$-vector and λ_* is an r-vector. Because $\nabla h(z_*) = 0$, Bz_* must also vanish, and we have

$$\begin{aligned} \nabla f(x_*) &= A^{\mathrm{T}} \lambda_* \\ &= \mathrm{J}_g(x_*)^{\mathrm{T}} \lambda_*. \end{aligned} \tag{6.51}$$

Thus, at the optimum, the gradient of the objective function is a linear combination of the columns of the Jacobian of the constraints. The elements of the linear combination vector λ_* are called *Lagrange multipliers.*

The condition expressed in (6.51) implies that the objective function cannot be reduced any further without violating the constraints.

We can see this in a simple example with equality constraints:

$$\begin{aligned} \min_x \quad & f(x) = 2x_1 + x_2 \\ \text{s.t. } & g(x) = x_1^2 - x_2 = 1. \end{aligned}$$

In this example the objective function is linear, and the single equality constraint is quadratic. The optimum is $x_* = (-1, 0)$. The gradient of $f(x)$ is $\nabla f(x) = (2, 1)$, that of $g(x)$ is $\nabla g(x) = (2x_1, -1)$, and $\nabla g(x_*) = (-2, -1)$. As we see in Figure 6.10 at the optimum,

$$\begin{aligned} \nabla f(x_*) &= -\nabla g(x_*) \\ &= -\mathrm{J}_g(x_*)^{\mathrm{T}}. \end{aligned}$$

The Lagrangian Function

The relationship between the gradient of the objective function and the Jacobian of the constraint function, motivates the definition of the *Lagrangian function*:

$$L(x, \lambda) = f(x) + \lambda^{\mathrm{T}}(g(x) - b), \tag{6.52}$$

where λ is an m-vector, the elements of which are the Lagrange multipliers.

The derivatives of the Lagrangian function can be analyzed in a manner similar to the analysis of the derivatives of the objective function to determine necessary and sufficiency conditions for a minimum subject to equality constraints.

Linear Programming

The basic linear program, which is often written as

$$\begin{aligned} \min_x \; & z = c^{\mathrm{T}}x \\ \text{s.t.} \quad & x \geq 0 \\ & Ax \leq b, \end{aligned} \tag{6.53}$$

is a problem over a dense domain. A solution to the problem, however, occurs at a vertex of the polytope formed by the constraints. (The polytope may be unbounded; that is, it may have “open” sides.) Because this is a finite set, the solution can be determined by inspecting a finite number of possibilities. It is in this sense that the linear programming problem is similar to other combinatorial optimization problems.

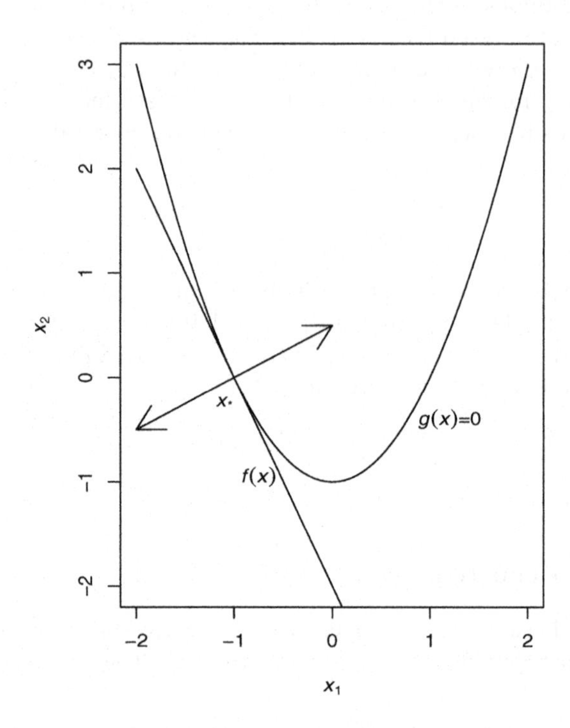

Fig. 6.10. Linear Objective and Quadratic Equality Constraint

The linear programming problem is generally most easily solved by a *simplex method*, which steps through the vertices efficiently.

More efficient methods for very large-scale linear programs are based on interior-point methods (see Griva, Nash, and Sofer, 2008, for a description). An interior-point method may proceed along interior points until the algorithm appears to slow, and then move to a vertex at some step and switch over to a simplex algorithm for the final iterations toward the solution x_*. The interior-point method uses a barrier function to proceed through the dense interior of the feasible region. This approach treats the problem a combinatorial optimization problem only in the latter stages.

Linear programming is a good example of how a specialized algorithm can perform very differently for some variation of the underlying optimization problem.

Special formulations of the simplex method make very significant differences in the speed of the solution. The problem of fitting a linear regression under the criterion of least absolute values is a linear programming problem,

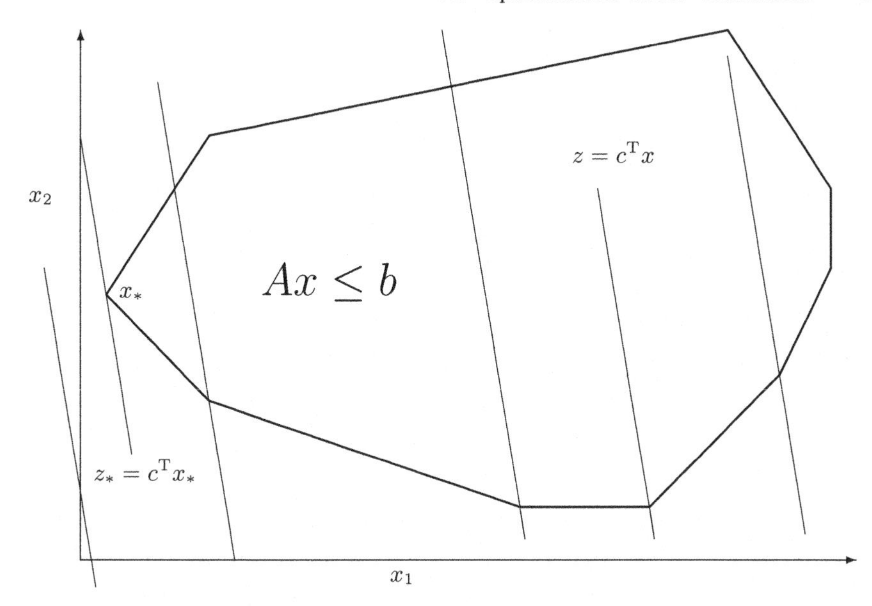

Fig. 6.11. A Linear Programming Problem. The Parallel Lines Are in the Direction of the Coefficient Vector c

but its solution is much more efficient when the simplex method is accelerated by taking into account its special structure. (See Kennedy and Gentle, 1980, Chapter 11, for a description of the modified linear programming methods applied to the L_1 and L_∞ fitting problems.)

An important variation of linear programming is integer programming, in which the decision variables are restricted to be integers. In mixed integer programming some variables are restricted to be integers and others are not.

General Constrained Optimization over Dense Domains

Inequality constraints present significant challenges in optimization problems. The extent of the difficulty depends on the type of the constraint. The simplest constraints are "box constraints", or simple bounds on the variables. Next are linear constraints of the form $l \leq Ax \leq u$. Finally, general nonlinear constraints are the most complicated.

As in other cases of optimization over dense domains, we will usually assume that the objective function is twice differentiable in all variables. We will only indicate some of the general approaches.

When there are both equality and inequality constraints, it is more convenient for the discussion to write the equality constraints explicitly as equalities, rather than as a pair of inequalities in the form of problem (6.44):

$$\begin{aligned} \min_x \quad & f(x) \\ \text{s.t. } & g_1(x) = b_1, \\ & g_2(x) \leq b_2. \end{aligned} \tag{6.54}$$

For any feasible point all equality constraints are active, while the any of the inequality constraints $g_2(x) \leq b_2$ may or may not be active.

The following well-known theorem is proved in many texts on optimization, such as Griva, Nash, and Sofer (2008).

Let $L(x, \lambda)$ be the Lagrangian, and let x_* be a solution to problem (6.54). If the gradients of the active constraints at x_*, $\nabla g_2^{(a)}(x_*)$, are linearly independent, then there exists λ_* such that

$$\nabla_x L(x_*, \lambda_*) = 0,$$

and for all active constraints, $g_2^{(a)}$ with corresponding $\lambda^{(a)}$,

$$\lambda_*^{(a)} \leq 0$$

and

$$\lambda_*^{(a)} g_2^{(a)}(x_*) = 0.$$

These necessary conditions are called the Karush-Kuhn-Tucker conditions, or just Kuhn-Tucker conditions. The Karush-Kuhn-Tucker conditions allow identification of potential solutions. These conditions, together with sufficient conditions involving second derivatives of $L(x, \lambda)$, form the basis for a variety of algorithms for constrained optimization of differentiable functions.

Another approach to solving constrained problems is to formulate a sequence of simpler problems that converges to problem of interest. The method is often called the sequential unconstrained minimization technique (SUMT). A possible problem arises in this approach if the behavior of the objective function is different outside the feasible region from its behavior when the constraints are satisfied.

Quadratic Objective Function with Linear Inequality Constraints

A common form of the general constrained optimization problem (6.44) has a quadratic objective function and linear inequality constraints:

$$\begin{aligned} \min_x \; & c^{\mathrm{T}} x + x^{\mathrm{T}} C x \\ \text{s.t.} \quad & Ax \leq b. \end{aligned} \tag{6.55}$$

This is called a quadratic programming problem. If C is positive semidefinite, the problem is particularly simple, and there are efficient methods for solving a quadratic programming problem that make use of the fact that if x_* is a solution, then there exists λ_* such that

$$2Cx_* + A^{\mathrm{T}}\lambda_* = c^{\mathrm{T}}. \tag{6.56}$$

A number of algorithms based on sequential quadratic programming problems are used for more general constrained optimization problems. As in the unconstrained sequences, the violations of the constraints are built into the objective functions of later stages.

The fact that the sequence of approximate problems does not maintain feasibility of the solution to the original problem can be a major disadvantage. In some cases the objective function may not even be defined outside of the feasible region.

Constrained Combinatorial Optimization

Constraints in combinatorial optimization problems are usually handled by restricting the mechanism that generates new points to generate only feasible points. In a simulated annealing method, for example, the feasibility of each potential state r is determined prior to the acceptance/rejection step.

6.5 Computations for Least Squares

One of the most common problems in applications in statistics and data analysis is the least squares problem. The usual context is in fitting the model

$$\mathrm{E}(Y_i) = f(x_i, \theta_*), \tag{6.57}$$

given observations (x_i, y_i).

For any given θ, we form the residuals

$$r_i(\theta) = y_i - f(x_i, \theta).$$

We will assume that $f(\cdot)$ is a smooth function and θ is an m-vector. Letting y be the n-vector of observations, we can write the least squares objective function as

$$s(\theta) = \left(r(\theta)\right)^{\mathrm{T}} r(\theta). \tag{6.58}$$

The gradient and the Hessian for a least squares problem have special structures that involve the Jacobian of the residuals, $\mathrm{J}_r(\theta)$. The gradient of s is

$$\nabla s(\theta) = 2\left(\mathrm{J}_r(\theta)\right)^{\mathrm{T}} r(\theta). \tag{6.59}$$

Taking derivatives of $\nabla s(\theta)$, we see that the Hessian of s can be written in terms of the Jacobian of r and the individual residuals:

$$\mathrm{H}_s(\theta) = 2\left(\mathrm{J}_r(\theta)\right)^{\mathrm{T}} \mathrm{J}_r(\theta) + 2\sum_{i=1}^{n} r_i(\theta)\mathrm{H}_{r_i}(\theta). \tag{6.60}$$

In the vicinity of the solution $\widehat{\theta}$, the residuals $r_i(\theta)$ should be small, and so $\mathrm{H}_s(\theta)$ may be approximated by neglecting the second term:

$$\mathrm{H}_s(\theta) \approx 2\left(\mathrm{J}_r(\theta)\right)^{\mathrm{T}} \mathrm{J}_r(\theta).$$

Using equation (6.59) and this approximation for equation (6.60) in the gradient descent equation (6.36), we have the system of equations

$$\left(\mathrm{J}_r(\theta^{(k-1)})\right)^{\mathrm{T}} \mathrm{J}_r(\theta^{(k-1)})\, d^{(k)} \;=\; -\left(\mathrm{J}_r(\theta^{(k-1)})\right)^{\mathrm{T}} r(\theta^{(k-1)}) \qquad (6.61)$$

that is to be solved for $d^{(k)}$, where

$$d^{(k)} \propto \theta^{(k)} - \theta^{(k-1)}.$$

It is clear that the solution $d^{(k)}$ is a descent direction; that is, if $\nabla s(\theta^{(k-1)}) \neq 0$, then

$$\begin{aligned}(d^{(k)})^{\mathrm{T}}\nabla s(\theta^{(k-1)}) &= -\left(\left(\mathrm{J}_r(\theta^{(k-1)})\right)^{\mathrm{T}} d^{(k)}\right)^{\mathrm{T}} \left(\mathrm{J}_r(\theta^{(k-1)})\right)^{\mathrm{T}} d^{(k)} \\ &< 0.\end{aligned}$$

The update step is determined by a line search in the appropriate direction:

$$\theta^{(k)} - \theta^{(k-1)} = \alpha^{(k)} d^{(k)}.$$

The method just described that uses the Gramian matrix formed from the Jacobian, rather than the Hessian, is called the *Gauss-Newton algorithm.* (The method is also sometimes called the "modified Gauss-Newton algorithm" because many years ago no damping was used in the Gauss-Newton algorithm, and $\alpha^{(k)}$ was taken as the constant 1. Without an adjustment to the step, the Gauss-Newton method tends to overshoot the minimum in the direction $d^{(k)}$.)

In practice, rather than a full search to determine the best value of $\alpha^{(k)}$, we just consider the sequence of values $1, \frac{1}{2}, \frac{1}{4}, \ldots$ and take the largest value so that $s(\theta^{(k)}) < s(\theta^{(k-1)})$. The algorithm terminates when the change is small.

If the residuals are not small, that is, if the Gramian is not a good approximation of the Hessian, or if $\mathrm{J}_r(\theta^{(k)})$ is poorly conditioned, the Gauss-Newton method can perform very poorly.

If the condition is poor, one possibility is to add a conditioning matrix to the coefficient matrix in equation (6.61). A simple choice is $\lambda^{(k)} I_m$, and the equation for the update becomes

$$\left(\left(\mathrm{J}_r(\theta^{(k-1)})\right)^{\mathrm{T}} \mathrm{J}_r(\theta^{(k-1)}) + \lambda^{(k)} I_m\right) d^{(k)} \;=\; -\left(\mathrm{J}_r(\theta^{(k-1)})\right)^{\mathrm{T}} r(\theta^{(k-1)}),$$

where I_m is the $m \times m$ identity matrix and $\lambda^{(k)}$ is nonnegative. A better choice may be an $m \times m$ scaling matrix, $S^{(k)}$, that takes into account the variability in the columns of $\mathrm{J}_r(\theta^{(k-1)})$; hence, we have for the update equation

$$\left(\left(J_r(\theta^{(k-1)})\right)^{\mathrm{T}} J_r(\theta^{(k-1)}) + \lambda^{(k)} \left(S^{(k)}\right)^{\mathrm{T}} S^{(k)} \right) d^{(k)} \\ = - \left(J_r(\theta^{(k-1)})\right)^{\mathrm{T}} r(\theta^{(k-1)}). \tag{6.62}$$

The basic requirement for the matrix $\left(S^{(k)}\right)^{\mathrm{T}} S^{(k)}$ is that it improve the condition of the coefficient matrix. There are various ways of choosing this matrix. One is to transform the matrix $\left(J_r(\theta^{(k-1)})\right)^{\mathrm{T}} J_r(\theta^{(k-1)})$ so that it has 1's along the diagonal (this is equivalent to forming a correlation matrix from a variance-covariance matrix), and to use the scaling vector to form $S^{(k)}$. The nonnegative factor $\lambda^{(k)}$ can be chosen to control the extent of the adjustment. The sequence $\lambda^{(k)}$ must go to 0 for the algorithm to converge.

Equation (6.62) can be thought of as a Lagrange multiplier formulation of the constrained problem,

$$\begin{aligned} &\min_x \; \tfrac{1}{2} \left\| J_r(\theta^{(k-1)})x + r(\theta^{(k-1)}) \right\| \\ &\text{s.t.} \qquad \left\| S^{(k)} x \right\| \leq \delta_k. \end{aligned} \tag{6.63}$$

The Lagrange multiplier $\lambda^{(k)}$ is zero if $d^{(k)}$ from equation (6.61) satisfies $\|d^{(k)}\| \leq \delta_k$; otherwise, it is chosen so that $\left\| S^{(k)} d^{(k)} \right\| = \delta_k$.

Use of an adjustment such as in equation (6.62) in a Gauss-Newton algorithm is called the *Levenberg-Marquardt algorithm*. It is probably the most widely used method for nonlinear least squares.

Iteratively Reweighted Least Squares

In the weighted least squares problem, we have the objective function equation from page 69:

$$s_w(\theta) = \sum_{i=1}^{n} w_i \left(r_i(\theta)\right)^2.$$

The weights add no complexity to the problem, and the Gauss-Newton methods of the previous section apply immediately, with

$$\tilde{r}(\theta) = Wr(\theta),$$

where W is a diagonal matrix containing the weights.

The simplicity of the computations for weighted least squares suggests a more general usage of the method. Suppose that we are to minimize some other L_p norm of the residuals r_i, as in equation (1.156) on page 67. The objective function can be written as

$$s_p(\theta) = \sum_{i=1}^{n} \frac{1}{|y_i - f(\theta)|^{2-p}} \left(y_i - f(\theta)\right)^2, \tag{6.64}$$

so long as $y_i - f(\theta) \neq 0$.

This leads to an iteration on the least squares solutions. Beginning with $y_i - f(\theta^{(0)}) = 1$, we form the recursion that results from the approximation

$$s_p(\theta^{(k)}) \approx \sum_{i=1}^{n} \frac{1}{\left|y_i - f(\theta^{(k-1)})\right|^{2-p}} \left(y_i - f(\theta^{(k)})\right)^2. \tag{6.65}$$

Hence, we solve a weighted least squares problem, and then form a new weighted least squares problem using the residuals from the previous problem.

The method using the recursion (6.65) is called *iteratively reweighted least squares*, or IRLS. The iterations over the residuals are outside the loops of iterations to solve the least squares problems, so in nonlinear least squares, IRLS results in nested iterations.

There are some problems with the use of reciprocals of powers of residuals as weights. The most obvious problem arises from very small residuals. This is usually handled by use of a fixed large number as the weight.

Iteratively reweighted least squares can also be applied to other norms,

$$s_\rho(\theta) = \sum_{i=1}^{n} \rho\left(y_i - f(\theta)\right).$$

The weights at the kth step are just $\rho(y_i - f(\theta^{(k-1)}))/(y_i - f(\theta^{(k-1)}))^2$.

The approximations for the updates may not be as good as for L_p norms. No matter what norm is used, very small residuals can cause problems.

6.6 Computations for Maximum Likelihood

Although methods based on the maximum of the likelihood function require strong assumptions about the underlying probability distributions, they are widely used in statistics and data analysis. Instead of the model of only the expectation (6.57), $\mathrm{E}(Y_i) = f(x_i, \theta_*)$, as for approaches based on least squares, for maximum likelihood, we must have a model of the PDF. We assume it to be of a given form,

$$p_{Y_i}(y_i \mid f(x_i, \theta_*)). \tag{6.66}$$

Again, the objective is to determine an estimate of θ_*.

We should be aware of the strength of the assumptions we must make about the probability distribution. The assumption underlying the maximum likelihood approach is stronger than an assumption about expected values, which underlies approaches based on minimizing residuals.

Given the PDF, we form the likelihood $L(\theta\,;\,y)$ or the log-likelihood $l_L(\theta\,;\,y)$ as describe beginning on page 44. The maximum likelihood estimate of θ_* is

$$\underset{\theta}{\arg\max}\, l_L(\theta\,;\,y).$$

If the likelihood is twice differentiable and if the range does not depend on the parameter, Newton's method (see equation (6.36)) could be used to solve the optimization problem. Newton's equation

$$\mathrm{H}_{l_L}(\theta^{(k-1)}\,;\,y)\,d^{(k)} = \nabla l_L(\theta^{(k-1)}\,;\,y) \tag{6.67}$$

is used to determine the step direction in the kth iteration. A quasi-Newton method, as we mentioned on page 269, uses a matrix $\widetilde{H}_{l_L}(\theta^{(k-1)})$ in place of the Hessian $\mathrm{H}_{l_L}(\theta^{(k-1)})$. At this point, we should remind the reader:

The form of a mathematical expression and the way the expression should be evaluated in actual practice may be quite different.

There are many additional considerations for the numerical computations, and the expressions below, such as equations (6.68), (6.70), and (6.71), rarely should be used directly in a computer program.

The optimization problem can be solved by Newton's method, equation (6.29) on page 266, or by a quasi-Newton method. (We should first note that this is a maximization problem, so the signs are reversed from our previous discussion of a minimization problem.)

A common quasi-Newton method for optimizing $l_L(\theta\,;\,y)$ is *Fisher scoring*, in which the Hessian in Newton's method is replaced by its expected value. The expected value can be replaced by an estimate, such as the sample mean. The iterates then are

$$\theta^{(k)} = \theta^{(k-1)} - \left(\widetilde{E}(\theta^{(k-1)})\right)^{-1} \nabla l_L(\theta^{(k-1)}\,;\,y), \tag{6.68}$$

where $\widetilde{E}(\theta^{(k-1)})$ is an estimate or an approximation of

$$\mathrm{E}\Big(\mathrm{H}_{l_L}(\theta^{(k-1)} \mid Y)\Big), \tag{6.69}$$

which is itself an approximation of $\mathrm{E}_\theta(\mathrm{H}_{l_L}(\theta \mid Y))$. By equation (1.167) on page 72, this is the negative of the Fisher information matrix *if* the differentiation and expectation operators can be interchanged. (This is one of the "regularity conditions" we alluded to earlier.) The most common practice is to take $\widetilde{E}(\theta^{(k-1)})$ as the Hessian evaluated at the current value of the iterations on θ; that is, as $\mathrm{H}_{l_L}(\theta^{(k-1)}\,;\,y)$. This is called the *observed* information matrix.

In the case of a covariate x_i where we have $\mu = x_i(\theta)$, another quasi-Newton method may be useful. The Hessian in equation (6.67) is replaced by

$$\left(X(\theta^{(k-1)})\right)^{\mathrm{T}} K(\theta^{(k-1)})\, X(\theta^{(k-1)}), \tag{6.70}$$

where $K(\theta^{(k-1)})$ is a positive definite matrix that may depend on the current value $\theta^{(k-1)}$. (Again, think of this in the context of a regression model, but not necessarily linear regression.) This method is called the *Delta algorithm* because of its similarity to the delta method for approximating a variance-covariance matrix (described on page 50).

Maximization over Subvectors

In some cases, when θ is a vector, the optimization problem can be solved by alternating iterations on the elements of θ. In this approach, with $\theta = (\theta_i, \theta_j)$, iterations based on equations such as (6.67) are

$$\widetilde{H}_{l_L}\left(\theta_i^{(k-1)}\ ;\ \theta_j^{(k-1)}, y\right) d_i^{(k)} = \nabla l_{l_L}\left(\theta_i^{(k-1)}\ ;\ \theta_j^{(k-1)}, y\right), \tag{6.71}$$

where d_i is the update direction for θ_i, and θ_j is considered to be constant in this step. In the next step, the indices i and j are exchanged. This is componentwise optimization. For some objective functions, the optimal value of θ_i for fixed θ_j can be determined in closed form. In such cases, componentwise optimization may be the best method.

Sometimes, we may be interested in the MLE of θ_i given a fixed value of θ_j, so the iterations do not involve an interchange of i and j as in componentwise optimization. Separating the arguments of the likelihood or log-likelihood function in this manner leads to what is called *profile likelihood*, or *concentrated likelihood.*

As a purely computational device, the separation of θ into smaller vectors makes for a smaller optimization problem for which the number of computations is reduced by more than a linear amount. The iterations tend to zigzag toward the solution, so convergence may be quite slow. If, however, the Hessian is block diagonal, or almost block diagonal (with sparse off-diagonal submatrices), two successive steps of the alternating method are essentially equivalent to one step with the full θ. The rate of convergence would be the same as that with the full θ. Because the total number of computations in the two steps is less than the number of computations in a single step with a full θ, however, the method may be more efficient in this case. EM methods, which we discuss next, are special cases of this general approach.

EM Methods for Maximum Likelihood

EM methods alternate between updating $\theta^{(k)}$ by maximization of a likelihood and use of conditional expected values. This method is called the *EM method* because the alternating steps involve an expectation and a maximization.

The EM methods can be explained most easily in terms of a random sample that consists of two components, one observed and one unobserved or missing. A simple example of missing data occurs in life-testing, when, for example, a number of electrical units are switched on and the time when each fails is recorded. In such an experiment, it is usually necessary to curtail the recordings prior to the failure of all units. The failure times of the units still working are unobserved. The data are said to be *right censored.* The number of censored observations and the time of the censoring obviously provide information about the distribution of the failure times.

The missing data can be missing observations on the same random variable that yields the observed sample, as in the case of the censoring example; or the missing data can be from a different random variable that is related somehow to the random variable observed.

Many common applications of EM methods do involve missing-data problems, but this is not necessary. Often, an EM method can be constructed based on an artificial "missing" random variable to supplement the observable data.

Let $Y = (U, V)$, and assume that we have observations on U but not on V. We wish to estimate the parameter θ, which figures in the distribution of both components of Y. An EM method uses the observations on U to obtain a value of $\theta^{(k)}$ that increases the likelihood and then uses an expectation based on V that increases the likelihood further.

Let $L_c(\theta \; ; \; u, v)$ and $l_{L_c}(\theta \; ; \; u, v)$ denote, respectively, the likelihood and the log-likelihood for the complete sample. The likelihood for the observed U is

$$L(\theta \; ; \; u) = \int L_c(\theta \; ; \; u, v) \, \mathrm{d}v,$$

and $l_L(\theta \; ; \; u) = \log L(\theta \; ; \; u)$. The EM approach to maximizing $L(\theta \; ; \; u)$ has two alternating steps. The first one begins with a value $\theta^{(0)}$. The steps are iterated until convergence.

- E step : compute $q^{(k)}(\theta) = \mathrm{E}_{V|u,\theta^{(k-1)}}\left(l_{L_c}(\theta \mid u, V)\right)$.
- M step : determine $\theta^{(k)}$ to maximize $q^{(k)}(\theta)$, subject to any constraints on acceptable values of θ.

The sequence $\theta^{(1)}, \theta^{(2)}, \ldots$ converges to a local maximum of the observed-data likelihood $L(\theta \; ; \; u)$ under fairly general conditions. The EM method can be very slow to converge, however.

As is usual for estimators defined as solutions to optimization problems, we may have some difficulty in determining the statistical properties of the estimators. There are various ways that we might estimate the variance-covariance matrix using computations that are part of the EM steps. The most obvious method is to use the gradient and Hessian of the complete-data log-likelihood, $l_{L_c}(\theta \; ; \; u, v)$.

It is interesting to note that under certain assumptions on the distribution, the iteratively reweighted least squares method discussed on page 294 can be formulated as an EM method (see Dempster, Laird, and Rubin, 1980).

For a simple example of the EM method, see Exercise 6.12, in which the problem in Dempster, Laird, and Rubin (1977) is described. As a further example of the EM method, consider an experiment described by Flury and Zoppè (2000). It is assumed that the lifetime of light bulbs follows an exponential distribution with mean θ. To estimate θ, n light bulbs were tested until they all failed. Their failure times were recorded as $u_1, \ldots, u_n$. In a separate experiment, m bulbs were tested, but the individual failure times were not recorded. Only the number of bulbs, r, that had failed at time t was recorded.

The missing data are the failure times of the bulbs in the second experiment, $v_1, \ldots, v_m$. We have

$$l_{L_c}(\theta\ ;\ u, v) = -n(\log\theta + \bar{u}/\theta) - \sum_{i=1}^{m}(\log\theta + v_i/\theta).$$

The expected value, $\mathrm{E}_{V|u,\theta^{(k-1)}}$, of this is

$$q^{(k)}(\theta) = -(n+m)\log\theta - \frac{1}{\theta}\left(n\bar{u} + (m-r)(t+\theta^{(k-1)}) + r(\theta^{(k-1)} - th^{(k-1)})\right),$$

where $h^{(k-1)}$ is given by

$$h^{(k-1)} = \frac{\mathrm{e}^{-t/\theta^{(k-1)}}}{1 - \mathrm{e}^{-t/\theta^{(k-1)}}}.$$

The kth M step determines the maximum, which, given $\theta^{(k-1)}$, occurs at

$$\theta^{(k)} = \frac{1}{n+m}\left(n\bar{u} + (m-r)(t+\theta^{(k-1)}) + r(\theta^{(k-1)} - th^{(k-1)})\right). \qquad (6.72)$$

Starting with a positive number $\theta^{(0)}$, equation (6.72) is iterated until convergence.

This example is interesting because if we assume that the distribution of the light bulbs is uniform, $\mathrm{U}(0, \theta)$ (such bulbs are called "heavybulbs"!), the EM algorithm cannot be applied. As we have pointed out above, maximum likelihood methods must be used with some care whenever the range of the distribution depends on the parameter. In this case, however, there is another problem. It is in computing $q^{(k)}(\theta)$, which does not exist for $\theta < \theta^{(k-1)}$.

Notes and Further Reading

Rootfinding

The problem I have called "solving equations" or "finding roots" is often associated with the keyword "zero". (In the IMSL Library, the routines for "finding zeros" were in Chapter Z.)

General Methods for Optimization

Griva, Nash, and Sofer (2008) provide a comprehensive coverage of the basic ideas and methods of optimization in dense domains. They present the methods in the form of what they call a General Optimization Algorithm, which consists of two steps, an optimality convergence test, and a step that improves the current solution. In this chapter, I have described the second step as itself consisting of two steps: finding a new possible point, and the deciding whether or not to accept that point. Many stochastic methods may accept a new point even when the current solution is better.

For differentiable objective functions, the first and second derivatives can be used to move toward an optimum and to decide when a local optimum has been achieved. Conn, Scheinberg, and Vicente (2009) describe various algorithms for optimization in dense domains that do not require derivatives of the objective function.

Methods for Specialized Optimization Problems

One of the most widely-encountered specialized optimization problems is the linear programming problem and related problems in network optimization. Griva, Nash, and Sofer (2008) describe methods for such problems.

Stochastic Optimization and Evolutionary Methods

Stochastic optimization is discussed in some detail by Spall (2004). De Jong (2006) describes the basic ideas of evolutionary computation and how the methods can be used in a variety of optimization problems.

Optimization for Statistical Applications

Many of the relevant details of numerical optimization for statistical applications are discussed by Rustagi (1994) and by Gentle (2009).

The EM method for optimization is covered in some detail by Ng, Krishnan, and McLachlan (2004). The EM method itself was first described and analyzed systematically by Dempster, Laird, and Rubin (1977).

Numerical Software for Optimization

Most of the comprehensive scientific software packages such as the IMSL Libraries, Matlab, and R have functions or separate modules for solution of systems of nonlinear equations and for optimization.

The R function `uniroot` (which is `zbrent` in the IMSL Libraries) is based on an algorithm of Richard Brent that uses a combination of linear interpolation, inverse quadratic interpolation, and bisection to find a root of a univariate function in an interval whose endpoints evaluate to values with different signs. The R function `polyroot` (which is `zpolrc` or `zpolcc` in the IMSL Libraries) is based on the Traub-Jenkins algorithm to find the roots of a univariate polynomial.

It is difficult to design general-purpose software for optimization problems because the problems tend to be somewhat specialized and different solution methods are necessary for different problems; hence, there are several specialized software packages for optimization. Some address general optimization problems for continuous nonlinear functions, with or without constraints. There are several packages for linear programming. These often also handle quadratic programming problems, as well as other variations, such as mixed integer problems and network problems.

Another reason it is difficult to design general-purpose software for optimization problems is because the formulation of the problems in simple computer interfaces is difficult.

The need for an initial guess may also complicate the design of optimization software, especially for the unsophisticated user. The software would be hardpressed to decide on a reasonable starting value, however. Sometimes an obvious default such as $x^{(0)} = 0$ will work, and there are some software packages that will choose such a starting value if the user does not supply a value. Most packages, however, require the user to input a starting value.

Many of the standard routines for optimization use derivative-free methods. For optimization of univariate functions, in R the function `optimize`, based on an algorithm of Richard Brent, uses a combination of golden section search and successive parabolic interpolation, for optimization of a univariate function. The IMSL routine `uvmif`, based on a method of Mike Powell, uses a safeguarded interpolation, and tends to be somewhat more robust.

For optimization of multivariate functions, in R the function `nlm` uses a Newton-type method either with a user-supplied gradient and Hessian or with numerically-approximated derivatives along with a simple bisection line search. The R function `optim` uses a method that the user can choose, including Nelder-Mead.

There is a wide range of software for least squares problems. Most of the general-purpose software includes special routines for least squares. Packages for statistical data analysis often include functions for nonlinear least squares. For example, in the IMSL Libraries the routine `rnlin` performs least squares fits of general models and in R the function `nls` performs the computations for nonlinear least squares regression.

Because the development of a mathematical model that can be communicated easily to the computer is an important, but difficult aspect of optimization problems, there are packages that implement modeling languages, and many of the general packages accept optimization problems expressed in these languages.

It is also possible for a user to access computational servers for optimization over the internet, so that the user client does not need to run the software. The site is

`http://www-neos.mcs.anl.gov/`

Hans Mittelmann maintains a useful guide to non-commercial optimization software at

`http://plato.la.asu.edu/guide.html`

This website also provides additional items such as benchmarks and testbeds, annotated bibliography, and a glossary for optimization.

Exercises

6.1. Apply Aitken's Δ^2-extrapolation to equation (6.5) to obtain equation (6.6).

6.2. Apply Algorithm 6.1 to equation (6.4) and collect data similar to the bisection iterations shown in Table 6.1.

6.3. Use a plain Newton's method to construct a linearly convergent sequence $\{x_n\}$ that converges slowly to the multiple root $x = 1$ of the function $f(x) = x^3 - 3x + 2$. Then use Aitken acceleration to construct $\{\tilde{x}_n\}$, which converges faster to the root $x = 1$. Use Newton's method and Steffensen's acceleration method to find numerical approximations to the multiple root, starting with $x_0 = 1$. Compare the number of iterations for the two methods.

6.4. Use a plain Newton's method to construct a linearly convergent sequence $\{x_n\}$ that converges slowly to the multiple root $x = \sqrt{2}$ of the function $f(x) = \sin(x^2 - 2)(x^2 - 2)$. Then use Aitken acceleration to construct $\{\tilde{x}_n\}$, which converges faster to the root $x = \sqrt{2}$. Use Newton's method and Steffensen's acceleration method to find numerical approximations to the multiple root, starting with $x_0 = 1$. Compare the number of iterations for the two methods.

6.5. Bisection method.
Write a program module to implement the bisection method to find a root of a given function, which is input together with values that bracket a root, and an epsilon as the stopping criterion. Your program should check that the two starting values are legitimate.
Use your bisection program to determine the first zero of the Bessel function of the first kind, of order 0:

$$J_0(x) = \frac{1}{\pi}\int_0^{\pi} \cos(x \sin\theta)\,\mathrm{d}\theta.$$

(This function is available in Matlab, `besselj`; in PV-Wave, `beselj`; in the IMSL Library, `bsj0/dbsj0`; and in the Unix math library, `j0`.)

6.6. Newton's method.
Write a program module similar to that of Exercise 6.5 to implement Newton's method to find a root of a given function, which is input together with its derivative, a starting value, and two stopping criteria: an epsilon and a maximum number of iterations.

a) Observe the performance of the method on the function

$$f(x) = x^3 - 14x^2 + 68x - 115,$$

which is the function used in the examples in this chapter. Start with $x_0^{(0)} = 9$. Print $x_0^{(k)}$ to 10 digits, and observe the number of correct digits at each iteration until the solution is accurate to 10 digits. Produce a table similar to Table 6.1 on page 248. What is the rate of convergence?

b) Now observe the performance of the method on the function

$$f(x) = x^3 - 15x^2 + 75x - 125,$$

whose solution is also 5. Again start with $x_0^{(0)} = 9$. What is the rate of convergence? What is the difference?

6.7. Secant method.
Write a program module similar to that of Exercise 6.5 to implement the secant method to find a root of a given function, which is input together with two starting values, and two stopping criteria: an epsilon and a maximum number of iterations. Observe the performance of the method on the function

$$f(x) = x^3 - 14x^2 + 68x - 115.$$

Produce a table similar to Table 6.1 on page 248.

6.8. Regula falsi method.
Write a program module similar to that of Exercise 6.5 to implement the regula falsi method to find a root of a given function, which is input together with two starting values and two stopping criteria: an epsilon and a maximum number of iterations. Your program should check that the two starting values are legitimate. Observe the performance of the method on the function

$$f(x) = x^3 - 14x^2 + 68x - 115.$$

Produce a table similar to Table 6.1 on page 248.

6.9. Compare the performance of the four methods in Exercises 6.5 through 6.8 and that of the bisection method for the given polynomial.
Summarize your findings in a clearly-written report. Consider such things as rate of convergence and ease of use of the method.

6.10. Consider the function

$$f(x) = x_1^2 + 5x_2^2,$$

whose minimum obviously is at $(0, 0)$.

a) Plot contours of f. (You can do this easily in R, S-Plus or Matlab, for example.)
b) In the steepest descent method, determine the first 10 values of $\alpha^{(k)}$, $f(x^{(k)})$, $\nabla f(x^{(k)})$, and $x^{(k)}$, starting with $x^{(0)} = (5, 1)$. For the step length, use the optimal value (equation (6.25), page 263).
c) Plot contours of the scaled quadratic model (6.34) of f at the point $(5, 1)$.
d) Repeat Exercise 6.10b using Newton's method. (How many steps does it take?)

6.11. Show that the rank-one update of equation (6.39), page 270, results in a matrix $B^{(k+1)}$ that satisfies the secant condition (6.37).

6.12. Assume a random sample $y_1, \ldots, y_n$ from a gamma distribution with parameters α and β. (Refer to Exercise 1.20 on page 78.)

a) Write a function in a language such as R, Matlab, or Fortran that accepts a sample of size n and computes the least squares estimator of α and β and an approximation of the variance-covariance matrix using both expression (1.160) and expression (1.161).

b) Try out your program in Exercise 6.12a by generating a sample of size 500 from a gamma(2,3) distribution and then computing the estimates. (The sample can be generated by `rgamma` in R or S-Plus and by `rngam` in IMSL.)

c) Write a function in a language such as R, Matlab, or Fortran that accepts a sample of size n and computes the maximum likelihood estimator of α and β and computes an approximation of the variance-covariance matrix using expression (1.168), page 73.

d) Try out your program in Exercise 6.12c by computing the estimates from an artificial sample of size 500 from a gamma(2,3) distribution.

6.13. Dempster, Laird, and Rubin (1977) consider the multinomial distribution with four outcomes, that is, the multinomial with probability function,

$$p(x_1, x_2, x_3, x_4) = \frac{n!}{x_1!x_2!x_3!x_4!}\pi_1^{x_1}\pi_2^{x_2}\pi_3^{x_3}\pi_4^{x_4},$$

with $n = x_1 + x_2 + x_3 + x_4$ and $1 = \pi_1 + \pi_2 + \pi_3 + \pi_4$. They assumed that the probabilities are related by a single parameter, θ:

$$\begin{aligned}
\pi_1 &= \frac{1}{2} + \frac{1}{4}\theta \\
\pi_2 &= \frac{1}{4} - \frac{1}{4}\theta \\
\pi_3 &= \frac{1}{4} - \frac{1}{4}\theta \\
\pi_4 &= \frac{1}{4}\theta,
\end{aligned}$$

where $0 \leq \theta \leq 1$. (This model goes back to an example discussed by Fisher, 1925, in *Statistical Methods for Research Workers.*) Given an observation (x_1, x_2, x_3, x_4), the log-likelihood function is

$$l(\theta) = x_1 \log(2 + \theta) + (x_2 + x_3)\log(1 - \theta) + x_4 \log(\theta) + c$$

and

$$\mathrm{d}l(\theta)/\mathrm{d}\theta = \frac{x_1}{2+\theta} - \frac{x_2 + x_3}{1-\theta} + \frac{x_4}{\theta}.$$

The objective is to estimate θ.

a) Determine the MLE of θ. (Just solve a simple polynomial equation.) Evaluate the estimate using the data that Dempster, Laird, and Rubin used: $n = 197$ and $x = (125, 18, 20, 34)$.
b) Although the optimum is easily found as in the previous part of this exercise, it is instructive to use Newton's method (as in equation (6.29) on page 266). Write a program to determine the solution by Newton's method, starting with $\widehat{\theta^{(0)}} = 0.5$.
c) Write a program to determine the solution by scoring (which is the quasi-Newton method given in equation (6.68) on page 295), again starting with $\widehat{\theta^{(0)}} = 0.5$.
d) Write a program to determine the solution by the EM algorithm, again starting with $\widehat{\theta^{(0)}} = 0.5$.
e) How do these methods compare? (Remember, of course, that this is a particularly simple problem.)

7

Generation of Random Numbers

Monte Carlo simulation is a core technology in computational statistics. Monte Carlo methods require numbers that appear to be realizations of random variables. Obtaining these numbers is the process called "generation of random numbers".

Our objective is usually not to generate a truly random sample. Deep understanding of the generation process and strict reproducibility of any application involving the "random" numbers is more important. We often emphasize this perspective by the word "pseudorandom", although almost anytime we use a phrase similar to "generation of random numbers", we refer to "pseudorandom" numbers.

The quality of a process for random number generation is measured by the extent to which the sample generated appears, from every imaginable perspective, to be a random sample (that is, i.i.d.) from a given probability distribution. Some methods of random number generation are better than others.

7.1 Randomness of Pseudorandom Numbers

The initial step in random number generation is to obtain a sequence that appears to be independent realizations from a uniform distribution over the open interval $(0, 1)$. We denote this distribution by $\mathrm{U}(0, 1)$.

While mathematically there is no difference in a continuous distribution over $[0, 1] \subset \mathbb{R}$ and one over $(0, 1) \subset \mathbb{R}$, there is a difference in a distribution over $[0, 1] \subset \mathbb{F}$ and over $(0, 1) \subset \mathbb{F}$. Because of the computations we may perform with samples that appear to be from $\mathrm{U}(0, 1)$, we must make sure that we exclude the zero-probability events of 0 and 1. (See Exercise 2.9a and its solution on page 677 for a different situation, which may superficially appear to be the same as this.)

J.E. Gentle, *Computational Statistics*, Statistics and Computing,
DOI: 10.1007/978-0-387-98144-4_7,

Generation of Pseudorandom Numbers from a Uniform Distribution

There are several methods for generating uniform numbers. Most of these are sequential congruential methods; that is, methods in which if a subsequence of length j of positive numbers $u_{k-1}, \ldots u_{k-j}$ is given, the next value in the sequence is

$$u_k = f(u_{k-1}, \ldots u_{k-j}) \bmod m \tag{7.1}$$

for some function f and some positive number m and with u_k chosen so that $0 \leq u_k < m$. In this recursion, j is often chosen as 1.

It is clear that if the subsequence $u_{k-1}, \ldots u_{k-j}$ ever occurs again, the next value in the sequence will always be the same; hence, it is clear that on a computer, for any f, the sequence will be periodic in $\mathbb{F}$ or $\mathbb{Z}$ because those sets are finite. In practice, however, the form of f is such that the sequence will also be periodic within $\mathbb{R}$. The *period* is an important property of a random number generator. Obviously the period must be great enough that we do not expect to exhaust it in applications.

A simple instance of equation (7.1) in which m and the u_i are integers is

$$u_k = a u_{k-1} \bmod m \quad \text{with } 0 < u_k < m. \tag{7.2}$$

This is called a linear congruential generator. Because the u_i are integers, it is clear that the period cannot exceed $m - 1$.

Another type of modular reduction scheme works at the bit level, doing circular shifts and additions of selected bits to the popped bit. One class of such methods is called a generalized feedback shift register (GFSR) method in which numbers between 0 and 1 are formed by successively circularly shifting the bits in a fixed-size register while adding a bit from a fixed location within the register. (This is the "feedback".) After a fixed number of circular shifts with the feedback, the bits in a fixed subset of the register are selected to represent a random number. The process continues, with a new random number being delivered after each fixed number of feedback circular shifts.

One of the best of the current methods for generating $\mathrm{U}(0,1)$ is the Mersenne twister described in Matsumoto and Nishimura (1998). This generator "twists" the terms in a sequence from a GFSR by a matrix multiplication. It is called "Mersenne" because the period is a Mersenne prime (which is a prime of the form $2^p - 1$, where p is a prime). One widely-used form of the Mersenne twister that was constructed by Matsumoto and Nishimura is called MT19937. It has a period of $2^{19937} - 1$ and can be implemented so as to execute very fast.

A problem in any random sampling, of course, is that a particular sample might not very well reflect the distribution of interest. Use of a pseudorandom number generator can yield "good" or "bad" samples. Another approach to random number generation is not to try to simulate randomness, but rather to ensure that any generated sequence is more like a "good" random sample.

Sequences generated in this way are called "quasirandom" numbers. We will not describe this approach here.

A sequence of pseudorandom numbers generated by a computer program is determined by a *seed*; that is, an initial state of the program. In the generator of equation (7.2), the seed is x_0. The seed for the generator of equation (7.1) is the initial sequence of j positive numbers. A given seed generates the same sequence of pseudorandom numbers every time the program is run. The ability to control the sequence is important because this allows the experimenter to reproduce results exactly and also to combine experiments that use pseudorandom numbers.

There are various algorithms for generating pseudorandom numbers, and various computer programs that implement these algorithms. Some algorithms and programs are better than others. Statistical tests for randomness applied to samples of pseudorandom numbers generated by good random number generators yield results consistent with hypotheses of randomness. The pseudorandom numbers simulate random samples. A large set of statistical tests for random number generators is TESTU01, developed by L'Ecuyer and Simard (2007). The tests can be run at three different levels.

Although the algorithms for random number generation seem fairly simple, there are a number of issues that must be taken into account when implementing these algorithms in computer programs. Rather than writing code from scratch, it is generally better to use existing computer code. In Section 7.6 we describe available software in Fortran or C and in R or S-Plus.

7.2 Generation of Nonuniform Random Numbers

Samples from other distributions are generated by using transformations of sequences of a stream of U(0, 1) random numbers. We will briefly describe some of these methods below. These techniques are sometimes called "sampling methods".

To generate a realization of a random variable, X, with any given distribution, we seek a transformation of one or more independent U(0, 1) random variables, $U_1, \ldots, U_k$,

$$X = f(U_1, \ldots, U_k),$$

such that X has the desired distribution. In some cases, the transformation f may be a simple transformation of a single uniform variable. For example, to obtain a standard exponential random variable, the transformation

$$X = -\log(U)$$

yields one exponential for each uniform variable. In other cases, the transformation may involve multiple stages in which we first transform a set of uniform variables to a set of variables with some other joint distribution and then identify a marginal or conditional distribution that corresponds to the desired distribution.

The transformations must be done with care and must respect the non-randomness in the underlying uniform generator.

For example, for a double exponential distribution, with density

$$p(x) = \frac{1}{2}\mathrm{e}^{-|x|},$$

consider the simple method:
Generate U_1 and U_2; set $X = \log(U_1)$; then if $U_2 > 0.5$, set $X = -X$.

Often such mathematically correct transformations must be performed with special care on the computer. In this example, if the uniform stream is from a linear congruential generator with a relatively small multiplier, the method will yield a stream of double exponentials in which all extreme values are positive. (Because if U_1 is very small, U_2 will be also.)

Inverse CDF Method

If X is a scalar random variable with a continuous cumulative distribution function (CDF) P_X, then the random variable

$$U = P_X(X)$$

has a $\mathrm{U}(0,1)$ distribution.

This fact provides a very simple relationship with a uniform random variable U and a random variable X with CDF P_X, namely,

$$X = P_X^{-1}(U), \tag{7.3}$$

where the inverse of the CDF exists. Use of this straightforward transformation is called the *inverse CDF* technique. The log transformation mentioned above that yields an exponential random variable uses the inverse CDF.

For a discrete random variable, although the inverse of the CDF does not exist, the inverse CDF method can still be used. The value of the discrete random variable is chosen as the smallest value within its countable range such that the CDF is no less than the value of the uniform variate.

For a multivariate random variable, the inverse CDF method yields a level curve in the range of the random variable; hence, the method is not directly useful for multivariate random variables. Multivariate random variates can be generated using the inverse CDF method first on a univariate marginal and then on a sequence of univariate conditionals.

Acceptance/Rejection Methods

Acceptance/rejection methods for generating realizations of a random variable X make use of realizations of another random variable Y whose PDF g_Y is similar to the PDF of X, p_X. The random variable Y is chosen so that we can easily generate realizations of it and so that its density g_Y can be scaled to majorize p_X using some constant c; that is, so that $cg_Y(x) \geq p_X(x)$ for all x. The density g_Y is called the *majorizing* density, and cg_Y is called the majorizing function. The majorizing density is also called the "proposal density". The density of interest, p_X, is called the "target density". The support of the target density must be contained in the support of the majorizing density; for densities with infinite support, the majorizing density must likewise have infinite support. In the case of infinite support, it is critical that the majorizing density not approach zero faster than the target density.

Acceptance/rejection methods can also be used for discrete random variables. We use the term "probability density" to include a probability mass function, and all of the discussion in this section applies equally to probability functions and probability densities.

Unlike the inverse CDF method, acceptance/rejection methods apply immediately to multivariate random variables.

Algorithm 7.1 The Acceptance/Rejection Method to Convert Uniform Random Numbers

1. Generate y from the distribution with density function g_Y.
2. Generate u from a uniform (0,1) distribution.
3. If $u \leq p_X(y)/cg_Y(y)$, then
   3.a. take y as the desired realization;

   otherwise,

   3.b. return to step 1. ∎

It is easy to see that Algorithm 7.1 produces a random variable with the density p_X. Let Z be the random variable delivered. For any x, because Y (from the density g) and U are independent, we have

$$\begin{aligned}
\Pr(Z \leq x) &= \Pr\left(Y \leq x \mid U \leq \frac{p_X(Y)}{cg_Y(Y)}\right) \\
&= \frac{\int_{-\infty}^{x} \int_0^{p_X(t)/cg_Y(t)} g_Y(t)\,\mathrm{d}s\,\mathrm{d}t}{\int_{-\infty}^{\infty} \int_0^{p_X(t)/cg_Y(t)} g_Y(t)\,\mathrm{d}s\,\mathrm{d}t} \\
&= \int_{-\infty}^{x} p_X(t)\,\mathrm{d}t,
\end{aligned}$$

which is the CDF corresponding to p_X. (Differentiating this quantity with respect to x yields $p_X(x)$.) Therefore, Z has the desired distribution.

It is easy to see that the random variable corresponding to the number of passes through the steps of Algorithm 7.1 until the desired variate is delivered has a geometric distribution. This random variable is a measure of the inefficiency of the algorithm. Because both cg_Y and p_X are densities, it is easy to see that the expected value of this random variable is c. (See Exercise 7.3.)

A straightforward application of the acceptance/rejection method is very simple. For distributions with finite support, the density g can always be chosen as a uniform. For example, to generate deviates from a beta distribution with parameters α and β — that is, the distribution with density,

$$p(x) = \frac{1}{\mathrm{B}(\alpha, \beta)} x^{\alpha-1}(1-x)^{\beta-1}, \quad \text{for } 0 \leq x \leq 1,$$

where $\mathrm{B}(\alpha, \beta)$ is the complete beta function — we could use a uniform majorizing density, as shown in Figure 7.1.

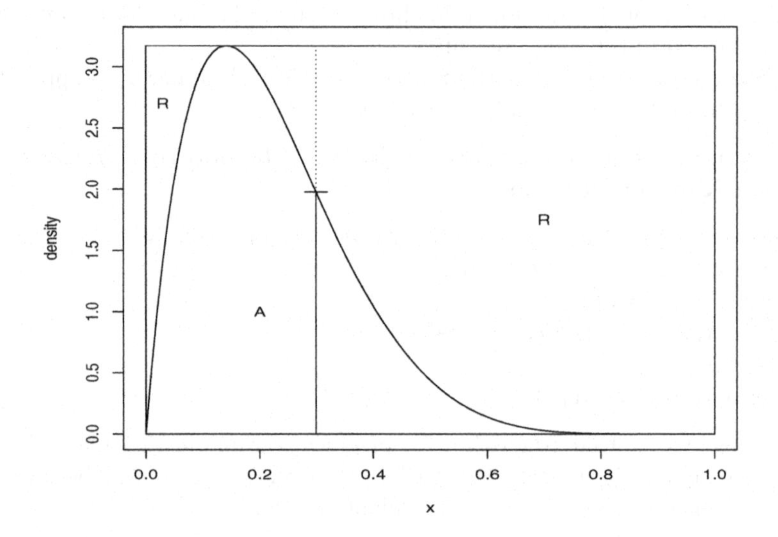

Fig. 7.1. Beta (2, 7) Density with a Uniform Majorizing Density

The value of c by which we scale the uniform density should be as small as possible to minimize the frequency with which we reject the candidate points. This requires determination of the maximum value of the beta density, which we can compute very easily in S-Plus or R just by evaluating the density at the mode:

```
xmode <- (alpha-1.)/(alpha+beta-2.)
dmax  <- xmode^(alpha-1.)*(1-xmode)^(beta-1)*
           gamma(alpha+beta) / (gamma(alpha)*gamma(beta))
```

To generate deviates from the beta using the uniform majorizing density, we could write the following R statements:

```
y<-runif(1000)
```

```
x<-na.omit(ifelse(
          runif(1000)<=dbeta(y,alpha,beta)/dmax, y, NA))
```

Of course, in these statements, the number of beta variates delivered in x will not be known a priori; in fact, the number will vary with different executions of the statements. Instead of using a program that holds all the values in vectors, we generally form an explicit loop in the program to obtain a given number of deviates.

Considering the large area between a scaled uniform density and the beta density shown in Figure 7.1, it is clear that the uniform is not a very efficient density to use for the majorizing density, even though it is extremely easy to use. When a random uniform point falls in the areas marked "R", the point is rejected; when it falls in the area marked "A", the point is accepted. Only 1 out of dmax (≈ 3.18) will be accepted.

As another example just for illustration, consider the use of a normal with mean 0 and variance 2 as a majorizing density for a normal with mean 0 and variance 1, as shown in Figure 7.2. A majorizing density like this whose shape more closely approximates that of the target density is more efficient. The problem in this case, obviously, is that if we could generate deviates from the $\mathrm{N}(0, 2)$ distribution, we could generate ones from the $\mathrm{N}(0, 1)$ distribution.

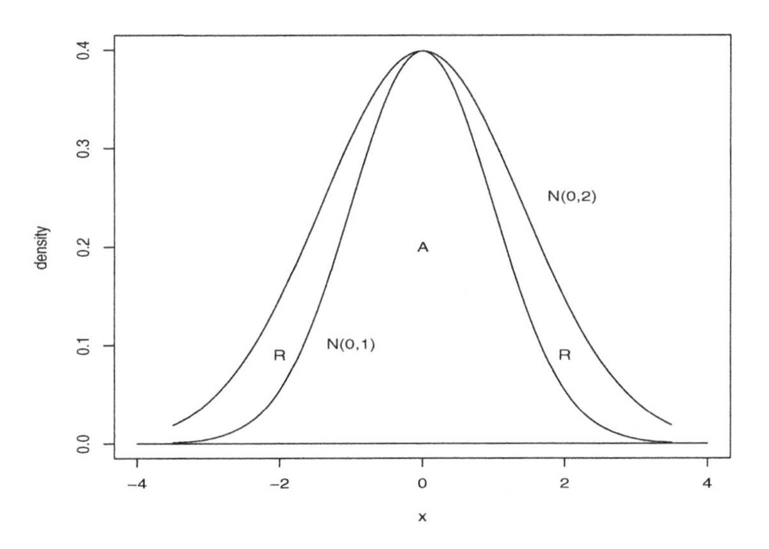

Fig. 7.2. Normal (0, 1) Density with a Normal (0, 2) Majorizing Density

The value of c required to make the density of $\mathrm{N}(0, 2)$ majorize that of $\mathrm{N}(0, 1)$ is $\sqrt{2}$. Hence, one out of ≈ 1.41 candidate points will be accepted.

Although the acceptance/rejection method can be used for multivariate random variables, in that case the majorizing distribution must also be multivariate. For higher dimensions, another problem is the relationship of the rejection region to the acceptance region. In the one-dimensional case, as shown in Figure 7.2, the acceptance region is the area under the lower curve, and the rejection region is the thin shell between the two curves. In higher dimensions, even a thin shell contains most of the volume, so the rejection proportion would be high. See Section 16.7, page 573, and Exercise 16.10, page 583.

Use of Conditional Distributions

Sometimes, the density of interest, p_X, can be represented as a marginal density of some joint density, p_{XY}, that has tractable conditional densities, $p_{X|Y}$ and $p_{Y|X}$. If we can generate realizations from the conditional distributions, observations on X can often be generated as a discrete-time Markov process whose elements have densities

$$p_{Y_i|X_{i-1}}, \quad p_{X_i|Y_i}, \quad p_{Y_{i+1}|X_i}, \quad p_{X_{i+1}|Y_{i+1}}, \ldots. \tag{7.4}$$

This is possible if the distribution of the X_i in the sequence converges to that of X. (A note on terminology: The term "Markov chain" is often restricted to a Markov process with a countable state space. If the support of X or Y is continuous, the state space of the bivariate sequence $\{(X_i, Y_i)\}$ is uncountable. Current terminology in random number generation for such a process, however, is "Markov chain". There are some differences, and Tierney, 1994, 1996, discusses some of the additional complexities arising from a continuous state space that are relevant to the use of such processes in random number generation.)

The *transition kernel* of X_i in the Markov chain is

$$p_{X_i|X_{i-1}}(x_i|x_{i-1}) = \int p_{X_i|Y_i}(x_i|y)\, p_{Y_i|X_{i-1}}(y|x_{i-1})\, \mathrm{d}y.$$

Starting with X_0 and stepping through the transitions, we have

$$p_{X_i|X_0}(x|x_0) = \int p_{X_i|X_{i-1}}(x|t)\, p_{X_{i-1}|X_0}(t|x_0)\, \mathrm{d}t. \tag{7.5}$$

As $i \to \infty$, the density in equation (7.5) converges to p_X under very mild regularity conditions on the densities p_{X_i} and $p_{Y_i|X_i}$. (Existence and absolute continuity are sufficient; see, for example, Nummelin, 1984.) The problem is analogous to the more familiar one involving a discrete-state Markov chain, where convergence is assured if all of the entries in the transition matrix $T_{X|X} = T_{Y|X}T_{X|Y}$ are positive.

The usefulness of this method for random number generation depends on identifying a joint density with conditionals that are easy to simulate.

For example, if the distribution of interest is a standard normal for the random variable X, and Y is a random variable conditionally uniform over $(0, \mathrm{e}^{-X^2})$, the joint density

$$p_{XY}(x, y) = \frac{1}{\sqrt{2\pi}} \frac{1}{\mathrm{e}^{-x^2/2}}, \quad \text{for } -\infty < x < \infty, \; 0 < y < \mathrm{e}^{-x^2},$$

has a marginal density corresponding to the distribution of interest, and it has simple conditionals. The conditional distribution of $Y|X$ is $\mathrm{U}(0, \mathrm{e}^{-X^2})$, and the conditional of $X|Y$ is $\mathrm{U}(-\sqrt{-\log Y}, \sqrt{-\log Y})$. Starting with x_0 in the range of X, we generate y_1 as a uniform conditional on x_0, then x_1 as a uniform conditional on y_1, and so on.

The auxiliary variable Y that we introduce just to simulate X is called a "latent variable". Use of conditional distributions to generate random variables in this way is called *Gibbs sampling*, which we consider again on page 317.

A chain of conditional distributions can also be used for discrete random variables. In that case, the Markov process is a discrete-state Markov chain, and the analysis is even simpler.

Conditional distributions can also be used for a multivariate random variable, and in fact that is one of the most important applications of the method. We discuss the Gibbs algorithm further, for generation of multivariate random variables, on page 317.

7.3 Acceptance/Rejection Method Using a Markov Chain

A discrete-time Markov chain is the basis for several schemes for generating random numbers, either continuous or discrete, and multivariate as well as univariate. The differences in the various methods using Markov processes come from differences in the transition kernel. Sometimes, the transition kernel incorporates an acceptance/rejection decision. The elements of the chain can be accepted or rejected in such a way as to form a different chain whose stationary distribution is the distribution of interest. Simulation methods that make use of a Markov chain to generate samples are called Markov chain Monte Carlo, or MCMC, methods. The interest is not in the sequence of the Markov chain itself. Methods based on Markov chains are *iterative* because several steps must be taken before the stationary distribution is achieved. In practice, it is very difficult to determine the length of the "burn-in" period, that is, to determine when a stationary distribution has been achieved.

For a distribution with density p, the *Metropolis algorithm* or *Metropolis random walk* introduced by Metropolis et al. (1953), generates a random walk and performs an acceptance/rejection based on p evaluated at successive steps in the walk. In the simplest version, the walk moves from the point y_i to the point $y_{i+1} = y_i + s$, where s is a realization from $\mathrm{U}(-a, a)$, and accepts y_{i+1} if

$$p(y_{i+1}) \,/\, p(y_i) \geq u, \tag{7.6}$$

where u is an independent realization from $\mathrm{U}(0, 1)$.

If the range of the distribution is finite, the random walk is not allowed to go outside of the range.

Hastings (1970) developed an algorithm that is based on a transition kernel with a more general acceptance/rejection decision. The *Metropolis-Hastings sampler* to generate deviates from a distribution with density p_X uses deviates from a Markov chain with a completely different density, $g_{Y_{t+1}|Y_t}$. The conditional density $g_{Y_{t+1}|Y_t}$ is chosen so that it is easy to generate deviates from it, and realizations from this distribution are selectively chosen as realizations from the distribution with density p_X.

Algorithm 7.2 Metropolis-Hastings Algorithm

0. Set $i = 0$, and choose x_i in the support of p.
1. Generate y from the density $g_{Y_{t+1}|Y_t}(y|x_i)$.
2. Set r:

$$r = p_X(y) \frac{g_{Y_{t+1}|Y_t}(x_i|y)}{p_X(x_i) g_{Y_{t+1}|Y_t}(y|x_i)}.$$

3. If $r \geq 1$, then
   3.a. set $x_{i+1} = y$;
   otherwise
   3.b. generate u from the uniform(0,1) distribution and
      if $u < r$, then
         3.b.i. set $x_{i+1} = y$,
      otherwise
         3.b.ii. set $x_{i+1} = x_i$.
4. Set $i = i + 1$ and go to step 1. ∎

The r in step 2 is called the *Hastings ratio*, and step 3 is called the "Metropolis rejection". The conditional density, $g_{Y_{t+1}|Y_t}(\cdot|\cdot)$, is called the "proposal density" or the "candidate generating density". Notice that because the majorizing function contains p_X as a factor, we only need to know p_X to within a constant of proportionality. This is an important characteristic of the Metropolis algorithms, including the random walk that only uses the ratio in (7.6).

We can illustrate the use of the Metropolis-Hastings algorithm in using a Markov chain in which the density of X_{t+1} is normal with a mean of X_t and a variance of σ^2. Let us use this density to generate a sample from a standard normal distribution (that is, a normal with a mean of 0 and a variance of 1). We start with x_0, chosen arbitrarily. We take logs and cancel terms in the expression for r. The output from an R implementation of the method is shown in Figure 7.3. Notice that the values descend very quickly from the starting value, which would be a very unusual realization of a standard normal.

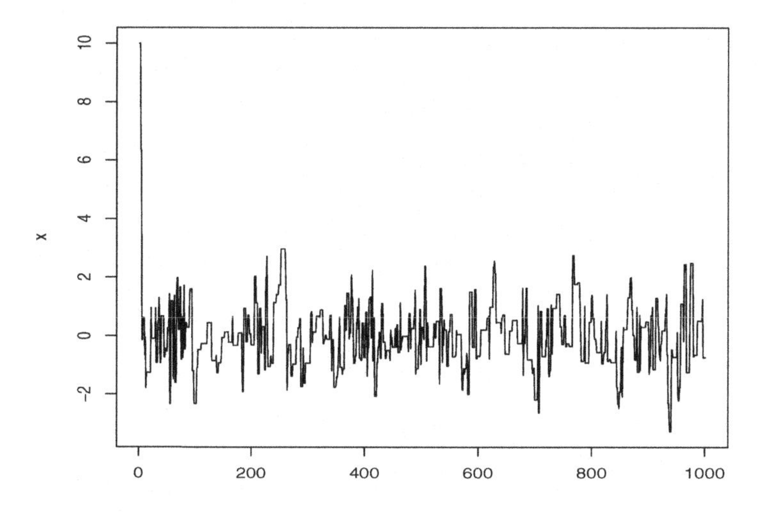

Fig. 7.3. Sequential Output from a N(0, 1) Distribution Using a Markov Chain, $\mathrm{N}(X_t, \sigma^2)$

In practice, we generally cannot expect such a short burn-in period. Notice also the horizontal line segments where the underlying Markov chain did not advance.

7.4 Generation of Multivariate Random Variates

For multivariate distributions with a very large number of variables, the standard acceptance/rejection method is difficult to apply because it is difficult to determine a usable majorizing density. In addition, the acceptance/rejection method is not very efficient because the rejection rate becomes higher in higher dimensions, as we mentioned when discussing Figure 7.2.

The most common ways of generating multivariate random variates are by use of either i.i.d. (independent, identically distributed) univariates followed by a transformation or else by a sequence of conditional univariates.

Transformations Based on the Variance-Covariance Matrix

If $Y_1, \ldots, Y_d$ is a sequence of i.i.d. univariate random variables with variance 1, the variance-covariance matrix of the random d-vector Y composed of those elements is the identity I_d. Assume that the mean of Y is 0. This is without loss of generality because the mean can always be adjusted by an addition.

Consider the random d-vector X, where $X = AY$ for the nonsingular matrix A. The variance-covariance matrix of this transformed random variable is AA^{T}. Suppose that we want to determine a transformation of i.i.d. random variables with unit variances that yields a random variable with variance-covariance matrix Σ. If Y is the vector of the i.i.d. random variables, and A is a matrix such that $AA^{\mathrm{T}} = \Sigma$, then $X = AY$ is the transformation. The matrix A could be either the Cholesky factor or the square root of Σ, for example (see Gentle, 2007, Section 5.9).

This transformation is a very good way of generating multivariate normal random variables. For other multivariate distributions, however, its usefulness is more limited.

Sequences of Conditional Distributions

The other common way of generating a multivariate random number is by use of a sequence of univariate random numbers from conditional univariate distributions that combine to yield the desired multivariate distribution.

Again, for the multivariate normal distribution, this is a simple method. For example, consider a multivariate normal with mean of 0 and variance-covariance matrix Σ, with elements σ_{ij}. If

X_1 is generated as $\mathrm{N}(0, \sigma_{11})$,
X_2 is generated as $\mathrm{N}(\sigma_{12}X_1/\sigma_{11}, \; \sigma_{22} - \sigma_{12}^2/\sigma_{11})$,
and so on,

then

$$X = (X_1, X_2, \ldots)$$

has a multivariate normal distribution with variance-covariance matrix Σ. Some other multivariate distributions can also be easily generated by a sequence of conditional distributions, but for many distributions, the method may be considerably more complicated.

Covariances or correlations are the natural way to define multivariate normal distributions, but for other distributions, copulas may be more appropriate for expressing the associations (see page 32). If marginal distributions are given, a multivariate distribution with bivariate associations expressed through copulas can be formed by use of the inverse CDF method applied both to a marginal CDF and a conditional CDF, using equation (1.82). Suppose we have two marginal CDFs P_{X_1} and P_{X_2}, and a joint CDF defined by the copula $C(P_{X_1}, P_{X_2})$. In notation following that of equation (1.82), let C_u be the partial derivative of C with respect to its first argument. We then generate independent uniforms U_1 and U_2, and set $V = C_u^{-1}(U_2)$. Next we form our desired deviates (X_1, X_2) as $X_1 = P_{X_1}^{-1}(U_1)$ and $X_2 = P_{X_2}^{-1}(V)$. The same idea can be extended to more than two random variables, working first with uniforms as above and then transforming all of the uniforms by the inverses of the marginal CDFs. In Exercise 7.5 you are asked to use copulas to generate bivariate distributions.

Gibbs Sampling

In some cases, it is possible to reduce the problem to a sequence that begins with a univariate marginal distribution and then builds up the random vector by conditional distributions that include the generated elements one at a time. This is possible by decomposing the multivariate density into a marginal and then a sequence of conditionals:

$$p_{X_1 X_2 X_3 \cdots X_d} = p_{X_1|X_2 X_3 \cdots X_d} \cdot p_{X_2|X_3 \cdots X_d} \cdots p_{X_d}.$$

In other cases, we may have a full set of conditionals:

$$p_{X_i|\{X_j; j \neq i\}}.$$

In this case, we can sample from a Markov process by updating individual random variables given the values of the other random variables at a previous time, in the same way as the process (7.4) on page 312. This iterative technique is called "Gibbs sampling". It was introduced by Geman and Geman (1984) and further developed by Gelfand and Smith (1990), among others.

The Gibbs sampler begins with an arbitrary starting point, $x_1^{(0)}, x_2^{(0)}, \ldots, x_d^{(0)}$; generates $x_1^{(1)}$ from knowledge of $p_{X_1^{(1)}|x_2^{(0)},\ldots,x_d^{(0)}}$; generates $x_2^{(1)}$ from knowledge of $p_{X_2^{(1)}|x_1^{(1)},x_3^{(0)},\ldots,x_d^{(0)}}$; and so on.

The process is then iterated in this systematic fashion to get $x_1^{(2)}, x_2^{(2)}, \ldots, x_d^{(2)}$, and so on. A full iteration requires generation of d random variables. Because of the arbitrary starting point, the iterations may not immediately yield deviates from the target distribution.

Geman and Geman showed that $(X_1^{(i)}, X_2^{(i)}, \ldots, X_d^{(i)})$ converges in distribution to $(X_1, X_2, \ldots, X_d)$ so that each component individually converges. This result does not depend on the conditional generations at each iteration being done in the same order.

Algorithm 7.3 Gibbs Sampling Method

0. Set $k = 0$ and choose $x^{(0)}$.
1. Generate $x_1^{(k+1)}$ conditionally on $x_2^{(k)}, x_3^{(k)}, \ldots, x_d^{(k)}$,
   Generate $x_2^{(k+1)}$ conditionally on $x_1^{(k+1)}, x_3^{(k)}, \ldots, x_d^{(k)}$,
   ...
   Generate $x_{d-1}^{(k+1)}$ conditionally on $x_1^{(k+1)}, x_2^{(k+1)}, \ldots, x_d^{(k)}$,
   Generate $x_d^{(k+1)}$ conditionally on $x_1^{(k+1)}, x_2^{(k+1)}, \ldots, x_{d-1}^{(k+1)}$.
2. If convergence has occurred, then
   2.a. deliver $x = x^{(k+1)}$;
   otherwise,
   2.b. set $k = k + 1$, and go to step 1. ∎

Gibbs sampling can be extremely slow to converge. Furthermore, it is often difficult to determine when convergence has occurred (see the discussion beginning on page 420). The convergence is slower when the correlations among the variables are larger (see Exercise 7.6).

Probability Densities Known Only Proportionally

It is often easy to specify a model up to a constant of proportionality. For example, let t be a vector-valued statistic over some sample space and let

$$h(x) = \mathrm{e}^{\langle t(x), \theta \rangle},$$

where $\langle t(x), \theta \rangle$ denotes the dot product of $t(x)$ and θ. This specifies a family of densities

$$f_\theta(x) = \frac{1}{c(\theta)} \mathrm{e}^{\langle t(x), \theta \rangle},$$

where

$$c(\theta) = \int \mathrm{e}^{\langle t(x), \theta \rangle} \, \mathrm{d}\mu(x).$$

In general, if h is a nonnegative integrable function that is not zero almost everywhere, a probability density p can be defined by normalizing h:

$$p(x) = \frac{1}{c} h(x),$$

where

$$c = \int h(x) \, \mathrm{d}\mu(x).$$

We may know h but not c, and c may not be easy to evaluate, especially if h is multivariate. Some Markov chain Monte Carlo methods are particularly useful in dealing with such multivariate distributions that have densities known up to a constant of proportionality. In the ratio (7.6) in the Metropolis random walk or in the Hastings ratio in Algorithm 7.2, the constant c would not be required; h alone could be used to simulate realizations from p.

There are many problems in Bayesian inference in which densities are known only up to a constant of proportionality. In such problems h is the likelihood times the prior. Normalizing h—that is, determining the integral c—may be difficult, but Markov chain Monte Carlo methods allow simulations of realizations from the posterior without knowing c.

7.5 Data-Based Random Number Generation

Often we have a set of data and wish to generate pseudorandom variates from the same data-generating process that yielded the given data. How we do this depends on how much we know or what assumptions we make about the data-generating process. At one extreme, we may assume full knowledge of the data-generating process; for example, we may assume that the given set of data came from a normal distribution with known mean and variance. In this case, we use the well-known techniques for generating pseudorandom variates from a $\mathrm{N}(\mu, \sigma^2)$ distribution.

A slightly weaker assumption is that the data came from a normal distribution, but we do not know the mean or variance. In this case, a simple approach may be to use the given data to estimate the mean and standard deviation and then proceed as if the estimates were the true values. (Notice that if we want the process to be unbiased, we cannot use the square root of the sample variance as the estimate of the standard deviation.) Use of a parametric model, such as the normal distribution, with given or estimated values of the parameters, is a *parametric* approach. Because the parameters are estimated from the data, we call it the *empirical parametric method.* In Section 11.3 on page 424, we discuss the use of estimated parameters in bootstrap simulations for statistical inference such as hypothesis testing or setting confidence bounds. The empirical parametric method is also sometimes called a "parametric bootstrap".

For many of the parametric families shown in Tables B.1 and B.2 beginning on page 660 there are standard methods of generating random variables. The empirical parametric method is appropriate when the parameters can be estimated from a given dataset.

In addition to the standard parametric families shown in Tables B.1 and B.2, there are some general families of probability distributions that are very useful in data-based random number generation because they cover wide ranges of shapes and have a variety of interesting properties that are controlled by a few parameters. Some, such as Tukey's generalized lambda distribution are designed to be particularly simple to simulate. We discuss these families of distributions in Section 14.2.

A strong assumption that does not involve parameters in the usual sense is that the given data resulted from a discrete data-generating process (that is, one that can yield only a countable set of distinct values). In this case, we would generate pseudorandom variates by sampling (or "resampling") the given set of data.

Even for a given sample of univariate data from a continuous data-generating process, the ECDF could be used in place of the CDF in a standard inverse CDF method, as described on page 308. We discuss direct use of the ECDF in Sections 11.2 and 11.4. The ECDF defines a distribution with a finite range $[y_{(1)}, y_{(n)}]$ corresponding to the smallest and largest order statistics of the data. Instead of using the ECDF to generate random data, we may choose to use the probabilities associated with the empirical quantiles, as discussed on page 62, if they can be determined.

For a given sample of multivariate data, our pseudorandom samples must capture probabilities of general regions. The correlations in the given sample must be replicated.

As we have mentioned above, it is not practical to use the inverse CDF method directly for multivariate distributions.

Taylor and Thompson (1986) suggest a different way that avoids the step of estimating a density. The method has some of the flavor of density estimation; however, in fact it is essentially equivalent to fitting a density with a normal kernel. It uses the m nearest neighbors of a randomly selected point; m is a *smoothing parameter*. The method is particularly useful for multivariate data. Suppose that the given sample is $\{x_1, x_2, \ldots, x_n\}$ (the xs are vectors). A random vector deviate is generated by the steps given in Algorithm 7.4.

Algorithm 7.4 Thompson–Taylor Data-Based Simulation

1. Randomly choose a point, x_j, from the given sample.
2. Identify the m nearest neighbors of x_j (including x_j), $x_{j_1}, x_{j_2}, \ldots, x_{j_m}$, and determine their mean, $\bar{x}_j$.
3. Generate a random sample, $u_1, u_2, \ldots, u_m$, from a uniform distribution with lower bound $\frac{1}{m} - \sqrt{\frac{3(m-1)}{m^2}}$ and upper bound $\frac{1}{m} + \sqrt{\frac{3(m-1)}{m^2}}$.
4. Deliver the random variate

$$z = \sum_{k=1}^{m} u_k (x_{j_k} - \bar{x}_j) + \bar{x}_j.$$

■

The limits of the uniform weights and the linear combination for z are chosen so that the expected value of the ith element of a random variable Z that yields z is the ith element of the sample mean of the xs, $\bar{x}_i$; that is,

$$\mathrm{E}(Z_i) = \bar{x}_i.$$

(The subscripts in these expressions refer to the elements of the data vectors rather than to the element of the sample.) Likewise, the variance and covariance of elements of Z are close to the sample variance and covariance of the elements of the given sample. If $m = 1$, they would be exactly the same. For $m > 1$, the variance is slightly larger because of the variation due to the random weights. The exact variance and covariance, however, depend on the distribution of the given sample because the linear combination is of nearest points. The routine `rndat` in the IMSL Libraries implements this method.

7.6 Software for Random Number Generation

Random number generators are widely available in a variety of software packages. Although the situation may not be as dire as when Park and Miller (1988) stated, "good ones are hard to find", the user must be careful in selecting a random number generator.

Basic Uniform Generators

Some programming languages, such as C, Fortran, and Ada 95, provide built-in random number generators. In C, the generator is the function `rand()` in stdlib.h. This function returns an integer in the range 0 through RAND_MAX, so the result must be normalized to the range $(0,1)$. (The scaling should be done with care. It is desirable to have uniform numbers in $(0,1)$ rather than $[0,1]$.) The seed for the C random number generator is set in srand().

In Fortran, the generator is the subroutine random_number, which returns $\mathrm{U}(0,1)$ numbers. (The user must be careful, however; the generator may yield either a 0 or a 1.) The seed can be set in the subroutine random_seed. The design of the Fortran module as a subroutine yields a major advantage over the C function in terms of efficiency. (Of course, because Fortran has the basic advantage of arrays, the module could have been designed as an array function and would still have had an advantage over the C function.)

A basic problem with the built-in generator of C, Fortran, and Ada 95 is lack of portability. The standards do not specify the algorithm. The bindings are portable, but none of these generators will necessarily generate the same sequence on different platforms.

Other Distributions

Given a uniform random number generator, it is usually not too difficult to generate variates from other distributions. For example, in Fortran, the inverse CDF technique for generating a random deviate from a Bernoulli distribution with parameter π can be implemented by the code in Figure 7.4.

```
integer, parameter      :: n  = 100  ! INITIALIZE THIS
real, parameter (pi)    :: pi = .5   ! INITIALIZE THIS
real, dimension (n)     :: uniform
real, dimension (n)     :: bernoulli
call random_number (uniform)
where (uniform .le. pi)
       bernoulli = 1.0
elsewhere
       bernoulli = 0.0
endwhere
```

Fig. 7.4. A Fortran Code Fragment to Generate n Bernoulli Random Deviates with Parameter π

Implementing one of the simple methods to convert a uniform deviate to that of another distribution may not be as efficient as a special method for the target distribution, and those special methods may be somewhat complicated. The IMSL Libraries and S-Plus and R have a number of modules that use efficient methods to generate variates from several of the more common distributions. Matlab has a basic uniform generator, rand, and a standard normal generator, randn. The Matlab Statistics Toolbox also contains generators for several other distributions.

A number of Fortran or C programs are available in collections published by *Applied Statistics* and by *ACM Transactions on Mathematical Software*. These collections are available online at `statlib` and `netlib`, respectively. See page 692 in the bibliography for more information.

The freely distributed GNU Scientific Library (GSL) contains several C functions for random number generation. There are several different basic uniform generators in the library. Utility functions in the library allow selection of a uniform generator for use by the functions that generate nonuniform numbers. In addition to a number of newer uniform generators, including quasirandom number generators, there are basic uniform generators that yield output sequences that correspond (or almost correspond) to legacy generators provided by various systems developers, such as the IBM `RANDU` and generators associated with various Unix distributions. The random number generators in GSL can be accessed from R by use of the `gsl` package.

Information about the GNU Scientific Library, including links to sites from which source code can be obtained, is available at

`http://www.gnu.org/software/gsl/`

The Guide to Available Mathematical Software, or GAMS (see the Bibliography) can be used to locate special software for various distributions.

The User Interface for Random Number Generators

Software for random number generation must provide a certain amount of control by the user, including the ability to:

- set or retrieve the seed;
- select seeds that yield separate streams;
- possibly select the method from a limited number of choices.

Whenever the user invokes a random number generator for the first time in a program or session, the software should not require the specification of a seed but should allow the user to set it if desired. If the user does not specify the seed, the software should use some mechanism, such as accessing the system clock, to form a "random" seed. On a subsequent invocation of the random number generator, unless the user specifies a seed, the software should use the last value of the seed from the previous invocation. This means that the routine for generating random numbers must produce a "side effect"; that is, it changes something other than the main result. It is a basic tenet of software engineering that side effects must be carefully noted. At one time, side effects were generally to be avoided. In object-oriented programming, however, objects may encapsulate many entities, and as the object is acted upon, any of the components may change. Therefore, in object-oriented software, side effects are to be expected. In object-oriented software for random number generation, the state of the generator is an object.

Another issue to consider in the design of a user interface for a random number generator is whether the output is a single value (and an updated seed) or an array of values. Although a function that produces a single value as the C function `rand()` is convenient to use, it can carry quite a penalty in execution time because of the multiple invocations required to generate an array of random numbers. It is generally better to provide both single- and multivalued procedures for random number generation, especially for the basic uniform generator.

Random Number Generation in IMSL Libraries

For doing Monte Carlo studies, it is usually better to use a software system with a compilable programming language, such as Fortran or C. Not only do such systems provide more flexibility and control, but the programs built in the compiler languages execute faster. To do much work in such a system, however, a library or routines both to perform the numerical computations in the inner loop of the Monte Carlo study and to generate the random numbers driving the study are needed.

The IMSL Libraries contain a large number of routines for random number generation. The libraries are available in both Fortran and C, each providing the same capabilities and with essentially the same interface within the two languages. In Fortran the basic uniform generator is provided in both function and subroutine forms.

The uniform generator allows the user to choose among seven different algorithms: a linear congruential generator with modulus of $2^{31} - 1$ and with three choices of multiplier, each with or without shuffling, and the generalized feedback shift generator described by Fushimi (1990), which has a period of $2^{521} - 1$. The multipliers that the user can choose are the "minimal standard" one of Park and Miller (1988), which goes back to Lewis, Goodman, and Miller (1969) and two of the "best" multipliers found by Fishman and Moore (1982, 1986).

The user chooses which of the basic uniform generators to use by means of the Fortran routine `rnopt` or the C function `imsls_random_option`. For whatever choice is in effect, that form of the uniform generator will be used for whatever type of pseudorandom events are to be generated. The states of the generators are maintained in a `common` block (for the simple congruential generators, the state is a single seed; for the shuffled generators and the GFSR generator, the state is maintained in a table). There are utility routines for setting and saving states of the generators and a utility routine for obtaining a seed to skip ahead a fixed amount.

There are routines to generate deviates from most of the common distributions. Most of the routines are subroutines but some are functions. The algorithms used often depend on the values of the parameters to achieve greater efficiency. The routines are available in both single and double precision. (Double precision is more for the purpose of convenience for the user than it is for increasing accuracy of the algorithm.)

A single-precision IMSL Fortran subroutine for generating from a specific distribution has the form

`rn`*name* (*number*, *parameter_1*, *parameter_2*, ..., *output_array*)

where "*name*" is an identifier for the distribution, "*number*" is the number of random deviates to be generated, "*parameter_i*" are parameters of the distribution, and "*output_array*" is the output argument with the generated deviates. The Fortran subroutines generate variates from standard distributions, so location and scale parameters are not included in the argument list. The subroutine and formal arguments to generate gamma random deviates, for example, are

```
rngam (nr, a, r)
```

where `a` is the shape parameter (α) of the gamma distribution. The other parameter in the common two-parameter gamma distribution (usually called β) is a scale parameter. The deviates produced by the routine `rngam` have a scale parameter of 1; hence, for a scale parameter of `b`, the user would follow the call above with a call to a BLAS routine:

```
sscal (nr, b, r, 1)
```

Identifiers of distributions include those shown in Tables B.1 and B.2 beginning on page 660. In addition to the ones shown in those tables there are IMSL random number generators for random two-way tables, exponential mixtures, correlation matrices, points on a circle or sphere, order statistics from a normal or uniform, an ARMA process, and a nonhomogeneous Poisson process.

For general distributions, the IMSL Libraries provide routines for an alias method and for table lookup, for either discrete or continuous distributions. The user specifies a discrete distribution by providing a vector of the probabilities at the mass points and specifies a continuous distribution by giving the values of the cumulative distribution function at a chosen set of points. In the case of a discrete distribution, the generation can be done either by an alias method or by an efficient table lookup method. For a continuous distribution, a cubic spline is first fit to the given values of the cumulative distribution function, and then an inverse CDF method is used to generate the random numbers from the target distribution. Another routine uses the Thompson-Taylor data-based scheme (Taylor and Thompson, 1986) to generate deviates from an unknown population from which only a sample is available.

Other routines in the IMSL Libraries generate various kinds of time series, random permutations, and random samples. The routine `rnuno`, which generates order statistics from a uniform distribution, can be used to generate order statistics from other distributions.

All of the IMSL routines for random number generation are available in both Fortran and C. The C functions have more descriptive names, such as random_normal. Also, the C functions may allow specification of additional arguments, such as location and scale parameters. For example, random_normal has optional arguments IMSLS_MEAN and IMSLS_VARIANCE.

Controlling the State of the Generators

Figure 7.5 illustrates the way to save the state of an IMSL generator and then restart it. The functions to save and to set the seed are rnget and rnset.

```
    call rnget (iseed)        ! save it
    call rnun (nr, y)         ! get sample, analyze, etc.
 ...
    call rnset (iseed)        ! restore seed
    call rnun (nr, yagain)    ! will be the same as y
```

Fig. 7.5. Fortran Code Fragment to Save and Restart a Random Sequence Using the IMSL Library

In a library of numerical routines such as the IMSL Libraries, it is likely that some of the routines will use random numbers in regular deterministic computations, such as an optimization routine generating random starting points. In a well-designed system, before a routine in the system uses a random number generator in the system, it will retrieve the current value of the seed if one has been set, use the generator, and then reset the seed to the former value. IMSL subprograms are designed this way. This allows the user to control the seeds in the routines called directly.

Random Number Generation in R and S-Plus

Both R and S-Plus provides some choices for the basic type of random number generator, but R and S-Plus do not use the same random number generators. *Monte Carlo studies conducted using one system cannot reliably be reproduced exactly in the other system.*

In R, the function RNGkind can be used to choose the type of the generator. The default currently is the Mersenne twister MT19937.

Random number generation in either R or S-Plus is done with basic functions of the form

r*name* (*number* [, *parameters*])

where "*name*" is an identifier for the distribution, "*number*" is the number of random deviates to be generated, which can be specified by an array argument, in which case the number is the number of elements in the array, and "*parameters*" are parameters of the distribution, which may or may not be required.

For distributions with standard forms, such as the normal, the parameters may be optional, in which case they take on default values if they are not specified. For other distributions, such as the gamma or the t, there are required parameters. Optional parameters are both positional and keyword.

For example, the normal variate generation function is

```
rnorm (n, mean=0, sd=1)
```

so

`rnorm (n)`	yields n normal (0,1) variates
`rnorm (n, 100, 10)`	yields n normal (100,100) variates
`rnorm (n, 100)`	yields n normal (100,1) variates
`rnorm (n, sd=10)`	yields n normal (0,100) variates

(Note that R and S-Plus consider one of the parameters of the normal distribution to be the standard deviation or the scale rather than the variance, as is more common.)

For the gamma distribution, at least one parameter (the shape parameter) is required. The function reference

```
rgamma (100,5)
```

generates 100 random numbers from a gamma distribution with a shape parameter of 5 and a scale parameter of 1 (a standard gamma distribution).

Identifiers of distributions include those shown in Tables B.1 and B.2 beginning on page 660.

The function `sample` generates a random sample with or without replacement. Sampling with replacement is equivalent to generating random numbers from a (finite) discrete distribution. The mass points and probabilities can be specified in optional arguments:

```
xx <- sample(massp, n, replace=T, probs)
```

Order statistics in R and S-Plus can be generated using the beta distribution and the inverse distribution function. For example, 10 maximum order statistics from normal samples of size 30 can be generated by

```
x <- qnorm(rbeta(10,30,1))
```

Controlling the State of the Generators

Both R and S-Plus use an object called `.Random.seed` to maintain the state of the random number generators. In R, `.Random.seed` also maintains an indicator of which of the basic uniform random number generators is the current choice. Anytime random number generation is performed, if `.Random.seed` does not exist in the user's working directory, it is created. If it exists, it is used to initiate the pseudorandom sequence and then is updated after the sequence is generated. Setting a different working directory will change the state of the random number generator.

The function `set.seed(i)` provides a convenient way of setting the value of the `.Random.seed` object in the working directory to one of a fixed number of values. The argument `i` is an integer between 0 and 1023, and each value represents a state of the generator, which is "far away" from the other states that can be set in `set.seed`.

To save the state of the generator, just copy `.Random.seed` into a named object, and to restore, just copy the named object back into `.Random.seed`, as in Figure 7.6.

```
oldseed <- .Random.seed  # save it
y <- runif(1000)        # get sample, analyze, etc.
...
.Random.seed <- oldseed  # restore seed
yagain <- rnorm(1000)    # will be the same as y
```

Fig. 7.6. Code Fragment to Save and Restart a Random Sequence Using R or S-Plus

A common situation is one in which computations for a Monte Carlo study are performed intermittently and are interspersed with other computations, perhaps broken over multiple sessions. In such a case, we may begin by setting the seed using the function `set.seed(i)`, save the state after each set of computations in the study, and then restore it prior to resuming the computations, similar to the code shown in Figure 7.7.

```
set.seed(10)               # set seed at beginning of study
... # perform some computations for the Monte Carlo study
MC1seed <- .Random.seed  # save the generator state
... # do other computations
.Random.seed <- MC1seed  # restore seed
... # perform some computations for the Monte Carlo study
MC1seed <- .Random.seed  # save the generator state
```

Fig. 7.7. Starting and Restarting Monte Carlo Studies in S-Plus or R

The built-in functions in S-Plus that use the random number generators have the side effect of changing the state of the generators, so the user must be careful in Monte Carlo studies where the computational nuclei, such as `ltsreg` for robust regression, for example, invoke an S-Plus random number generator. In this case, the user must retrieve the state of the generator prior to calling the function and then reset the state prior to the next invocation of a random number generator.

To avoid the side effect of changing the state of the generator, when writing a function in R or S-Plus, the user can preserve the state upon entry to the function and restore it prior to exit. The assignment

```
.Random.seed <- oldseed
```

in Figure 7.6, however, does not work if it occurs within a user-written function in R or S-Plus. Within a function, the assignment must be performed by the `<<-` operator. A well-designed R or S-Plus function that invokes a random number generator would have code similar to that in Figure 7.8.

```
oldseed <- .Random.seed    # save seed on entry
...
.Random.seed <<- oldseed  # restore seed on exit
return(...)
```

Fig. 7.8. Saving and Restoring the State of the Generator within an S-Plus or R Function

Monte Carlo in R and S-Plus

Explicit loops in R or S-Plus execute slowly. In either package, it is best to use array arguments for functions rather than to loop over scalar values of the arguments. Consider, for example, the problem of evaluating the integral

$$\int_0^2 \log(x+1)x^2(2-x)^3\,\mathrm{d}x.$$

This could be estimated in a loop as follows:

```
# First, initialize n.
uu <- runif(n, 0, 2)
eu <- 0
for (i in 1:n) eu <- eu + log(uu[i]+1)*uu[i]^2*(2-uu[i])^3
eu <- 2*eu/n
```

A much more efficient way, without the `for` loop, but still using the uniform, is

```
uu <- runif(n, 0, 2)
eu <- 2*sum(log(uu+1)*uu^2*(2-uu)^3)/n
```

Alternatively, using the beta density as a weight function, we have

```
eb <- (16/15)*sum(log(2*rbeta(n,3,4)+1))/n
```

(Of course, if we recognize the relationship of the integral to the beta distribution, we would not use the Monte Carlo method for integration.)

For large-scale Monte Carlo studies, an interpretive language such as S-Plus or R may require an inordinate amount of running time. These systems are very useful for prototyping Monte Carlo studies, but it is often better to do the actual computations in a compiled language such as Fortran or C.

Notes and Further Reading

There are a number of books and review papers on random number generation. I am most familiar with Gentle (2003). Chapter 1 in that book has an extensive discussion of recursive methods for generating sequences of U(0, 1) random numbers; Chapter 2 addresses quality of random number generators and methods of testing their quality; Chapter 3 discusses quasirandom numbers; and Chapters 4 and 5 describe methods of transforming a uniform sequence into a sequence from a given distribution. Section 4.14 in that book describes methods for generating random variates to simulate a general multivariate distribution.

L'Ecuyer (2004) gives an overview of random number generation, with an emphasis on the basic uniform generators and testing their quality.

The use of Markov chains to form a "proposal" distribution has become a very useful tool in Bayesian statistical analyses. Random number generation using a stationary Markov chain majorizing density for applications in Bayesian analyses is discussed and illustrated extensively in Albert (2007) and Marin and Robert (2007).

In Appendix B, for the standard distributions, we give the root name of the R/S-Plus and IMSL functions for generating deviates from those distributions.

Exercises

7.1. Prove that if X is a random variable with an absolutely continuous distribution function P_X, the random variable $P_X(X)$ has a U(0, 1) distribution.

7.2. Acceptance/rejection methods.
   a) Give an algorithm to generate a normal random deviate using the acceptance/rejection method with the double exponential density as the majorizing density. After you have obtained the acceptance/rejection test, try to simplify it.
   b) What would be the problem with using a normal density to make a majorizing function for the double exponential distribution (or using a half-normal for an exponential)?
   c) Write a program to generate bivariate normal deviates with mean $(0, 0)$, variance $(1, 1)$, and correlation ρ. Use a bivariate product double

exponential density as the majorizing density. Now, set $\rho = 0.5$ and generate a sample of 1,000 bivariate normals. Compare the sample statistics with the parameters of the simulated distribution.

7.3. Acceptance/rejection methods.
Let T be the number of passes through the steps of the algorithm until a variate is accepted.
   a) Determine the mean and variance of T for the method described in Algorithm 7.1.
   b) Consider a modification of the acceptance/rejection method given in Algorithm 7.1, in which steps 1 and 2 are reversed and the branch in step 3 is back to the new step 2; that is:
      1. Generate u from a uniform (0,1) distribution.
      2. Generate y from the distribution with density function g_Y.
      3. If $u \leq p_X(y)/cg_Y(y)$, then take y as the desired realization; otherwise, return to step 2.

      Is this a better method? Determine the mean and variance of T for this method. (This method was suggested by Sibuya, 1961.)

7.4. Use the Metropolis-Hastings algorithm (page 314) to generate a sample of standard normal random variables. Use as the candidate generating density, $g(x|y)$, a double exponential density in x with mean y; that is, $g(x|y) = \frac{1}{2}e^{-|x-y|}$. Experiment with different burn-in periods and different starting values. Plot the sequences generated. Test your samples for goodness-of-fit to a normal distribution. (Remember that they are correlated.) Experiment with different sample sizes.

7.5. Let Y and Z have marginal distributions as exponential random variables with parameters α and β respectively.
   a) Consider a joint distribution of Y and Z defined by a Gumbel copula (equation (1.86), page 34). Write an algorithm to generate a random pair (Y, Z).
   b) Consider a joint distribution of Y and Z difined by a Gaussian copula (equation (1.83)). Write an algorithm to generate a random pair (Y, Z).
   c) Write a program to implement your algorithm in Exercise 7.5b. (For any serious random number generation, you should use a compiled language such as Fortran or C, but for this, you can use any language. The point of the question is more important than the programming.) Now generate 10,000 bivariate exponentials defined by the Gaussian copula with $\rho = 0.5$. Compute an estimate of the correlation coefficient of Y and Z. Compare this with Exercise 1.7 on page 75. (Although the correlation coefficient may not have much meaning for bivariate exponentials, it is still defined in the usual way.)

7.6. Consider the use of Gibbs sampling to generate samples from a bivariate normal distribution. Let the means be 0, the variances be 1, and the correlation be ρ. Both conditional distributions have the same form, which is given in the discussion of the use of marginal/conditional distributions on page 316. Let $\rho = 0, 0.2, 0.5, 0.8$. Generate samples with varying lengths of burn-in and assess the fidelity of the samples by computing summary statistics and by plots of your samples. Describe the efficiency of Gibbs sampling for this problem.

Part III

Methods of Computational Statistics

Introduction to Part III

The field of computational statistics includes a set of statistical methods that are computationally intensive. These methods may involve looking at data from many different perspectives and looking at various subsets of the data. Even for moderately sized datasets, the multiple analyses may result in a large number of computations. Statistical methods may be computationally intensive also because the dataset is extremely large. With the ability to collect data automatically, ever-larger datasets are available for analysis.

Viewing data from various perspectives often involves transformations such as projections onto multiple lower-dimensional spaces. Interesting datasets may consist of subsets that are different in some important way from other subsets of the given data. The identification of different subsets and the properties that distinguish them is computationally intensive because of the large number of possible combinations.

Another type of computationally intensive method useful in a wide range of applications involves simulation of the data-generating process. Study of many sets of artificially generated data helps to understand the process that generates real data. This is an exciting method of computational statistics because of the inherent possibilities of unexpected discoveries through experimentation.

Monte Carlo experimentation is the use of simulated random numbers to estimate some functional of a probability distribution. In simple applications of Monte Carlo, a problem that does not naturally have a stochastic component may be posed as a problem with a component that can be identified with an expectation of some function of a stochastic variable. The problem is then solved by estimating the expected value by use of a simulated sample from the distribution of a random variable. In such applications, Monte Carlo methods are similar to other methods of numerical analysis.

Monte Carlo methods differ from other methods of numerical analysis, however, in yielding an *estimate* rather than an *approximation*. The "numerical error" in a Monte Carlo estimate is due to a pseudovariance associated with a pseudorandom variable; but the numerical error in standard numerical analysis is associated with approximations, including discretization, truncation, and roundoff.

Monte Carlo methods can also be used to make inferences about parameters of models and to study random processes. In statistical inference, real data are used to estimate parameters of models and to study random processes assumed to have generated the data. Some of the statistical methods discussed in Part III use simulated data in the analysis of real data. There are several ways this can be done.

If the simulated data are used just to estimate one or more parameters, rather than to study the probability model more generally, we generally use the term *Monte Carlo* to refer to the method. Whenever simulated data are used in the broader problem of studying the complete process and building models, the method is often called *simulation*. This distinction between a simulation method and a Monte Carlo method is by no means universally employed; and we will sometimes use the terms "simulation" and "Monte Carlo" synonymously.

In either simulation or Monte Carlo, an actual dataset may be available; but it may be supplemented with artificially generated data. The term "*resampling*" is related to both "simulation" and "Monte Carlo", and some authors use it synonymously with one or both of the other terms. In this text, we generally use the term "resampling" to refer to a method in which random subsamples are generated from a given dataset; that is, there is no additional artificially generated data.

In the chapters in Part III, we discuss the general methods of computational statistics. These include:

- graphical methods;
- projection and other methods of transforming data and approximating functions;
- Monte Carlo methods and simulation;
- randomization and use of subsets of the data;
- bootstrap methods.

Some of the chapters in Part III have close correspondence to chapters in Part II. The methods of Chapter 9 rely heavily on Chapter 5; those of Chapter 10, on Chapter 4; and those of Chapter 11, on Chapter 7. Of course, the basic methods of random number generation (Chapter 7) underlie many of the methods of computational statistics.

8

Graphical Methods in Computational Statistics

One of the first steps in attempting to understand data is to visualize it. Visualization of data and information provides a wealth of tools that can be used in detecting features, in discovering relationships, and finally in retaining the knowledge gained.

Graphical displays have always been an important part of statistical data analysis, but with the continuing developments in high-speed computers and high-resolution devices, the usefulness of graphics has greatly increased. Higher resolution makes for a more visually pleasing display, and occasionally it allows features to be seen that could not be distinguished otherwise. The most important effects of the computer on graphical methods in statistics, however, arise from the ease and speed with which graphical displays can be produced, rather than from the resolution. Rapid production of graphical displays has introduced motion and articulated projections and sections into statistical graphics. Such graphical methods are important tools of computational statistics. The multiple views are tools of discovery, not just ways of displaying a set of data that has already been analyzed. Although the power of graphical displays has greatly increased, some of the most useful graphs are the simple ones, as illustrated in Section 1.1, and they should not be ignored just because we can do more impressive things.

Proper design of a graphical display depends on the context of the application and the purpose of the graphics, whether it is for the analyst to get a better understanding of the data or to present a picture that conveys a message. Our emphasis in the following discussion is on methods useful in exploratory graphics.

One thing that is lagging in the statistical literature is the use of color in graphical displays. The simple mechanics used in producing popular magazines are yet to be incorporated in the production of learned journals. Journals available in electronic form do not have these production problems, and it is likely that the paper versions of many journals will be discontinued before the production of color is mastered.

J.E. Gentle, *Computational Statistics*, Statistics and Computing,
DOI: 10.1007/978-0-387-98144-4_8,

Data of three or fewer dimensions can be portrayed on a two-dimensional surface fairly easily, but for data of higher dimensions, various transformations must be employed. The simplest transformations are just projections onto two dimensions, but transformations of points into other geometric objects may often reveal salient features. It is useful to have multiple views of the data in which graphical objects are linked by color or some other visual indicator. This linking is called *brushing*. An interactive graphics program may allow interesting sets points to be "roped" by the data analyst using a pointing device to draw a curve around the representations of the observations. When the analyst associates a particular color or other identifying attribute with given observations in one view, those same observations are endowed with the same attribute simultaneously in the other views. The analyst may also wish to magnify the region of the graph containing these special observations or perform other transformations selectively.

The number of variables and the number of observations may determine the way that graphical displays are constructed. If the number of observations is large we may first make a few plots of samples of the full dataset. Even for multivariate data, some initial plots of single variables may be useful. A preliminary 4-plot for each variable on a dataset can be a useful, automatic part of almost any analysis of data (see page 9).

Most often we are interested in graphical representations of a dataset, but we can distinguish three basic types of objects that a graphical display may represent:

- discrete data;
- mathematical functions;
- geometrical objects.

The graphical elements that represent discrete data are simple plotting symbols: dots, circles, and so on. The graphical elements that represent functions or geometrical objects may be curves or surfaces. Because of the discrete nature of the picture elements (pixels) of a graphical display, both continuous functions and geometrical objects must be converted to discrete data to produce a graph. Hence, beginning with either functions or geometrical objects, we arrive at the task of graphing discrete data. The data, at the lowest level, correspond to adjacent pixels.

The Graphical Coordinate System

The basic activity in producing a graphical display is to translate data represented in a "world coordinate system" into a representation in a graphics "device coordinate system". Positions in the world coordinate system are represented in the computer in its floating-point scheme. Positions in the device coordinate system correspond to picture elements, or "pixels", which can appear as black or white dots or dots of various colors. The graphical display is the pointillistic image produced by the pixels. The type of the coordinate system used, cartesian or homogeneous, may depend on the types of transformations on the data to be performed in the graphical analysis.

For display on a flat surface, such as a sheet of paper, the graphical coordinate system is necessarily two-dimensional. This means that the data in a world coordinate system, which may be multidimensional, must be projected onto a two-dimensional system. An orthogonal projection of the points onto the two-dimensional system are just the two-tuples of coordinates of the points in the graphical coordinate system. If the world coordinate system of the data is two-dimensional, this projection does not sacrifice information. If, however, the data in the world coordinate system are three-dimensional, the relative orientation of the two coordinate systems becomes important because the amount of information conveyed in a given two-dimensional system is usually less than the information available in the full coordinate system. This orientation corresponds to the angle from which we view the three-dimensional data. Not only is the angle important, the distance from which the three-dimensional data are viewed is important. For objects that are not transparent, this distance may determine which object can be seen from the given angle. In any event, the distance determines the perspective, which provides a sense of depth. These ideas of the viewing angle and the "eye position" for a two-dimensional perspective extends to data of any dimension.

Images

The images themselves are usually constructed in one of two ways: as a raster or as a vector. A raster image is a fixed set of pixels. It is resolution-dependent, so if it is displayed at a higher resolution or its size is increased, jagged edges may appear. A vector image is made up of mathematically defined lines and curves. The definitions do not depend on the resolution. Modifications to the image, such as moving it or changing its size, are relatively simple and scalable because they are made to the mathematical definition.

Displays of Large Data Sets

As the number of observations increases, information should increase. A problem with many graphical displays, however, is that large amounts of data result in too dense a graph, and there may actually be a loss of information. Data points are overplotted. There is too much "ink" and too little information. If there were no low-dimensional structure in the data shown in Figure 9.4 on page 379 in Chapter 9, for example, the plot, which represents only 1,000 points, would just be an almost solid blob.

When overplotting occurs for only a relatively small number of points, and especially if the data are not continuous (that is, data points occur only on a relatively coarse lattice), the overplotting problem can be solved by jittering, which is the process of plotting the data at nearby points rather than at the exact point ("exact" subject to the resolution of the plot).

For large datasets we may associate each observation with a single pixel. The number of observations in a dataset may be even larger than the number of pixels, in which case we must locally smooth the data.

For very large datasets where the number of observations is greater than the number of pixels, it is clear that some other approach must be used to represent the data. It is necessary either to sample the data, to smooth the data and plot the smoothed function, and/or to plot a representation of the density of the data. Gray scale or relative sizes of the plotting symbols can also be effective for representing the data density, especially for univariate or bivariate data. It can also be used in two-dimensional projections of multivariate data.

Smoothing and graphing of data are activities that go well together. Smoothing provides a better visual display, and conversely, the display of the smoothed data provides a visual assessment of smoothing.

Datasets that are large because of the number of variables present a more difficult problem. The nature of the variables may allow special representations that are useful for that type of data.

Data Analysis and Human Perception

Visual perception by humans is a complex process. We may identify three fairly distinct aspects. The first is the physical and physiological, the optics, the retinal photoreceptors and their responses, and the matching of colors. The second phase is the representation, that is, the analysis of images in the neural retina and the visual cortex. This involves sensitivity to and recognition of patterns, requiring multiresolution of images. The third aspect is the interpretation of information in the visual representation. Perception of color, motion, and depth plays an important role in all of these phases.

Although color can be very useful in the visual representation of data, poorly chosen colors are a major distraction in statistical graphics. Many statisticians and other data analysts are aware of the importance of carefully chosen colors for enhancing visual displays. There are two issues that need emphasis, however. One is the differences in the effects of color in different media. A color scheme that appears very useful on one computer monitor may not be appropriate on other monitors or on printed media. Another problem arises from the not insignificant proportion of persons who are color-blind but in all other ways are normally-sighted. Color-blindness usually involves only two colors, often complementary ones. The most common type of color-blindness in America is red-green; that is, the inability to distinguish red and green. For normally-sighted persons, however, these two colors show up well and are easily distinguished; hence, they are obvious choices for use in color graphics.

Other senses, such as hearing and touch, may be usefully employed in coming to a better understanding of a set of data.

Immersive techniques in which data are used to simulate a "virtual reality" may help in understanding complicated data. Such systems consist of various projectors, mirrors, and speakers; eyeglasses with alternating shutters; and user controls for feedback.

8.1 Smoothing and Drawing Lines

In typical applications, the observed data represent points along some continuous range of possibilities. Hence, although we begin with discrete data, we wish to graph a continuous function. We assume a continuous function as an underlying model of the process that generated the data. The process of determining a continuous function from the data is called *smoothing*.

Smoothing is often an integral process of graphing discrete data. A smooth curve helps us to visualize relationships and trends.

Graphing Continuous Functions

There are two common situations in statistics that lead to approximation and estimation of functions. Sometimes, one of the variables in a dataset is modeled as a stochastic function of the other variables, and a model of the form

$$y \approx f(x) \tag{8.1}$$

is used. The "dependent" or "response" variable y is related to the variable x by an unknown function f.

In another type of situation, the variable x is assumed to be a realization of a random variable X, and we are interested in the probability density function

$$p_X(x). \tag{8.2}$$

In both of these cases, x may be a vector.

In the former case, in which relationships of variables are being modeled as in model (8.1), the dataset consists of pairs (y_i, x_i). A smooth curve or surface that represents an estimate or an approximation of f helps us to understand the relationship between y and x. Fitting this curve smoothes the scatter plot of y_i versus x_i. There are, of course, several ways of smoothing the data, as discussed from various perspectives in Chapters 4, 10, and 17.

Bézier Curves

In graphical applications and in geometric modeling, Bézier curves are used extensively because they are quickly computed. Bézier curves are smooth curves in two dimensions that connect two given points with a shape that depends on points in between. For a given set of points in two dimensions, $p_0, p_1, \ldots, p_n$, called control points, Bézier curves are required to satisfy two conditions:

1. The two endpoints p_0 and p_n must be interpolated.

2. The rth derivatives at p_0 and p_n are determined by r adjacent points to produce a smooth curve. The first derivative at p_0, for example, is the line determined by p_0 and p_1.

These conditions obviously do not uniquely determine the curves.

The Bézier curve is determined by the set of points $\{p_0, \ldots, p_n\}$ (that is, in two dimensions, $p_i = (x, y)$), defined parametrically by

$$p(u) = \sum_{i=0}^{n} p_i B_{i,n}(u), \tag{8.3}$$

where $u \in [0, 1]$ and $B_{i,n}(u)$ is the Bernstein polynomial,

$$B_{i,n}(u) = \frac{n!}{i!(n-i)!} u^i (1-u)^{n-i} \quad \text{for} \quad u \in [0, 1]. \tag{8.4}$$

For example,

$$\begin{aligned} B_{0,3}(u) &= (1-u)^3, \\ B_{1,3}(u) &= 3u(1-u)^2, \\ B_{2,3}(u) &= 3u^2(1-u), \\ B_{3,3}(u) &= u^3. \end{aligned}$$

The Bernstein polynomial $B_{i,n}$ is proportional to the PDF of a standard beta distribution with parameters $\alpha = i+1$ and $\beta = n-i+1$. (They are essentially the same as the beta weight function used in defining the Jacobi polynomials on page 170, except those beta weights were over the interval $[-1, 1]$. The standard beta is over $[0, 1]$.) The Bernstein polynomials are not orthogonal polynomials.

For $n+1$ control points, $(x_0, y_0), (x_1, y_1), \ldots, (x_n, y_n)$, we use the series of Bernstein polynomials $B_{i,n}(u)$ for $i = 0 \ldots n$.

Note that because of the form of the Bernstein polynomials, the sequence of points could be reversed without changing the curve.

In Figure 8.1, we show four Bézier curves of various degrees. In the top left panel, there are three control points, $(0, 5), (10, 4), (9, 2)$ and the quadratic Bernstein polynomials are used. In the top right panel, an additional control point is inserted between the second and third: $(0, 5), (10, 4), (12, 3), (9, 2)$ and the cubic Bernstein polynomials are used. Notice, how the additional point pulls the curve farther to the right, so that the curve doubles back onto itself.

In the bottom panels of Figure 8.1, we form closed Bézier curves by making the first and last control points the same. In the lower left panel there are two control points (points 1 and 2) that pull the cubic curve out into a narrow loop. In the lower right panel, two more control points have been added and the quintic curve forms a more open loop.

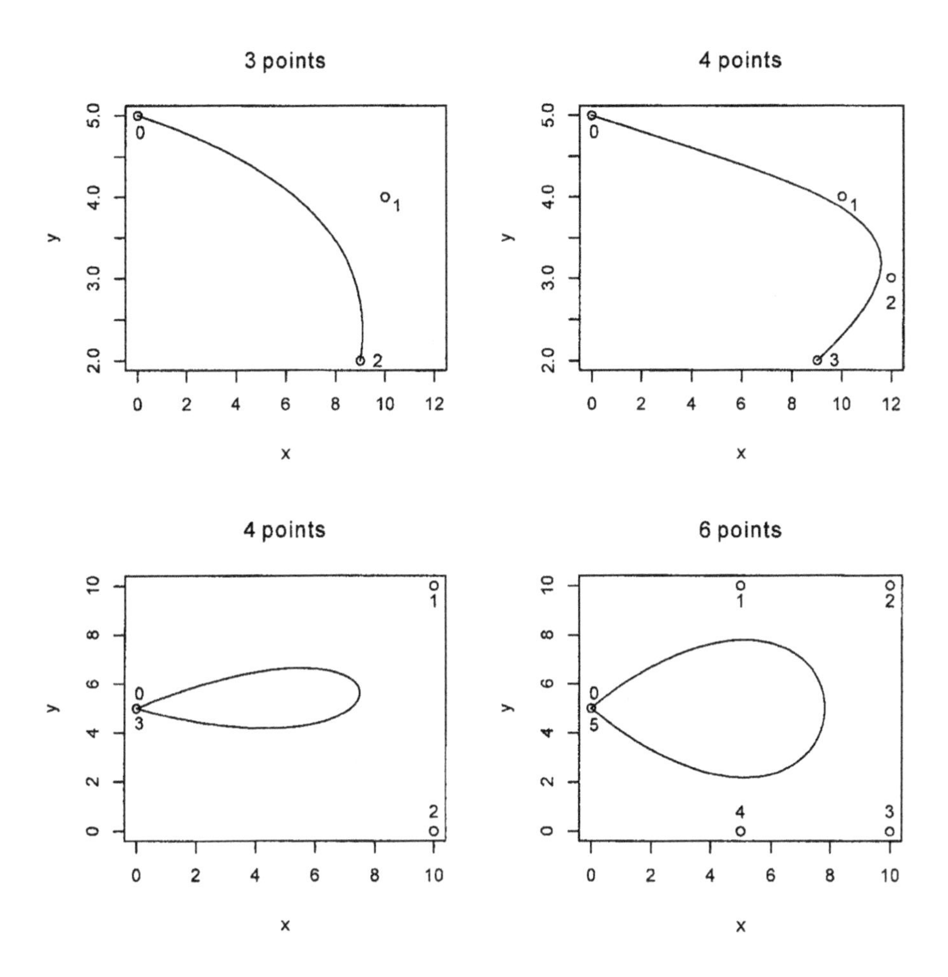

Fig. 8.1. Bézier Curves and Control Points

Bézier curves are widely used in graphics because they can be computed quickly. They can be implemented in interactive graphics software to allow the user to adjust curves smoothly. They are also used by graphic designers to define characters or logos that are immediately scalable.

Continuous Densities

If no particular variable in a multivariate dataset is considered a dependent variable, we may be interested in the probability density function p that describes the distribution of a multivariate random variable. A histogram is one representation of the probability density. The histogram is a method of smoothing data, but the histogram itself can be smoothed in various ways, or, alternatively, other smooth estimates of the density can be computed.

Fitting models using observed data is an important aspect of statistical data analysis. Distributional assumptions may be used to arrive at precise statements about parameters in the model. Smoothing for graphical displays is generally less formal. The purpose is to help us to visualize relationships and distributions without making formal inferences about models.

8.2 Viewing One, Two, or Three Variables

Plots of one or two variables are easy to construct and often easy to interpret. Plots of three variables can use some of the same techniques as for one or two variables. For datasets with more variables, it is often useful to look at the variables one, two, or three at a time, or to look at projections of all variables into a two- or three-dimensional subspace.

One of the most important properties of data is the shape of its distribution, that is, a general characterization of the density of the data. The density of the data is measured by a nonnegative real number. The density is thus an additional variable on the dataset. The basic tool for looking at the shape of the distribution of univariate data is the *histogram*. A histogram is a graph of the counts or the relative frequency of the data within contiguous regions called bins. Graphs such as histograms that represent the density have one more dimension than the dimension of the original dataset.

A *scatter plot*, which is just a plot of the points on cartesian axes representing the variables, is useful for showing the distribution of two-dimensional data. The dimension of the scatter plot is the same as the dimension of the data. In a scatter plot, data density is portrayed by the density of the points in the plot.

We use the phrases “two-dimensional” and “three-dimensional” in reference to graphical displays to refer to the dimension of the space that the display depicts in a cartesian system. Thus, the dimension of a scatter plot of either two or three variables is the same as the dimension of the data, although in either case the actual display either on a monitor or on paper is two-dimensional. In statistical displays, we often are interested in an additional dimension that represents the distribution or density of the data. As noted above, plots representing densities have one more dimension than the data.

Histograms and Variations

A histogram is a presentation, either graphical or tabular, of the counts of binned or discrete data. The vertical axis in a histogram may be the counts (frequencies) in the bins or may be proportions representing densities such that the total area adds up to 1.

The formation of bins for grouping data is one of the most fundamental aspects of visualizing and understanding data. The bins in a histogram generally all have the same width, but this is not necessary; sometimes, if there are only a small number of observations over a wide range, the bins over that range can be made wider to smooth out the roughness of the variation in the small counts.

The number of bins in a histogram can markedly affect its appearance, especially if the number of observations is small. In Figure 8.2, four histograms are shown, each of the same dataset. The data are a pseudorandom sample of size 30 from a gamma distribution with shape parameter 3 and scale parameter 10. In this simple example, we get different pictures of the overall shape of the data depending on the number of bins. It is worthwhile to consider this example of a small dataset because the same issues may arise even in very large datasets in high dimensions. In large datasets in high dimension, we also sometimes have "small sample" problems, especially when we focus on "slices" of the data.

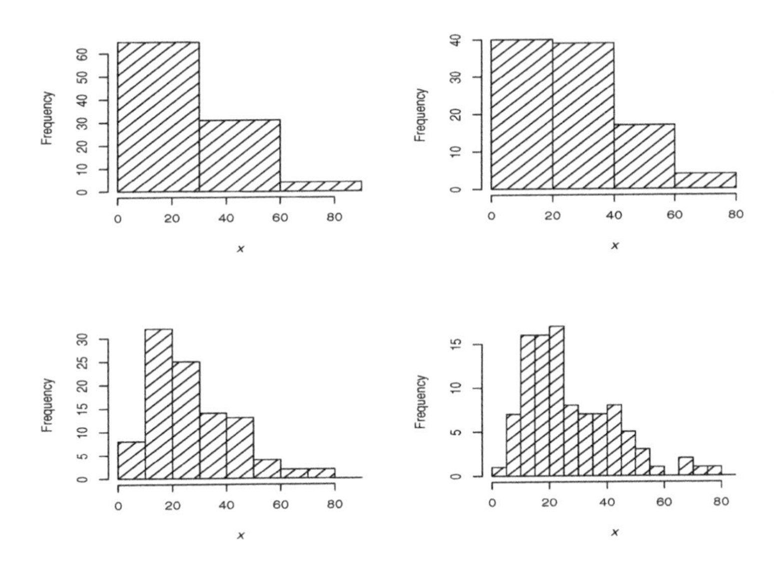

Fig. 8.2. Histograms of the Same Data with Different Fixed Bin Sizes (Data from a Gamma Distribution)

More bins give a rougher appearance of the histogram. Either too few or too many bins can obscure structure in the data. In Figure 8.2, when only three or four bins are used, the curvature of the density is not apparent; conversely, when twelve bins are used, it is difficult to ascertain any simple pattern in the data. In general, the number of bins should be greater for a larger number of observations. A simple rule for the approximate number of bins to use is

$$1 + \log_2 n,$$

where n is the number of observations.

Figure 8.3 shows the same data as used in Figure 8.2 and with the same cutpoints for the bins as the histogram with seven bins, except that some bins have been combined. The appearance of the histogram is smoother and, in fact, is closer to the appearance expected of a histogram of a sample from a gamma distribution with a shape parameter of 3. (Notice that this is a density histogram, rather than a frequency histogram, as in Figure 8.2. The actual counts are less relevant when the bin widths are variable.)

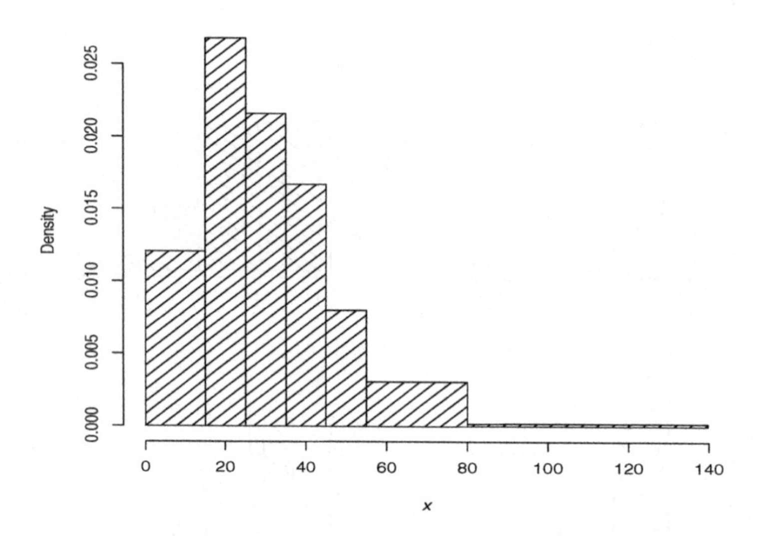

Fig. 8.3. Density Histogram of the Gamma Data with Bins of Variable Widths

Exploring the data by using histograms with different bin widths is useful in understanding univariate data. The objective is not to match some known or desired distribution but rather to get a better view of the structure of the data. Obviously, the histogram of a random sample from a particular distribution may never look like the density of that distribution. Exploration of the data for gaining a better understanding of it is not to be confused with manipulation of the data for presentation purposes, which is a common objective in statistical graphics.

The appearance of a histogram can also be affected by the location of the cutpoints of the bins. The cutpoints of the histogram in the lower left of Figure 8.2, which has seven bins, are

$$0, 10, 20, 30, 40, 60, 70.$$

Shifting the cutpoints by 2, so that the cutpoints are

$$2, 12, 22, 32, 42, 62, 72,$$

results in the histogram in the upper right-hand corner of Figure 8.4, and further shifts in the cutpoints result in the other histograms in that figure. Notice the changes in the apparent structure in the data.

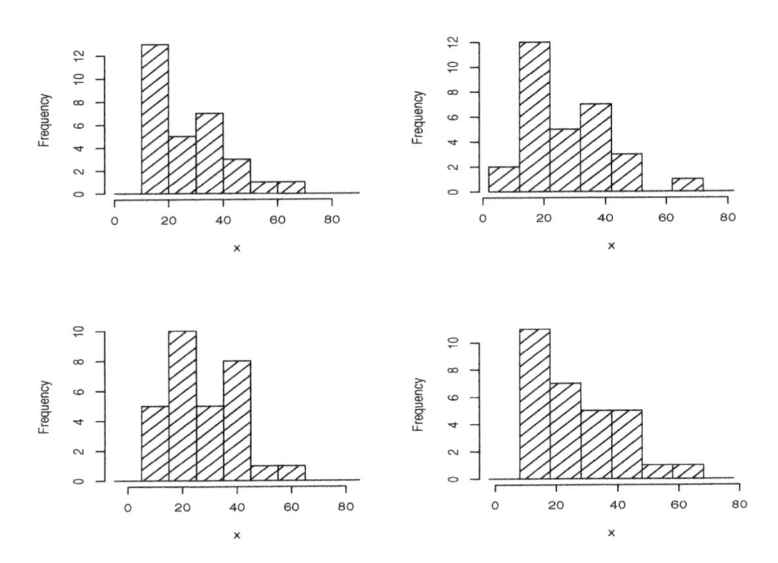

Fig. 8.4. Histograms of the Same Set of Gamma Data with Different Locations of Bins

The use of histograms with different widths and locations of bins is an example of a method of computational statistics. In the analysis of a dataset, we should consider a number of different views of the same data. The emphasis is on exploration of the data rather than on confirmation of hypotheses or graphical presentation.

In the toy example that we considered, we could attribute the problems to the relatively small sample size and so perhaps decide that the problems are not very relevant. These same kinds of problems, however, can occur in very large datasets if the number of variables is large.

The Empirical Cumulative Distribution Function and q-q Plots

The empirical cumulative distribution function, or ECDF, is one of the most useful summaries of a univariate sample. The ECDF is a step function, with a saltus of $\frac{1}{n}$ at each point in a sample of size n. A variation of the ECDF, the broken-line ECDF, with lines connecting the points, is often more useful. A plot of the broken-line ECDF is shown in the graph on the left-hand side in Figure 8.5.

Another variation of an ECDF plot is one that is flipped or folded at some point of interest, such as the median. Such a plot is called a *mountain plot.* It is often easier to see certain properties, such as symmetry, in a mountain plot. A folded ECDF plot, or mountain plot, is shown on the right in Figure 8.5.

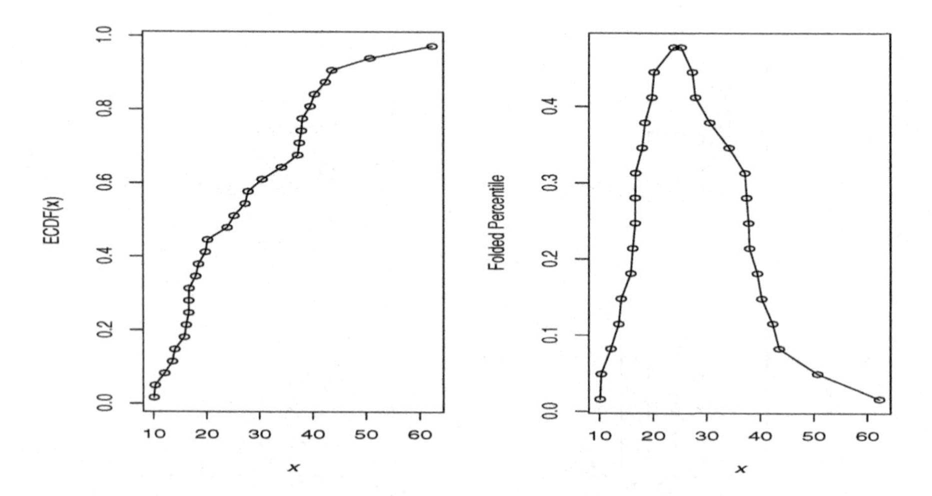

Fig. 8.5. Plots of a Broken-Line ECDF and a Folded ECDF or Mountain Plot of Data from a Gamma Distribution

The plot of the ECDF provides a simple comparison of the sample with the uniform distribution. If the sample were from a uniform distribution, the broken-line ECDF would be close to a straight line, and the folded ECDF would be close to an isosceles triangle. The ECDF of a unimodal sample is concave. The ECDF of a multimodal sample is convex over some intervals. The plots of the sample of gamma variates in Figure 8.5 show a skewed, unimodal pattern.

A sample can be compared to some other distribution very easily by a transformation of the vertical axis so that it corresponds to the cumulative distribution function of the given distribution. If the vertical axis is transformed in this way, a broken-line ECDF plot of a sample from that distribution would be close to a straight line. A plot with a transformed vertical axis is called a probability plot.

A related plot is the quantile-quantile plot or q-q plot. In this kind of plot, the quantiles or "scores" of the reference distribution are plotted against the sorted data. The $1/n^{\text{th}}$ quantile is plotted against the first order statistic in the sample of size n, and so on. As we mentioned on page 62, the probabilities associated with the empirical quantiles depend on the underlying distribution, and in any event, are difficult to work out.

However the probability is chosen, the p_k^{th} *quantile* (or "population quantile") is the value, x_{p_k}, of the random variable, X, such that

$$\Pr(X \leq x_{p_k}) = p_k.$$

In the case of the normal distribution, this value is also called the p_k^{th} *normal score.*

A q-q plot with a vertical axis corresponding to the quantiles of a gamma distribution with a shape parameter of 4 is shown on the left-hand side in Figure 8.6. This plot was produced by the R statement

```
plot(qgamma(ppoints(length(x)),4),sort(x))
```

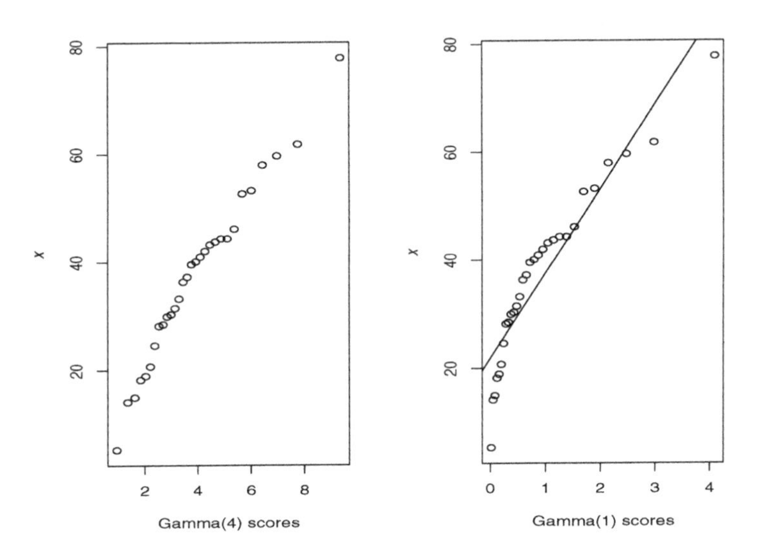

Fig. 8.6. Quantile-Quantile Plot for Comparing the Sample to Gamma Distributions

If the relative values of the sample quantiles correspond closely to the distribution quantiles, the points fall along a straight line, as in the plot on the left-hand side in Figure 8.6. The data shown were generated from a gamma distribution with a shape parameter of 4. When the sample quantiles are compared with the quantiles of a gamma distribution with a shape parameter of 1, as in the plot on the right-hand side in Figure 8.6, the extremes of the sample do not match the quantiles well. The pattern that we observe for the smaller observations (that is, that they are below a straight line that fits most of the data) is characteristic of data with a heavier left tail than the reference distribution to which it is being compared. Conversely, the larger observations, being below the straight line, indicate that the data have a lighter right tail than the reference distribution.

The sup absolute difference between the ECDF and the reference CDF is the *Kolmogorov distance*, which is the basis for the Kolmogorov test (and the Kolmogorov-Smirnov test) for distributions. The Kolmogorov distance, however, does poorly in measuring differences in the tails of the distribution. A q-q plot, on the other hand, generally is very good in revealing differences in the tails.

An important property of the q-q plot is that its shape is independent of the location and the scale of the data. In Figure 8.6, the sample is from a gamma distribution with a scale parameter of 10, but the distribution quantiles are from a population with a scale parameter of 1.

For a random sample from the distribution against whose quantiles it is plotted, the points generally deviate most from a straight line in the tails. This is because of the larger variability of the extreme order statistics. Also, because the distributions of the extreme statistics are skewed, the deviation from a straight line is in a specific direction (toward lighter tails) more than half of the time (see Exercise 8.2, page 368).

The ECDF is most useful for univariate data. Plots based on the ECDF for a multivariate dataset are generally difficult to interpret.

Representation of the Third Dimension

A three-dimensional plot on a two-dimensional surface is sometimes called a "perspective plot". Important characteristics of a perspective plot are the viewing angle and the assumed position of the eye that is viewing the plot. In a perspective plot, the simulated location of the eye determines the viewing angle or line of sight and also affects the perspective.

A perspective plot attempts to give the appearance of a three-dimensional space. It may consist of the individual three-dimensional data points, or, if one of the coordinate variables is considered to be a function of the other two, that variable usually is made to correspond to the vertical coordinate in the display and the display itself is of a surface. We show a perspective plot on page 382, which was produced by the R function `persp`.

Another simple way of reducing a three-dimensional display to a two-dimensional one is by use of contour lines, or contour bands (usually of different colors). A contour line represents a path over which the values in the dimension not represented are constant.

Contours represent one extra dimension, so three-dimensional data are often represented on a two-dimensional surface in a *contour plot.* A contour plot is especially useful if one variable is a "dependent" variable (that is, one for which we are interested in its relationship or dependence on the two other variables). In a contour plot, lines or color bands are used to represent regions over which the values of the dependent variable are constant.

For representing three-dimensional data in which one variable is a dependent variable, an *image plot* is particularly useful. An image plot is a plot of three-dimensional data in which one dimension is represented by color or by a gray scale.

Image plots are especially useful in identifying structural dependencies. They are often used when the two "independent" variables are categorical. In such cases, the ordering of the categories along the two axes has a major effect on the appearance of the plot. Figure 8.7 shows four image plots of the same set of data representing gene expression activity for 500 genes from cells from 60 different locations in a human patient. In the plot in the upper left, the cells and genes are arbitrarily positioned along the axes, whereas in the other plots there has been an attempt to arrange the cells and/or genes along their respective axes to discover patterns that may be present.

Exploration with image plots is a powerful tool for discovering structure and relationships. Clustering methods discussed in Chapter 16 can be used to suggest orderings of the categorical variables. A "clustered image map" can be particularly useful in detecting structural differences. See the programs and data at

http://discover.nci.nih.gov

Both contour plots and image plots can be effective for large datasets in which overplotting would be a serious problem in other types of plots.

Contour plots are produced by the `contour` function in both S-Plus and R, and image plots are produced by the `image` function in both packages.

Other methods of representing the third dimension use color or some other noncartesian property, such as discussed for general multivariate data beginning on page 359. Another approach for three-dimensional data is to simulate a visual perception of depth.

Rendering

Some methods of representing three-dimensional data attempt to simulate a geometric object. These methods include *direct volume rendering* and *surface rendering.* In either case, an important consideration is the point, called the viewpoint, from which the space is viewed. In some cases, the viewpoint may be a pair of points corresponding to the viewer's two eyes.

In direct volume rendering, we attempt to display the density of material or data along imaginary rays from the viewer's eyes through the dataset. This

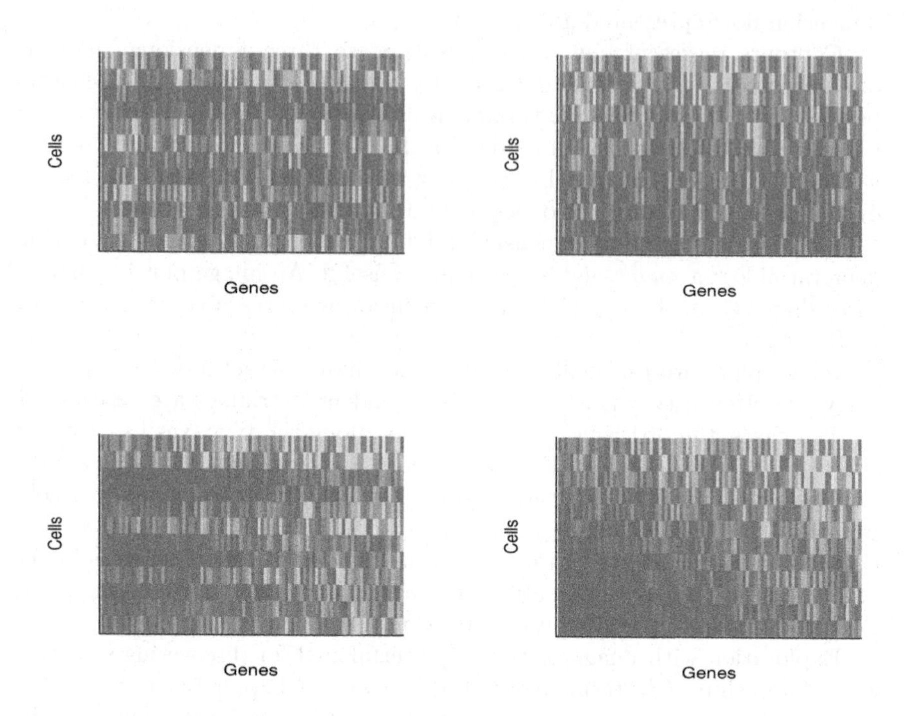

Fig. 8.7. Image Plots with Different Orderings of the Categorical Variables

procedure is called ray tracing. The volume is rendered by assigning values to voxels (three-dimensional equivalents of pixels). In one form, called binary voxel volume rendering, the first object along the ray from the viewer's eye results in an opaque voxel that hides objects farther along the ray. Binary rendering is often done by "z-buffering". In this technique, a value representing the depth of each point (discretized as a pixel) is stored, and the minimum such depth (discretized) from the viewpoint is stored in a "z-buffer". (It is called this because in graphics for three dimensions, a cartesian coordinate system (x, y, z) is commonly used, and by custom the plane of the display is the x-y plane, or that plane moved slightly.) A point is hidden if its depth is greater than the value in the z-buffer for that direction. In another form, called semitransparent volume rendering, a transparency is assigned to a voxel based on the value of the parameter that is being rendered. This allows visualization of the interior of an object. This technique is widely used in medical imaging.

The simplest method of representing a surface is by use of a "wire frame", which is a grid deformed to lie on the surface. Generally, in a wire frame, the grid lines on regions of surfaces that would be obscured by other surfaces are not shown. This is usually done by z-buffering. An example of a wire frame surface is shown in Figure 9.5 on page 382. Sometimes, however, it is useful to show the hidden lines in a wire frame, and with different line types, it is possible to show lines that would be hidden but make clear that those lines are in the background. It is often useful to combine a wire frame with a contour plot on the flat surface representing the plane of two variables.

Other ways of depicting the surface, which use a continuous representation, require consideration of the surface texture and the source of light that the surface reflects.

Stereograms

A perception of depth or a third dimension in a two-dimensional display can be induced by use of two horizontally juxtaposed displays of the same set of points and a mechanism to cause one of the viewer's eyes to concentrate on one display and the other eye to concentrate on the other display. The stereo pair may be in different colors, and the viewer uses glasses with one lens of one color and the other lens of the other color (the colors are chosen so that a lens of one color cancels the other color). Such a pair is called an *anaglyph*. Another type of stereo pair, called a *stereogram*, requires the viewer to defocus each separate view and fuse the two views into a single view. This fusion can be aided by special glasses that present one view to one eye and the other view to the other eye. Many people can perform the fusion without the aid of glasses by focusing their eyes on a point either beyond the actual plane of the display or in front of the plane. Either way works. The perspective in one way is the reverse of the perspective in the other way. For some displays and for some people, one way is easier than the other. In either type of stereo pair, the features on the two displays are offset in such a way as to appear in perspective.

Figure 8.8 shows a stereogram of data with three variables, x, y, and z. The stereoscopic display is formed by two side-by-side plots of x and y in which x is on the horizontal axes and z determines the depth. The perception of depth occurs because the values of x are offset by an amount proportional to the depth. The depth at the ith point is

$$d_i = c \cdot \left(z_{\max} - z_i\right) \frac{x_{\max} - x_{\min}}{z_{\max} - z_{\min}}. \tag{8.5}$$

The choice of c in equation (8.5) depends on the separation of the displays and on the units of measurement of the variables. The left-hand plot is of the vectors $x - d$ and y, and the right-hand plot is of $x + d$ and y. If the eyes are focused behind the plane of the paper, the vertex that is lowest on the graphs in Figure 8.8 is at the front of the image; if the eyes are focused in front of the plane of the paper (that is, if the eyes are crossed) the vertex that is highest on the graphs in Figure 8.8 is at the front of the image.

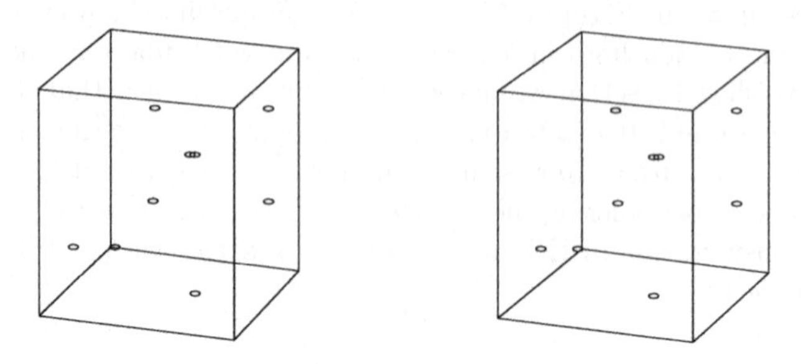

Fig. 8.8. Trivariate Data in a Stereogram

Although statisticians have experimented with anaglyphs and stereograms for many years, these visual devices have generally had more value as entertainment than as effective as tools of discovery. They may be useful, however, for displaying features of the data that are already known.

Changing the Viewing Angle

The perception of depth can also be induced by rotating a single three-dimensional scatter plot on a monitor. The perception of a third dimension ceases when the movement stops, however. One way of preserving the motion and hence the depth perception while still maintaining the viewing angle is to rock the scatter plot (that is, to rotate the scatter plot back and forth through a small angle).

In addition to the induced perception of depth, rotations, because they give different perspectives on the data, often prompt discoveries of structure in the data. For example, observations that are outliers in a direction that does not correspond to the axis of any single variable may become apparent when a three-dimensional scatter plot is rotated. We discuss rotation transformations in some detail in Chapter 9 beginning on page 375.

Different people have differing abilities to interpret graphical displays. The stereoscopic devices discussed above are not very useful for people with vision in only one eye and for the growing number of people who use artificial lenses to provide near vision in one eye and distance vision in the other.

Contours in Three Dimensions

Contour plots in three dimensions are surfaces that represent constant values of a fourth variable. Methods for drawing contours for three-dimensional data are similar to those for two-dimensional data; a three-dimensional mesh is formed, and the values of the function or of a fourth variable at the lattice points of the mesh are used to decide whether and how a contour surface may cut through the mesh volume element. The most widely used method for drawing contours in three dimensions is "the marching cubes" method. (See Schroeder, Martin, and Lorensen, 1996, for a good description of the method.)

Contours of three-dimensional data can also be represented in stereograms using the same offsets as in equation (8.5). See Scott (2004, Section 4.4) for examples.

8.3 Viewing Multivariate Data

Graphical displays are essentially limited to the two dimensions of a computer monitor or a piece of paper. Various devices that simulate depth allow visualization of a third dimension using just the two-dimensional surface. The simplest such devices are reference objects, such as color saturation or perspective lines, that provide a sense of depth. Because of our everyday experience in a three-dimensional world with an atmosphere, visual clues such as diminished color or converging lines readily suggest distance from the eye. Other, more complicated mechanisms making use of the stereo perspective of the two human eyes may suggest depth more effectively, as we discuss on page 353. There are not many situations in which these more complicated devices provide more understanding of data than would be available by looking at various two-dimensional views. They are more fun, however.

There are basically two ways of displaying higher-dimensional data on a two-dimensional surface. One is to use multiple two-dimensional views, each of which relates points in a cartesian plane with a projection of the points in higher dimensions. The number of separate two-dimensional views of d-dimensional data that convey the full information of the original data is $\mathrm{O}(d^2)$. The projections do not have to be orthogonal, of course, so the number of projections is uncountable.

The other way is to use graphical objects that have characteristics other than just cartesian coordinates that are associated with values of variables in the data. These graphical objects may be other geometric mappings, or they may be icons or glyphs whose shapes are related to specific values of variables. The number of graphical objects is n, the number of observations, so some of these methods of displaying high-dimensional data are useful only if the number of observations is relatively small.

Projections

Numeric data can easily be viewed two variables at a time using scatter plots in two-dimensional cartesian coordinates. Each scatter plot represents a projection from a given viewing angle. The projection may also be scaled to simulate an eye position.

An effective way of arranging these two-dimensional scatter plots of multidimensional data is to lay them out in a square (or triangular) pattern. All scatter plots in one row of the pattern have the same variable on the vertical axis, and all scatter plots in one column of the pattern have the same variable on the horizontal axis, as shown in Figure 8.9. In this case, the various two-dimensional graphical coordinate systems correspond directly to orthogonal projections of the world coordinate system in which the data are represented.

Each view is a two-dimensional projection of a multidimensional scatter plot, so all observations are represented in each view. This arrangement is sometimes called a scatter plot matrix, or "SPLOM". One is shown in Figure 8.9. The plot shows pairwise relationships among the variables and also that the observations fall into distinct groups. (The plot in Figure 8.9 was produced by the R function `pairs`. The plot uses the "Fisher iris data". This is a relatively small dataset that has been widely studied. The data are four measurements on each of 50 iris plants from each of three species. The data are available in S-Plus and R as `iris`.)

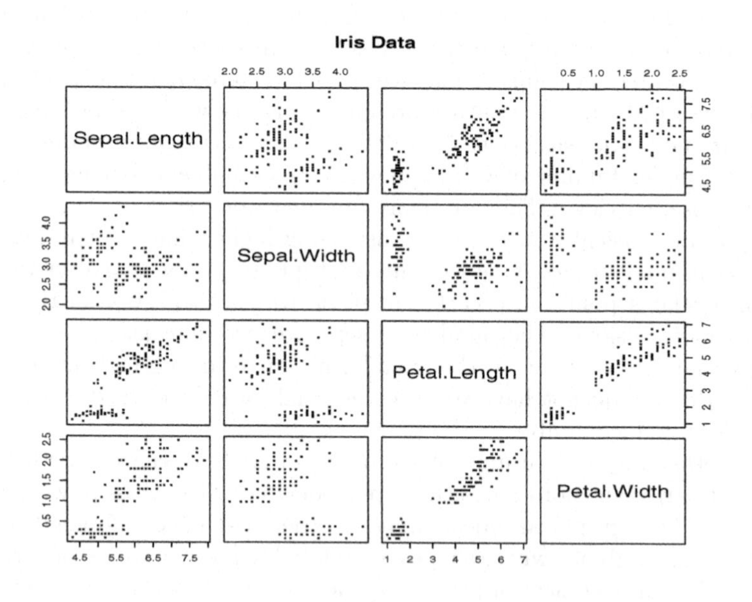

Fig. 8.9. Scatter Plot Matrix of Fisher Iris Data

The two-dimensional scatter plots represent the data in the $n \times 2$ matrix, $X_{jk} = X[e_j|e_k]$, where $[e_j|e_k]$ is a $d \times 2$ matrix and e_i is the ith unit column vector of length d. More general two-dimensional projections of the d-dimensional data may also be useful. The $n \times d$ matrix X that contains the data is post-multiplied by a $d \times d$ projection matrix of rank 2, and then the data are plotted on cartesian axes that form an orthogonal basis of the two-dimensional space. (A projection matrix is any real symmetric idempotent matrix. The projection matrix used in the formation of X_{jk} above, if $j < k$, is the $d \times d$ matrix consisting of all zeros except for ones in the (j, j) and (k, k) positions. We will encounter projection matrices again in Chapter 9.)

The scatter plots in a SPLOM as described above are plots of unadjusted marginal values of the individual variables. Davison and Sardy (2000) suggest use of a partial scatter plot matrix (that is, scatter plots of residuals from linearly adjusting each variable for all others except the other one in the current scatter plot). This has the advantage of showing more precisely the relationship between the two variables, because they are conditioned on the other variables that may be related. Davison and Sardy also suggest forming a scatter plot matrix using the marginal values in one half (say, the plots above the diagonal) and using the adjusted values or residuals in the other half of the matrix.

A sequence of projections is useful in identifying interesting properties of data. In a "grand tour", a plane is moved through a d-dimensional scatter plot. As the plane moves through the space, all points are projected onto the plane in directions normal to the plane. We discuss the grand tour on page 363. Projection pursuit, as discussed in Section 16.5, is another technique for successively considering lower-dimensional projections of a dataset in order to identify interesting features of the data.

Projections cannot reveal structure of a higher dimension than the dimension of the projection. Consider, for example, a sphere, which is a three-dimensional structure. A two-dimensional projection of the sphere is the same as that of a ball; that is, the projection does not reveal the hollow interior.

A matrix of image plots can be useful in simultaneously exploring the relationship of a response variable to pairs of other variables. The plot in Figure 8.10 is an image plot matrix, or IMPLOM, showing values of a response variable in gray scale at various combinations of levels of three independent variables. In Figure 8.10, the ordering of each of the independent variables is the same above the diagonal as below it, but this is not necessary. Orderings can be compared by using different ones in the two images that correspond to the same two independent variables.

Conditioning Plots

Another type of lower-dimension view of a dataset is provided by a *section*. A section of a space of d dimensions is the intersection of the space with a lower-dimensional object. The lower-dimensional object is often chosen as a hyperplane.

A two-dimensional section of a three-dimensional sphere can reveal the hollow interior if the section passes through the sphere. The ability of a section to determine a feature depends on where the section is relative to the feature.

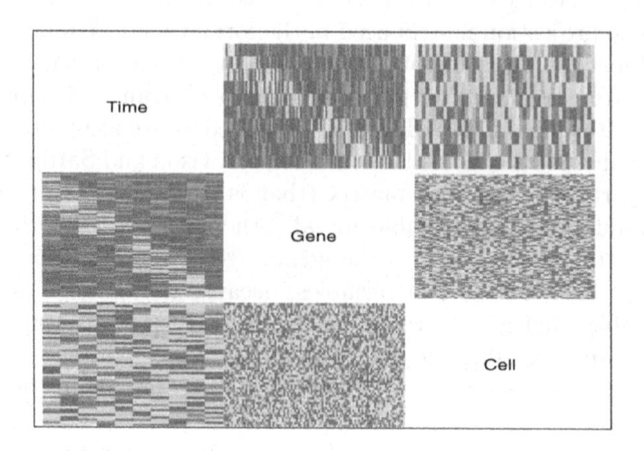

Fig. 8.10. Matrix of Image Plots

Projections and sections can be used together to help in identifying structure in data.

Sectioning leads to another way of forming an interesting lower-dimensional dataset, which is to restrict the full dataset to its intersection with a given lower-dimensional subset. That intersection has the dimensions of the given lower-dimensional subset. The idea is to look at specific slices of the data, one at a time. For example, if a plane is passed through a trivariate space, the result is a plane. In such a case, however, the resulting dataset may be quite sparse. If the lower-dimensional subset is determined by specific values of nominal variables, however, the intersection may contain a meaningful proportion of the observations. In another approach, the subset used in the intersection may represent a range of values, and if the range is relatively narrow, it may be the case that a projection preserves most of the information. This approach yields a conditional dataset, and the displays of the conditional datasets are called called conditioning plots, or coplots.

When a multivariate dataset has different subsets of interest, perhaps determined by the values of a nominal variable, it is convenient to view the data in separate panels corresponding to the different subsets. Such a multipaneled graphic is called a "casement display".

In a conditioning plot, the overall graphical display is divided into two parts, a panel for the conditioning slices and a set of *dependence panels* showing the bivariate relationship between the dependent variable and the panel variable, at each level of the conditioning slices.

Conditioning is also called splitting or nesting. The "trellis" displays of S-Plus are designed to do this. The trellis displays are implemented in the Lattice system in R; see Sarkar (2008). The `coplot` function in both S-Plus and R produces conditioning plots.

In conditioning plots, even if the dataset is very large, if the dimensionality is high, the sizes of the dataset in each panel of the display may be relatively small.

Noncartesian Displays

Noncartesian displays are often developed to aid in identification of specific features. The box plot is very useful in seeing the overall shape of the data density and in highlighting univariate outliers.

One way of dealing with multivariate data is to represent each observation as a more complicated object than just a point. The values of the individual variables that make up the observation are represented by some aspect of the iconic object. For some kinds of displays, each observation takes up a lot of space, so the use of those techniques is generally limited to datasets with a small number of observations. Other displays use curves to represent observations, and those kinds of displays can be used with larger datasets, although at some point they may suffer from extreme overplotting.

A limitation to the usefulness of such noncartesian displays is that the graphical elements may not have any natural correspondence to the variables. This sometimes makes the plot difficult to interpret until the viewer has established the proper association between the graphical features and the variables. Even so, it often is not easy to visualize the data using the icons or curves.

Glyphs and Icons

Various kinds of glyphs and icons can be used to represent multivariate data. In many cases, a separate glyph or icon is used for each observation. One of the most typical is a star diagram, of which there are several variations. In a star diagram, for each observation of d-dimensional data, rays pointing from a central point in d equally spaced directions represent the values of the variables.

An issue in noncartesian displays is how to represent magnitudes. In one variation of the star diagram, the lengths of the rays correspond to the magnitude of the value of the variable. In another variation, sometimes called "snowflakes", the lengths of the rays are linearly scaled individually so that the minimum of any variable in the dataset becomes 0 and the maximum becomes 1.

Once the rays are drawn, their endpoints are connected. Once a mental association is made between a direction and the variable that corresponds to it, one can get a quick idea of the relative values of the variables in a given observation.

Chernoff (1973) suggested the use of stylized human faces for representing multivariate data. Each variable is associated with some particular feature of the face, such as height, width, shape of the mouth, and so on. Because of our visual and mental ability to process facial images rapidly, it is likely that by viewing observations represented as faces, we can identify similarities among observations very quickly.

As with star diagrams, the features of Chernoff faces can be associated with the variables and their values in different ways, for example, the area of face may correspond to the first variable, the shape of face to the second, the length of nose to the third, and so on.

In star diagrams, each variable is represented in the same way: as a ray. The stars have different appearances based only on the order in which the variables are assigned to rays. In faces, however, the variables correspond to very different features of the diagram, so there are many more differences in appearance that can result from the same dataset. Perhaps for this reason, faces are not used nearly as often as stars.

Anderson (1957) proposed use of a glyph consisting of a circle and from one to seven rays emanating from its top. Variations of these glyphs yield "feather plots", "compass plots", and "rose plots". Of course there are several other types of icons and glyphs that can be used for representing multivariate data.

Stars, faces, and other kinds of icons that individually represent a single observation are not very useful if the number of observations is more than 20 or 30. The usefulness of such graphics, however, results from their ability to represent 20 or 30 variables.

Parallel Coordinates: Points Become Broken Line Segments

Another type of curve for representing a multivariate observation is a piecewise linear curve joining the values of the variables on a set of parallel axes, each of which represents values of a given variable. This type of plot is called a "parallel coordinates plot". A parallel coordinates plot is similar to a nomogram.

With parallel coordinates, a point in the multidimensional space becomes a curve (a broken line) in a two-dimensional space. The scales on the horizontal lines in a parallel coordinates plot are generally scaled linearly to cover the range of the value in the sample.

Parallel coordinates help to identify relationships between variables. Pairwise positive correlations between variables represented by adjacent coordinate lines result in the line segments of the observations having similar slopes, whereas negative correlations yield line segments with different slopes. Correlations between variables on adjacent coordinate lines are most easily recognized. If the columns in the dataset are reordered, the adjacency pattern in the parallel coordinates plot changes, so other relationships may be observed.

Observations that are similar tend to have lines that have similar tracks. Parallel coordinates plots are therefore useful in identifying groups in data. If a variable on the dataset indicates group membership, all of the lines will go through the points on that coordinate that represent the (presumably small number of) groups. The visual impact is enhanced by placing this special coordinate either at the top or the bottom of the parallel coordinates plot (although if this special coordinate is placed in the middle, sometimes the groups are more obvious, because there may be fewer line crossings).

In Figure 8.11 we show the parallel coordinates plot of four artificially-generated points from a 7-variate normal distribution. The points were from independent distributions except for two, which had a correlation of 0.9. The two lines that track each other very closely represent those two points.

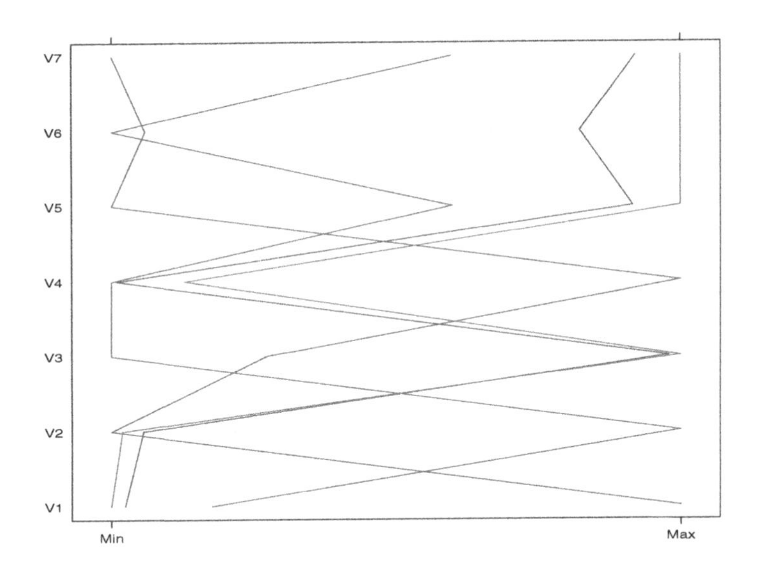

Fig. 8.11. Parallel Coordinates Plot

For large numbers of observations with little structure, parallel coordinates plots will suffer from too much overplotting. If there is low-dimensional structure, however, a parallel coordinates plot may help to uncover it. Figure 9.4 on page 379 in Chapter 9, for example, shows a parallel coordinates plot of data that have been rotated so a to show more interesting structure. The same plot in the unrotated coordinates, would just be a mass of ink.

Trigonometric Series: Points Become Curves

Another way of depicting a multivariate observation is with a curve in two dimensions. One type of curve is built by a sum of trigonometric functions. Plots of this type are called "Fourier series curves", or "Andrews curves" after David Andrews. An Andrews curve representing the point

$$x = (x_1, x_2, \ldots, x_d)$$

is

$$s(t) = x_1/\sqrt{2} + x_2 \sin t + x_3 \cos t + x_4 \sin(2t) + x_5 \cos(2t) + \ldots. \qquad (8.6)$$

Each observation in a data set yields one Andrews curve. The curves can be useful in identifying observations that are similar to each other. If the number of observations is large, however, the amount of overplotting generally becomes excessive.

In Figure 8.12 we show the Andrews curves for the same four artificially-generated points used in Figure 8.11. The points were from independent distributions except for points 3 and 4, which were generated from a distribution with correlation of 0.9. The lines representing those points are seen to track each other very closely.

As t goes from 0 to 2π, the curve traces out a full period, so the plot is shown just over that range. It is often useful, however, to make plots of two full periods of these periodic curves.

In Andrews curves, the value of the first feature (variable) determines the overall height of the curve; hence Andrews curves are very dependent on the order of the variables. It is generally a good idea to reorder the variables, so that the ones of most interest occur before others.

Andrews curves are also sometimes plotted in polar coordinates, resulting in a star-shaped figure with four arms.

Rotations and Dynamical Graphics

When a cluster of points or a surface in three dimensions is rotated (or alternatively, when the viewing direction is changed), patterns and structures in the data may be more easily recognized. Changing the eye position, that is, the distance from which a projection of the data is viewed, can also help to identify structure.

Rotations are orthogonal transformations that preserve norms of the data vectors and angles between the vectors. The simplest rotation to describe is that of a plane defined by two coordinates about the other principal axes. Such a rotation changes two elements of a vector that represents cartesian coordinates in the plane that is rotated and leaves all the other elements, representing the other coordinates, unchanged. A rotation matrix, introduced

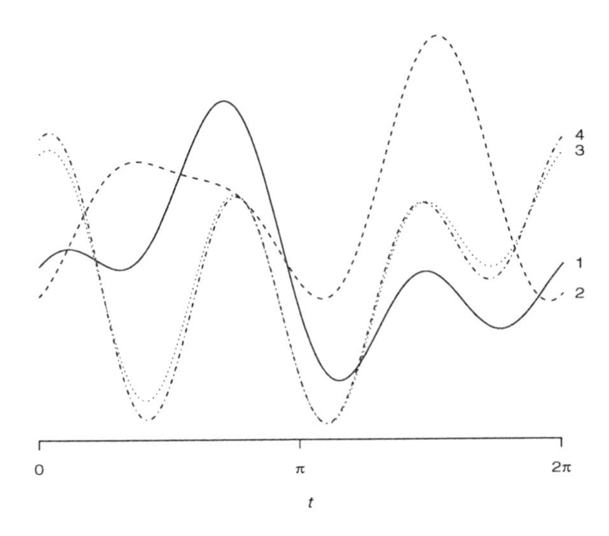

Fig. 8.12. Fourier Curves Plot

on page 233 and shown in a general form in equation (9.4) on page 377, is the same as an identity matrix with four elements changed.

A generalized rotation matrix, Q, can be built as a product of $(d^2 - d)/2$ such Q_{ij} simple rotation matrices,

$$Q = Q_{12}Q_{13} \cdots Q_{1d}Q_{23}Q_{24} \cdots Q_{2d} \cdots Q_{d-1,d}.$$

Rotating a plot in all directions, along with various projections, is called a "grand tour". In one method of performing a grand tour, the angles for the rotations are taken as

$$t\phi_{ij} \bmod 2\pi, \tag{8.7}$$

where the ϕ_{ij} are fixed constants that are linearly independent over the integers; that is, if for any set of integers $k_{12}, k_{13}, \ldots, k_{d-1,d}$,

$$\left(\sum_{i=1}^{d-1} \sum_{j=i+1}^{d} k_{ij}\phi_{ij} \right) \bmod 2\pi = 0,$$

then all $k_{ij} = 0$. As t is stepped over time, the angle in the rotation matrix Q_{ij} is taken as $t\phi_{ij}$ and the generalized rotation matrix is computed and applied to the dataset.

The rotated datasets can be viewed in various ways. In the most common grand tour, the point cloud is projected onto a representation of a three-dimensional cartesian system. In the grand tour, the data points appear to be in continuous motion on the computer monitor. The motion of the system, or equivalently, the apparent continuous rotations of the data, provide a perception of the third dimension.

Rotated datasets can also be viewed using parallel coordinates or Andrews curves. Structure appears in various ways in these plots. A hyperplane, for example, appears as a point on a parallel coordinate axis that corresponds to the coefficients of the hyperplane.

After we consider general rotations in Chapter 9 we show an example on page 379 of how a dataset can be rotated before plotting. You are asked to develop systematic rotation and plotting methods for producing Andrews curves and parallel coordinates plots in Exercises 9.4 and 9.5.

There is another way of "touring the data" by using the Andrews curves, $s(t)$, of equation (8.6) on page 362. These representations have a natural dynamic quality. The variable t in $s(t)$ of equation (8.6) can be interpreted as "time" and varied continuously.

Wegman and Shen (1993) modify and generalize the Andrews curves as

$$\begin{aligned} r(t) &= x_1\sin(\omega_1 t) + x_2\cos(\omega_1 t) + x_3\sin(\omega_2 t) + x_4\cos(\omega_2 t) + \ldots \\ &= (a(t))^{\mathrm{T}} x, \end{aligned} \tag{8.8}$$

where the vector $a(t)$ is chosen as

$$a(t) = \big(\sin(\omega_1 t),\ \cos(\omega_1 t),\ \sin(\omega_2 t),\ \cos(\omega_2 t),\ \ldots\big).$$

Wegman and Shen (1993) then consider an orthogonal linear combination, $q(t) = (b(t))^{\mathrm{T}}x$, where

$$b(t) = \big(\cos(\omega_1 t),\ -\sin(\omega_1 t),\ \cos(\omega_2 t),\ -\sin(\omega_2 t),\ \ldots\big). \tag{8.9}$$

They define a two-dimensional "pseudo grand tour" as the plots of $r(t)$ and $q(t)$ as t varies continuously. For the pseudo grand tour, they suggest defining $a(t)$ and $b(t)$ so that each has an even number of elements (if d is odd, the data vector x can be augmented with a 0 as its last element) and then normalizing both $a(t)$ and $b(t)$. They also recommend centering the data vectors about 0.

If the ω's are chosen so that ω_i/ω_j is irrational for all i and j not equal (i and j range from 1 to $\lceil d/2 \rceil$), a richer set of orientations of the data are encountered when t is varied. The generalized curves are not periodic in this case. Specialized graphics software often provides interaction that allows "guided tours" using controlled rotations. In a guided tour, the data analyst, using knowledge of the dataset or information provided by previous views of the data, actively decides which portions of the data space are explored. This, of course, is an instance in which concepts of real numbers do not have an analogue in $\mathbb{F}$. (There are no irrational numbers in $\mathbb{F}$; see Section 2.2.)

Projection pursuit, as discussed in Section 16.5 on page 564, can be used to determine the rotations in any grand tour using either the standard cartesian coordinates or the other types of displays.

Notes and Further Reading

There is a wealth of statistical literature on graphics, and visualization is being used ever more widely in data analysis. Developments in statistical graphics are reported in several journals, most notably, perhaps, *Journal of Computational and Graphical Statistics*, as well as in unrefereed conference proceedings and newsletters, such as *Statistical Computing & Graphics Newsletter*, published quarterly by the Statistical Computing and the Statistical Graphics sections of the American Statistical Association. Many of the advances in computer graphics are reported at the annual ACM SIGGRAPH Conference. The proceedings of these conferences, with nominal refereeing, are published as *Computer Graphics, ACM SIGGRAPH xx Conference Proceedings* (where "xx" is a two-digit representation of the year).

Hardware and Low-Level Software for Graphics

Hardware for graphics includes the computational engine, input devices, and various display devices. Rapid advances are being made for almost all types of graphics hardware, and the advances in the quality and capabilities of the hardware are being accompanied by decreases in the costs of the equipment. The image is the result of the collage of pixels displayed on these devices.

Software for graphics often interacts very closely with the hardware, taking advantage of specific design features of the hardware.

Because a graph may require many computations to produce lines and surfaces that appear smooth, the speed of the computational engine is very important. The appropriate pixels must be identified and set to the proper value to make the entire graph appear correctly to the human eye. The need for computer speed is even greater if the object being displayed is to appear to move smoothly.

A typical computer monitor has a rectangular array of approximately one to two million pixels. (Common monitors currently are 1,280 by 1,024, 1,600 by 1,200, or 1,920 by 1,200.) This is approximately 100 pixels per inch. This resolution allows arbitrary curves to appear fairly smooth. Whether graphical displays can respond to real-time motion depends on how fast the computer can perform the calculations necessary to update 10^6 pixels fast enough for the latency period of the human eye.

Color is determined by the wavelength of light. Violet is the shortest and red the longest. (Here, "color" is assumed to refer to something visible.) White color is a mixture of waves of various lengths.

Most light sources generate light of various wavelengths. Our perception of color depends on the mix of wavelengths. A single wavelength in light is a highly saturated single color. Multiple wavelengths reduce the saturation and affect the overall hue and lightness. Multiple wavelengths are perceived as a different single color that is a combination of the individual colors.

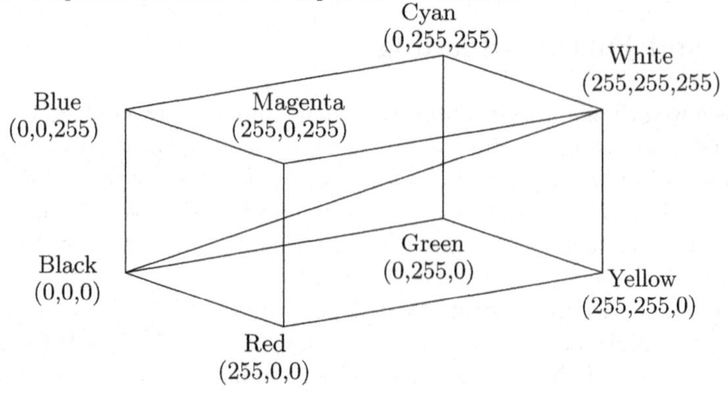

Fig. 8.13. RGB Color Cube

A given color can be formed by combining up to three basic colors. Red, green, and blue are often used as the basic colors. Colors can be combined additively using different light sources or in a subtractive way by filters or absorptive surfaces.

To specify a given color or mix of colors and other characteristics of the light, we use a *color system.* Color values are defined in the given color system and then used by the software and hardware to control the colors on the output device. Different systems use different combinations of values to describe the same color and the same overall effect. The common color systems are RGB (red, green, blue), CMY (cyan, magenta, yellow), HLS (hue, lightness, saturation), and HSV (hue, saturation, value).

The RGB color system uses a vector of three elements ranging in value from 0 to 255. The system can be illustrated as in Figure 8.13 by a cube whose sides are 255 units long. Three corners correspond to the primary colors of red, green, and blue; three corners correspond to the secondary colors of cyan, magenta, and yellow; and two corners correspond to black and white. Each color is represented by a point within or on the cube. The point $(255, 255, 255)$ represents an additive mixture of the full intensity of each of the three primary colors. Points along the main diagonal are shades of gray because the intensity of each of the three primaries is equal.

Digital display devices represent each component of an RGB color coordinate in binary as an integer in the range of 0 to $2^n - 1$, for some n. Each displayable color is an RGB coordinate triple of n-bit numbers, so the total number of representable colors is 2^{3n}, including black and white. An m-bit pixel can represent 2^m different colors. If m is less than $3n$, a *color translation table* (or just *color table*) with 2^m entries is used to map color triples to values of the pixels.

Low-Level Software

Software for producing graphics must interact very closely with the display devices. Because the devices vary in their capabilities, the approach generally taken is to design and produce the software at various levels so that graphics software at one level will interface with different software at another level in different environments.

The lowest-level software includes the *device drivers*, which are programs in the machine language of the particular display device. The next level of software provides the primitive graphics operations, such as illuminating a pixel or drawing a line. There have been a number of efforts to standardize the interface for this set of graphics primitives. The Open Graphics Library, or OpenGL, is a library of primitive graphics functions developed by Silicon Graphics, Inc. It was standardized by the OpenGL Architecture Review Board (1992), and it is now controlled by the Khronos Group, which is an industry consortium. For each of these sets of standard graphics functions there are bindings for Fortran, C, and C++. Glaeser and Stachel (1999) describe the use of OpenGL in a C++ graphics system called Open Geometry.

Software for Graphics Applications

There are a number of higher-level graphics systems ranging from Fortran, C, or Java libraries to interactive packages that allow modification of the display as it is being produced. Many graphics packages have a number of preconstructed color tables from which the user can select to match the colors a given device produces to the desired colors.

Gnuplot is an interactive plotting package that provides a command-driven interface for making a variety of data- and function-based graphs. The system is primarily for two-dimensional graphics, but there are some three-dimensional plotting capabilities. The graphs produced can be exported into a number of formats. The package is freeware and is commonly available on both Unix/Linux systems and MS Windows.

Xfig is a graphics package for Unix/Linux windowing systems (X11) that provides capabilities for the basic objects of vector graphics, including lines and various curves such as Bézier curves.

Advanced Visual Systems, Inc., develops and distributes a widely used set of graphics and visualization systems, AVS5 and AVS/Express together with various associated products. These systems run on most of the common platforms.

The Visualization Toolkit, or vtk, developed by Schroeder, Martin, and Lorensen (2004), is an object-oriented system that emphasizes three-dimensional graphics. The software manual also has good descriptions of the algorithms implemented.

Because making graphical displays is generally not an end in itself, graphics software is usually incorporated into the main application software. Software systems for statistical data analysis, such as S-Plus, R, and SAS, have extensive graphics capabilities. Some of the graphical capabilities in S-Plus and R are similar. Most features in one package are available in the other package, but there are differences in how the two packages interact with the operating system, and this means that there are some differences in the way that graphics files are produced. The function `expression` in R is a useful feature for producing text containing mathematical notation or Greek letters. The function can be used in most places that expect text, such as `xlab`. For example,

```
main = expression(paste("Plot of ",
                        Gamma(x)," versus",hat(beta) x^hat(gamma)))
```

produces the main title

Plot of $\Gamma(x)$ versus $\widehat{\beta}\, x^{\widehat{\gamma}}$

The actual appearance is device dependent and in any event is unlikely to have the beauty of a display produced by TEX.

There is also a very useful system in R, called Grid Graphics, that facilitates layout design; see Murrell (2006). A higher-level system for visualization of multivariate data, called Lattice, has been built on Grid Graphics; see Sarkar (2008).

Cook and Swayne (2008) describe a system for interactive graphics called GGobi that is integrated with R.

Wilkinson (2004) provides a unified structure for the production of meaningful graphical displays from data in tabular or matrix form.

Exercises

8.1. Generate a sample of size 200 of pseudorandom numbers from a mixture of two univariate normal distributions. Let the population consist of 80% from a $\mathrm{N}(0,1)$ distribution and 20% from a $\mathrm{N}(3,1)$ distribution. Plot the density of this mixture. Notice that it is bimodal. Now plot a histogram of the data using nine bins. Is it bimodal? Choose a few different numbers of bins and plot histograms. (Compare this with Exercise 15.9 of Chapter 15 on page 512.)

8.2. Generate a sample of pseudorandom numbers from a normal (0,1) distribution and produce a quantile plot of the sample against a normal (0,1) distribution, similar to Figure 8.6 on page 349. Do the tails of the sample seem light? (How can you tell?) If they do not, generate another sample and plot it. Does the erratic tail behavior indicate problems with the random number generator? Why might you expect often (more than 50% of the time) to see samples with light tails?

8.3. Stereoscopic displays.

   a) Show that equation (8.5) on page 353, is the offset, in the plane of the display, for each eye in viewing points at a depth z. *Hint:* Draw rays representing the lines of sight through the plane of the graph.

b) Using any graphics software, reproduce the stereogram in Figure 8.14 that represents a cone resting on its base being viewed from above. (If the eyes focus in front of the plane of the paper, the view of the cone is from below.)

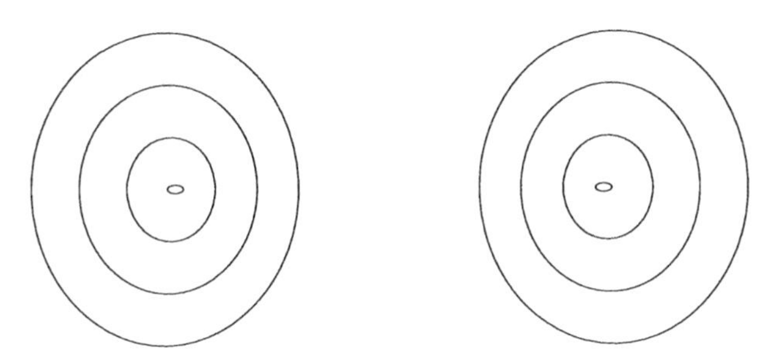

Fig. 8.14. A Cone Resting on Its Base

8.4. Write a program for the simple linear congruential random number generator

$$x_i \equiv 259x_{i-1} \bmod 2^{15}.$$

Generate a sequence of length 1,008. Look for structure in triples of sequential values, (x_i, x_{i+1}, x_{i+2}), by plotting two-dimensional sections of a three-dimensional scatter plot.

8.5. Plot the ellipse $x^2 + 4y^2 = 5$ in cartesian coordinates. Now, plot it in parallel coordinates. What is the shape of the parallel coordinates plot of an ellipse?

8.6. Generate 1,000 pseudorandom 4-variate normal deviates with mean 0 and the identity as the variance-covariance matrix. Now, delete from the dataset all deviates whose length is less than 2. This creates a dataset with a "hole" in it. Try to find the hole using various graphic displays.

8.7. Generate 100 pseudorandom trivariate normal variates with mean 0 and variance-covariance matrix

$$\begin{bmatrix} 1.00 & -.90 & .90 \\ -.90 & 1.81 & -1.71 \\ .90 & -1.71 & 2.62 \end{bmatrix}.$$

The Cholesky factor of the variance-covariance matrix is

$$\begin{bmatrix} 1.00 & & \\ -.90 & 1.00 & \\ .90 & -.90 & 1.00 \end{bmatrix}.$$

a) Plot the data using parallel coordinates. What shapes result from the correlations?
b) Plot the data using Andrews curves. What shapes result from the correlations?

8.8. Program a modification of parallel coordinates in which there is a common scale for all coordinates (that is, one in which a vertical line would pass through the same value on all parallel coordinate lines). Plot the Fisher iris data in a display in which all coordinates have the same scale and compare it to a display in which the coordinates have their original scales. Now, try some other datasets. How would you recommend that the scales on the parallel coordinate lines be constructed? What are the advantages and disadvantages of a fixed scale for all lines?
Summarize your findings in a clearly-written report.

9

Tools for Identification of Structure in Data

In recent years, with our increased ability to collect and store data, have come enormous datasets. These datasets may consist of billions of observations and millions of variables. Some of the classical methods of statistical inference, in which a parametric model is studied, are neither feasible nor relevant for analysis of these datasets. The objective is to identify interesting structures in the data, such as clusters of observations, or relationships among the variables. Sometimes, the structures allow a reduction in the dimensionality of the data.

Many of the classical methods of multivariate analysis, such as principal components analysis, factor analysis, canonical correlations analysis, and multidimensional scaling, are useful in identifying interesting structures. These methods generally attempt to combine variables in such a way as to preserve information yet reduce the dimension of the dataset. Dimension reduction generally carries a loss of some information. Whether the lost information is important is the major concern in dimension reduction.

Another set of methods for reducing the complexity of a dataset attempts to group observations together, combining observations, as it were.

In the following we will assume that an observation consists of a vector $x = (x_1, \ldots, x_m)$. In most cases, we will assume that $x \in \mathbb{R}^m$. In statistical analysis, we generally assume that we have n observations, and we use X to denote an $n \times m$ matrix in which the rows correspond to observations.

In practice, it is common for one or more of the components of x to be measured on a nominal scale; that is, one or more of the variables represents membership in some particular group within a countable class of groups. We refer to such variables as "categorical variables". Although sometimes it is important to make finer distinctions among types of variables (see Stevens, 1946, who identified nominal, ordinal, interval, and ratio types), we often need to make a simple distinction between variables whose values can be modeled by $\mathbb{R}$ and those whose values essentially indicate membership in a particular group. We may represent the observation x as being composed of these two types, "real" or "numerical", and "categorical":

J.E. Gentle, *Computational Statistics*, Statistics and Computing,
DOI: 10.1007/978-0-387-98144-4_9,

$$x = (x^{\mathrm{r}},\ x^{\mathrm{c}}).$$

In the following, we often use the phrase "numerical data" to indicate that each element of the vector variable takes on values in $\mathbb{R}$, that the relevant operations of $\mathbb{R}$ are available, and that the properties of the reals apply.

Major concerns for methods of identifying structure are the number of computations and amount of storage required.

In this chapter, we introduce some of the tools that are used for identifying structure in data. There are two distinct tools: transformations of data, and internal measures of structure. Although these two topics are to some extent independent, transformations may change the internal measures or may help us to use them more effectively. Transformations also play important roles in exploration of data, as in the graphical displays discussed in Chapter 8.

In Chapter 16, using the tools covered in this chapter, we discuss various methods of exploring data.

Linear Structure and Other Geometric Properties

Numerical data can conveniently be represented as geometric vectors. We can speak of the length of a vector, or of the angle between two vectors, and relate these geometric characteristics to properties of the data. We will begin with definitions of a few basic terms.

The *Euclidean length* or just the *length* of an n-vector x is the square root of the sum of the squares of the elements of the vector. We generally denote the Euclidean length of x as $\|x\|_2$ or just as $\|x\|$:

$$\|x\| = \left(\sum_{i=1}^{n} x_i^2\right)^{1/2}.$$

The Euclidean length is a special case of a more general real-valued function of a vector called a "norm", which is defined on page 13.

The *angle* θ between the vectors x and y is defined in terms of the cosine by

$$\cos(\theta) = \frac{\langle x, y\rangle}{\sqrt{\langle x, x\rangle\langle y, y\rangle}}.$$

(See Figure 9.1.)

Linear structures in data are the simplest and often the most interesting. Linear relationships can also be used to approximate other more complicated structures.

Flats

The set of points x whose components satisfy a linear equation

$$b_1x_1 + \cdots b_dx_d = c$$

is called a flat. Such linear structures often occur (approximately) in observational data, leading to a study of the linear regression model,

$$x_d = \beta_0 + \beta_1x_1 + \cdots + \beta_mx_m + \epsilon.$$

A flat through the origin, that is, a set of points whose components satisfy

$$b_1x_1 + \cdots b_dx_d = 0,$$

is a vector space. Such equations allow simpler transformations, so we often transform regression models into the form

$$x_d - \bar{x}_d = \beta_1(x_1 - \bar{x}_1) + \cdots + \beta_m(x_m - \bar{x}_m) + \epsilon.$$

The data are centered to correspond to this model.

9.1 Transformations

Transformations of data often give us a better perspective on its structure, and may allow us to use simpler models of the structure. Nonlinear transformations, such as logarithmic transformations, may allow us to use a linear model of the relationships among variables. In this section, however, we will focus mainly on linear transformations.

Linear Transformations

Linear transformations play a major role in analyzing numerical data and identifying structure.

A linear transformation of the vector x is the vector Ax, where A is a matrix with as many columns as the elements of x. If the number of rows of A is different, the resulting vector has a dimension different from x.

Orthogonal Transformations

An important type of linear transformation is an orthogonal transformation, that is, a transformation in which the matrix of the transformation, Q, is square and has the property that

$$Q^{\mathrm{T}}Q = I,$$

where Q^{T} denotes the transpose of Q, and I denotes the identity matrix.

If Q is orthogonal, for the vector x, we have

$$\|Qx\| = \|x\|. \tag{9.1}$$

(This is easily seen by writing $\|Qx\|$ as $\sqrt{(Qx)^{\mathrm{T}}Qx}$, which is $\sqrt{x^{\mathrm{T}}Q^{\mathrm{T}}Qx}$.) Thus, we see that orthogonal transformations preserve Euclidean lengths.

If Q is orthogonal, for vectors x and y, we have

$$\langle Qx, Qy\rangle = (Qx)^{\mathrm{T}}(Qy) = x^{\mathrm{T}}Q^{\mathrm{T}}Qy = x^{\mathrm{T}}y = \langle x, y\rangle,$$

hence,

$$\arccos\left(\frac{\langle Qx, Qy\rangle}{\|Qx\|_2\,\|Qy\|_2}\right) = \arccos\left(\frac{\langle x, y\rangle}{\|x\|_2\;\|y\|_2}\right). \tag{9.2}$$

Thus, we see that orthogonal transformations preserve angles.

Geometric Transformations

In many important applications of linear algebra, a vector represents a point in space, with each element of the vector corresponding to an element of a coordinate system, usually a cartesian system. A set of vectors describes a geometric object. Algebraic operations are geometric transformations that rotate, deform, or translate the object. Although these transformations are often used in the two or three dimensions that correspond to the easily perceived physical space, they have similar applications in higher dimensions.

Important characteristics of these transformations are what they leave *unchanged* (that is, their *invariance properties*). We have seen, for example, that an orthogonal transformation preserves lengths of vectors (equation (9.1)) and angles between vectors (equation (9.2)). A transformation that preserves lengths and angles is called an *isometric transformation*. Such a transformation also preserves areas and volumes.

Another isometric transformation is a *translation*, which for a vector x is just the addition of another vector:

$$\tilde{x} = x + t.$$

A transformation that preserves angles is called an *isotropic transformation*. An example of an isotropic transformation that is not isometric is a uniform scaling or dilation transformation, $\tilde{x} = ax$, where a is a scalar.

The transformation $\tilde{x} = Ax$, where A is a diagonal matrix with not all elements the same, does not preserve angles; it is an *anisotropic* scaling.

Another anisotropic transformation is a *shearing transformation*, $\tilde{x} = Ax$, where A is the same as an identity matrix except for a single row or column that has a one on the diagonal but possibly nonzero elements in the other positions; for example,

$$\begin{bmatrix} 1 & 0 & a_1 \\ 0 & 1 & a_2 \\ 0 & 0 & 1 \end{bmatrix}.$$

Although they do not preserve angles, both anisotropic scaling and shearing transformations preserve parallel lines. A transformation that preserves parallel lines is called an *affine transformation.* Preservation of parallel lines is equivalent to preservation of collinearity, so an alternative characterization of an affine transformation is one that preserves collinearity. More generally, we can combine nontrivial scaling and shearing transformations to see that the transformation Ax for any nonsingular matrix A is affine. It is easy to see that addition of a constant vector to all vectors in a set preserves collinearity within the set, so a more general affine transformation is $\tilde{x} = Ax + t$ for a nonsingular matrix A and a vector t.

All of these transformations are *linear transformations* because they preserve straight lines. A *projective transformation,* which uses the homogeneous coordinate system of the projective plane, preserves straight lines but does not preserve parallel lines. These transformations are very useful in computer graphics.

The invariance properties are summarized in Table 9.1.

Table 9.1. Invariance Properties of Linear Transformations

Transformation	Preserves
general	lines
affine	lines, collinearity
shearing	lines, collinearity
scaling	lines, angles (and, hence, collinearity)
translation	lines, angles, lengths
rotation	lines, angles, lengths
reflection	lines, angles, lengths

Rotations

Two major tools in seeking linear structure are rotations and projections of the data matrix X. Rotations and projections of the observations are performed by postmultiplication of X by special matrices. In this section, we briefly review these types of matrices for use in multivariate data analysis.

The simplest rotation of a vector can be thought of as the rotation of a plane defined by two coordinates about the other principal axes. Such a rotation changes two elements of all vectors in that plane and leaves all of the other elements, representing the other coordinates, unchanged. This rotation can be described in a two-dimensional space defined by the coordinates being changed, without reference to the other coordinates.

Consider the rotation of the vector x through the angle θ into $\tilde{x}$. The length is preserved, so we have $\|\tilde{x}\| = \|x\|$. Referring to Figure 9.1, we can write

$$\tilde{x}_1 = \|x\| \cos(\phi + \theta),$$
$$\tilde{x}_2 = \|x\| \sin(\phi + \theta).$$

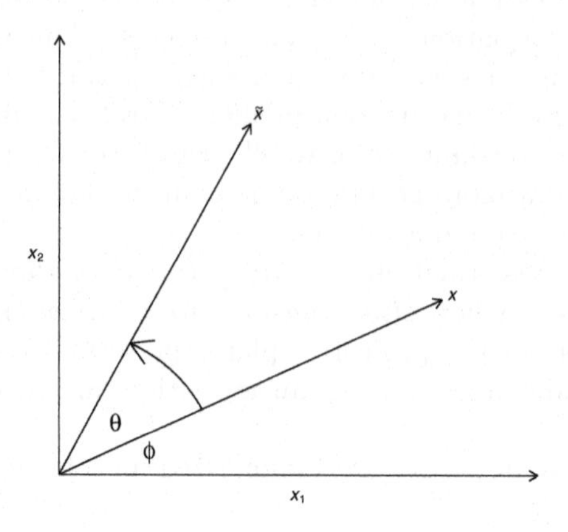

Fig. 9.1. Rotation of x

Now, from elementary trigonometry, we know that

$$\begin{aligned}\cos(\phi+\theta) &= \cos\phi\cos\theta - \sin\phi\sin\theta,\\ \sin(\phi+\theta) &= \sin\phi\cos\theta + \cos\phi\sin\theta.\end{aligned}$$

Because $\cos\phi = x_1/\|x\|$ and $\sin\phi = x_2/\|x\|$, we can combine these equations to get

$$\begin{aligned}\tilde{x}_1 &= x_1\cos\theta - x_2\sin\theta,\\ \tilde{x}_2 &= x_1\sin\theta + x_2\cos\theta.\end{aligned} \tag{9.3}$$

Hence, multiplying x by the orthogonal matrix

$$\begin{bmatrix}\cos\theta & -\sin\theta\\ \sin\theta & \cos\theta\end{bmatrix}$$

performs the rotation of x.

This idea easily extends to the rotation of a plane formed by two coordinates about all of the other (orthogonal) principal axes. The $m\times m$ orthogonal matrix

$$
Q_{pq}(\theta) = \begin{bmatrix}
1\,0 \cdots 0 & 0 & 0 \cdots 0 & 0 & 0 \cdots 0 \\
0\,1 \cdots 0 & 0 & 0 \cdots 0 & 0 & 0 \cdots 0 \\
& \ddots & & & \\
0\,0 \cdots 1 & 0 & 0 \cdots 0 & 0 & 0 \cdots 0 \\
0\,0 \cdots 0 & \cos\theta & 0 \cdots 0 & \sin\theta & 0 \cdots 0 \\
0\,0 \cdots 0 & 0 & 1 \cdots 0 & 0 & 0 \cdots 0 \\
& & \ddots & & \\
0\,0 \cdots 0 & 0 & 0 \cdots 1 & 0 & 0 \cdots 0 \\
0\,0 \cdots 0 & -\sin\theta & 0 \cdots 0 & \cos\theta & 0 \cdots 0 \\
0\,0 \cdots 0 & 0 & 0 \cdots 0 & 0 & 1 \cdots 0 \\
& & & & \ddots \\
0\,0 \cdots 0 & 0 & 0 \cdots 0 & 0 & 0 \cdots 1
\end{bmatrix}, \tag{9.4}
$$

in which p and q denote the rows and columns that differ from the identity, rotates the data vector x_i through an angle of θ in the plane formed by the pth and q^{th} principal axes of the m-dimensional cartesian coordinate system. This rotation can be viewed equivalently as a rotation of the coordinate system in the opposite direction. The coordinate system remains orthogonal after such a rotation. In the matrix XQ, all of the observations (rows) of X have been rotated through the angle θ.

How a rotation can reveal structure can be seen in Figures 9.2 and 9.3. In the original data, there do not appear to be any linear relationships among the variables. After applying a rotation about the third axis, however, we see in the scatter plot in Figure 9.3 a strong linear relationship between the first and third variables of the rotated data.

Rotations of the data matrix provide alternative views of the data. There is usually nothing obvious in the data to suggest a particular rotation; however, dynamic rotations coupled with projections that are plotted and viewed as they move are very useful in revealing structure.

Figure 9.4 shows 1,000 points in three-space that were generated by a random number generator called `RANDU`. The data have been rotated by postmultiplication by a 3×3 orthogonal matrix whose second column is proportional to $(9, -6, 1)$. We see from the figure that there are exactly 15 planes in that rotation. (See Gentle, 2003, page 18, for further discussion of data generated by `RANDU`. Because of the sharp indentations between the axes in the plot in Figure 9.4, we may conclude that there are strong negative correlations between these orthogonal linear combinations of the original data. This is a further indication that the random number generator is not a good one. In Exercise 9.3, you are asked to consider similar problems.) You are asked to develop systematic rotation and plotting methods for producing Andrews curves and parallel coordinates plots in Exercises 9.4 and 9.5.

A rotation of any plane can be formed by successive rotations of planes formed by two principal axes. Furthermore, any orthogonal matrix can be written as the product of a finite number of rotation matrices; that is, any

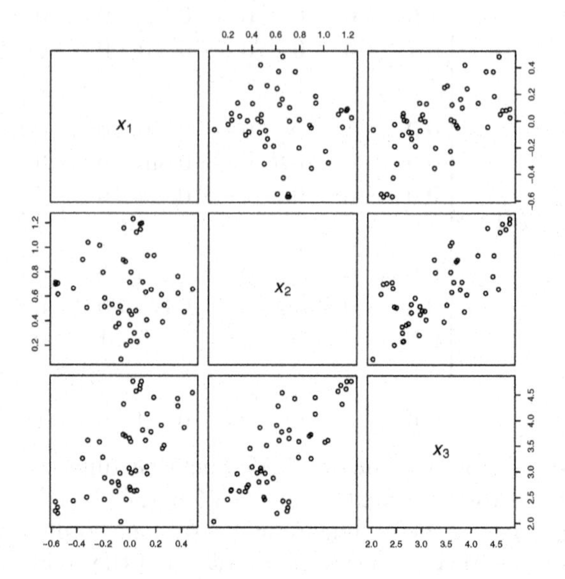

Fig. 9.2. Scatter Plot Matrix of Original Data

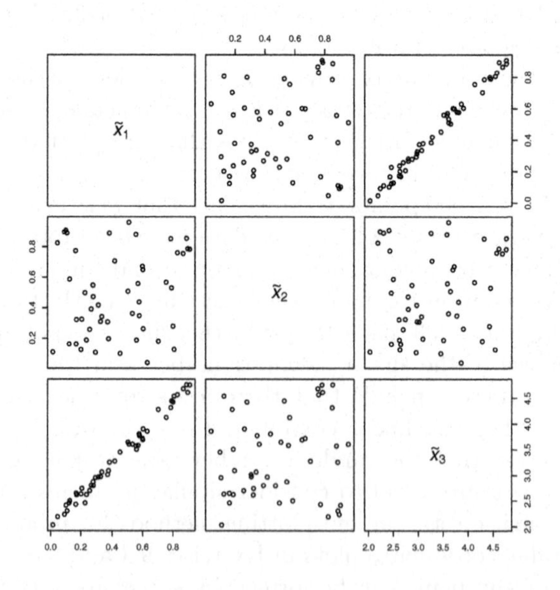

Fig. 9.3. Scatter-Plot Matrix of Rotated Data

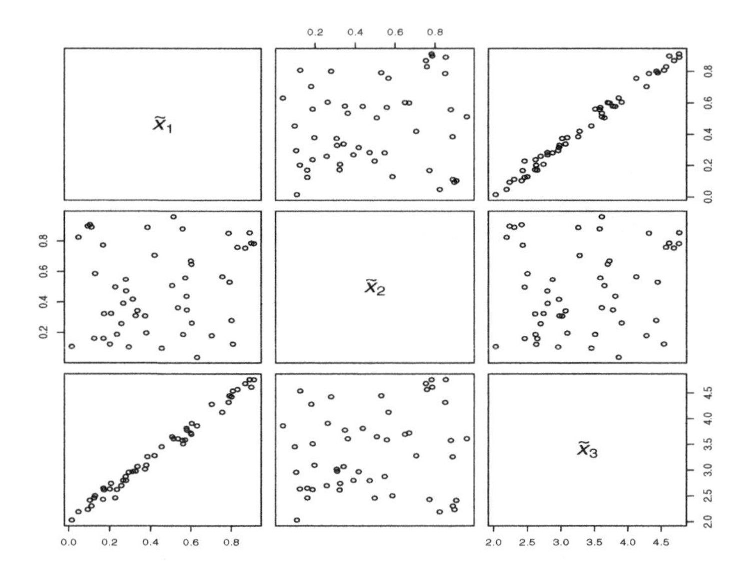

Fig. 9.4. 1,000 Triples from RANDU Rotated onto Interesting Coordinates

orthogonal transformation can be viewed as a sequence of rotations. These rotations, which are sometimes called Givens rotations, play an important role in numerical linear algebra (see Gentle, 2007, Chapter 7). It should be noted that the computations for Givens rotations can be subject to severe numerical inaccuracies. As with many numerical problems accurate computations are not nearly as simple as they may appear (see Gentle, 2007, page 185).

Projections

Another way of getting useful alternative views of the data is to project the data onto subspaces. A symmetric idempotent matrix P *projects* vectors onto the subspace spanned by the rows (or columns) of P. Except for the identity matrix, a projection matrix is of less than full rank; hence, it projects a full-rank matrix into a space of lower dimension. Although we may only know that the rows of the data matrix X are in $\mathbb{R}^m$, the rows of XP are in the subspace spanned by the rows of P. It may be possible to identify relationships and structure in this space of lower dimension that are obscured in the higher-dimensional space.

Projection transformations are often performed by rotating a given orthogonal coordinate system into a new orthogonal coordinate system in which one or more of the axes are chosen to reveal some aspect of the data, such a different groups in the data, as in linear discrimination (which we will discuss on page 623). The coordinates corresponding to any subset of the new set of coordinate axes immediately represent the projection of the data onto the subspace defined by that set of axes.

Translations

Translations are relatively simple transformations involving the addition of vectors. Rotations and other geometric transformations such as shearing involve multiplication by an appropriate matrix, as we have seen. In applications where several geometric transformations are to be made, it would be convenient if translations could also be performed by matrix multiplication. This can be done by using *homogeneous coordinates.*

Homogeneous coordinates, which form the natural coordinate system for projective geometry, have a very simple relationship to cartesian coordinates. The point with cartesian coordinates $(x_1, x_2, \ldots, x_d)$ is represented in homogeneous coordinates as $(x_0^{\mathrm{h}}, x_1^{\mathrm{h}}, x_2^{\mathrm{h}}, \ldots, x_d^{\mathrm{h}})$, where, for arbitrary x_0^{h} not equal to zero, $x_1^{\mathrm{h}} = x_0^{\mathrm{h}} x_1$, $x_2^{\mathrm{h}} = x_0^{\mathrm{h}} x_2$, and so on.

Each value of x_0^{h} corresponds to a hyperplane in the ordinary cartesian coordinate system. The special plane $x_0^{\mathrm{h}} = 0$ does not have a meaning in the cartesian system. It corresponds to a hyperplane at infinity in the projective geometry.

Because the point is the same, the two different symbols represent the same thing, and we have

$$(x_1, x_2, \ldots, x_d) = (x_0^{\mathrm{h}}, x_1^{\mathrm{h}}, x_2^{\mathrm{h}}, \ldots, x_d^{\mathrm{h}}). \tag{9.5}$$

Alternatively, of course, the hyperplane coordinate may be added at the end, and we have

$$(x_1, x_2, \ldots, x_d) = (x_1^{\mathrm{h}}, x_2^{\mathrm{h}}, \ldots, x_d^{\mathrm{h}}, x_0^{\mathrm{h}}). \tag{9.6}$$

An advantage of the homogeneous coordinate system is that we can easily perform translations by matrix multiplications. We can effect the translation $\tilde{x} = x + t$ by first representing the point x as $(1, x_1, x_2, \ldots, x_d)$ and then multiplying by the $(d + 1) \times d$ matrix

$$T = \begin{bmatrix} 1 & 0 & \cdots & 0 \\ t_1 & 1 & \cdots & 0 \\ & & \cdots & \\ t_d & 0 & \cdots & 1 \end{bmatrix}.$$

We will use the symbol x^{h} to represent the vector of corresponding homogeneous coordinates:

$$x^{\mathrm{h}} = (1, x_1, x_2, \ldots, x_d).$$

The translated point can be represented as $\tilde{x} = Tx^{\mathrm{h}}$.

We must be careful to distinguish the point x from the vector of coordinates that represents the point. In cartesian coordinates, there is a natural correspondence, and the symbol x representing a point may also represent the vector $(x_1, x_2, \ldots, x_d)$. The vector of homogeneous coordinates of the result Tx^{h} corresponds to the vector of cartesian coordinates of $\tilde{x}$, $(x_1 + t_1, x_2 + t_2, \ldots, x_d + t_d)$.

Homogeneous coordinates are used extensively in computer graphics not only for the ordinary geometric transformations but also for projective transformations. A standard use of homogeneous coordinates is in mapping three-dimensional graphics to two dimensions. The perspective plot function `persp` in R, for example, produces a 4×4 matrix for projecting three-dimensional points represented in homogeneous coordinates onto two-dimensional points in the displayed graphic. R uses homogeneous coordinates in the form of equation (9.6) rather than equation (9.5). If the matrix produced is T and if a^{h} is the representation of a point (x_a, y_a, z_a) in homogeneous coordinates, in the form of equation (9.6), then $a^{\mathrm{h}}T$ yields transformed homogeneous coordinates that correspond to the projection onto the two-dimensional coordinate system of the graphical display. Consider the graph in Figure 9.5. The wire frame plot, which is of the standard bivariate normal density, was produced by the following simple R statements.

```
x <- seq(-3,3,.1)
y <- seq(-3,3,.1)
f <- function(x,y){dnorm(x)*dnorm(y)}
z <- outer(x,y,f)
persp(x,y,z,theta=-30,phi=30,zlab="p(x,y)",ylab="y",xlab="x"
) -> trot
```

The angles `theta` and `phi` are the azimuthal and latitudinal viewing angles, respectively, in degrees. The matrix `trot` is the rotation matrix that will carry a point in the three-dimensional space that was plotted, which is represented in homogeneous coordinates, into the two-dimensional plane on which the plot is displayed.

Now, suppose we want to plot a single point on the surface, say the point corresponding to $x = 0$ and $y = -1$. We compute the corresponding z value, represent the vector in homogeneous coordinates, with the coordinate of the hyperplane being 1, rotate it onto the two-dimensional plane on which the plot is displayed using `trot`, and finally plot the point. The following simple R statements do this, and the point is seen in Figure 9.5.

```
x1 <- 0
y1 <- -1
z1 <- f(x1,y1)
tp <- cbind(x1,y1,z1,1)%*%trot
p1 <- tp[1,1]/tp[1,4]
p2 <- tp[1,2]/tp[1,4]
text(p1,p2,"*",cex=2)
```

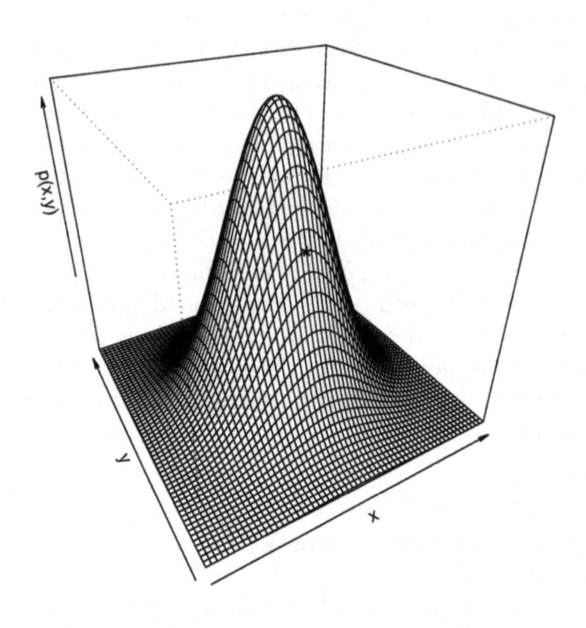

Fig. 9.5. Illustration of the Use of Homogeneous Coordinates to Locate Three-Dimensional Points on a Two-Dimensional Graph

General Transformations of the Coordinate System

Although the transformations that we have discussed above can be thought of either as transforming the data within a fixed coordinate system or as transforming the coordinate system, the coordinate system itself remains essentially a cartesian coordinate system. Homogeneous coordinates correspond in a simple way to cartesian coordinates, as we see in equation (9.5).

We can make more general transformations of the coordinate system that can be useful in identifying structure in the data. Two kinds of coordinate transformations especially useful in graphical displays are parallel coordinates, which we discuss on page 360, and Fourier curves, which we discuss on page 362.

Polar coordinates are useful in a variety of applications. They are particularly simple for bivariate data, but they can be used in any number of dimensions. The point

$$x = (x_1, x_2, \ldots, x_d)$$

is represented in polar coordinates by a length,

$$r = \|x\|,$$

and $d-1$ angles, $\theta_1, \ldots, \theta_{d-1}$. There are various ways that the relationships among the cartesian coordinates and the polar coordinates could be defined. One way is given by Kendall (1961). The relationships among the coordinates are given by

$$\begin{aligned}
x_1 &= r\cos\theta_1 \quad \cdots \quad \cos\theta_{d-2}\cos\theta_{d-1} \\
x_2 &= r\cos\theta_1 \quad \cdots \quad \cos\theta_{d-2}\sin\theta_{d-1} \\
&\;\;\vdots \\
x_j &= r\cos\theta_1\cdots\cos\theta_{d-j}\sin\theta_{d-j+1} \\
&\;\;\vdots \\
x_{d-1} &= r\cos\theta_1\sin\theta_2 \\
x_d &= r\sin\theta_1,
\end{aligned} \tag{9.7}$$

where

$$-\pi/2 \leq \theta_j \leq \pi/2, \quad \text{for} \quad j = 1, 2, \ldots, d-2,$$

and

$$0 \leq \theta_{d-1} \leq 2\pi.$$

In a variation of this definition, the sines and cosines are exchanged, with an appropriate change in the limits on the angles. In this variation, for $d=2$, we have the usual polar coordinates representation; and for $d=3$, we have what is sometimes called the spherical coordinates representation.

9.2 Measures of Similarity and Dissimilarity

There are many ways of measuring the similarity or dissimilarity between two observations or between two variables. For numerical data, the most familiar measures of similarity are covariances and correlations.

Dissimilarities in numerical data are generally distances of some type. The dissimilarity or distance function is often a *metric* (see page 14).

Other measures of dissimilarity can often be useful. Nonmetric functions, such as ones allowing ties and that do not obey the triangle inequality, can also be used for defining dissimilarity, especially in applications in which there is some noise or in which there is some subjectivity in the data. Distance measures defined on a finite set of points, $x_1, x_2, \ldots, x_n$, may be required to satisfy the "ultrametric" inequality:

$$\Delta(x_i, x_k) \leq \max_j\bigl(\Delta(x_i, x_j), \Delta(x_j, x_k)\bigr),$$

instead of just the triangle inequality. Ultrametric distances are sometimes used as dissimilarity measures in clustering applications.

Other measures of both similarity and dissimilarity must be used for categorical data or for mixed data (that is, for data consisting of some numerical variables and categorical variables),

$$x = (x^{\mathrm{r}},\, x^{\mathrm{c}}).$$

The measures may involve ratings of judges, for example. The measures may not be metrics.

In some cases, it is useful to allow distance measures to be asymmetric. If $d(x_i, x_j)$ represents the cost of moving from point x_i to point x_j it may be the case that $d(x_i, x_j) \neq d(x_j, x_i)$. If the distance represents a perceptual difference, it may also be the case that $d(x_i, x_j) \neq d(x_j, x_i)$. Sullivan (2002) has developed a theory for asymmetric measures of dissimilarity, and explored their use in clustering and other applications.

Similarities: Covariances and Correlations

Measures of similarity include covariances, correlations, rank correlations, and cosines of the angles between two vectors. Any measure of dissimilarity, such as the distances discussed in the next section, can be transformed into a measure of similarity by use of a decreasing function, such as the reciprocal. For example, whereas the cosine of the angle formed by two vectors can be considered a measure of similarity, the sine can be considered a measure of dissimilarity.

Although we can consider similarities/dissimilarities between either columns (variables) or rows (observations), in our common data structures, we often evaluate covariances and correlations between columns and distances among rows. We speak of the covariance or the correlation between columns or between variables. The covariance between a column (variable) and itself is its variance.

For an $n \times m$ data matrix X, we have the $m \times m$ *variance-covariance matrix* (or just the *covariance matrix*):

$$S = \begin{bmatrix} s_{11} & s_{12} & \cdots & s_{1m} \\ s_{21} & s_{22} & \cdots & s_{2m} \\ \vdots & \vdots & \vdots & \vdots \\ s_{m1} & s_{m2} & \cdots & s_{mm} \end{bmatrix}, \tag{9.8}$$

where

$$s_{jk} = s_{kj} = \frac{\sum_{i=1}^{n}(x_{ij} - \bar{x}_j)(x_{ik} - \bar{x}_k)}{n-1}. \tag{9.9}$$

If $\overline{X}$ is the matrix in which each column consists of the mean of the corresponding column of X, we see that

$$S = \frac{1}{n-1}(X - \overline{X})^{\mathrm{T}}(X - \overline{X}). \tag{9.10}$$

The matrix S is therefore nonnegative definite. The matrix $X - \overline{X}$ is called the "centered data matrix"; each column sums to 0.

Assuming none of the variables is constant, the correlation is often a more useful measure because it is scaled by the variances. For an $n \times m$ data matrix, the $m \times m$ *correlation matrix* is

$$R = \begin{bmatrix} 1 & r_{12} & \cdots & r_{1m} \\ r_{12} & 1 & \cdots & r_{2m} \\ \vdots & \vdots & \vdots & \vdots \\ r_{1m} & r_{2m} & \cdots & 1 \end{bmatrix}, \tag{9.11}$$

where

$$r_{jk} = r_{kj} = \frac{s_{jk}}{\sqrt{s_{jj}s_{kk}}}; \tag{9.12}$$

that is,

$$R = \left(\mathrm{diag}(\sqrt{s_{11}}, \sqrt{s_{22}}, \ldots, \sqrt{s_{mm}})\right)^{-1} S \left(\mathrm{diag}(\sqrt{s_{11}}, \sqrt{s_{22}}, \ldots, \sqrt{s_{mm}})\right)^{-1}.$$

The data matrix X together with either S or R is a complete graph in which the columns of X constitute the vertices.

Notice that covariances and correlations are based on the L_2 norm. They are sometimes called "product-moment" covariances and correlations.

Because the concepts of covariance and correlation are also used to refer to properties of random variables, we sometimes refer to the quantities that we have defined above as "sample covariance" or "sample correlation" to distinguish them from the "population" quantities of abstract variables.

There are variations of these such as rank correlations and robust covariances. Rank correlations are computed by first replacing the elements of each column of X by the ranks of the elements within the column and then computing the correlation as above. Robust covariances and correlations are computed either by using a different measure than the L_2 norm or by scaling of the covariance matrix based on an expectation taken with respect to a normal (or Gaussian) distribution. ("Robustness" usually assumes a normal or Gaussian distribution as the reference standard.) See page 395 for a specific robust alternative to S.

Similarities When Some Variables Are Categorical

If all of the variables are measured on a scale that can be modeled as a real number, covariances and/or correlations or similar measures are the obvious choice for measuring the similarity between two points, x_j and x_k. If, however, some of the variables are categorical variables, that is, if the generic x can be represented in the notation introduced earlier,

$$x = (x^{\mathrm{r}},\ x^{\mathrm{c}}),$$

a different measure of similarity must be chosen.

Sometimes, the values of the categorical variables represent such different situations that it does not make sense to consider similarities between observations from different groups. In such a case, the similarity between

$$x_j = (x_j^{\mathrm{r}},\, x_j^{\mathrm{c}})$$

and

$$x_k = (x_k^{\mathrm{r}},\, x_k^{\mathrm{c}})$$

may be measured by the function

$$\begin{aligned} s(x_j, x_k) &= \frac{\sum_{i=1}^{n}(x_{ij}^{\mathrm{r}} - \bar{x}_j^{\mathrm{r}})(x_{ik}^{\mathrm{r}} - \bar{x}_k^{\mathrm{r}})}{n-1}, \quad \text{if } x_j^{\mathrm{c}} = x_k^{\mathrm{c}}, \\ &= 0, \quad \text{otherwise.} \end{aligned} \tag{9.13}$$

Instead of requiring an exact match of the categorical variables, we can allow some degrees of similarity between observations with different values of their categorical variables. One way would be by using the count of how many variables within x_j^{c} and x_k^{c} agree. Such a simple count can be refined to take into account the number of possible values each of the categorical variables can assume. The measure can also be refined by incorporating some measure of the similarity of different classes.

Similarities among Functional Observations

Interest-bearing financial instruments such as bonds or U.S. Treasury bills have prices that depend on the spot or current interest rate and so-called forward rates at future points in time. (A forward rate at time t_1 for a future time t_2 can be thought of as the value of cash or a riskless security at time $t_2 > t_1$ discounted back to time t_1.) The forward rates depend on, among other things, the investors' perception of future spot or actual rates. At any point, a set of forward rates together with the spot rate determine the "yield curve" or the "term structure" for a given financial instrument:

$$r(t).$$

Observational data for measuring and comparing term structures consist of functions for a set of securities measured at different time points.

Another example of observations that are functions are the measurements on various units of individual features of developing organisms taken over time. For example, the observational unit may be a developing organism, the features may be gene expressions, and the data elements may be measures of these expressions taken at fixed times during the development of the organism. The observations on feature j may consist of measurements $(x_{j1}, x_{j2}, \ldots, x_{jm})$ taken at times $t_1, t_2, \ldots, t_m$. The overall patterns of the measurements may be of interest. The underlying model is a continuous function,

$$x(t).$$

The observation on each feature is a discrete function, evaluated at discrete points in its time domain.

Consider, for example, the three observations

$$x_1 = (1,\ 2,\ 1),$$

$$x_2 = (1,\ 2,\ 3),$$

and

$$x_3 = (4,\ 8,\ 4).$$

Because of the obvious patterns, we may wish to consider x_1 and x_3 more similar than are x_1 and x_2.

There are several ways to define a similarity measure to capture this kind of relationship. A very simple one in this case is the relative changes over time. We may first of all augment the existing data with measures of changes. In the example above, taking a simplistic approach of just measuring changes and scaling them, and then augmenting the original vectors, we have

$$\tilde{x}_1 = \left(1,\ 2,\ 1,\ \middle|\ 1,\ -\frac{1}{2}\right),$$

$$\tilde{x}_2 = \left(1,\ 2,\ 3,\ \middle|\ 1,\ \ \frac{1}{2}\right),$$

and

$$\tilde{x}_3 = \left(4,\ 8,\ 4,\ \middle|\ 1,\ -\frac{1}{2}\right).$$

After transforming the data in this way, we may employ some standard similarity measure, possibly one that downweights the first three elements of each observation.

Another approach is to fit a smoothing curve to each observational vector and then form a new vector by evaluating the smoothing curve at fixed points. A standard similarity measure would then be applied to the transformed vectors.

There are many issues to consider when comparing curves. Whereas the data-generating process may follow a model $x(t)$, the data are of the form $x_i(t_{ij})$. In the model, the variable t (usually "time") may not be measured in an absolute sense, but rather may be measured relative to a different starting point for each observational unit. Even when this shift is taken into consideration two responses that are similar overall may not begin at the same relative time; that is, one observational unit may follow a model $x(t)$ and another $x(t + \delta)$. To proceed with the analysis of such data, it is necessary to *register* the data (that is, to shift the data to account for such differences in the time). More generally, two observational units may follow the same functional process under some unknown transformation of the independent variable:

$$x_1(t) = x_2(h(t)).$$

Unraveling this transformation is a more difficult process of registration.

We may want to base similarity among observations on some more general relationship satisfied by the observations. Suppose, for example, that a subset of some bivariate data lies in a circle. This pattern may be of interest, and we may want to consider all of the observations in the subset lying in the circle to be similar to one another and different from observations not lying in the circle.

Many such similarity measures depend on the context (that is, on a subset of variables or observations, not just on the relationship between two variables or two observations). Similarities defined by a context are of particular use in pattern recognition.

Similarities between Groups of Variables

We may want to combine variables that have similar values across all observations into a single variable, perhaps a linear combination of some of the original variables. This is an objective of the methods discussed in Sections 16.3 and 16.4.

The general problem of studying linear relationships between two sets of variables is addressed by the method of *canonical correlations*. We will not pursue that topic here.

Dissimilarities: Distances

There are several ways of measuring dissimilarity. One measure of dissimilarity is distance, and there are several ways of measuring distance. Some measures of distance between two points are based only on the elements of the vectors defining those two points. These distances, which are usually defined by a commutative function, are useful in a homogeneous space. Other measures of distance may be based on a structure imposed by a set of observations.

In a homogeneous space, there are several commonly used measures of distance between two observations. Most of these are based on some *norm* of the difference between the two numeric vectors representing the observations. A norm of the difference between two vectors is a metric, as we have observed in Chapter 1.

Some of the commonly used measures of distance between observations of numerical data represented in the vectors x_i and x_k are the following:

- Euclidean distance, the root sum of squares of differences:

$$\|x_i - x_k\|_2 \tag{9.14}$$

  or

$$\left(\sum_{j=1}^{m}(x_{ij}-x_{kj})^2\right)^{1/2}.$$

The Euclidean distance is sometimes called the L_2 norm.

- maximum absolute difference:

$$\|x_i - x_k\|_\infty \tag{9.15}$$

or

$$\max_j |x_{ij} - x_{kj}|.$$

- Manhattan distance, the sum of absolute differences:

$$\|x_i - x_k\|_1 \tag{9.16}$$

or

$$\sum_{j=1}^{m} |x_{ij} - x_{kj}|.$$

- Minkowski or L_p distance:

$$\|x_i - x_k\|_p \tag{9.17}$$

or

$$\left(\sum_{j=1}^{m} |x_{ij} - x_{kj}|^p\right)^{1/p}.$$

The L_p distance is the L_p norm of the difference in the two vectors. Euclidean distance, maximum difference, and Manhattan distance are special cases, with $p = 2$, $p \to \infty$, and $p = 1$, respectively.

- Canberra distance (from Lance and Williams, 1966):

$$\sum_{j=1}^{m} \frac{|x_{ij} - x_{kj}|}{|x_{ij}| + |x_{kj}|}, \tag{9.18}$$

as long as $|x_{ij}| + |x_{kj}| \neq 0$; otherwise, 0 (sometimes normalized by m to be between 0 and 1).

- correlation-based distances:

$$f(r_{ik}).$$

The correlation between two vectors r_{ik} (equation (9.12)) can also be used as a measure of dissimilarity. Values close to 0 indicate small association. The absolute value of the correlation coefficient is a decreasing function in what is intuitively a dissimilarity, so a distance measure based on it, $f(r_{ik})$, should be a decreasing function of the absolute value. Two common choices are

$$1 - |r_{ik}|$$

and

$$1 - r_{ik}^2.$$

- distances based on angular separation:

$$\frac{x_i^{\mathrm{T}} x_k}{\|x_i\|_2 \, \|x_k\|_2} \tag{9.19}$$

or

$$\frac{\sum_{j=1}^m x_{ij}x_{kj}}{\sqrt{\sum_{j=1}^m x_{ij}^2 \sum_{j=1}^m x_{kj}^2}}.$$

This measure of angular separation is the cosine of the angle; hence, it is a decreasing function in what is intuitively a dissimilarity. Other quantities, such as the sine of the angle, can be used instead. For centered data, the angular separation is the same as the correlation of equation (9.12).

For categorical data, other measures of distance must be used. For vectors composed of zeros and ones, for example, there are two useful distance measures:

- Hamming distance: the number of bits that are different in the two vectors;
- binary difference: the proportion of non-zeros that two vectors do not have in common (the number of occurrences of a zero and a one, or a one and a zero divided by the number of times at least one vector has a one).

Lance and Williams (1967a, 1967b, and 1968) provide a general framework for definitions of distances and discuss the differences in the measures in cluster analysis.

Notice that generally the *distances* are between the *observations*, whereas the *covariances* discussed above are between the *variables.*

The distances are elements of the $n \times n$ dissimilarity matrix,

$$D = \begin{bmatrix} 0 & d_{12} & d_{13} & \cdots & \cdots & d_{1n} \\ d_{21} & 0 & d_{23} & \cdots & \cdots & d_{2n} \\ \vdots & \vdots & \vdots & \vdots & \vdots & \vdots \\ \vdots & \vdots & \vdots & \vdots & \vdots & \vdots \\ \vdots & \vdots & \vdots & \vdots & \vdots & \vdots \\ d_{n1} & d_{n2} & d_{n3} & \cdots & \cdots & 0 \end{bmatrix}. \tag{9.20}$$

All of the distance measures discussed above are metrics; in particular, they satisfy $\Delta(x_1, x_2) = \Delta(x_2, x_1)$ for all $x_1, x_2 \in \mathrm{IR}^m$. This means, among other things, that any dissimilarity matrix D, in which the elements correspond to those measures, is symmetric.

The data matrix X together with D is a complete graph. In this graph, the rows of X constitute the vertices.

The measures of distance listed above are appropriate in a homogeneous space in which lengths have the same meaning in all directions. A scaling of the units in any of the cardinal directions (that is, a change of scale in the measurement of a single variable) may change the distances. In many applications, the variables have different meanings. Because many statistical techniques give preferential attention to variables with larger variance, it is often useful to scale all variables to have the same variance. Sometimes, it is more useful to scale the variables so that all have the same range.

Notice that the angular separation, as we have defined it, is based on the L_2 norm. A transformation that preserves L_2 distances and angles is called an "isometric transformation". If Q is an orthogonal matrix, the Euclidean distance between Qx_i and Qx_k and the angular separation between those two vectors are the same as the distance and angle between x_i and x_k. Hence, an orthogonal matrix is called an isometric matrix because it preserves Euclidean distances and angles.

Other Dissimilarities Based on Distances

The various distance measures that we have described can be used to define dissimilarities in other ways. For example, we may define the distance from x_j to x_k, $d^{\mathrm{R}}(x_j, x_k)$, as the rank of an ordinary distance d_{jk} in the set of all distances d_{ji}. If x_k is the point closest to x_j, then $d^{\mathrm{R}}(x_j, x_k) = 1$. This type of dissimilarity depends on the "direction"; that is, in general,

$$d^{\mathrm{R}}(x_j, x_k) \neq d^{\mathrm{R}}(x_k, x_j).$$

A distance measure such as $d^{\mathrm{R}}(\cdot, \cdot)$ is dependent on the neighboring points, or the "context".

If we think of the distance between two points as the cost or effort required to get from one point to another, the distance measure often may not be symmetric. (It is therefore not a metric.) Common examples in which distances measured this way are not symmetric arise in anisotropic media under the influence of a force field (say, electrical or gravitational) or in fluids with a flow (see Exercise 9.11).

Dissimilarities in Anisometric Coordinate Systems: Scaling and Sphering Data

If the elements of the observation vectors represent measurements made on different scales, it is usually best to scale the variables so that all have the same variance or else have the same range. A scaling of the data matrix X so that all columns have a variance of 1 is achieved by postmultiplication by a diagonal matrix whose elements are the square roots of the diagonal elements of S in equations (9.8):

$$X_{\mathrm{N}} = X\mathrm{diag}(\sqrt{s_{ii}}). \qquad (9.21)$$

We refer to X_{N} or any data matrix whose columns have a variance of 1 as "scaled data" or "normalized data".

If the scaling is applied to centered data, we have the "standardized" data matrix:

$$X_{\mathrm{S}} = (X - \overline{X})\,\mathrm{diag}(\sqrt{s_{ii}}). \tag{9.22}$$

This scaling is what is done in computing correlations. The correlation matrix in equation (9.11) can be computed as $X_{\mathrm{S}}^{\mathrm{T}} X_{\mathrm{S}}/(n-1)$.

If there are relationships among the variables whose observations comprise the columns of X, and if there are more rows than columns (that is, $n > m$), it may be appropriate to perform an oblique scaling,

$$X_{\mathrm{W}} = (X - \overline{X})H, \tag{9.23}$$

where H is the Cholesky factor of S^{-1} (equation (9.8)); that is,

$$\begin{aligned} H^{\mathrm{T}}H &= (n-1)\big((X - \overline{X})^{\mathrm{T}}(X - \overline{X})\big)^{-1} \\ &= S^{-1}. \end{aligned}$$

(If the matrix S is not of full rank, the generalized inverse is used in place of the inverse. In any case, the matrix is nonnegative definite, so the decomposition exists.) The matrix X_{W} is a *centered and sphered* matrix. It is sometimes called a *white* matrix. The matrix is orthonormal; that is, $X_{\mathrm{W}}^{\mathrm{T}} X_{\mathrm{W}} = I$.

In general, a structure may be imposed on the space by $(X - \overline{X})^{\mathrm{T}}(X - \overline{X})$ or S. A very useful measure of the distance between two vectors is the *Mahalanobis squared distance*. The Mahalanobis squared distance between the ith and k^{th} observations, x_i and x_k (the ith and k^{th} rows of X) is

$$(x_i - x_k)^{\mathrm{T}} S^{-1} (x_i - x_k). \tag{9.24}$$

Notice that the Mahalanobis squared distance is the squared Euclidean distance after using $S^{-1/2}$ to scale the data. It is the squared Euclidean distance between rows in the X_{S} matrix above. It is often more natural to use the Mahalanobis distance, that is, the square root of expression (9.24), because it is a metric (see page 15).

There are other types of distance. Certain paths from one point to another can be specified. The distance can be thought of as the cost of getting from one node on a graph to another node. Although distances are usually considered to be symmetric (that is, the distance from point x_i to point x_k is the same as the distance from point x_k to point x_i), a more general measure may take into account fluid flow or elevation gradients, so the dissimilarity matrix would not be symmetric.

Another type of data that presents interesting variations for measuring dissimilarities or similarities is directional data, or circular data (that is, data that contain a directional component). The angular separation (9.19) measures this, of course, but often in directional data, one of the data elements is a plane angle. As the size of the angle increases, ultimately it comes close to a measure of 0. A simple example is data measured in polar coordinates. When one of the data elements is an angle, the component of the overall distance between two observations i and j attributable to their angles, θ_i and θ_j, could be taken as

$$d_{ij}^{\mathrm{d}} = 1 - \cos(\theta_i - \theta_j).$$

The directional component must be combined additively with a component due to Euclidean-like distances, d_{ij}^{r}. In polar coordinates, the radial component is already a distance, so d_{ij}^{r} may just be taken as the absolute value of the difference in the radial components r_i and r_j. The overall distance d_{ij} may be formed from d_{ij}^{d} and d_{ij}^{r} in various ways that weight the radial distance and the angle differently.

There are many examples, such as wind direction in meteorology or climatology, in which directional data arise.

Properties of Dissimilarities

A dissimilarity measure based on a metric conforms generally to our intuitive ideas of distance. The norm of the difference between two vectors is a metric, that is, if

$$\Delta(x_1, x_2) = \|x_1 - x_2\|,$$

then $\Delta(x_1, x_2)$ is a metric. Distance measures such as the L_p distance and the special cases of Euclidean distance, maximum difference, and Manhattan distance, which are based on norms of the difference between two vectors, have useful properties, such as satisfying the triangle inequality:

$$d_{ik} \leq d_{ij} + d_{jk}.$$

There are many different measures that may be useful in different applications.

Dissimilarities between Groups of Observations

In clustering applications, we need to measure distances between groups of observations. We are faced with two decisions. First, we must choose the distance metric to use, and then the points in the two groups between which we measure the distance. Any of the distance measures discussed above could be used.

Once a distance measure is chosen, the distance between two groups can be defined in several ways, such as the following;

- the distance between a central point, such as the mean or median, in one cluster and the corresponding central point in the other cluster;
- the minimum distance between a point in one cluster and a point in the other cluster;
- the largest distance between a point in one cluster and a point in the other cluster;

- the average of the distances between the points in one cluster and the points in the other cluster.

The average of all of the pairwise point distances is the most common type of measure used in some applications. This type of measure is widely used in genetics, where the distance between two populations is based on the differences in frequencies of chromosome arrangements (for example, Prevosti's distance) or on DNA matches or agreement of other categorical variables (for example, Sanghvi's distance).

Effects of Transformations of the Data

In the course of an analysis of data, it is very common to apply various transformations to the data. These transformations may involve various operations on real numbers, such as scaling a variable (multiplication), summing all values of a variable (addition), and so on. Do these kinds of operations have an effect on the results of the data analysis? Do they change the relative values of such things as measures of similarity and dissimilarity?

Consider a very simple case in which a variable represents length, for example. The actual data are measurements such as 0.11 meters, 0.093 meters, and so on. These values are recorded simply as the real numbers 0.11, 0.093, and so on. In analyzing the data, we may perform certain operations (summing the data, squaring the data, and so on) in which we merely assume that the data behave as real numbers. (Notice that 0.11 is a real number but 0.11 meters is not a real number; 0.11 meters is a more complicated object.) After noting the range of values of the observations, we may decide that millimeters would be better units of measurement than meters. The values of the variable are then scaled by 1,000. Does this affect any data analysis we may do?

Although, as a result of scaling, the mean goes from approximately μ (for some value μ) to $1,000\mu$, and the variance goes from σ^2 (for some value σ) to $1,000,000\sigma^2$, the scaling certainly should not affect any analysis that involves that variable alone.

Suppose, however, that another variable in the dataset is also length and that typical values of that variable are 1,100 meters, 930 meters, and so on. For this variable, a more appropriate unit of measure may be kilometers. To change the unit of measurement results in dividing the data values by 1,000. The differential effects on the mean and variance are similar to the previous effects when the units were changed from meters to millimeters; the effects on the means and on the variances differ by a factor of 1,000. Again, the scaling certainly should not affect any analysis that involves that variable alone.

This scaling, however, does affect the relative values of measures of similarity and dissimilarity. Consider, for example, the Euclidean distance between two observations, $x_1 = (x_{11}, x_{12})$ and $x_2 = (x_{21}, x_{22})$. The squared distance prior to the scaling is

$$(x_{11} - x_{21})^2 + (x_{12} - x_{22})^2.$$

Following the scaling, it is

$$10^6(x_{11} - x_{21})^2 + 10^{-6}(x_{12} - x_{22})^2.$$

The net effect depends on the relative distances between x_1 and x_2 as measured by their separate components.

As we mention above, an orthogonal transformation preserves Euclidean distances and angular separations; that is, it is an isometric transformation. An orthogonal transformation also preserves measures of similarity based on the L_2 norm. An orthogonal transformation, however, does not preserve other measures of similarity or distance.

Outlying Observations and Robust Measures

Many methods of data analysis may be overly affected by observations that lie at some distance from the other observations. Using a least squares criterion for locating the center of a set of observations, for example, can result in a "central point" that is outside of the convex hull of all of the data except for just one observation. As an extreme case, consider the mean of 100 univariate observations, all between 0 and 1 except for one outlying observation at 100. The mean of this set of data is larger than 99% of the data.

An outlier may result in one row and column in the dissimilarity matrix D having very large values compared to the other values in the dissimilarity matrix. This is especially true of dissimilarities based on the L_2 norm. Dissimilarities based on other norms, such as the L_1 norm, may not be as greatly affected by an outlier.

Methods of data analysis that are not as strongly affected by outlying observations are said to be "robust". (There are various technical definitions of robustness, which we will not consider here.) The variance-covariance matrix S in equation (9.8), because it is based on squares of distances from unweighted means, may be strongly affected by outliers. A robust alternative is

$$S_{\mathrm{R}} = (s_{\mathrm{R}jk}), \tag{9.25}$$

where the $s_{\mathrm{R}jk}$ are robust alternatives to the s_{jk} in equation (9.9).

There are various ways of defining the $s_{\mathrm{R}jk}$. In general, they are formed by choosing weights for the individual observations to decrease the effect of outlying points; for example,

$$s_{\mathrm{R}jk} = \frac{\sum_{i=1}^n w_i^2 (x_{ij} - \bar{x}_{\mathrm{R}j})(x_{ik} - \bar{x}_{\mathrm{R}k})}{\sum_{i=1}^n w_i^2 - 1}, \tag{9.26}$$

where

$$\bar{x}_{\mathrm{R}j} = \sum_{i=1}^n w_i x_{ij} \Big/ \sum_{i=1}^n w_i, \tag{9.27}$$

for a given function ω,

$$w_i = \omega(d_i)/d_i, \tag{9.28}$$

and

$$d_i = (x_i - \bar{x}_{\mathrm{R}})^{\mathrm{T}} S_{\mathrm{R}}^{-1}(x_i - \bar{x}_{\mathrm{R}}). \tag{9.29}$$

(In this last expression, x_i represents the m-vector of the ith observation, and $\bar{x}_{\mathrm{R}}$ represents the m-vector of the weighted means. These expressions are circular and require iterations to evaluate them.)

The function ω is designed to downweight outlying observations. One possibility, for given constants b_1 and b_2, is

$$\begin{aligned} \omega(d) &= d \qquad\qquad\qquad\qquad \text{if } d \leq d_0 \\ &= d_0 \mathrm{e}^{-\frac{1}{2}(d-d_0)^2/b_2^2} \text{ if } d > d_0, \end{aligned} \tag{9.30}$$

where $d_0 = \sqrt{m} + b_1/\sqrt{2}$.

Instead of defining the "center" as a weighted mean as in equation (9.27), we may use some other measure of the center, such as a median, a geometric mean, or a harmonic mean. The effect of outlying observations on these measures is different. Similarity measures based on these measures may be more robust to outlying observations. Some comparisons are given by Sebe, Lew, and Huijsmans (2000).

Collinear Variables

A problem of a different type arises when the variables are highly correlated. In this case, the covariance matrix S and the correlation matrix R, which are based on the L_2 norm, are both ill-conditioned. The ranking transformation mentioned on page 385 results in a correlation matrix that is better conditioned.

Depending on the application, some type of regularization may be useful when the variables are highly correlated. We consider some regularization methods in Chapter 17.

Multidimensional Scaling: Determining Observations that Yield a Given Distance Matrix

Given an $n \times n$ distance matrix such as D in equation (9.20), could we reconstruct an $n \times m$ data matrix X that yields D for some metric $\Delta(\cdot, \cdot)$? The question, of course, is constrained by m (that is, by the number of variables). The problem is to determine the elements of rows of X such that

$$\begin{aligned} \tilde{d}_{ij} &= \Delta(x_i, x_j) \\ &\approx d_{ij}. \end{aligned}$$

This is called *multidimensional scaling.*

The approximation problem can be stated precisely as an optimization problem to minimize

$$\frac{\sum_i \sum_j f(\tilde{d}_{ij} - d_{ij})}{\sum_i \sum_j f(d_{ij})},$$

where $f(\cdot)$ is some function that is positive for nonzero arguments and is monotone increasing in the absolute value of the argument, and $f(0) = 0$. An obvious choice is $f(t) = t^2$. Clarkson and Gentle (1986) describe an alternating iteratively reweighted least squares algorithm to compute the minimum when f is of the form $f(t) = |t|^p$. If the distances in D do not arise from a metric, they discussed ways of transforming the dissimilarities so that the least squares approach would still work.

The larger the value of m, of course, the closer the $\tilde{d}_{ij}$ will be to the d_{ij}. If $m \ll n$ and the approximations are good, significant data reduction is achieved.

There are various modifications of the basic multidimensional scaling problem, and software programs are available for different ones. The S-Plus and R function `cmdscale` performs computations for multidimensional scaling when the dissimilarities are Euclidean distances. (In R, `cmdscale` is in the `mva` package.)

Notes and Further Reading

Because many of the similarity and dissimilarity measures are based on least squares approaches, they may be sensitive to heavy tailed distributions or to samples with outliers. Ammann (1989) and (1993) discusses ways of robustifying the measures prior to such analyses as principal components. Amores, Sebe, and Radeva (2006) discuss robust measures of distance for use in nearest neighbor classification.

Exercises

9.1. Determine the rotation matrix that transforms the vector $x = (5, 12)$ into the vector $\tilde{x} = (0, 13)$.

9.2. Reproduce the surface shown in the wire frame of Figure 9.5, and then add a circular band around the surface centered at $(0, 0)$, and with radius 1.

9.3. Write a program for the simple linear congruential random number generator

$$x_i \equiv 35x_{i-1} \bmod 2^{15}.$$

Generate a sequence of length 1008. Look for structure in d-tuples of sequential values, $(x_i, x_{i+1}, \ldots, x_{i+d-1})$, for $3 \leq d \leq 1005$.

Compare this with the output of the similar generator in Exercise 8.4 where we consider $d = 3$.

a) Use a program that plots points in three dimensions and rotates the axes. Now, look for structure in higher dimensions. *Hint:* All of the points lie on the hyperplanes

$$x_{i+3} - 9x_{i+2} + 27x_{i+1} - 27x_i = j,$$

where j is an integer.

b) Examine this structure using parallel coordinates.

9.4. Develop a grand tour in parallel coordinates plots. Apply it to data generated by the random number generator in Exercise 9.3.

9.5. Develop a grand tour in Andrews curves. Apply it to data generated by the random number generator in Exercise 9.3. What is the difference in the grand tour and the pseudo grand tour in Andrews curves discussed on page 364?

9.6. Data exploration.

a) Generate 25 numbers independently from U(0, 1), and form five five-dimensional vectors, x_i, from them by taking the first five, the second five, and so on. Now, using Gram-Schmidt orthogonalization, form a set of orthogonal vectors y_i from the x's. You now have two multivariate datasets with very different characteristics. See if you can discover the difference graphically using either cartesian or noncartesian displays.

b) Now, generate five n-dimensional vectors, for n relatively large. Do the same thing as in the previous part. (Now, you can decide what is meant by "n relatively large".) In this exercise, you have two datasets, each with five variables and n observations. Your graphical displays have been of the *variables* instead of the *observations*.

c) Now, use the two datasets of the previous part and graph them in the traditional way using displays in which the upper-level graphical objects are the n observations.

Summarize your findings in a clearly-written report.

9.7. Consider the relative interpoint distances between the three 3-vectors

$$x_1 = (x_{11}, x_{12}, x_{13}),$$

$$x_2 = (x_{21}, x_{22}, x_{23}),$$

and

$$x_3 = (x_{31}, x_{32}, x_{33}).$$

For each of the other distance measures listed on pages 389 and 390, give specific values (small integers) for the x_{ij} such that, for the Euclidean distance, the distance between the first and second, d_{12}, is less than the distance between the second and third, d_{23}, but for that other distance measure, $d_{12} > d_{23}$. For the Hamming and binary distances, use the binary representations of the elements.

9.8. Show that the Mahalanobis distance (9.24), on page 392, between any two observations is nonnegative.

9.9. Show that the norm of the difference between two vectors is a metric; that is, if

$$\Delta(x_1, x_2) = \|x_1 - x_2\|,$$

$\Delta(x_1, x_2)$ is a metric.

9.10. a) Show that all of the distance measures in equations (9.14) through (9.18), as well as the Hamming distance and the binary difference, are metrics.

b) Which of those distance measures are based on norms?

c) Why are the correlation-based distances, including equation (9.19), not metrics?

9.11. Consider a two-dimensional surface with an orthogonal coordinate system over which there is a fluid flow with constant velocity $f = (f_1, f_2)$. Suppose that an object can move through the fluid with constant velocity v with respect to the fluid and measured in the same units as f. (The magnitude of the velocity is $\|v\| = \sqrt{v_1^2 + v_2^2}$.) Assume that $\|v\| > \|f\|$.

a) Define a distance measure, d, over the surface such that for two points x_i and x_j, the distance from x_i to x_j is proportional to the time required for the object to move from x_i to x_j.

b) Compare your distance measure with those listed on page 388.

c) What properties of a norm does your distance measure possess?

9.12. Consider the problem of a dissimilarity measure for two-dimensional data represented in polar coordinates, as discussed on page 392. One possibility, of course, is to transform the data to cartesian coordinates and then use any of the distance measures for that coordinate system. Define a dissimilarity measure based on d_{ij}^{d} and d_{ij}^{r}. Is your measure a metric?

9.13. Given two n-vectors, x_1 and x_2, form a third vector, x_3, as $x_3 = a_1x_1 + a_2x_2 + \epsilon$, where ϵ is a vector of independent $\mathrm{N}(0,1)$ realizations. Although the matrix $X = [x_1\ x_2\ x_3]$ is in $\mathbb{R}^{n\times 3}$, the linear structure, even obscured by the noise, implies a two-dimensional space for the data matrix (that is, the space $\mathbb{R}^{n\times 2}$).

a) Determine a rotation matrix that reveals the linear structure. In other words, determine matrices Q and P such that the rotation XQ followed by the projection $(XQ)P$ is a noisy line in two dimensions.

b) Generate x_1 and x_2 as realizations of a $\mathrm{U}(0,1)$ process and x_3 as $5x_1 + x_2 + \epsilon$, where ϵ is a realization of a $\mathrm{N}(0,1)$ process. What are Q and P from the previous question?

9.14. Given the distance matrix

$$D = \begin{bmatrix} 0 & 4.34 & 4.58 & 7.68 & 4.47 \\ 4.34 & 0 & 1.41 & 4.00 & 4.36 \\ 4.58 & 1.41 & 0 & 5.10 & 5.00 \\ 7.68 & 4.00 & 5.10 & 0 & 6.56 \\ 4.47 & 4.36 & 5.00 & 6.56 & 0 \end{bmatrix},$$

where the elements are Euclidean distances, determine a $5 \times m$ matrix with a small value of m that has a distance matrix very close to D.

10

Estimation of Functions

An interesting problem in statistics, and one that is generally difficult, is the estimation of a continuous function, such as a probability density function or a nonlinear regression model. The statistical properties of an estimator of a function are more complicated than statistical properties of an estimator of a single parameter or of a countable set of parameters.

In Chapter 4 we discussed ways of numerically *approximating* functions. In this brief chapter we will discuss ways of statistically *estimating* functions. Many of these methods are based on approximation methods such as orthogonal systems, splines, and kernels discussed in Chapter 4. The PDF decomposition plays an important role in the estimation of functions.

We will discuss the properties of an estimator in the general case of a real scalar-valued function over real vector-valued arguments (that is, a mapping from $\mathbb{R}^d$ into $\mathbb{R}$). One of the most common situations in which these properties are relevant is in nonparametric probability density estimation, which we discuss in Chapter 15. (In that application, of course, we do not have to do a PDF decomposition.) The global statistical properties we discuss in Section 10.3 are the measures by which we evaluate probability density estimators.

First, we say a few words about notation. We may denote a function by a single letter, f, for example, or by the function notation, $f(\cdot)$ or $f(x)$. When $f(x)$ denotes a function, x is merely a placeholder. The notation $f(x)$, however, may also refer to the value of the function at the point x. The meaning is usually clear from the context.

Using the common "hat" notation for an estimator, we use $\widehat{f}$ or $\widehat{f}(x)$ to denote the estimator of f or of $f(x)$. Following the usual terminology, we use the term "estimator" to denote a random variable, and "estimate" to denote a realization of the random variable.

The hat notation is also used to denote an estimate, so we must determine from the context whether $\widehat{f}$ or $\widehat{f}(x)$ denotes a random variable or a realization of a random variable.

J.E. Gentle, *Computational Statistics*, Statistics and Computing,
DOI: 10.1007/978-0-387-98144-4_10,

The estimate or the estimator of the value of the function at the point x may also be denoted by $\widehat{f}(x)$. Sometimes, to emphasize that we are estimating the ordinate of the function rather than evaluating an estimate of the function, we use the notation $\widehat{f(x)}$. In this case also, we often make no distinction in the notation between the realization (the estimate) and the random variable (the estimator). We must determine from the context whether $\widehat{f}(x)$ or $\widehat{f(x)}$ denotes a random variable or a realization of a random variable. In most of the following discussion, however, the hat notation denotes a random variable. Its distribution depends on the underlying random variable that yields the sample from which the estimator is computed.

The usual optimality properties that we use in developing a theory of estimation of a finite-dimensional parameter must be extended for estimation of a general function. As we will see, two of the usual desirable properties of point estimators, namely unbiasedness and maximum likelihood, cannot be attained globally or in general by estimators of functions.

There are many similarities in *estimation* of functions and *approximation* of functions, but we must be aware of the fundamental differences in the two problems. Estimation of functions is similar to other estimation problems: We are given a sample of observations; we make certain assumptions about the probability distribution of the sample; and then we develop estimators. The estimators are random variables, and how useful they are depends on properties of their distribution, such as their expected values and their variances.

Approximation of functions is an important aspect of numerical analysis. Functions are often approximated to interpolate functional values between directly computed or known values. Functions are also approximated as a prelude to quadrature. In this chapter, we will often approximate a function as a step in the statistical estimation of the function.

In the problem of function estimation, we may have observations on the function at specific points in the domain, or we may have indirect measurements of the function, such as observations that relate to a derivative or an integral of the function. In either case, the problem of function estimation has the competing goals of providing a good fit to the observed data and predicting values at other points. In many cases, a smooth estimate satisfies this latter objective. In other cases, however, the unknown function itself is not smooth. Functions with different forms may govern the phenomena in different regimes. This presents a very difficult problem in function estimation, but it is one that we will not consider in any detail here.

There are various approaches to estimating functions. Maximum likelihood (see page 70) has limited usefulness for estimating functions because in general the likelihood is unbounded. A practical approach is to assume that the function is of a particular form and estimate the parameters that characterize the form. For example, we may assume that the function is exponential, possibly because of physical properties such as exponential decay. We may then use various estimation criteria, such as least squares, to estimate the parameter. An extension of this approach is to assume that the function is a mixture of other functions. The mixture can be formed by different functions over different domains or by weighted averages of the functions over the whole domain. Estimation of the function of interest involves estimation of various parameters as well as the weights.

Another approach to function estimation is to represent the function of interest as a linear combination of basis functions, that is, to represent the function in a series expansion. The basis functions are generally chosen to be orthogonal over the domain of interest, and the observed data are used to estimate the coefficients in the series. We discuss the use of basis functions beginning on page 18 and again on page 161.

It is often more practical to estimate the function value at a given point. (Of course, if we can estimate the function at any given point, we can effectively have an estimate at all points.) One way of forming an estimate of a function at a given point is to take the average at that point of a filtering function that is evaluated in the vicinity of each data point. The filtering function is called a kernel, and the result of this approach is called a kernel estimator. We discussed use of kernels in approximation of functions in Section 4.5. Kernel methods have limited use in function approximation, but they are very useful in function estimation. We briefly discuss the use of kernel filters in function estimation on page 406, but we discuss those kinds of methods more fully in the context of probability density function estimation in Section 15.3, beginning on page 499.

In the estimation of functions, we must be concerned about the properties of the estimators at specific points and also about properties over the full domain. Global properties over the full domain are often defined in terms of integrals or in terms of suprema or infima.

10.1 General Approaches to Function Estimation

The theory of statistical estimation is based on probability distributions. In order to develop an estimation procedure, we need to identify random variables, and make some assumptions about their distributions. In the case of statistical estimation of a function, this may involve decomposing the function of interest so as to have a factor that is a PDF. This PDF decomposition is a preliminary step in function estimation.

Once a random variable and a probability distribution are identified, methods for estimation of functions often parallel the methods of approximation of functions as in Chapter 4.

Function Decomposition and Estimation of the Coefficients in an Orthogonal Expansion

In the following, we will work with functions that are square-integrable over some domain D; that is, functions in $L^2(D)$.

The estimation approach follows the approximation approach discussed in Sections 4.2 and 4.3. It begins, however, with a PDF decomposition.

We decompose the function of interest to have a factor that is a probability density function, say

$$f(x) = g(x)p(x), \tag{10.1}$$

where

$$\int_D p(x)\mathrm{d}x = 1$$

and $p(x) > 0$ on D; that is, p is a PDF and the distribution has support D.

This PDF decomposition is important, because now we can introduce the expectation of a random variable. We expand the function as in equation (4.25) on page 162, and then from equation (4.26), we have

$$\begin{aligned} c_k &= \langle f, q_k \rangle \\ &= \int_D q_k(x)g(x)p(x)\mathrm{d}x \\ &= \mathrm{E}(q_k(X)g(X)), \end{aligned} \tag{10.2}$$

where X is a random variable whose probability density function is p.

If we have a random sample, $x_1, \ldots, x_n$, from the distribution with density p, an estimator of c_k is

$$\widehat{c}_k = \frac{1}{n}\sum_{i=1}^{n} q_k(x_i)g(x_i). \tag{10.3}$$

It is clear that this estimator is unbiased for the expectation in equation (10.2), if we assume that the expectation is finite.

The series estimator of the function for all x using the truncated series approximation, as in equation (4.27), therefore is

$$\widehat{f}(x) = \frac{1}{n}\sum_{k=0}^{j}\sum_{i=1}^{n} q_k(x_i)g(x_i)q_k(x) \tag{10.4}$$

for some truncation point j. *Note that this estimator assumes a random sample from a known distribution over the domain of* f.

The random sample, $x_1, \ldots, x_n$, may be an observed dataset, or it may be the output of a random number generator.

For univariate function, the basis functions in the expansion above are often chosen from the standard series of univariate orthogonal polynomials, such as the Legendre, Laguerre, or Hermite polynomials. (See Table 4.1 on page 170.)

Use of Splines

The approach to function estimation that we pursued in the previous section makes use of a finite subset of an infinite basis set which often consists of polynomials of degrees $p = 0, 1, \ldots$. This approach yields a smooth estimate $\widehat{f}(x)$. The polynomials in $\widehat{f}(x)$, however, cause oscillations that may be undesirable. This is because of the approximation used prior to the estimation. The approximation oscillates a number of times one less than the highest degree of the polynomial used. Also, if the function being approximated has quite different shapes in different regions of its domain, the global approach of using the same polynomials over the full domain may not be very effective.

Another approach is to subdivide the interval over which the function is to be approximated and then on each subinterval use polynomials with low degree. The approximation at any point is a sum of one or more piecewise polynomials. Even with polynomials of very low degree, if we use a large number of subintervals, we can obtain a good approximation to the function. Zero-degree polynomials, for example, would yield a piecewise constant function that could be very close to a given function if enough subintervals are used. Using more and more subintervals, of course, is not a very practical approach. Not only is the approximation a rather complicated function, but it may be discontinuous at the interval boundaries. We can achieve smoothness of the approximation by imposing continuity restrictions on the piecewise polynomials and their derivatives. This is the approach in *spline* approximation and smoothing, which we discussed in Section 4.4.

As described on page 178, there are three types of spline basis functions commonly used:

- truncated power functions (or just power functions),
- B-splines,
- "natural" polynomial splines.

Some basis functions for various types of splines over the interval $[-1, 1]$ are shown in Figure 4.5 on page 181.

Smoothing Splines

Smoothing splines are generally more useful in function estimation than are interpolating splines. The individual points may be subject to error, so the approximating spline may not go through any of the given points. In this usage, the splines are evaluated at each abscissa point, and the ordinates are fitted by some criterion (such as least squares) to the spline.

The choice of knots is a difficult problem. One approach is to include the knots as decision variables in the optimization problem for determining the fit. Other approaches are to add (pre-chosen) knots in a stepwise manner or to use a regularization method (addition of a component to the fitting optimization objective function that increases for roughness or for some other undesirable characteristic of the fit).

In an important type of application of splines, we have assumed a linear relationship similar to equation (4.62) on page 179, but with an error term:

$$y = \sum_{k=1}^{j} c_k b_k(x) + \epsilon. \tag{10.5}$$

If we have a sample $(y_1, x_1), \ldots, (y_n, x_n)$, we first evaluate each of the spline basis functions b_k at each x_i, yielding $\tilde{x}_{ik}$, where

$$\tilde{x}_{ik} = b_k(x_i). \tag{10.6}$$

The observations on the dependent variable y_i are then fit to the spline function values by choosing appropriate values of c_k. The fit can be based on any of the criteria that we discussed in Section 1.8. A least squares fit is most commonly used; that is, the c_k are chosen to minimize

$$\sum_{i=1}^{n} \left(y_i - \sum_{k=1}^{j} c_k \tilde{x}_{ik} \right)^2 . \tag{10.7}$$

Kernel Methods

An approach to function approximation discussed in Section 4.5 is to use a *filter* or *kernel* function to provide local weighting of the observed data. This approach ensures that at a given point the observations close to that point influence the estimate at the point more strongly than more distant observations. A standard method in this approach is to convolve the observations with a unimodal function that decreases rapidly away from a central point. This function is the filter or the kernel. A kernel has two arguments representing the two points in the convolution, but we typically use a single argument that represents the distance between the two points.

The univariate kernel functions equations (4.65) through (4.67) are often used in function estimation. These are the uniform,

$$K_{\mathrm{u}}(t) = 1/(2\lambda)\mathrm{I}_{[-\lambda,\lambda]}(t),$$

the quadratic,

$$K_{\mathrm{q}}(t) = 3/(\lambda^2(6 - 2\lambda))(\lambda - t^2)\mathrm{I}_{[-\lambda,\lambda]}(t),$$

and the normal,

$$K_{\mathrm{n}}(t) = \frac{1}{\sqrt{2\pi}}\mathrm{e}^{-(t/\lambda)^2/2},$$

each with a smoothing parameter λ. Kernel methods are often used in the estimation of probability density functions which we discuss in Chapter 15, and in Section 15.3, we will discuss some more kernels.

In kernel methods, the locality of influence is controlled by a smoothing parameter or a window around the point of interest. The choice of the size of the window is the most important issue in the use of kernel methods. In practice, for a given choice of the size of the window, the argument of the kernel function is transformed to reflect the size. In general, the transformation is accomplished using a positive definite matrix, V, whose determinant measures the volume (size) of the window.

To estimate the function f at the point x, we first form a PDF decomposition of f, as in equation (10.1),

$$f(x) = g(x)p(x),$$

where p is a probability density function. In the multivariate case, for a given set of data, $x_1, \ldots, x_n$, and a given scaling transformation matrix V, the kernel estimator of the function at the point x is

$$\widehat{f(x)} = (n|V|)^{-1} \sum_{i=1}^{n} g(x) K\left(V^{-1}(x - x_i)\right). \tag{10.8}$$

In the univariate case, the size of the window is just the width h. The argument of the kernel is transformed to s/h, so the function that is convolved with the function of interest is $K(s/h)/h$. The univariate kernel estimator is

$$\widehat{f(x)} = \frac{1}{nh} \sum_{i=1}^{n} g(x) K\left(\frac{x - x_i}{h}\right).$$

10.2 Pointwise Properties of Function Estimators

The statistical properties of an estimator of a function at a given point are analogous to the usual statistical properties of an estimator of a scalar parameter. The statistical properties involve expectations or other properties of random variables. In the following, when we write an expectation, $\mathrm{E}(\cdot)$, or a variance, $\mathrm{V}(\cdot)$, the expectations are usually taken with respect to the (unknown) distribution of the underlying random variable. Occasionally, we may explicitly indicate the distribution by writing, for example, $\mathrm{E}_p(\cdot)$, where p is the density of the random variable with respect to which the expectation is taken.

Bias

The bias of the estimator of a function value at the point x is

$$\mathrm{E}\left(\widehat{f}(x)\right) - f(x).$$

If this bias is zero, we would say that the estimator is unbiased at the point x. If the estimator is unbiased at every point x in the domain of $\widehat{f}$, we say that the estimator is pointwise unbiased. Obviously, in order for $\widehat{f}(\cdot)$ to be pointwise unbiased, it must be defined over the full domain of f.

Variance

The variance of the estimator at the point x is

$$\mathrm{V}\left(\widehat{f}(x)\right) = \mathrm{E}\left(\left(\widehat{f}(x) - \mathrm{E}\left(\widehat{f}(x)\right)\right)^2\right).$$

Estimators with small variance are generally more desirable, and an optimal estimator is often taken as the one with smallest variance among a class of unbiased estimators.

Mean Squared Error

The mean squared error, MSE, at the point x is

$$\mathrm{MSE}\left(\widehat{f}(x)\right) = \mathrm{E}\left(\left(\widehat{f}(x) - f(x)\right)^2\right). \tag{10.9}$$

The mean squared error is the sum of the variance and the square of the bias:

$$\begin{aligned}\mathrm{MSE}\left(\widehat{f}(x)\right) &= \mathrm{E}\left(\left(\widehat{f}(x)\right)^2 - 2\widehat{f}(x)f(x) + (f(x))^2\right) \\ &= \mathrm{V}\left(\widehat{f}(x)\right) + \left(\mathrm{E}\left(\widehat{f}(x)\right) - f(x)\right)^2. \end{aligned} \tag{10.10}$$

Sometimes, the variance of an unbiased estimator is much greater than that of an estimator that is only slightly biased, so it is often appropriate to compare the mean squared error of the two estimators. In some cases, as we will see, unbiased estimators do not exist, so rather than seek an unbiased estimator with a small variance, we seek an estimator with a small MSE.

Mean Absolute Error

The mean absolute error, MAE, at the point x is similar to the MSE:

$$\mathrm{MAE}\left(\widehat{f}(x)\right) = \mathrm{E}\left(\left|\widehat{f}(x) - f(x)\right|\right). \tag{10.11}$$

It is more difficult to do mathematical analysis of the MAE than it is for the MSE. Furthermore, the MAE does not have a simple decomposition into other meaningful quantities similar to the MSE.

Consistency

Consistency of an estimator refers to the convergence of the expected value of the estimator to what is being estimated as the sample size increases without bound. A point estimator T_n, based on a sample of size n, is consistent for θ if

$$T_n - \theta \to 0 \quad \text{as } n \to \infty.$$

The convergence is stochastic, of course, so there are various types of convergence that can be required for consistency. The most common kind of convergence considered is weak convergence, or convergence in probability.

In addition to the type of stochastic convergence, we may consider the convergence of various measures of the estimator. In general, if m is a function (usually a vector-valued function that is an elementwise norm), we may define consistency of an estimator T_n in terms of m if

$$\mathrm{E}(m(T_n - \theta)) \to 0. \tag{10.12}$$

For an estimator, we are often interested in *weak convergence in mean square* or *weak convergence in quadratic mean*, so the common definition of consistency of T_n is

$$\mathrm{E}\left((T_n - \theta)^{\mathrm{T}}(T_n - \theta)\right) \to 0,$$

where the type of convergence is convergence in probability. Consistency defined by convergence in mean square is also called L_2 consistency.

If convergence does occur, we are interested in the rate of convergence. We define rate of convergence in terms of a function of n, say $r(n)$, such that

$$\mathrm{E}(m(T_n - \theta)) = \mathrm{O}(r(n)).$$

A common form of $r(n)$ is n^α, where $\alpha < 0$. For example, in the simple case of a univariate population with a finite mean μ and finite second moment, use of the sample mean $\bar{x}$ as the estimator T_n, and use of $m(z) = z^2$, we have

$$\begin{aligned} \mathrm{E}(m(\bar{x} - \mu)) &= \mathrm{E}\left((\bar{x} - \mu)^2\right) \\ &= \mathrm{MSE}(\bar{x}) \\ &= \mathrm{O}\left(n^{-1}\right). \end{aligned}$$

See Exercise 10.1, page 414.

In the estimation of a function, we say that the estimator $\widehat{f}_n$ of the function f is *pointwise consistent* if

$$\mathrm{E}\left(\widehat{f}_n(x)\right) \to f(x) \tag{10.13}$$

for every x the domain of f. Just as in the estimation of a parameter, there are various kinds of pointwise consistency in the estimation of a function. If the convergence in expression (10.13) is in probability, for example, we say that the estimator is weakly pointwise consistent. We could also define other kinds of pointwise consistency in function estimation along the lines of other types of consistency.

10.3 Global Properties of Estimators of Functions

Often, we are interested in some measure of the statistical properties of an estimator of a function over the full domain of the function. The obvious way of defining statistical properties of an estimator of a function is to integrate the pointwise properties discussed in the previous section.

Statistical properties of a function estimator, such as the bias of the estimator, are often defined in terms of a norm of the function.

For comparing $\widehat{f}(x)$ and $f(x)$, the L_p norm of the error is

$$\left(\int_D \left|\widehat{f}(x) - f(x)\right|^p \mathrm{d}x\right)^{1/p}, \tag{10.14}$$

where D is the domain of f. The integral may not exist, of course. Clearly, the estimator $\widehat{f}$ must also be defined over the same domain.

Three useful measures are the L_1 norm, also called the *integrated absolute error*, or IAE,

$$\mathrm{IAE}(\widehat{f}\,) = \int_D \left|\widehat{f}(x) - f(x)\right| \mathrm{d}x, \tag{10.15}$$

the square of the L_2 norm, also called the *integrated squared error*, or ISE,

$$\mathrm{ISE}(\widehat{f}\,) = \int_D \left(\widehat{f}(x) - f(x)\right)^2 \mathrm{d}x, \tag{10.16}$$

and the L_∞ norm, the *sup absolute error*, or SAE,

$$\mathrm{SAE}(\widehat{f}\,) = \sup\left|\widehat{f}(x) - f(x)\right|. \tag{10.17}$$

The L_1 measure is invariant under monotone transformations of the coordinate axes, but the measure based on the L_2 norm is not. See Exercise 4.1 on page 199.

The L_∞ norm, or SAE, is the most often used measure in general function approximation. In statistical applications, this measure applied to two cumulative distribution functions is the *Kolmogorov distance.* The measure is not so useful in comparing densities and is not often used in density estimation.

Other measures of the difference in $\widehat{f}$ and f over the full range of x are the Kullback-Leibler measure,

$$\int_D \widehat{f}(x) \log\left(\frac{\widehat{f}(x)}{f(x)}\right) \mathrm{d}x,$$

and the Hellinger distance,

$$\left(\int_D \left(\widehat{f}^{1/p}(x) - f^{1/p}(x)\right)^p \mathrm{d}x\right)^{1/p}.$$

For $p = 2$, the Hellinger distance is also called the Matusita distance.

Integrated Bias and Variance

We now want to develop global concepts of bias and variance for estimators of functions. Bias and variance are statistical properties that involve expectations of random variables. The obvious global measures of bias and variance are just the pointwise measures integrated over the domain. In the case of the bias, of course, we must integrate the absolute value, otherwise points of negative bias could cancel out points of positive bias.

The estimator $\widehat{f}$ is pointwise unbiased if

$$\mathrm{E}\left(\widehat{f}(x)\right) = f(x) \quad \text{for all } x \in \mathrm{IR}^d.$$

Because we are interested in the bias over the domain of the function, we define the *integrated absolute bias* as

$$\mathrm{IAB}\left(\widehat{f}\right) = \int_D \left|\mathrm{E}\left(\widehat{f}(x)\right) - f(x)\right| \,\mathrm{d}x \tag{10.18}$$

and the *integrated squared bias* as

$$\mathrm{ISB}\left(\widehat{f}\right) = \int_D \left(\mathrm{E}\left(\widehat{f}(x)\right) - f(x)\right)^2 \mathrm{d}x. \tag{10.19}$$

If the estimator is unbiased, both the integrated absolute bias and integrated squared bias are 0. This, of course, would mean that the estimator is pointwise unbiased almost everywhere. Although it is not uncommon to have unbiased estimators of scalar parameters or even of vector parameters with a countable number of elements, it is not likely that an estimator of a function could be unbiased at almost all points in a dense domain. ("Almost" means all except possibly a set with a probability measure of 0.)

The *integrated variance* is defined in a similar manner:

$$\begin{aligned}\mathrm{IV}\left(\widehat{f}\right) &= \int_D \mathrm{V}\left(\widehat{f}(x)\right) \mathrm{d}x \\ &= \int_D \mathrm{E}\left(\left(\widehat{f}(x) - \mathrm{E}\left(\widehat{f}(x)\right)\right)^2\right) \mathrm{d}x.\end{aligned} \tag{10.20}$$

Integrated Mean Squared Error and Mean Absolute Error

As we suggested above, global unbiasedness is generally not to be expected. An important measure for comparing estimators of functions is, therefore, based on the mean squared error.

The *integrated mean squared error* is

$$\begin{aligned}\mathrm{IMSE}\left(\widehat{f}\right) &= \int_D \mathrm{E}\left(\left(\widehat{f}(x) - f(x)\right)^2\right) \mathrm{d}x \\ &= \mathrm{IV}\left(\widehat{f}\right) + \mathrm{ISB}\left(\widehat{f}\right) \end{aligned} \tag{10.21}$$

(compare equations (10.9) and (10.10)).

If the expectation integration can be interchanged with the outer integration in the expression above, we have

$$\begin{aligned}\mathrm{IMSE}\left(\widehat{f}\right) &= \mathrm{E}\left(\int_D \left(\widehat{f}(x) - f(x)\right)^2 \mathrm{d}x\right) \\ &= \mathrm{MISE}\left(\widehat{f}\right),\end{aligned}$$

the *mean integrated squared error.* We will assume that this interchange leaves the integrals unchanged, so we will use MISE and IMSE interchangeably.

Similarly, for the *integrated mean absolute error*, we have

$$\begin{aligned}\mathrm{IMAE}\left(\widehat{f}\right) &= \int_D \mathrm{E}\left(\left|\widehat{f}(x) - f(x)\right|\right) \mathrm{d}x \\ &= \mathrm{E}\left(\int_D \left|\widehat{f}(x) - f(x)\right| \mathrm{d}x\right) \\ &= \mathrm{MIAE}\left(\widehat{f}\right),\end{aligned}$$

the *mean integrated absolute error.*

Mean SAE

The *mean sup absolute error*, or MSAE, is

$$\mathrm{MSAE}\left(\widehat{f}\right) = \int_D \mathrm{E}\left(\sup\left|\widehat{f}(x) - f(x)\right|\right) \mathrm{d}x. \tag{10.22}$$

This measure is not very useful unless the variation in the function f is relatively small. For example, if f is a density function, $\widehat{f}$ can be a "good" estimator, yet the MSAE may be quite large. On the other hand, if f is a cumulative distribution function (monotonically ranging from 0 to 1), the MSAE may be a good measure of how well the estimator performs. As mentioned earlier, the SAE is the *Kolmogorov distance.* The Kolmogorov distance (and, hence, the SAE and the MSAE) does poorly in measuring differences in the tails of the distribution.

Large-Sample Statistical Properties

The pointwise consistency properties are extended to the full function in the obvious way. In the notation of expression (10.12), consistency of the function estimator is defined in terms of

$$\int_D \mathrm{E}\left(m\left(\widehat{f}_n(x) - f(x)\right)\right) \mathrm{d}x \to 0,$$

where m is some function, usually a norm or a power of a norm.

The estimator of the function is said to be *mean square consistent* or L_2 *consistent* if the MISE converges to 0; that is,

$$\int_D \mathrm{E}\left(\left(\widehat{f}_n(x) - f(x)\right)^2\right) \mathrm{d}x \;\to\; 0. \tag{10.23}$$

If the convergence is weak, that is, if it is convergence in probability, we say that the function estimator is weakly consistent; if the convergence is strong, that is, if it is convergence almost surely or with probability 1, we say the function estimator is strongly consistent.

The estimator of the function is said to be L_1 *consistent* if the mean integrated absolute error (MIAE) converges to 0; that is,

$$\int_D \mathrm{E}\left(\left|\widehat{f}_n(x) - f(x)\right|\right) \mathrm{d}x \;\to\; 0. \tag{10.24}$$

As with the other kinds of consistency, the nature of the convergence in the definition may be expressed in the qualifiers "weak" or "strong".

As we have mentioned above, the integrated absolute error is invariant under monotone transformations of the coordinate axes, but the L_2 measures are not. As with most work in L_1, however, derivation of various properties of IAE or MIAE is more difficult than for analogous properties with respect to L_2 criteria.

If the MISE converges to 0, we are interested in the rate of convergence. To determine this, we seek an expression of MISE as a function of n. We do this by a Taylor series expansion.

In general, if $\widehat{\theta}$ is an estimator of θ, the Taylor series for $\mathrm{ISE}(\widehat{\theta})$, equation (10.16), about the true value is

$$\mathrm{ISE}\left(\widehat{\theta}\right) = \sum_{k=0}^{\infty} \frac{1}{k!}\left(\widehat{\theta} - \theta\right)^k \mathrm{ISE}^{k'}(\theta), \tag{10.25}$$

where $\mathrm{ISE}^{k'}(\theta)$ represents the kth derivative of ISE evaluated at θ.

Taking the expectation in equation (10.25) yields the MISE. The limit of the MISE as $n \to \infty$ is the *asymptotic mean integrated squared error*, AMISE. One of the most important properties of an estimator is the order of the AMISE.

In the case of an unbiased estimator, the first two terms in the Taylor series expansion are zero, and the AMISE is

$$\mathrm{V}(\widehat{\theta})\,\mathrm{ISE}''(\theta)$$

to terms of second order.

Other Global Properties of Estimators of Functions

There are often other properties that we would like an estimator of a function to possess. We may want the estimator to weight given functions in some particular way. For example, if we know how the function to be estimated, f, weights a given function r, we may require that the estimate $\widehat{f}$ weight the function r in the same way; that is,

$$\int_D r(x)\widehat{f}(x)\mathrm{d}x = \int_D r(x)f(x)\mathrm{d}x.$$

We may want to restrict the minimum and maximum values of the estimator. For example, because many functions of interest are nonnegative, we may want to require that the estimator be nonnegative.

We may want to restrict the variation in the function estimate. This can be thought of as the "roughness" of the function (see page 151). Often, in function estimation, we may seek an estimator $\widehat{f}$ such that its roughness (by some definition) is small.

Notes and Further Reading

Function estimation of course is closely related to the problem in numerical analysis of function approximation, which is the topic of Chapter 4. "Estimation" in this case means statistical estimation; that is, the use of observed data to make inferences about the objects that define a function. In simpler cases, these objects are just parameters in a given parametric representation of the function. In more interesting cases, the function is not specified parametrically. The result of the estimation procedure is not a mathematical expression of a functional form; rather, it is an algorithm that takes as input an argument of the function and produces the corresponding estimated value of the function. Extensive discussions of methods of function estimation are available in Ramsay and Silverman (2002, 2005) and Efromovich (1999).

This chapter has surveyed general methods of function estimation and properties of function estimators. The most common kind of function that we estimate in statistical applications is the probability density function. That is the topic of Chapters 14 and 15.

Exercises

10.1. Consider the problem of estimating μ and σ (the mean and standard deviation) in a normal distribution. For estimators in a sample of size n, we will use the sample mean, $\bar{y}_n$, and the sample standard deviation, s_n. Assume that

$$\text{MSE}(\bar{y}_n) = \text{O}(n^{\alpha})$$

and

$$\text{MSE}(s_n) = \text{O}(n^{\beta}).$$

Perform a Monte Carlo experiment to estimate α and β. Plot your data on log-log axes, and use least squares to estimate α and β. Now, derive the exact values for α and β and compare them with your estimates.

10.2. Formally derive equation (10.21) using equations (10.20), and (10.19).

10.3. Some problems in function estimation are relatively easy. Consider the problem of estimation of

$$f(t) = \alpha + \beta t,$$

for $t \in [0, 1]$. Suppose, for $t_1, \ldots, t_n$ we observe $f(t_1) + \epsilon_1, \ldots, f(t_n) + \epsilon_n$, where the ϵ_i are independent realizations from a $\text{N}(0, \sigma^2)$ distribution. As we know, a good estimator of $f(t)$ is $\widehat{f}(t) = \widehat{\alpha} + \widehat{\beta}t$, where $\widehat{\alpha}$ and $\widehat{\beta}$ are the least squares estimators from the data. Determine $\text{MISE}(\widehat{f})$.

10.4. Consider the $\text{U}(0, \theta)$ distribution, with θ unknown. The true probability density is $p(x) = 1/\theta$ over $(0, \theta)$ and 0 elsewhere. Suppose we have a sample of size n and we estimate the density as $\hat{p}(x) = 1/x_{(n)}$ over $(0, x_{(n)})$ and 0 elsewhere, where $x_{(n)}$ is the maximum order statistic. The density of the distribution of $X_{(n)}$ is $nx_{(n)}^{n-1}\theta^{-n}$ over $(0, \theta)$ and 0 elsewhere.

a) Determine (that is, write an explicit expression for) the integrated squared bias, ISB, of $\hat{p}(x)$.
b) Determine the integrated squared error, ISE, of $\hat{p}(x)$.
c) Determine the mean integrated squared error, MISE, of $\hat{p}(x)$.
d) Determine the asymptotic (as $n \to \infty$) mean integrated squared error, AMISE, of $\hat{p}(x)$.

11

Monte Carlo Methods for Statistical Inference

Monte Carlo methods are experiments. Monte Carlo experimentation is the use of simulated random numbers to estimate some functional of a probability distribution. A problem that does not have a stochastic component can sometimes be posed as a problem with a component that can be identified with an expectation of some function of a random variable. This is often done by means of a PDF decomposition. The problem is then solved by estimating the expected value by use of a simulated sample from the distribution of the random variable.

Monte Carlo methods use random numbers, so to implement a Monte Carlo method it is necessary to have a source of random numbers. On the computer, we generally settle for *pseudorandom* numbers, that is, numbers that *appear to be* random but are actually deterministic. Generation of pseudorandom numbers is the topic of Chapter 7.

Often, our objective is not to simulate random sampling directly, but rather to estimate a specific quantity related to the distribution of a given sample. In this case, we may want to ensure that a chosen sample closely reflects the distribution of the population we are simulating. Because of random variation, a truly random sample or a pseudorandom sample that simulates a random sample would not necessarily have this property. Sometimes, therefore, we generate a *quasirandom* sample, which is a sample constrained to reflect closely the distribution of the population we are simulating, rather than to exhibit the variability that would result from random sampling. Because in either case we proceed to treat the samples as if they were random, we will refer to both pseudorandom numbers and quasirandom numbers as "random numbers", except when we wish to emphasize the "pseudo" or "quasi" nature.

In this chapter, we discuss various ways random numbers are used in statistical inference. Monte Carlo methods are also used in many of the techniques described in other chapters.

J.E. Gentle, *Computational Statistics*, Statistics and Computing,
DOI: 10.1007/978-0-387-98144-4_11,

11.1 Monte Carlo Estimation

The general objective in Monte Carlo simulation is to estimate some characteristic of a random variable X. Often, the objective is to calculate the expectation of some function g of X.

We begin by reviewing some of the material covered in Section 4.7.

Estimation of a Definite Integral

Monte Carlo inference, as for statistical inference generally, can be formulated as estimation of either a definite integral

$$\theta = \int_D f(x)\mathrm{d}x \tag{11.1}$$

or, given the integral θ, of a domain D, or of a function f that satisfies certain optimality conditions. If the integral can be evaluated in closed form, there is no need for Monte Carlo methods. If D is of only one or two dimensions, there are several good, straightforward numerical quadrature methods available to solve the problem. For domains of higher dimension, Monte Carlo estimation is sometimes the best method for the quadrature.

Function Decomposition

If the function f is decomposed to have a factor that is a probability density function, say

$$f(x) = g(x)p(x), \tag{11.2}$$

where

$$\int_D p(x)\mathrm{d}x = 1$$

and $p(x) \geq 0$, then the integral θ is the expectation of the function g of the random variable with probability density p; that is,

$$\theta = \mathrm{E}(g(X)) = \int_D g(x)p(x)\mathrm{d}x. \tag{11.3}$$

Notice that this PDF decomposition is a standard method in statistical estimation; we identify a random variable, a probability distribution, and finally an expectation. Compare this with the development leading up to equation (10.2) on page 404 for estimating a function.

With a random sample $x_1, \ldots, x_m$ from the distribution with probability density p, an estimate of θ is

$$\widehat{\theta} = \frac{\sum g(x_i)}{m}. \tag{11.4}$$

We use this technique in many settings in statistics. There are three steps:

1. Decompose the function of interest to include a probability density function as a factor;
2. identify an expected value;
3. use a sample (simulated or otherwise) to estimate the expected value.

The PDF decomposition is not unique, of course, and sometimes a particular decomposition is more useful than another. In the Monte Carlo application, it is necessary to be able to generate random numbers easily from the distribution with the given density. As we will see in the discussion of importance sampling on page 426, there are other considerations for efficient Monte Carlo estimation.

We should note here that the use of Monte Carlo procedures for numerical quadrature is rarely the best method for lower-dimensional integrals. Use of Newton-Cotes or Gaussian quadrature, as discussed in Chapter 4, is usually better. For higher-dimensional integrals, however, Monte Carlo quadrature is often a viable alternative.

Estimation of the Variance

A Monte Carlo estimate usually has the form of the estimator of θ in equation (11.4). An estimate of the variance of this estimator is

$$\widehat{\mathrm{V}}(\widehat{\theta}) = \frac{\sum \left(g(x_i) - \overline{g(x)}\right)^2}{m(m-1)}. \tag{11.5}$$

This is because the elements of the set of random variables $\{g(X_i)\}$, on which we have observations $\{g(x_i)\}$, are (assumed to be) independent and thus to have zero correlations.

Estimating the Variance Using Batch Means

If the $g(X_i)$ do not have zero correlations, as may be the case when the X_i are from a Markov process, the estimator (11.5) has an expected value that includes the correlations; that is, it is biased for estimating $\mathrm{V}(\widehat{\theta})$. This situation arises often in simulation. In many processes of interest, however, observations are "more independent" of observations farther removed within the sequence than they are of observations closer to them in the sequence. A common method for estimating the variance in a sequence of nonindependent observations, therefore, is to use the means of successive subsequences that are long enough that the observations in one subsequence are almost independent of the observations in another subsequence. The means of the subsequences are called "batch means".

If $G_1, \ldots, G_b, G_{b+1}, \ldots, G_{2b}, G_{2b+1}, \ldots, G_{kb}$ is a sequence of random variables such that the correlation of G_i and G_{i+b} is approximately zero, an estimate of the variance of the mean, $\overline{G}$, of the $m = kb$ random variables can be developed by observing that

$$
\begin{aligned}
\mathrm{V}(\overline{G}) &= \mathrm{V}\left(\frac{1}{m}\sum G_i\right) \\
&= \mathrm{V}\left(\frac{1}{k}\sum_{j=1}^{k}\left(\frac{1}{b}\sum_{i=(j-1)b+1}^{jb} G_i\right)\right) \\
&\approx \frac{1}{k^2}\sum_{j=1}^{k}\mathrm{V}\left(\frac{1}{b}\sum_{i=(j-1)b+1}^{jb} G_i\right) \\
&\approx \frac{1}{k}\mathrm{V}(\overline{G}_b),
\end{aligned}
$$

where $\overline{G}_b$ is the mean of a batch of length b. If the batches are long enough, it may be reasonable to assume that the means have a common variance. An estimator of the variance of $\bar{G}_b$ is the standard sample variance from k observations, $\bar{g}_1, \bar{g}_2, \ldots, \bar{g}_k$:

$$\frac{\sum(\bar{g}_j - \bar{g})^2}{k-1}.$$

Hence, the batch-means estimator of the variance of $\overline{G}$ is

$$\widehat{\mathrm{V}}(\overline{G}) = \frac{\sum(\bar{g}_j - \bar{g})^2}{k(k-1)}. \tag{11.6}$$

This batch-means variance estimator should be used if the Monte Carlo study yields a stream of nonindependent observations, such as in a time series or when the simulation uses a Markov chain. The size of the subsamples should be as small as possible and still have means that are independent. A test of the independence of the $\overline{G}_b$ may be appropriate to help in choosing the size of the batches.

Batch means are useful in variance estimation whenever a Markov chain is used in the generation of the random deviates.

Convergence of Iterative Monte Carlo and Mixing of the Markov Chain

In ordinary Monte Carlo simulation, estimation relies on the fact that for independent, identically distributed variables $X_1, X_2, \ldots$ from the distribution P of X,

$$\frac{1}{n}\sum_{i=1}^{n} g(X_i) \to \mathrm{E}(g(X))$$

almost surely as n goes to infinity. This convergence is a simple consequence of the law of large numbers in the case of i.i.d. random variables. In Monte Carlo simulation, a random number generator simulates an independent stream. When X is multivariate or a complicated stochastic process, however, it may be difficult to simulate independent realizations.

The mean of a sample from an irreducible Markov chain $X_1, X_2, \ldots$ that has P as its equilibrium distribution also converges to the desired expectation. For this fact to have relevance in applications, the finite sampling from a Markov chain in the application must be concentrated in the equilibrium distribution; that is, the burn-in sample must not dominate the results. We mention below some methods for assessing convergence of MCMC samples to the stationary distribution, but it is not easy to determine when the Markov chain has begun to resemble its stationary distribution.

Once convergence to the stationary distribution is achieved, however, subsequent iterations are from that distribution; that is, they do not depend on the starting point. In Gibbs sampling, if

$$X_1, \ldots, X_{i-1},\ X_{i+1}, \ldots, X_d$$

have the marginal stationary distribution and X_i is given a new realization from the correct conditional distribution given the rest, then all of them still have the correct joint distribution.

In MCMC we must be concerned with more than just the length of a burn-in period and convergence to the stationary distribution, however. We must also be concerned with the *mixing* of the Markov chain, that is, how independently states X_i and X_{i+k} behave. Rapid mixing of the chain (meaning X_i and X_{i+k} are "relatively independent" for small k) ensures that the regions in the state-space will be visited in relatively small sequences with a frequency similar to long-term frequencies in the stationary distribution.

Some of the most important issues in MCMC concern the rate of convergence, that is, the length of the burn-in, and how fast the sampler mixes. These issues are more difficult to assess for multivariate distributions, but it is for multivariate distributions that MCMC is most important. The burn-in can often be much longer than a quick analysis might lead us to expect.

Various diagnostics have been proposed to assess convergence. A general approach to assess convergence is to use multiple simultaneous simulations of the chain and compare the output of the simulations. Large differences in the output would indicate that one or more of the simulations is in the burn-in phase. A related approach using only a single simulation is to inspect and compare separate subsequences or blocks of the output. Large differences in relatively long blocks would indicate that convergence has not occurred.

The results of a method for assessing convergence may strongly indicate that convergence *has not* occurred, but they cannot strongly indicate that convergence *has* occurred. Different methods may be more or less reliable in different settings, but no single method is completely dependable. In practice, the analyst generally should use several different methods and conclude that convergence has occurred only if no method indicates a lack of convergence.

There are many possibilities for assessing convergence of MCMC methods, but unfortunately, the current methodology is not sufficiently reliable to allow decisions to be made on the basis of any standard set of tests (see Chib, 2004, page 96). The careful analyst chooses and performs various ad hoc assessments, often based on exploratory graphics.

Monte Carlo, Iterative Monte Carlo, and Simulation

Convergence of the Monte Carlo estimator $\frac{1}{n}\sum_{i=1}^{n} g(X_i)$ to its expectation $\mathrm{E}(g(X))$ is not the only issue. If the constant c is such that

$$g(c) = \mathrm{E}(g(X)),$$

a random number generator that yields $x_i = c$ in each iteration would yield a very good estimate of $\mathrm{E}(g(X))$. Usually, however, our objectives in using Monte Carlo include obtaining other estimates or assessing the behavior of a random process that depends on the distribution P of X. The degenerate generator yielding $x_i = c$ would not provide these other results. Although it may not be efficient, sometimes it is very important to simulate the underlying random process with all of its variability.

Whenever a correlated sequence such as a Markov chain is used, variance estimation must be performed with some care. In the more common cases of positive autocorrelation, the ordinary variance estimators are negatively biased. The method of batch means or some other method that attempts to account for the autocorrelation should be used.

If the noniterative approach is possible, it is to be preferred. There are many situations in which an MCMC method is easy to devise but performs very poorly. See Robert (1998) for an example of such a problem.

11.2 Simulation of Data from a Hypothesized Model: Monte Carlo Tests

One of the most straightforward methods of computational inference is the *Monte Carlo test.* Barnard (1963) suggested use of Monte Carlo methods to estimate quantiles of a test statistic, T, under the null hypothesis. In Barnard's Monte Carlo test, m random (or pseudorandom) samples of the same size as the given sample are generated under the null hypothesis, and the test statistic is computed from each sample. This yields a sample of test statistics, $t_1^*, \ldots, t_m^*$. The ECDF, P_m^*, of the sample of test statistics is used as an estimate of the CDF of the test statistic, P_T; and the critical region for the test or the p-value of the observed test statistic can be estimated from P_m^*.

An estimate of the p-value of the observed test statistic can be taken as the proportion of the number of simulated values that exceed the observed value.

If the distribution of the test statistic is continuous, and r is the number that exceed the observed value,

$$r/m$$

is an unbiased estimate of the p-value. Because this quantity can be 0, we usually use

$$\frac{r+1}{m+1}$$

as an estimate of the p-value associated with the upper tail of the test statistic. This is also the simple empirical quantile if, as under the null hypothesis, the observed value is from the same distribution. For test statistics with discrete distributions, we must estimate the probability of the observed value, and allocate that proportionally to the rejection and acceptance regions.

The expected power of a Monte Carlo test can be quite good even for relatively small values of m. In simple situations (testing means, for example), $m = 99$ may be a good choice. This allows the p-value to be expressed simply in two decimal places. In more complicated situations (inference concerning higher moments or relationships between variables), a value of $m = 999$ may be more appropriate. The p-value resulting from a Monte Carlo test is an estimate based on a sample of size m, so in general the larger m is, the better the estimate. In practical applications, it is not likely that a decision, other than to gather additional data, would be made based on more than two significant digits in a p-value.

To use a Monte Carlo test, the distribution of the random component in the assumed model must be known, and it must be possible to generate pseudorandom samples from that distribution under the null hypothesis. Notice that a Monte Carlo test is based on an *estimate* of a critical value of the test statistic rather than on an *approximation* of it.

In many applications of statistics, there is no simple model of the phenomenon being studied. If a simple approximation is chosen as the model, subsequent decisions rely on the adequacy of the approximation. On the other hand, if a more realistic model is chosen, the distributions of the statistics used in making inferences are intractable. The common approach is to approximate the distributions using asymptotic approximations. Monte Carlo tests provide an alternative; the distributional properties can be estimated by simulation. Computational inference can replace asymptotic inference. (Of course, in many complicated models, both approaches may be used.) If the sample size is not compatible with the order of the asymptotic approximation, an inferential procedure using Monte Carlo methods is clearly better than one using the approximation.

The ECDF of the simulated test statistic provides us with more information about the test statistic than just the critical values. It allows us to make other inferences about the distribution of the test statistic under the null hypothesis, such as an estimate of the variance of the test statistic, its symmetry, and so on.

An obvious problem with a Monte Carlo test is that the null hypothesis, together with underlying assumptions, must fully specify the distribution at least up to any pivotal quantity used in the test. In Chapter 13, we discuss Monte Carlo methods that involve *resampling* from the given sample; hence, a complete specification of an underlying distribution is not necessary.

11.3 Simulation of Data from a Fitted Model: "Parametric Bootstraps"

Instead of using the hypothesized value of the parameter, another approach in computational inference is to use an estimate of the parameter from the sample. In a similar manner as in the previous section, we can simulate samples from the fitted model to obtain a sample of test statistics $t_1^*, \ldots, t_m^*$. Again, the ECDF, P_m^*, of the sample of test statistics can be used as an estimate of the CDF of the test statistic, P_T; and critical regions for a test, p-values of the observed test statistic, or other properties of the distribution of the test statistic can be estimated from P_m^*. In this case, of course, the distributional properties are not those that hold under a particular hypothesis; rather they are the properties under a model whose parameters correspond to values fitted from the data. This kind of approach to statistical inference is sometimes called a *parametric bootstrap.*

In the parametric bootstrap, the CDF of the population of interest, P, is assumed known up to a finite set of parameters, θ. An estimate of the CDF is P with θ replaced by an estimate $\widehat{\theta}$ obtained from the given sample. Hence, the first step is to obtain estimates of the parameters that characterize the distribution within the assumed family. After this, the procedure is to generate m random samples each of size n from the estimated distribution, and for each sample, compute an estimator T_j^* of the same functional form as the original estimator T. The distribution of the T_j^*'s is used to make inferences about the CDF of T. The estimate of the CDF of T can be used to test hypotheses about θ, using the observed value of T from the original sample. If $f(T, \theta)$ is a pivotal quantity when the distribution of T is known, the estimate of the CDF of T can be used to form confidence intervals for θ.

11.4 Random Sampling from Data

Some statistical methods involve formation of subsets of the data or randomization of the data. The number of subsets or permutations can be very large. For this reason, in the application of such methods, rather than using all possible subsets or all possible permutations, we generally resort to generating random samples of subsets or permutations. Some methods we discuss in Chapters 12 and 13 necessitate use of Monte Carlo sampling.

The discrete uniform population defined by the data is a useful surrogate for the population from which the data were drawn. Properties of the discrete population are used in making an inference about the "real" population. The analysis of the discrete population is often facilitated by drawing samples from it. This is, in effect, a *resampling* of the given data, which is a sample from the "real" population. Some of the bootstrap methods discussed in Chapter 13 use Monte Carlo procedures in this way.

Many other statistical methods involve sampling from the data. For example, in survey sampling, the dataset often includes incomplete records or missing data. In the missing-data problem, we think of the full dataset as being represented by an $n \times d$ matrix Y (that is, n observations, each of which contains d elements), of which a certain portion, Y^{mis}, is actually not observed. The missing portion together with the observed portion, Y^{obs}, constitute the full dataset. For analyzing the data and providing descriptive statistics of the population, it is often desirable to fill in the missing data using complete records as "donors" to impute the missing data. There are various approaches to this problem. In one approach, called *multiple imputation*, m simulated values, $Y_1^{\text{mis}*}, \ldots, Y_m^{\text{mis}*}$, of the missing data are generated from an appropriate population, and the complete datasets, $Y_1^*, \ldots, Y_m^*$, are analyzed. This procedure provides a measure of the uncertainty due to the missing data. (In order for this approach to be valid, the simulated missing data must come from an appropriate distribution. See Rubin, 1987, or Schafer, 1997, for discussions of the properties the distribution must have.) Because multiple imputation only simulates from the missing data portion of the dataset and because the simulation variance is likely to be relatively small compared to the overall sampling variance, the value of m does not need to be large. A value of $m = 3$ is often adequate in multiple imputation.

11.5 Reducing Variance in Monte Carlo Methods

Monte Carlo methods involve a inferences from random (or pseudorandom) samples. The usual principles of inference apply. We seek procedures with small (generally zero) bias and small variance. As with other methods for statistical inference, various procedures with differing bias and variance are available. In sampling from artificially generated random numbers on the computer, just as in taking observations of other events, an objective is to devise a sampling plan that will yield estimators with small variance. It is often possible to modify a procedure to reduce the bias or the variance. There are a number of ways of reducing the variance in Monte Carlo sampling.

As with any statistical estimation procedure, an objective is to choose an estimator and/or a sampling design that will have a small, possibly minimum, variance. The first principle in achieving this objective is to remove or reduce sampling variation wherever possible. This principle is *analytic reduction*. An example that has been considered in the literature (see Ripley, 1987, for example) is the estimation of the probability that a Cauchy random variable is larger than 2; that is, the evaluation of the integral

$$\int_2^\infty \frac{1}{\pi(1+x^2)} \mathrm{d}x.$$

This integral can be transformed analytically to

$$\int_0^{1/2} \frac{y^{-2}}{\pi(1+y^{-2})} \mathrm{d}y,$$

and the variance of a simple estimator of the latter integral using a sample from $\mathrm{U}(0, \frac{1}{2})$ is only about one-thousandth the variance of a simple estimator of the former integral using a sample from a Cauchy distribution. Inspection of the original integral, however, reveals that the antiderivative of the integrand is the arctangent. If reasonable software for evaluating trigonometric functions is available, one should not estimate the integral in the original problem. The rule is *do not resort to Monte Carlo methods unnecessarily.*

Importance Sampling

Given the integral $\int_D f(x)\mathrm{d}x$, there may be a number of ways that we can decompose f into g and a probability density function p. This PDF decomposition determines the variance of our estimator $\widehat{\theta}$. The intuitive rule is to sample more heavily where $|f|$ is large. This principle is called *importance sampling.* As we do following equation (11.1) on page 418, we write the integral as

$$\begin{aligned}\theta &= \int_D f(x)\,\mathrm{d}x \\ &= \int_D \frac{f(x)}{p(x)} p(x)\,\mathrm{d}x.\end{aligned}$$

where $p(x)$ is a probability density over D. The density $p(x)$ is called the *importance function.* The objective in importance sampling is to use an optimal PDF decomposition. We will now proceed to determine *the* optimal PDF decomposition.

From a sample of size m from the distribution with density p, we have the estimator

$$\widehat{\theta} = \frac{1}{m} \sum \frac{f(x_i)}{p(x_i)}. \tag{11.7}$$

It is clear that $\widehat{\theta}$ is unbiased for θ (assuming, of course, that the integral exists). The variance of this estimator is

$$\mathrm{V}(\widehat{\theta}) = \frac{1}{m} \mathrm{V}\left(\frac{f(X)}{p(X)}\right), \tag{11.8}$$

where the variance is taken with respect to the distribution of the random variable X with density $p(x)$. The variance of the ratio can be expressed as

$$\mathrm{V}\left(\frac{f(X)}{p(X)}\right) = \mathrm{E}\left(\frac{f^2(X)}{p^2(X)}\right) - \left(\mathrm{E}\left(\frac{f(X)}{p(X)}\right)\right)^2. \tag{11.9}$$

The objective in importance sampling is to choose p so this variance is minimized. Because

$$\left(\mathrm{E}\left(\frac{f(X)}{p(X)}\right)\right)^2 = \left(\int_D f(x)\,\mathrm{d}x\right)^2, \tag{11.10}$$

the choice involves only the first term in the expression above for the variance. By Jensen's inequality, we have a lower bound on that term:

$$\begin{aligned}\mathrm{E}\left(\frac{f^2(X)}{p^2(X)}\right) &\geq \left(\mathrm{E}\left(\frac{|f(X)|}{p(X)}\right)\right)^2 \\ &= \left(\int_D |f(x)|\,\mathrm{d}x\right)^2.\end{aligned} \tag{11.11}$$

That bound is obviously achieved when

$$p(x) = \frac{|f(x)|}{\int_D |f(x)|\,\mathrm{d}x}. \tag{11.12}$$

This is the optimal PDF decomposition.

Of course, if we knew $\int_D |f(x)|\,\mathrm{d}x$, we would probably know $\int_D f(x)\,\mathrm{d}x$ and would not even be considering a Monte Carlo procedure to estimate the integral. In practice, for importance sampling, we generally seek a probability density p that is nearly proportional to $|f|$ (that is, such that $|f(x)|/p(x)$ is nearly constant).

Control Variates

Another way of reducing the variance, just as in ordinary sampling, is to use covariates, or control variates, as they are often called in Monte Carlo sampling. Any variable that is correlated with the variable of interest has potential value as a control variate. The control variate is useful if it is easy to generate and if it has properties that are known or that can be computed easily.

As an example, consider a method of using control variates to reduce the variance in Monte Carlo tests in two-way contingency tables described by Senchaudhuri, Mehta, and Patel (1995). An $r \times c$ contingency table can be thought of as an $r \times c$ matrix, A, whose nonnegative integer elements a_{ij} represent the counts in the cells of the table. For such tables, we may be interested in patterns of values in the cells. Specifically, we ask whether, given the marginal totals

$$a_{\bullet j} = \sum_{i=1}^{r} a_{ij}$$

and

$$a_{i\bullet} = \sum_{j=1}^{c} a_{ij},$$

the cells are independent. (Here, we use the "dot notation" for summation: $a_{\bullet j}$ is the sum of the counts in the jth column, for example, and $a_{\bullet\bullet}$ is the grand total. Also, below we use the notation a_{*j} to represent the vector that is the jth column.) There are several statistical tests that address this question or aspects of it under various assumptions. The test statistic is some function of the observed table, $T(A)$. The objective is to compute the p-value of the observed value of the test statistic. The distributions of most test statistics for this problem are very complicated, so either an approximation is used or a Monte Carlo test is performed.

A Monte Carlo test involves generation of a large number of random tables that satisfy the null hypothesis and for each table computing the test statistic to determine if it is more extreme than that of the observed table. The problem with the Monte Carlo test, however, is the large amount of computation involved. Both generation of the tables and computation of the test statistic are tedious.

Senchaudhuri et al. suggested use of an additional statistic related to the test statistic of interest to serve as a control variate. The auxiliary statistic, which is easy to compute, has a known mean.

The relationship between the auxiliary statistic and the test statistic relies on a "separability property" of the test statistic that allows the test statistic to be written as

$$T(A) = \sum_{j=1}^{c} T_j(a_{*j}),$$

where a_{*j} is the jth column of A. Given this representation, form $\widetilde{T}_j(a_{*j})$ as the contribution $T_j(a_{*j})$ rounded to p digits,

$$\widetilde{T}_j(a_{*j}) = \lfloor T_j(a_{*j}) \times 10^p + 0.5 \rfloor \times 10^{-p},$$

and form the new test statistic as

$$\widetilde{T}(A) = \sum_{j=1}^{c} \widetilde{T}_j(a_{*j}). \tag{11.13}$$

As an example, consider Pearson's test, a common test for lack of dependence between the rows and columns that uses the test statistic

$$T(A) = \sum_{i=1}^{r} \sum_{j=1}^{c} \frac{(a_{ij} - a_{i\bullet}a_{\bullet j}/a_{\bullet\bullet})^2}{a_{i\bullet}a_{\bullet j}/a_{\bullet\bullet}}.$$

The distribution of this statistic is complicated, but under the null hypothesis, the statistic has an asymptotic chi-squared distribution with $(r-1)(c-1)$ degrees of freedom. The exact significance level of an observed value of this test statistic is often determined by Monte Carlo methods.

Pearson's test statistic has the separability property, so the procedure described above can be used.

Identification of appropriate control variates and other techniques of variance reduction often requires some ingenuity. The techniques are often ad hoc.

Replication of the Inherent Variance

Although generally we want to construct Monte Carlo estimators with small variance, we must be aware that sometimes it is very important to simulate the underlying random process in such a way that variances of the Monte Carlo stream are representative of variances within the process.

Acceleration of Markov Chain Monte Carlo Methods

In MCMC, not only are we concerned with the variance of estimators, but also with the convergence of the chain. Until the stationary distribution is reached, the Monte Carlo sample may not be representative of the target distribution. Even after the stationary distribution is reached, the Monte Carlo samples must be large enough that the Markov chain moves through the state-space sufficiently to ensure that the sample is representative. In many practical applications, MCMC is extremely slow.

The efficiency of MCMC may vary depending on the approach. In some cases, the Gibbs algorithm is more efficient because there is no rejection step as there is in the Metropolis-Hastings method. The setup step in the Gibbs method is more complicated, however, and unless the sampling from the one-dimensional conditionals is very fast, there may be simple Metropolis-Hastings methods that will run faster.

The progress of Markov chain used in Monte Carlo can sometimes be assessed and possibly accelerated by running sequences in parallel. There are various possibilities for exchanging information from one sequence to another, possibly the rates of change of the different sequences. When one sequence seems more stable than another, it may be advantageous to restart other sequences at the state of the more stable sequence.

11.6 Software for Monte Carlo

Monte Carlo studies typically require many repetitive computations, which are usually implemented through looping program-control structures. Some higher-level languages do not provide efficient looping structures. For this reason, it is usually desirable to conduct moderate- to large-scale Monte Carlo studies using a lower-level language such as C or Fortran. Some higher-level languages provide the capability to produce compiled code, which will execute faster. If Monte Carlo studies are to be conducted using an interpretive language, and if the production of compiled code is an option, that option should be chosen for the Monte Carlo work.

Controlling the Seeds in Monte Carlo Studies

There are three reasons why the user must be able to control the seeds in Monte Carlo studies: for testing of the program, for use of blocks in Monte Carlo experiments, and for combining results of Monte Carlo studies.

In the early phases of programming for a Monte Carlo study, it is very important to be able to test the output of the program. To do this, it is necessary to use the same seed from one run of the program to another.

Controlling seeds in a parallel random number generator is much more complicated than in a serial generator. Performing Monte Carlo computations in parallel requires some way of ensuring the independence of the parallel streams.

In Section 7.6, we describe how to control the seed in IMSL and R random number generators.

Notes and Further Reading

Monte Carlo Tests

Several applications of Monte Carlo tests are reported in the literature. Many involve spatial distributions of species of plants or animals. Manly (2006), for example, describes several uses of Monte Carlo tests in biology, some of which, for example, are based on interpoint distances to assess randomness in a spatial distribution. (In Exercise 11.7, you are asked to devise a Monte Carlo test for spatial independence.) Zhu (2005) provides a very comprehensive coverage of theory and applications of Monte Carlo tests.

Hope (1968) and Marriott (1979) studied the power of the test and found that the power of Monte Carlo tests can be quite good even for relatively small values of m. Hall and Titterington (1989) compared the power of Monte Carlo tests with that of tests that use asymptotic approximations. The results obviously depend on the sample size of the observations. Some asymptotic approximations become fairly good for samples as small as 20, while others require sample sizes in the hundreds.

Senchaudhuri, Mehta, and Patel (1995) describe Monte Carlo tests in contingency tables. Ziff (2006) develops a Monte Carlo test for a type of lattice in chemical bond systems. Brigo and Liinev (2005) and Dufour and Khalaf (2002) describe Monte Carlo tests for hypotheses regarding financial models. Forster, McDonald, and Smith (1996) describe conditional Monte Carlo tests based on Gibbs sampling in log-linear and logistic models. Besag and Clifford (1989, 1991) describe randomized significance tests with exact p-values using Markov chain Monte Carlo methods. Dufour (2006) discusses Monte Carlo tests in the presence of nuisance parameters.

Markov Chain Monte Carlo

MCMC has been one of the most active areas of statistical research since around 1990. Chib (2004) provides a summary of much of the work. The most important unsettled issue is the assessment of convergence. Robert (1998) provides an interesting example for the evaluation of various convergence assessment techniques that have been proposed.

The use of Markov chains to form a "proposal" distribution and the applications in Bayesian analyses that depend on this approach are discussed and illustrated extensively in Albert (2007) and Marin and Robert (2007).

Other Applications of Monte Carlo

In this chapter we have been concerned with the use of Monte Carlo as a statistical method for inference. An important and common use of Monte Carlo is to evaluate and compare other statistical methods for inference. This type of use is the subject of Appendix A.

Exercises

11.1. Monte Carlo integration. Use Monte Carlo to evaluate each of the following integrals:

a)
$$\int_0^1 x^2 \mathrm{d}x$$

b)
$$\int_0^1 \int_{-2}^2 x^2 \cos(xy) \mathrm{d}x \mathrm{d}y$$

c)
$$\int_0^\infty \frac{3}{4} x^4 \mathrm{e}^{-x^3/4} \mathrm{d}x$$

11.2. Use Monte Carlo methods to study least squares/normal drift. Let $\mu = 0$ and $\sigma^2 = 1$, and generate a sample of size 100 from a $\mathrm{N}(\mu, \sigma^2)$ distribution. From this sample, compute $\bar{y}^{(1)}$ and $\bar{s}^{2(1)}$. Now let $\mu = \bar{y}^{(1)}$ and $\sigma^2 = \bar{s}^{2(1)}$; generate a sample of size 100 and compute $\bar{y}^{(2)}$ and $\bar{s}^{2(2)}$. Continue in this way, generating the sequences $\{\bar{y}^{(k)}\}$ and $\{\bar{s}^{2(k)}\}$. Describe these stochastic processes.

11.3. Assume that we have a sample, $x_1, \ldots, x_n$ from a $\mathrm{N}(\mu, \sigma^2)$ distribution, and we wish to test the hypothesis

$$\mathrm{H}_0 : \mu = 0$$

versus

$$\mathrm{H}_1 : \mu \neq 0.$$

a) Describe a Monte Carlo test for this hypothesis. *Hint:* Use the standard test statistic for this situation.
b) How would you study the power of this test?

11.4. Suppose a random sample of size n is taken from what is believed to be a double exponential distribution, that is, a distribution with density

$$f(y) = \frac{1}{2\theta} \mathrm{e}^{-|y|/\theta}.$$

All you have available from the sample are the mean m and the second and fourth central moments:

$$m_2 = \sum (y_i - m)^2 / n,$$

$$m_4 = \sum (y_i - m)^4 / n.$$

Describe a test, at the 0.05 level of significance, of the hypothesis that the original sample came from a distribution with that density.

11.5. Consider a common application in statistics: Three different treatments are to be compared by applying them to randomly selected experimental units. This, of course, usually leads us to "analysis of variance" using a model such as $y_{ij} = \mu + \alpha_i + e_{ij}$ with the standard meanings of these symbols and the usual assumptions about the random component e_{ij} in the model. Suppose that instead of the usual assumptions, we assume that the e_{ij} have independent and identical double exponential distributions centered on zero.
a) Describe how you would perform a Monte Carlo test instead of the usual AOV test. Be clear in stating the alternative hypothesis.
b) Is the Monte Carlo test that you described nonparametric? Describe some other computer-intensive test that you could use even if you make no assumptions about the distribution of the e_{ij}.

11.6. It is often said in statistical hypothesis testing that "if the sample size is large enough, any test will be significant".
Is that true?
Study this by Monte Carlo methods. Use a two-sample t test for equality of means on some data you generate from $\mathrm{N}(\mu, \sigma^2)$ and $\mathrm{N}(\mu + \delta, \sigma^2)$. Set $\delta = 0$, and generate various sizes of samples n_1 and n_2. Does the size of your test increase as the sample sizes increase? (It should not, despite the quote from the folklore.) What about the power? (It should, of course; that is, the sensitivity of the test to δ should increase.) There are, of

course, valid reasons to be aware of the increasing chance of rejecting the null hypothesis because of a large sample size. The analyst must balance statistical significance with practical significance.

11.7. A biologist studying the distribution of maple trees within New England forests is interested in how the trees tend to cover areas within the forests. The question is whether mature trees tend to be at maximal spatial separation or tend to cluster by some spatial measure (perhaps due to prevailing winds or other environmental factors). The biologist plotted a given area in a forest of mature maple trees and located each tree in a rectangular coordinate system. The area contained 47 trees.

a) Describe how the biologist might construct a Monte Carlo test of the null hypothesis that the trees are randomly distributed. There are several possibilities for constructing a Monte Carlo test. (There are also alternative procedures that do not involve Monte Carlo tests.) The test statistic might be various functions of the cartesian coordinates of the trees (such as distances between trees).

b) Now, to be specific and to change the problem to be much smaller, suppose that in a given area on which a coordinate system has been imposed in such a way that the area constitutes a unit square, trees were observed at the following 5 points:

(.2, .3)
(.8, .3)
(.3, .7)
(.4, .5)
(.7, .9)

Using this random sample, develop and perform a Monte Carlo test based on interpoint distances that the positions of the trees are independent of one another. (In Chapter 12, we will discuss a randomization test for this hypothesis, and we will consider the problem again in Exercise 12.1.)

11.8. Suppose that we want to use Monte Carlo methods to compare the variances of two estimators, T_1 and T_2. In a simple approach to the problem, we wish to estimate the sign of $\mathrm{V}(T_1) - \mathrm{V}(T_2)$. Suppose that it is known that both of the estimators are unbiased. Why is it better to compute the Monte Carlo estimate as $\mathrm{V}(T_1 - T_2)$ rather than as $\mathrm{V}(T_1) - \mathrm{V}(T_2)$?

12

Data Randomization, Partitioning, and Augmentation

Although subsampling, resampling, or otherwise rearranging a given dataset cannot increase its information content, these procedures can sometimes be useful in extracting information. Randomly rearranging the observed dataset, for example, can give an indication of how unusual the dataset is with respect to a given null hypothesis. This idea leads to randomization tests.

There are many useful procedures for data analysis that involve partitioning the original sample. Using subsets of the full sample, we may be able to get an estimate of the bias or the variance of the standard estimator or test statistic without relying too heavily on the assumptions that led to that choice of estimator or test statistic. It is often useful to partition the dataset into two parts and use the data in the "training set" or "estimation set" to arrive at a preliminary estimate or fit and then use the data in the "validation set" or "test set" to evaluate the fit. This kind of approach is particularly appropriate when we are unsure of our model of the data-generating process. In actual applications, of course, we are always at least somewhat unsure of our model. If the full dataset is used to fit the model, we are very limited in the extent to which we can validate the model.

No matter what we do with a given dataset, we are still left with uncertainty about the relevance of that dataset to future modeling problems. Prediction requires some assumption about the model of the data-generating processes, both the one that yielded the given data and the unseen one for which inferences are to be made. The variance of predictors is called *prediction error* or *generalization error*. Obviously, since this involves unseen data and unknown scenarios, there is no way of measuring or estimating this kind of error with any confidence. The use of partitions of the given dataset, however, is one way of getting some feel for the generalization error for scenarios that are somewhat similar to the one that gave rise to the observed dataset.

Subsets of the data can be formed systematically or they can be formed as random samples from the given dataset. Sometimes the given dataset is viewed as a set of mass points of a finite distribution whose distribution function is the same as the empirical distribution function of the given dataset. In this

J.E. Gentle, *Computational Statistics*, Statistics and Computing,
DOI: 10.1007/978-0-387-98144-4_12,

case, the data partitioning may be done in such a way that observations may occur multiple times in the various partitions. In most cases, when we speak of "sets" or "subsets", we mean "multisets" (that is, collections in which items may occur more than once and are distinguished in each occurrence).

12.1 Randomization Methods

The basic idea of randomization tests is to compare an observed configuration of outcomes with all possible configurations. The randomization procedure does not depend on assumptions about the underlying probability distribution, so it is usable in a wide range of applications. When a hypothesis of interest does not have an obvious simple test statistic, a randomization test may be useful. It is important to note, however, that the procedure does depend on the overall data-generating process. The data collection process and the sampling design must be respected in any randomization of the data. For clinical trials, for example, it is unlikely that randomization methods could be used in the analysis.

In straightforward applications of randomization tests, the null hypothesis for the test is that all outcomes are equally likely, and the null hypothesis is rejected if the observed outcome belongs to a subset that has a low probability under the null hypothesis but a relatively higher probability under the alternative hypothesis.

A simple example of a randomization test is a test of whether the means of two data-generating processes are equal. The decision would be based on observations of two samples of results using the two treatments. There are several statistical tests for this null hypothesis, both parametric and nonparametric, that might be used. Most tests would use either the differences in the means of the samples, the number of observations in each sample that are greater than the overall mean or median, or the overall ranks of the observations in one sample. Any of these test statistics could be used as a test statistic in a randomization test. Consider the difference in the two sample means, for example. Without making any assumptions about the distributions of the two populations, the significance of the test statistic (that is, a measure of the extremeness of the observed difference) can be estimated by considering all configurations of the observations among the two treatment groups. This is done by computing the same test statistic for each possible arrangement of the observations, and then ranking the observed value of the test statistic within the set of all computed values.

More precisely, consider two samples,

$$x_1, \ldots, x_{n_1},$$

$$y_1, \ldots, y_{n_2},$$

for which we want to test the equality of the respective population means. We choose the unscaled test statistic $t_0 = \bar{x} - \bar{y}$. Now consider a different configuration of the same set of observations,

$$y_1, x_2, \ldots, x_{n_1},$$

$$x_1, y_2, \ldots, y_{n_2},$$

in which an observation from each set has been interchanged with one from the other set. The same kind of test statistic, namely the difference in the sample means, is computed. Let t_1 be the value of the test statistic for this combination. Now, consider a different configuration in which other values of the original samples have been switched. Again, compute the test statistic. Continuing this way through the full set of x's, we would eventually obtain $\binom{n_1 + n_2}{n_2}$ different configurations and a value of the test statistic for each one of these artificial samples. Without making any assumptions about the distribution of the random variable corresponding to the test statistic, we can consider the set of computed values to be a realization of a random sample from that distribution under the null hypothesis. The empirical "significance" of the value corresponding to the observed configuration could then be computed simply as the rank of the observed value in the set of all values.

In Exercise 11.7 on page 433, we considered the problem of deciding whether the locations of trees within a field were randomly distributed. This question can be addressed by a Monte Carlo test, as suggested in the exercise, or by a randomization test (or by several other methods).

Mead (1974) described a randomization test based on counts within nested grids. If the field is divided into four quadrants as shown by the solid lines in Figure 12.1, and then each quadrant is divided as shown by the dashed lines, the uniformity of the distribution can be assessed by comparing the counts within the two levels of gridding.

Mead's test is based on a ratio of measures of variation of counts within the larger grids to variation of counts within the smaller grids. Let n_{ij} be the count of observations within the jth grid cell of the ith large cell, and let $\bar{n}_{i\bullet}$ be the mean within the ith large cell and $\bar{n}_{\bullet\bullet}$ be the overall mean, using the common AOV notation. There are various measures that could be used. One measure of overall variation among the small grids is the total sum of squares,

$$t = \sum_i \sum_j (n_{ij} - \bar{n}_{\bullet\bullet})^2,$$

and a measure of variation among the small grids within the larger grids is the sum of squares,

$$w = 4 \sum_i (\bar{n}_{i\bullet} - \bar{n}_{\bullet\bullet})^2.$$

A test statistic is

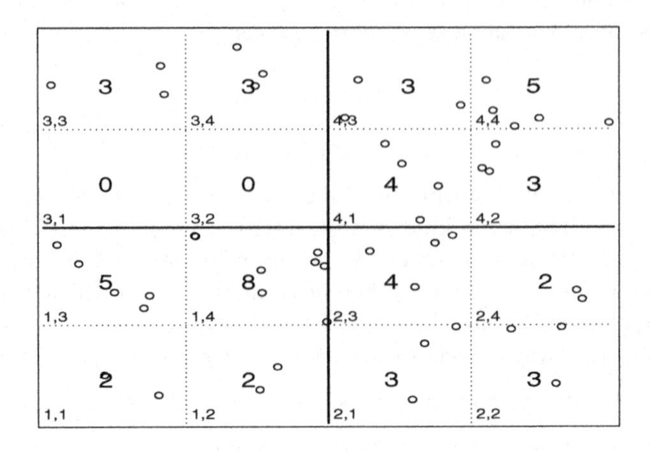

Fig. 12.1. Quadrants and Subquadrants i, j with Cell Counts for a Randomization Test

$$Q = w/t. \tag{12.1}$$

Notice the similarity of Q to the F statistic in ANOVA. The numerator in both is the pooled among-sums-of-squares, but in Q the denominator is the total-sum-of-squares, rather than the error-sum-of-squares.

Clearly, values of Q that are either very small (close to 0) or very large (close to 1) indicate nonuniformity. In the example shown in Figure 12.1, $Q = 0.309$.

Although an approximation to the distribution of Q under the null hypothesis may be possible, a randomization approach can provide an accurate value for the significance level of the observed statistic. We consider values of Q computed by all possible arrangements of the small grid cells into the large grid. The total number of arrangements of the 16 counts and the ways of organizing them into 4 quadrants with 4 subquadrants is $16!/(4!)^4$, which is a very large number. (Of course, in our small example, because of the duplication of counts, the effective number of arrangements is much smaller.) Instead of computing Q for all possible arrangements, we may consider only a random sample of the arrangements. In this sampling for the randomization test, we generally do sampling with replacement because of the bookkeeping involved otherwise.

In this example, in a sample of 5,000 arrangements, the observed value of Q was near the second quartile, so this partitioning provided no evidence of nonrandomness. Mead's test, however, proceeds to further subdivisions. Each of the quadrants in Figure 12.1 could be divided into quadrants and subquadrants, and Q statistics could be computed for each as shown in Figure 12.2.

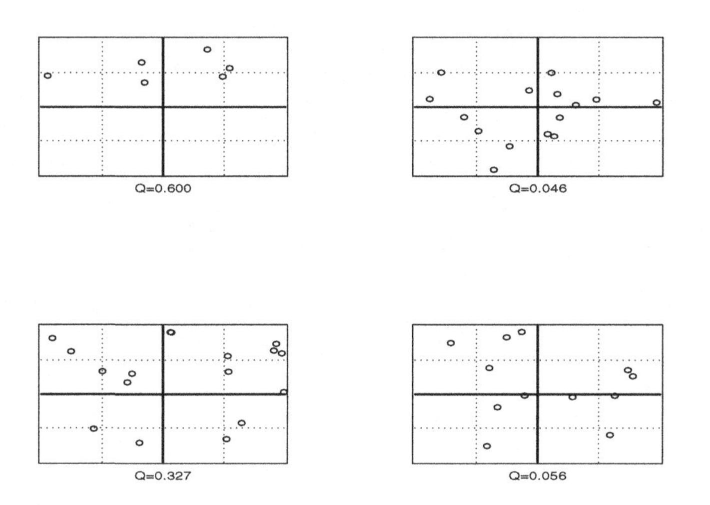

Fig. 12.2. Further Subdivision into Quadrants and Subquadrants

There are various ways to handle the statistics for the smaller quadrants. One way is to take the average of the individual values as a single value for Q. There are many ways of doing the randomization that results in an unbiased test. It is difficult to make statements about the omnibus power of this test or any other test for randomness.

If tests from different levels of subdivision are combined, some kind of Bonferroni bound may be used on the significance level. Randomization, however, can allow us to avoid complications of multiple tests if the statistics computed in each randomization are the same as those computed on the observed sample.

Mead's test may have some difficulties caused by edge effects that occur because of the original square (or rectangular) outline we impose on the field. It may be advisable to compute the test statistic for different grids that have just been shifted slightly.

In many applications of randomization methods, as in the one above, because there may be a very large number of all possible configurations, it may be necessary to sample randomly from the possible configurations rather than considering them all. When a sample of the configurations is used, the test is sometimes called an "approximate randomization test".

12.2 Cross Validation for Smoothing and Fitting

Cross validation methods are used with a variety of statistical procedures, such as regression, time series analysis, and density estimation, that yield a "fit" at each observation.

Consider the problem of fitting Y using X; that is, the problem of determining a function $g_{X,Y}(x)$ such that $Y \sim g_{X,Y}(X)$. For a given point, (x_0, y_0), how well does $g_{X,Y}(x_0)$ match y_0? The goodness of the match probably depends on whether the point (x_0, y_0) was used in determining the function g. If that point is used in fitting, then it is likely that $g_{X,Y}(x_0)$ is closer to y_0 than it would be if the point (x_0, y_0) is not used in the fitting. Our interest, of course, is in how well our fitted model $g_{X,Y}(x)$ would perform at new points; that is, how useful the model is in *prediction.*

Let $R(y, g)$ be a measure of the error between an observed value y and the predicted value g. (Often, R is the L_2 norm of the difference of the observed and the predicted. In the univariate case, this is just the square, $(y - g)^2$.) We are interested in the expected value of this error. More precisely, we are interested in the expected value with respect to the conditional distribution of Y given X; that is,

$$\mathrm{E}_{P_{Y|X}}\Big(R(Y_0,\, g_{X,Y}(x_0))\Big). \tag{12.2}$$

Of course, we do not know $P_{Y|X}$.

The fitted function $g_{X,Y}(x)$ provides an estimate of the conditional distribution of Y given X, $\widehat{P}_{Y|X}$, so we can evaluate the expectation with respect to this distribution. We could estimate it as the sample average:

$$\mathrm{E}_{\widehat{P}_{Y|X}}\Big(R(Y_0,\, g_{X,Y}(x_0))\Big) = \frac{1}{n}\sum_{i=1}^{n} R(y_i,\, g_{X,Y}(x_i)). \tag{12.3}$$

This quantity, which is easy to compute, is the "apparent error". It is typically smaller than the true error at any point x_0 whether or not a point corresponding to x_0 was in the dataset used in fitting $g_{X,Y}$. The fit is usually chosen to minimize a sum such as in equation (12.3).

We may arrive at a better estimate if we partition the dataset into two parts, say S_1 and S_2, and use the data in the training set or "estimation set" S_1 to get the fit $g_{1X,Y}$ and then use the data in the "validation set" or "test set" S_2 to estimate the error:

$$\mathrm{E}_{\widehat{P}_{1Y|X}}\Big(R(Y_0,\, g_{X,Y}(X_0))\Big) = \frac{1}{\#(S_1)}\sum_{i\in S_2} R(y_i,\, g_{1X,Y}(x_i)). \tag{12.4}$$

This quantity is likely to be larger than the quantity in equation (12.3).

We can also get an estimate after exchanging the roles of S_1 and S_2 and then combine the estimates:

$$\mathrm{E}_{\widetilde{P}_{Y|X}}\Big(R(Y_0,\, g_{X,Y}(X_0))\Big) = \frac{1}{n}\left(\sum_{i\in S_2} R(y_i,\, g_{1X,Y}(x_i)) + \sum_{i\in S_1} R(y_i,\, g_{2X,Y}(x_i))\right). \tag{12.5}$$

This is an old idea. An extension and related idea is balanced half-sampling, which is a technique that is often used in finite-population sampling.

This kind of data partitioning is the idea behind cross validation. Instead of dividing the sample in half, we could form multiple partial datasets with overlap. One way would be to leave out just one observation at a time. The idea is to hold out one (or more) observation(s) at a time, apply the basic procedure, and compare the fitted value with the observed value. In K-fold cross validation, the sample is divided into K approximately equal-sized subsets, and each subset is used to get a measure of the prediction error by using the fit from all of the rest of the data. The average from the K subsets is then taken as the estimate of the prediction error.

Cross validation can be useful in model building. In regression model building, the standard problem is, given a set of potential regressors, to choose a relatively small subset that provides a good fit to the data. Standard techniques include "stepwise" regression and all best subsets. One of the main problems in model building is overfitting. In regression models, the more independent variables are included in the model, the better the fit unless the observations on the added variables do not increase the rank of the coefficient matrix. This is a simple consequence of having more decision variables in the optimization problem that is used to fit the model. Various statistics, such as the adjusted R^2 or C_p, can be used to determine when the improvement of the fit due to an additional variable is "worthwhile". These statistics are based on penalties for the number of variables in the model.

Allen (1971, 1974) suggested a cross-validation "prediction sum of squares", PRESS, to aid in variable selection in full-rank linear models that are fit using a least squares criterion. PRESS is similar to the sums in equation (12.5) except that instead of just one partition into two sets, the dataset is partitioned n times into an estimation set with $n-1$ observation and a test set with only one observation. In the linear model $y = X\beta$, where y is an n-vector of observations, X is an $n \times m$ matrix of corresponding observations, and β is an m-vector of parameters, using the notation $\widehat{\beta}_{-j}$ to denote the least squares estimate of β based on all but the jth observation, PRESS is defined as

$$\sum_{j=1}^{n}(y_j - x_j^{\mathrm{T}}\widehat{\beta}_{-j})^2, \tag{12.6}$$

where x_j^{T} is the jth row of X (that is, the one that was not used in computing $\widehat{\beta}_{-j}$).

As more variables are added to the model, PRESS may decrease initially and then begin to increase when overfitting occurs. It is important to understand that variable selection and model fitting are very different from inferential statistical procedures. Our usual concerns about bias, power, significance levels, and so on just do not apply. The problem is because to make statements about such things we must have a model, which we do not have if we are building a model.

Although computation of PRESS involves n regression fits, when efficient updating and downdating techniques are used, the additional amount of computation is only about twice the amount of computation to do a single regression fit.

In fitting models to data, it is often appropriate to limit the range of influence of observations. For example, in estimation of a probability density function, observations at one extreme of the distribution may provide very little information about the shape of the density in a different range of the distribution. The range of influence of observations in a statistical procedure is often controlled by "smoothing" parameters. A smoothing parameter may be the width of an interval to use in constructing a frequency function, for example. Another example of a smoothing parameter is the number of knots in splines. Cross validation is a common method of selecting smoothing parameters. Further examples of smoothing parameters include such things as bin widths in histograms or window sizes for various kernel methods.

12.3 Jackknife Methods

Jackknife methods make use of systematic partitions of a dataset to estimate properties of an estimator computed from the full sample. Quenouille (1949, 1956) suggested the technique to estimate the bias of an estimator. John Tukey coined the term "jackknife" to refer to the method, and showed that the method is also useful in estimating the variance of an estimator.

Suppose that we have a random sample, $Y_1, \ldots, Y_n$, from which we compute a statistic T as an estimator of a parameter θ in the population from which the sample was drawn. In the jackknife method, we partition the given dataset into r groups, each of size k. (For simplicity, we will assume that the number of observations n is kr.)

Now, we remove the jth group from the sample and compute the estimator from the reduced sample. Let $T_{(-j)}$ denote the estimator computed from the sample with the jth group of observations removed. (This sample is of size $n-k$.) The estimator $T_{(-j)}$ has properties similar to those of T. For example, if T is unbiased, so is $T_{(-j)}$. If T is not unbiased, neither is $T_{(-j)}$; its bias, however, is likely to be different.

The mean of the $T_{(-j)}$,

$$\overline{T}_{(\bullet)} = \frac{1}{r}\sum_{j=1}^{r} T_{(-j)}, \tag{12.7}$$

can be used as an estimate of θ. The $T_{(-j)}$ can also be used in some cases to obtain more information about the estimator T from the full sample. (For the case in which T is a linear functional of the ECDF, then $\overline{T}_{(\bullet)} = T$, so the systematic partitioning of a random sample will not provide any additional information.)

Consider the weighted differences in the estimate for the full sample and the reduced samples:

$$T_j^* = rT - (r-1)T_{(-j)}. \tag{12.8}$$

The T_j^* are called "pseudovalues". (If T is a linear functional of the ECDF and $k = 1$, then $T_j^* = T(x_j)$; that is, it is the estimator computed from the single observation, x_j.) We call the mean of the pseudovalues the "jackknifed" T and denote it as $\mathrm{J}(T)$:

$$\begin{aligned} \mathrm{J}(T) &= \frac{1}{r}\sum_{j=1}^{r} T_j^* \\ &= \overline{T}^*. \end{aligned} \tag{12.9}$$

We can also write $\mathrm{J}(T)$ as

$$\mathrm{J}(T) = T + (r-1)\left(T - \overline{T}_{(\bullet)}\right)$$

or

$$\mathrm{J}(T) = rT - (r-1)\overline{T}_{(\bullet)}. \tag{12.10}$$

In most applications of the jackknife, it is common to take $k = 1$, in which case $r = n$. It has been shown that this choice is optimal under certain assumptions about the population (see Rao and Webster, 1966).

Jackknife Variance Estimate

Although the pseudovalues are not independent (except when T is a linear functional), we treat them as if they were independent, and use $\mathrm{V}(\mathrm{J}(T))$ as an estimator of the variance of T, $\mathrm{V}(T)$. The intuition behind this is simple: a small variation in the pseudovalues indicates a small variation in the estimator. The sample variance of the mean of the pseudovalues can be used as an estimator of $\mathrm{V}(T)$:

$$\widehat{\mathrm{V}(T)}_{\mathrm{J}} = \frac{\sum_{j=1}^{r}\left(T_j^* - \mathrm{J}(T)\right)^2}{r(r-1)}. \tag{12.11}$$

(Notice that when T is the mean and $k = 1$, this is the standard variance estimator.) From expression (12.11), it may seem more natural to take $\widehat{\mathrm{V}(T)}_{\mathrm{J}}$ as an estimator of the variance of $\mathrm{J}(T)$, and indeed it often is.

A variant of this expression for the variance estimator uses the original estimator T:

$$\frac{\sum_{j=1}^{r}(T_j^* - T)^2}{r(r-1)}. \tag{12.12}$$

There are several methods of estimating or approximating the variances of estimators, including the delta method we discussed on page 50, which depends on knowing the variance of some simpler statistic; Monte Carlo methods, discussed in Chapter 11; and bootstrap methods, discussed in Chapter 13. How good any of these variance estimates are depends on the estimator T and on the underlying distribution. Monte Carlo studies indicate that $\widehat{\mathrm{V}(T)}_{\mathrm{J}}$ is often conservative; that is, it often overestimates the variance (see Efron, 1982, for example). The alternate expression (12.12) is greater than or equal to $\widehat{\mathrm{V}(T)}_{\mathrm{J}}$, as is easily seen; hence, it is an even more conservative estimator (see Exercises 12.4 and 12.5).

Jackknife Bias Correction

In the following, for simplicity, we will consider the group sizes to be 1; that is, we assume that $r = n$. As we mentioned above, this is the most common case in practice, and it has certain optimality properties.

Suppose that we can represent the bias of T as a power series in n^{-1}; that is,

$$\begin{aligned}\mathrm{Bias}(T) &= \mathrm{E}(T) - \theta \\ &= \sum_{q=1}^{\infty} \frac{a_q}{n^q},\end{aligned} \tag{12.13}$$

where the a_q do not involve n. If all $a_q = 0$, the estimator is unbiased. If $a_1 \neq 0$, the order of the bias is n^{-1}. (Such an estimator is sometimes called "second-order accurate". "First-order" accuracy implies a bias of order $n^{-1/2}$.)

Using the power series representation for the bias of T, we see that the bias of the jackknife estimator is

$$\begin{aligned}\mathrm{Bias}(\mathrm{J}(T)) &= \mathrm{E}(\mathrm{J}(T)) - \theta \\ &= n(\mathrm{E}(T) - \theta) - \frac{n-1}{n}\sum_{j=1}^{n}\mathrm{E}(T_{(-j)} - \theta) \\ &= n\sum_{q=1}^{\infty}\frac{a_q}{n^q} - (n-1)\left(\sum_{q=1}^{\infty}\frac{a_q}{(n-1)^q}\right) \\ &= a_2\left(\frac{1}{n} - \frac{1}{n-1}\right) + a_3\left(\frac{1}{n^2} - \frac{1}{(n-1)^2}\right) + \ldots \\ &= -a_2\left(\frac{1}{n(n-1)}\right) + a_3\left(\frac{1}{n^2} - \frac{1}{(n-1)^2}\right) + \ldots;\end{aligned} \tag{12.14}$$

that is, the bias of the jackknife estimator, $\text{Bias}(\text{J}(T))$, is at most of order n^{-2}. If $a_q = 0$ for $q = 2, \ldots$, the jackknife estimator is unbiased.

This reduction in the bias is a major reason for using the jackknife. Any explicit analysis of the bias reduction, however, depends on a representation of the bias in a power series in n^{-1} with constant coefficients. This may not be possible, of course.

From

$$\text{E}(\text{J}(T)) - \theta = \text{E}(T) - \theta + (n-1)\left(\text{E}(T) - \frac{1}{n}\sum_{j=1}^{n}\text{E}(T_{(-j)})\right),$$

we have the jackknife estimator of the bias in T,

$$B_\text{J} = (n-1)\left(\overline{T}_{(\bullet)} - T\right), \tag{12.15}$$

and the jackknife bias-corrected estimator of θ,

$$T_\text{J} = nT - (n-1)\overline{T}_{(\bullet)}. \tag{12.16}$$

Higher-Order Bias Corrections

Suppose that we pursue the bias correction to higher orders by using a second application of the jackknife. The pseudovalues are

$$T_j^{**} = n\text{J}(T) - (n-1)\text{J}(T_{(-j)}). \tag{12.17}$$

Assuming the same series representations for the bias as before, a second-order jackknife estimator,

$$\text{J}^2(T) = \frac{n^2\text{J}(T) - (n-1)^2\sum_{j=1}^{n}\text{J}(T)_{(-j)}/n}{n^2 - (n-1)^2}, \tag{12.18}$$

is unbiased to order $\text{O}(n^{-3})$.

There are two major differences between this estimator and the first-order jackknifed estimator. For the first-order jackknife, $\text{J}(T)$ differs from T by a quantity of order n^{-1}; hence, if T has variance of order n^{-1} (as we usually hope), the variance of $\text{J}(T)$ is asymptotically the same as that of T. In other words, the bias reduction carries no penalty in increased variance. This is not the case for higher-order bias correction of $\text{J}^2(T)$.

The other difference is that in the bias expansion,

$$\text{E}(T) - \theta = \sum_{q=1}^{\infty} a_q/n^q,$$

if $a_q = 0$ for $q \geq 2$, then the first-order jackknifed estimator is unbiased. For the second-order jackknifed estimator, even if $a_q = 0$ for $q \geq 3$, the estimator may not be unbiased. Its bias is

$$\text{Bias}(\text{J}^2(T)) = \frac{a_2}{(n-1)(n-2)(2n-1)}; \tag{12.19}$$

that is, it is still of order n^{-3}.

The Generalized Jackknife

Schucany, Gray, and Owen (1971) suggested a method of systematically reducing the bias by combining higher-order jackknives. First, consider two biased estimators of θ, T_1 and T_2. Let

$$w = \frac{\mathrm{Bias}(T_1)}{\mathrm{Bias}(T_2)}.$$

Now, consider the estimator

$$T_w = \frac{T_1 - wT_2}{1 - w}. \tag{12.20}$$

We have

$$\begin{aligned}
\mathrm{E}(T_w) &= \frac{1}{1-w}\mathrm{E}(T_1) - \frac{w}{1-w}\mathrm{E}(T_2) \\
&= \frac{1}{1-w}(\theta + \mathrm{Bias}(T_1)) - \frac{w}{1-w}(\theta + \mathrm{Bias}(T_2)) \\
&= \theta,
\end{aligned}$$

so this weighted combination of the estimators is unbiased.

Now, consider the biases of the jackknifed estimators, from equations (12.14) and (12.19),

$$\mathrm{Bias}(\mathrm{J}(T)) = -\frac{a_2}{n(n-1)} + \mathrm{O}(n^{-3})$$

and

$$\mathrm{Bias}(\mathrm{J}^2(T)) = -\frac{a_2}{(n-1)(n-2)(2n-1)} + \mathrm{O}(n^{-3}),$$

and let

$$w = \frac{\mathrm{Bias}(\mathrm{J}(T))}{\mathrm{Bias}(\mathrm{J}^2(T))}.$$

Notice that if $w = (n-1)/n$, then the jackknife estimator,

$$nT - (n-1)\overline{T}_{(-j)},$$

is unbiased. This suggests a different second-order jackknife instead of the one in equation (12.18). Schucany, Gray, and Owen (1971) therefore set

$$\begin{aligned}
w &= \frac{\frac{1}{n(n-1)}}{\frac{1}{(n-1)(n-2)}} \\
&= \frac{n-2}{n}
\end{aligned}$$

and take

$$\mathrm{J}^2(T) = \frac{n}{2}\mathrm{J}(T) - \frac{n-2}{2}\sum_{j=1}^{n} T^*_{(-j)}/n \tag{12.21}$$

as the second-order jackknifed estimator.

So, generalizing, and writing $T_1 = T$ and $T_2 = \overline{T}_{(-j)}$, we jackknife T_1 by the ratio of the determinants

$$\mathrm{J}(T_1) = \frac{\begin{vmatrix} T_1 & T_2 \\ 1/n & 1/(n-1) \end{vmatrix}}{\begin{vmatrix} 1 & 1 \\ 1/n & 1/(n-1) \end{vmatrix}}.$$

Suppose that for two estimators, T_1 and T_2, we can express the biases as

$$\mathrm{E}(T_1) - \theta = f_1(n)b$$

and

$$\mathrm{E}(T_2) - \theta = f_2(n)b.$$

We define the generalized jackknife of T_1 as

$$\begin{aligned} \mathrm{J}(T_1) &= \frac{\begin{vmatrix} T_1 & T_2 \\ f_1(n) & f_2(n) \end{vmatrix}}{\begin{vmatrix} 1 & 1 \\ f_1(n) & f_2(n) \end{vmatrix}} \\ &= \frac{1}{1-w}T_1 - \frac{w}{1-w}T_2, \end{aligned} \tag{12.22}$$

where

$$w = \frac{f_1(n)}{f_2(n)}.$$

The higher-order generalized jackknife estimators can be developed by writing the bias of the jth estimator as

$$\mathrm{E}(T_j) - \theta = \sum_{i=1}^{\infty} f_{ij}(n)b_i$$

for $j = 1, \ldots, k+1$. Then

$$\mathrm{J}(T_k) = \frac{\begin{vmatrix} T_1 & T_2 & \cdots & T_{k+1} \\ f_{11}(n) & f_{12}(n) & \cdots & f_{1,k+1}(n) \\ & & \vdots & \\ f_{k1}(n) & f_{k2}(n) & \cdots & f_{k,k+1}(n) \end{vmatrix}}{\begin{vmatrix} 1 & 1 & \cdots & 1 \\ f_{11}(n) & f_{12}(n) & \cdots & f_{1,k+1}(n) \\ & & \vdots & \\ f_{k1}(n) & f_{k2}(n) & \cdots & f_{k,k+1}(n) \end{vmatrix}},$$

where $T_2, \ldots, T_{k+1}$ are the means of the estimators of the successive jackknifed estimators from the reduced samples.

The generalized jackknife reduces the order of the bias by $1/n$ in each application and if all terms beyond the kth in the expansion of the bias are zero, then $J(T_k)$ is unbiased. The variance of the jackknifed estimator may increase, however.

The Delete-k Jackknife

Although as we mentioned earlier it has been shown that deleting one observation at a time is optimal under certain assumptions about the population, it does not lead to a consistent procedure for some estimators that are not differentiable functions of the sample. An example, considered by Efron and Tibshirani (1993), is the jackknife estimate of the variance of the sample median. Because in leaving out one observation at a time, the median of the reduced samples will only take on at most two different values, the jackknife procedure cannot lead to a good estimate of the variance. This is obviously the case, no matter how large is the sample size.

In a practical sense, the consistency is not the issue. The jackknife does not perform well in finite samples in this case. (Note that "performing well" in real applications may not be very closely related to consistency.)

Instead of deleting a single observation, we can form pseudo-observations by deleting k observations. This leads to the "delete-k jackknife". The delete-k jackknifed estimator may be consistent for a wider range of estimators than is the delete-one jackknife. The delete-k jackknifed estimator is consistent in the case of estimation of the variance of the median under certain conditions. The asymptotics require that k also gets large. For the median, the requirements are $n^{1/2}/k \to 0$ and $n - k \to \infty$; see Efron and Tibshirani (1993). (See also Shao and Tu, 1995, for further discussion of properties of the delete-k jackknife.)

In the delete-k jackknife, if each subset of size k is deleted, this could lead to a large number of pseudo-observations; we have the number of combinations of k items from a set of n items. The large number of pseudo-observations can be accommodated by random sampling, however; that is, we do not form the exact mean of all pseudo-observations.

Notes and Further Reading

The idea of partitioning data is an old one. It may be done prior to collecting data and, in this case, the statistician may use principles of experimental design to enhance the power of the procedure. "Half samples" have been used for many years in sampling finite populations. The immediate purpose is to get better estimates of variances, but a more general purpose is to assess the validity of assumptions and the quality of estimates.

The use of a training set and a test set has been standard procedure in classification and machine learning for years. Various ways of selecting and using multiple training sets have also been proposed; see Amit and Geman (1997) and Breiman (2001), for example. We will also discuss some of these methods, called boosting, bagging, arcing, and random forests, again in Chapter 17.

In building a model of a data-generating process, generally the more complicated we allow the model to become, the better the model will appear to fit a given set of data. Rather than using some measure of the goodness-of-fit that depends on all of the data at once, such as an R-squared, it is much more sensible to use a criterion such as PRESS that depends on the ability of the model to fit data that were not included in the fitting process. The latter approach uses partitions of the data. Picard and Berk (1990) discuss and give examples of various uses of data partitioning in statistical inference.

Randomization tests have been used on small datasets for a long time. R. A. Fisher's famous "lady tasting tea" experiment (Fisher, 1935) used a randomization test. Because such tests can require very extensive computations, however, their use has been limited until recently. Edgington and Onghena (2007) give an extensive description of randomization tests and their applications.

Data from spatial point processes are particularly difficult to analyze. There are no really effective tests to distinguish a completely random distribution of locations from various models of relationships among the locations. This is one reason that the methods of statistical inference for such processes are often based on randomization or Monte Carlo methods. Møller and Wasgepetersen (2004) discuss various models of spatial point processes and methods of simulating data from the models.

The generalized jackknife was developed by Gray and Schucany (1972). See their paper and Sharot (1976) for further discussions of it.

Exercises

12.1. Consider again the problem of Exercise 11.7 on page 433, in which we must decide whether the locations of trees within a field were randomly distributed. In that exercise, you were to develop a Monte Carlo test, possibly based on distances between pairs of trees.

  a) Write a program in Fortran, C, or a higher-level language to compute the randomization test statistic of Mead, allowing three levels of nested grids.
  b) Design and conduct a small Monte Carlo study to compare the Monte Carlo test of Exercise 11.7 with the Mead randomization test. The important issue here is the omnibus alternative hypothesis (that the distribution is not randomly uniform). There are many ways that the distribution could be nonrandom, and it is possible that the relative

performance of the tests is dependent on the nature of the nonrandomness. Define and study some different types of nonrandomness. One simple type you may include is due to a neighborhood exclusion, in which one tree exerts an inhibition on other trees within a neighborhood of radius r. Obviously, the larger r is, the less random is the distribution.

12.2. PRESS.

a) Write a program in Fortran, C, or a higher-level language that computes PRESS efficiently.

b) Generate n observations according to the polynomial model

$$y_i = 1 + x_i + x_i^2 + e_i,$$

where e_i is from a normal distribution with a variance of 1. Let $n = 50$, and let x_i be from a standard normal distribution. Compute PRESS for each of the models:

$$\begin{aligned} y_i &= \beta_0 + \beta_1 x_i + e_i, \\ y_i &= \beta_0 + \beta_1 x_i + \beta_2 x_i^2 + e_i, \\ y_i &= \beta_0 + \beta_1 x_i + \beta_2 x_i^2 + \beta_3 x_i^3 + e_i. \end{aligned}$$

12.3. For $r = n$, show that the jackknife variance estimate, V_{J} (equation (12.11), page 443), can be expressed as

$$\frac{n-1}{n} \sum_{j=1}^{n} \left(T_{(-j)} - \overline{T}_{(\bullet)}\right)^2.$$

12.4. Show that

$$V_{\mathrm{J}} \leq \frac{\sum_{j=1}^{n} (T_j^* - T)^2}{n(n-1)}.$$

12.5. The statistic

$$b_2 = \frac{\sum (y_i - \bar{y})^4}{\left(\sum (y_i - \bar{y})^2\right)^2}$$

is sometimes used to decide whether a least squares estimator is appropriate (otherwise, a robust method may be used). What is the jackknife estimate of the standard deviation of b_2?

Design and conduct a Monte Carlo study of the performance of the jackknife estimator of the standard deviation of b_2 in two specific cases: a normal distribution and a double exponential distribution. In each case, use only one sample size, $n = 100$, but for each case, use both $k = 1$ and $k = 5$ (nonoverlapping partitions).

Summarize your findings in a clearly-written report. Be specific about the basis on which you assess the performance of the jackknife estimator.

12.6. Jackknife bias reduction. Assume that $Y_1, \ldots, Y_n$ are i.i.d.

a) Consider $M_2 = \sum Y_i^2/n$ as an estimator of the second raw population moment, $\mu_2 = \mathrm{E}(Y^2)$. What is the jackknife bias-reduced estimator of μ_2? Is it unbiased?
b) Consider $M_3 = \sum Y_i^3/n$ as an estimator of the third raw population moment, $\mu_3 = \mathrm{E}(Y^3)$. What is the jackknife bias-reduced estimator of μ_3? Is it unbiased?

13

Bootstrap Methods

Resampling methods involve the use of many samples, each taken from a single sample that was taken from the population of interest. Inference based on resampling makes use of the conditional sampling distribution of a new sample (the "resample") drawn from a given sample. Statistical functions on the given sample, a finite set, can easily be evaluated. Resampling methods therefore can be useful even when very little is known about the underlying distribution.

A basic idea in bootstrap resampling is that, because the observed sample contains all the available information about the underlying population, the observed sample can be considered *to be* the population; hence, the distribution of any relevant test statistic can be simulated by using random samples from the "population" consisting of the original sample.

Suppose that a sample $y_1, \ldots, y_n$ is to be used to estimate a population parameter, θ. For a statistic T that estimates θ, as usual, we wish to know the sampling distribution so as to correct for any bias in our estimator or to set confidence intervals for our estimate of θ. The sampling distribution of T is often intractable in applications of interest.

A basic bootstrapping method formulated by Efron (1979) uses the discrete distribution represented by the sample to study the unknown distribution from which the sample came. The basic tool is the empirical cumulative distribution function. The ECDF is the CDF of the finite population that is used as a model of the underlying population of interest.

The functional of the CDF that defines a parameter defines a plug-in estimator of that parameter when the functional is applied to the ECDF. A functional of a population distribution function, $\Theta(P)$, defining a parameter θ can usually be expressed as

$$\begin{aligned} \theta &= \Theta(P) \\ &= \int g(y) \, \mathrm{d}P(y). \end{aligned} \tag{13.1}$$

The plug-in estimator T is the same functional of the ECDF:

J.E. Gentle, *Computational Statistics*, Statistics and Computing,
DOI: 10.1007/978-0-387-98144-4_13,

$$\begin{aligned} T &= T(P_n) \\ &= \Theta(P_n) \\ &= \int g(y)\,\mathrm{d}P_n(y). \end{aligned} \tag{13.2}$$

(In both of these expressions, we are using the integral in a general sense. In the second expression, the integral is a finite sum. It is also a countable sum in the first expression if the random variable is discrete. Note also that we use the same symbol to denote the functional and the random variable.) Various properties of the distribution of T can be estimated by use of "bootstrap samples", each of the form $\{y_1^*, \ldots, y_n^*\}$, where the y_i^*'s are chosen from the original y_i's with replacement.

We define a *resampling vector*, p^*, corresponding to each bootstrap sample as the vector of proportions of the elements of the original sample in the given bootstrap sample. The resampling vector is a realization of a random vector P^* for which nP^* has an n-variate multinomial distribution with parameters n and $(1/n, \ldots, 1/n)$. The resampling vector has random components that sum to 1. For example, if the bootstrap sample $(y_1^*, y_2^*, y_3^*, y_4^*)$ happens to be the sample (y_2, y_2, y_4, y_3), the resampling vector p^* is

$$(0,\ 1/2,\ 1/4,\ 1/4).$$

The bootstrap replication of the estimator T is a function of p^*, $T(p^*)$. The resampling vector can be used to estimate the variance of the bootstrap estimator. By imposing constraints on the resampling vector, the variance of the bootstrap estimator can be reduced.

The *bootstrap principle* involves repeating the process that leads from a population CDF to an ECDF. Taking the ECDF P_n to be the CDF of a population, and resampling, we have an ECDF for the new sample, $P_n^{(1)}$. (In this notation, we could write the ECDF of the original sample as $P_n^{(0)}$.) The difference is that we know more about $P_n^{(1)}$ than we know about P_n. Our knowledge about $P_n^{(1)}$ comes from the simple discrete uniform distribution, whereas our knowledge about P_n depends on knowledge (or assumptions) about the underlying population.

The bootstrap resampling approach can be used to derive properties of statistics, regardless of whether any resampling is done. Most common uses of the bootstrap involve computer simulation of the resampling; hence, bootstrap methods are usually instances of computational inference.

13.1 Bootstrap Bias Corrections

For an estimator T that is the same functional of the ECDF as the parameter is of the CDF, the problem of bias correction is to find a functional f_T that allows us to relate the distribution function of the sample P_n to the population distribution function P, that is, such that

$$\mathrm{E}(f_T(P, P_n) \mid P) = 0. \tag{13.3}$$

Correcting for the bias is equivalent to finding b that solves the equation

$$\begin{aligned} f_T(P, P_n) &= \Theta(P_n) - \Theta(P) - b \\ &= T(P_n) - T(P) - b \end{aligned}$$

so that f_T has zero expectation with respect to P.

Using the bootstrap principle, we look for $f_T^{(1)}$ so that

$$\mathrm{E}\left(f_T^{(1)}(P_n, P_n^{(1)}) \;\middle|\; P_n \right) = 0, \tag{13.4}$$

where $P_n^{(1)}$ is the empirical cumulative distribution function for a sample from the discrete distribution formed from the original sample.

We know more about the items in equation (13.4) than those in equation (13.3), so we now consider the simpler problem of finding b_1 so that

$$\mathrm{E}\Big(T(P_n^{(1)}) - T(P_n) - b_1 \;\Big|\; P_n \Big) = 0.$$

We can write the solution as

$$b_1 = T(P_n) - \mathrm{E}\Big(T(P_n^{(1)}) \;\Big|\; P_n \Big). \tag{13.5}$$

An estimator with less bias is therefore

$$T_1 = 2T(P_n) - \mathrm{E}\Big(T(P_n^{(1)}) \;\Big|\; P_n \Big). \tag{13.6}$$

Suppose, for example, that

$$\begin{aligned} \theta &= \Theta(P) \\ &= \int y \, \mathrm{d}P(y), \end{aligned}$$

and we wish to estimate θ^2. From a random sample of size n, the plug-in estimator of θ is

$$\begin{aligned} \Theta(P_n) &= \int y \, \mathrm{d}P_n(y) \\ &= \bar{y}. \end{aligned}$$

A candidate estimator for θ^2 is the square of the sample mean, that is, $\bar{y}^2$. Because P_n completely defines the sample, we can represent the estimator as a functional $T(P_n)$, and we can study the bias of T by considering the problem of estimating the square of the mean of a discrete uniform distribution with mass points $y_1, \ldots, y_n$. We do this using a single sample of size n from this distribution, $y_1^*, \ldots, y_n^*$. For this sample, we merely work out the expectation of $(\sum y_i^*/n)^2$. You are asked to complete these computations in Exercise 13.2.

In general, to correct the bias, we must evaluate

$$\mathrm{E}\Big(T(P_n^{(1)}) \,\Big|\, P_n\Big) \tag{13.7}$$

in equation (13.6). We may be able to compute $\mathrm{E}\Big(T(P_n^{(1)}) \mid P_n\Big)$, as in the simple example above, or we may have to resort to Monte Carlo methods to estimate it.

The Monte Carlo estimate is based on m random samples each of size n, taken with replacement from the original sample. This is a nonparametric procedure. Specifically, the basic nonparametric Monte Carlo bootstrap procedure for bias correction is

- take m random samples each of size n, *with replacement* from the given set of data, the original sample $y_1, \ldots, y_n$;
- for each sample, compute an estimate T^{*j} of the same functional form as the original estimator T.

The mean of the T^{*j}, $\overline{T}^*$, is an unbiased estimator of $\mathrm{E}\Big(T(P_n^{(1)}) \mid P_n\Big)$.

The distribution of T^{*j} is related to the distribution of T. The variability of T about θ can be assessed by the variability of T^{*j} about T (see below), and the bias of T can be assessed by the mean of $T^{*j} - T$.

Notice that in the bootstrap bias correction we use an estimate of the bias of bootstrap estimators $T^{*j} - T$ of the estimate T from the original sample. This estimate may not be independent of the true estimand. Another problem with our standard approach may arise in a multiparameter case when the bias of individual estimators depends on the bias of other estimators. There is no universal approach to addressing these problems, but often ad hoc methods can be developed.

13.2 Bootstrap Estimation of Variance

From a given sample $y_1, \ldots, y_n$, suppose that we have an estimator $T(y)$. The estimator T^* computed as the same function T, using a bootstrap sample (that is, $T^* = T(y^*)$), is a *bootstrap observation* of T.

The bootstrap estimate of some function of the estimator T is a plug-in estimate that uses the empirical distribution P_n in place of P. This is the bootstrap principle, and this bootstrap estimate is called the *ideal bootstrap*.

For the variance of T, for example, the ideal bootstrap estimator is the variance $\mathrm{V}(T^*)$. This variance, in turn, can be estimated from bootstrap samples. The bootstrap estimate of the variance, then, is the sample variance of T^* based on the m samples of size n taken from P_n:

$$\begin{aligned}\widehat{\mathrm{V}}(T) &= \widehat{\mathrm{V}}(T^*) \\ &= \frac{1}{m-1}\sum(T^{*j} - \overline{T}^*)^2, \end{aligned} \tag{13.8}$$

where T^{*j} is the jth bootstrap observation of T. This, of course, can be computed by Monte Carlo methods by generating m bootstrap samples and computing T^{*j} for each.

If the estimator of interest is the sample mean, for example, the bootstrap estimate of the variance is $\widehat{\mathrm{V}}(Y)/n$, where $\widehat{\mathrm{V}}(Y)$ is an estimate of the variance of the underlying population. (This is true no matter what the underlying distribution is, as long as the variance exists.) The bootstrap procedure does not help in this situation.

13.3 Bootstrap Confidence Intervals

As in equation (1.128) on page 55, a method of forming a confidence interval for a parameter θ is to find a pivotal quantity that involves θ and a statistic T, $f(T, \theta)$, and then to rearrange the terms in a probability statement of the form

$$\Pr\Big(f_{(\alpha/2)} \;\leq\; f(T, \theta) \;\leq\; f_{(1-\alpha/2)}\Big) \;=\; 1 - \alpha. \tag{13.9}$$

When distributions are difficult to work out, we may use bootstrap methods for estimating and/or approximating the percentiles, $f_{(\alpha/2)}$ and $f_{(1-\alpha/2)}$.

Basic Intervals

For computing confidence intervals for a mean, the pivotal quantity is likely to be of the form $T - \theta$. The simplest application of the bootstrap to forming a confidence interval is to use the sampling distribution of $T^* - T_0$ as an approximation to the sampling distribution of $T - \theta$; that is, instead of using $f(T, \theta)$, we use $f(T^*, T_0)$, where T_0 is the value of T in the given sample. The percentiles of the sampling distribution determine $f_{(\alpha/2)}$ and $f_{(1-\alpha/2)}$ in the expressions above. If we cannot determine the sampling distribution of $T^* - T_0$, we can easily estimate it by Monte Carlo methods.

For the case $f(T, \theta) = T - \theta$, the probability statement above is equivalent to

$$\Pr\Big(T - f_{(1-\alpha/2)} \;\leq\; \theta \;\leq\; T - f_{(\alpha/2)}\Big) \;=\; 1 - \alpha. \tag{13.10}$$

The $f_{(\pi)}$ may be estimated from the percentiles of a Monte Carlo sample of $T^* - T_0$.

Bootstrap-t Intervals

Methods of inference based on a normal distribution often work well even when the underlying distribution is not normal. A useful approximate confidence interval for a location parameter can often be constructed using as a template the familiar confidence interval for the mean of a normal distribution,

$$\left(\overline{Y} - t_{(1-\alpha/2)}\, s/\sqrt{n}, \quad \overline{Y} - t_{(\alpha/2)}\, s/\sqrt{n}\right),$$

where $t_{(\pi)}$ is a percentile from the Student's t distribution, and s^2 is the usual sample variance.

A confidence interval for any parameter constructed in this pattern is called a *bootstrap-t interval.* A bootstrap-t interval has the form

$$\left(T - \widehat{t}_{(1-\alpha/2)}\sqrt{\widehat{\mathrm{V}}(T)}, \quad T - \widehat{t}_{(\alpha/2)}\sqrt{\widehat{\mathrm{V}}(T)}\right), \tag{13.11}$$

where $\widehat{t}_{(\pi)}$ is the estimated percentile from the studentized statistic,

$$\frac{T^* - T_0}{\sqrt{\widehat{\mathrm{V}}(T^*)}}.$$

For many estimators T, no simple expression is available for $\widehat{\mathrm{V}}(T)$. The variance could be estimated using a bootstrap and equation (13.8). This bootstrap nested in the bootstrap to determine $\widehat{t}_{(\pi)}$ increases the computational burden multiplicatively.

If the underlying distribution is normal and T is a sample mean, the interval in expression (13.11) is an exact $(1-\alpha)100\%$ confidence interval of shortest length. If the underlying distribution is not normal, however, this confidence interval may not have good properties. In particular, it may not even be of size $(1-\alpha)100\%$. An asymmetric underlying distribution can have particularly deleterious effects on one-sided confidence intervals. Exercise 1.14 on page 76, provides some insight as to why this is the case.

If the estimators T and $\widehat{\mathrm{V}}(T)$ are based on sums of squares of deviations, the bootstrap-t interval performs very poorly when the underlying distribution has heavy tails. This is to be expected, of course. Bootstrap procedures can be no better than the statistics used.

Bootstrap Percentile Confidence Intervals

Given a random sample $(y_1, \ldots, y_n)$ from an unknown distribution with CDF P, we want an interval estimate of a parameter, $\theta = \Theta(P)$, for which we have a point estimator, T.

A bootstrap estimator for θ is T^*, based on the bootstrap sample $(y_1^*, \ldots, y_n^*)$. Now, if $G_{T^*}(t)$ is the distribution function for T^*, then the exact upper $1-\alpha$

confidence limit for θ is the value $t^*_{(1-\alpha)}$, such that $G_{T^*}(t^*_{(1-\alpha)}) = 1 - \alpha$. This is called the *percentile upper confidence limit.* A lower limit is obtained similarly, and an interval is based on the lower and upper limits.

In practice, we generally use Monte Carlo and m bootstrap samples to estimate these quantities. The probability-symmetric bootstrap percentile confidence interval of size $(1 - \alpha)100\%$ is thus

$$\left(t^*_{(\alpha/2)},\quad t^*_{(1-\alpha/2)}\right),$$

where $t^*_{(\pi)}$ is the $[\pi m]^{\text{th}}$ order statistic of a sample of size m of T^*. (Note that we are using T and t, and hence T^* and t^*, to represent estimators and estimates in general; that is, $t^*_{(\pi)}$ here does not refer to a percentile of the Student's t distribution.) This percentile interval is based on the ideal bootstrap and may be estimated by Monte Carlo simulation.

Confidence Intervals Based on Transformations

Suppose that there is a monotonically increasing transformation g and a constant c such that the random variable

$$W = c(g(T^*) - g(\theta)) \tag{13.12}$$

has a symmetric distribution about zero. Here $g(\theta)$ is in the role of a mean and c is a scale or standard deviation.

Let H be the distribution function of W, so

$$G_{T^*}(t) = H\big(c(g(t) - g(\theta))\big) \tag{13.13}$$

and

$$t^*_{(1-\alpha/2)} = g^{-1}\big(g(t^*) + w_{(1-\alpha/2)}/c\big), \tag{13.14}$$

where $w_{(1-\alpha/2)}$ is the $(1-\alpha/2)$ quantile of W. The other quantile $t^*_{(\alpha/2)}$ would be determined analogously.

Instead of approximating the ideal interval with a Monte Carlo sample, we could use a transformation to a known W and compute the interval that way. Use of an exact transformation g to a known random variable W, of course, is just as difficult as evaluation of the ideal bootstrap interval. Nevertheless, we see that forming the ideal bootstrap confidence interval is equivalent to using the transformation g and the distribution function H.

Because transformations to approximate normality are well-understood and widely used, in practice, we generally choose g as a transformation to normality. The random variable W above is a standard normal random variable, Z. The relevant distribution function is Φ, the normal CDF. The normal approximations have a basis in the central limit property. Central limit approximations often have a bias of order $\mathrm{O}(n^{-1})$, however, so in small samples, the percentile intervals may not be very good.

Correcting the Bias in Intervals Due to Bias in the Estimator or to Lack of Symmetry

It is likely that the transformed statistic $g(T^*)$ in equation (13.12) is biased for the transformed θ, even if the untransformed statistic is unbiased for θ. We can account for the possible bias by using the transformation

$$Z = c(g(T^*) - g(\theta)) + z_0,$$

and, analogous to equation (13.13), we have

$$G_{T^*}(t) = \Phi\big(c(g(t) - g(\theta)) + z_0\big).$$

The bias correction z_0 is $\Phi^{-1}\big(G_{T^*}(t)\big)$.

Even when we are estimating θ directly with T^* (that is, g is the identity), another possible problem in determining percentiles for the confidence interval is the lack of symmetry of the distribution about z_0. We would therefore need to make some adjustments in the quantiles instead of using equation (13.14) without some correction.

Rather than correcting the quantiles directly, we may adjust their levels. For an interval of confidence $(1 - \alpha)$, instead of $(t^*_{(\alpha/2)},\ t^*_{(1-\alpha/2)})$, we take

$$\left(t^*_{(\alpha_1)},\ t^*_{(\alpha_2)}\right),$$

where the adjusted probabilities α_1 and α_2 are determined so as to reduce the bias and to allow for the lack of symmetry.

As we often do, even for a nonnormal underlying distribution, we relate α_1 and α_2 to percentiles of the normal distribution.

To allow for the lack of symmetry—that is, for a scale difference below and above z_0—we use quantiles about that point. Efron (1987), who developed this method, introduced an "acceleration", a, and used the distance $a(z_0 + z_{(\pi)})$. Using values for the bias correction and the acceleration determined from the data, Efron suggested the quantile adjustments

$$\alpha_1 = \Phi\left(\widehat{z}_0 + \frac{\widehat{z}_0 + z_{(\alpha/2)}}{1 - \widehat{a}(\widehat{z}_0 + z_{(\alpha/2)})}\right)$$

and

$$\alpha_2 = \Phi\left(\widehat{z}_0 + \frac{\widehat{z}_0 + z_{(1-\alpha/2)}}{1 - \widehat{a}(\widehat{z}_0 + z_{(1-\alpha/2)})}\right).$$

Use of these adjustments to the level of the quantiles for confidence intervals is called the accelerated bias-corrected, or "BC$_a$", method. This method automatically takes care of the problems of bias or asymmetry resulting from transformations that we discussed above.

Note that if $\widehat{a} = \widehat{z}_0 = 0$, then $\alpha_1 = \Phi(z_{(\alpha)})$ and $\alpha_2 = \Phi(z_{(1-\alpha)})$. In this case, the BC$_a$ is the same as the ordinary percentile method.

The problem now is to estimate the acceleration a and the bias correction z_0 from the data.

The bias-correction term z_0 is estimated by correcting the percentile near the median of the m bootstrap samples:

$$\widehat{z}_0 = \Phi^{-1}\left(\frac{1}{m}\sum_j \mathrm{I}_{(-\infty,T]}\big(T^{*j}\big)\right).$$

The idea is that we approximate the bias of the median (that is, the bias of a central quantile) and then adjust the other quantiles accordingly.

Estimating a is a little more difficult. The way we proceed depends on the form the bias may take and how we choose to represent it. Because one cause of bias may be skewness, Efron (1987) adjusted for the skewness of the distribution of the estimator in the neighborhood of θ. The skewness is measured by a function of the second and third moments of T. We can use the jackknife to estimate those moments. The expression is

$$\widehat{a} = \frac{\sum\big(\mathrm{J}(T) - T_{(i)}\big)^3}{6\Big(\sum\big(\mathrm{J}(T) - T_{(i)}\big)^2\Big)^{3/2}}. \tag{13.15}$$

There may be a bias that results from other departures from normality, such as heavy tails. This adjustment does nothing for this kind of bias.

Bootstrap-t and BC$_a$ confidence intervals may be used for inference concerning the difference in the means of two groups. For moderate and approximately equal sample sizes, the coverage of BC$_a$ intervals is often closer to the nominal confidence level, but for samples with very different sizes, the bootstrap-t intervals are often better in the sense of coverage frequency.

Because of the variance of the components in the BC$_a$ method, it generally requires relatively large numbers of bootstrap samples. For location parameters, for example, we may need $m = 1,000$.

A delta method approximation (equation (1.124)) for the standard deviation of the estimator may also be useful for bootstrap confidence intervals. Terms in the Taylor series expansions are used for computing $\widehat{a}$ and $\widehat{z}_0$ rather than using bootstrap estimates for these terms. As with the usual BC$_a$ method, however, there may be a bias that results from other departures from normality, such as heavy tails.

13.4 Bootstrapping Data with Dependencies

When there are relationships among the variables originally sampled, resampling methods must preserve these relationships.

In analyzing data in a regression model,

$$y = X\beta + \epsilon,$$

we may resample the (y_i, x_i) observations (note x_i is a vector), or we may attempt to resample the ϵ_i. The former approach is generally less efficient. The latter approach uses the fitted $\widehat{\beta}$ to provide a set of residuals, which are then resampled and added to the $x_i^{\mathrm{T}}\widehat{\beta}$ to obtain y_i^*. This approach is more efficient, but it relies more strongly on the assumption that the distribution of ϵ is the same in all regions of the model.

Another common case in which dependencies do not allow a straightforward application of the bootstrap is in time series data, or data with serial correlations. This problem can often be addressed by forming subsequences in batches whose summary statistics are (almost) independent. The method of batch means described on page 419 in Chapter 11 is one way of doing this.

Bootstrapping is critically dependent on reproducing the variance in the original population. A correlated sample will not reproduce this variance, so the first step in bootstrapping data with serial correlations, such as time series data, is to model out the dependencies as well as any serial trend. This can be done with various time series models. The simplest model assumes that residuals from the means of disjoint blocks of data are essentially independent (batch means). A linear model or a higher-degree polynomial model may be useful for removal of trends. Diagnostic plots based on histograms of block means can be used to assess the success of blocking schemes.

In more complicated problems with dependencies, such as the problem of selection of variables in a regression model, bootstrapping is rarely useful. Other methods, such as cross validation, that provide comparative measures must be used. Bootstrapping does not provide such measures.

13.5 Variance Reduction in Monte Carlo Bootstrap

Monte Carlo bootstrap estimators have two sources of variation: one is due to the initial sampling, and the other is due to the bootstrap sampling.

Jackknife After Bootstrap

The first problem, of course, is to estimate the variance of the bootstrap estimator. One way of estimating the variance is to use a jackknife. The brute force way would be to do n separate bootstraps on the original sample with a different observation removed each time.

A more computationally efficient way, called jackknife-after-bootstrap, was suggested by Efron (1992). The procedure is to store the indices of the sample included in each bootstrap sample (an $n \times m$ matrix) and then, for each bootstrap sample that does not contain a given element y_j of the original sample, treat that bootstrap sample as if it had been obtained from an original sample from which y_j had been omitted. The two bootstrap samples do indeed have the same distribution; that is, the distribution of a bootstrap sample conditioned on not containing y_j is the same as the unconditional distribution of a bootstrap sample from a given sample that does not contain y_j.

This procedure would have problems, of course, if it so happened that for a given y_j, every bootstrap sample contained y_j. Efron (1992) shows that the probability of this situation is extremely small, even for n as small as 10 and m as small as 20. For larger values of m relative to n, the probability is even lower.

Efron and Tibshirani (1993) report on a small Monte Carlo study that indicates that the jackknife-after-bootstrap tends to overestimate the variance of the bootstrap estimator, especially for small values of m. Efron and Tibshirani attribute this to the overestimation by the jackknife of the resampling variance caused by using the same set of m bootstrap samples to obtain the n jackknife estimates. The jackknife-after-bootstrap should only be used for large values of m, where "large" is subject to user discretion but generally is of the order of 1,000.

The Bootstrap Estimate of the Bias of a Plug-In Estimator

The Monte Carlo estimate of the bootstrap estimate of the bias can be improved if the estimator whose bias is being estimated is a plug-in estimator.

Consider the resampling vector, $p^{*0} = (1/n, \ldots, 1/n)$.

Such a resampling vector corresponds to a permutation of the original sample. If the estimator is a plug-in estimator, then its value is invariant to permutations of the sample; and, in fact,

$$T(p^{*0}) = T(P_n),$$

so the Monte Carlo estimate of the bootstrap estimate of the bias can be written as

$$\sum_{j=1}^{m} s(y_1^{*j}, \ldots, y_n^{*j})/m \; - \; T(p^{*0}).$$

Instead of using $T(p^{*0})$, however, we can increase the precision of the Monte Carlo estimate by using the mean of the individual p^*'s actually obtained:

$$\sum s(y_1^{*j}, \ldots, y_n^{*j})/m \; - \; T(\bar{p}^*),$$

where

$$\bar{p}^* = \sum p^{*j}/m.$$

Notice that for an unbiased plug-in estimator (e.g., the sample mean), this quantity is 0.

If the objective in Monte Carlo experimentation is to estimate some quantity, just as in any estimation procedure, we want to reduce the variance of our estimator (while preserving its other good qualities).

The basic idea is usually to reduce the problem analytically as far as possible and then to use Monte Carlo methods on what is left.

Beyond that general reduction principle, in Monte Carlo experimentation, there are several possibilities for reducing the variance, as discussed in Section 11.5 on page 425. The two main types of methods are judicious use of an auxiliary variable and use of probability sampling. Auxiliary variables may be:

- control variates (any correlated variable, either positively or negatively correlated);
- antithetic variates (in the basic uniform generator);
- regression covariates.

Probability sampling is:

- stratified sampling in the discrete case;
- importance sampling in the continuous case.

Balanced Resampling

Another way of reducing the variance in Monte Carlo experimentation is to constrain the sampling so that some aspects of the samples reflect precisely some aspects of the population.

We may choose to constrain $\bar{p}^*$ to equal p^{*0}. This makes $T(\bar{p}^*) = T(p^{*0})$ and hopefully makes $\sum s(y_1^{*j}, \ldots, y_n^{*j})/m$ closer to its expected value while preserving its correlation with $T(\bar{p}^*)$. This is called *balanced resampling*.

Hall (1990) has shown that the balanced-resampling Monte Carlo estimator of the bootstrap estimator has a bias $\mathrm{O}(m^{-1})$ but that the reduced variance generally more than makes up for it.

Notes and Further Reading

Standard references on the bootstrap are Efron and Tibshirani (1993) and Davison and Hinkley (1997). Each of these texts has associated software. A library of R and S-Plus software called `bootstrap` is used with the Efron/Tibshirani text, and a library called `boot` developed by A. J. Canty is associated in the Davison/Hinkley text. The R program `boot.ci` in `boot` computes BC_a confidence intervals and `abc.ci` computes ABC confidence intervals.

Chernick (2008) provides a very extensive bibliography on the bootstrap. The volume edited by LePage and Billard (1992) contains a number of articles that describe situations requiring special care in the application of the bootstrap. Politis and Romano (1992, 1994) describe the "stationary bootstrap" and other methods of blocking to overcome serial dependencies.

Manly (2006) discusses many applications of bootstrap methods, especially in biology.

Exercises

13.1. Let $S = \{Y_1, \ldots, Y_n\}$ be a random sample from a population with mean μ, variance σ^2, and distribution function P. Let $\widehat{P}$ be the empirical distribution function. Let $\overline{Y}$ be the sample mean for S. Let $S^* = \{Y_1^*, \ldots, Y_n^*\}$ be a random sample taken with replacement from S. Let $\overline{Y}^*$ be the sample mean for S^*.

a) Show that

$$\mathrm{E}_{\widehat{P}}(\overline{Y}^*) = \overline{Y}.$$

b) Show that

$$\mathrm{E}_P(\overline{Y}^*) = \mu.$$

c) Note that in the questions above there was no replication of the bootstrap sampling. Now, suppose that we take m samples S_j^*, compute $\overline{Y}_j^*$ for each, and compute

$$V = \frac{1}{m-1} \sum_j \left(\overline{Y}_j^* - \overline{\overline{Y}^*} \right)^2.$$

Derive $\mathrm{E}_{\widehat{P}}(V)$.

d) Derive $\mathrm{E}_P(V)$.

13.2. Use equation (13.5) to determine the bootstrap bias correction for the square of the sample mean as an estimate of the square of the population mean.

13.3. Show that the bootstrap estimate of the bias of the sample second central moment is $\sum(y_i - \bar{y})^2/n^2$. (Notice that here the y's are used to denote the realization of the random sample rather than the random sample.)

13.4. Show that, for the sample mean, both the bootstrap estimate of the bias and the Monte Carlo estimate of the bootstrap estimate of the bias using the mean resampling vector are 0. Is this also true for the ordinary Monte Carlo estimate?

13.5. The questions in Exercise 10.4 on page 415 assume we know $p(x)$. How would you use the bootstrap to estimate the bias in $X_{(n)}$ for θ (knowing only that θ is the upper bound on the range, but not knowing the distribution is uniform)?
Is the bootstrap very reliable in this case? Why or why not?

13.6. Consider a bootstrap estimate of the variance of an estimator T. Show that the estimate from a bootstrap sample of size m has the same expected value as the ideal bootstrap estimator but that its variance is greater than or equal to that of the ideal bootstrap estimate. (This is the variance of the variance estimator. Also, note that the expectation and variance of these random variables should be taken with respect to the true distribution, not the empirical distribution.)

13.7. Conduct a Monte Carlo study of confidence intervals for the variance in a normal distribution. (This is similar to a study reported by Schenker, 1985.) Use samples of size 20, 35, and 100 from a normal distribution with variance of 1. Use the nonparametric percentile, the BCa, and the ABC methods to set 90% confidence intervals, and estimate the coverage probabilities using Monte Carlo methods. You can use the R library `boot` developed by A. J. Canty (see Davison and Hinkley, 1997) to compute the confidence intervals. The library is available from `statlib`. Use bootstrap $m = 1,000$ and 1,000 Monte Carlo replications. Prepare a two-way table of the estimated coverage percentages:

n	Percentile	BCa	ABC
20	_%	_%	_%
35	_%	_%	_%
100	_%	_%	_%

Summarize your findings in a clearly-written report. Explain the difference in the confidence intervals for the mean and for the variance in terms of pivotal quantities.

13.8. Let $(y_1, y_2, ..., y_{20})$ be a random sample from an exponential distribution with mean $\theta = 1$. Based on a Monte Carlo sample size of 400 and bootstrap sizes of 200, construct a table of percentages of coverages of 95% confidence intervals based on:
- a standard normal approximation;
- a nonparametric percentile method;
- the BCa method;
- the nonparametric ABC method.

13.9. Assume that we have a random sample, $Y_1, \ldots, Y_n$ from a gamma distribution with shape parameter α and scale parameter 1.
a) Describe how you would use a parametric bootstrap to set a 95% lower one-sided confidence interval for the standard deviation, $\sqrt{\alpha}$.
b) Carefully describe how you would perform a Monte Carlo test at the 0.05 significance level for

$$\mathrm{H}_0 : \sqrt{\alpha} \leq 10$$

versus

$$\mathrm{H}_1 : \sqrt{\alpha} > 10.$$

c) What is the relationship (if any) between the answers to the previous two questions?
d) What estimator would you use for $\sqrt{\alpha}$? Your estimator is probably biased. Describe how you would use the jackknife to reduce the bias of the estimator.
e) Describe how you would use the jackknife to estimate the variance of your estimator.

f) Now, assume that you have the same sample as before, but you do not assume a particular form of the distribution. Describe how you would use a nonparametric bootstrap to set a 95% two-sided confidence interval for the standard deviation. (Use any type of nonparametric bootstrap confidence interval you wish.) Clearly specify the interval limits. Is your interval symmetric in any sense?

13.10. Assume a sample of size n. Write a program to generate m resampling vectors p^{*j} so that $\bar{p}^* = (1/n, \ldots, 1/n)$.

Part IV

Exploring Data Density and Relationships

Introduction to Part IV

A canonical problem in statistics is to gain understanding of a given random sample,

$$y_1, \ldots, y_n,$$

in order to understand better the data-generating process that yielded the data. The specific objective is to make inferences about the population from which the random sample arose. In many cases, we wish to make inferences only about some finite set of parameters, such as the mean and variance, that describe the population. In other cases, we want to predict a future value of an observation. Sometimes, the objective is more difficult: We want to estimate a *function* that characterizes the distribution of the population. The cumulative distribution function (CDF) or the probability density function (PDF) provides a complete description of the population, so we may wish to estimate these functions.

In the simpler cases of statistical inference, we assume that the form of the CDF P is known and that there is a parameter, $\theta = \Theta(P)$, of finite dimension that characterizes the distribution within that assumed family of forms. An objective in such cases may be to determine an estimate $\widehat{\theta}$ of the parameter θ. The parameter may completely characterize the probability distribution of the population, or it may just determine an important property of the distribution, such as its mean or median. If the distribution or density function is assumed to be known up to a vector of parameters, the complete description is provided by the parameter estimate. For example, if the distribution is assumed to be normal, then the form of P is known. It involves two parameters, the mean μ and the variance σ^2. The problem of completely describing the distribution is merely the problem of estimating $\theta = (\mu, \sigma^2)$. In this case, the estimates of the CDF, $\widehat{P}$, and the density, $\widehat{p}$, are the normal CDF and density with the estimate of the parameter, $\widehat{\theta}$, plugged in.

If no assumptions, or only weak assumptions, are made about the form of the distribution or density function, the estimation problem is much more difficult. Because the distribution function or density function is a characterization from which all other properties of the distribution could be determined, we expect the estimation of the function to be the most difficult type of statistical inference. "Most difficult" is clearly a heuristic concept and here may mean that the estimator is most biased, most variable, most difficult to compute, most mathematically intractable, and so on.

Estimators such as $\widehat{\theta}$ for the parameter θ or $\widehat{p}$ for the density p are usually random variables; hence, we are interested in the statistical properties of these estimators. If our approach to the problem treats θ and p as fixed (but unknown), then the distribution of $\widehat{\theta}$ and $\widehat{p}$ can be used to make informative statements about θ and p. Alternatively, if θ and p are viewed as realizations of random variables, then the distribution of $\widehat{\theta}$ and $\widehat{p}$ can be used to make informative statements about conditional distributions of the parameter and the function, given the observed data.

Although the CDF in some ways is more fundamental in characterizing a probability distribution (it always exists and is defined the same for both continuous and discrete distributions), the probability density function is more familiar to most data analysts. Important properties such as skewness, modes, and so on can be seen more readily from a plot of the probability density function than from a plot of the CDF. We are therefore usually more interested in estimating the density, p, than the CDF, P. Some methods of estimating the density, however, are based on estimates of the CDF. The simplest estimate of the CDF is the empirical cumulative distribution function, the ECDF, which is defined as

$$P_n(y) = \frac{1}{n} \sum_{i=1}^{n} \mathrm{I}_{(-\infty,y]}(y_i).$$

(See page 669 for the definition and properties of the indicator function $\mathrm{I}_S(\cdot)$ in the ECDF.) As we have seen on page 59, the ECDF is pointwise unbiased for the CDF.

The derivative of the ECDF, the empirical probability density function (EPDF),

$$p_n(y) = \frac{1}{n} \sum_{i=1}^{n} \delta(y - y_i),$$

where δ is the Dirac delta function, is just a series of spikes at points corresponding to the observed values. It is not very useful as an estimator of the probability density. It is, however, unbiased for the probability density function at any point.

In the absence of assumptions about the form of the density p, the estimation problem may be computationally intensive. A very large sample is usually required in order to get a reliable estimate of the density. The goodness of the estimate depends on the dimension of the random variable. Heuristically, the higher the dimension, the larger the sample required to provide adequate representation of the sample space.

Density estimation generally has more modest goals than the development of a mathematical expression that yields the probability density function p everywhere. Although we may develop such an expression, the purpose of the estimate is usually a more general understanding of the population:

- to identify structure in the population, its modality, tail behavior, skewness, and so on;
- to classify the data and to identify different subpopulations giving rise to it; or
- to make a visual presentation that represents the population density.

There are several ways to approach the probability density estimation problem. In a parametric approach mentioned above, the parametric family of distributions, such as a normal distribution or a beta distribution, is assumed. The density is estimated by estimating the parameters of the distribution and substituting the estimates into the expression for the density. In a nonparametric approach, only very general assumptions about the distribution are made. These assumptions may only address the shape of the distribution, such as an assumption of unimodality or an assumption of continuity or other degrees of smoothness of the density function. There are various semiparametric approaches in which, for example, parametric assumptions may be made only over a subset of the range of the distribution, or, in a multivariate case, a parametric approach may be taken for some elements of the random vector and a nonparametric approach for others. Another approach is to assume a more general family of distributions, perhaps characterized by a differential equation, for example, and to fit the equation by equating properties of the sample, such as sample moments, with the corresponding properties of the equation.

In the case of parametric estimation, we have a complete estimate of the density (that is, an estimate at all points). In nonparametric estimation, we generally develop estimates of the ordinate of the density function at specific points. After the estimates are available at given points, a smooth function can be fitted.

Chapters 14 and 15 in this part address the problem of estimating a probability density function, either parametrically or nonparametrically. The probability density of a data-generating process is one of the most important determinants of the structure in data, and it is by studying some aspects of that structure that we make inferences about the probability density.

Chapter 16 considers the problems of finding other structure in the data. Many of the methods we discuss are sensitive to artificial structure, which, similar to artificial ill-conditioning that we discussed on page 208, is structure that can be removed by univariately scaling the data. Scaling has implications not only for numerical computations; it also affects the results of many multivariate analyses, even if the computations are exact.

It is now common to search through datasets and compute summary statistics from various items that may indicate relationships that were not previously recognized. The individual items or the relationships among them may not have been of primary interest when the data were originally collected. This process of prowling through large datsets is sometimes called *data mining* or *knowledge discovery in databases* (KDD). (The names come and go with current fads; there is very little of substance indicated by use of different names. The meaning of "large" in the phrase "large datasets" becomes ever more restrictive as the capacity of computer systems grows.) The objective is to discover characteristics of the data that may not be expected based on the existing theory. In the language of the database literature, the specific goals of data mining are:

- classification of observations;
- linkage analysis;
- deviation detection;

and finally

- predictive modeling.

Of course, the first three of these are the objectives of any exploratory statistical data analysis. Data mining is exploratory data analysis (EDA) applied to large datasets. An objective of an exploratory analysis is often to generate hypotheses, and exploratory analyses are generally followed by more formal confirmatory procedures. The explorations in massive datasets must be performed without much human intervention. Searching algorithms need to have some means of learning and adaptively improving. This will be a major area of research for some time.

Predictive modeling uses inductive reasoning rather than the more common deductive reasoning, which is much easier to automate.

In the statistical classification of observations, the dataset is partitioned recursively. The partitioning results in a classification tree, which is a decision tree, each node of which represents a partition of the dataset. The decision at each node is generally based on the values of a single variable at a time.

The partitioning can also be based on linear combinations of the variables. This is sometimes called "oblique partitioning" because the partitions are not parallel to the axes representing the individual variables. Seeking good linear combinations of variables on which to build oblique partitions is a much more computationally intensive procedure than just using single variables.

Linkage analysis is often the most important activity of data mining. In linkage analysis, relationships among different variables are discovered and analyzed. This step follows partitioning and is the interpretation of the partitions that were formed.

It is also important to identify data that do not fit the patterns that are discovered. The deviation of some subsets of the data often makes it difficult to develop models for the remainder of the data.

In Chapter 17 we consider building models that express asymmetric relationships between variables and then making inferences about those models.

14

Estimation of Probability Density Functions Using Parametric Models

The problem addressed in this chapter is the estimation of an unknown probability density $p(y)$. The way of doing this that we discuss in this chapter is to *approximate* the unknown distribution by use of familiar parametric functions, and then to estimate the parameters in these functions. While mechanically this is essentially point estimation, there are differences in the objectives. We are more concerned about the estimator of the function $\widehat{p}$ or the value of the function at the point y, $\widehat{p(y)}$, than we are about the point estimators of some parameters.

Parametric statistical procedures involve inference about the parameters of a model. In this chapter, although we use parametric models, we can view the methods as *nonparametric*, in the sense that the role of the parameters is to serve as tuning constants so as to have a density function that corresponds to the density of the observed data. The phrase "nonparametric density estimation", however, is generally reserved for methods such as we discuss in Chapter 15. In Section 15.6, however, we consider use of parametric models in more obviously nonparametric methods. While we can identify certain procedures that are "parametric", the classification of other statistical procedures is less clear. "Semiparametric" is sometimes used, but it is generally not a very useful term for describing a statistical procedure.

There are some relatively simple standard distributions that have proven useful for their ability to model the distribution of observed data from many different areas of application. The normal distribution is a good model for symmetric, continuous data from various sources. For skewed data, the lognormal and gamma distributions often work very well. Discrete data are often modeled by the Poisson or binomial distributions. Distributions such as these are *families of distributions* that have various numbers of parameters to specify the distribution completely. To emphasize that the density is dependent on parameters, we may write the density as $p(y \mid \theta)$, where θ may be a vector. Several of the standard parametric families are shown in Tables B.1 and B.2 beginning on page 660.

J.E. Gentle, *Computational Statistics*, Statistics and Computing,
DOI: 10.1007/978-0-387-98144-4_14,

A standard way of estimating a density is to identify appropriate characteristics, such as symmetry, modes, range, and so on, choose some well-known parametric distribution that has those characteristics, and then estimate the parameters of that distribution. For example, if the density is known or assumed to be zero below some point, to be unimodal, and to extend without limit along the positive axis, a three-parameter gamma distribution with density

$$p(y \mid \alpha, \beta, \gamma) = \frac{1}{\Gamma(\alpha)\beta^{\alpha}}(y-\gamma)^{\alpha-1}\mathrm{e}^{-(y-\gamma)/\beta}, \quad \text{for } \gamma \leq y,$$

may be used to model the data. The three parameters α, β, and γ are then estimated from the data.

If the probability density of interest has a finite range, a beta distribution may be used to model it, and if it has an infinite range at both ends, a normal distribution, a Student's t distribution, or a stable distribution may be a useful approximation.

14.1 Fitting a Parametric Probability Distribution

Fitting a parametric density to a set of data is done by estimating the parameters. The estimate of the density, $\widehat{p}(y)$, is formed by substitution of the estimate of the parameters:

$$\widehat{p}(y) = p(y \mid \widehat{\theta}). \tag{14.1}$$

There are several ways of estimating the parameters, and for more complicated models there are many issues that go beyond just estimating the parameters. Many of the methods of fitting the model involve minimization of residuals. To fit a parametric probability density, the most common ways of estimating the parameters are maximum likelihood, matching moments, and matching quantiles.

Maximum Likelihood Methods

The method of maximum likelihood involves the use of a *likelihood function* that comes from the joint density for a random sample. If $p(y \mid \theta)$ is the underlying density, the joint density is just $\prod_i p(y_i \mid \theta)$. The likelihood is a function of the parameter θ:

$$L(\theta;\, y_1, \ldots, y_n) = \prod_i p(y_i \mid \theta).$$

Note the reversal in roles of variables and parameters.

The mode of the likelihood (that is, the value of θ for which L attains its maximum value) is the *maximum likelihood estimate* of θ for the given data, y. The data, which are realizations of the variables in the density function, are considered as fixed and the parameters are considered as variables of the optimization problem in maximum likelihood methods.

Fitting by Matching Moments

Because many of the interesting distributions are uniquely determined by a few of their moments, another method of estimating the density is just to determine parameters of a given family so that the population moments (or model moments) match the sample moments. In some distributions, the parameter estimates derived from matching moments are the same as the maximum likelihood estimates. In general, we would expect the number of moments that can be matched exactly to be the same as the number of parameters in the family of distributions being used.

Fitting by Matching Quantiles

The moments of distributions with infinite range may exhibit extreme variability. This is particularly true of higher-order moments. For that reason it is sometimes better to fit distributions by matching population quantiles with sample quantiles. In general, we would expect the number of quantiles that can be matched exactly to be the same as the number of parameters in the family of distributions being used. A quantile plot may be useful in assessing the goodness of the fit.

14.2 General Families of Probability Distributions

Instead of using some parametric family with relatively limited flexibility of shape, a more general parametric family may be defined. The parameters of these families are generally viewed as tuning parameters to control such things as skewness or kurtosis. These kinds of general distributions are often used in simulation. The general families of distributions are useful in modeling an observed set of data in order to simulate observations from a similar population.

The parameters of these distributions generally do not have any intrinsic meaning. We usually estimate them for the sole purpose of *approximating* the distribution of the population that gave rise to an observed dataset. The most common ways of estimating the parameters in these families is by use of sample moments and known relationships between the population moments and the parameters of the distribution, and by matching quantiles of the sample and those of the family of distributions.

The data are standardized before fitting the parameters. The fitted distribution is then translated and scaled to match the sample mean and standard deviation.

Pearson Curves

Because a probability density is a derivative of a function, a family of differential equations may be used to model the density. The Pearson family of distributions is based on a differential equation whose parameters allow a wide range of characteristics.

The univariate *Pearson family* of probability density functions is developed from the probability function of a hypergeometric random variable, Y,

$$\begin{aligned} p(y) &= \Pr(Y = y) \\ &= \frac{\binom{N\pi}{y}\binom{N(1-\pi)}{n-y}}{\binom{N}{n}}. \end{aligned}$$

The difference equation at y is

$$\begin{aligned} \frac{\Delta p}{p} &= \frac{\Pr(Y = y) - \Pr(Y = y - 1)}{\Pr(Y = y)} \\ &= 1 - \frac{y(N(1-\pi) - n + y)}{(N\pi - y + 1)(n - y + 1)} \\ &= \frac{y - \frac{(n+1)(N\pi+1)}{N+2}}{-\frac{(n+1)(N\pi+1)}{N+2} + \frac{N\pi+n+2}{N+2}y - \frac{1}{N+2}y^2} \\ &= \frac{(y-a)}{b + cy + dy^2}, \end{aligned}$$

where we have introduced the parameters a, b, c, and d. The associated differential equation is

$$\frac{\mathrm{d}(\log p(y))}{\mathrm{d}y} = \frac{(y-a)}{b + cy + dy^2}$$

or

$$(b + cy + dy^2)\,\mathrm{d}p = (y - a)p\,\mathrm{d}y. \tag{14.2}$$

The solution of the differential equation $(b + cy + dy^2)\,\mathrm{d}p = (y - a)p\,\mathrm{d}y$ depends on the roots of $(b + cy + dy^2) = 0$. Certain common parametric families correspond to particular situations. For example, if the roots are real and of opposite sign (that is, if $c^2 - 4bd > 0$ and $|c^2 - 4bd| > |b|$), then the corresponding density function is that of a beta distribution. This is called a Pearson Type I distribution.

If the roots are real, but of the same sign, the distribution is called a Pearson Type VI distribution. A familiar member of this family is the beta distribution of the second kind, which can be obtained from a common beta distribution by means of the transformation of a common beta random variable Y as $X = Y/(1 - Y)$.

Although Karl Pearson distinguished eleven types and subtypes based on the differential equation, the only other one that corresponds to a common distribution is the Pearson Type III, in which $d = 0$. After a translation of the origin, this is the gamma distribution.

The usefulness of the Pearson family of distributions arises from its ability to model many types of empirical data. By multiplying both sides of equation (14.2) by y^k, for $k = 1, 2, 3, 4$, and integrating (by parts), we can express the parameters a, b, c, d in terms of the first four moments. For modeling the distribution of observed data, the first four sample moments are used to determine the parameters of the Pearson system.

The Johnson Family

The *Johnson family* of distributions is based on transformations to an approximate normal distribution. There are three distributional classes, called S_B, S_L, and S_U, based on the form of the transformation required to cause a given random variable Y to have an approximate normal distribution. Each type of transformation has four parameters. If Z is the random variable that has an approximate normal distribution, the transformations that define the distributional classes are

$$S_B : Z = \gamma + \eta \log\left(\tfrac{Y-\epsilon}{\lambda+\epsilon-Y}\right), \text{ for } \epsilon \leq Y \leq \epsilon + \lambda, \tag{14.3}$$

$$S_L : \quad Z = \gamma + \eta \log\left(\tfrac{Y-\epsilon}{\lambda}\right), \quad \text{for } \epsilon \leq Y, \tag{14.4}$$

$$S_U : Z = \gamma + \eta \sinh^{-1}\left(\tfrac{Y-\epsilon}{\lambda}\right), \text{ for } -\infty \leq Y \leq \infty, \tag{14.5}$$

where $\eta, \lambda > 0$, and γ and ϵ are unrestricted.

An attractive property of the Johnson family is the ability to match it to empirical quantiles.

The Burr Family of Distributions

The *Burr family* of distributions (Burr, 1942, and Burr and Cislak, 1968) is defined by the distribution function

$$P(y) = \frac{1}{1 + \exp(-G(y))}, \tag{14.6}$$

where

$$G(y) = \int_{-\infty}^{y} g(t)\,\mathrm{d}t, \tag{14.7}$$

and $g(y)$ is a nonnegative integrable function (a scaled density function). There are many forms that the Burr distribution function can take; for example,

$$P(y) = 1 - \frac{1}{(1 + y^{\alpha})^{\beta}} \quad \text{for } 0 \leq y; 0 < \alpha, \beta. \tag{14.8}$$

The Tukey Lambda Family of Distributions

John Tukey introduced a general symmetric distribution with a single parameter, called λ, for fitting a given set of data. The *lambda family* of distributions is described by its inverse distribution function,

$$\begin{aligned} P^{-1}(u) &= (u^\lambda - (1-u)^\lambda)/\lambda \quad \text{for } \lambda \neq 0 \\ &= \log(u) - \log(1-u) \quad \text{for } \lambda = 0. \end{aligned} \tag{14.9}$$

Ramberg and Schmeiser (1974) extended this distribution to accommodate various amounts of skewness and kurtosis. The *generalized lambda family* of distributions is also defined by its inverse distribution function,

$$P^{-1}(u) = \lambda_1 + \frac{u^{\lambda_3} - (1-u)^{\lambda_4}}{\lambda_2}. \tag{14.10}$$

Least squares can be used to fit moments or to match quantiles to determine values of the λ parameters that fit a given set of data well.

The lambda and the generalized lambda distributions are particularly useful in simulation because the percentiles can be taken as uniform random variables. The λ variates are directly generated by the inverse CDF method.

14.3 Mixtures of Parametric Families

Rather than trying to fit an unknown density $p(y)$ to a single parametric family of distributions, it may be better to fit it to a finite mixture of densities and to represent it as

$$p(y) \approx \sum_{j=1}^{m} \omega_j p_j(y \mid \theta_j), \tag{14.11}$$

where $\sum_{j=1}^{m} \omega_j = 1$. Such a linear combination provides great flexibility for approximating many distributions, even if the individual densities $p_j(y \mid \theta_j)$ are from a restricted class. For example, even if the individual densities are all normals, a skewed distribution can be approximated by a proper choice of the ω_j.

The use of mixtures for density estimation involves *choice* of the number of terms m and of the component families of densities $p_j(y \mid \theta_j)$ and *estimation* of the weights ω_j and the parameters θ_j. The mixture density estimator is

$$\widehat{p}_M(y) = \sum_{j=1}^{\widehat{m}} \widehat{\omega}_j p_j(y \mid \widehat{\theta}_j). \tag{14.12}$$

Here, we have written the number of terms as $\widehat{m}$ because we can think of it as an estimated value under the assumption that the true density is a finite mixture of the p_j.

The choice of the number of terms is usually made adaptively; that is, after an initial arbitrary choice, terms are added or taken away based on the apparent fit of the data.

The process of fitting a mixture of probability densities to a given dataset involves what is called "model-based clustering". Each observation is assigned to a particular distribution, and each set of observations from a given distribution is a cluster.

A standard way of fitting a mixture of distributions if the densities have known forms is by maximum likelihood estimation of the mixing weights and of the individual parameters. Given a random sample $Y_1, \ldots, Y_n$ from a mixture distribution, if the particular distribution from which each given observation arose, the MLE of the mixing parameter ω, is the vector of counts of the individual memberships normalized by the sample size. In the more common case of modeling mixture distributions, we do not know the memberships of the individual observations. Instead, we introduce an auxiliary unobserved variable, D, to represent the distribution from 1 to m to which the observation belongs. Our data then consists of the pairs (Y_i, D_i), in which D is missing. This is a classic setup for an EM method, and indeed that is the standard way of fitting mixture distributions by maximum likelihood.

We will illustrate the EM fitting in a simple example. The general EM method is described beginning on page 296 in Chapter 6.

A two-component normal mixture model can be defined by two normal distributions, $\mathrm{N}(\mu_1, \sigma_1^2)$ and $\mathrm{N}(\mu_2, \sigma_2^2)$, and the probability that the random variable (the observable) arises from the first distribution is w.

The parameter in this model is the vector $\theta = (w, \mu_1, \sigma_1^2, \mu_2, \sigma_2^2)$. (Note that w and the σ's have the obvious constraints.)

The PDF of the mixture is

$$p(y; \theta) = \omega p_1(y; \mu_1, \sigma_1^2) + (1 - \omega) p_2(y; \mu_2, \sigma_2^2),$$

where $p_j(y; \mu_j, \sigma_j^2)$ is the normal PDF with parameters μ_j and σ_j^2. (I am just writing them this way for convenience; p_1 and p_2 are actually the same parameterized function of course.)

In the standard formulation with $C = (X, U)$, X represents the observed data, and the unobserved U represents class membership.

Let $U = 1$ if the observation is from the first distribution and $U = 0$ if the observation is from the second distribution.

The unconditional $\mathrm{E}(U)$ is the probability that an observation comes from the first distribution, which of course is ω.

Suppose we have n observations on X, $x_1, \ldots, x_n$.

Given a provisional value of θ, we can compute the conditional expected value $\mathrm{E}(U|x)$ for any realization of X. It is merely

$$\mathrm{E}(U|x, \theta^{(k)}) = \frac{\omega^{(k)} p_1(x; \mu_1^{(k)}, \sigma_1^{2^{(k)}})}{p(x; \omega^{(k)}, \mu_1^{(k)}, \sigma_1^{2^{(k)}}, \mu_2^{(k)}, \sigma_2^{2^{(k)}})}.$$

The M step is just the familiar MLE of the parameters:

$$\omega^{(k+1)} = \frac{1}{n} \sum \mathrm{E}(U|x_i, \theta^{(k)})$$

$$\mu_1^{(k+1)} = \frac{1}{n\omega^{(k+1)}} \sum q^{(k)}(x_i, \theta^{(k)}) x_i$$

$$\sigma_1^{2(k+1)} = \frac{1}{n\omega^{(k+1)}} \sum q^{(k)}(x_i, \theta^{(k)})(x_i - \mu_1^{(k+1)})^2$$

$$\mu_2^{(k+1)} = \frac{1}{n(1-\omega^{(k+1)})} \sum q^{(k)}(x_i, \theta^{(k)}) x_i$$

$$\sigma_2^{2(k+1)} = \frac{1}{n(1-\omega^{(k+1)})} \sum q^{(k)}(x_i, \theta^{(k)})(x_i - \mu_2^{(k+1)})^2.$$

(Recall that the MLE of σ^2 has a divisor of n, rather than $n-1$.)

The estimation procedure proceeds by iterations over k from the starting value of $\theta^{(0)}$.

14.4 Statistical Properties of Density Estimators Based on Parametric Families

The use of a parametric family to estimate a probability density depends on two things. One is an *approximation*. The unknown probability density is approximated by some standard probability density. Of course, this is what is always done in statistics; some probability model is used as an approximation for some unknown data-generating process. There is a slight difference in the present case, however. In the usual case, we *assume* the probability model *is* the data-generating process. (Of course, we know that we are assuming this, and if pressed, we might just say that the assumed probability model is an approximation.) In the case of probability density estimation by the means discussed in this chapter, however, we begin explicitly with an approximation; we assume that the probability model that we are using is an approximation. Now, the second thing that our estimator depends on is our fit of the approximant; that is, on our estimators of the parameters of the approximating probability model.

The statistical properties of parametric density estimators depend on the properties of the estimators of the parameters. The properties also depend on the approximation, that is, on the parametric family chosen to use in the estimation. The true density is a function of the parameter, and the estimator of the density is a function of an estimator of the parameter, so, as we discussed on page 49, properties of the estimator such as unbiasedness do not carry over to the density estimator. This applies to the pointwise properties of the density estimator and so obviously applies to global properties.

The use of a parametric family of densities for estimating an unknown density will result in good estimators if the unknown density is a member of that parametric family. If this is not the case, the density estimator is not robust to many types of departures from the assumed distribution. Use of a symmetric family, for example, would not yield a good estimator for a skewed distribution.

We generally assess the performance of statistical procedures based on assumed probability models. Both the pointwise and global properties of the estimator are difficult to assess because of the dependence on the *approximating* distribution (not the *assumed* distribution). Of course, this is not so different from what we always try to do in statistical applications: assess the correspondence of the observed data to our assumptions about its distribution. The cross validation and jackknife methods discussed in Chapter 12 can be used to estimate the combined effects of estimation and model selection.

Notes and Further Reading

General Families of Distributions

Many of the common distributions arise from first principles. For example, an exponential distribution results from the definition of a Poisson process; a normal distribution arises from axioms about transformations or, from a different perspective, from central limit theorems. Other distributions, such as a gamma or a t are derived from those basic distributions.

Definitions of general, flexible families of distributions are motivated by modeling applications. The parameters of these families act as tuning parameters that control skewness and kurtosis or other general properties of a distribution.

For the Johnson family of distributions, Chou et al. (1994) identify the distributional classes and the appropriate transformations for a number of well-known distributions. Slifker and Shapiro (1980) describe a method for selection of the particular Johnson family based on ratios of quantiles of the density to be fitted. Chou et al. (1994) give a method for fitting Johnson curves using quantiles. Devroye (1986) describes a method for simulating variates from a Johnson family.

Albert, Delampady, and Polasek (1991) defined another family of distributions that is very similar to the lambda distributions with proper choice of the parameters. The family of distributions of Albert, Delampady, and Polasek is particularly useful in Bayesian analysis with location-scale models.

Mixture Distributions

Solka, Poston, and Wegman (1995) describe visualization methods to accompany EM methods for estimation of the mixture parameter.

Everitt and Hand (1981) provide a good general discussion of the use of finite mixtures for representing distributions. Solka et al. (1998) describe how fitting mixtures to observed data can provide insight about the underlying structure of the data. Roeder and Wasserman (1997) describe the use of mixtures in Bayesian density estimation.

Exercises

14.1. Consider the $\mathrm{U}(0, \theta)$ distribution. The maximum likelihood estimator of θ is the maximum order statistic, $x_{(n)}$, in a sample of size n. This estimator, which is biased, provides a parametric estimator of the function $p(x) = 1/\theta$, which is the probability density function corresponding to $\mathrm{U}(0, \theta)$:

$$\widehat{p}_P(x) = 1/x_{(n)}, \quad \text{for } 0 < x < x_{(n)}.$$

a) Determine the ISE of $\widehat{p}_P(x) = 1/x_{(n)}$. Remember that the domain is $(0, \theta)$.
b) Determine the MISE of $\widehat{p}_P(x) = 1/x_{(n)}$ (with respect to $\mathrm{U}(0, \theta)$).
c) The bias in $x_{(n)}$ can be reduced by taking $\widehat{\theta} = cx_{(n)}$ for some $c > 1$. Determine the value of c that minimizes the MISE.

Notice that this exercise and the following ones do not address the properties of the estimator of the density that gave rise to the sample; they only concern the properties of the estimator of the *given* density, using the *given* sample and assuming that it came from the given density. See Exercise 14.7

14.2. Repeat Exercise 14.1 for the IAE and the MIAE instead of the ISE and the MISE.

14.3. Repeat Exercise 14.1 for the SAE and the MSAE.

14.4. For the U(0,1) distribution, compute the SAE and the MSAE of the ECDF as an estimator of the CDF.

14.5. Consider the gamma(α, β) distribution, with probability density,

$$p(x) = \frac{1}{\Gamma(\alpha)\beta^\alpha} x^{\alpha-1} \mathrm{e}^{-x/\beta} \quad \text{for } 0 \le x \le \infty.$$

a) Generate a random sample of size 100 from a gamma(2,3) distribution.
b) Using your sample, compute the MLE of α and β in a gamma(α, β) distribution.
c) Using as $\widehat{p}$ the density above with the MLE substituted for the parameters, plot your estimated density and superimpose on it a histogram of the random sample you generated.
d) Using a histogram with 10 bins, compute an estimate of the integrated error of your estimate.

14.6. Consider a mixture of three normal densities:

$$p(x) = \frac{\omega_1}{\sqrt{2\pi}\sigma_1} e^{-(x-\mu_1)^2/2\sigma_1^2} + \frac{\omega_2}{\sqrt{2\pi}\sigma_2} e^{-(x-\mu_2)^2/2\sigma_2^2} + \frac{\omega_3}{\sqrt{2\pi}\sigma_3} e^{-(x-\mu_3)^2/2\sigma_3^2}.$$

a) Given a random sample $x_1, x_2, \ldots, x_n$, where n is greater than 9, from this population, what is the likelihood function of $\omega_i, \mu_i, \sigma_i$?

b) With

$$\begin{array}{ccc} \mu_1 = 0 & \mu_2 = 2 & \mu_3 = 6 \\ \sigma_1^2 = 1 & \sigma_2^2 = 4 & \sigma_3^2 = 9 \\ \omega_1 = 0.5 & \omega_2 = 0.4 & \omega_3 = 0.1 \end{array}$$

generate a random sample of size 1000 from this mixture.

c) Using your sample, compute the MLE of $\omega_i, \mu_i, \sigma_i$. (This is a rather complicated optimization problem.)

d) Now, consider your sample to be bivariate random variables, (X, J), where X is as before and J with range $\{1, 2, 3\}$ is an indicator variable corresponding to the population in the mixture from which X arises. Write out the likelihood for the bivariate sample, and recognizing that J is missing in the sample, use an EM method to compute the MLE of $\omega_i, \mu_i, \sigma_i$. Do you get the same results as using a direct computational approach?

e) Using $\widehat{p}$ as the mixture above with the MLE substituted for the parameters, plot your estimated density and superimpose on it a histogram of the random sample you generated.

f) Using a histogram with 10 bins, compute an estimate of the integrated error of your estimate.

14.7. Suppose we have a set of data from an unknown distribution. We use a Tukey's lambda distribution with $\lambda = 0.14$ as an approximation to the density of the standardized sample.

a) Suppose the sample is actually from a $\mathrm{N}(0, 1)$ distribution. What is the MISE of the estimated (that is, the approximated) density?

b) Suppose the sample is actually from a $\mathrm{N}(\mu, \sigma^2)$ distribution. What is the MISE of the estimated (that is, the approximated) density?

c) Suppose the sample is actually from a standard Cauchy distribution. What is the MISE of the estimated (that is, the approximated) density?

The setting $\lambda = 0.14$ is the best fit (by matching of quantiles) of Tukey's lambda distribution to a normal distribution. The setting $\lambda = -1$ is the best fit (by same criterion) of Tukey's lambda distribution to a Cauchy distribution.

15

Nonparametric Estimation of Probability Density Functions

Estimation of a probability density function is similar to the estimation of any function, and the properties of the function estimators that we have discussed are relevant for density function estimators. A density function $p(y)$ is characterized by two properties:

- it is nonnegative everywhere;
- it integrates to 1 (with the appropriate definition of "integrate").

In this chapter, we consider several nonparametric estimators of a density; that is, estimators of a general nonnegative function that integrates to 1 and for which we make no assumptions about a functional form other than, perhaps, smoothness.

It seems reasonable that we require the density estimate to have the characteristic properties of a density:

- $\widehat{p}(y) \geq 0$ for all y;
- $\int_{\mathrm{IR}^d} \widehat{p}(y)\,\mathrm{d}y = 1$.

A probability density estimator that is nonnegative and integrates to 1 is called a *bona fide* estimator.

Rosenblatt (1956) showed that no unbiased bona fide estimator can exist for all continuous p. Rather than requiring an unbiased estimator that cannot be a bona fide estimator, we generally seek a bona fide estimator with small mean squared error or a sequence of bona fide estimators $\widehat{p}_n$ that are asymptotically unbiased; that is,

$$\mathrm{E}_p(\widehat{p}_n(y)) \to p(y) \quad \text{for all } y \in \mathrm{IR}^d \text{ as } n \to \infty.$$

15.1 The Likelihood Function

Suppose that we have a random sample, $y_1, \ldots, y_n$, from a population with density p. Treating the density p as a variable, we write the likelihood functional as

J.E. Gentle, *Computational Statistics*, Statistics and Computing,
DOI: 10.1007/978-0-387-98144-4_15,

$$L(p;\, y_1, \ldots, y_n) = \prod_{i=1}^{n} p(y_i).$$

The *maximum likelihood method* of estimation obviously cannot be used directly because this functional is unbounded in p. We may, however, seek an estimator that maximizes some modification of the likelihood. There are two reasonable ways to approach this problem. One is to restrict the domain of the optimization problem. This is called *restricted maximum likelihood*. The other is to *regularize* the estimator by adding a penalty term to the functional to be optimized. This is called *penalized maximum likelihood*.

We may seek to maximize the likelihood functional subject to the constraint that p be a bona fide density. If we put no further restrictions on the function p, however, infinite Dirac spikes at each observation give an unbounded likelihood, so a maximum likelihood estimator cannot exist, subject only to the restriction to the bona fide class. An additional restriction that p be Lebesgue-integrable over some domain D (that is, $p \in L^1(D)$) does not resolve the problem because we can construct sequences of finite spikes at each observation that grow without bound.

We therefore must restrict the class further. Consider a finite dimensional class, such as the class of step functions that are bona fide density estimators. We assume that the sizes of the regions over which the step function is constant are greater than 0.

For a step function with m regions having constant values, $c_1, \ldots, c_m$, the likelihood is

$$\begin{aligned} L(c_1, \ldots, c_m;\, y_1, \ldots, y_n) &= \prod_{i=1}^{n} p(y_i) \\ &= \prod_{k=1}^{m} c_k^{n_k}, \end{aligned} \tag{15.1}$$

where n_k is the number of data points in the kth region. For the step function to be a bona fide estimator, all c_k must be nonnegative and finite. A maximum therefore exists in the class of step functions that are bona fide estimators.

If v_k is the measure of the volume of the kth region (that is, v_k is the length of an interval in the univariate case, the area in the bivariate case, and so on), we have

$$\sum_{k=1}^{m} c_k v_k = 1.$$

We incorporate this constraint together with equation (15.1) to form the Lagrangian,

$$L(c_1, \ldots, c_m) + \lambda \left(1 - \sum_{k=1}^{m} c_k v_k\right).$$

Differentiating the Lagrangian function and setting the derivative to zero, we have at the maximum point $c_k = c_k^*$, for any λ,

$$\frac{\partial L}{\partial c_k} = \lambda v_k.$$

Using the derivative of L from equation (15.1), we get

$$n_k L = \lambda c_k^* v_k.$$

Summing both sides of this equation over k, we have

$$nL = \lambda,$$

and then substituting, we have

$$n_k L = nLc_k^* v_k.$$

Therefore, the maximum of the likelihood occurs at

$$c_k^* = \frac{n_k}{nv_k}.$$

The restricted maximum likelihood estimator is therefore

$$\begin{aligned} \widehat{p}(y) &= \frac{n_k}{nv_k}, \quad \text{for } y \in \text{region } k, \\ &= 0, \qquad \text{otherwise.} \end{aligned} \tag{15.2}$$

Instead of restricting the density estimate to step functions, we could consider other classes of functions, such as piecewise linear functions. For given subsets S_i of densities for which a maximum likelihood estimator exists, Grenander (1981) developed estimators based on sequences of such subsets, each containing its predecessor. The sequence is called a sieve, so the approach is called the method of sieves. Such sequences have been constructed that yield standard estimators of the density, such as histogram estimators and orthogonal series estimators (see Banks, 1989, for example).

We may also seek other properties, such as smoothness, for the estimated density. One way of achieving other desirable properties for the estimator is to use a penalizing function to modify the function to be optimized. Instead of the likelihood function, we may use a penalized likelihood function of the form

$$L_{\mathrm{p}}(p;\, y_1, \ldots, y_n) = \prod_{i=1}^{n} p(y_i) \mathrm{e}^{-\mathcal{T}(p)},$$

where $\mathcal{T}(p)$ is a transform that measures some property that we would like to minimize. For example, to achieve smoothness, we may use the transform $\mathcal{R}(p)$

of equation (4.7) on page 151 in the penalizing factor. To choose a function $\hat{p}$ to maximize $L_{\mathrm{p}}(p)$ we could use some finite series approximation to $\mathcal{T}(\hat{p})$.

For densities with special properties there may be likelihood approaches that take advantage of those properties. For example, for nonincreasing densities, Grenander (1981) suggested use of the slope of the least concave majorant of the ECDF.

15.2 Histogram Estimators

Let us assume finite support D, and construct a fixed partition of D into a grid of m nonoverlapping bins T_k. (We can arbitrarily assign bin boundaries to one or the other bin.) Let v_k be the volume of the kth bin (in one dimension, v_k is a length and in this simple case is often denoted h_k; in two dimensions, v_k is an area, and so on). The number of such bins we choose, and consequently their volumes, depends on the sample size n, so we sometimes indicate that dependence in the notation: $v_{n,k}$. For the sample $y_1, \ldots, y_n$, the histogram estimator of the probability density function is defined as

$$\begin{aligned}\widehat{p}_H(y) &= \sum_{k=1}^{m} \frac{1}{v_k} \frac{\sum_{i=1}^{n} \mathrm{I}_{T_k}(y_i)}{n} \mathrm{I}_{T_k}(y), \quad \text{for } y \in D,\\ &= 0, \quad \text{otherwise.}\end{aligned}$$

The histogram is the restricted maximum likelihood estimator (15.2).

Letting n_k be the number of sample values falling into T_k,

$$n_k = \sum_{i=1}^{n} \mathrm{I}_{T_k}(y_i),$$

we have the simpler expression for the histogram over D,

$$\widehat{p}_H(y) = \sum_{k=1}^{m} \frac{n_k}{n v_k} \mathrm{I}_{T_k}(y). \tag{15.3}$$

As we have noted already, this is a bona fide estimator:

$$\widehat{p}_H(y) \geq 0$$

and

$$\begin{aligned}\int_{\mathrm{IR}^d} \widehat{p}_H(y) \mathrm{d}y &= \sum_{k=1}^{m} \frac{n_k}{n v_k} v_k\\ &= 1.\end{aligned}$$

Although our discussion generally concerns observations on multivariate random variables, we should occasionally consider simple univariate observations. One reason why the univariate case is simpler is that the derivative is a scalar function. Another reason why we use the univariate case as a model is because it is easier to visualize. The density of a univariate random variable is two-dimensional, and densities of other types of random variables are of higher dimension, so only in the univariate case can the density estimates be graphed directly.

In the univariate case, we assume that the support is the finite interval $[a, b]$. We partition $[a, b]$ into a grid of m nonoverlapping bins $T_k = [t_{n,k}, t_{n,k+1})$ where

$$a = t_{n,1} < t_{n,2} < \ldots < t_{n,m+1} = b.$$

The univariate histogram is

$$\widehat{p}_H(y) = \sum_{k=1}^{m} \frac{n_k}{n(t_{n,k+1} - t_{n,k})} \mathrm{I}_{T_k}(y). \tag{15.4}$$

If the bins are of equal width, say h (that is, $t_k = t_{k-1} + h$), the histogram is

$$\widehat{p}_H(y) = \frac{n_k}{nh}, \quad \text{for } y \in T_k.$$

This class of functions consists of polynomial splines of degree 0 with fixed knots, and the histogram is the maximum likelihood estimator over the class of step functions. Generalized versions of the histogram can be defined with respect to splines of higher degree. Splines with degree higher than 1 may yield negative estimators, but such histograms are also maximum likelihood estimators over those classes of functions.

The histogram as we have defined it is sometimes called a "density histogram", whereas a "frequency histogram" is not normalized by the n.

Some Properties of the Histogram Estimator

The histogram estimator, being a step function, is discontinuous at cell boundaries, and it is zero outside of a finite range. As we have seen (page 344 and Figure 8.4 on page 347), it is sensitive both to the bin size and to the choice of the origin.

An important advantage of the histogram estimator is its simplicity, both for computations and for analysis. In addition to its simplicity, as we have seen, it has two other desirable global properties:

- It is a bona fide density estimator.
- It is the unique maximum likelihood estimator confined to the subspace of functions of the form

$$\begin{aligned} g(t) &= c_k, \text{ for } t \in T_k, \\ &= 0, \text{ otherwise,} \end{aligned}$$

and where $g(t) \geq 0$ and $\int_{\cup_k T_k} g(t)\,\mathrm{d}t = 1$.

Pointwise and Binwise Properties

Properties of the histogram vary from bin to bin. From equation (15.3), the expectation of the histogram estimator at the point y in bin T_k is

$$\mathrm{E}(\widehat{p}_H(y)) = \frac{p_k}{v_k}, \tag{15.5}$$

where

$$p_k = \int_{T_k} p(t)\,\mathrm{d}t \tag{15.6}$$

is the probability content of the kth bin.

Some pointwise properties of the histogram estimator are the following:

- The **bias** of the histogram at the point y within the kth bin is

$$\frac{p_k}{v_k} - p(y). \tag{15.7}$$

  Note that the bias is different from bin to bin, even if the bins are of constant size. The bias tends to decrease as the bin size decreases. We can bound the bias if we assume a regularity condition on p. If there exists γ such that for any $y_1 \neq y_2$ in an interval

$$|p(y_1) - p(y_2)| < \gamma\|y_1 - y_2\|,$$

  we say that p is Lipschitz-continuous on the interval, and for such a density, for any ξ_k in the kth bin, we have

$$\begin{aligned} |\mathrm{Bias}(\widehat{p}_H(y))| &= |p(\xi_k) - p(y)| \\ &\leq \gamma_k\|\xi_k - y\| \\ &\leq \gamma_k v_k. \end{aligned} \tag{15.8}$$

- The **variance** of the histogram at the point y within the kth bin is

$$\begin{aligned} \mathrm{V}\big(\widehat{p}_H(y)\big) &= \mathrm{V}(n_k)/(nv_k)^2 \\ &= \frac{p_k(1-p_k)}{nv_k^2}. \end{aligned} \tag{15.9}$$

  This is easily seen by recognizing that n_k is a binomial random variable with parameters n and p_k. Notice that the variance decreases as the bin size increases. Note also that the variance is different from bin to bin. We can bound the variance:

$$\mathrm{V}(\widehat{p}_H(y)) \leq \frac{p_k}{nv_k^2}.$$

  By the mean-value theorem, we have $p_k = v_k p(\xi_k)$ for some $\xi_k \in T_k$, so we can write

$$\mathrm{V}(\widehat{p}_H(y)) \leq \frac{p(\xi_k)}{nv_k}.$$

Notice the tradeoff between bias and variance: *as h increases the variance, equation (15.9), decreases, but the bound on the bias, equation (15.8), increases.*

- The **mean squared error** of the histogram at the point y within the kth bin is

$$\mathrm{MSE}\big(\widehat{p}_H(y)\big) = \frac{p_k(1-p_k)}{nv_k^2} + \left(\frac{p_k}{v_k} - p(y)\right)^2. \tag{15.10}$$

For a Lipschitz-continuous density, within the kth bin we have

$$\mathrm{MSE}\big(\widehat{p}_H(y)\big) \leq \frac{p(\xi_k)}{nv_k} + \gamma_k^2 v_k^2. \tag{15.11}$$

We easily see that the histogram estimator is L_2 pointwise consistent for a Lipschitz-continuous density if, as $n \to \infty$, for each k, $v_k \to 0$ and $nv_k \to \infty$. By differentiating, we see that the minimum of the bound on the MSE in the kth bin occurs for

$$h^*(k) = \left(\frac{p(\xi_k)}{2\gamma_k^2 n}\right)^{1/3}. \tag{15.12}$$

Substituting this value back into MSE, we obtain the order of the optimal MSE at the point x,

$$\mathrm{MSE}^*\big(\widehat{p}_H(y)\big) = \mathrm{O}\big(n^{-2/3}\big).$$

Asymptotic MISE (or AMISE) of Histogram Estimators

Global properties of the histogram are obtained by summing the binwise properties over all of the bins.

The expressions for the integrated variance and the integrated squared bias are quite complicated because they depend on the bin sizes and the probability content of the bins. We will first write the general expressions, and then we will assume some degree of smoothness of the true density and write approximate expressions that result from mean values or Taylor approximations. We will assume rectangular bins for additional simplification. Finally, we will then consider bins of equal size to simplify the expressions further.

First, consider the integrated variance for a histogram with m bins,

$$\begin{aligned}
\mathrm{IV}(\widehat{p}_H) &= \int_{\mathrm{IR}^d} \mathrm{V}(\widehat{p}_H(t))\,\mathrm{d}t \\
&= \sum_{k=1}^{m} \int_{T_k} \mathrm{V}(\widehat{p}_H(t))\,\mathrm{d}t \\
&= \sum_{k=1}^{m} \frac{p_k - p_k^2}{nv_k} \\
&= \sum_{k=1}^{m} \left(\frac{1}{nv_k} - \frac{\sum p(\xi_k)^2 v_k}{n} \right) + \mathrm{o}(n^{-1})
\end{aligned}$$

for some $\xi_k \in T_k$, as before. Now, taking $\sum p(\xi_k)^2 v_k$ as an approximation to the integral $\int (p(t))^2\,\mathrm{d}t$, and letting $\mathcal{S}$ be the functional that measures the variation in a square-integrable function of d variables,

$$\mathcal{S}(g) = \int_{\mathrm{IR}^d} (g(t))^2\,\mathrm{d}t, \tag{15.13}$$

we have the integrated variance,

$$\mathrm{IV}(\widehat{p}_H) \approx \sum_{k=1}^{m} \frac{1}{nv_k} - \frac{\mathcal{S}(p)}{n}, \tag{15.14}$$

and the asymptotic integrated variance,

$$\mathrm{AIV}(\widehat{p}_H) = \sum_{k=1}^{m} \frac{1}{nv_k}. \tag{15.15}$$

The measure of the variation, $\mathcal{S}(p)$, is a measure of the roughness of the density because the density integrates to 1.

Now, consider the other term in the integrated MSE, the integrated squared bias. We will consider the case of rectangular bins, in which $h_k = (h_{k_1}, \ldots, h_{k_d})$ is the vector of lengths of sides in the kth bin. In the case of rectangular bins, $v_k = \Pi_{j=1}^d h_{k_j}$.

We assume that the density can be expanded in a Taylor series, and we expand the density in the kth bin about $\bar{t}_k$, the midpoint of the rectangular bin. For $\bar{t}_k + t \in T_k$, we have

$$p(\bar{t}_k + t) = p(\bar{t}_k) + t^{\mathrm{T}} \nabla p(\bar{t}_k) + \frac{1}{2} t^{\mathrm{T}} \mathrm{H}_p(\bar{t}_k) t + \cdots, \tag{15.16}$$

where $\mathrm{H}_p(\bar{t}_k)$ is the Hessian of p evaluated at $\bar{t}_k$.

The probability content of the kth bin, p_k, from equation (15.6), can be expressed as an integral of the Taylor series expansion:

$$
\begin{aligned}
p_k &= \int_{\bar{t}_k + t \in T_k} p(\bar{t}_k + t)\, \mathrm{d}t \\
&= \int_{-h_{kd}/2}^{h_{kd}/2} \cdots \int_{-h_{k1}/2}^{h_{k1}/2} \left(p(\bar{t}_k) + t^{\mathrm{T}} \nabla p(\bar{t}_k) + \ldots \right) \mathrm{d}t_1 \cdots \mathrm{d}t_d \\
&= v_k p(\bar{t}_k) + \mathrm{O}\left(h_{k*}^{d+2} \right),
\end{aligned}
\tag{15.17}
$$

where $h_{k*} = \min_j h_{kj}$. The bias at a point $\bar{t}_k + t$ in the kth bin, after substituting equations (15.16) and (15.17) into equation (15.7), is

$$
\frac{p_k}{v_k} - p(\bar{t}_k + t) = -t^{\mathrm{T}} \nabla p(\bar{t}_k) + \mathrm{O}\left(h_{k*}^2 \right).
$$

For the kth bin the integrated squared bias is

$$
\begin{aligned}
&\mathrm{ISB}_k(\widehat{p}_{\mathrm{H}}) \\
&= \int_{T_k} \left(\left(t^{\mathrm{T}} \nabla p(\bar{t}_k) \right)^2 - 2\mathrm{O}\left(h_{k*}^2 \right) t^{\mathrm{T}} \nabla p(\bar{t}_k) + \mathrm{O}\left(h_{k*}^4 \right) \right) \mathrm{d}t \\
&= \int_{-h_{kd}/2}^{h_{kd}/2} \cdots \int_{-h_{k1}/2}^{h_{k1}/2} \sum_i \sum_j t_{ki} t_{kj} \nabla_i p(\bar{t}_k) \nabla_j p(\bar{t}_k)\, \mathrm{d}t_1 \cdots \mathrm{d}t_d + \mathrm{O}\left(h_{k*}^{4+d} \right).
\end{aligned}
\tag{15.18}
$$

Many of the expressions above are simpler if we use a constant bin size, v, or $h_1, \ldots, h_d$, where, if the bins are rectangular, $v = h_1 \cdots h_d$. In the case of constant bin size, the asymptotic integrated variance in equation (15.15) becomes

$$
\mathrm{AIV}\left(\widehat{p}_H \right) = \frac{m}{nv}.
\tag{15.19}
$$

In this case, the integral in equation (15.18) simplifies as the integration is performed term by term because the cross-product terms cancel, and the integral is

$$
\frac{1}{12} (h_1 \cdots h_d) \sum_{j=1}^{d} h_j^2 \left(\nabla_j p(\bar{t}_k) \right)^2.
\tag{15.20}
$$

This is the asymptotic squared bias integrated over the kth bin.

When we sum the expression (15.20) over all bins, the $\left(\nabla_j p(\bar{t}_k) \right)^2$ become $\mathcal{S}\left(\nabla_j p \right)$, and we have the asymptotic integrated squared bias,

$$
\mathrm{AISB}\left(\widehat{p}_H \right) = \frac{1}{12} \sum_{j=1}^{d} h_j^2 \mathcal{S}\left(\nabla_j p \right).
\tag{15.21}
$$

Combining the asymptotic integrated variance, equation (15.19), and squared bias, equation (15.21), for the histogram with rectangular bins of constant size, we have

$$\mathrm{AMISE}(\widehat{p}_H) = \frac{m}{n(h_1 \cdots h_d)} + \frac{1}{12} \sum_{j=1}^{d} h_j^2 \mathcal{S}(\nabla_j p). \tag{15.22}$$

As we have seen before, smaller bin sizes increase the variance but decrease the squared bias.

Bin Sizes

As we have mentioned and have seen by example, the histogram is very sensitive to the bin sizes, both in appearance and in other properties. Equation (15.22) for the AMISE assuming constant rectangular bin size is often used as a guide for determining the bin size to use when constructing a histogram. This expression involves $\mathcal{S}(\nabla_j p)$ and so, of course, cannot be used directly. Nevertheless, differentiating the expression with respect to h_j and setting the result equal to zero, we have the bin width that is optimal with respect to the AMISE,

$$h_{j*} = \mathcal{S}(\nabla_j p)^{-1/2} \left(6 \prod_{i=1}^{d} \mathcal{S}(\nabla_i p)^{1/2} \right)^{\frac{1}{2+d}} n^{-\frac{1}{2+d}}. \tag{15.23}$$

Substituting this into equation (15.22), we have the optimal value of the AMISE

$$\frac{1}{4} \left(36 \prod_{i=1}^{d} \mathcal{S}(\nabla_i p)^{1/2} \right)^{\frac{1}{2+d}} n^{-\frac{2}{2+d}}. \tag{15.24}$$

Notice that the optimal rate of decrease of AMISE for histogram estimators is $\mathrm{O}(n^{-\frac{2}{2+d}})$. Although histograms have several desirable properties, this order of convergence is not good compared to that of some other bona fide density estimators, as we will see in later sections.

The expression for the optimal bin width involves $\mathcal{S}(\nabla_j p)$, where p is the unknown density. An approach is to choose a value for $\mathcal{S}(\nabla_j p)$ that corresponds to some good general distribution. A "good general distribution", of course, is the normal with a diagonal variance-covariance matrix. For the d-variate normal with variance-covariance matrix $\Sigma = \mathrm{diag}(\sigma_1^2, \ldots, \sigma_d^2)$,

$$\mathcal{S}(\nabla_j p) = \frac{1}{2^{d+1} \pi^{d/2} \sigma_j^2 |\Sigma|^{1/2}}.$$

For a univariate normal density with variance σ^2,

$$\mathcal{S}(p') = 1/(4\sqrt{\pi}\sigma^3)$$

(Exercise 4.2b on page 200), so the optimal constant one-dimensional bin width under the AMISE criterion is

$$3.49\sigma n^{-1/3}.$$

In practice, of course, an estimate of σ must be used. The sample standard deviation s is one obvious choice. Freedman and Diaconis (1981a) proposed using a more robust estimate of the scale based on the sample interquartile range, r. The sample interquartile range leads to a bin width of $2rn^{-1/3}$.

The AMISE is essentially an L_2 measure. The L_∞ criterion — that is, the sup absolute error (SAE) of equation (10.17) — also leads to an asymptotically optimal bin width that is proportional to $n^{-1/3}$. Based on that criterion, Freedman and Diaconis (1981b) derived the rule

$$1.66s\left(\frac{\log n}{n}\right)^{1/3},$$

where s is an estimate of the scale.

One of the most commonly used rules is for the number of bins rather than the width. Assume a symmetric binomial model for the bin counts, that is, the bin count is just the binomial coefficient. The total sample size n is

$$\sum_{k=0}^{m-1}\binom{m-1}{k} = 2^{m-1},$$

and so the number of bins is

$$m = 1 + \log_2 n.$$

Bin Shapes

In the univariate case, histogram bins may vary in size, but each bin is an interval. For the multivariate case, there are various possibilities for the shapes of the bins. The simplest shape is the direct extension of an interval, that is a hyperrectangle. The volume of a hyperrectangle is just $v_k = \prod h_{kj}$. There are, of course, other possibilities; any tessellation of the space would work. The objects may or may not be regular, and they may or may not be of equal size. Regular, equal-sized geometric figures such as hypercubes have the advantages of simplicity, both computationally and analytically. In two dimensions, there are three possible regular tessellations: triangles, squares, and hexagons.

For two dimensions, hexagons are slightly better than squares and triangles with respect to the AMISE (see Exercise 15.6 on page 511). Binning in hexagons can be accomplished using two staggered tessellations by rectangles, as indicated in Figure 15.1 (see also Exercise 15.7). The lattice determining one tessellation is formed by the points at the centers of the rectangles of the other tessellation. For nonrectangular tessellations, different data structures are necessary for efficient processing.

Various other tessellations may also work well, especially adaptive tessellations. We discuss tessellations for use in clustering data beginning on page 528.

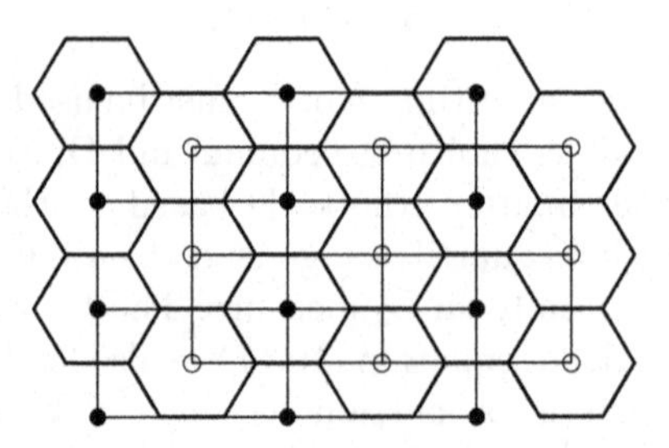

Fig. 15.1. Regular Hexagonal Tiling of 2-Space with Two Rectangular Lattices Superimposed

For hyperrectangles of constant size, the univariate theory generally extends fairly easily to the multivariate case. The histogram density estimator is

$$\widehat{p}_H(y) = \frac{n_k}{nh_1h_2\cdots h_d}, \quad \text{for } y \in T_k,$$

where the h's are the lengths of the sides of the rectangles. The variance within the kth bin is

$$\mathrm{V}(\widehat{p}_H(y)) = \frac{np_k(1-p_k)}{(nh_1h_2\cdots h_d)^2}, \quad \text{for } y \in T_k,$$

and the integrated variance is

$$\mathrm{IV}(\widehat{p}_H) \approx \frac{1}{nh_1h_2\cdots h_d} - \frac{\mathcal{S}(f)}{n}.$$

Other Density Estimators Related to the Histogram

There are several variations of the histogram that are useful as probability density estimators. The most common modification is to connect points on the histogram by a continuous curve. A simple way of doing this in the univariate case leads to the *frequency polygon*. This is the piecewise linear curve that connects the midpoints of the bins of the histogram. The endpoints are usually zero values at the midpoints of two appended bins, one on either side.

The *histospline* is constructed by interpolating knots of the empirical CDF with a cubic spline and then differentiating it. More general methods use splines or orthogonal series, such as we discuss in Section 15.5, to fit the histogram.

As we have mentioned and have seen by example, the histogram is somewhat sensitive in appearance to the location of the bins, even for a fixed width of the bins. To overcome the problem of location of the bins, the estimate of the density at a given point can be taken as the average of several histograms with equal bin widths but different bin locations. This is called the *average shifted histogram*, or ASH. It also has desirable statistical properties, and it is computationally efficient in the multivariate case. See Scott (2004) for further discussions of ASH.

15.3 Kernel Estimators

Kernel methods are probably the most widely used technique for building nonparametric probability density estimators. They are best understood by developing them as a special type of histogram. The difference is that the bins in kernel estimators are centered at the points at which the estimator is to be computed. The problem of the choice of location of the bins in histogram estimators does not arise.

Rosenblatt's Histogram Estimator; Kernels

For the one-dimensional case, Rosenblatt (1956) defined a histogram that is shifted to be centered on the point at which the density is to be estimated. Given the sample $y_1, \ldots, y_n$, Rosenblatt's histogram estimator at the point y is

$$\widehat{p}_R(y) = \frac{\#\{y_i \text{ s.t. } y_i \in (y - h/2, \quad y + h/2]\}}{nh}. \tag{15.25}$$

This histogram estimator avoids the ordinary histogram's constant-slope contribution to the bias. This estimator is a step function with variable lengths of the intervals that have constant value.

Rosenblatt's centered histogram can also be written in terms of the ECDF:

$$\widehat{p}_R(y) = \frac{P_n(y + h/2) - P_n(y - h/2)}{h}, \tag{15.26}$$

where, as usual, P_n denotes the ECDF. As seen in this expression, Rosenblatt's estimator is a centered finite-difference approximation to the derivative of the empirical cumulative distribution function (which, of course, is not differentiable at the data points). We could, of course, use the same idea and form other density estimators using other finite-difference approximations to the derivative of P_n.

Another way to write Rosenblatt's shifted histogram estimator over bins of length h is

$$\widehat{p}_R(y) = \frac{1}{nh} \sum_{i=1}^{n} K\left(\frac{y - y_i}{h}\right), \tag{15.27}$$

where $K(t) = K_{\mathrm{u}}(t)$ is the uniform or "boxcar" kernel of equation (4.65) with smoothing parameter $\lambda = 1/2$. Notice, however, that the role of λ in our earlier formulation of kernel functions is effectively taken over by the h; hence, we will consider it to be the smoothing parameter in this kind of density estimator. Other values of the smoothing parameter or kernel functions could be used, and equation (15.27) is the general form of the univariate kernel probability density estimator.

The estimator extends easily to the multivariate case. In the general kernel estimator, we usually use a more general scaling of $y - y_i$,

$$V^{-1}(y - y_i),$$

for some positive-definite matrix V. The determinant of V^{-1} scales the estimator to account for the scaling within the kernel function. The general kernel estimator is given by

$$\widehat{p}_K(y) = \frac{1}{n|V|}\sum_{i=1}^{n} K\left(V^{-1}(y - y_i)\right), \tag{15.28}$$

where the function K is the *kernel*, and V is the *smoothing matrix*. The determinant of the smoothing matrix is exactly analogous to the bin volume in a histogram estimator. The univariate version of the kernel estimator is the same as Rosenblatt's estimator (15.27), but in which a more general function K is allowed.

In practice, V is usually taken to be constant for a given sample size, but, of course, there is no reason for this to be the case, and indeed it may be better to vary V depending on the number of observations near the point y. The dependency of the smoothing matrix on the sample size n and on y is often indicated by the notation $V_n(y)$.

Properties of Kernel Estimators

The appearance of the kernel density estimator depends to some extent on the support and shape of the kernel. Unlike the histogram estimator, the kernel density estimator may be continuous and even smooth.

It is easy to see that if the kernel satisfies

$$K(t) \geq 0, \tag{15.29}$$

and

$$\int_{\mathrm{IR}^d} K(t)\,\mathrm{d}t = 1 \tag{15.30}$$

(that is, if K is a density), then $\widehat{p}_K(y)$ is a bona fide density estimator.

There are other requirements that we may impose on the kernel either for the theoretical properties that result or just for their intuitive appeal. It also seems reasonable that in estimating the density at the point y, we would want to emphasize the sample points near y. This could be done in various ways, but one simple way is to require

$$\int_{\mathrm{IR}^d} tK(t)\,\mathrm{d}t = 0. \tag{15.31}$$

In addition, we may require the kernel to be symmetric about 0.

For multivariate density estimation, the kernels are usually chosen as a radially symmetric generalization of a univariate kernel. Such a kernel can be formed as a product of the univariate kernels. For a product kernel, we have for some constant σ_K^2,

$$\int_{\mathrm{IR}^d} tt^{\mathrm{T}} K(t)\,\mathrm{d}t = \sigma_K^2 I_d, \tag{15.32}$$

where I_d is the identity matrix of order d. We could also impose this as a requirement on any kernel, whether it is a product kernel or not. This makes the expressions for bias and variance of the estimators simpler. The spread of the kernel can always be controlled by the smoothing matrix V, so sometimes, for convenience, we require $\sigma_K^2 = 1$.

In the following, we will assume the kernel satisfies the properties in equations (15.29) through (15.32).

Pointwise Properties

The pointwise properties of the kernel estimator are relatively simple to determine because the estimator at a point is merely the sample mean of n independent and identically distributed random variables. The expectation of the kernel estimator (15.28) at the point y is the convolution of the kernel function and the probability density function,

$$\begin{aligned}\mathrm{E}\left(\widehat{p}_K(y)\right) &= \frac{1}{|V|}\int_{\mathrm{IR}^d} K\left(V^{-1}(y-t)\right) p(t)\,\mathrm{d}t \\ &= \int_{\mathrm{IR}^d} K(u)p(y-Vu)\,\mathrm{d}u, \end{aligned} \tag{15.33}$$

where $u = V^{-1}(y-t)$ (and, hence, $\mathrm{d}u = |V|^{-1}\mathrm{d}t$).

If we approximate $p(y - Vu)$ about y with a three-term Taylor series, using the properties of the kernel in equations (15.29) through (15.32) and using properties of the trace, we have

$$\begin{aligned}\mathrm{E}\left(\widehat{p}_K(y)\right) &\approx \int_{\mathrm{IR}^d} K(u)\left(p(y) - (Vu)^{\mathrm{T}}\nabla p(y) + \frac{1}{2}(Vu)^{\mathrm{T}}\mathrm{H}_p(y)Vu\right)\mathrm{d}u \\ &= p(y) - 0 + \frac{1}{2}\mathrm{trace}\left(V^{\mathrm{T}}\mathrm{H}_p(y)V\right). \end{aligned} \tag{15.34}$$

To second order in the elements of V (that is, $\mathrm{O}(|V|^2)$), the bias at the point y is therefore

$$\frac{1}{2}\mathrm{trace}\left(VV^{\mathrm{T}}\mathrm{H}_p(y)\right). \tag{15.35}$$

Using the same kinds of expansions and approximations as in equations (15.33) and (15.34) to evaluate $\mathrm{E}\left((\widehat{p}_K(y))^2\right)$ to get an expression of order

$\mathrm{O}(|V|/n)$, and subtracting the square of the expectation in equation (15.34), we get the approximate variance at y as

$$\mathrm{V}\left(\widehat{p}_K(y)\right) \approx \frac{p(y)}{n|V|} \int_{\mathrm{IR}^d} (K(u))^2 \, \mathrm{d}u,$$

or

$$\mathrm{V}\left(\widehat{p}_K(y)\right) \approx \frac{p(y)}{n|V|} \mathcal{S}(K). \tag{15.36}$$

Integrated Properties

Integrating the variance at each point y, because p is a density, we have

$$\mathrm{AIV}\big(\widehat{p}_K\big) = \frac{\mathcal{S}(K)}{n|V|}, \tag{15.37}$$

and integrating the square of the asymptotic bias in expression (15.35), we have

$$\mathrm{AISB}\big(\widehat{p}_K\big) = \frac{1}{4} \int_{\mathrm{IR}^d} \left(\mathrm{trace}\left(V^{\mathrm{T}} \mathrm{H}_p(y) V\right)\right)^2 \mathrm{d}y. \tag{15.38}$$

These expressions are much simpler in the univariate case, where the smoothing matrix V is the smoothing parameter or window width h. We have a simpler approximation for $\mathrm{E}\left(\widehat{p}_K(y)\right)$ than that given in equation (15.34),

$$\mathrm{E}\left(\widehat{p}_K(y)\right) \approx p(y) + \frac{1}{2} h^2 p''(y) \int_{\mathrm{IR}} u^2 K(u) \, \mathrm{d}u,$$

and from this we get a simpler expression for the AISB. After likewise simplifying the AIV, we have

$$\mathrm{AMISE}\big(\widehat{p}_K\big) = \frac{\mathcal{S}(K)}{nh} + \frac{1}{4} \sigma_K^4 h^4 \mathcal{R}(p), \tag{15.39}$$

where we have left the kernel unscaled (that is, $\int u^2 K(u) \, \mathrm{d}u = \sigma_K^2$).

Minimizing this with respect to h, we have the optimal value of the smoothing parameter

$$\left(\frac{\mathcal{S}(K)}{n \sigma_K^4 \mathcal{R}(p)}\right)^{1/5}. \tag{15.40}$$

Substituting this back into the expression for the AMISE, we find that its optimal value in this univariate case is

$$\frac{5}{4} \mathcal{R}(p) (\sigma_K \mathcal{S}(K))^{4/5} n^{-4/5}. \tag{15.41}$$

The AMISE for the univariate kernel density estimator is thus $\mathrm{O}(n^{-4/5})$. Recall that the AMISE for the univariate histogram density estimator is $\mathrm{O}(n^{-2/3})$ (expression (15.24) on page 496).

We see that the bias and variance of kernel density estimators have similar relationships to the smoothing matrix that the bias and variance of histogram estimators have. As the determinant of the smoothing matrix gets smaller (that is, as the window of influence around the point at which the estimator is to be evaluated gets smaller), the bias becomes smaller and the variance becomes larger. This agrees with what we would expect intuitively.

Choice of Kernels

On page 183, we listed three common kernels, the uniform, $K_{\mathrm{u}}(t)$, which is the one used in Rosenblatt's shifted histogram estimator (15.27), the normal, $K_{\mathrm{n}}(t) = \frac{1}{\sqrt{2\pi}}\mathrm{e}^{-t^2/2}$, and the quadratic,

$$K_{\mathrm{q}}(t) = 0.75(1 - t^2)\mathrm{I}_{[-1,1]}(t). \tag{15.42}$$

The quadratic kernel is also called the "Epanechnikov" kernel, because Epanechnikov (1969) showed that it yields the optimal rate of convergence of the MISE. As it turns out, however, the kernel density estimator is not very sensitive to the form of the kernel.

Kernels with finite support (that is, compact kernels) are generally easier to work with. In the univariate case, a useful general form of a compact kernel is

$$K(t) = \kappa_{rs}(1 - |t|^r)^s\mathrm{I}_{[-1,1]}(t), \tag{15.43}$$

where

$$\kappa_{rs} = \frac{r}{2\mathrm{B}(1/r, s+1)}, \quad \text{for } r > 0,\ s \geq 0,$$

and $\mathrm{B}(a, b)$ is the complete beta function. This general form leads to several simple specific cases:

- for $r = 1$ and $s = 0$, it is the uniform or rectangular kernel;
- for $r = 1$ and $s = 1$, it is the triangular kernel;
- for $r = 2$ and $s = 1$ ($\kappa_{rs} = 3/4$), it is the quadratic or Epanechnikov kernel;
- for $r = 2$ and $s = 2$ ($\kappa_{rs} = 15/16$), it is the "biweight" kernel.

If $r = 2$ and $s \to \infty$, we have the Gaussian kernel (with some rescaling), which, of course, is not compact.

As mentioned above, for multivariate density estimation, the kernels are often chosen as a product of the univariate kernels. The product Epanechnikov kernel, for example, is

$$K(t) = \frac{d+2}{2c_d}(1 - t^{\mathrm{T}}t)\mathrm{I}_{(t^{\mathrm{T}}t \leq 1)}, \tag{15.44}$$

where

$$c_d = \frac{\pi^{d/2}}{\Gamma(d/2+1)}.$$

We have seen that the AMISE of a kernel estimator (that is, the sum of equations (15.37) and (15.38)) depends on $\mathcal{S}(K)$ and the smoothing matrix V. As we mentioned above, the amount of smoothing (that is, the window of influence) can be made to depend on σ_K. We can establish an approximate equivalence between two kernels, K_1 and K_2, by choosing the smoothing matrix to offset the differences in $\mathcal{S}(K_1)$ and $\mathcal{S}(K_2)$ and in σ_{K_1} and σ_{K_2}.

Although the kernel may be from a parametric family of distributions, in kernel density estimation, we do not estimate those parameters; hence, the kernel method is a nonparametric method.

Computation of Kernel Density Estimators

If the estimate is required at one point only, it is simplest just to compute it directly. If the estimate is required at several points, it is often more efficient to compute the estimates in some regular fashion.

If the estimate is required over a grid of points, a fast Fourier transform (FFT) can be used to speed up the computations. Silverman (1982) describes an FFT method using a Gaussian kernel. He first takes the discrete Fourier transform of the data (using a histogram on 2^k cells) and then inverts the product of that and the Fourier transform of the Gaussian kernel, $\exp(-h^2s^2/2)$.

15.4 Choice of Window Widths

An important problem in nonparametric density estimation is to determine the smoothing parameter, such as the bin volume, the smoothing matrix, the number of nearest neighbors, or other measures of locality. In kernel density estimation, the window width has a much greater effect on the estimator than the kernel itself does.

An objective is to choose the smoothing parameter that minimizes the MISE. We often can do this for the AMISE, as in equation (15.23) on page 496. It is not as easy for the MISE. The first problem, of course, is just to estimate the MISE.

In practice, we use cross validation with varying smoothing parameters and alternate computations between the MISE and AMISE.

In univariate density estimation, the MISE has terms such as $h^\alpha \mathcal{S}(p')$ (for histograms) or $h^\alpha \mathcal{S}(p'')$ (for kernels). We need to estimate the roughness of a derivative of the density.

Using a histogram, a reasonable estimate of the integral $\mathcal{S}(p')$ is a Riemann approximation,

$$\begin{aligned}\widehat{\mathcal{S}}(p') &= h\sum\left(\widehat{p}'(t_k)\right)^2\\ &= \frac{1}{n^2h^3}\sum(n_{k+1}-n_k)^2,\end{aligned}$$

where $\widehat{p}'(t_k)$ is the finite difference at the midpoints of the kth and $(k+1)^{\text{th}}$ bins; that is,

$$\widehat{p}'(t_k) = \frac{n_{k+1}/(nh) - n_k/(nh)}{h}.$$

This estimator is biased. For the histogram, for example,

$$\mathrm{E}(\widehat{\mathcal{S}}(p')) = \mathcal{S}(p') + 2/(nh^3) + \ldots.$$

A standard estimation scheme is to correct for the $2/(nh^3)$ term in the bias and plug this back into the formula for the AMISE (which is $1/(nh) + h^2\mathcal{S}(r')/12$ for the histogram).

We compute the estimated values of the AMISE for various values of h and choose the one that minimizes the AMISE. This is called *biased cross validation* because of the use of the AMISE rather than the MISE.

These same techniques can be used for other density estimators and for multivariate estimators, although at the expense of considerably more complexity. The sampling variability in cross validation, however, makes it of limited value in selecting the window width.

15.5 Orthogonal Series Estimators

A continuous real function $p(x)$, integrable over a domain D, can be represented over that domain as an infinite series in terms of a complete spanning set of real orthogonal functions $\{q_k\}$ over D:

$$p(x) = \sum_k c_k q_k(x). \tag{15.45}$$

The orthogonality property allows us to determine the coefficients c_k in the expansion (15.45):

$$c_k = \langle q_k, p \rangle. \tag{15.46}$$

Approximation using a truncated orthogonal series can be particularly useful in estimation of a probability density function because the orthogonality relationship provides an equivalence between the coefficient and an expected value. Expected values can be estimated using observed values of the random variable and the approximation of the probability density function. Assume that the probability density function p is approximated by an orthogonal series $\{q_k\}$:

$$p(y) = \sum_k c_k q_k(y).$$

From equation (15.46), we have

$$\begin{aligned} c_k &= \langle q_k, p \rangle \\ &= \int_D q_k(y) p(y) \mathrm{d}y \\ &= \mathrm{E}(q_k(Y)), \end{aligned} \tag{15.47}$$

where Y is a random variable whose probability density function is p.

The c_k can therefore be unbiasedly estimated by

$$\widehat{c}_k = \frac{1}{n}\sum_{i=1}^{n} q_k(y_i)w(y_i).$$

Notice that we do not even have to do a PDF decomposition, as we do for a general function in equation (10.1) on page 404.

The orthogonal series estimator is therefore

$$\widehat{p}_S(y) = \frac{1}{n}\sum_{k=0}^{j}\sum_{i=1}^{n} q_k(y_i)q_k(y) \tag{15.48}$$

for some truncation point j.

Without some modifications, this generally is not a good estimator of the probability density function. It may not be smooth, and it may have infinite variance. The estimator may be improved by shrinking the $\widehat{c}_k$ toward the origin. This could be done by formulating an objective function that consists of a weighted average of the estimated AMISE and $\|\widehat{c}_k\|$. (This is similar to shrinkage estimators in the linear model, such as ridge regression models, that we discuss beginning on page 607.)

The number of terms in the finite series approximation also has a major effect on the statistical properties of the estimator. Having more terms is not necessarily better.

One useful property of orthogonal series estimators, however, is that the convergence rate is independent of the dimension of the random variable. This may make orthogonal series methods more desirable for higher-dimensional problems.

There are several standard orthogonal series that could be used, as we discuss in Sections 4.2 and 4.3. These two most commonly used series are the Fourier and the Hermite.

A Fourier series is commonly used for distributions with bounded support. In those cases, it yields estimators with generally better properties than estimators based on the Hermite series. In the bounded support cases, an orthogonal series estimator based on Jacobi polynomials may be even better, however, due to the flexibility of choosing the two shape parameters. Tarter, Freeman, and Hopkins (1986) gave a Fortran program for computing probability density estimates based on the Fourier series.

For distributions with unbounded support, the Hermite polynomials are most commonly used. Some of the "natural" Hermite polynomials are shown in equation (4.53) on page 174.

15.6 Other Methods of Density Estimation

There are several other methods of probability density estimation. Most of them are modifications of the ones we have discussed. A simple example is to form a density estimator as a smooth function interpolating or approximating a histogram. This can be done with splines or Bézier curves, for example. Either the midpoint or one of the vertices of the top of each cell could be used.

Another variation is to fit a spline or Bézier curve to a cumulative histogram; that is, to a grouped ECDF, and then differentiate the spline or Bézier curve.

Any estimator based on an approximating function must be constrained to be nonnegative, of course, and the issue of an integral of 1 may not be easily settled.

Some methods work only in the univariate case, whereas others can be applied in multivariate density estimation.

All of the nonparametric methods of density estimation involve decisions such as window width, number of mixtures, number of terms in an expansion, and so on. All of these quantities can be thought of as smoothing parameters. There are various rules for making these decisions in an asymptotically optimal fashion for some known distributions. (See Stone, 1984, for window selection rules for kernel density estimators.) Absent assumptions about the nature of the true distribution, it is difficult to decide on the extent of smoothing. Much of the current work is on developing adaptive methods in which these choices are made based on the data.

As we mentioned earlier, it may be reasonable to vary the smoothing parameter in the kernel density estimator in such a way that the locality of influence is smaller in areas of high density and larger in sparser regions. In this case, the general kernel density estimator of equation (15.28) would become

$$\widehat{p}_K(y) = \frac{1}{n}\sum_{i=1}^{n}\frac{1}{|V_i|}K\left(V_i^{-1}(y - y_i)\right). \tag{15.49}$$

The V_i would likely be chosen to be constant over various regions. The variable smoothing parameter would be selected adaptively in the various regions, possibly using the estimated AMISE in a manner similar to biased cross validation mentioned above.

Mixtures and Kernel Methods

In Chapter 14 we discussed the use of mixtures of parametric models as probability density estimators. We can combine that approach with the use of kernels. We choose a set of weighting functions $f_1, f_2, \ldots, f_m$, such that $\sum f_j(x) = 1$, and associated smoothing matrices V_j, and form the estimator

$$\widehat{p}_F(y) = \frac{1}{n}\sum_{i=1}^{n}\sum_{j=1}^{m}\frac{f_j(y_i)}{|V_j|}K\left(V_i^{-1}(y - y_i)\right). \tag{15.50}$$

This is sometimes called a filtered kernel density estimator.

If the kernel function is the standard d-variate normal density $\phi_d(t)$ and the filtering functions are weighted normal densities $\pi_j \phi_d(t \,|\, \mu_j, \Sigma_j)$, we can express the filtered kernel density estimator as

$$\widehat{p}_F(y) = \frac{1}{n} \sum_{i=1}^{n} \sum_{j=1}^{m} \frac{\pi_j \phi_d(t \,|\, \mu_j, \Sigma)}{h \left| \Sigma_j^{1/2} \right| f_\bullet(y_i)} \phi_d \left(\Sigma_j^{-1/2} (y - y_i)/h \right),$$

where $f_\bullet(t) = \sum_{j=1}^{m} \pi_j \phi_d(t \,|\, \mu_j, \Sigma_j)$. We now have the choices in the estimator as m, π_j, μ_j, and Σ_j. This, of course, gives us more flexibility, but it makes the adaptive selection of the tuning parameters more difficult.

Another approach is to alternate between nonparametric filtered kernel estimators and parametric mixture estimators composed of the same number of terms (filter functions or component densities). The estimator is computed iteratively by beginning with a mixture estimator $\widehat{p}_M^{(1)}(y)$ of the form in equation (14.12) on page 480 and a filtered kernel estimator $\widehat{p}_F^{(1)}(y)$ of the form in equation (15.50) above. A new mixture estimator $\widehat{p}_M^{(2)}(y)$ is chosen as the mixture estimator closest to $\widehat{p}_F^{(1)}(y)$ (in some norm). The new mixture estimator is used to determine the choices for a new filtered kernel estimator. (In general, these are the f_j and the V_j in equation (15.50). For normal filters and kernels, they are the variances.) The process is continued until

$$\left\| \widehat{p}_M^{(k+1)}(y) - \widehat{p}_M^{(k)}(y) \right\|$$

is small.

The method of alternating kernel and mixture estimators can easily be used for multivariate density estimation.

Comparisons of Methods

Nonparametric probability density estimation involves the fundamental tradeoff between a spike at each observation (that is, no smoothing) and a constant function over the range of the observations (that is, complete smoothing) ignoring differences in relative frequencies of observations in various intervals. It is therefore not surprising that the comparison of methods is not a trivial exercise.

One approach for comparing methods of estimation is to define broad classes of densities and to evaluate the performance of various estimators within those classes. One way of forming interesting families of distributions is to use a mixture of normal densities $\phi(y \,|\, \mu_j, \sigma_j^2)$. (You are asked to do this in Exercise 15.9.) The triangular density, $p(y) = (1-|x|)_+$, can also be used to form mixtures of densities with interesting shapes. Three useful special cases are the following.

- claw density

$$p(y) = \frac{1}{10}\Big(5\phi(y\,|\,0,1) + \phi(y\,|\,-1,0.1) + \phi(y\,|\,-0.5,0.1) + \phi(y\,|\,0,0.1) + \phi(y\,|\,0.5,0.1) + \phi(y\,|\,1,0.1)\Big);$$

- smooth comb density

$$p(y) = \frac{32}{63}\phi\left(y\,|\,-\frac{31}{21},\frac{32}{63}\right) + \frac{16}{63}\phi\left(y\,|\,\frac{17}{21},\frac{16}{63}\right) + \frac{8}{63}\phi\left(y\,|\,\frac{41}{21},\frac{8}{63}\right) + \frac{4}{63}\phi\left(y\,|\,\frac{53}{21},\frac{4}{63}\right) + \frac{2}{63}\phi\left(y\,|\,\frac{59}{21},\frac{2}{63}\right) + \frac{1}{63}\phi\left(y\,|\,\frac{62}{21},\frac{1}{63}\right);$$

- saw-tooth density

$$g(y) = p(y+9) + p(y+7) + \cdots p(y-7) + p(y-9),$$

  where p is the triangular density.

These densities cover a wide range of shapes.

Notes and Further Reading

Nonparametric estimation of probability density functions was a topic for much research during the last twenty years of so of the twentieth century. Much of the research focused on the univariate problem, of course. While some of the methods extend to higher dimensions, the issues of structures become much more important because they ultimately involve all possible subsets of the variables. The increasing number of subsets, which is of exponential order, is one manifestation of the curse of dimensionality. The focus is generally restricted to relationships within subsets of size two.

In all methods of nonparametric probability density function estimation, there is some kind of smoothing parameter.

Scott (2004) summarizes the theory and methods of nonparametric probability density function estimation, with an emphasis on histogram and kernel methods. He also discusses visualization of the densities, which is an important aid in choosing the fineness of the structure that should be conveyed by the estimate.

Bin Shapes

Newman and Barkema (1999) discuss data structures for working with hexagonal grids and a related type of grid formed by a Kagomé lattice, which is a tessellation composed of hexagons and twice as many triangles, in which the hexagons meet at their vertices and spaces between three hexagons are form the triangles.

General discussions of tessellations are given by Conway and Sloane (1999) (particularly Chapter 2 of that book) and by Okabe et al. (2000). Conway and Sloane (1982) give algorithms for binning data into lattices of various types for dimensions from 2 to 8.

Smoothing Functions

The orthogonal series estimators generally are based on a smoothing of a histogram; thus, the smoothing parameter is the bin size. In addition to the orthogonal series we discussed, other orthogonal systems can be used in density estimation. Walter and Ghorai (1992) describe the use of wavelets in density estimation, and discuss some of the problems in their use. Vidakovic (2004) also discusses density estimation and other applications of wavelets.

Kooperberg and Stone (1991) describe use of splines to smooth the histogram. Kim et al. (1999) propose use of the cumulative histogram and fitting it using Bézier curves.

Exercises

15.1. Use Monte Carlo methods to study the performance of the histogram density estimator $\widehat{p}_H(x)$ using univariate normal data. Generate samples of size 500 from a N(0, 1) distribution. Use a Monte Carlo sample size of 100.
  a) Choose three different bin sizes. Tell how you chose them.
  b) For each bin size, estimate the variance of $\widehat{p}(0)$.
  c) For each bin size, compute the average MISE.

  Summarize your findings in a clearly-written report.

15.2. Use Monte Carlo methods to study the performance of the histogram density estimator $\widehat{p}_H(x)$ using a simple but nonstandard univariate distribution that has density

$$
\begin{aligned}
p(x) &= 3x \quad \text{for} \quad 0.0 \leq x < 0.5 \\
&= 3 - 3x \quad \text{for} \quad 0.5 \leq x < 1.0 \\
&= x - 1 \quad \text{for} \quad 1.0 \leq x < 1.5 \\
&= 2 - x \quad \text{for} \quad 1.5 \leq x < 2.0 \\
&= 0 \quad \text{otherwise.}
\end{aligned}
$$

  The programs you will need to write are incremental; that is, the program written in one question may be used in another question.
  a) Generation of random variates.
     i. Describe a method for generating random variables from this distribution.
     ii. Write a program to generate random variables from this distribution.

b) Use Monte Carlo methods to estimate the variance of $\widehat{p}_H(1)$ for a fixed sample size of 500, using three different bin widths: 0.2, 0.3, and 0.4. Use a Monte Carlo sample size of 100. If a cutpoint corresponds to $x = 1$, use the average of the two bins. (Notice that this is the variance of the density estimator at one point.)
Now, work out the true variance of $\widehat{p}_H(1)$ for this distribution when a bin width of 0.4 is used. (Tell what bin this point is in. It depends on how you set the cutpoints.)

c) For each bin size in the previous question, compute an estimate of the MISE.

d) Now, investigate the order of the MISE in the sample size for a given bin width that is dependent on the sample size. Choose the bin width as $0.84n^{-1/3}$.
   i. First, show that this bin width is optimal for the AMISE. (The coefficient is given to two decimal places.)
   ii. Use Monte Carlo methods to study the order of the MISE; that is, estimate α in AMISE $= \mathrm{O}(n^\alpha)$, for the given bin width sequence. Use sample sizes of 128, 256, 512, and 1,024. Compute the MISE at each sample size, and plot the MISE versus n on log-log axes. Estimate α using least squares.

15.3. Derive the variance of the histogram (equation (15.9) on page 492).

15.4. Derive the bin size that achieves the lower bound of the MSE for the histogram (equation (15.12), page 493).

15.5. Let

$$
\begin{aligned}
p(y) &= \frac{1}{\Gamma(\alpha)\beta^\alpha} y^{\alpha-1} \mathrm{e}^{-y/\beta} \quad \text{for } 0 \leq y, \\
&= 0 \quad \text{elsewhere.}
\end{aligned}
$$

(This is the probability density function for a gamma distribution.) Determine $\mathcal{S}(p')$, as in equation (15.13) on page 494.

15.6. Bivariate histograms over any of the three regular tessellations are relatively easy to construct. (The triangular and square tessellations are straightforward. See Exercise 15.7 for the hexagonal tessellation.) For a random sample from a density $p(y_1, y_2)$, show that a bivariate histogram has AMISE of the form

$$
\frac{1}{nh^2} + ch^2\big(\mathcal{S}(p_{y_1}) + \mathcal{S}(p_{y_2})\big),
$$

where h^2 is the area of the bin, $p_{y_1} = \partial p(y_1, y_2)/\partial y_1$ and $p_{y_2} = \partial p(y_1, y_2)/\partial y_2$. Notice that the number of bins is different for the three bin shapes. Determine the value of c for each of the regular tessellations. Which tesselation has the best AMISE, given equal bin sizes (but unequal numbers of bins)?

15.7. Write a program to count the bivariate observations for regular hexagonal bins. The input to your program is a set of observations, the bin definitions, and the bin counts from prior sets of observations. The output is the updated array of bin counts. Use the fact that the centers of the hexagons are the lattice points of two rectangular tilings of the plane. See Figure 15.1 on page 498.

15.8. Frequency polygons.

a) Show that the univariate frequency polygon is a bona fide estimator.

b) Suppose that it is known that the true density is zero below a certain bound (for example, suppose that the nature of the phenomenon being modeled requires the data to be nonnegative, so the support is the positive half line). Derive an expression for the integrated squared bias of the frequency polygon density estimator. (The main thing to determine is the order of the bias in the bin width h.)

c) For the density estimation problem when a fixed bound is known, suggest a modification to the frequency polygon that might reduce the bias. (Reflection is one possibility, but you may want to suggest a different approach.) Derive an expression for the integrated squared bias of your modified frequency polygon density estimator. Show whether your modified estimator is a bona fide estimator.

15.9. Use a random number generator to generate a sample of size 1,000 from a mixture of two univariate normal distributions to study the performance of a histogram density estimator and a kernel estimator. Among other things, the object of the study will be to assess the ability of the two types of estimators to identify mixtures. In each case, the width of the bins must be decided on empirically in such a way that the resulting density estimator is visually smooth (that is, so that it is not very jagged), yet may have more than one "hump". (Of course, since you know the density, you can "cheat" a little on this.) Let the population consist of a fraction π from a $\mathrm{N}(0,1)$ distribution and a fraction $1-\pi$ from a $\mathrm{N}(\delta,1)$ distribution. Let π take the values 0.5, 0.7, and 0.9. Let δ take two positive values, δ_1 and δ_2.

a) For each value of π, choose δ_1 and δ_2 so that for δ_1 the distribution is unimodal and for δ_2 the distribution is bimodal. (δ is nested in π.) Choose δ_2 so that the minimum of the density between the two modes is at least 0.05. (Marron and Wand, 1992, describe general types of bimodal and multimodal families of distributions useful in assessing the performance of density estimators.)

b) For each of the six combinations of π and δ, choose a sequence of bin widths, compute the estimates, and by visual inspection of plots of the estimates, choose an "optimal" bin size.

The functions `hist` and `density` in R can be used to compute the density estimates.

15.10. Using the univariate analogue of equation (15.34) on page 501, derive the AMISE for the kernel density estimator given in expression (15.39).

15.11. Suppose that we have a random sample as follows:

$$-1.8,\ -1.2,\ -.9,\ -.3,\ -.1,\ .1,\ .2,\ .4,\ .7,\ 1.0,\ 1.3,\ 1.9.$$

(*Be aware that this sample is too small for any serious density estimation!*)

a) Compute the kernel density estimate at the point 0 using a normal kernel.
b) Compute the kernel density estimate at the point 0 using the Epanechnikov kernel.
c) Compute a smoothed kernel density estimate over the range $(-2, 2)$ using the Epanechnikov kernel.
d) Compute the orthogonal series estimator of the probability density using Hermite polynomials in equation (15.48) and truncating at $j = 4$.

15.12. Given a random sample $y_1, \ldots, y_n$ from an unknown population, the basic problems in density estimation are to estimate $\Pr(Y \in S)$ for a random variable Y from that population, to estimate the density at a specific point, $p(y_0)$, or to estimate the density function p or the distribution function P at all points.
Suppose instead that the problem is to generate random numbers from the unknown population.

a) Describe how you might do this. There are several possibilities that you might want to explore. Also, you should say something about higher dimensions.
b) Suppose that we have a random sample as follows:

$$-1.8,\ -1.2,\ -.9,\ -.3,\ -.1,\ .1,\ .2,\ .4,\ .7,\ 1.0,\ 1.3,\ 1.9.$$

Generate a random sample of size 5 from the population that yielded this sample. There are obviously some choices to be made.
Describe your procedure in detail, and write a computer program to generate the random sample.

15.13. Consider another problem related to density estimation:
Given a random sample $y_1, \ldots, y_n$ from an unknown population, estimate the mode of that population.

a) Describe how you might do this. Again, there are several possibilities that you might want to explore, and your solution will be evaluated on the basis of how you address the alternative possibilities and how you select a specific procedure to use. (You might also want to say something about higher dimensions.)
b) Suppose that we have a random sample as follows:

$$-1.8,\ -1.2,\ -.9,\ -.3,\ -.1,\ .1,\ .2,\ .4,\ .7,\ 1.0,\ 1.3,\ 1.9.$$

Estimate the mode of the population that yielded this sample. Describe your procedure in detail, and write a computer program to estimate the mode.

15.14. Consider the three density estimators

$$\widehat{p}_1(y) = \frac{P_n(y) - P_n(y-h)}{h},$$

$$\widehat{p}_2(y) = \frac{P_n(y+h) - P_n(y)}{h},$$

$$\widehat{p}_3(y) = \frac{P_n(y+\frac{h}{2}) - P_n(y-\frac{h}{2})}{h},$$

where P_n is the empirical distribution function based on a random sample of size n. For each,

a) putting the estimator in the form of a kernel density estimator, write out the kernel;
b) compute the integrated variance; and
c) compute the integrated squared bias.

15.15. The L_p error of bona fide density estimators.

a) Show that the L_1 error is less than or equal to 2.
b) Let g be a monotone, continuous function, and consider the random variable, $Z = g(Y)$. Show that the L_1 error is invariant to this change of variable. (Apply the same function g to the elements of the sample of y's.)
c) By an example, show that the L_2 error has no bound.

15.16. A common way of fitting a parametric probability density function to data is to use estimates of the parameters that yield moments of the fitted density that match the sample moments. The second and higher moments used in this method of estimation are usually taken as the central moments. This means that if $\widehat{p}$ is the density estimated from the sample $y_1, \ldots, y_n$, then

$$\mathrm{E}_{\widehat{p}}(Y) = \bar{y}$$

and, in the univariate case,

$$\mathrm{E}_{\widehat{p}}(Y^r) = \frac{1}{n}\sum(y_i - \bar{y})^r$$

for $r = 2, \ldots$.

a) For the univariate histogram estimator, $\widehat{p}_H$, and a sample $y_1, \ldots, y_n$, how does $\mathrm{E}_{\widehat{p}_H}(Y^r)$ compare to the rth sample moment?
b) For the univariate kernel estimator, $\widehat{p}_K$, with a rectangular kernel, how does $\mathrm{E}_{\widehat{p}_K}(Y^r)$ compare to the rth sample moment?

15.17. Make a plot of each of the test densities shown on page 508.

16

Statistical Learning and Data Mining

A major objective in data analysis is to identify interesting features or structure in the data. In this chapter, we consider the use of some of the tools and measures discussed in Chapters 9 and 10 to identify interesting structure. The graphical methods discussed in Chapter 8 are also very useful in discovering structure, but we do not consider those methods further in the present chapter.

There are basically two ways of thinking about "structure". One has to do with *counts* of observations. In this approach, patterns in the *density* are the features of interest. We may be interested in whether the density is multimodal, whether it is skewed, whether there are holes in the density, and so on. The other approach seeks to identify relationships among the variables. The two approaches are related in the sense that if there are relationships among the variables, the density of the observations is higher in regions in which the relationships hold. Relationships among variables are generally not exact, and the relationships are identified by the higher density of observations that exhibit the approximate relationships.

An important kind of pattern in data is a relationship to time. Often, even though data are collected at different times, the time itself is not represented by a variable on the dataset. A simple example is one in which the data are collected sequentially at roughly equal intervals. In this case, the index of the observations may serve as a surrogate variable. Consider the small univariate dataset in Table 16.1, for example.

A static view of a histogram of these univariate data, as in Figure 16.1, shows a univariate bimodal dataset. Figure 16.2, however, in which the data are plotted against the index (by rows in Table 16.1), shows a completely different structure. The data appear to be sinusoidal with an increasing frequency. The sinusoidal responses at roughly equal sampling intervals result in a bimodal static distribution, which is the structure seen in the histogram.

Interesting structure may also be groups or clusters of data based on some measure of similarity, as discussed in Section 9.2 beginning on page 383. When there are separate groups in the data, but the observations do not contain

J.E. Gentle, *Computational Statistics*, Statistics and Computing,
DOI: 10.1007/978-0-387-98144-4_16,

Table 16.1. Dataset with Two Interesting Structures

0.85	0.89	0.94	0.95	0.99	1.00	0.96	0.94	0.97	0.90
0.84	0.71	0.57	0.43	0.29	0.08	-0.09	-0.30	-0.49	-0.72
-0.83	-0.88	-1.02	-0.94	-0.95	-0.78	-0.60	-0.38	-0.04	0.26
0.55	0.77	0.97	1.04	0.91	0.69	0.41	0.02	-0.37	-0.70
-0.96	-1.01	-0.87	-0.50	-0.06	0.44	0.79	0.99	0.95	0.59
0.10	-0.46	-0.87	-0.95	-0.77	-0.22	0.36	0.89	1.03	0.66

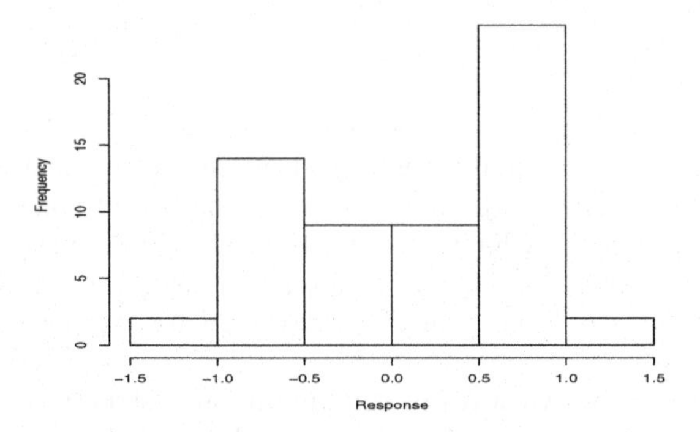

Fig. 16.1. Histogram of the Data in Table 16.1

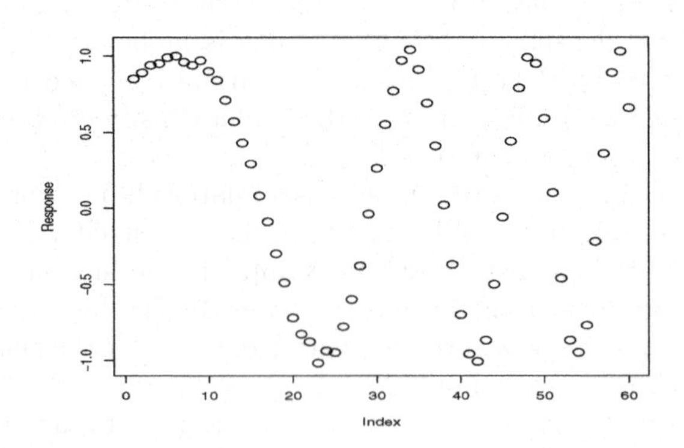

Fig. 16.2. Data in Table 16.1 Plotted against Its Index

an element or an index variable representing group membership, identifying nearby elements or clusters in the data requires some measure of similarity (or, equivalently, of dissimilarity).

Figure 16.3 shows four different bivariate datasets, each of which consists of two clusters. The criteria that distinguish the clusters are different in the datasets. In Figure 16.3(a), the clusters are defined by proximity; the points in each cluster are closer to the respective cluster centroid than they are to the centroid of the other cluster.

In Figures 16.3(b), 16.3(c), and 16.3(d), the definitions of the clusters are somewhat more difficult. The clusters are defined by characteristics of the clusters themselves (that is, by structures that the clusters individually exhibit). These clusters are sometimes called "conceptual clusters"; the points are members of a cluster because of some concept or holistic characteristic of the set of points, such as lying close to a straight line.

The plots in Figure 16.3 also illustrate one of the problems in the identification of clusters: In some cases, although the clusters are obvious, there are a few individual observations that could apparently belong to either cluster.

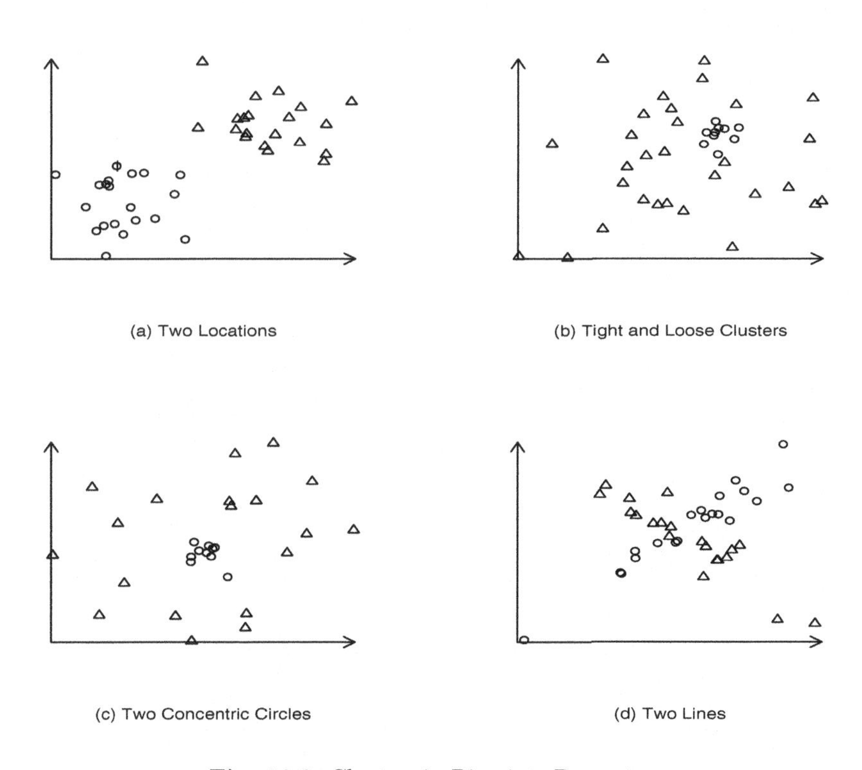

Fig. 16.3. Clusters in Bivariate Datasets

Figure 16.4 shows two different two-dimensional datasets (that is, datasets containing two variables) whose members fall almost within one-dimensional manifolds.

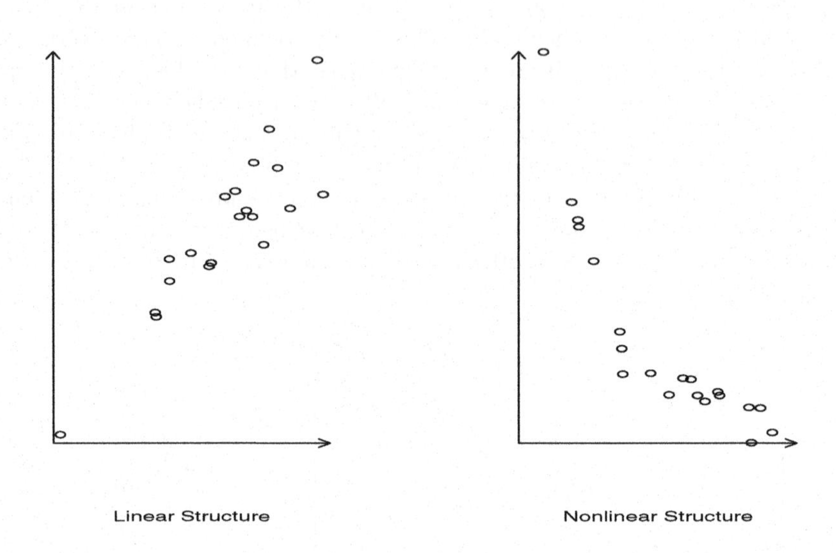

Fig. 16.4. Relational Structures in Bivariate Datasets

We may occasionally wish to identify any of the types of groups or structures shown in Figures 16.2, 16.3, and 16.4, but we will concentrate in this chapter on identifying the types of clusters shown in the first graph in Figure 16.3 (that is, clusters whose centroids are different).

Although we often assume that the data space is a subspace of $\mathbb{R}^m$, a data space may be more general. Data, for example, may be character strings such as names. The more general types of data may be mapped from the original data space to a "feature space", which is a subspace of $\mathbb{R}^m$. The variables may be measured on different scales; they may, of course, represent completely different phenomena, so measurement scales cannot be made the same. One way of reconciling the measurements, however, is to standardize the data using the transformation (9.22) on page 392,

$$X_{\mathrm{S}} = (X - \overline{X})\,\mathrm{diag}(1/\sqrt{s_{ii}}),$$

where $\overline{X}$ is the matrix whose constant columns contain the means of the corresponding columns of X, and $\sqrt{s_{ii}}$ is the sample standard deviation of the ith column of X.

We may be interested in finding the *nearest neighbors* of a given observation based on their similarity; or, alternatively, we may be interested in identifying all observations within a given degree of closeness to a given observation. This problem is called a "proximity search".

In the following sections, we consider general problems in multivariate data analysis. Our emphasis will be on exploratory analysis, and the main goals will be to identify clusters in data and to determine lower-dimensional structures in multidimensional data. We will use the methods and measurements that we discussed in Section 9.2 beginning on page 383.

Interesting structure may involve clusters of data, or it may be the result of the data lying on or near a space of reduced dimension. Interesting structure may also be defined generically as properties of the data that differ from expected properties if the data were a random sample from a multivariate normal distribution or from some other standard distribution. The normal (or Gaussian) distribution lies at the heart of many methods of data analysis. The heuristic definition of structure as a departure from normality can be motivated by the fact that most randomly selected low-dimensional projections of any high-dimensional dataset will appear similar to a random sample from a multivariate normal distribution (see Diaconis and Freedman, 1984).

The usual objective in cluster analysis is to divide the observations into groups that are close to each other or are more homogeneous than the full set of observations. An observation may consist of categorical variables that may (or may not) specify the class to which the observation belongs. In general, as we discuss on page 385, if the ith observation can be represented as

$$x_i = (x_i^{\text{r}},\ x_i^{\text{c}}), \tag{16.1}$$

where the subvector x_i^{c} represents values of the categorical variables, we may wish to handle the x_i^{c} component separately. In Figure 16.3, for example, suppose that each observation consists of values for three variables, x_1 and x_2 as shown and a third variable that represents group membership that corresponds to the symbol in the graphical depiction. In that case, the classes may already be defined, or we may want to allow the possibility that observations with different values of the categorical variable nevertheless belong to the same class. In most of the following, we will assume that none of the variables are categorical.

16.1 Clustering and Classification

Identifying groups of similar observations in a dataset is an important step in making sense of the data and in understanding the phenomena represented by the data. Clustering, classification, and discrimination are terms that describe this activity, which lies at the crossroads of a number of traditional disciplines, including statistics, computer science, artificial intelligence, and electrical engineering. Classification is sometimes called *statistical learning* or *machine learning*, especially in the more engineering-oriented disciplines. As is often the case when scientific methods are developed within diverse areas, there are several slight variations of theory and methodology, which are sometimes described as "statistical", "inductive", and so on. The slight variations lead to a variety of terms to describe the methods, and there is generally scant theory to support optimality of one method over another. The various approaches to clustering and classification also lead to the use of terms such as "hypothesis", "bias", and "variance" that have different meanings from their technical statistical definitions.

Clustering and classification make use of a wide range of statistical techniques, both descriptive methods utilizing simple summary statistics and graphics and methods of fitting equations to data. Statistical techniques in clustering and classification often emphasize uncertainty and the importance of dealing with noise in the data. A good general reference on clustering and classification, generally from a statistical perspective, is Gordon (1999). Hastie, Tibshirani, and Friedman (2009) discuss classification using terminology from both the statistics and machine learning disciplines.

The first step in forming groups is to develop a definition of the groups. This may be based on similarities of the observations or on closeness of the observations to one another.

Clustering

Cluster analysis is generally exploratory. It seeks to determine what groups are present in the data. If the groups are known from some training set, "discriminant analysis" seeks to understand what makes the groups different and then to provide a method of classifying observations into the appropriate groups. When discriminant analysis is used to "train" a clustering method, we refer to the procedure as "supervised" classification. Discriminant analysis is mechanically simpler than cluster analysis. Clustering is "unsupervised" classification. We will discuss classification in Chapter 17.

Because of the large number of possibilities for grouping a set of data into clusters, we generally must make some decisions to simplify the problem. One way is to decide a priori on the number of clusters; this is done in K-means clustering, discussed below. Another way is to do recursive clustering; that is, once trial clusters are formed, observations are not exchanged from one cluster to another. Two pairs of observations that are in different clusters at one stage of the clustering process would never be split so that at a later stage one member of each pair is in one cluster and the other member of each pair is in a different cluster.

There are two fundamentally different approaches to recursive clustering. One way is to start with the full dataset as a single group and, based on some reasonable criterion, partition the dataset into two groups. This is called divisive clustering. The criterion may be the value of some single variable; for example, any observation with a value of the third variable larger than 5 may be placed into one group and the other observations placed in the other group. Each group is then partitioned based on some other criterion, and the partitioning is continued recursively. This type of divisive clustering or partitioning results in a classification tree, which is a decision tree each node of which represents a partition of the dataset.

Another way of doing recursive clustering is to begin with a complete clustering of the observations into singletons. Initially, each cluster is a single observation, and the first multiple-unit cluster is formed from the two closest observations. This agglomerative, bottom-up approach is continued so that at each stage the two nearest clusters are combined to form one bigger cluster.

K-Means Clustering

The objective in K-means clustering is to find a partition of the observations into a preset number of groups, k, that minimizes the variation within each group. The variation of the jth variable in the gth group is measured by the within sum-of-squares,

$$s^2_{j(g)} = \frac{\sum_{i=1}^{n_g} \left(x_{ij(g)} - \bar{x}_{j(g)}\right)^2}{n_g - 1}, \tag{16.2}$$

where n_g is the number of observations in the gth group, and $\bar{x}_{j(g)}$ is the mean of the jth variable in the gth group. Assuming there are m variables in each observation, there are m such quantities.

A measure of the overall variation within each group is some combination of the individual $s^2_{j(g)}$. As we have suggested, in most multivariate analyses, it is best to scale or normalize the data prior to the analysis (see page 391). If the X data have been normalized, then an appropriate measure for the variation within a group is just the the sum of the within sum-of-squares for each variable, $\sum_{j=1}^{m} \sum_{i=1}^{n_g} (x_{ij(g)} - \bar{x}_{j(g)})^2$.

Now, to state more precisely the objective in K-means clustering, it is to find a partition of the observations into a preset number of groups k that minimizes, over all groups, the total of the linear combinations of the within sum-of-squares for all variables. For linear combinations with unit coefficients, this quantity is

$$w = \sum_{g=1}^{k} \sum_{j=1}^{m} \sum_{i=1}^{n_g} \left(x_{ij(g)} - \bar{x}_{j(g)}\right)^2. \tag{16.3}$$

Determining the partitioning to minimize this quantity is a computationally intensive task.

In practice, we seek a local minimum (that is, a solution such that there is no single switch of an observation from one group to another group that will decrease the objective). Even the procedure used to achieve the local minimum is rather complicated. Hartigan and Wong (1979) give an algorithm (and Fortran code) for performing the clustering. Their algorithm forms a set of initial trial clusters and then transfers observations from one cluster to another while seeking to decrease the quantity in equation (16.3). Simulated annealing can also be used to do K-means clustering.

Most of the algorithms for K-means clustering will yield different results if the data are presented in a different order. The algorithms, such as simulated annealing, that use techniques that depend on random numbers may yield different results on different runs with the same data in the same order.

In whatever method is used for K-means clustering, it is necessary to choose initial points, and then trial points to move around. The points can be chosen randomly or chosen arbitrarily. A random choice gives preference to points in dense regions, which is consistent with an underlying concept of K-means clustering in which the inner two sums in expression (16.3) are similar to an expected value with respect to a distribution from which the observed data constitute a random sample.

The clustering depends on the variability of the variables. It is generally best to scale the variables in order for the clustering to be sensible because the larger a variable's variance, the more impact it will have on the clustering. See page 533, however, for further discussion of the issue of scaling variables prior to clustering.

Choosing the Number of Clusters

A major issue is how many clusters should be formed. The question of the number of groups must generally be addressed in an ad hoc manner. Most algorithms form nonempty clusters, so the number of clusters is pre-specified. Minimizing the within-groups sum-of-squares, w, in equation (16.3) leads to exactly k clusters except in an extreme case of multiple observations with the same values, which yields multiple solutions. For a fixed k, w is function of the assignments of observations to groups, indicated in the equation by the subscript $ij(g)$.

If k is also made a decision variable in the optimization problem of minimizing w, that is, if w is also a function of k, then $w(k,g)$ is minimized by increasing k until each n_g is 1. Hence, it is not appropriate simply to make k a variable in the objective function.

We may modify the objection function $w(k,g)$ with a penalty for the number of groups. The simplest penalty is just division by the degrees of freedom $n-k$. Even if we normalize w by the degrees of freedom, however, the minimum value occurs at a point where k is fairly large.

In addition to homogeneity of the observations within each group, we also seek heterogeneity of the groups; hence, intuitively, an F-like statistic as a ratio of the between-group dissimilarity to the within-group dissimilarity could be used to indicate the goodness of a given clustering. We define a "pseudo F":

$$\widetilde{F}_k = \frac{b/(k-1)}{w/(n-k)}, \tag{16.4}$$

where b is the between-groups sum-of-squares,

$$b = \sum_{g=1}^{k} \sum_{j=1}^{m} \left(\bar{x}_{j(g)} - \bar{x}_j \right)^2,$$

and w is the pooled within-groups sum-of-squares of equation (16.3). This measure is also called the Calinski-Harabasz index after Calinski and Harabasz (1974) who suggested its use as a stopping criterion in hierarchical clustering (see below). The larger that ratio, the better the clustering at any fixed number of clusters. In general, however, the within-groups sum-of-squares of equation (16.3) will decrease more quickly than the between-groups sum-of-squares will decrease as more groups are formed, so the pseudo F favors a value of k that is too large.

A practical approach to the problem is to vary k, as $k_0, k_0 + 1, \ldots$, computing $\widetilde{F}_k$, which will initially increase fairly rapidly, and to choose the value of k as the point at which $\widetilde{F}_{k+1} - \widetilde{F}_k$ is relatively small. This is similar to the type of approach in a different context; that is, choosing the number of principal components, which we consider on page 554.

Hierarchical Clustering

It is useful to consider a hierarchy of clusterings from a single large cluster to a large number of very small clusters. Hierarchical clustering yields these alternative clusterings.

The results of a hierarchical clustering can be depicted as a tree, as shown in Figure 16.5. Each point along the bottom of the tree may correspond to a single observation. Nodes higher up in the diagram represent successively larger groups. The number of clusters depends on the level in the tree, as indicated in the plot.

The vertical axis, which is not shown in Figure 16.5, corresponds to a distance metric, which could be based on any of the measures of distance described beginning on page 388. The actual units of the vertical axis must be interpreted in that context.

Many of the algorithms for hierarchical clustering will yield different results if the data are presented in a different order.

Agglomerative Hierarchical Clustering

In agglomerative hierarchical clustering, we first begin with a large number of clusters, generally as many as the number of observations, so that each cluster consists of a single observation, and then we combine clusters that are nearest to each other.

To define distances between groups, we must first consider from what points within the groups to measure distance. As we mentioned on page 393, there are several ways of doing this. One way is to measure the distance between a central point in one group, such as the mean or median of the group,

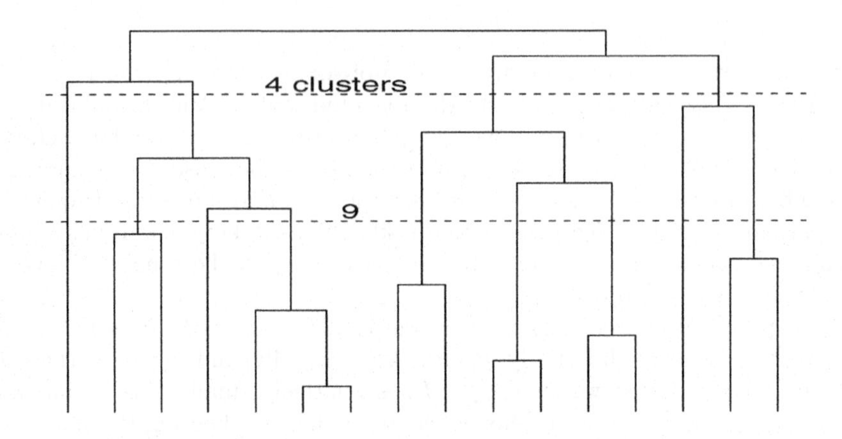

Fig. 16.5. A Cluster Tree. Each Leaf Represents an Observation or a Group of Observations

and the corresponding central point in the other group. These methods often do not work very well in hierarchical clustering. In agglomerative hierarchical clustering, the distance between two clusters is usually chosen in one of the following three ways.

- The minimum distance between a point in the first cluster and a point in the second cluster. Using this criterion results in what is sometimes called "single linkage" clustering.
- The distance between clusters is the average of the distances between the points in one cluster and the points in the other cluster.
- The largest distance between a point in one cluster and a point in the other cluster. Using this criterion results in what is sometimes called "complete linkage" clustering.

In addition to the choice of the two points to define the distance, different distance metrics can be chosen. Any of the distances described beginning on page 388 could be used, and in a given situation, one may be more appropriate than another. Most clustering methods use an L_2 metric. Other metrics will produce different clusters, and it may be appropriate to consider the clustering produced by various metrics.

By changing the distance metric and the clustering method, several different cluster trees can be created from a single dataset. Which one is more appropriate depends on the situation. For example, if the data has outliers in one or more dimensions, an L_1 metric might yield clusters that appear more reasonable.

We can see the differences in hierarchical clustering with different distance measures between clusters using a simple example that consists of five observations with the distance matrix

$$D = \left[\begin{array}{c|cccc} & 2 & 3 & 4 & 5 \\ \hline 1 & 4.34 & 4.58 & 7.68 & 4.47 \\ 2 & & 1.41 & 4.00 & 4.36 \\ 3 & & & 5.10 & 5.00 \\ 4 & & & & 6.56 \end{array}\right]. \tag{16.5}$$

Using either type of distance measure, the first cluster is formed from observations 2 and 3 because 1.41 is the minimum in any case. The subsequent clusters are different in the three methods, as shown in Figure 16.6 by the matrices that contain distances between clusters.

$$\left[\begin{array}{c|cccc} & 2 & 3 & 4 & 5 \\ \hline 1 & 4.34 & 4.58 & 7.68 & 4.47 \\ 2 & & \underline{1.41} & 4.00 & 4.36 \\ 3 & & & 5.10 & 5.00 \\ 4 & & & & 6.56 \end{array}\right]$$

single linkage ("connected") | average | complete linkage ("compact")

$$\left[\begin{array}{c|ccc} & 2,3 & 4 & 5 \\ \hline 1 & 4.34 & 7.68 & 4.47 \\ 2,3 & & \underline{4.00} & 4.36 \\ 4 & & & 6.56 \end{array}\right] \quad \left[\begin{array}{c|ccc} & 2,3 & 4 & 5 \\ \hline 1 & \underline{4.46} & 7.68 & 4.47 \\ 2,3 & & 4.55 & 4.68 \\ 4 & & & 6.56 \end{array}\right] \quad \left[\begin{array}{c|ccc} & 2,3 & 4 & 5 \\ \hline 1 & 4.58 & 7.68 & \underline{4.47} \\ 2,3 & & 5.10 & 5.00 \\ 4 & & & 6.56 \end{array}\right]$$

$$\left[\begin{array}{c|cc} & 2,3,4 & 5 \\ \hline 1 & \underline{4.34} & 4.47 \\ 2,3,4 & & \underline{4.36} \end{array}\right] \quad \left[\begin{array}{c|cc} & 4 & 5 \\ \hline 1,2,3 & 6.12 & \underline{4.58} \\ 4 & & 6.56 \end{array}\right] \quad \left[\begin{array}{c|cc} & 2,3 & 4 \\ \hline 1,5 & \underline{4.58} & 7.68 \\ 2,3 & & 5.10 \end{array}\right]$$

$$\Big(5, 1, (4, (2, 3))\Big) \qquad \Big(4, \big(5, (1, (2, 3))\big)\Big) \qquad \Big(4, ((2, 3), (1, 5))\Big)$$

Fig. 16.6. Hierarchical Clustering Using Three Different Methods

In this example, we have carried the clustering to a single final cluster. The clusters at any intermediate stage except the first are different. Thus, in complete linkage, for example, after the cluster with observations 2 and 3 is formed, a separate cluster with observations 1 and 5 is formed; then, these two clusters are grouped into a cluster with four observations, and finally observation 4 is added to form a single cluster.

Another agglomerative hierarchical clustering method proceeds by forming the clusters in such a way that each new cluster leads to a minimum increase in the total within-cluster sums of squares, equation (16.2). Beginning with all clusters consisting of single observations, this total is 0. The closest two points are combined to form the first cluster with two elements. In the example in Figure 16.6, this would be observations 2 and 3, the same as in all of the other methods. Assuming that the distances are Euclidean distances in D in equation (16.5), the increase in the sum of squares is $1.41^2/2$. This is sometimes called Ward's method, from Ward (1963).

Figure 16.7 shows the cluster trees that result from each method of clustering. The lengths of the vertical lines indicate the closeness of the clusters that are combined. In each tree, for example, the first level of combination (between observations 2 and 3) occurred at a measure of 1.41, as shown on the vertical scale. In the connected linkage, as shown in the tree on the left-hand side, the second step was to add observation 4 to the cluster containing observations 2 and 3. This combination occurred at a measure of 4.00. On the other hand, in the compact linkage, as shown in the tree on the right-hand side, the cluster containing observations 2 and 3 was unchanged and a second cluster was formed between observations 1 and 5 at a measure of 4.47.

The cluster trees in Figure 16.7 were produced with the following R commands:

```
plclust(hclust(D,method="single"))
plclust(hclust(D,method="average"))
plclust(hclust(D,method="complete"))
plclust(hclust(D,method="ward"))
```

The height h_{ik} at which the two observations i and k enter the same cluster is a measure of the closeness of the two observations. For any reasonable dissimilarity matrix and any reasonable method of linkage, the heights will satisfy the ultrametric inequality

$$h_{ik} \leq \max_j(h_{ij}, h_{kj}).$$

This property is trivially satisfied by the linkages illustrated in Figures 16.6 and 16.7, as may be easily checked.

Consideration of the heights in a hierarchical clustering may suggest the appropriate number of clusters. For example, the relative heights separating clusters in the tree in the upper left in Figure 16.7 indicates that there may be four clusters: (5), (1), (4), and (2,3), three of which are singletons. (Remember that this is a toy dataset!) The tree on the lower left indicates that there may be three clusters: (4), (2,3), and (1,5).

The R function agnes also does agglomerative hierarchical clustering and provides more information about the levels at which clusters are combined.

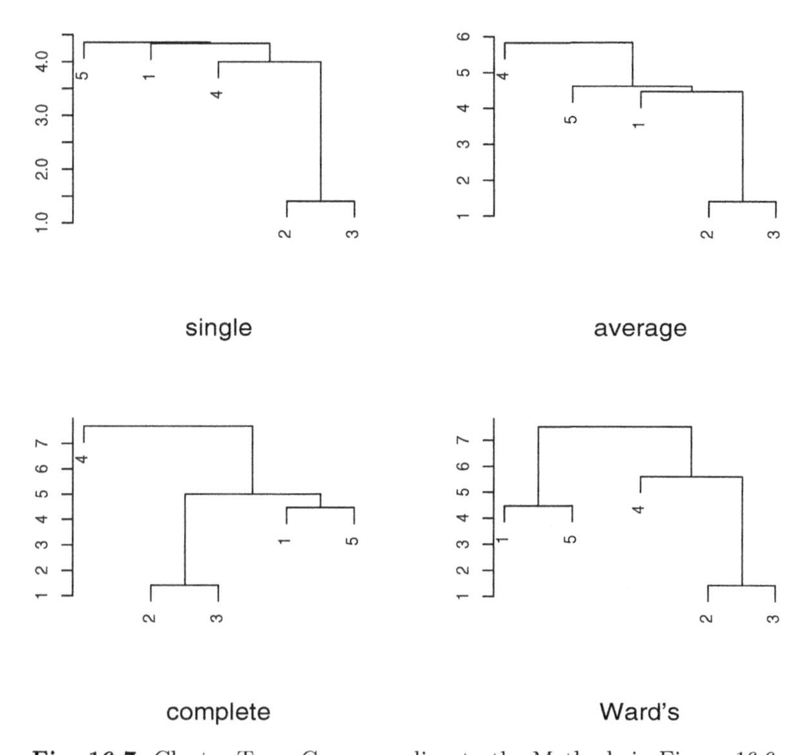

Fig. 16.7. Cluster Trees Corresponding to the Methods in Figure 16.6

It is important to note the computational and storage burdens in agglomerative hierarchical clustering that begins with individual observations. The size of the distance matrix (D in the example above) is of order $O(n^2)$.

Model-Based Hierarchical Clustering

In the general clustering problem, we may assume that the data come from several distributions, and our problem is to identify the distribution from which each observation arose. Without further restrictions, this problem is ill-posed; no solution is any better than any other. We may, however, impose the constraint that the distributions be of a particular type. We may then formulate the problem as one of fitting the observed data to a mixture of distributions of the given type. The problem posed thusly is similar to the problem of density estimation using parametric mixtures, as we discuss in Section 14.3 beginning on page 480. The R function `mclust` performs model-based clustering.

Divisive Hierarchical Clustering

Most hierarchical clustering schemes are agglomerative; that is, they begin with no clusters and proceed by forming ever-larger clusters. In divisive hierarchical clustering, we begin with a single large cluster and successively divide the clusters into smaller ones.

Kaufman and Rousseeuw (1990) have described a divisive hierarchy in which clusters are divided until each cluster contains only a single observation. At each stage, the cluster with the largest dissimilarity between any two of its observations is selected to be divided. To divide the selected cluster, the observation with the largest average dissimilarity to the other observations of the selected cluster is used to define a "splinter group". Next, observations that are closer to the splinter group than to their previous groups are assigned to the splinter group. This is continued until all observations have been assigned to a single cluster. The result is a hierarchical clustering. The R function `diana` in the `cluster` package determines clusters by this method.

Other Divisive Clustering Schemes

Because of the computational time required in agglomerative clustering or global partitioning such as by K-means, for large datasets, simpler methods are sometimes more useful. A recursive partitioning scheme can be efficient. One simple recursive method groups the observations into hyperrectangular regions based on the medians of the individual variables. In the first step of the median-split divisive scheme, the n observations are divided into two sets of $n/2$ based on the median of the variable with the largest range. The subsequent steps iterate that procedure. At any stage, the number of observations in all clusters is nearly equal. This procedure and the motivation for it are closely related to the k-d-tree (see page 547). A related scheme uses the mean rather than the median. This scheme is less intensive computationally. It does not have the property of almost equal-size clusters, however.

Clustering and Classification by Space Tessellations

Groups in data can naturally be formed by partitioning a space in which the data are represented. If the data are represented in a cartesian coordinate system, for example, the groupings can be identified by polytopes that fill the space. Groups are separated by simple planar structures.

Groups may be formed by the values of only a subset of the variables. A simple example is data in which one or more variables represent geographic location. Clusters may be defined based on location, either by methods such as we have discussed above or by regions that tessellate the space. The tessellations may be preassigned regions, perhaps corresponding to administrative or geographical boundaries.

More interesting tessellations can be constructed from data. Let T be a set of points, possibly all of the observations in a dataset, a random sample of the data, or some statistics formed from subsets of the data. The set T may be used to form a tessellation that defines groups, and the tessellation may be used to classify additional data. Formation of a subset of observations for classification is a form of dimension reduction, which, as we have noted with caveats, is one of the general approaches for understanding multivariate data.

A simple and useful tessellation constructed from a dataset is the *Dirichlet tessellation*, or the *Voronoi tessellation*. (The names are synonymous.) This tiling forms regions containing single points of T in such a way that all points within a region are closer to the given point than they are to any other point in T. The points in T that form the tessellation are called *generators*. The points on a boundary in a Dirichlet tessellation are equidistant to two points in T. This type of tessellation generalizes to higher dimensions.

The set of edges of the polygons (or faces of polyhedra or hyperpolyhedra) on which points are equidistant to two points in T is called the *Voronoi diagram*. The Dirichlet tessellation determined by a set of six points is shown in Figure 16.8.

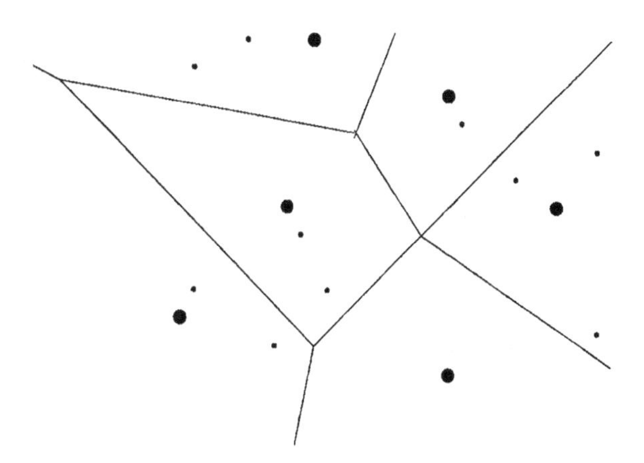

Fig. 16.8. A Dirichlet Tessellation in a Plane Formed by Six Generator Points

The other points shown in Figure 16.8 are clustered with respect to the tessellation formed by the given six points.

A unique set of simplices is obtained by joining all nodes that share an edge in the Voronoi diagram. The set of triangles (which are simplices in two dimensions) formed by the Dirichlet tessellation in Figure 16.8 is shown in

Figure 16.9. This construction is called a Delaunay triangulation. The triangulation is also a tessellation and is sometimes called a Delaunay tessellation. The Voronoi diagram and the Delaunay triangulation are duals of each other; one determines the other.

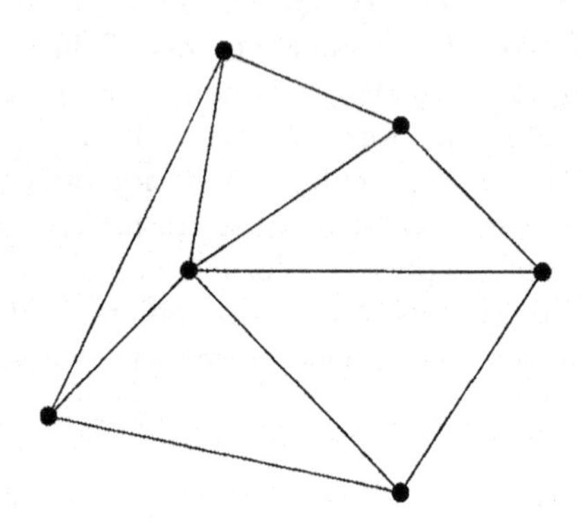

Fig. 16.9. A Delaunay Triangulation

The Dirichlet tessellation or Voronoi diagram and the Delaunay triangulation have many interesting properties. One important property of the Delaunay triangulation is that it is the unique triangulation that maximizes the minimum angle in a grid formed from a fixed set of vertices. This property is easy to see in two dimensions. (See Figure 16.10 for an example of another triangulation that obviously lacks this property when compared to Figure 16.9.) This property makes the Delaunay triangulation very useful in various fields of scientific computation. For example, it is a good way to form a set of solution points for the numerical solution of partial differential equations. (This is an “unstructured grid”.)

Another property of the Voronoi diagram and the associated Delaunay triangulation in two dimensions is that a circle centered at a point where three Voronoi tiles meet and that passes through the vertices of the Delaunay triangle enclosing that point will not contain a vertex of any other triangle. (There are possible degeneracies when multiple points are collinear, but the property still holds when more than three tiles meet.) This property also holds in higher dimensions for spheres and hyperspheres.

The Bowyer-Watson algorithm exploits this property for computing a Delaunay triangulation (Bowyer, 1981, Watson, 1981). Starting with $d+1$ points

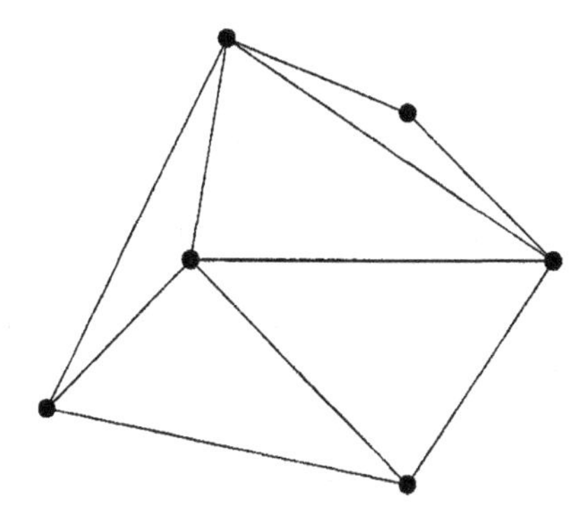

Fig. 16.10. Another Triangulation

and a simplex, the algorithm proceeds by recursive insertion of nodes. For each new node:

1. Find any simplices whose circumscribed hyperspheres include the new node.
2. Create a cavity by eliminating these simplices (if there are any).
3. Create the new set of simplices by connecting the new point to the nodes that define this cavity.

This triangulation is relatively simple to implement in two dimensions, as in the simple application of unstructured grids for the numerical solution of partial differential equations. O'Rourke (1998) and Lee (1999a, 1999b) provide general descriptions of computational methods for Delaunay triangulations as well as other problems in computational geometry. Various programs for performing the tessellations and other computations in d dimensions are available at

www.geom.umn.edu/software/download/

Renka (1997) gives an algorithm for computing the Delaunay tessellation on the surface of a sphere.

A special type of Voronoi tessellation is one in which the generators are the centroids of the regions of the tessellation. This is called a *centroidal Voronoi tessellation.* A centroidal Voronoi tessellation with k regions can be formed by an iterative routine in which a random set of k generators is chosen, the Voronoi tessellation is determined, and the centroids of the regions are taken as generators for a new tessellation. The generators, which were the centroids of the previous tessellation, will not in general be the centroids of the new tessellation, so the iterations are continued until a stopping criterion is achieved. See Lloyd (1982) for descriptions and properties of the process. Kieffer (1983) proved convergence of the process for a fairly restricted class of problems. General convergence properties are open questions.

A tessellation of a finite point set, T, can be defined in terms of a tiling over a continuous region. The points within a given tile form the finite set of points within a given tessellation of the set T. A K-means clustering is a centroidal Voronoi tessellation of a finite point set in which the means of the clusters are the generators.

A minimal spanning tree (see page 538) can also be used to cluster by tessellations, as shown in Figure 16.18 on page 547.

Meanings of Clusters; Conceptual Clustering

Identification of clusters in a dataset is usually only a small part of a larger scientific investigation. Another small step toward the larger objective of understanding the phenomenon represented by the data is to characterize the groups in the data by simple descriptions in terms of ranges of individual variables. For the case of hierarchical clustering, decision trees may be an effective way of describing the clustering. A decision tree can be expressed as a set of conjunctive rules that can aid in understanding the phenomenon being studied. The rules that define classes can be formulated in terms of either numerical or categorical variables. We consider this way of summarizing a clustered dataset again on page 623.

If the intent of an analysis is interpretation or understanding of the phenomenon that gave rise to the data, simplicity of the description of clusters has great value, even if it is achieved at some cost in accuracy. If, on the other hand, the objective is an ad hoc classification of the observations in a given dataset, simplicity is not important, and often an algorithmic "black box" is a more effective classifier (see Breiman, 2001).

We could formulate the clustering problem so that there are unobserved categorical variables whose values range over a predetermined set. In this situation, the observation x may be represented as

$$x = (x^{\mathrm{r}},\, x^{\mathrm{c}}), \tag{16.6}$$

as we have discussed on page 519, but we may not observe x^{c} directly. We may know, however, that $x^{\mathrm{c}} \in \mathcal{C}$, where $\mathcal{C}$ is some given set of characteristics.

In another variation, we may have a set of characteristics of interest, $\mathcal{C}$, and wish to assign to each observation a variable x^{c} that takes a value in our previously identified set $\mathcal{C}$. The set $\mathcal{C}$ may consist of rules or general characteristics. The characteristics may be dependent on the context, as we discussed on page 388. The set of characteristics of interest may include such things as "lies on a straight line with negative slope", for example. The data represented by triangles in the graph in the lower left of Figure 16.3 on page 517 would have this characteristic, so all of those points are similar in that respect.

This kind of similarity cannot be identified by considering only pairwise similarity measurements. Michalski (1980) and Michalski and Stepp (1983) described methods for clustering using sets of characteristics, or "concepts". They called this approach conceptual clustering.

Fuzzy Clustering

Fuzzy set theory has been applied to clustering, as it has to most problems that have a component of uncertainty. Instead of observations being grouped into definite or "crisp" clusters, they are given membership probabilities. The membership probability of the ith observation in the gth group is u_{ig}. The memberships satisfy

$$0 \leq u_{ig} \leq 1$$

and

$$\sum_{g=1}^{k} u_{ig} = 1 \quad \text{for all } i = 1, \ldots, n.$$

The quantity analogous to equation (16.3) in standard K-means clustering is

$$\sum_{g=1}^{k} \sum_{j=1}^{m} \sum_{i=1}^{n} u_{ig}^2 \left(x_{ij} - \bar{x}_{j(g)} \right)^2, \tag{16.7}$$

where, as before, $\bar{x}_{j(g)}$ is the mean of the jth element of the vectors x_i that are in the gth group. Because group membership is a weight, however,

$$\bar{x}_{j(g)} = \frac{\sum_{i=1}^{n} u_{ig}^2 x_{ij}}{\sum_{i=1}^{n} u_{ig}^2}.$$

Clustering and Transformations of the Data

As we discuss on page 394, transformations on the data may change the relative values of measures of similarity. This, of course, affects any method of analysis that depends on measures of similarity. A severe limitation of clustering results from the dependence of the clusters on the scaling of the data. In many data-analytic procedures, we perform various linear transformations on the data, with predictable results on the analysis. For example, we often perform a simple univariate standardization of the data by subtracting the sample mean and dividing by the sample standard deviation. For the typical data matrix X whose columns represent variables and whose rows represent multivariate observations, we may standardize each variable by subtracting the column mean from each value in the column and dividing by the standardization of the column. Doing this, however, affects the clustering, as seen in Figure 16.11.

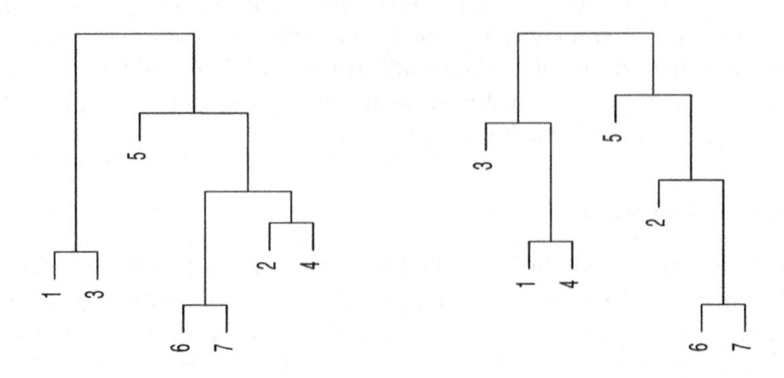

Fig. 16.11. Cluster Trees; Raw Data and Standardized Data

The cluster trees in Figure 16.11 were produced with the following R commands that first create a matrix of seven observations with five variables and perform hierarchical clustering (using largest distances) and then standardize the data univariately and perform the same hierarchical clustering. (As mentioned in Section 7.6, the same seed and `rnorm` function in R do not produce the same data as in S-Plus. Also, there is no guarantee that the S-Plus code executed on a different computer system will produce the same data.)

```
set.seed(3)
x <- matrix(rnorm(35),ncol=5)
plclust(hclust(dist(x)),axes=F,sub="",xlab="",ylab="")
#  univariate standardization
standard <- function(vec) (vec-mean(vec))/sqrt(var(vec))
y<-apply(x,2,standard)
plclust(hclust(dist(y)),axes=F,sub="",xlab="",ylab="")
```

The dependence of the clustering on transformations of the data results from the effect on the distance measures discussed in Section 9.2 beginning on page 383. Whether one variable is measured in grams or kilograms affects the relative distance of any one observation to the other observations. If all variables in the dataset are of the same type (mass, say), it is easy to measure them all in the same units; if some are of one type and some are of another type, decisions on units are not as easy. These decisions, however, affect the results of clustering. As we mentioned in Section 1.1 scaling induces artificial structure. Cluster analysis is sensitive to artificial structure.

We also observe effects of transformations of the data on other structures in the data, such as some we discuss in later sections.

Transformations are useful in finding other types of structure in data, as we will see in later sections. Even in the identification of clusters, transformations can help. The two clusters called "concentric circles" in Figure 16.3 on page 517 could be identified easily using any of the clustering methods discussed in the section if the data were centered and then transformed to polar coordinates. (Other methods for this kind of structure may involve "slicing" the data.)

Clustering of Variables

There is a basic duality between the m "variables" and the n "observations" of the dataset X. We have been discussing clustering of observational units. Clustering observations is done by measures of distance, possibly scaled by S. Consider reversing the roles of variables and observations. Suppose that we wish to cluster the variables (that is, we wish to know which variables have values that are strongly related to each other).

The relative values of the variables provide information on how similar or dissimilar the observations are; conversely, the relative values of the multivariate observations provide information on the similarity of the variables. Clustering of variables is conceptually and mechanically the same as clustering of observations. Instead of an $n \times n$ matrix of dissimilarities between observations, such as D in equation (9.20), we would use an $m \times m$ matrix of association between variables, such as S in equation (9.8) or R in equation (9.11) on page 385.

There is one obvious difference in the variance-covariance matrix or the correlation matrix and the dissimilarity matrix: Covariances and correlations can be positive or negative. Positive and negative covariances or correlations of the same magnitude, however, represent the same degree of association between the variables, so instead of S or R, similar matrices with all elements replaced by their absolute values are more useful for clustering variables. See Soffritti (1999) for some comparisons of various ways of using these and other measures of association for clustering variables.

Comparing Clusterings

As we have seen, various methods of clustering yield different results, and, furthermore, the same method yields different results if the data have been transformed. Which clustering is best cannot in general be determined by analysis of data with no context. The purpose of the clustering, after all, is to develop a better understanding of a phenomenon of which the data measure various aspects. Nevertheless, it is instructive to develop numerical measures of the agreement (or, equivalently, disagreement) of different clusterings of the same dataset.

A two-way contingency table can be used to represent agreement of two clusterings. (A p-way contingency table could be used to represent agreement of p clusterings.) If the classes of one clustering are denoted as $C_{11}, \ldots, C_{1k_1}$ and those of a second clustering as $C_{21}, \ldots, C_{2k_2}$, a two-way table of the numbers of units falling in the cells is constructed, as shown.

	C_{11}	$\ldots$	C_{1k_1}	
C_{21}	n_{11}	$\ldots$	n_{1k_1}	$n_{1\bullet}$
$\vdots$		$\ddots$		$\vdots$
C_{2k_2}	n_{k_21}	$\ldots$	$n_{k_2k_1}$	$n_{k_2\bullet}$
	$n_{\bullet 1}$	$\ldots$	$n_{\bullet k_2}$	n

The labeling of the clusters is arbitrary. (In a classification problem, the clusters correspond to classes, which are usually known and fixed, given the data.)

From the cluster trees shown in Figure 16.11, there appear to be two obvious clusters in the first clustering and three clusters in the second clustering. If we identify the clusters from left to right in each tree (so that, for example, the first cluster in the first tree contains the points 2, 5, and 7, and the first cluster in the second tree contains the single point 3), we would have the table below.

	C_{11}	C_{12}	
C_{21}	0	1	1
C_{22}	3	0	3
C_{23}	0	3	3
	3	4	7

The marginal totals are the counts for the corresponding clusters. The numbers in the cells indicate the extent of agreement of the two clusterings. Perfect agreement would yield, first of all, $k_1 = k_2$, and, secondly, a table in which each column and each row contains only one nonzero value.

Rand (1971) suggested a measure of the agreement of two clusterings by considering the number of pairs of points that are in common clusters. Of the total of $\binom{n}{2}$ pairs of points, each pair may be:

1. in the same cluster in both clusterings;
2. in different clusters in both clusterings;
3. in the same cluster in one clustering but in different clusters in the other clustering.

Both the first and second events indicate agreement of the clusterings, and the third indicates disagreement. Rand's statistic is a count of the number of pairs of the first and second types divided by the total number of pairs. This statistic is obviously in the interval $[0, 1]$, and a value of 0 indicates total disagreement and a value of 1 complete agreement. Rand gave a method of computing the total number of pairs of the first and second types by subtracting the count of the number of the third type from the total:

$$\binom{n}{2} - \frac{1}{2}\left(\sum_i n_{i\bullet}^2 - 2\sum_i\sum_j n_{ij}^2 + \sum_j n_{\bullet j}^2\right).$$

This can be seen by expanding $(\sum_i \sum_j n_{ij})^2$. Rand's statistic therefore is

$$R = 1 - \frac{\sum_i n_{i\bullet}^2 - 2\sum_i\sum_j n_{ij}^2 + \sum_j n_{\bullet j}^2}{n(n-1)}. \tag{16.8}$$

For the two clusterings shown in Figure 16.11, with two clusters in the first clustering and three clusters in the second, we see that the count of agreements is 18; hence, Rand's statistic is 6/7.

Statistical significance may be determined in terms of the distribution of such a statistic given random clusterings. Hubert and Arabie (1985) studied and modified Rand's measure to account for the expected values of random clusterings. Their statistic is

$$R_{HA} = \frac{\sum_i\sum_j \binom{n_{ij}}{2} - \sum_i \binom{n_{i\bullet}}{2}\sum_j \binom{n_{\bullet j}}{2}/\binom{n}{2}}{\left(\sum_i \binom{n_{i\bullet}}{2} + \sum_j \binom{n_{\bullet j}}{2}\right)/2 - \sum_i \binom{n_{i\bullet}}{2}\sum_j \binom{n_{\bullet j}}{2}/\binom{n}{2}}, \tag{16.9}$$

where the sums over i go to k_1 and the sums over j go to k_2. In practice, the statistical significance of such a comparison of two observed clusterings would be approximated by use of a few randomly formed clusterings.

An interesting problem that has received very little attention is to develop methods for drawing cluster trees to facilitate comparisons of clusterings.

There have been and will continue to be a multitude of Monte Carlo studies assessing the performance of various clustering procedures in various settings. The difficulty in making a simple statement about the performance of clustering methods arises from the multitude of possible clustering patterns.

Computational Complexity of Clustering

The task of identifying an unknown number of clusters that are distinguished by unknown features is an exceedingly complex problem. In practical clustering methods, there are generally trade-offs between how clusters are defined and the algorithm used to find the clusters. In the hierarchical clustering algorithms, the algorithm dominates the approach to the problem. In those hierarchical clustering methods, the definition of clusters at any level is merely what results from a specified algorithm. After the $\mathrm{O}(mn^2)$ computations to compute the distance matrix, the algorithm requires only $\mathrm{O}(n)$ computations. As we have discussed, identification of clusters may involve concepts and iterations requiring human interactions. Even with a fixed algorithm-based approach, however, the method is computationally intensive.

K-means clustering begins with a reasonable definition of clusters, assuming a known number of clusters. Even with the simplifying assumption that the number of clusters is known, the definition of clusters requires a very computationally intensive algorithm. Just to compute the objective function (16.3) on page 521 for a given trial clustering requires kmn computations. A trial clustering is defined by a permutation of the n data elements together with a choice of k nonnegative integers n_g such that $\sum n_g = n$. Clearly, the number of computations required to satisfy the definition of clusters, even under the assumption of a known number of clusters, is often not acceptable.

Development of clustering algorithms that are feasible for large datasets is an important current activity.

16.2 Ordering and Ranking Multivariate Data

For univariate data, many nonparametric methods and methods of comparing distributions use ranks of the data. Many of these methods, such as q-q plots, tests for equality of distributions of two samples, and nonparametric regression can be extended to multivariate data once there are meaningful ways of ranking the data, such as by minimal spanning trees. The rankings can also be used for systematically exploring graphical projections of the data such as in a grand tour.

The concept of order or rank within a multivariate dataset can be quite complicated. (See Barnett, 1976, and Eddy, 1985, for general discussions of the problem.) The simple approach of defining a "sort key" that consists of a priority ordering of the variables to use in ranking the data is not very useful except in simple, well-structured datasets.

Ranking of a multivariate dataset means among other things that we identify exactly one "first" and one "last" in the data. We may decide that the first and last are both on the "outside" boundary of the data, and so from one extreme to the other, the rankings pass through the central portion of the data. This *extreme ranking* is the natural way we rank univariate ordinal data, which, of course, has a natural "smallest" and "largest" element. In the multivariate case, there may be natural extremes, but sometimes which extreme is smallest and which is largest may be arbitrary. Another way of thinking of the first and last is in terms of "most central" first (or last), and "farthest out" as last (or first). This is a *radial ranking*, and is rarely used in ordering univariate data, but for multivariate data it may be the most useful way of ranking the data.

In this section, we will describe three approaches to ranking multivariate data. They are based on minimal spanning trees, on convex hulls, and on data clusters. Each general approach may yield different rankings depending on how the extremes are identified.

Minimal Spanning Trees

A *spanning tree* for a graph is a tree subgraph that contains all nodes of the given graph. A spanning tree is not necessarily rooted. A useful graph of observations is the spanning tree whose edges have the least total distance. This is called a *minimal spanning tree*, or MST. It is obvious that the number of edges in a minimal spanning tree would be one less than the number of nodes. A minimal spanning tree may not be unique.

A minimal spanning tree can be used to determine an ordering of the data so that the total distance between successive observations is least. This ranking, which would start at one extreme of the data and move to the other extreme, is in the spirit of the definition of the tree. Another ranking is a radial ranking in which the next element in the ranking has the smallest maximum distance to any of the remaining (unranked) elements.

Kruskal (1956) gave the method shown in Algorithm 16.1 for forming a minimal spanning tree. The set of all edge distances should first be put into a minimum heap so that the updating can proceed more rapidly.

Algorithm 16.1 Formation of a Minimal Spanning Tree T from a Connected Graph with n Nodes and with Edge Distances in H

0. Set $T = \emptyset$ (the empty set), and set $k = 0$.
1. Choose the edge e from H with shortest distance.
2. Remove e from H.
3. If e does not create a cycle in T, then add e to T and set $k = k + 1$.
4. If $k < n - 1$, go to step 1. ∎

It is easy to see that the problem of determining an MST is $O(n^2)$ and Algorithm 16.1 is of that order. This is prohibitive for large datasets. Bentley and Friedman (1978) described an algorithm to approximate an MST that is $O(n \log n)$.

A tree that connects observations with nearby ones helps us to understand the distribution of the observations and to identify clusters of observations and outlying observations. The number of edges in the longest path starting at any node is called the *eccentricity* of that node or observation. The node most distant from a given node is called an *antipode* of the node, and the path between a node with greatest eccentricity and its antipode is called a *diameter* of the tree. The length of such a path is also called the diameter. (This word also carries both meanings in the familiar context of a circle.) A node with minimum eccentricity is called a *center* node or a *median.*

Minimal spanning trees have a variety of uses in multivariate analysis. A related problem is to determine the shortest path between two given nodes. An algorithm to determine the shortest path, called Dijkstra's algorithm, is described in Horowitz, Sahni, and Rajasekaran (1998), for example.

In the following few pages, we will use a simple bivariate dataset for illustrations. The dataset is shown in Table 16.2. A plot of the data and a minimal spanning tree for this simple bivariate dataset is shown in Figure 16.12. In most cases, when we display this dataset we will suppress the scales on the axes.

The median of the tree shown in the right-hand plot in Figure 16.12 is observation number 6 (as labeled in the left-hand plot in Figure 16.12).

Table 16.2. Dataset for Illustrations

Obs. Number	x_1	x_2
1	10	66
2	19	52
3	8	88
4	37	25
5	66	75
6	53	55
7	89	76
8	73	91
9	21	32
10	12	23
11	29	41
12	86	65
13	91	81
14	42	23
15	36	38
16	90	85

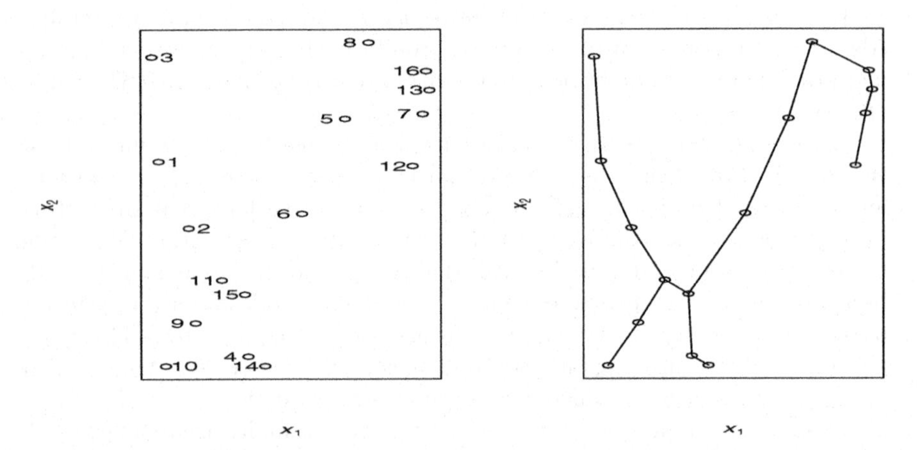

Fig. 16.12. A Scatter Plot of the Bivariate Dataset (Table 16.2) and a Minimal Spanning Tree for It

Although we cannot produce a useful visual graph of an MST in higher dimensions, the concept carries through, and the same algorithm applies.

Ranking Data Using Minimal Spanning Trees

Friedman and Rafsky (1979a, 1979b) defined a method for sorting multivariate observations based on a minimal spanning tree. The procedure is to define a starting node at an endpoint of a tree diameter and then to proceed through the tree in such a way as to visit any shallow subtrees at a given node before proceeding to the deeper subtrees. For the tree shown in the right-hand plot in Figure 16.12, for example, if we choose to begin on the right of the tree, the next seven nodes are in the single path from the first one. Finally, at the eighth node, we choose the shallow subtree for the ninth and tenth nodes. The eleventh node is then the other node connected to the eighth one. This ordering is shown in Figure 16.13.

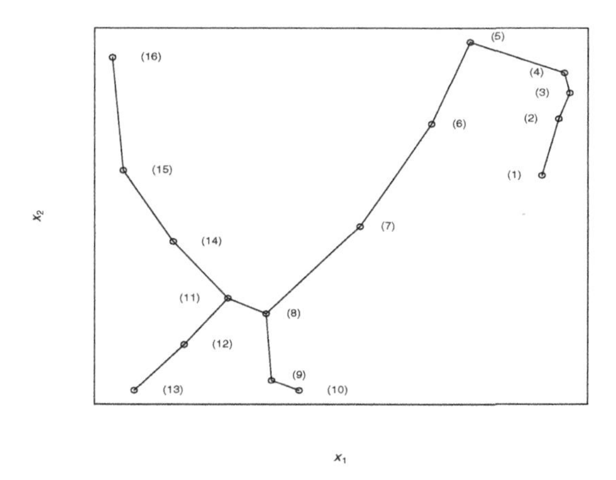

Fig. 16.13. Extreme Ranking of the Dataset Using the MST

A radial ranking of the dataset is shown in Figure 16.14. The first element is the most central, as determined by the interpoint distances of the MST, and the last element is the most extreme. Notice, as expected, the most extreme radially is one of the extremes in the extreme ranking of Figure 16.13. In that case, it was chosen as the "first" element. The "last" element in Figure 16.13 is the next to last in the radial ranking, as we should expect if the "first" element in Figure 16.13 is the last in the radial ranking. There is a certain amount of arbitrariness in "first" and "last".

There are several other ways of using a graph to define a sequence or ranking of multivariate data. A connected acyclic graph that contains only one path would be a reasonable possibility. (A cyclic path that includes every node exactly once is called a "Hamiltonian circuit".) Such a graph with minimal diameter is the solution to the traveling salesperson problem. The traveling

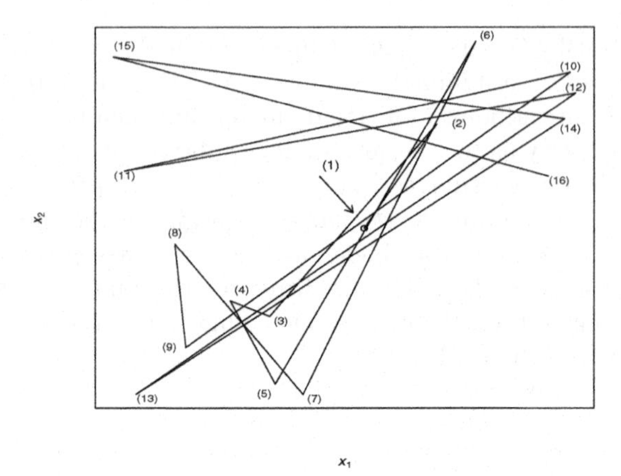

Fig. 16.14. Radial Ranking of the Dataset Using the MST

salesperson problem is computationally more complex than determining the minimal spanning tree.

Ranking Data Using Convex Container Hulls

Another way to order data is by convex hull peeling. The idea behind convex hull peeling, which is due to John Tukey, is that the convex hull of a dataset identifies the extreme points, and the most extreme of these is the one whose removal from the dataset would yield a much "smaller" convex hull of the remaining data. In two dimensions, convex hull peeling takes as the most extreme observation the one on the convex hull with the smallest angle. Next, the convex hull of all remaining points is determined, and the second most extreme observation is the one on this convex hull with the smallest angle. This process continues until a total ordering of all observations is achieved. The first few steps are shown in Figure 16.15. The ordering by convex hull peeling tends to move around the edges of the set of points, often similar to the radial ordering in a minimal spanning tree. Notice that the three largest observations identified by convex hull peeling are all among the six largest identified by the MST radial ordering shown in Figure 16.14.

Various programs for computing convex hulls and other problems in computational geometry are available at the site,

www.geom.umn.edu/software/download/

The convex hull of a two-dimensional dataset is particularly easy to compute.

Instead of a convex hull, formed by planes, we may consider a smoothed figure such as an ellipsoid with minimum volume. Hawkins (1993a), Cook,

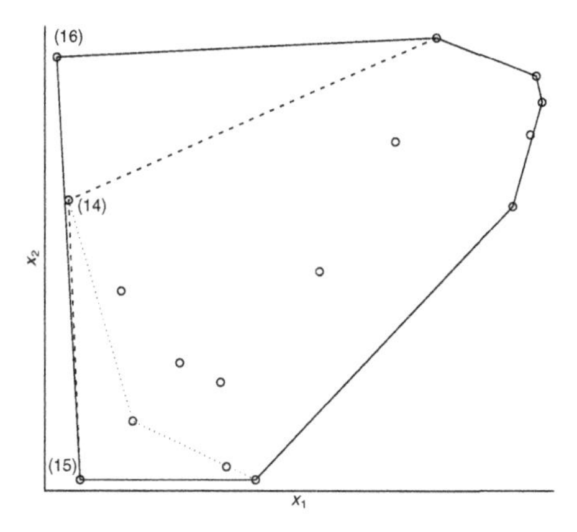

Fig. 16.15. Ordering of the Dataset (Table 16.2) by Peeling the Convex Hull

Hawkins, and Weisberg (1993), and Woodruff and Rocke (1993) give algorithms for computing the minimum-volume ellipsoid. The algorithms in the last reference include various heuristic combinatorial algorithms, such as simulated annealing, genetic algorithms, and tabu searches. The minimum-volume ellipsoid that contains a given percentage of the data provides another way of ordering the points in a multivariate dataset. It would be expected to produce a very similar ranking as that resulting from peeling a convex hull. The convex hull algorithms are generally faster than those for minimum-volume ellipsoids, and so would probably be a better choice for very large datasets (that is, datasets with very large numbers of observations or very large numbers of variables).

Ranking Data Using Location Depth

A peeled convex hull or an ellipsoid containing a given percentage of data provides an ordering of the data from the outside in. Another approach to ordering data is to begin in the inside — that is, at the densest part of the data — and proceed outward. For bivariate data, John Tukey introduced the concept of *halfspace location depth* for a given point x_c relative to the dataset X whose rows are in $\mathbb{R}^2$. The halfspace location depth, $d_{\text{hsl}}(x_c, X)$, is the smallest number of x_i contained in any closed halfplane whose boundary passes through x_c. The halfspace location depth is defined for datasets X whose rows are in $\mathbb{R}^m$ by immediate extension of the definition for the bivariate case. Figure 16.16 shows some halfspaces defined by lines, together with the counts of points on either of the lines.

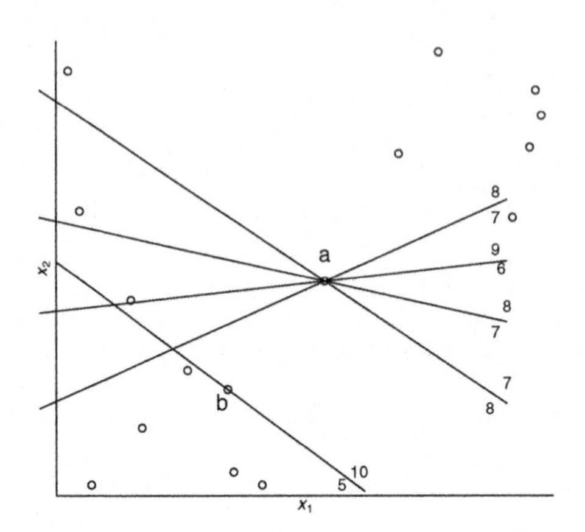

Fig. 16.16. Halfplanes that Define Location Depths in the Dataset (Table 16.2)

There are 15 possible pairs of counts for each point in Figure 16.16 (no three points in this dataset are collinear). Some halfplanes for which point A lies on the boundary contain as few as six points (one such halfplane is shown in the figure), but all contain at least six points. Thus,

$$d_{\text{hsl}}(a, X) = 6.$$

Of all the points in this dataset, a has the greatest halfspace location depth. The halfspace location depth of point b, $d_{\text{hsl}}(b, X)$, for example, is 4. The halfplane shown in Figure 16.16 with point b on the boundary contains five points. A clockwise rotation of that boundary line yields a halfplane containing four points.

The halfspace location depth provides another way of ordering the data. There are generally many ties in this ordering.

Rousseeuw and Ruts (1996) provide an algorithm for computing the halfspace location depth for bivariate data. (See also Ruts and Rousseeuw, 1996, who discuss contours of regions with equal location depth.)

Ordering by location depth emphasizes the interior points, whereas convex hull peeling emphasizes the outer points. The outer points have a halfspace location depth of 0, and generally the first few points removed in convex hull peeling, such as points 15 and 16 in Figure 16.15, have a location depth of 0. The point that is removed next, however, has a location depth of 1, whereas there are other points in the dataset with location depths of 0.

If a single point has a greater location depth than any other point in the dataset, the point is called the *depth median.* If multiple points have the largest location depth of any in the dataset, the depth median is the centroid of all such points.

The depth median of the dataset shown in Figure 16.16 is the point labeled "a". That is the same as observation number 6 as labeled in the left-hand plot in Figure 16.12 (page 540), which was the median of the minimal spanning tree shown in the right plot in Figure 16.12.

Determination of the depth median is computationally intensive. Rousseeuw and Ruts (1998) give an algorithm for computing the depth median in a bivariate dataset, and Struyf and Rousseeuw (2000) give an approximate algorithm for higher-dimensional data.

There are other ways of defining data depth. One approach is to define a measure of distance of depth based on maximal one-dimensional projections. This measure can be used to order the data, and it is also useful as an inverse weight for robust estimators of location and scale. It is computationally intensive, and most methods in use depend on sampling of the data.

Ordering by Clustering

Clustering also provides a way of ordering or, especially, of partially ordering data. The ordering that arises from clustering, whether divisive or agglomerative, however, depends on local properties, and a global ordering is difficult to identify. A hierarchical clustering of the data in Table 16.2 is shown in Figure 16.17. The ordering, or partial ordering, would be from left to right (or from right to left) along the leaves of the tree. Comparison of the cluster tree with the scatter plot in Figure 16.12 (page 540) shows how nearby points are grouped first. In this dataset, the points closest together are on the periphery of the data cloud. For a cloud of points that is concentrated around the median, as is perhaps more common, the central points would be grouped first.

Clustering by Ordering

The ordering of the data by the minimal spanning tree can be used to cluster the data. The minimal spanning tree shown in Figure 16.12 can be used to form the clusters indicated in Figure 16.18. The clusters are formed by a tessellation formed by boundaries perpendicular to the longer edges in the MST.

As it turns out, the four clusters shown in the MST correspond to the four clusters formed by the hierarchical clustering shown in Figure 16.17. It is not always the case that a clustering can be formed by simple cuts of the

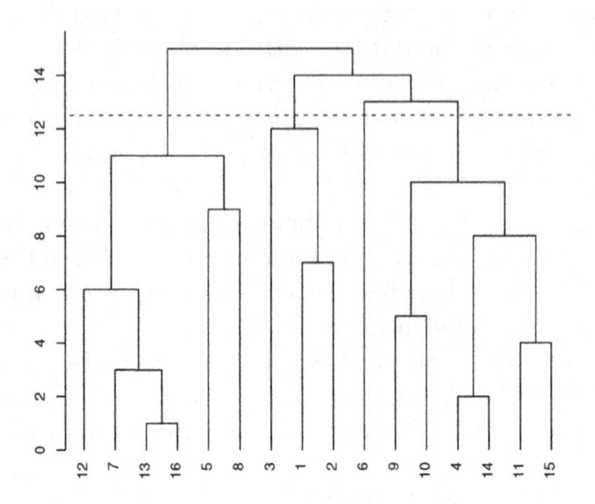

Fig. 16.17. Hierarchical Clustering of the Dataset (Table 16.2)

branches of a minimal spanning tree to correspond to the clustering formed by a particular hierarchical algorithm.

Nearest Neighbors and k-d-Trees

For a given observation, x_i, we may want to find its "nearest neighbor", x_k, where we define a nearest neighbor as one for which some function, $f(x_i, x_k)$, is minimized. For example, f may be the square of the Euclidean distance,

$$\sum_{j=1}^{m}(x_{ij} - x_{kj})^2.$$

To search bivariate data for a point that is close in Euclidean distance to a given point, a *quad tree* is useful (see Knuth, 1973).

For the more general problem of finding nearest neighbors in multivariate data, a *k-d-tree* developed by Friedman, Bentley, and Finkel (1977) may be more useful. A k-d-tree is a multivariate form of a B-tree; see Bentley and Friedman (1979).

Consider an $n \times m$ data matrix X in which columns represent variables and rows represent observations. A k-d-tree for X is defined by two arrays, v, which contains indicators of the variables to be used as discriminators, and p, which contains values of the corresponding variables to be used in forming partitions. Let b be the maximum bucket size (that is, the largest number of elements to be left at a terminal node).

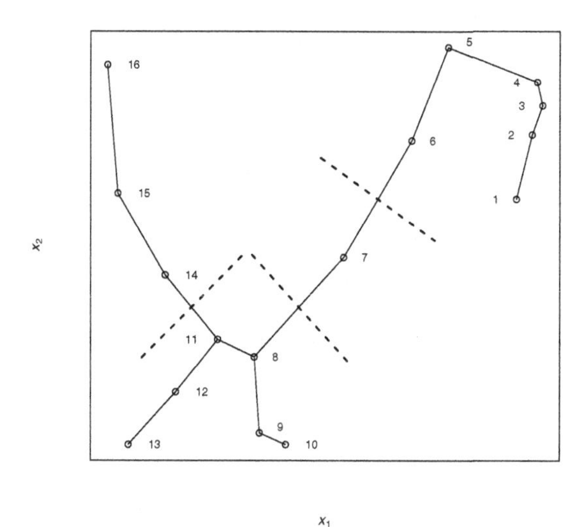

Fig. 16.18. Clustering Using the MST of the Dataset (Table 16.2)

Algorithm 16.2 Formation of a k-d-Tree

0. Set $l = 1$ and $h = n$.
1. Let $k = \lfloor (l + h)/2 \rfloor$.
2. Let v_k be the column number with maximum spread.
3. Let p_k be the median in the range $[l, h]$ of the v_k^{th} column.
4. Interchange the rows of X so that all rows of X with values in the v_k^{th} column less than or equal to p_k occur before (or at) the kth element.
5. If $k - l > b$, then form a submatrix with $h = k$.
   If $h - k - l > b$, then form a submatrix with $l = k + 1$ (with h as in step 4).
   Process steps 1 through 4 for each submatrix formed and then return to step 5. ∎

Although trees are used often at a lower level, there is not much software available at the user level to form trees. The IMSL Fortran routine `QUADT` builds a k-d-tree and the routine `NGHBR` uses a k-d-tree to find nearest neighbors.

Murtagh (1984) provides a comparative review of algorithms for computing nearest neighbors.

Ordering and Ranking of Transformed Data

Minimal spanning trees depend on relative distances between points, so, as we would expect, the minimal spanning tree and any ordering based on it may be different if the data are transformed. Likewise, of course, orderings based on clustering may be changed by transformations of the data.

An important property of convex hulls and the depth of data is that they are not affected by affine transformations.

16.3 Linear Principal Components

In addition to clusters and orderings in data, other types of interesting structure are lower-dimensional relationships in the data. The information in observations that consist of m components may be adequately represented by transformed observations consisting of a smaller set of k components. This reduction in the dimension of the data may allow a reduction in the storage requirements, but, more importantly, it may help in understanding the data. Dimension reduction is useful in identifying structure in the data and also in discovering properties that the data do not measure directly. We may wish to extract features in the data that have been obscured by measurement error or other sources of noise.

A basic tenet of data analysis is that variation provides information, and an important approach in statistics is the analysis of variation. When many variables are present, however, it is often difficult to identify individual effects, so it may be useful to reduce the number of variables.

Another basic tenet is that covariance among a set of variables reduces the amount of information that the variables contain. We therefore seek to combine variables in such a way that their covariance is reduced or, more generally, that they are independent. For normal variables, of course, zero covariance is equivalent to independence.

The basic problem therefore is to transform the observed m-vectors x into k-vectors $\tilde{x}$ that, as a set, exhibit almost as much variation as the original set and are mutually independent or "almost" so.

Because of differences in the meanings of the variables, it is best first to standardize the data using the transformation (9.22)

$$X_{\mathrm{S}} = (X - \overline{X})\,\mathrm{diag}(1/\sqrt{s_{ii}}),$$

where $\overline{X}$ is the matrix whose constant columns contain the means of the corresponding columns of X, and $\sqrt{s_{ii}}$ is the sample standard deviation of the ith column of X.

There are various ways of combining a set of variables into a smaller set (that is, of transforming m-vectors into k-vectors). One of the simplest methods is to use linear transformations. If the linear transformations result in new variables that are orthogonal (that is, if they have zero sample correlation), and if the data are multivariate normal, then the new variables are independent. The linear combinations are called "principal components", "empirical orthogonal functions" or "EOF" (especially in meteorology and atmospheric research), "latent semantic indices" (especially in information retrieval), "factors" (especially in the social sciences), Karhunen-Loève transforms (especially in signal analysis), or "independent components" (also in signal analysis).

There are some differences among and within these methods. The differences have to do with assumptions about probability distributions and with the nature of the transformations. In factor analysis, which we discuss in Section 16.4, a rather strong stochastic model guides the analysis. In independent components analysis, rather than seeking only orthogonality, which yields zero correlations, and hence independence for normal data, transformations may be performed to yield zero cross moments of higher order. (Correlations are cross moments of second order.) Independent component analysis, therefore, may involve nonlinear transformations. Any of the methods mentioned above may also utilize nonlinear combinations of the observed variables.

Linear *principal components analysis* (PCA) is a technique for data reduction by constructing linear combinations of the original variables that account for as much of the total variation in those variables as possible. We discuss this method first.

The Probability Model Underlying Principal Components Analysis

Linear principal components is a method of "decorrelating" the elements of a vector random variable. The method depends on the variances of the individual elements, so it is generally best to perform transformations as necessary so that all elements have the same variance. In addition and without loss of generality, it is convenient to subtract the mean of each element of the random variable. The transformed vector is thus standardized so that the mean of each element is 0 and the variance of each element is 1.

Consider an m-vector random variable Y with variance-covariance matrix Σ, which has 1's along the diagonal. We will refer to the elements of the random variable as "variables". We seek a transformation of Y that produces a random vector whose elements are uncorrelated; that is, we seek a matrix W with m columns such that $\mathrm{V}(WY)$ is diagonal. (Here, $\mathrm{V}(\cdot)$ is the variance.) Now,

$$\mathrm{V}(WY) = W\Sigma W^{\mathrm{T}},$$

so the matrix W must be chosen so that $W\Sigma W^{\mathrm{T}}$ is diagonal.

The obvious solution is to decompose Σ:

$$\Sigma = W^{\mathrm{T}}\Lambda W. \tag{16.10}$$

The spectral decomposition of the variance-covariance matrix is

$$\Sigma = \sum_{k=1}^{m} \lambda_k w_k w_k^{\mathrm{T}}, \tag{16.11}$$

with the eigenvalues λ_k indexed so that $0 \leq \lambda_m \leq \cdots \leq \lambda_1$ and with the w_k orthonormal; that is,

$$I = \sum_k w_k w_k^{\mathrm{T}}.$$

Now, consider the random variables

$$\widetilde{Y}_{(k)} = w_k^{\mathrm{T}} Y,$$

which we define as the *principal components* of Y.

The first principal component, $\widetilde{Y}_{(1)}$, is the projection of Y in the direction in which the variance is maximized; the second principal component, $\widetilde{Y}_{(2)}$, is the projection of Y in an orthogonal direction with the largest variance; and so on.

It is clear that the variance of $\widetilde{Y}_{(k)}$ is λ_k and that the $\widetilde{Y}_{(k)}$ are uncorrelated; that is, the variance-covariance matrix of the random vector $(\widetilde{Y}_{(1)}, \ldots, \widetilde{Y}_{(m)})$ is $\mathrm{diag}(\lambda_1, \ldots, \lambda_m)$. Heuristically, the kth principal component accounts for the proportion

$$\frac{\lambda_k}{\sum \lambda_j}$$

of the "total variation" in the original random vector Y.

The linear combinations $\widetilde{Y}_{(k)}$ that correspond to the largest eigenvalues are most interesting. If we consider only the ones that account for a major portion of the total variation, we have reduced the dimension of the original random variable without sacrificing very much of the potential explanatory value of the probability model. Thus, using only the p largest eigenvalues, instead of the m-vector Y, we form the transformation matrix W as

$$W = \begin{bmatrix} w_1^{\mathrm{T}} \\ w_2^{\mathrm{T}} \\ \vdots \\ w_p^{\mathrm{T}} \end{bmatrix}.$$

This produces the p-vector $\widetilde{Y} = (\widetilde{Y}_{(1)}, \ldots, \widetilde{Y}_{(p)})$.

The matrix

$$\Sigma_p = \sum_{k=1}^{p} \lambda_k w_k w_k^{\mathrm{T}} \tag{16.12}$$

is the variance-covariance matrix of $\widetilde{Y}$.

Eckart and Young (1936) proved an interesting fact about Σ_p as an approximation to Σ. It is the matrix of rank p closest to Σ as measured by the Frobenius norm,

$$\|\Sigma - \Sigma_p\|_{\mathrm{F}}.$$

Although all of the statements above are true for any distribution for which the first two moments exist, the properties of the principal components are even more useful if the underlying random variable Y has a multivariate normal distribution. In this case, the principal components vector $\widetilde{Y}$ also has a multivariate normal distribution, and the elements of $\widetilde{Y}$ are independent.

Principal Components Analysis of Data

In the basic multivariate data structure of X, we often consider the rows to be realizations of some multivariate random variable, such as Y in the discussion above. Because of differences in the meanings of the variables in the data matrix X, it is best first to standardize the data using the transformation (9.22) on page 392:

$$X_{\mathrm{S}} = (X - \overline{X})\,\mathrm{diag}(1/\sqrt{s_{ii}}).$$

In the following, we will assume that this has been done. We will assume that X has been standardized and not continue to use the notation X_{S}.

Using S as an estimate of Σ, we can perform a principal components analysis of the data that follows the same techniques as above for a random variable. We determine the spectral decomposition of S just as we did for Σ in equation (16.11):

$$S = \sum_j \hat{\lambda}_j \hat{w}_j \hat{w}_j^{\mathrm{T}}. \tag{16.13}$$

The principal components of the vector of observed variables x are

$$\tilde{x}_{(j)} = \hat{w}_j^{\mathrm{T}} x. \tag{16.14}$$

Corresponding to the generic data vector x is the generic vector of principal components,

$$\tilde{x} = (\tilde{x}_{(1)}, \ldots, \tilde{x}_{(m)}). \tag{16.15}$$

For each observation x_i, we can compute a value of the principal components vector, $\tilde{x}_i$. From the spectral decomposition that yields the principal components, it is easy to see that the sample variance-covariance matrix of the principal components is diagonal.

Figure 16.20 shows the marginal distributions of the elliptical data cloud shown in Figure 16.19 in the original coordinates x_1 and x_2 and in the coordinates of the principal components z_1 and z_2.

The first principal component is the hyperplane that minimizes the orthogonal distances to the hyperplane as discussed on page 610, and illustrated in Figure 17.3.

The principal components are transformations of the original system of coordinate axes. It is difficult to relate any of the new axes to the old axes, however. To aid in their interpretability, Hausman (1982) suggests a constrained PCA in which each principal component is approximated by a linear combination of the original axes with coefficients of the combination being $+1$,

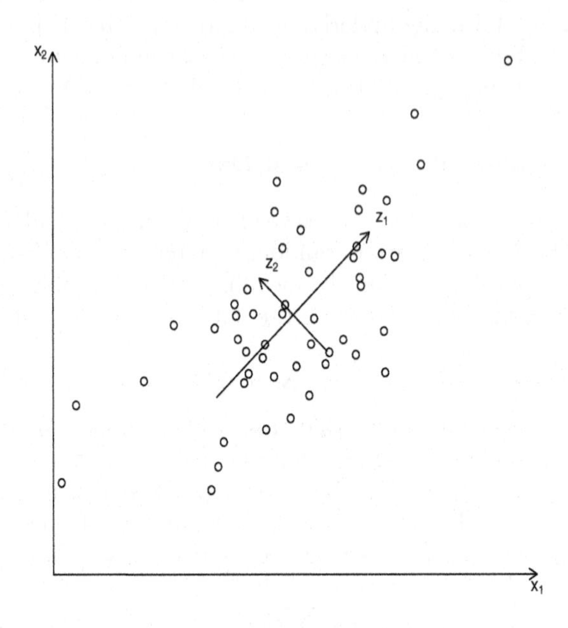

Fig. 16.19. Principal Components of Some Bivariate Data

−1, or 0. Presumably, a combination in which the original variables are just added or subtracted is easier to interpret in terms of the quantities measured by the original variables.

In the same spirit, Vines (2000) and Jolliffe and Uddin (2000) describe methods to determine “simple components”. In Vines’s approach, “simplicity preserving” transformations — that is, linear combinations with integer coefficients (but not just +1, −1, or 0) — are applied sequentially in such a way that the combination with the largest variance has a larger variance than that of the original variable with the largest variance. The allowable combinations result from a fixed set of coefficients or from a rule that defines allowable coefficients. After fixing that combination as the first simple component, another simple component is chosen similarly. There are several different criteria that could be applied to determine the order of all of the transformations, and more studies are needed to evaluate various ways of proceeding. Of course, underlying any decision on the exact algorithm is the question of what is a “simple” component.

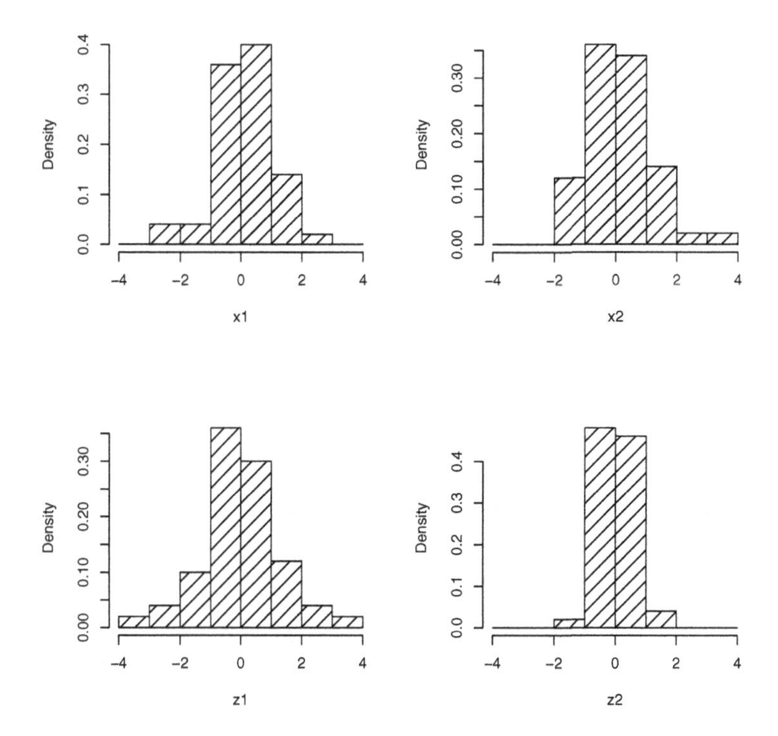

Fig. 16.20. Univariate Histograms of the Original Coordinates and the Principal Components

In the method of Jolliffe and Uddin (2000), instead of choosing a priori a set of allowable combinations, a penalty function is used to push the combinations, which are determined sequentially, toward simplicity. Jolliffe, Trendafilov, and Uddin (2003) carried this idea further by using a regularization similar to that in lasso (see page 608) to drive some of the $\hat{w}_j$ to zero, so that the corresponding original variables are no longer in the picture.

Dimension Reduction by Principal Components Analysis

We can reduce the dimension of the data by considering the transformed variables $\tilde{x}_{(i)}$, each of which is a vector formed using the eigenvectors corresponding only to the p largest eigenvalues. As before, we form the $p \times m$ transformation matrix $\widehat{W}$,

$$\widehat{W} = \begin{bmatrix} \hat{w}_1^{\mathrm{T}} \\ \hat{w}_2^{\mathrm{T}} \\ \vdots \\ \hat{w}_p^{\mathrm{T}} \end{bmatrix}.$$

For the ith observation x_i, this produces the p-vector $\tilde{x}_i = (\tilde{x}_{i(1)}, \ldots, \tilde{x}_{i(p)})$. The application of this transformation matrix is often called the (discrete) Karhunen-Loève transform in signal analysis after independent work by K. Karhunen and M. Loève in the late 1940s or the Hotelling transform after the work of H. Hotelling in the 1930s.

Choosing p

The question arises, naturally, of how to choose p. This is the question of how much we can reduce the dimensionality of the original dataset. A simple approach that is often employed is to choose p as the number of the ranked eigenvalues just prior to a large gap in the list. For example, if $m = 6$ and the eigenvalues are 10.0, 9.0, 3.0, 2.5, 2.1, and 2.0, a logical choice of p may be 2, because of the large decrease after the second eigenvalue. A plot of these ordered values or of the values scaled by their total may be useful in identifying the point at which there is a large dropoff in effect. Such a plot, called a scree plot, is shown as the left-hand plot in Figure 16.21. The scree plot can be either a line plot as in the figure or a bar chart in which the heights of the bars represent the relative values of the eigenvalues. The key feature in a scree plot is an "elbow", if one exists. A plot of the accumulated "total variation" accounted for by the principal components, as shown in the right-hand plot in Figure 16.21, may also be useful.

The effect of each of the original variables (the elements of x) on each principal component is measured by the correlation between the variable and the principal component. This is called the "component loading" of the variable on the principal component. The component loading of the jth variable on the kth principal component is the correlation

$$\frac{w_{kj}\sqrt{\hat{\lambda}_k}}{\sqrt{s_{jj}}}.$$

(Note that w_{kj} is the jth element of the kth eigenvector.)

Principal Components and Transformations of the Data

As we mentioned at the outset, variation provides information. Variables with large sample variances will tend to predominate in the first principal component. Consider the extreme case in which the variables are uncorrelated (that

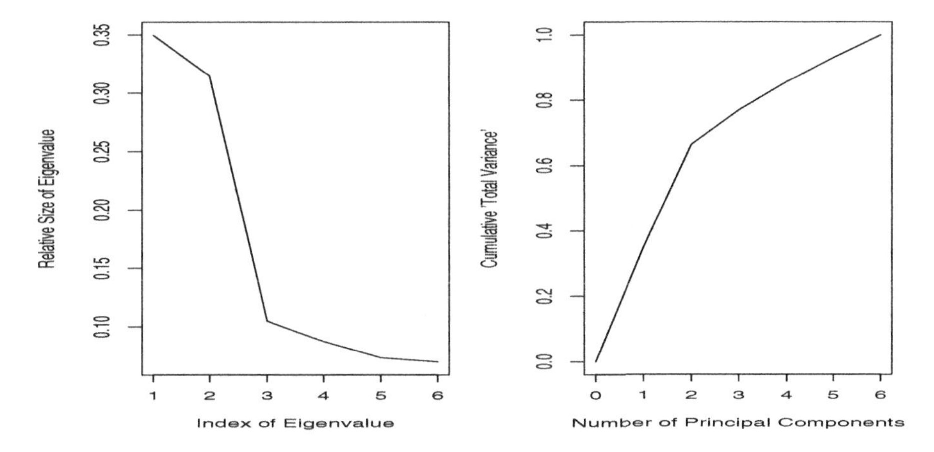

Fig. 16.21. Scree Plot of Scaled Eigenvalues and Plot of Proportion of "Total Variance"

is, in which S is diagonal). The principal components are determined exactly by the variances, from largest to smallest. This is a natural and desirable consequence of the fact that variation provides information. In principal components analysis, the relative variation from one variable to another is the important factor in determining the rankings of the components. It is difficult, however, to measure the relative variation from one variable to another. The variance of a variable depends on the units of measurement. Suppose that one variable represents linear measurements in meters. If, for some reason, the unit of measurement is changed to centimeters, the effect of that variable in determining the principal components will increase one hundredfold.

The component loadings can help in understanding the effects of data reduction through principal components analysis. Notice that the component loadings are scaled by the square root of the variance. Another approach to scaling problems resulting from the choice of unit of measurement is to use the correlation matrix, R (see equation (9.11)), rather than the variance-covariance matrix. The correlations result from scaling the covariances by the square roots of the variances. The obvious should be noted, however: *The principal components resulting from the use of* R *are not the same as those resulting from the use of* S.

Change of units of measurement is just one kind of simple scaling transformation. Transformations of any kind are likely to change the results of a multivariate analysis, as we see, for example, in the case of clustering on page 533. As we mentioned in Section 1.1 scaling induces artificial structure. Principal component analysis is sensitive to artificial structure.

Principal Components of Observations

Just as in Section 16.1, on page 535, we observed the basic symmetry between the "variables" and the "observations" of the dataset X, we can likewise reverse their roles in principal components analysis. Suppose, for example, that the observational units are individual persons and the variables are responses to psychological tests. Principal components analysis as we have described it would identify linear combinations of the scores on the tests. These principal components determine relationships among the test scores. If we replace the data matrix X by its transpose and proceed with a principal components analysis as described above, we identify important linear combinations of the observations that, in turn, identify relationships among the observational units. In the social sciences, a principal components analysis of variables is called an "R-Type" analysis and the analysis identifying relationships among the observational units is called "Q-Type".

In the usual situation, as we have described, the number of observations, n, is greater than the number of variables, m. If X has rank m, then the variance-covariance matrix and the correlation matrix are of full rank. In a reversal of the roles of observations and variables, the corresponding matrix would not be of full rank. Of course, the analysis could proceed mechanically as we have described, but the available information for identifying meaningful linear combinations of the observations would be rather limited. This problem could perhaps be remedied by collecting more data on each observational unit (that is, by defining and observing more variables).

Principal Components Directly from the Data Matrix

Formation of the S or R matrix emphasizes the role that the sample covariances or correlations play in principal component analysis. However, there is no reason to form a matrix such as $(X - \overline{X})^{\mathrm{T}}(X - \overline{X})$, and indeed we may introduce significant rounding errors by doing so.

The singular value decomposition (SVD) of the $n \times m$ matrix $X - \overline{X}$ yields the square roots of the eigenvalues of $(X - \overline{X})^{\mathrm{T}}(X - \overline{X})$ and the same eigenvectors. (The eigenvalues of $(X - \overline{X})^{\mathrm{T}}(X - \overline{X})$ are $(n-1)$ times the eigenvalues of S.) We will assume that there are more observations than variables (that is, that $n > m$). In the SVD of the centered data matrix, we write

$$X - \overline{X} = UAV^{\mathrm{T}},$$

where U is an $n \times m$ matrix with orthogonal columns, V is an $m \times m$ orthogonal matrix, and A is an $m \times m$ diagonal matrix with nonnegative entries, called the singular values of $X - \overline{X}$.

The spectral decomposition in terms of the singular values and outer products of the columns of the factor matrices is

$$X - \overline{X} = \sum_{i=1}^{m} \sigma_i u_i v_i^{\mathrm{T}}. \tag{16.16}$$

The vectors u_i, called the "left eigenvectors" or "left singular vectors" of $X - \overline{X}$, are the same as the eigenvectors of S in equation (16.13). The vectors v_i, the "right eigenvectors", are the eigenvectors that would be used in a Q-type principal components analysis. The reduced-rank matrix that approximates $X - \overline{X}$ is

$$\widetilde{X}_p = \sum_{i=1}^{p} \sigma_i u_i v_i^{\mathrm{T}} \tag{16.17}$$

for some $p < \min(n, m)$.

Computational Issues

For the eigenanalysis computations in PCA, if the sample variance-covariance matrix S is available, it is probably best to proceed with the decomposition of it as in equation (16.13). Because the interest is generally only in the largest eigenvalues, the power method (see Gentle, 2007, Section 7.2) may be the best method to use. If S is not available, there is generally no reason to compute it just to perform PCA. The computations to form S are $\mathrm{O}(m^3)$. Not only do these computations add significantly to the overall computational burden, but, as we see from inequality (5.18), on page 208, S is more poorly conditioned than X (or $X - \overline{X}$). The SVD decomposition (16.16) is therefore the better procedure.

Artificial neural nets have been proposed as a method for computing the singular values in equation (16.16). A study by Nicole (2000), however, indicates that neural nets may not perform very well for identifying any but the first principal component.

PCA for Clustering

An objective of principal components analysis is to identify linear combinations of the original variables that are useful in accounting for the variation in those original variables. This is effectively a clustering of the variables. For many purposes, these derived features carry a large amount of the information that is available in the original larger set of variables. For other purposes, however, the principal component may completely lose the relevant information. For example, the information carried by the smaller set of features identified in PCA may be useless in clustering the observations. Consider the bivariate dataset in Figure 16.22. There are two clusters in this dataset, each of which appears to be a sample from an elliptical bivariate normal distribution with a small positive correlation. The two clusters are separated by a displacement of the mean of the second component. The two principal components are shown in Figure 16.22. As would be expected, the first principal component is in the direction of greater spread of the data, and the second principal component is orthogonal to the first.

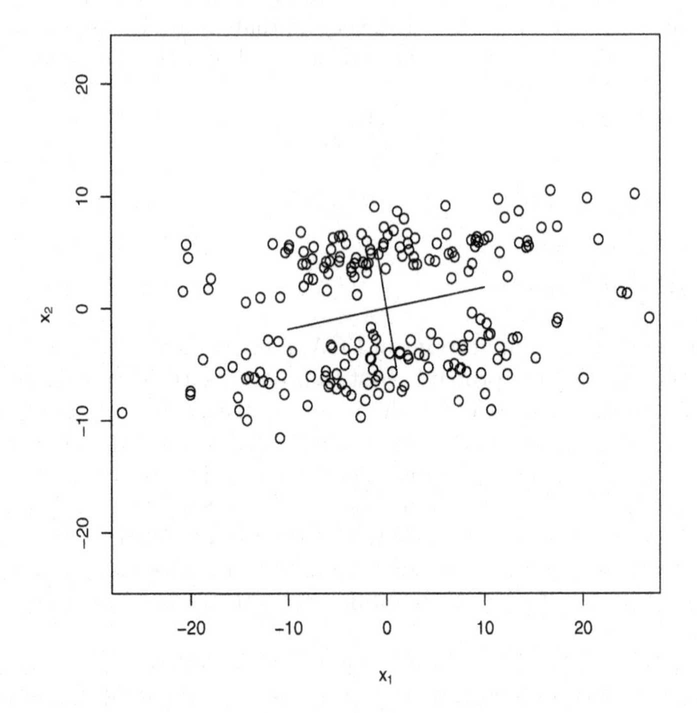

Fig. 16.22. Principal Components of a Bivariate Dataset with Two Clusters

The first principal component contains no information about the clusters in the data. Figure 16.23 shows histograms of the data projected onto the two principal components. The second principal component carries information about the two clusters, but the first principal component appears to be a single normal sample.

Principal components analysis emphasizes the direction of maximum variation. If the main source of variation in the dataset is the variation between clusters, then PCA will identify the clusters. This is not always the case, and the principal component may not lie in the most informative direction. Other techniques, such as projection pursuit (see Section 16.5), seek projections of the data in directions that exhibit other interesting structure, such as the bimodality in the direction of the second principal component in this example.

Robustness of Principal Components

As we have mentioned above, outlying observations or (nearly) collinear variables can present problems in data analysis. Principal components is one way

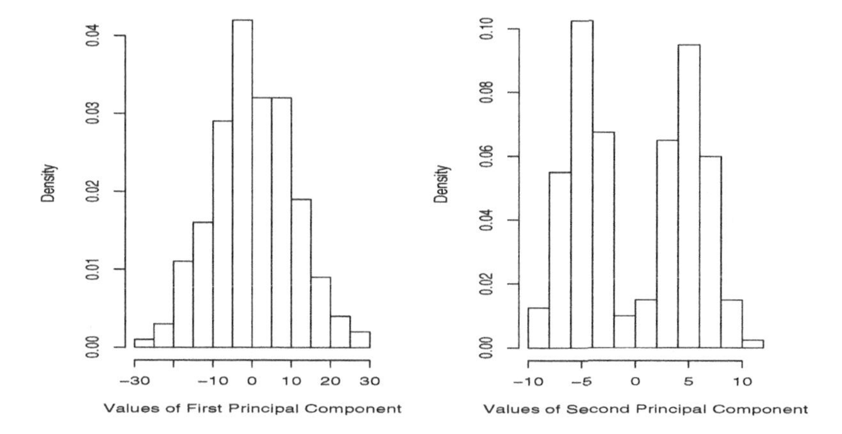

Fig. 16.23. Histograms of Projections of Bivariate Dataset onto the Two Principal Components

of dealing with collinear variables. These variables have large correlations among themselves. The dimension of the space of principal components will likely be reduced so that all variables that are collinear with each other are replaced by a single variable.

Outlying observations, however, may have a major effect on the principal components analysis. The first few principal components are very sensitive to these outlying observations. If the outliers were not present, or if they were perturbed slightly, a different set of the first few principal components would likely result.

There are generally two ways of dealing with outlying observations. One way is to identify the outliers and remove them temporarily. Another way is to use methods that are not much affected by outliers. Use of a robust sample variance-covariance, S_{R}, in equation 9.25 on page 395 will yield principal components that are less affected by outliers than those resulting from the usual sample variance-covariance, S.

If outliers can be identified and removed temporarily, a standard analysis can be performed. This identification, removal, and analysis procedure can be applied in stages. The major problem, of course, is that as extreme observations are removed, the variability in the dataset is reduced, so other, valid observations are more likely to appear as outliers. In general, a data analyst must assume that every observation carries useful information, and no observation must be discarded until its information is incorporated into the analysis.

For purposes of PCA, outliers can be identified in a preliminary step using a clustering procedure or even by using Q-type principal components analysis.

Caroni (2000) describes a way of using the robust PCA based on S_{R} to identify outliers.

16.4 Variants of Principal Components

Factor Analysis

Factor analysis is mechanically similar to principal components analysis. The main differences involve the probability model. Our discussion of factor analysis will be brief.

The Probability Model Underlying Factor Analysis

In factor analysis, we begin with a model that relates a centered m-vector random variable Y (observable) to an underlying, unobservable k-vector random variable, whose elements are called "factors". The factors have a mean of 0. In this model, the observable vector Y consists of linear combinations of the factors plus an independent random vector of "unique errors", which is modeled by a random variable with a mean of 0. The unique errors are independent of the factors. Now, letting F represent the vector of factors and E represent the errors, we have

$$Y - \mu = \Gamma F + E, \tag{16.18}$$

where μ is the mean of Y and Γ is an $m \times k$ fixed (but unknown) matrix, called the "factor loadings" matrix. Generally, the number of factors is less than the number of the observable variables. In some applications, such as in psychology, the factors may be related to some innate characteristics that are manifested in observable behavior.

We denote the variance-covariance matrix of Y by Σ, that of F by Σ_F, and that of E by Ψ, which is diagonal by the assumptions in the model. We therefore have the relationship

$$\Sigma = \Gamma \Sigma_F \Gamma^{\mathrm{T}} + \Psi.$$

Now, if we let $\widetilde{\Gamma} = \Gamma \Sigma_F^{\frac{1}{2}}$ and $\widetilde{F} = (\Sigma_F^{\frac{1}{2}})^{-1} F$, we have

$$\Sigma = \widetilde{\Gamma}\widetilde{\Gamma}^{\mathrm{T}} + \Psi.$$

Hence, a model equivalent to equation (16.18) is one in which we assume that the underlying factors have the identity as their variance-covariance matrix, and so we have

$$\Sigma = \Gamma \Gamma^{\mathrm{T}} + \Psi. \tag{16.19}$$

The diagonal elements of Ψ are called the *specific variances* of the factors and the diagonal elements of $\Gamma\Gamma^{\mathrm{T}}$ are called the *commonalities* of the factors.

The transformations above that indicate that $\Gamma\Gamma^{\mathrm{T}}$ can be used instead of $\Gamma\Sigma_F\Gamma^{\mathrm{T}}$ raise the issue of more general transformations of the factors, leading to an indeterminacy in the analysis.

If we decompose $\Sigma - \Psi$ as we did in PCA with Σ in equation (16.10) on page 549, (with Δ replacing Λ) we have

$$\Sigma - \Psi = W^{\mathrm{T}} \Delta W. \tag{16.20}$$

The factor-loading matrix therefore is

$$\Gamma = W^{\mathrm{T}} \Delta^{\frac{1}{2}}. \tag{16.21}$$

Factor Analysis of Data

In practical applications of factor analysis, we must begin with a chosen value of k, the number of factors. This is similar to choosing the number of principal components in PCA, and there are some ways of adaptively choosing k, but the computational approaches that we discuss below assume a fixed value for k. As usual, we consider the rows of the data matrix X to be realizations of a multivariate random variable. In factor analysis, the random variable has the same relationships to other random variables as Y above; hence, the observation x (a row of X) is related to the realization of two other random variables, f and e, by

$$x - \bar{x} = \Gamma f + e.$$

The objective in factor analysis is to estimate the parameters in the model (16.18)—that is, the factor loadings, Γ, and the variances, Σ and Ψ, in equation (16.19). There are several methods for estimating these parameters. In one method, the estimation criterion is *least squares* of the sum of the differences in the diagonal elements of Σ and S, that is, minimize the function g:

$$g(\Gamma, \Psi) = \mathrm{trace}\big((S - \Sigma)^2\big). \tag{16.22}$$

This criterion leads to the *principal factors method.* The minimization proceeds by first choosing a value $\widehat{\Psi}^{(0)}$ and then performing a decomposition similar to that in principal components, except that instead of decomposing the sample variance-covariance matrix S, an eigenanalysis of $S - \widehat{\Psi}^{(0)}$ as suggested by equation (16.20) is performed:

$$S - \widehat{\Psi}^{(0)} = \left(\widehat{W}^{(0)}\right)^{\mathrm{T}} \widehat{\Delta}^{(0)} \widehat{W}^{(0)}. \tag{16.23}$$

This yields the value for Γ, analogous to equation (16.21):

$$\widehat{\Gamma}^{(0)} = \left(\widehat{W}^{(0)}\right)^{\mathrm{T}} \left(\widehat{\Delta}^{(0)}\right)^{\frac{1}{2}}.$$

Next, the minimization problem (16.22) is solved for Ψ with the fixed value of $\widehat{\Gamma}^{(0)}$, that is,

$$\min g\big(\widehat{\Gamma}^{(0)}, \Psi\big). \tag{16.24}$$

Newton's method is usually used to solve this problem, leading to $\widehat{\Psi}^{(1)}$. The steps are then repeated; that is, $S - \widehat{\Psi}^{(1)}$ is decomposed, leading to

$$\widehat{\Gamma}^{(1)} = \left(\widehat{W}^{(1)}\right)^{\mathrm{T}} \left(\widehat{\Delta}^{(1)}\right)^{\frac{1}{2}},$$

which is used in the next application of Newton's method to solve $\min g\big(\widehat{\Gamma}^{(1)}, \Psi\big)$. Convergence criteria are usually chosen based on norms of the change in the estimates from one iteration to the next.

A simple method that is often used to get started is to take $\widehat{\psi}_{jj}^{(0)}$ as

$$\left(1 - \frac{k}{2m}\right)\left(S_{jj}^{-1}\right)^{-1},$$

where S_{jj}^{-1} is the jth diagonal element of S^{-1} if S is full rank; otherwise, take $\widehat{\psi}_{jj}^{(0)}$ as $s_{jj}/2$.

The factors derived using the principal factors method (that is, the linear combinations of the original variables) are the same as would be obtained in ordinary PCA if the variance of the noise (the unique errors) were removed from the variance-covariance of the observations prior to performing the PCA.

Another common method for estimating Γ, Σ, and Ψ uses the *likelihood criterion* that results from the asymptotic distributions. Using the negative of the log of the likelihood, we have the minimization problem,

$$\min l(\Gamma, \Psi) = \min\Big(\log\big|\Sigma^{-1}S\big| - \mathrm{trace}\big(\Sigma^{-1}S\big)\Big). \tag{16.25}$$

This criterion results in the method of maximum likelihood. In this method, we require that S be positive definite.

Solution of the minimization problem (16.25) is also done in iterations over two stages, as we did in the least squares method above. First, we choose a starting value $\widehat{\Psi}^{(0)}$. Its square root, $\big(\widehat{\Psi}^{(0)}\big)^{\frac{1}{2}}$, is symmetric. We then decompose $\big(\widehat{\Psi}^{(0)}\big)^{\frac{1}{2}} S^{-1} \big(\widehat{\Psi}^{(0)}\big)^{\frac{1}{2}}$ as in equation (16.23):

$$\left(\widehat{\Psi}^{(0)}\right)^{\frac{1}{2}} S^{-1} \left(\widehat{\Psi}^{(0)}\right)^{\frac{1}{2}} = \left(\widehat{W}^{(0)}\right)^{\mathrm{T}} \widehat{\Delta}^{(0)}\, \widehat{W}^{(0)}. \tag{16.26}$$

Using the relationship (16.19) and equation (16.21), we get a value for Γ:

$$\widehat{\Gamma}^{(0)} = \left(\widehat{\Psi}^{(0)}\widehat{W}^{(0)}\right)^{\mathrm{T}} \left(\widehat{\Delta}^{(0)} - I\right)^{\frac{1}{2}}.$$

Next, the minimization problem (16.25) is solved for Ψ with the fixed value of $\widehat{\Gamma}^{(0)}$. This problem may be rather ill-conditioned, and the convergence can be rather poor. The transformations

$$\theta_j = \psi_{jj}$$

can help. (The ψ_{jj} are the only variable elements of the diagonal matrix Ψ.) Hence, the minimization problem is

$$\min l\left(\widehat{\Gamma}^{(0)}, \theta\right). \tag{16.27}$$

An advantage of the maximum likelihood method is that it is independent of the scales of measurement. This results from the decomposition of $\widehat{\Psi}^{\frac{1}{2}} S^{-1} \widehat{\Psi}^{\frac{1}{2}}$ in equation (16.26). Suppose that we make a scale transformation on the random variable, Y, in equation (16.18); that is, we form $T = YD$, where D is a fixed diagonal matrix with positive entries. The resulting variance-covariance matrix for the unique errors, Ψ_T, is $D\Psi D^{\mathrm{T}}$. Likewise, the corresponding sample variance-covariance matrix, S_T, is DSD^{T}. The matrix to be decomposed as in equation (16.26) is

$$\begin{aligned}
\left(\widehat{\Psi}_T^{(0)}\right)^{\frac{1}{2}} S_T^{-1} \left(\widehat{\Psi}_T^{(0)}\right)^{\frac{1}{2}} &= \left(D\widehat{\Psi}^{(0)} D^{\mathrm{T}}\right)^{\frac{1}{2}} \left(DSD^{\mathrm{T}}\right)^{-1} \left(D\widehat{\Psi}^{(0)} D^{\mathrm{T}}\right)^{\frac{1}{2}} \\
&= \left(\widehat{\Psi}^{(0)}\right)^{\frac{1}{2}} D^{\mathrm{T}} \left(D^{\mathrm{T}}\right)^{-1} S^{-1} D^{-1} D \left(\widehat{\Psi}^{(0)}\right)^{\frac{1}{2}} \\
&= \left(\widehat{\Psi}^{(0)}\right)^{\frac{1}{2}} S^{-1} \left(\widehat{\Psi}^{(0)}\right)^{\frac{1}{2}},
\end{aligned}$$

which is the same as the one for the untransformed data.

Other common methods for factor analysis include generalized least squares, image analysis (of two different types), and alpha factor analysis.

The methods for factor analysis begin with the computation of the sample variance-covariance matrix S or the sample correlation matrix R. As we noted in the case of PCA, the results are different, just as the results are generally different following any transformation of the data.

Note that the model (16.18) does not define the factors uniquely; any rotation of the factors would yield the same model. In principal components analysis, a similar indeterminacy could also occur if we allow an arbitrary basis for the PCA subspace defined by the chosen k principal components.

The factors are often rotated to get a basis with some interesting properties. A common criterion is parsimony of representation, which roughly means that the matrix has few significantly nonzero entries. This principle has given rise to various rotations, such as the varimax, quartimax, and oblimin rotations.

Factor analysis is often applied to grouped data under a model with the same factors in each group, called a common factor model.

In general, because of the stronger model, factor analysis should be used with more caution than principal components analysis.

Latent Semantic Indexing

An interesting application of the methods of principal components, called *latent semantic indexing*, is used in matching keyword searches with documents. The method begins with the construction of a term-document matrix, X, whose rows correspond to keywords, whose columns correspond to documents (web pages, for example), and whose entries are the frequencies of occurrences of the keywords in the documents. A singular value decomposition is performed on X (or on $X - \overline{X}$) as in equation (16.16), and then a reduced-rank matrix $\widetilde{X}_p$ is defined, as in equation (16.17). A list of keywords is matched to documents by representing the keyword list as a vector, q, of 0's and 1's corresponding to the rows of X. The vector $\widetilde{X}_p^{\mathrm{T}} q$ is a list of scores for the documents. Documents with larger scores are those deemed relevant for the search.

A semantic structure for the set of documents can also be identified by $\widetilde{X}_p$. Semantically nearby documents are mapped onto the same singular vectors.

A variation of latent semantic indexing is called probabilistic latent semantic indexing, or nonnegative-part factorization. This approach assumes a set of hidden variables whose values in the matrix H correspond to the columns of X by a *nonnegative matrix factorization*,

$$X = WH,$$

where W is a matrix with nonnegative elements.

The relationship of the model in probabilistic latent semantic indexing to the standard latent semantic indexing model is similar to the differences in factor analysis and principal components analysis.

Linear Independent Components Analysis

Independent components analysis (ICA) is similar to principal components analysis and factor analysis. Both PCA and ICA have nonlinear extensions. In linear PCA and ICA, the objective is to find a linear transformation W of a random vector Y so that the elements of WY have small correlations. In linear PCA, the objective then is to find W so that $\mathrm{V}(WY)$ is diagonal, and, as we have seen, this is simple to do. If the random vector Y is normal, then 0 correlations imply independence. The objective in linear ICA is slightly different; instead of just the elements of WY, attention may be focused on chosen transformations of this vector, and instead of small correlations, independence is the goal. Of course, because most multivariate distributions other than the normal are rather intractable, in practice small correlations are usually the objective in ICA. The transformations of WY are often higher-order sample moments. The projections that yield diagonal variance-covariance matrices are not necessarily orthogonal.

We discuss independent components analysis further in Section 16.6.

16.5 Projection Pursuit

The objective in projection pursuit is to find "interesting" projections of multivariate data. Interesting structure in multivariate data may be identified by analyzing projections of the data onto lower-dimensional subspaces. The projections can be used for optimal visualization of the clustering structure of the data or for density estimation or even regression analysis. The approach is related to the visual approach of the grand tour (page 363). Reduction of dimension is also an important objective, especially if the use of the projections is in visualization of the data. Projection pursuit requires a measure of the "interestingness" of a projection.

Diaconis and Freedman (1984) showed that a randomly selected projection of a high-dimensional dataset onto a low-dimensional space will tend to appear similar to a sample from a multivariate normal distribution with that lower dimension. This result, which may be thought of as a central limit theorem for projections, implies that a multivariate normal dataset is the least "interesting". A specific projection of the given dataset, however, may reveal interesting features of the dataset. In projection pursuit, therefore, the objective is to find departures from normality in linear projections of the data.

Departures from normality may include such things as skewness and "holes" in the data, or multimodality. The projection whose histogram is shown on the right-hand side of Figure 16.23 on page 559 exhibits a departure from normality, whereas the histogram on the left-hand side appears to be of normal univariate data. The projections are of the same dataset.

The Probability Model Underlying Projection Pursuit

Consider an m-vector random variable Y. In general, we are interested in a k-dimensional projection of Y, say $A^{\mathrm{T}}Y$, such that the random variable $A^{\mathrm{T}}Y$ is very different from a k-variate normal distribution.

Because all one-dimensional marginals of a multivariate normal are normal, and cross products of normals are multivariate normal, we will concentrate on one-dimensional projections of Z. For a given m-variate random variable Y, our problem is to find $Z = a^{\mathrm{T}}Y$ such that the scalar random variable Z is "most different" from a normal random variable. Two-dimensional projections are also of particular interest, especially in graphics. In the following, we will discuss just the one-dimensional projections.

The structure of interest (that is, a departure from normality) can be considered separately from the location, variances, and covariances of the vector Y; therefore, we will assume that $\mathrm{E}(Y) = 0$ and $\mathrm{V}(Y) = I$. Prior to applying projection pursuit to data, we center and sphere the data so that the sample characteristics are consistent with these assumptions.

To quantify the objectives in projection pursuit, we need a measure, or index, of the departure from normality.

Projection Indexes for the Probability Model

One way to quantify departure from normality is to consider the probability density function of the projected variable and compare it to the probability density function ϕ of a standard normal random variable. For a given random variable Y and $Z = a^{\mathrm{T}}Y$, if the PDF of Z is p, the question is how different is p from ϕ. This is similar to the problem in function approximation. Whereas in function approximation, the Chebyshev norm is generally of most interest, in seeking a function that is "different", an L_2 norm,

$$H(a) = \int_{-\infty}^{\infty} (p(z) - \phi(z))^2 \mathrm{d}z, \tag{16.28}$$

may be more appropriate as a measure of the difference, because it is not just the difference at a single point.

We consider projections The objective in projection pursuit is to find an a that maximizes this norm. It has become common practice in the literature on projection pursuit to name indexes of departure from normality by the type of orthogonal polynomials used in approximating the index. The index in expression (16.28) is called the *Hermite index* because Hermite polynomials are appropriate for approximation over the unbounded domain (see Table 4.1 on page 170). It is also called *Hall's index* because it was studied by Hall (1989).

For a given a, Friedman (1987) proposed first mapping $Z = a^{\mathrm{T}}Y$ into $[-1, 1]$ by the transformation

$$R_a = 2\Phi(Z) - 1, \tag{16.29}$$

where Φ is the CDF of a standard normal distribution. If p_Z is the probability density of Z, then the probability density of R_a is

$$p_{R_a}(r) = \frac{\frac{1}{2} p_Z\left(\Phi^{-1}\left(\frac{r+1}{2}\right)\right)}{\phi\left(\Phi^{-1}\left(\frac{r+1}{2}\right)\right)}.$$

If Z has a normal distribution with a mean of 0 and variance of 1, R_a has a uniform distribution over $(-1,\ 1)$ and so has a constant density of $\frac{1}{2}$. (This is the idea behind the inverse CDF method of random number generation.) Hence, the problem is to find a such that the density, p_R, of R_a is very different from $\frac{1}{2}$. The relevant L_2 norm is

$$L(a) = \int_{-1}^{1} \left(p_{R_a}(r) - \frac{1}{2}\right)^2 \mathrm{d}r,$$

which simplifies to

$$L(a) = \int_{-1}^{1} p_{R_a}^2(r) \mathrm{d}r - \frac{1}{2}. \tag{16.30}$$

This norm, which is a scalar function of a and a functional of p_{R_a}, is sometimes called the *Legendre index* because Legendre polynomials are natural approximating series of orthogonal polynomials for functions over finite domains (see Table 4.1, on page 170).

Cook, Buja, and Cabrera (1993) suggested another index based on the L_2 norm in equation (16.28) being weighted with the normal density:

$$H_{\mathrm{n}}(a) = \int_{-\infty}^{\infty} (p(z) - \phi(z))^2 \phi(z) \mathrm{d}z. \tag{16.31}$$

Cook, Buja, and Cabrera call this the *natural Hermite index*. The index is evaluated by expanding both $p(z)$ and $\phi(z)$ in the Hermite polynomials that are orthogonal with respect to $\mathrm{e}^{-x^2/2}$ over $(-\infty, \infty)$. (These are not the standard Hermite polynomials, but they are the ones most commonly used by statisticians because the weight function is proportional to the normal density.) These Hermite polynomials are the H_k^{e} in equations (4.53) on page 174. The index is called "natural" because the difference in p and ϕ is weighted by the normal density. The natural Hermite index has some desirable invariance properties for two-dimensional projections. See Cook, Buja, and Cabrera (1993) for a discussion of these properties.

Various other measures of departure from normality are possible; in fact, almost any goodness-of-fit criterion could serve as the basis for a projection index.

The normal distribution is a member of a class of distributions that are elliptically symmetric. We could extend the concept of "interesting" structure to be ones that lack an elliptical symmetry. Nason (2001) suggested use of a circular multivariate t distribution following sphering of the data. Nason defined three indices similar to those in equations (16.28) and (16.31), but based on departures from an m-variate t distribution,

$$\int_{\mathbb{R}^m} (p(z) - t_{\nu,m}(z))^2 (t_{\nu,m}(z))^\alpha \mathrm{d}z,$$

where $t_{\nu,m}(z)$ is the density function of a spherical m-variate t distribution with ν degrees of freedom, and α is 0, as in equation (16.28) or 1, as in equation (16.31). As Nason points out and confirms empirically, a procedure based on an index of departure from a multivariate t distribution is likely to be more robust to a small number of outlying observations than would a procedure based on a normal distribution.

Projection Pursuit in Data

We now consider one-dimensional projection pursuit in a given set of data X (the familiar $n \times m$ matrix in our data analysis paradigm). For each projection a, we *estimate* the projection index associated with a under the assumption that the rows in X are independent realizations of a random variable Y. The vector Xa contains independent realizations of the scalar random variable $Z = a^{\mathrm{T}}Y = Y^{\mathrm{T}}a$.

The question is how similar the distribution of Z is to a normal distribution. The problem with measures of departure from normality is the difficulty in estimating the terms.

To estimate the projection index, we must *approximate* an integral. As we suggested above, the indexes lend themselves to approximation by standard series of orthogonal polynomials.

For $L(a)$, expanding one factor of $p^2_{R_a}$ in equation (16.30) in Legendre polynomials and leaving the other unexpanded, we have

$$L(a) = \int_{-1}^{1} \left(\sum_{k=0}^{\infty} c_k P_k(r) \right) p_{R_a}(r)\,\mathrm{d}r - \frac{1}{2},$$

where P_k is the kth Legendre polynomial.

Using equations (4.26) and (4.44), we have the Legendre coefficients

$$c_k = \frac{2k+1}{2} \int_{-1}^{1} P_k(r) p_{R_a}(r)\,\mathrm{d}r$$

for the expansion.

Substituting this into the expression above, because of the orthogonality of the P_k, we have

$$L(a) = \frac{1}{2} \sum_{k=0}^{\infty} (2k+1) \Big(\mathrm{E}(P_k(R_a)) \Big)^2 - \frac{1}{2}, \tag{16.32}$$

where the expectation E is taken with respect to the distribution of the random variable R_a. Each term in equation (16.32) is an expectation and therefore can be estimated easily from a random sample. The sample mean is generally a good estimate of an expectation; hence, for the kth term, from the original observations x_i, the projection a, and the normal CDF transformation, we have

$$\begin{aligned} \widehat{\mathrm{E}}\left(P_k(R_a)\right) &= \frac{1}{n} \sum_{i=1}^{n} P_k(r_i) \\ &= \frac{1}{n} \sum_{i=1}^{n} P_k(2\Phi(a^{\mathrm{T}} x_i) - 1). \end{aligned}$$

A simple estimate of the squared expectation is just the square of this quantity.

Obviously, in practice, we must use a finite approximation to the infinite expansion of p_{R_a}. After terminating the expansion at j, we have the truncated Legendre projection index, $L_j(a)$,

$$L_j(a) = \frac{1}{2} \sum_{k=0}^{j} (2k+1) \left(\mathrm{E}\left(P_k(R_a)\right)\right)^2 - \frac{1}{2}. \tag{16.33}$$

The approximation in equation (16.33) can be estimated easily from the sample:

$$\widehat{L}_j(a) = \frac{1}{2n^2}\sum_{k=0}^{j}(2k+1)\left(\sum_{i=1}^{n} P_k(2\Phi(a^{\mathrm{T}}x_i)-1)\right)^2 - \frac{1}{2}. \tag{16.34}$$

This expression is easily evaluated. The first six Legendre polynomials are shown in equation (4.43) on page 171. We usually use the recurrence relationship, equation (4.45), in computing the truncated Legendre index, especially if we are using more than three or four terms.

The problem now is to determine

$$\max_a \widehat{L}_j(a).$$

Scaling of a is not relevant, so we may restrict a so that the sum of its elements is some given value, such as 1. In general, this is not an easy optimization problem. There are local minima. Use of an optimization method such as Newton's method may require multiple starting points. An optimization method such as simulated annealing may work better.

After both $p(z)$ and $\phi(z)$ are expanded in the Hermite polynomials, the natural Hermite index of equation (16.31) reduces to

$$\sum_{k=0}^{\infty}(d_k - b_k)^2,$$

where the d_k are the coefficients of the expansion of $p(z)$ and the b_k are the coefficients of the expansion of $\phi(z)$. The b_k can be evaluated analytically. They are, for $k = 0, 1, \ldots$,

$$\begin{aligned} b_{2k} &= \frac{(-1)^k((2k)!)^{1/2}}{2^{2k+1}k!\sqrt{\pi}} \\ b_{2k+1} &= 0. \end{aligned}$$

The d_k can be represented as expected values, and are estimated from the data in a similar manner as done for the Legendre index above. The estimates are given by

$$\hat{d}_k = \sum_{i=1}^{n} H_k^{\mathrm{e}}(x_i)\phi(x_i). \tag{16.35}$$

The index is the truncated series

$$\sum_{k=0}^{j}(\hat{d}_k - b_k)^2.$$

The first six Hermite polynomials are shown in equation (4.53) on page 174. We usually use the recurrence relationship, equation (4.54), in computing the truncated Hermite index, especially if we are using more than three or four terms.

Although the more terms retained in the orthogonal series expansions, the better is the approximation, it is not necessarily the case that the better-approximated index is more useful. Hall (1989) develops an asymptotic theory that suggests that the optimal choice of j depends on the sample size and type of index. (He considers both the norm in expression (16.28) and that in equation (16.30).) Sun (1992) suggests the choice of j between 3 and 6 inclusive, with smaller values being chosen for smaller numbers of observations and smaller values being chosen for larger values of the dimension m. Cook, Buja, and Cabrera (1993) found that the discrimination ability of the index for different values of j depends on the nature of the nonnormality in the data.

Friedman (1987) addresses the statistical significance of $L(a)$ (that is, the question of whether the projection using random data is significantly different from a projection of normal data). He gives a method for computing a p-value for the projection.

Exploratory Projection Pursuit

The most important use of projection pursuit is for initial exploratory analysis of multivariate datasets.

Different indexes may be useful in identifying different kinds of structure. The Legendre index is very sensitive to outliers in the data. If identification of the outliers is of specific interest, this may make the index useful. On the other hand, if the analysis should be robust to outliers, the Legendre index would not be a good one. The Laguerre-Fourier index, which is based on an expansion in Laguerre polynomials, is particularly useful in identifying clusters in the data.

Example

As an example of projection pursuit, consider the simple dataset shown in Figure 16.22. The obvious nonnormal structure in that dataset is the existence of two groups. Performing principal components analysis on the data resulted in a first principal component that completely obscured this structure (see Figure 16.23). As it turned out, the second principal component identified the structure very well.

In projection pursuit, the first step is generally to sphere the data. The result of the sphering is to make the variances in all directions almost equal. A "principal" component is no longer very relevant. As we have emphasized, scaling or sphering or any other transformation of the data is likely to have an effect on the results. Because sphering obscures important structure, whether or not to sphere the data in projection pursuit is subject to question.

The sphered data corresponding to those in Figure 16.22 are shown in Figure 16.24. (See Exercise 16.7 for the data.)

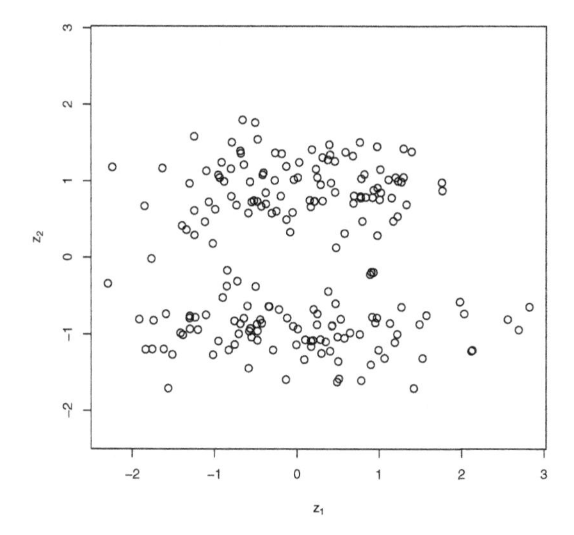

Fig. 16.24. The Sphered Bivariate Dataset with Two Clusters

After sphering the data, the plot does not show the correlations or the elliptical scatter of Figure 16.22. The principal components would be along the axes. The sphered data do, however, continue to exhibit a bimodality, which is a departure from normality.

Now, consider the Legendre index for the principal components — that is, for the projections $a_1 = (1, 0)$ and $a_2 = (0, 1)$. Using equation (16.34) with $j = 5$, we obtain $\widehat{L}_5(a_1) = 1.03$ and $\widehat{L}_5(a_2) = 1.66$. Clearly, the projection onto z_2 exhibits the greater nonnormality.

In most applications of projection pursuit, of course, we have datasets of higher dimension.

Computational Issues

Projection pursuit involves not only the computation of an index but the optimization of the index as a function of the linear combination vector. This approach is therefore computationally intensive.

The optimization problem is characterized by many local maxima. Rather than being interested in a global maximum, in data analysis with projection pursuit, we are generally interested in inspecting several projections, each of which exhibits an interesting structure — that is, some locally maximal departure from normality as measured by a projection index. This also adds to the computational intensity.

16.6 Other Methods for Identifying Structure

Structure in data is defined in terms of transformations of the data. In PCA, for example, the linear structure that is identified consists of a set of one-dimensional linear projections that is ordered by the norm of the projection of the centered dataset. In projection pursuit, the linear structure is also a set of projections, but they are ordered by their deviation from normality.

Nonlinear structure is generally much more difficult to detect. One approach is to generalize the methods of PCA. Girard (2000) describes a nonlinear PCA based on manifolds instead of linear projections. Hastie and Stuetzle (1989) discuss the generalization of principal (linear) components to principal curves.

Independent Components Analysis

In PCA, the objective is to determine components (that is, combinations of the original variables) that have zero correlations. If the data are normally distributed, zero correlations imply independence. If the data are not normally distributed, independence does not follow. Independent components analysis (ICA) is similar to PCA except that the objective is to determine combinations of the original variables that are independent. In ICA, moments of higher order than two are used to determine base vectors that are statistically as independent as possible. PCA is often used as a preprocessing step in ICA.

The probability model underlying independent components analysis assumes the existence of k *independent* data-generating processes that yield an observable n-vector through an unknown mixing process. Because many applications of ICA involve a time series in the observed x, we often express the model as

$$x(t) = As(t),$$

where A is a mixing matrix. The problem of uncovering $s(t)$ is sometimes called *blind source separation.* Let x_i for $i = 1, \ldots, m$ be measured signals and s_j for $j = 1, \ldots, k$ be independent components (ICs) with the zero mean. The basic problem in ICA is to estimate the mixing matrix A and determine the components s in

$$x = As.$$

Independent components analysis is similar to principal components analysis and factor analysis, but much of the research on ICA has been conducted without reference to PCA and factor analysis. Both PCA and ICA have nonlinear extensions. In linear PCA and ICA, the objective is to find a linear transformation W of a random vector Y so that the elements of WY have small correlations. In linear PCA, the objective then is to find W so that $\mathrm{V}(WY)$ is diagonal, and, as we have seen, this is simple to do. If the random vector Y is normal, then 0 correlations imply independence. The objective in linear ICA is slightly different; instead of just the elements of WY, attention may be focused on chosen transformations of this vector, and instead of small correlations, independence is the goal. Of course, because most multivariate distributions other than the normal are rather intractable, in practice, small correlations are usually the objective in ICA. The transformations of WY are often higher-order sample moments. The projections that yield diagonal variance-covariance matrices are not necessarily orthogonal.

In the literature on ICA, which is generally in the field of signal processing, either a "noise-free ICA model", similar to the simple PCA model, or a "noisy ICA model", similar to the factor analysis model, is used. Most of the research has been on the noise-free ICA model.

16.7 Higher Dimensions

The most common statistical datasets can be thought of as rows, representing observations, and columns, representing variables. In traditional multiple regression and correlation and other methods of multivariate analysis, there are generally few conceptual hurdles in thinking of the observations as ranging over a multidimensional space. In multiple regression with m regressors, for example, it is easy to visualize the hyperplane in $m + 1$ dimensions that represents the fit $\widehat{y} = X\widehat{\beta}$. It is even easy to visualize the projection of the n-dimensional vector that represents a least-squares fit.

Many properties of one- and two-dimensional objects (lines and planes) carry over into higher-dimensional space just as we would expect.

Although most of our intuition is derived from our existence in a three-dimensional world, we generally have no problem dealing with one- or two-dimensional objects. On the other hand, it can be difficult to view a 3-D world from a two-dimensional perspective. The delightful fantasy, *Flatland*, written by Edwin Abbott in 1884, describes the travails of a two-dimensional person (one "A. Square") thrown into a three-dimensional world. (See also Stewart, 2001, *Flatterland, Like Flatland Only More So.*) The small book by Kendall (1961), *A Course in the Geometry of n Dimensions*, gives numerous examples in which common statistical concepts are elucidated by geometrical constructs.

There are many situations, however, in which our intuition derived from the familiar representations in one-, two-, and three-dimensional space leads us completely astray. This is particularly true of objects whose dimensionality is greater than three, such as volumes in higher-dimensional space. The problem is not just with our intuition, however; it is indeed the case that some properties do not generalize to higher dimensions. Exercise 16.11 in this chapter illustrates such a situation.

The *shape* of a dataset is the total information content that is invariant under translations, rotations, and scale transformations. Quantifying the shape of data is an interesting problem.

Data Sparsity in Higher Dimensions

We measure space both linearly and volumetrically. The basic cause of the breakdown of intuition in higher dimensions is that the relationship of linear measures to volumetric measures is exponential in dimensionality. The cubing we are familiar with in three-dimensional space cannot be used to describe the relative sizes of volumes (that is, the distribution of space). Volumes relative to the linear dimensions grow very rapidly. There are two consequences of this. One is that the volumes of objects with interior holes, such as thin boxes or thin shells, are much larger than our intuition predicts. Another is that the density of a fixed number of points becomes extremely small.

The density of a probability distribution decreases as the distribution is extended to higher dimensions by an outer product of the range. This happens fastest going from one dimension to two dimensions but continues at a decreasing rate for higher dimensions. The effect of this is that the probability content of regions at a fixed distance to the center of the distribution increases; that is, outliers or isolated data points become more common. This is easy to see in comparing a univariate normal distribution with a bivariate normal distribution. If $X = (X_1, X_2)$ has a bivariate normal distribution with mean 0 and variance-covariance matrix $\mathrm{diag}(1, 1)$,

$$\Pr(|X_1| > 2) = 0.0455,$$

whereas

$$\Pr(\|X\| > 2) = 0.135.$$

The probability that the bivariate random variable is greater than two standard deviations from the center is much greater than the probability that the univariate random variable is greater than two standard deviations from the center. We can see the relative probabilities in Figures 16.25 and 16.26. The area under the univariate density that is outside the central interval shown is relatively small. It is about 5% of the total area. The volume under the bivariate density in Figure 16.26 beyond the circle is relatively greater than the volume within the circle. It is about 13% of the total volume. The percentage increases with the dimensionality (see Exercises 16.9 and 16.10).

The consequence of these density patterns is that an observation in higher dimensions is more likely to appear to be an outlier than one in lower dimensions.

Volumes of Hyperspheres and Hypercubes

It is interesting to compare the volumes of regular geometrical objects and observe how the relationships of volumes to linear measures change as the number of dimensions changes. Consider, for example, that the volume of a sphere of radius a in d dimensions is

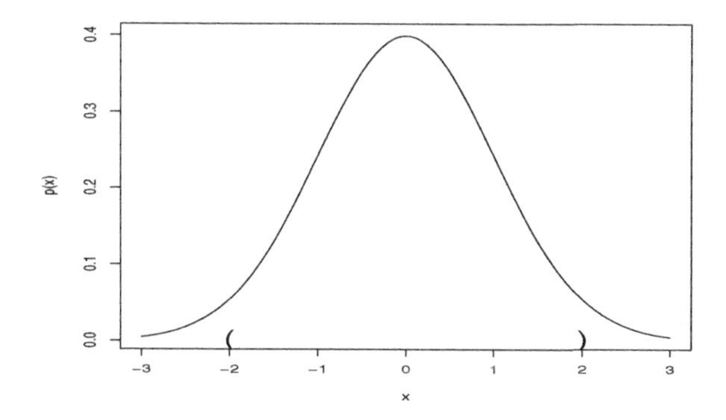

Fig. 16.25. Univariate Extreme Regions

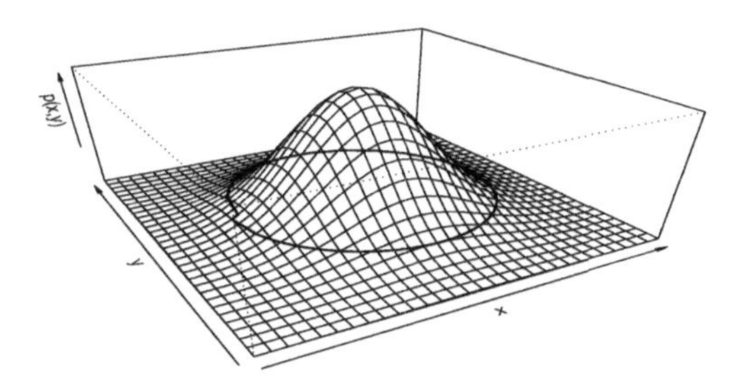

Fig. 16.26. Bivariate Extreme Regions

$$\frac{a^d \pi^{d/2}}{\Gamma(1 + d/2)}.$$

The volume of a superscribed cube is $(2a)^d$. Now, compare the volumes. Consider the ratio

$$\frac{\pi^{d/2}}{d 2^{d-1} \Gamma(d/2)}.$$

For $d = 3$ (Figure 16.27), this is 0.524; for $d = 7$, however, it is 0.037. As the number of dimensions increases, more and more of the volume of the cube is in the corners.

For two objects of different sizes but the same shape, with the smaller one centered inside the larger one, we have a similar phenomenon of the content of the interior object relative to the larger object. The volume of a thin shell

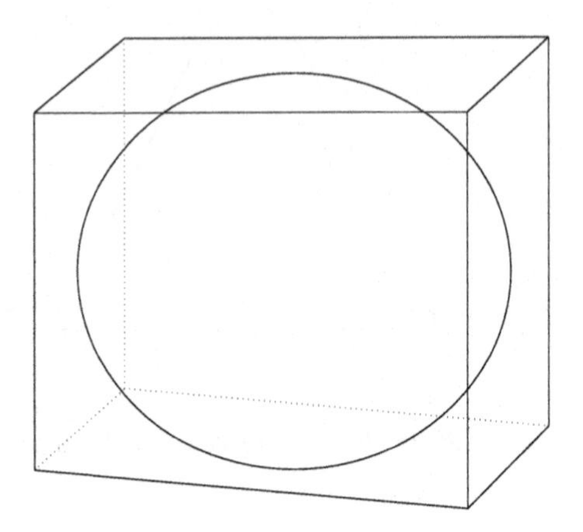

Fig. 16.27. A Superscribed Cube

as the ratio of the volume of the outer figure (sphere, cube, whatever) is

$$\frac{V_d(r) - V_d(r - \epsilon)}{V_d(r)} = 1 - \left(1 - \frac{\epsilon}{r}\right)^d.$$

As the number of dimensions increases, more and more of the volume of the larger object is in the outer thin shell. This is the same phenomenon that we observed above for probability distributions. In a multivariate distribution whose density is the product of identical univariate densities (which is the density of a simple random sample), the relative probability content within extreme regions becomes greater as the dimension increases.

The Curse of Dimensionality

The computational and conceptual problems associated with higher dimensions have often been referred to as "the curse of dimensionality". How many dimensions cause problems depends on the nature of the application. Golub and Ortega (1993) use the phrase in describing the solution to the diffusion equation in three dimensions, plus time as the fourth dimensions of course.

In higher dimensions, not only do data appear as outliers, but they also tend to lie on lower dimensional manifolds. This is the problem sometimes called "multicollinearity". The reason that data in higher dimensions are multicollinear, or more generally, concurve, is that the number of lower dimensional manifolds increases very rapidly in the dimensionality: The rate is 2^d.

Whenever it is possible to collect data in a well-designed experiment or observational study, some of the problems of high dimensions can be ameliorated. In computer experiments, for example, Latin hypercube designs can be useful for exploring very high dimensional spaces.

Data in higher dimensions carry more information in the same number of observations than data in lower dimensions. Some people have referred to the increase in information as the "blessing of dimensionality". The support vector machine approach in fact attempts to detect structure in data by mapping the data to higher dimensions.

Tiling Space

As we have mentioned in previous chapters, tessellations of the data space are useful in density estimation and in clustering and classification. Generally, regular tessellations, or tilings (objects with the same shapes), are used. Regular tessellations are easier both to form and to analyze.

Regular tessellations in higher dimensional space have counterintuitive properties. As an example, consider tiling by hypercubes, as illustrated in Figure 16.28 for squares in two dimensions.

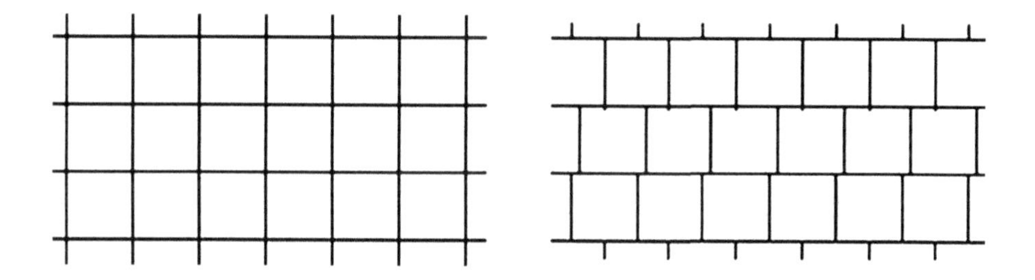

Fig. 16.28. Hypercube (Square) Tilings of 2-Space

The tiling on the left-hand side in Figure 16.28 is a lattice tiling. In both tilings, we see that each tile has an entire side in common with at least one adjacent tile. This is a useful fact when we use the tiles as bins in data analysis, and it is always the case for a lattice tiling. It is also always the case in two dimensions. (To see this, make drawings similar to those in Figure 16.28.) In fact, in lower dimensions (up to six dimensions for sure), tilings by hypercubes of equal size always have the property that some adjacent tiles have an entire face (side) in common. It is an open question as to what number of dimensions ensures this property, but the property definitely does not hold in ten dimensions, as shown by Peter Shor and Jeff Lagarias. (See *What's*

Happening in the Mathematical Sciences, Volume 1, American Mathematical Society, 1993, pages 21–25.)

Notes and Further Reading

Statistical learning has been one of the most rapidly growing areas of statistics during the past several years. Indicating the importance of this field, the American Statistical Association in 2009 formed a new section, Statistical Learning & Data Mining, and began a new journal, *Statistical Analysis and Data Mining*.

Recursive Partitioning

Many of the statistical learning methods that we have discussed are based on recursively partitioning the dataset. Trees or other graphs are used to maintain the status during the recursion. General methods as well as specific applications of recursive partitions are summarized by Zhang (2004). We encounter some of these methods again in Chapter 17.

Tesselations

Ash et al. (1988), Aurenhammer (1991), and Okabe et al. (2000) discuss many interesting properties of the Dirichlet tessellation or Voronoi diagram and the Delaunay triangulation that hold in d dimensions. Du, Faber, and Gunzburger (1999) discuss additional properties as well as applications.

Conceptual Clustering and Fuzzy Clustering

Dale (1985) provides a comparison of conceptual clustering with other clustering methods.

Hathaway and Bezdek (1988) discuss the performance of algorithms that identify fuzzy clusters by minimizing the quantity (16.7) above. Seaver, Triantis, and Reeves (1999) describe an algorithm for fuzzy clustering that begins with a stage of hard clustering and in the second step treats cluster membership as fuzzy. Rousseeuw (1995) discusses concepts of fuzzy clustering, and Laviolette et al. (1995) discuss some general issues in fuzzy methods.

Data Depth

Maronna and Yohai (1995) and Zuo and Serfling (2000a, 2000b) describe properties of the measure of depth based on maximal one-dimensional projections. Liu (1990) defined the *simplicial location depth* as the proportion of data simplices (triangles formed by three observations in the bivariate case) that contain the given point. Rousseeuw and Hubert (1999) define and study a depth measure based on regression fits. Liu, Parelius, and Singh (1999) describe and compare measures of data depth and discuss various applications of data depth in multivariate analysis.

Nearest Neighbor Classification

Different distance measures used in clustering methods generally result in different clusters or classifications, especially in the presence of outlying observations. Robust measures of distance, such as discussed in Chapter 9 may be more appropriate. Amores, Sebe, and Radeva (2006) discuss an adaptive approach to measuring distance for use in nearest neighbor classification.

Dimension Reduction

One of the main objectives in exploratory statistical learning is to reduce the dimension of the observable data. This means finding manifolds of lower dimension that contain all or most of the interesting features of the original dataset. There are many ways to reduce the dimensionality, and we have discussed some in this chapter. Mizuta (2004) provides further discussion, as well as consideration of others beyond those discussed in this chapter.

Independent Components Analysis

In the literature on ICA, which is generally in the field of signal processing, either a "noise-free ICA model", similar to the simple PCA model, or a "noisy ICA model", similar to the factor analysis model, is used. Most of the research has been on the noise-free ICA model. The reader is referred to Comon (1994) for further descriptions of ICA and an iterative algorithm. Comon also discusses some of the similarities and differences between ICA and PCA. Hyvärinen, Karhunen, and Oja (2001) provide a comprehensive discussion of ICA.

Projection Pursuit

Cook, Buja, and Cabrera (1993) discuss two-dimensional projections and their applications in graphics. They also consider the use of the natural Hermite index as a measure of the non-normality of the projection.

Other kinds of measures of departure from normality can be contemplated. Almost any goodness-of-fit criterion could serve as the basis for a projection index. Posse (1990, 1995a, 1995b) suggests a projection index based on a chi-squared measure of departure from normality. It has an advantage of computational simplicity. Jones (see Jones and Sibson, 1987) suggested an index based on ratios of moments of a standard normal distribution, and Huber (1985) suggested indexes based on entropy (called Shannon entropy or differential entropy):

$$-\int_{\mathrm{IR}^m} p(z) \log p(z) \mathrm{d}z.$$

The entropy is maximized among the class of all random variables when the density p is the standard multivariate normal density (mean of zero and variance-covariance matrix equal to the identity). For any other distribution, the entropy is strictly smaller.

Morton (1992) suggests an "interpretability index" that gives preference to simple projections (that is, to linear combinations in which a has more zeros), and when comparing two combinations a_1 and a_2, the vectors are (nearly) orthogonal. This work anticipated similar attempts in PCA to determine approximate principal components that are "simple" (see Vines, 2000, and Jolliffe and Uddin, 2000).

Sun (1992, 1993) reports comparisons of the use of Friedman's Legendre index, $L(a)$, and Hall's Hermite index (16.28). Cook, Buja, and Cabrera (1993) give comparisons of these two indexes and the natural Hermite index, $H_{\mathrm{n}}(a)$. Posse (1990, 1995a, 1995b) and Nason (2001) also report comparisons of the various indices. The results of these comparisons were inconclusive; which index is better in identifying departures from normality (or from uninteresting structure) seems to depend on the nature of the nonnormality.

Cook et al. (1995) describe the use of projection pursuit in a grand tour. Cabrera and Cook (1992) discuss the relationship of projection pursuit to the fractal dimension of a dataset.

Exercises

16.1. Consider the clusterings of the toy dataset depicted in Figure 16.7 on page 527.
   a) How many clusters seem to be suggested by each?
   b) Compute Rand's statistic (16.8) on page 537 and the modified Rand statistic (16.9) to compare the clustering on the left-hand side with that in the middle. Assume four clusters in each.
   c) Compute Rand's statistic and the modified Rand statistic to compare the clustering in the middle with that on the right-hand side. Assume four clusters in the middle, and three clusters on the right-hand side.
   d) Compute Rand's statistic and the modified Rand statistic to compare the clustering in the middle with that on the right-hand side. Assume two clusters in the middle, and three clusters on the right-hand side.

16.2. a) Develop a combinatorial optimization algorithm, perhaps using simulated annealing or a genetic algorithm, to perform K-means clustering in such a way that less than k groups may be formed. The objective function would need to be penalized for the number of groups. Try a modification of expression (16.3),

$$\sum_{g=1}^{k}\sum_{j=1}^{m}\sum_{i=1}^{n_g}\left(x_{ij(g)}-\bar{x}_{j(g)}\right)^2+\alpha k,$$

where α is a tuning parameter. Its magnitude depends on the sizes of the sums of squares, which of course are unknown a priori. Write a program to implement your algorithm. In the program, α is an input parameter. Use your program to form five or fewer clusters of the data:

$$\begin{bmatrix} x_1 \\ x_2 \\ x_3 \\ x_4 \\ x_5 \\ x_6 \\ x_7 \\ x_8 \\ x_9 \\ x_{10} \\ x_{11} \\ x_{12} \\ x_{13} \\ x_{14} \\ x_{15} \\ x_{16} \end{bmatrix} = \begin{bmatrix} 1 & 1 \\ 1 & 2 \\ 1 & 3 \\ 1 & 4 \\ 2 & 1 \\ 2 & 2 \\ 2 & 3 \\ 2 & 4 \\ 3 & 1 \\ 3 & 2 \\ 3 & 3 \\ 3 & 4 \\ 4 & 1 \\ 4 & 2 \\ 4 & 3 \\ 4 & 4 \end{bmatrix}.$$

How many clusters do you get?

b) Using the given data, do K-means clustering for $k = 1, 2, 3, 4, 5$. For each number of clusters, compute the Calinski-Harabasz index (16.4) on page 522. How many clusters are suggested?

16.3. What happens in cluster analysis if the data are sphered prior to the analysis? Sphere the data used in Figure 16.11 (on page 534), and then do the hierarchical clustering. To do this, replace the statement

```
y <- apply(x,2,standard)
```

with

```
y <- x %*% solve(chol(var(x)))
```

The strange result in this case of clustering sphered data does not always occur, but the point is that sphering can have unexpected effects.

16.4. Consider the problem of identification of the "concentric circles" structure in Figure 16.3 on page 517. As we mentioned, representation of the data in polar coordinates provides one way of finding this structure. A dissimilarity measure consisting of an angular component d^{d}_{ij} and a radial component d^{r}_{ij} may be defined, as discussed on page 392 and in Exercise 9.12. Develop a procedure for clustering the concentric circles. Test your procedure using data similar to that in Figure 16.3, which can be generated in R by

```
x1<-rnorm(10)
x2<-rnorm(10)
x<-cbind(x1,x2)
y1<-rnorm(40,0,5)
y2<-rnorm(40,0,5)
sel<-y1^2+y2^2>25
y<-cbind(y1[sel],y2[sel])
```

16.5. Consider the problem of Exercise 9.13 on page 399; that is, given two n-vectors, x_1 and x_2, form a third vector x_3 as $x_3 = a_1x_1 + a_2x_2 + \epsilon$, where ϵ is a vector of independent $\mathrm{N}(0,1)$ realizations. Although the matrix $X = [x_1\ x_2\ x_3]$ is in $\mathrm{IR}^{n\times 3}$, the linear structure, even obscured by the noise, implies a two-dimensional space for the data matrix (that is, the space $\mathrm{IR}^{n\times 2}$). Generate x_1 and x_2 as realizations of a $\mathrm{U}(0,1)$ process, and x_3 as $5x_1 + x_2 + \epsilon$, where ϵ is a realization of a $\mathrm{N}(0,1)$ process. Do a principal components analysis of the data (perhaps using `prcomp` or `princomp` in R). Make a scree plot of the eigenvalues (perhaps using `plot.pcs.scree` in the R `bio3d` package or `screeplot` in S-Plus, which produces a bar plot, rather than a line plot as shown in Figure 16.21). How many principal components would you choose?

16.6. a) Write out the gradient and Hessian for the optimization problem (16.24) on page 562. Remember Ψ is a diagonal matrix.
b) Write out the gradient and Hessian for the optimization problem (16.27) on page 563.

16.7. The data shown in Figure 16.22 and used in the PCA and the projection pursuit examples were generated by the R commands

```
n <- 200
x <- rnorm(n)
y <- rnorm(n)
xx <- 10*x + y
yy <- 2*y +x
n2 <- n/2
yy[1:n2] <- yy[1:n2] + 5
yy[(n2+1):n] <- yy[(n2+1):n] - 5
```

a) Sphere this dataset and plot it. Your plot should look like that in Figure 16.24.
b) Determine the optimal projection a using the estimated truncated Legendre index with $j = 4$.

16.8. Indexes for projection pursuit.

a) Derive equation (16.35) on page 569 for use in the natural Hermite index, equation (16.31). (Compare equation (16.34).) See page 174 for the Hermite polynomials.

b) Determine the optimal projection of the data in Exercise 16.7 using the estimated truncated natural Hermite index with $j = 4$.

16.9. Let X be a standard 10-variate normal random variable (the mean is 0 and the variance-covariance is $\mathrm{diag}(1, 1, 1, 1, 1, 1, 1, 1, 1, 1)$). What is the probability that $\|X\| > 6$? In other words, what is the probability of exceeding six sigma?
Hint: Use polar coordinates. (Even then, the algebra is messy.)

16.10. Consider the use of a d-variate multivariate normal density as a majorizing density for another d-variate multivariate normal in an acceptance/rejection application. (See Figure 7.2 on page 311 and the discussion concerning it.) To be specific, let $d = 1,000$, let the majorizing density have a diagonal variance-covariance matrix with constant diagonal terms of 1.1, and let the target density also have a diagonal variance-covariance matrix but with constant diagonal terms of 1. (As mentioned in the discussion concerning Figure 7.2, this is just an illustrative example. This kind of majorizing density would not make sense for the given target because if we could generate from the majorizing density we could generate directly from the target.)

a) Determine the value of c in Algorithm 7.1 on page 309.

b) Determine the probability of acceptance.

16.11. In d dimensions, construct 2^d hyperspheres with centers at the points $(\pm 1, \ldots, \pm 1)$, and construct the hypercube with edges of length 2 that contains the unit hyperspheres. At the point $(0, \ldots, 0)$, construct the hypersphere that is tangent to the other 2^d spheres. In two dimensions, the spheres appear as

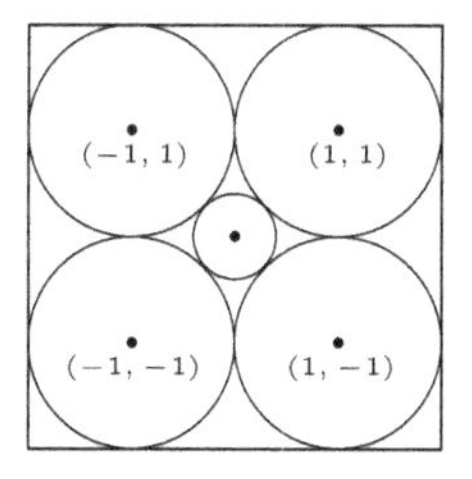

Is the interior hypersphere always inside the hypercube? (The answer is "No!") At what number of dimensions does the interior hypersphere poke outside the hypercube? (See *What's Happening in the Mathematical Sciences, Volume 1*, American Mathematical Society, 1993.)

16.12. Consider a cartesian coordinate system for IR^d, with $d \geq 2$. Let x be a point in IR^d_+ such that $\|x\|_2 = 1$ and x is equidistant from all axes of the coordinate system.

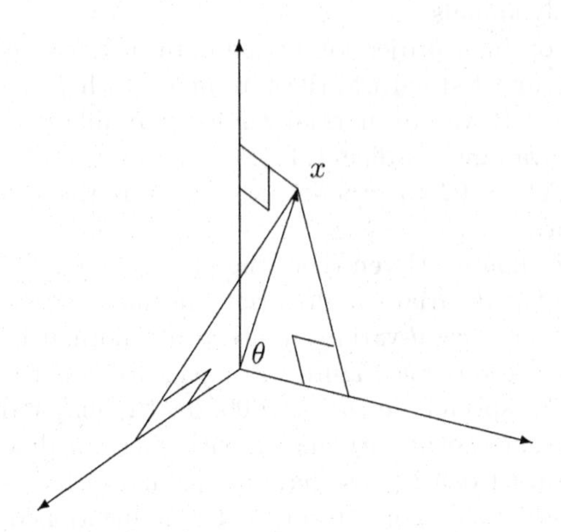

What is the angle between the line through x and any of the positive axes? *Hint:* for $d = 2$, the angles are $\pm\pi/4$.
What are the angles as $d \to \infty$?

17

Statistical Models of Dependencies

In the models and data-generating processes we have considered in previous chapters of Part IV, all of the variables or features were treated essentially in the same way. In this chapter, we consider models in which a subset of variables, often just one, is of special interest. This variable is the "response", and we seek to understand its dependence on the other variables. "Dependence" here refers to a stochastic relationship, not a causal one. By knowing the relationship of the other variables to the response variable, we may better understand the data-generating process, or we may be able to predict the response, given the values of the associated variables.

The models we consider in this chapter describe the stochastic behavior of one variable, Y, possibly a vector, as a function of other variables. Models of this type that express dependencies are called regression models if Y is a numeric variable or classification models if Y is a categorical variable. If Y is a numeric variable that takes on only a countable number of values, the model can be considered either a regression model (sometimes a "generalized model") or a classification model.

Another important type of dependency arises in sequentially sampled variables. The distribution of a random variable at time t_k depends on the realization of that random variable at times before t_k. There may also be covariates whose realizations affect the distribution of the variable of interest. A random process that possibly changes in time is called a *stochastic process*. Because change itself is of interest in such processes, the model is often expressed as a differential equation.

The development and use of a model is an iterative process that includes data collection and analysis. It requires looking at the data from various perspectives. The model embodies both knowledge and assumptions. The knowledge may result from first principles or from previous observations. A model can be strong (very specific) or weak (leaving open many possibilities).

If the range of possibilities in a model can be limited to a set of real numbers, the possibilities are represented by a *parameter*. Parametric statistical procedures involve inference about the parameters of a model. *Nonparamet-*

J.E. Gentle, *Computational Statistics*, Statistics and Computing,
DOI: 10.1007/978-0-387-98144-4_17,

ric methods in statistics also rely on models that usually contain parameters; the phrase "nonparametric" often refers to a method that does not specify a family of probability distributions except in a very general way.

Models

While some methods are referred to as "model-free", and the phrase "model-based approach" is sometimes used to describe a statistical method, implying that other, "non-model-based" approaches exist, in reality some model underlies all statistical analyses. The model is not immutable, however, and much of the effort of an analysis may go into developing and refining a model. In *exploratory data analysis*, or *EDA*, the model is quite weak. The patterns and other characteristics identified in an exploratory data analysis help to form a stronger model. In general, whenever the model is weak, a primary objective is usually to build a stronger model.

There are various types of models. They may have different purposes. A common form of a model is a mathematical equation or a system of equations. If the purpose of the model is to enhance the understanding of some phenomenon, there would be a large premium on simplicity of the model. If the model is very complicated, it may correspond very well to the reality being studied, but it is unlikely to be understandable. If its primary purpose is to aid understanding, an *equation model* should be relatively simple. It should not require an extensive period of time for scrutiny.

A model may be embedded in a computer program. In this case, the model itself is not ordinarily scrutinized; only its input and output are studied. The complexity of the model is not of essential consequence. Especially if the objective is prediction of a response given values of the associated variables, and if there is a large premium on making accurate predictions or classifications in a very short time, an *algorithmic model* may be appropriate. An algorithmic model prioritizes prediction accuracy. The details of the model may be very different from the details of the data-generating process being modeled. That is not relevant; the important thing is how well the output of the algorithmic model compares to the output of the data-generating process being modeled when they are given the same input.

Model Inference Using Data

Data analysis usually proceeds through some fairly standard steps. Before much is known about the process being investigated, the statistician may just explore the data to arrive at a general understanding of its structure. This may involve many graphical displays in various coordinate systems and under various projections. When more than one variable is present, relationships among the variables may be explored and models describing these relationships developed.

One aspect of the modeling process is to fit a tentative model to observed data. In a parametric model, this means determination of values of parameters in the model so that the model corresponds in some way to the observations. The criteria for the model to correspond to the observations may be based on some distributional motivation or merely on some heuristic measure of the correspondence of pairs of points to an equation. These criteria include the following.

- Some model moments match the corresponding sample moments (method of moments).
- Some norm of the vector of deviations of observed values from mean model values is minimized (least squares, for example).
- The joint probability density function evaluated at the observed values is maximized (maximum likelihood).
- Observations that are similar have similar responses in the model (homogeneity of classes).
- If the data are partitioned into two sets and a model is fit based on one set, the model fits the data in the other set well (training set and test set, or cross validation). In this case, the emphasis is on classification or prediction accuracy (based on an appropriate definition and a suitable quantification).

These criteria are not mutually exclusive, and some combination of them may be used. The criteria can be viewed purely as intuitive guidelines, or the stochastic components of the model may be modeled by some family of statistical probability distributions, and the distributional properties of the estimators under various assumptions can determine the approach. More formal methods of estimation as described in Chapter 1 may be used, especially if prior knowledge or beliefs are to be incorporated formally.

An important part of the modeling process is statistically testing of the correspondence of the available data to the model that has been fit. Depending on the type of model, the goodness-of-fit testing may or may not be a relatively straightforward process. We should also understand that goodness-of-fit tests generally are rather ineffective for addressing the basic question of what is the correct model.

The distributions of estimators under various assumptions may be quite difficult to work out. Rather than basing inference on asymptotic approximations, we can use computational inference for confidence levels or statements of probability about model parameters. Computational inference using simulated datasets can also be useful in assessing the fidelity of the evolving models to the observed data.

In this chapter we will consider statistical models of relationships among observable features. In Section 17.1, we briefly discuss models of dependencies in a general way. In Section 17.2, we discuss the incorporation of a probability distribution into the model. This gives us a basis for statistical inference. In the longer Section 17.3, we discuss the use of observational data to fit the model.

In Section 17.4, we discuss the use of observational data to fit a particular kind of model, namely a classification model. Finally, in Section 17.5, we discuss the use of transformations of data so that a model of a given form will fit it better. One-to-one transformations do not result in any information loss, and they often make observational data easier to understand.

17.1 Regression and Classification Models

In many applications, some subset of variables may be characterized as "dependent" on some other subset of variables; in fact, often there is just a single "dependent" variable, and our objective is to characterize its value in terms of the values of the other variables. (The word "dependent" is in quotes here because we do not wish necessarily to allow any connotation of causation or other stronger meanings of the word "dependence". In the following, we use "dependent" in this casual way but do not continue to emphasize that fact with quotation marks.) The dependent variable is often called the "response", and the other variables are often called "factors", "regressors", "independent variables", "carriers", "stimuli", or "covariates". (The word "independent" has some connotative difficulties because of its use in referring to a stochastic property, and some authors object to the use of "independent" here. Most choices of words have one kind of problem or another. A problem with "stimulus", for example, is the implication of causation. I am likely to use any one of these terms at various times. Fortunately, it does not matter; the meaning is always clear from the context.)

The asymmetric relationship between a random variable Y and a variable x may be represented as a black box that accepts x as input and outputs Y:

$$Y \leftarrow \boxed{\text{unknown process}} \leftarrow x. \tag{17.1}$$

The relationship might also be described by a statement of the form

$$Y \leftarrow f(x)$$

or

$$Y \approx f(x). \tag{17.2}$$

If f has an inverse, the model (17.2) appears symmetric. Even in that case, however, there is an asymmetry that results from the role of random variables in the model; we model the response as a random variable. We may think of $f(x)$ as a *systematic* effect and write the model with an additive adjustment, or error, as

$$Y = f(x) + E \tag{17.3}$$

or with a multiplicative error as

$$Y = f(x)\Delta, \tag{17.4}$$

where E and Δ are assumed to be random variables. (The "E" is the Greek uppercase epsilon.) We refer to these as "errors", although this word does not indicate a mistake. Thus, the model is composed of a *systematic component* related to the values of x and a *random component* that accounts for the indeterminacy between Y and $f(x)$. The relative contribution to the variability in the observed Y due to the systematic component and due to the random component is called the *signal to noise ratio.* (Notice that this is a nontechnical term here; we could quantify it more precisely in certain classes of models.)

In the case of the black-box model (17.1), both the systematic and random components are embedded in the box.

Models with multiplicative random effects are not as widely used. In the following, we will concentrate on models with additive random effects. In such models, E is also called the "residual".

Because the functional form f of the relationship between Y and x may contain a *parameter*, we may write the equation in the model as

$$Y = f(x;\theta) + E, \tag{17.5}$$

where θ is a parameter whose value determines a specific relationship within the family specified by f. In most cases, θ is a vector. In the usual linear regression model, for example, the parameter is a vector with two more elements than the number of elements in x,

$$Y = \beta_0 + x^{\mathrm{T}}\beta + E, \tag{17.6}$$

where $\theta = (\beta_0, \beta, \sigma^2)$.

A generalization of the linear model (17.6) is the *additive model*,

$$Y = \beta_0 + f_1(x_1, \beta_1) + \cdots + f_m(x_m, \beta_m) + E. \tag{17.7}$$

The specification of the distribution of the random component is a part of the model, and that part of the model can range from very general assumptions about the existence of certain moments or about the general shape of the density to very specific assumptions about the distribution. If the random component is additive, the mean, or more generally (because the moments may not exist) the appropriate location parameter, is assumed without loss of generality to be 0.

The model for the relationship between Y and x includes the equation (17.5) together with various other statements about Y and x such as the nature of the values that they may assume, statements about θ, and statements about the distribution of E. Thus, the *model* is

$$\begin{cases} Y = f(x;\theta) + E \\ \\ \text{additional statements about } Y, x, \theta, E. \end{cases} \tag{17.8}$$

In the following, for convenience, we will often refer to just the equation as the "model".

Another way of viewing the systematic component of equation (17.5) is as a conditional expectation of a random variable,

$$\mathrm{E}(Y|x;\theta) = f(x;\theta). \tag{17.9}$$

This formulation is very similar to that of equations (17.3) and (17.4). The main differences are that in the formulation of equation (17.9) we do not distinguish between an additive error and a multiplicative one, and we consider the error to be a random variable with finite first moment. In equations (17.3) and (17.4), or equation (17.5), we do not necessarily make these assumptions about E or Δ.

If we assume a probability distribution for the random component, we may write the model in terms of a probability density for the response in terms of the systematic component of the model,

$$p(y|x,\theta). \tag{17.10}$$

Cast in this way, the problem of statistical modeling is the same as fitting a probability distribution whose density depends on the values of a covariate.

Generalized Models

If the response can take on values only in a countable set, a model of the form

$$Y = f(x;\theta) + E$$

may not be appropriate, especially if the covariate x is continuous.

Suppose, for example, that the response is binary (0 or 1) representing whether or not a person has had a heart attack, and x contains various biometric measures such as blood pressure, cholesterol level, and so on. The expected value $\mathrm{E}(Y|x;\theta)$ is the probability that a person with x has had a heart attack. Even if a model such as

$$\mathrm{E}(Y|x;\theta) = f(x;\theta), \tag{17.11}$$

with continuous regressor x, made sense, it would not be clear how to fit the model to data or to make inferences about the model. A simple transformation of the response variable, $\tau(Y)$, does not improve the data-analysis problem; if Y is binary, so is $\tau(Y)$.

A problem with the model in this form is that the value of $f(x;\theta)$ must range between 0 and 1 for all (reasonable) values of x and θ. A function f could of course be constructed to have this range, but another way is to model a transformation of $\mathrm{E}(Y|x;\theta)$. This can often be done in a way that has a meaningful interpretation. We first let

$$\pi(x;\theta) = \mathrm{E}(Y|x;\theta), \tag{17.12}$$

which we generally write simply as π. If Y is binary with values 0 and 1 then π is just the probability that $Y = 1$.

Next, we introduce some function $g(\pi)$, that, it is hoped, relates to the data-generating process being studied. The function g is called the *link function.* The appropriate form of the link function depends on the nature of the probability distribution. Notice that the link function is not just a transformation of the observable variable Y, as is $\tau(Y)$ above. The link function is a transformation of the expected value of Y.

We now can form a "generalized model" that is more similar to the forms of the models for continuous responses than a form that models Y directly; that is,

$$g(\pi) \approx h(x;\theta). \tag{17.13}$$

In the case of Y being binary, letting $\pi = \mathrm{E}(Y|x;\theta)$, we may introduce the transformation

$$\begin{aligned} g(\pi) &= \log\left(\frac{\pi}{1-\pi}\right) \\ &= \mathrm{logit}(\pi). \end{aligned} \tag{17.14}$$

The *logit function* in equation (17.14) is the "odds ratio". The logit function is useful for a Bernoulli distribution (that is, a binary response).

In this kind of problem, we often form a generalized model such as

$$g(\pi) = \beta_0 + \beta_1 x_1 + \cdots + \beta_m x_m. \tag{17.15}$$

The generalized model formed in this way is called a logistic regression model.

For different distributions, other link functions may be more appropriate or useful. If Y is not binary, but takes on a countable number of values, the link function may be the log of the expected value of Y.

In many useful models, $h(x;\theta)$ in equation (17.13) is linear in θ. The resulting model, such as equation (17.15), is called a *generalized linear model.*

The analysis of generalized models is usually based on a maximum likelihood approach, and the elements of the analysis are often identified and organized in a manner that parallels the ANOVA of general linear models.

Classification Models

In the generic model for the classification problem, the variables are the pairs (Y, x), where Y is a categorical variable representing the subclass of the population to which the observation belongs.

The generalized models discussed above can be viewed as classification models. A generalized model yields the probability that an observation has a particular response, or that the observation is in a given category. A logit model as in equations (17.12), (17.14), and (17.15), is often used as a classification model for binary responses, designated arbitrarily as 0 and 1. The rounded predicted probability is the class a given observation is assigned to. The predicted probability can also be viewed as a "fuzzy classification".

In the classification problem, we often consider x to be a realization of a random variable X, and we represent the random variable pair as (G, X), where G is a discrete variable corresponding to the class of the pair. A classification rule, κ, is a mapping from $\mathcal{X}$, the space of X, to $\mathcal{G}$, the space of G. If we assume a particular distribution of the data, it may be possible to develop an optimal approach to the classification problem based on the distribution. In particular, if we have an expression for the conditional probability $p(g|x)$ that $G = g$, given $X = x$, then we can use a rule that assigns the value of G based on the largest conditional probability. This is called a *Bayes rule.* Given a set of distributional assumptions, a Bayes classification rule has the minimal expected misclassification rate.

In applications of classification models, we generally have a dataset with known values of x and G and are interested in predicting the values of G in another dataset with known values of x only.

Classification is similar to the clustering problem for the x's that we have discussed in Section 16.1 beginning on page 519, except that in classification models, the value of one of the variables indicates the cluster or group to which the observation belongs.

Generally, we do not assume a specific distribution but rather that the training set represents randomly chosen examples of the qualities for which we are trying to build a classification rule. This is sometimes called the *probably approximately correct*, or *PAC*, model of learning.

The starting point for studying the classification problem is classification into one of two groups. Bayes rules for the binary case are relatively simple to develop in a variety of scenarios. The multigroup problem is not as simple. One approach is a sequential one of assigning all unclassified observations to either the ith group or the group consisting of all other groups, and then continuing this process considering the observations in the group consisting of all other groups to be unclassified. This approach, however, does not necessarily yield an optimal classification, even if it is optimal at each stage.

As more data are collected, the properties of the groups may become known from the past training datasets, and future data can be classified in a supervised fashion. How well a classification scheme works can be assessed by observing the similarity of the new observations in each cluster. This process can also be applied to a single dataset by defining a subset of the data to be a training dataset. This type of cross validation is often useful in developing rules for classification.

Classification rules are often based on measures of distance to the means of the groups scaled by S, the sample variance-covariance matrix. The Mahalanobis distance of an observation x to the mean of the ith group, $\bar{x}_{(i)}$, is

$$(x - \bar{x}_{(i)})^{\mathrm{T}} S^{-1} (x - \bar{x}_{(i)}). \tag{17.16}$$

This is the basis for a linear discriminant function for classification, which we address on page 623.

An observation can be classified by computing its Euclidean distance from the group means projected onto a subspace defined by a subset of the canonical variates. The observation is assigned to the closest group. For two groups, this is easy; the discriminant for the observation x is just

$$\kappa(x) = \mathrm{sign}\big(x^{\mathrm{T}} S^{-1} (\bar{x}_{(1)} - \bar{x}_{(2)})\big). \tag{17.17}$$

A positive value of $\kappa(x)$ assigns x to the first group, and a negative value assigns it to the second group. Under certain assumptions (for example, normality), this is a Bayes rule. For more than two groups, it is a little more complicated. In general, a classification rule based on the Mahalanobis distance is

$$\kappa(x) = \arg\min_i \Big((x - \bar{x}_{(i)})^{\mathrm{T}} S^{-1} (x - \bar{x}_{(i)}) \Big). \tag{17.18}$$

We could represent the data in the classification problem as $x = (x^{\mathrm{r}},\ x^{\mathrm{c}})$ as in equation (16.6) on page 532. The x^{c} are categorical variables. If x^{c} is a vector, some elements may be unobserved. We may know that $x^{\mathrm{c}} \in \mathcal{C}$, where $\mathcal{C}$ is some given set of general characteristics or rules. The characteristics may be dependent on the context, as we discussed on page 388. The set of characteristics of interest may include "concepts" (that is, something more general than just a class index).

Models of Sequential Dependencies

A stochastic process is indexed by a time parameter, which may be a continuous variable over an interval or may be assumed to take on fixed values $\ldots, t_{k-1}, t_k, t_{k+1}, \ldots$. A model for a stochastic process may be written as

$$Y_{t_k} = f_{t_k}(x_{t_k}, y_{t_{k-1}}; \theta_{t_k}) + E_{t_k}. \tag{17.19}$$

The response variable in a stochastic process is often referred to as the *state* of the process.

In many applications of interest we can assume that f_{t_k}, θ_{t_k}, and the distribution of E_{t_k} do not change in time; that is, the model is *stationary.*

Another way of formulating a model of a stochastic process is to focus on the change in the dependent variable and to write a differential equation that represents the rate of change:

$$\frac{\mathrm{d}Y}{\mathrm{d}t} = g\big(x(t), Y(t); \theta(t)\big) + E(t). \tag{17.20}$$

In many cases, this is a natural way of developing a stochastic model from first principles of the theory underlying the phenomenon being studied. In other cases, such as financial applications, for the variable of interest there may be no obvious dependency on other variables. In financial analysis, the change in prices from day to day or from trade to trade is of interest, so the appropriate model is a differential equation.

Data and Models

The model (17.8) is expressed in terms of the general variables Y and x. The relationship for a particular pair of observed values of these variables may be written as

$$y_i = f(x_i; \theta) + \epsilon_i.$$

For a sample of n y_i's and x_i's, we may write the model as

$$y = f(X; \theta) + \epsilon, \tag{17.21}$$

where y is an n-vector, X is an $n \times m$ matrix of n observations on the m-vectors x_i, and ϵ is an n-vector representing a realization of the random error term E in the model equation (17.5).

The usual linear regression model would be written as

$$y = X\beta + \epsilon, \tag{17.22}$$

where X is understood to contain a column of 1's in addition to the columns of values of the covariates.

The problem in data analysis is to select the relevant factors x, the functional form f, the value of θ, and properties of the random component that best fit the data. Depending on the assumptions of the distribution of E, its variance may be of interest.

For given observations y and X, either a maximum likelihood approach or just a heuristic approach often leads to estimating β so as to minimize

$$\|y - Xb\|$$

with respect to the variable b.

If the form of the density is known, the theory of statistical inference can be used to assess properties of estimators or test procedures. Although the theory that allows identification of optimal procedures is interesting, the problem of model building is much more complicated than this. For each of the various models considered, however, it is useful to have simple theoretical guidelines for fitting a model, even if the model is tentative.

Accounting for an Intercept

Given a set of observations, the ith row of the system $Xb \approx y$ represents the linear relationship between y_i and the corresponding x's in the vector x_i:

$$y_i \approx b_1 x_{1i} + \cdots + b_m x_{mi}.$$

A different formulation of the relationship between y_i and the corresponding x's might include an intercept term:

$$y_i \approx \tilde{b}_0 + \tilde{b}_1 x_{1i} + \cdots + \tilde{b}_m x_{mi}.$$

There are two ways to incorporate this intercept term. One way is just to include a column of 1s in the X matrix, as in equation (17.22). Because we often prefer to think of the columns of X as representing covariates, we often take a different approach. We assume that the model is an exact fit for some set of values of y and the x's. If we assume that the model fits $y = 0$ and $x = 0$ exactly, we have a model without an intercept (that is, with a zero intercept). It may be a reasonable to assume that the model may have a nonzero intercept, but that it fits the means of the set of observations; that is, the equation is exact for $y = \bar{y}$ and $x = \bar{x}$, where the jth element of $\bar{x}$ is the mean of the jth column vector of X. (Students with some familiarity with the subject may think that this is a natural consequence of fitting the model. It may not be unless the model fitting is by ordinary least squares.)

If we require that the fitted equation be exact for the means (or if this happens naturally, as in the case of ordinary least squares), it is convenient to center each column of X by subtracting its mean from each element, and to form y_{c} as the vector $y - \bar{y}$. The matrix formed by centering all of the columns of a given matrix is called a centered matrix, and if the original matrix is X, we represent the centered matrix as X_{c}. If we represent the matrix whose i^{th} column is the constant mean of the ith column of X as $\overline{X}$,

$$X_{\text{c}} = X - \overline{X}. \tag{17.23}$$

Using the centered data provides two linear systems: a set of approximate equations in which the intercept is ignored,

$$y_{\text{c}} \approx X_{\text{c}} \beta_{\text{c}}, \tag{17.24}$$

and an equation that fits the point that is assumed to be satisfied exactly:

$$\bar{y} = \overline{X} \beta. \tag{17.25}$$

Transformations

Sometimes, the response of interest may not have a distribution that is amenable to analysis, but some transformation of the response variable may have a more tractable distribution. In such cases, rather than modeling the response Y, it may be preferable to model a transformation of the response, $\tau(Y)$. One reason for doing this is to remove dependence of various properties of the distribution of Y on x and θ. For example, if the variance of Y depends on $\mathrm{E}(Y)$, it may be desirable to consider a transformation $\tau(Y)$ whose variance does not change as the mean changes as a function of x and θ.

If statistics of the forms

$$\bar{y}|x;\theta$$

and

$$\sum \left(y_i - \bar{y}|x;\theta\right)^2$$

are to be used in data analysis, our objective might be to determine the transformation so that $\tau(Y)$ has a normal distribution because that is the only way that the statistics

$$\left.\overline{\tau(y)}\right| x;\theta$$

and

$$\sum \left(\tau(y_i) - \left.\overline{\tau(y)}\right| x;\theta\right)^2$$

would be independent. Other reasons for transforming the data may be to make some relationships linear, or else to make the variance of the error term constant. We describe various transformations beginning on page 629.

As we discuss on page 630, we may also consider transformations of the independent variable x. Transformations of the variable x in the clustering problem often result in quite different clusters, as we discuss on page 533. This effect of transformations can be exploited in the classification problem.

Exploration of various transformations and functional forms, using both numerical computations and graphical displays, is computationally intensive. Choice of a functional form involves selection of variables for inclusion in the model. Evaluation of subsets of potential variables is computationally intensive. We discuss some of the issues of variable selection and transformations of variables later.

Piecewise Models

As we have seen in Chapters 4 and 10, sometimes it is best to approximate a function differently over different domains. This can be done by use of splines, for example. Likewise, different statistical models of dependencies over different domains of the independent variables may be appropriate. Although a systematic component in a model that has a single global function is useful because of its simplicity, the data may not follow a single form of the systematic component very well.

One approach to developing a model that is piecewise smooth is to use polynomials over subintervals. We can impose smoothness constraints at the knots separating the subintervals. The difficult modeling problem is the choice of where to locate the knots. Obviously, the more knots, the better the model can fit any given set of data. On the other hand, the more knots, the more variation the model will exhibit. One approach is to add knots in a stepwise manner while monitoring the improvement in the fit of the model to the data (measured by some function of the residuals, as we discuss in Section 17.3, for example).

Overfitting

In any application in which we fit an overdetermined system,

$$y \approx Xb,$$

it is likely that the given values of X and y are only a sample (not necessarily a random sample) from some universe of interest. There is perhaps some error in the measurements. It is also possible that there is some other variable not included in the columns of X. In addition, there may be some underlying randomness that cannot be accounted for.

The fact that $y \neq Xb$ for all b results because the relationship is not exact. Whatever value of b provides the best fit (in terms of the criterion chosen) may not provide the best fit if some other equally valid values of X and y were used. The given dataset is fit optimally, but the underlying phenomenon of interest may not be modeled very well. The given dataset may suggest relationships among the variables that are not present in the larger universe of interest. Some element of the "true" b may be zero, but in the best fit for a given dataset, the value of that element may be significantly different from zero. Deciding on the evidence provided by a given dataset that there is a relationship among certain variables when indeed there is no relationship in the broader universe is an example of *overfitting*.

There are various approaches we may take to avoid overfitting, but there is no panacea. The problem is inherent in the process.

One approach to overfitting is *regularization*. In this technique, we restrain the values of b in some way. Minimizing $\|y - Xb\|$ may yield a b with large elements, or values that are likely to vary widely from one dataset to another. One way of "regularizing" the solution is to minimize also some norm of b. We will discuss this approach on page 607.

17.2 Probability Distributions in Models

Statistical inference (that is, estimation, testing, or prediction) is predicated on probability distributions. This means that the model (17.8) or (17.21) must include some specification of the distribution of the random component or the residual term, E. In statistical modeling, we often assume that the independent variables are fixed — not necessarily that there is no probability distribution associated with them, but that the set of observations we have are considered as given — and the random mechanism generating them is not of particular interest. This is another aspect of the asymmetry of the model. This distinction between what is random and what is not is a basic distinction between regression analysis and correlation analysis, although we do not wish to imply that this is a hard and fast distinction.

The probability distribution for the residual term determines a family of probability distributions for the response variable. This specification may be very complete, such as that $E \sim \mathrm{N}(0, \sigma^2 I)$, or it may be much less specific. If there is no a priori specification of the distribution of the residual term, the main point of the statistical inference may be to provide that specification. If the distribution of the residual has a first moment, it is taken to be 0; hence, the model (for an individual Y) can be written as

$$\mathrm{E}(Y) = f(x;\ \theta);$$

that is, the expected value of Y is the systematic effect in the model. If the first moment does not exist, the median is usually taken to be 0 so that the median of y is $f(x;\ \theta)$. More generally, we can think of the model as expressing a conditional probability density for Y:

$$p_Y(y) = p(y \mid f(x;\theta)). \tag{17.26}$$

Hierarchical Models

The basic general model (17.8) can be a component of a *hierarchical model:*

$$\begin{aligned} y &= f(x;\theta) + \epsilon, \\ x &= g(w;\tau) + \delta, \end{aligned} \tag{17.27}$$

or

$$\begin{aligned} y &= f(x;\theta) + \epsilon, \\ \theta &\sim D(\tau), \end{aligned} \tag{17.28}$$

where $D(\tau)$ is some distribution that may depend on a parameter, τ. Either of these models could be part of a larger hierarchy, of course.

Hierarchical models of the form (17.27) arise in various applications, such as population dynamics, where the components are often called "compartments", or in situations where the independent variables are assumed not to be observable or not to be measured without error.

Hierarchical models of the form (17.28) are useful in Bayesian statistics. In that case, we may identify the components of the model as various joint, marginal, and conditional distributions. For example, following the general outline presented on page 43, we may consider the joint distribution of the random variables Y and Θ (using an uppercase letter to emphasize that it is a random variable),

$$(Y, \Theta) \sim D_{Y,\Theta}(x, \tau),$$

the conditional distribution of Y given Θ,

$$Y \sim D_{Y|\theta}(f(x;\theta)),$$

or the conditional distribution of Θ given Y,

$$\Theta \sim D_{\Theta|y}(x, y, \tau). \tag{17.29}$$

In the analysis of data, the second component of model (17.28), which is a marginal distribution, which we rewrite as

$$\Theta \sim D_{\Theta}(\tau),$$

is called the *prior distribution*, and the conditional distribution of Θ given Y (17.29) is called the *posterior distribution*. The idea is that the conditional distribution represents knowledge of the "unknown parameter" that includes information from the observations y and the prior distribution of the parameter.

Probability Distributions in Models of Sequential Dependencies

There are two different ways to develop probability models for the dependent variable in a stochastic process as described by equation (17.19). In one approach, we consider a set of probability spaces indexed by t; that is, for each t, there is a different probability space that depends not only on t but also on the value of $Y(t - \epsilon)$.

In another approach, we define the outcome space to correspond to the sequence or path of values that can be assumed by Y_t. In a continuous stochastic process, the outcome space in the underlying probability space may be chosen to be the set of continuous mappings ("trajectories") of $[0, t]$ into $\mathbb{R}$. This approach is sometimes called the "canonical" setup.

The parameter space for these models includes time. The parameter that indexes time may be continuous, but it is often considered to be discrete.

We can develop a differential equation model of a stochastic process, as in equation (17.20), by starting with models of small changes. One of the simplest and most commonly used models developed in this way is called *Brownian motion*. In this model, the random variable, B_t, with a continuous index, t, has the following properties:

- the change ΔB_t during the small time interval Δt is

$$\Delta B_t = Z\sqrt{\Delta t},$$

  where Z is a random variable with distribution $N(0, 1)$;
- ΔB_{t_1} and ΔB_{t_2} are independent for $t_1 \neq t_2$.

From the definition, we see that $\mathrm{E}(\Delta B_t) = 0$ and $\mathrm{V}(\Delta B_t) = \Delta t$. Now, if $B_0 = 0$, and for the positive integer n we let $t = n\Delta t$, we have

$$B_t = \sum_{i=1}^{n} Z_i\sqrt{\Delta t}, \tag{17.30}$$

where the Z_i are i.i.d. $N(0, 1)$, so B_t has a $N(0, t)$ distribution.

In the limit as $\Delta t \to 0$, $\sqrt{\Delta t}$ becomes much larger than Δt, and the change in B_t (that is, ΔB_t) becomes relatively large compared to the change in t (that is, Δt). Therefore, as t moves through any finite interval:

- the expected length of the path followed by B_t is infinite; and
- the expected number of times B_t takes on any given value is infinite.

The limiting process is called a *Wiener process* and is denoted as $\mathrm{d}B_t$.

There are some useful generalizations of this process. One, called a *generalized Wiener process*, is a linear combination of a Wiener process and a constant rate in t. This can be represented by

$$\mathrm{d}Y_t = u\,\mathrm{d}t + v\,\mathrm{d}B_t, \tag{17.31}$$

where u and v are constants and $\mathrm{d}B_t$ is a Wiener process. (An equation in differentials of the form (17.31) is called a *Langevin equation.*) Another generalization is an *Ornstein-Uhlenbeck process*, which is similar to a generalized Wiener process except that the change in time depends on the state. The Langevin equation is

$$\mathrm{d}Y_t = uY_t\,\mathrm{d}t + v\,\mathrm{d}B_t. \tag{17.32}$$

A further generalization is an *Ito process*, which is similar to a generalized Wiener process except that both coefficients are functions of Y_t (and, hence, at least implicitly, of t). The Langevin-type equation for an Ito process is

$$\mathrm{d}Y_t(\omega) = u(Y_t(\omega), t)\,\mathrm{d}t + v(Y_t(\omega), t)\,\mathrm{d}B_t(\omega), \tag{17.33}$$

where B_t is a Brownian motion. This model is widely used for the price of a financial asset, such as a stock. The simple form is just

$$\mathrm{d}Y = \mu Y\,\mathrm{d}t + \sigma Y\,\mathrm{d}B_t, \tag{17.34}$$

where the drift parameter μ is the rate of return of the stock per unit of time, and σ is the "volatility" of the stock price. (In general terms, ignoring the ambiguities of continuous time, volatility is the standard deviation of the relative change in the price.) This version of an Ito process is called *geometric Brownian motion.* Variations on geometric Brownian motion that attempt to capture additional aspects of stock price behavior include discrete changes in μ or σ, resulting in a "jump process", and imposition of stochastic constraints on the magnitude of Y, yielding a "mean-reverting process".

17.3 Fitting Models to Data

Observational data help us to build a model. The model helps us to understand nature. Standard ways of developing our knowledge of nature involve estimation and tests of hypotheses — that is, statistical inference.

Inference about the model $y = f(X;\theta) + \epsilon$ involves estimation of the parameters θ, tests of hypotheses about θ, and inference about the probability distribution of ϵ. It may also involve further consideration of the model, the form of f, or other issues relating to the population, such as whether the population or the sample is homogeneous, whether certain observations are outliers, and so on.

The statistical characteristics of the estimator, such as its bias and variance, depend on how the model is fit (that is, on how the estimates are computed) and on the distribution of the random component. For a specific family of distributions of the random component and a specific form of f, it may be possible to determine estimators that are optimal with respect to some statistical characteristic such as mean squared error.

A unified approach to model inference involves a method of estimation that allows for statements of confidence and that provides the basis for the subsequent inference regarding the distribution of ϵ and the suitability of the model. In this section, we will be concerned primarily with methods of fitting the model rather than on the problem of statistical inference.

The Mechanics of Fitting

Fitting a model using data can be viewed simply as a mechanical problem of determining values of the parameters so that functional relationships expressed by the model are satisfied approximately by the given set of data. Fitting a model to data is often a step in statistical estimation, but estimation generally involves a deeper belief about the nature of the data. The data are realizations of a random variable whose distribution is related to or specified by the model. Statistical estimation also includes some assessment of the distribution of the estimator.

One of the first considerations is what approach to take in modeling. Is the objective to develop a model in the form of equations and statements about distributions of elements of the model, as in equation (17.8), or is it acceptable to have a black box of the form (17.1) together with an algorithm that accepts x and produces a good prediction of Y? Often, a set of rules is sufficient. Because there is no particular restriction on the complexity of the rules as there would be if we attempt to express the rules in a single equation, the black box together with a prediction algorithm performs best. A neural net, which can be quite complicated yet provides no insight into identifiable functions and parameters, often yields excellent predictions of the response for a given input x.

An additional consideration is whether the fit is to be global or local (that is, whether a single model describes the data over the full domain and all observations are used at once to fit the model, or whether different models apply in different domains and only "local" observations are used to fit the model within each domain).

On page 587, we listed five basic approaches for fitting models using data: method of moments, minimizing residuals, maximizing the likelihood, homogeneity of modeled classes, and predictive accuracy in partitioned data. Any of these approaches can be applied in fitting models that have a continuous-valued response. (We can assess class homogeneity if one or more of the covariates is a classification variable, or we may be able to discretize the response into meaningful groups.) If the model has a discrete-valued response, or if the purpose is classification, there are two possibilities. One is the use of a generalized model, which effectively makes the response continuous-valued and can yield a probability-based or fuzzy classification. Otherwise, the fitting problem can be addressed directly, and the class purity is the primary criterion. In the classification problem, the predictive accuracy in partitioned data is almost always considered in the model fitting. The dataset can be partitioned either randomly or systematically, as we discuss in Chapter 12. Whatever method is used to fit a model, it may be followed by some further steps to bring the model and the data into closer agreement. Individual observations that do not agree well with the fitted model may be treated specially in some way. Some outlying observations may be modified or even removed from the dataset, and then the model is refit using the cleaned data.

Estimation by Minimizing Residuals

Of the basic approaches for fitting models using data, listed on page 587, perhaps the most intuitive is fitting based on minimizing the residuals. This is the method most often used by data analysts and scientists to fit a curve to data without any assumptions about probability distributions.

For a given function f, the fit is accomplished by solving an optimization problem involving some function of the vector of residuals,

$$r = y - f(X; \theta),$$

where y is the vector of observed responses and X is the matrix of corresponding observations of the covariates. The decision variable in the optimization problem is θ.

Notice that the r_i are vertical distances as shown in Figure 17.1 for a simple linear model. Another way of measuring residuals in a model is indicated by the orthogonal residuals, d_i, shown in Figure 17.3.

For a given set of data $\{(y_i, x_i)\}$, the residuals r_i are functions of f and θ. Clearly, the space of functions from which to select f must be restricted in some way; otherwise, the problem is not well-defined. We generally restrict the function space to contain only relatively tractable functions, such as low-degree polynomials, often just linear functions, exponential functions, or functions that can be formed from common basis functions, as we discuss in Section 4.2 and in Chapter 10. Once a general form of f is chosen, the residuals are functions of θ, $r_i(\theta)$.

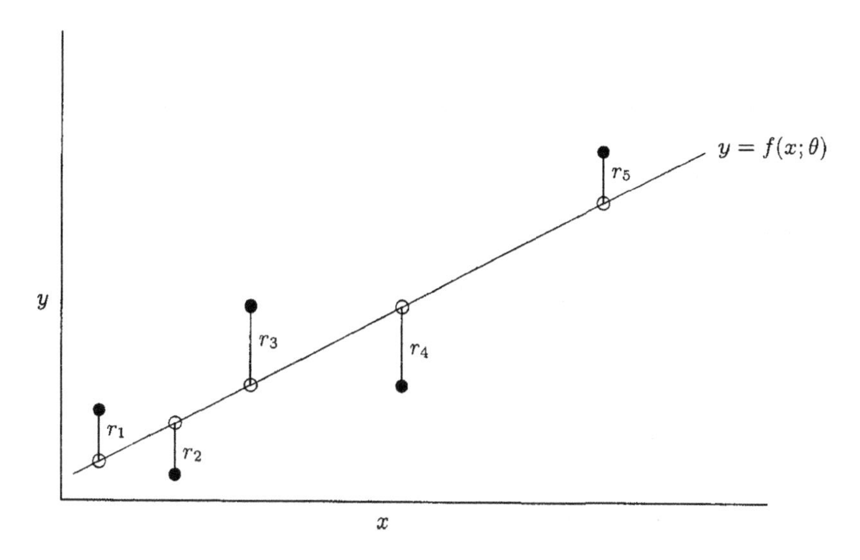

Fig. 17.1. Residuals

There are many reasonable choices as to how to minimize the $r_i(\theta)$. In general, we can minimize the sum

$$\sum \rho(r_i(\theta)), \tag{17.35}$$

where $\rho(t)$ is some nonnegative, nondecreasing function in $|t|$, at least near $t = 0$. A value of θ that minimizes the sum (17.35) is called an *M-estimator* because log-likelihood equations for common distributions often have a form similar to the negative of this sum. Most common choices of $\rho(\cdot)$ are such that the sum is a norm of the residual vector r:

$$\sum \rho(r_i(\theta)) = \|r\|. \tag{17.36}$$

The L_p norm is commonly used. For $p = 2$ this is least squares, for $p = 1$ this is least absolute values, and for $p \to \infty$ this is least maximum value (minimax).

For data from a normal distribution, least squares is the optimal minimal-residual criterion by various other criteria, such as maximum likelihood. It is, however, subject to strong effects of outliers or observations that have large (positive or negative) residuals. The least absolute values criterion, on the other hand, is not strongly affected by outlying observations.

In the simple case in which there are no covariates, that is, for the model of the form $y = \theta$, use of least absolute values leads to a fit that corresponds exactly to one of the observations. (It is the median, or in the case of a sample size of even number, either of the two central order statistics or any value in between.) In a linear regression model, a least absolute values fit also has this kind of discontinuous property. One fit always corresponds exactly to some "central" observations. Any of the other observations may be perturbed up to a certain amount without affecting the fit. Once the perturbation of a "noncentral" observation reaches a certain amount, however, that observation may become a central one that a newly-fitted model matches exactly.

The computations for determining the minimizer of (17.35) are discussed in Chapter 6. The most common method is some type of quasi-Newton algorithm. Another iterative method that is very simple to implement is one in which the individual elements of the m-vector θ are updated one at a time. Beginning with some starting value for $\theta^{(0)}$, we have a one-dimensional minimization problem, and letting j range over the indices of θ, we take

$$\theta_j^{(k)} = \arg\min_{\theta_j} \sum \rho \left(r_i \left(\theta_1^{(k-1)}, \ldots, \theta_j, \ldots, \theta_m^{(k-1)} \right) \right). \tag{17.37}$$

We then iterate on k until convergence. These kinds of iterations are especially useful in fitting an additive model such as equation (17.7) on page 589, in which

$$r_i = y_i - \beta_0 - \sum_{l \neq j} f_l(x_{li}, \beta_l) - f_j(x_{ji}, \beta_j). \tag{17.38}$$

In Chapter 1, in addition to outlining the basic quasi-Newton computations, we also discuss how those computations may yield estimates of the variance of the estimator of θ. Other methods of estimating the variance of estimators of θ are based on computational inference utilizing data partitioning (Chapter 12) or bootstrapping (Chapter 13).

Minimizing the sum (17.35) in the case of a generalized model may be quite difficult. If, however, we restrict attention to certain classes of link functions and to a simple form of ρ, such as a square, the computations are relatively stable, and a quasi-Newton method, a Gauss-Newton method, or iteratively reweighted least squares can be used. The optimization problem for estimation of the parameters in a generalized model is usually formulated as a likelihood to be maximized rather than a sum of functions of the residuals to be minimized. For the link function and ρ of simple forms, the MLE and the minimum-residual estimates are the same.

Least Squares Estimation in Linear Models

The most familiar example of fitting a statistical model to data uses the linear regression model (17.22):

$$y = X\beta + \epsilon.$$

The least-squares fit of this is exactly the same as the least-squares fit of the overdetermined system (5.54) $Xb \approx y$ on page 229 in Section 5.6. It is the solution to the normal equations,

$$X^{\mathrm{T}} X \widehat{\beta} = X^{\mathrm{T}} y,$$

which, as we determined in that section, is

$$\widehat{\beta} = X^+ y,$$

which, in the case of full-rank X, is

$$\widehat{\beta} = (X^{\mathrm{T}} X)^{-1} X^{\mathrm{T}} y. \tag{17.39}$$

As we pointed out in Section 5.6, fitting a model is not the same as statistical inference. Statistical inference requires a probability distribution on ϵ as part of the model.

If we ignore any distributions on X and β, or at least just state results conditionally on X and β, we have that if $\mathrm{E}(\epsilon) = 0$ then $\widehat{\beta}$ above is unbiased for β, that is, $\mathrm{E}(\widehat{\beta}) = \beta$, and $\mathrm{V}(\widehat{\beta}) = X^+ \mathrm{V}(\epsilon)(X^+)^{\mathrm{T}}$.

With an additional assumption that $\mathrm{V}(\epsilon) = \sigma^2 I$, for a constant σ^2, we have $\mathrm{V}(\widehat{\beta}) = X^+(X^+)^{\mathrm{T}}\sigma^2$, but more importantly, we have the Gauss-Markov theorem that $\widehat{\beta}$ is the best (in the sense of minimum variance of all linear combinations) linear unbiased estimator of β.

With the assumption that $\epsilon \sim \mathrm{N}_n(0, \sigma^2 I)$, we have distributional properties of $\widehat{\beta}$ as well as of $(y - X\widehat{\beta})^{\mathrm{T}}(y - X\widehat{\beta})$ that allow us to develop most powerful statistical tests of hypotheses regarding β.

Another interesting property of the least-squares estimator is the relationship between the model with the centered data, as in equation (17.24),

$$y_{\mathrm{c}} = X_{\mathrm{c}}\beta_{\mathrm{c}} + \epsilon,$$

and the model that may include an intercept, equation (17.22),

$$y = X\beta + \epsilon.$$

The least squares estimators are the same (except of course β_{c} does not contain a term corresponding to an intercept), and in any event, equation (17.25) is satisfied by the estimator

$$\bar{y} = \overline{X}\widehat{\beta}$$

without imposing this as a separate constraint. (Recall that fitting the equation by minimizing some other norm of the residuals does not ensure that the fit goes through the means.)

Variations on Minimizing Residuals

When data are contaminated with some (unidentified) observations from a different data-generating process, it is desirable to reduce or eliminate the effect of these contaminants. Even if all of the observations arise from the same data-generating process, if that process is subject to extreme variance, it may be desirable to reduce the effect of observations that lie far from the mean of the model. It is not possible, of course, to know which observations are which in either of these cases. Both situations can be addressed, however, by assuming that the bulk of the observations are close to the mean values of the model and fit the model allowing such observations to have larger relative effect on the fit. The objective is to obtain an estimator that is "robust" or resistant to contamination.

A useful variation on the sum (17.35) is

$$\sum w_i \rho(r_i(\theta)). \tag{17.40}$$

If the w_i are just given constants that do not depend on $\rho(r_i(\theta))$, the expression (17.40) is a simple weighted sum and presents no difficulties, either computational or inferential, beyond those of the unweighted sum (17.35). If, on the other hand, we want to choose smaller weights for the y's and x's that do not fit the model, w_i may be some function $w(x_i, \theta, \rho(r_i(\theta)))$. The problems, both computational and inferential, are more difficult in this case.

Another way of approaching this problem is to define a function ρ in the sum (17.35) that depends on y, x, and a given model determined by θ:

$$\rho(t; y, x, \theta) = \begin{cases} \rho_1(t) & \text{if } \theta \text{ provides a "good" fit for } x \text{ and } y, \\ \\ \rho_2(t) & \text{otherwise.} \end{cases} \tag{17.41}$$

This heuristic approach is appealing, but to carry it out would require some preliminary fits and some definition of what it means for "θ to provide a good fit for x and y". Once that meaning is quantified, this approach may be computationally intensive, but it is easily done. In one simple approach, for example, we could define a function ρ in the sum (17.35) that is a square near zero (small model residuals) and smoothly becomes an absolute value at some data-dependent distance from zero:

$$\rho(t; y, x, \theta) = \begin{cases} \frac{1}{2}t^2 & \text{if } |t| \le c, \\ \\ |t|c - \frac{1}{2}c^2 & \text{if } |t| > c, \end{cases} \tag{17.42}$$

where $c = c(y, x, \theta)$ is some constant that depends on the data. The $\frac{1}{2}$ factor is included to make the derivative continuous. (This form of ρ was suggested by Huber, and the resulting estimator of θ is called a *Huber estimator*. Various forms of $c(y, x, \theta)$ have been proposed and studied.)

We can easily modify this basic idea to define other estimators. Suppose that $\rho(t)$ is defined to be 0 for $|t| > c$. If for $|t| \le c$, $\rho(t) = t^2$, minimizing the sum (17.35) yields the least trimmed squares estimate; if for $|t| \le c$, $\rho(t) = |t|$, minimizing (17.35) yields the least trimmed absolute values estimate. Because of the dependence of c, computation of such estimators is more difficult. The basic approach consists of two-step iterations; at the jth iteration, set $c^{(j)} = c(y, x, \theta^{(j)})$, which determines $\rho^{(j)}$, and then determine $\theta^{(j+1)}$ as the solution to

$$\min_{\theta} \sum \rho^{(j)}(r_i(\theta)). \tag{17.43}$$

The iterations can be started with some value of $\theta^{(0)}$ that is computed without any trimming. If the functional form of $\rho(t)$ for $|t| \leq c$ is t^2 or some other simple form with second derivatives, quasi-Newton methods (see pages 269 and following) can be used to solve for $\theta^{(j+1)}$.

Instead of defining ρ_1 and ρ_2 in equation (17.41) based on the size of the residuals, we could define ρ based on order statistics of the residuals; that is, decrease the effects of the smallest and largest order statistics. This is the idea behind the commonly used univariate trimmed mean and winsorized mean statistics, in which the contributions to the estimator of a certain percentage of the smallest and largest order statistics are attenuated. If the percentage approaches 50%, the estimators become the median.

Following this same idea, we could either define weights in expression (17.40) or a ρ_1 in equation (17.41) to reduce the effects on the estimator of the observations whose residuals are the smallest and largest order statistics of all residuals. In the extreme case of eliminating the effect of all but one observation, we can write the optimization problem as

$$\min_{\theta} \mathrm{Med}(\rho(r_i(\theta))). \tag{17.44}$$

If $\rho(t) = t^2$, this yields the *least median of squares estimator.* Fitting the model by this criterion is obviously computationally intensive.

Regularized Solutions

Other variations on the basic approach of minimizing residuals involve some kind of regularization, which may take the form of an additive penalty on the objective function (17.35). Regularization often results in a shrinkage of the estimator toward 0. The general formulation of the problem then is

$$\min_{b}(\|y - Xb\|_r + \lambda\|b\|_b), \tag{17.45}$$

where λ is some appropriately chosen nonnegative number. The norm on the residuals, $\|\cdot\|_r$, and that on the solution vector b, $\|\cdot\|_b$, are often chosen to be the same. The weighting factor λ may be chosen adaptively, as various fits of the data are examined. Its effect depends on the relative sizes of the residuals and the regression coefficients (which depend on the scaling of the regressors). Its effect also depends on the number of observations and the number of regression variables.

If both norms are the L_2 norm, the fitting is called Tikhonov regularization in the applied mathematical literature. In the statistical literature, the fitting is called ridge regression, after terminology used in the description of the method by Hoerl (1962).

Ridge regression is often used in statistical modeling when there are large correlations among the independent variables. The ridge regression estimator for the model $y = X\beta + \epsilon$ is the solution of the normal equations $(X^{\mathrm{T}}X+\lambda I)\beta = X^{\mathrm{T}}y$. These normal equations correspond to the least squares approximation

$$\begin{pmatrix} y \\ \\ 0 \end{pmatrix} \approx \begin{bmatrix} X \\ \\ \sqrt{\lambda} I \end{bmatrix} \beta. \tag{17.46}$$

The choice of λ may be based on observing the changes in the parameter estimates, as λ is increased from 0. After an initial period of rapid change of the parameter estimates, the estimates generally stabilize and begin changing more slowly. The value of λ at which this appears to be happening may be chosen as the optimal value. Often, if there are large correlations among the independent variables, some of the parameter estimates may have signs that are different from what their marginal values would be (that is, if the other variables were not included in the model). A value of λ may be chosen whenever the signs of all of the parameter estimates are consistent with their marginal values (if such a value exists and if at that value, the estimated parameters provide a reasonable fit to the data).

As an indication of the form of ridge regression in a more general setting, we note that the Levenberg-Marquardt scaling matrix $S^{(k)}$ in the Gauss-Newton update (6.62) on page 293 can be thought of as a regularization of a standard Gauss-Newton step. The formulation as a constrained optimization problem in equation (6.63) emphasizes this view. The general ridge regression objective function can be formulated in a similar manner. In both the Gauss-Newton method with a scaling matrix and ridge regression, the computations can be performed efficiently by recognizing that the system is the normal equations for the least squares fit of

$$\begin{pmatrix} r(\theta^{(k)}) \\ \\ 0 \end{pmatrix} \approx \begin{bmatrix} \mathrm{J}_r(\theta^{(k)}) \\ \\ \sqrt{\lambda^{(k)}}\left(S^{(k)}\right) \end{bmatrix} p, \tag{17.47}$$

where p is the update step.

If $\|\cdot\|_r$ in expression (17.45) is the L_2 norm and $\|\cdot\|_b$ is the L_1 norm, the statistical method is called the "lasso". Because of the special properties of the L_1 norm (in particular, the discontinuity property alluded to above), increasing the weight of the L_1 norm leads to an exact zero for some coefficients. For this reason, the lasso can be used for selection of variables to be included in the model.

Figure 17.2 shows the contours of both $\|\cdot\|_r$ and $\|\cdot\|_b$ for a regression example with two variables. The norm used for $\|\cdot\|_r$ is the L_2 norm in both panels of Figure 17.2, and is shown in the heavier lines. In the left panel, the norm used for $\|\cdot\|_b$ is the L_2 norm, while in the right panel, the norm used for

$\|\cdot\|_b$ is the L_1 norm. The regularized fit in either case would result from minimizing a weighted sum of the two norms. The graph in the right panel seems to indicate that if the weight on $\|\cdot\|_b$ is large enough, the optimal b_2 would be 0. This is the variable selection aspect of the method mentioned above. There are several approaches to the variable selection problem in regression analysis. We will discuss this topic briefly again on page 613.

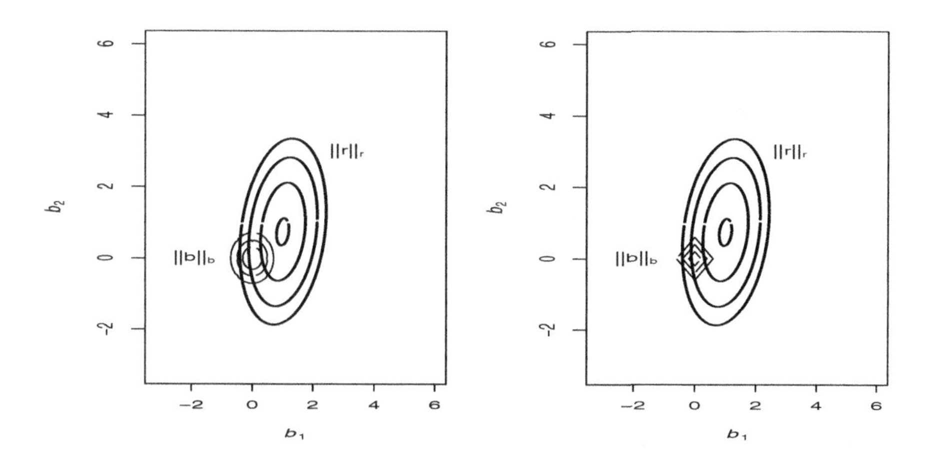

Fig. 17.2. Regularization Function

Comparisons of Estimators Defined by Minimum Residuals

It is easy to describe estimators determined by minimizing various functions of the residuals. What is difficult is understanding when to use which one. Because of the range of possibilities of forms of models and of distributions, it is also difficult to summarize what is known about the relative performance of various estimators.

Because of the difficulty of working out exact distributions, Monte Carlo methods are often used in comparing the performance of various estimators. An example is described in Appendix A that compares the power of a statistical hypothesis test for which the test statistic is computed by minimizing an L_2 norm with a test for which the test statistic is computed by minimizing an L_1 norm.

Orthogonal Residuals

In the models that we have considered so far, the error or residual is additive to the systematic part of the model. The method of fitting the model based on minimizing these additive residuals is appropriate both theoretically and from the intuitive perspective suggested by Figure 17.1.

The view in Figure 17.3 presents a different perspective, however. The d_i are orthogonal distances, and represent the shortest distances between observations and the model. This may suggest fitting the model by minimizing some function of the d_i's. An obvious choice for the function, just as for vertical distances, is the sum of the squares, and in that case, the fit is "orthogonal least squares". Fitting a model by minimizing the sum of the squared distances is called *orthogonal distance regression*, and the criterion is sometimes called *total least squares*.

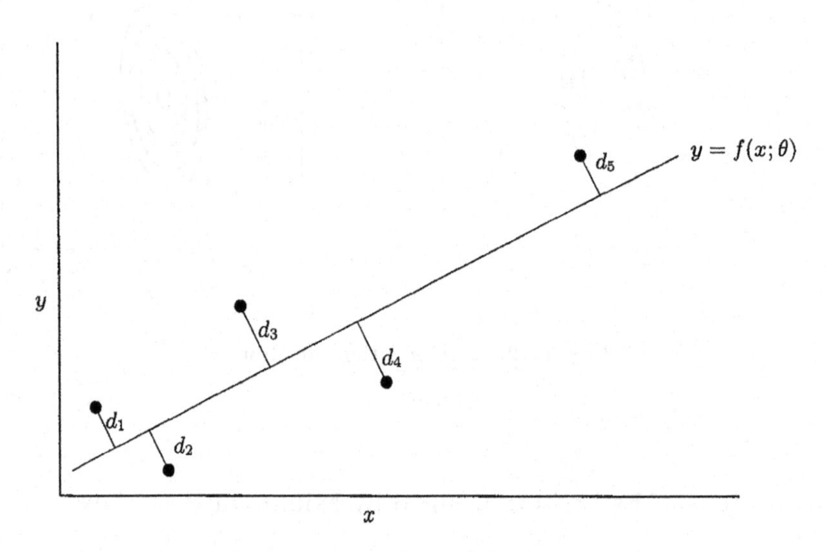

Fig. 17.3. Orthogonal Distances

Whether the vertical residuals in Figure 17.1 or the orthogonal residuals in Figure 17.3 are more interesting depends on the model. The orthogonal residuals criterion are sometimes suggested as appropriate for an *errors-in-variables* model, in which we assume that both y_i and x_i are observed with random error; that is, we observe

$$y_i + \epsilon_i \tag{17.48}$$

and

$$x_i + \delta_i. \tag{17.49}$$

A linear model

$$y_i \approx \beta_0 + \beta_1 x_i \tag{17.50}$$

that includes these errors on y_i and x_i is an errors-in-variables model. We might intuitively feel that it is appropriate to fit such a model by minimizing the orthogonal residuals.

Some reflection on this approach for estimation in an errors-in-variables model reveals two problems, however. The first is that minimization of the orthogonal residuals must be based on an assumption that the random errors have constant and equal variances. The assumption of constant variance is always problematic, but even ignoring that issue, are the y_i's and x_i's sufficiently similar that we can reasonably assume that they have equal variances? If not, then instead of a direction orthogonal to the model equation, we would want to consider the residuals in a direction that is scaled by the individual variances. (Recall the discussion on page 392 about scaling the Euclidean space using Mahalanobis distances.)

The second problem arises when we recall how we have been modeling relationships such as that shown by expression (17.50). If we just assume that the measurement error makes that expression an equality, we would have

$$y + E = \beta_0 + \beta_1(x + \Delta),$$

where E and Δ are random variables modeling the measurement errors. In the form that we have used to represent the data in the model, we would have

$$y = (X + \Delta)\beta + E$$

(assuming that E has a distribution symmetric about 0). This equation, however, is quite different from the models that we have been using, in which we had

$$\text{response} \quad = \quad \text{systematic component} \quad + \quad \text{random component}, \tag{17.51}$$

where the "random component" represented all "randomness", possibly inherent in the physical process, or possibly due to something else, but certainly not specifically measurement error. The random component in the context of the canonical model (17.51) is called "model error". In any event, the decision to fit the linear model by minimizing orthogonal residuals should not be based simply on a model of errors-in-variables.

Moving now from the statistical modeling issues, we will consider the mechanical problem of fitting a linear model such as in Figure 17.3, so as to minimize some norm of the orthogonal distances in the more general linear model in which there may be more than one covariate.

With n observations and with m covariates, we have the n-vector y and the $n \times m$ matrix X, which possibly includes a column of 1's, as on page 594, and the equation

$$(X + D)b = y + e, \tag{17.52}$$

where D is also in $\mathrm{IR}^{n\times m}$ (and we assume $n > m$).

In fitting the linear model as before only with adjustments to y, we determine b so as to minimize some norm of e. Likewise, with adjustments to both X and y, we seek b so as to minimize some norm of the matrix D and the vector e. There are obviously several ways to approach this. We could take norms of D and e separately and consider some weighted combination of the norms. Another way is to adjoin e to D and minimize some norm of the $n \times (m+1)$ matrix $[D|e]$. We may seek to minimize $\|[D|e]\|_{\mathrm{F}}$. This, of course, is the sum of squares of all elements in $[D|e]$. (That is why numerical analysts call the problem "total least squares".)

If it exists, the minimum of $\|[D|e]\|_{\mathrm{F}}$ is achieved at

$$b = -v_{2*}/v_{22}, \tag{17.53}$$

where

$$[X|y] = USV^{\mathrm{T}} \tag{17.54}$$

is the singular value decomposition (see equation (1.63) on page 28), and V is partitioned as

$$V = \begin{bmatrix} V_{11} & v_{*2} \\ v_{2*} & v_{22} \end{bmatrix}.$$

If D has some special structure, the problem of minimizing the orthogonal residuals may not have a solution. Golub and Van Loan (1980) show that a sufficient condition for a solution to exist is that the two smallest singular values s_m and s_{m+1} be such that $s_m > s_{m+1}$. (Recall that the s's in the SVD are nonnegative and they are indexed so as to be nonincreasing. If $s_m = s_{m+1}$, there may or may not be a solution.)

As in the usual case of weighted least squares, the orthogonal residuals can be weighted by premultiplication by a Cholesky factor of a weight matrix, as discussed on page 230.

If some norm other than the L_2 norm is to be minimized, an iterative approach must be used. Ammann and Van Ness (1988, 1989) describe an iterative method that is applicable to any norm, so long as a method is available to compute a value of b that minimizes the norm of the usual vertical distances. The method is simple. We first fit $y = Xb$, minimizing the vertical distances in the usual way; we then rotate y into $\tilde{y}$ and X into $\widetilde{X}$, so that the fitted plane is horizontal. Next, we fit $\tilde{y} = \widetilde{X}b$ and repeat. After continuing this way until the fits in the rotated spaces do not change from step to step, we adjust the fitted b back to the original unrotated space. Because of these rotations, if we assume that the model fits some point exactly, we must adjust y and X accordingly (see the discussion on page 594). In the following, we assume that the model fits the means exactly, so we center the data. We let m be the number of columns in the centered data matrix. (The centered matrix does not contain a column of 1s. If the formulation of the model $y = Xb$ includes an intercept term, then X is $n \times (m+1)$.)

Algorithm 17.1 Iterative Orthogonal Residual Fitting through the Means

0. Input stopping criteria, ϵ and $k_{\max}$.
   Set $k = 0$, $y_c^{(0)} = y_c$, $X_c^{(0)} = X_c$, and $D^{(0)} = I_{m+1}$.
1. Set $k = k + 1$.
2. Determine a value $b_c^{(k)}$ that minimizes the norm of $\left(y_c^{(k-1)} - X_c^{(k-1)} b_c^{(k)}\right)$.
3. If converged$(b_c^{(k)}, \epsilon, k_{\max})$, go to step 8.
4. Determine a rotation matrix $Q^{(k)}$ that makes the kth fit horizontal.
5. Transform the matrix $\left[y_c^{(k-1)}|X_c^{(k-1)}\right]$ $\left[y_c^{(k)}|X_c^{(k)}\right]$ by a rotation matrix:

$$\left[y_c^{(k)}|X_c^{(k)}\right] = \left[y_c^{(k-1)}|X_c^{(k-1)}\right] Q^{(k)}.$$

6. Transform $D^{(k-1)}$ by the same rotation: $D^{(k)} = D^{(k-1)}Q^{(k)}$.
7. Go to step 1.
8. For $j = 2, \ldots, m$, choose $b_j = d_{j,m+1}/d_{m+1,m+1}$ (So long as the rotations have not produced a vertical plane in the unrotated space, $d_{m+1,m+1}$ will not be zero.)
9. Compute $b_1 = \bar{y} - \sum_{j=2}^{k} b_j * \bar{x}_j$ (where $\bar{x}_j$ is the mean of the jth column of the original uncentered X). ∎

(Refer to page 244 for discussion of the converged$(\cdot)$ function.) An appropriate rotation matrix for Algorithm 17.1 is Q in the QR decomposition of

$$\begin{bmatrix} I_m & 0 \\ & \\ (b^{(k)})^{\mathrm{T}} & 1 \end{bmatrix}.$$

Note that forcing the fit to go through the means, as is done in Algorithm 17.1, is not usually done for norms other than the L_2 norm.

Variable Selection

If we start with a model such as equation (17.5),

$$Y = f(x; \theta) + E,$$

we are ignoring the most fundamental problem in data analysis: which variables *are really related* to Y, and *how are they related*?

We often begin with the premise that a linear relationship is at least a good approximation locally; that is, with restricted ranges of the variables. This leaves us with one of the most important tasks in linear regression analysis: selection of the variables to include in the model. There are many statistical issues that must be taken into consideration.

We first note that any measure of how well the model fits the given dataset will be better, or at least as good, the more variables we include in the model. This often leads to *overfitting*. Overfitting is to be avoided for three reasons. The first two are practical ones. Models with fewer variables are generally easier to understand. Models with fewer variables are usually easier to use in prediction because there are fewer variables that require data to be collected.

Some aspects of the statistical analysis involve tests of linear hypotheses. There is a major difference, however. Most of the theory of statistical hypothesis tests is based on knowledge of the *correct* model. The basic problem in variable selection is that we do not know the correct model. Most reasonable procedures to determine the correct model yield biased statistics. Some people attempt to circumvent this problem by recasting the problem in terms of a "full" model; that is, one that includes all independent variables that the data analyst has looked at. (Looking at a variable and then making a decision to exclude that variable from the model can bias further analyses.)

We generally approach the variable selection problem in linear models by writing the model with the data as

$$y = X_{\mathrm{i}}\beta_{\mathrm{i}} + X_{\mathrm{o}}\beta_{\mathrm{o}} + \epsilon, \tag{17.55}$$

where X_{i} and X_{o} are matrices that form some permutation of the columns of X. We rearrange the columns of X so that $X_{\mathrm{i}}|X_{\mathrm{o}} = X$, and β_{i} and β_{o} are vectors consisting of corresponding elements from β. We then consider the model

$$y = X_{\mathrm{i}}\beta_{\mathrm{i}} + \epsilon_{\mathrm{i}}. \tag{17.56}$$

In most cases, we will consider the vector y and the X matrices to be centered, as in equation (17.24). That means that we are not considering the intercept to be a variable that is selected; it is always in the model.

It is interesting to note that the least squares estimate of β_{i} in the model (17.56) is the same as the least squares estimate in the model

$$\hat{y}_{\mathrm{io}} = X_{\mathrm{i}}\beta_{\mathrm{i}} + \epsilon_{\mathrm{i}},$$

where $\hat{y}_{\mathrm{io}}$ is the vector of predicted values obtained by fitting the full model (17.55). An interpretation of this fact is that fitting the model (17.56) that includes only a subset of the variables is the same as using that subset to *approximate* the predictions of the full model. The fact itself can be seen from the normal equations associated with these two models. We have

$$X_{\mathrm{i}}^{\mathrm{T}}X(X^{\mathrm{T}}X)^{-1}X^{\mathrm{T}} = X_{\mathrm{i}}^{\mathrm{T}}. \tag{17.57}$$

This follows from the fact that $X(X^{\mathrm{T}}X)^{-1}X^{\mathrm{T}}$ is a projection matrix, and X_{i} consists of a set of columns of X (see page 26).

The problem in variable selection in the linear model is to move columns between X_{i} and X_{o}. There are many ways that we can do this systematically. One approach to the problem is to consider the best matrix X_{i} with one column, the best with two columns, and so on. This is called "all best" regressions, and there is an efficient algorithm for determining these. Another way, which is very simple, is called *forward selection.* We begin with the simple linear model

$$y \approx x_k \beta_k$$

(recall that the data are centered), and determine the j as

$$j = \arg\min_k \min_{b_k} \|y - x_k b_k\|_2. \tag{17.58}$$

The x_j for which this minimum is minimized becomes the first column in X_i. It is the column vector whose correlation with y is greatest in absolute value. The process continues by determining the next best variable, after y has been adjusted for x_j, and moving it from X_o to X_i until all variables are included in the model or until some stopping criterion is satisfied. At each step, the chosen variable is the one with the greatest correlation (in absolute value) with the current residuals, that is, with the values of y adjusted for all variables currently in the model.

In a variation of forward selection, we have *forward stepwise selection,* in which we consider the possibility of removing variables; that is, moving columns from X_i to X_o. Forward stepwise selection requires that some criterion be chosen for including the best variable in X_i; that is, we do not just identify the best variable, we decide whether it is good enough to enter the model. If it is not, we terminate the process. Likewise, forward stepwise selection requires that some criterion be chosen for removing the worst variable in X_i; that is, we do not just identify the worst variable, we decide whether it should be removed from the model. The criteria are generally based on the change in the residual L_2 norms:

$$\|y - X_\mathrm{i}^\mathrm{current} \beta_\mathrm{i}^\mathrm{current}\| - \min_{\beta_\mathrm{i}^\mathrm{new}} \|y - X_\mathrm{i}^\mathrm{new} \beta_\mathrm{i}^\mathrm{new}\|. \tag{17.59}$$

If a new variable is being added to $X_\mathrm{i}^\mathrm{current}$ we seek the maximum reduction; if a variable if being removed from $X_\mathrm{i}^\mathrm{current}$ we seek the minimum increase. The actual comparisons are often expressed in terms of a pseudo-F statistic.

The relevant computations for updating these solutions efficiently were discussed in Section 5.6, especially beginning on page 233.

Another interesting approach to variable selection is called least angle regression. This method begins by identifying the variable that is most correlated with y, just as in forward selection. This would be the same variable as x_j in equation (17.58), but we do not determine the optimal b_j. Instead we begin with $b_j = 0$ and increase it in the direction of the sign of the correlation between y and x_j. If $s_j = \pm 1$ with the same sign as that correlation, we let α_j increase by small step sizes h, that is, $\alpha_j = 0, h, 2h, \ldots$, and take $b_j = s_j \alpha_j$. As we do this, we form

$$r = y - b_j x_j,$$

and we compute the correlation between r and all the x's at each step. We want to identify the variable with the greatest correlation (in absolute value).

Initially x_j will have the greatest correlation with r. (That is the way j was selected at the beginning with $b_j = 0$.) At some point, the correlation between x_j and r becomes zero. (That is one of the properties of the residual vector in linear regression under least squares.) Suppose that at some value of b_j, x_k has greater correlation with r than x_j does. At that point, we put x_k in the model, and determine signs on b_j and b_k in the least squares fit of y to x_j and x_k, say s_j and s_k. We now begin changing b_j in the direction s_j from its current value and changing b_k from its current value (of 0) in the direction s_k. As we do this, we compute $r = y - b_j x_j - b_k x_k$, and we compute the correlation between r and all the x's at each step, and identify the next variable whose correlation is greatest. The correlation of both x_j and x_k with r will approach 0, so at some point we have a new variable to bring into the model.

We continue in this way until all variables have entered the model.

As we proceed through these steps, it is quite possible that the optimal values of some coefficients will change signs as new variables enter the model, hence the direction in which they are changed at each step may change. Depending on how far they are moved before another variable enters the model, the current values of the coefficients may change signs. (Recall that the optimal value and the current value of the coefficients may not be the same. The current value of each coefficient begins at 0.)

Software for least angle regression is available in the `lars` function in the R package `lars`.

As mentioned above, there are many difficult statistical issues in the variable selection problem. The exact methods of statistical inference generally do not apply (because they are based on a model, and we are trying to choose a model). In variable selection, as in any statistical analysis that involves the choice of a model, the effect of the given dataset may be greater than warranted, resulting in overfitting. One way of dealing with this kind of problem is to penalize the fitting criterion with some term that measures the complexity of the model. The regularized solutions (17.45) are a step in this direction. There are also criteria such as Mallow's C_p and the Akaike information criterion (AIC) that explicitly include the number of parameters to be fitted in the model. The AIC, for example, is

$$-2l^*(\beta_m) + 2m \tag{17.60}$$

where $l^*(\beta_m)$ is the maximum of the log-likelihood of the model with parameter β_m (see equation (1.110) on page 45) and m is the number of elements in β_m. As various β_m subvectors of the full vector β are used in the model, the one with minimum AIC may be chosen as the best. (Recall that a likelihood is defined in terms of a full specification of a probability distribution, which we have not done in most of this discussion. It is generally just taken as normal, hence the log-likelihood is just the negative of the residual sum of squares.) The R functions `step` and `stepAIC` (in `MASS`) compute the AIC for the regression models chosen by forward selection (not forward stepwise selection), as described above.

A general rule when a model or variables within a general type of model are to be chosen is to use part of the dataset for fitting and part for validation of the fit. There are many variations on exactly how to do this, but in general, cross validation is an important part of any analysis that involves building a model.

Assessing the performance of different variable selection methods is very difficult because of all of the differences from one scenario to another. These differences include overall signal to noise ratio, the extent of collinearity among the independent variables and a host of other factors. In Exercise 17.6 you are asked to design and conduct a Monte Carlo study to consider some of the issues in variable selection.

Local Fitting

Often, a single model of the form $Y \approx f(x;\theta)$ over the full range of interest either does not provide a very good fit or the form of f is overly complicated. If the functional form is complicated, it is unlikely that it provides insights into the relationship of Y and x. We may achieve a better fit and more accuracy in predictions if we abandon the global model. We could, of course, seek a piecewise model of the

$$Y \approx f_j(x;\theta_j) \quad \text{for } x \in R_j,$$

where R_j is some connected subregion of the region of interest. This may be a useful approach, but, of course, we are faced with the problem of determining the ranges R_j. In some applications, there may be first principles that suggest particular functional forms over certain ranges.

Two other ways of doing local fitting that we discuss in Chapter 10 (page 405) are by using splines or kernels. Rather than developing single functional forms, we can think of the problem as simply one of providing a rule that for a given x_0 provides a predicted $\widehat{Y}_0$. The rule may be expressed in the form of a regression tree (page 621) in which each terminal node is the predicted value of Y within the region defined by the path to the terminal node.

An even simpler approach is just to divide the range of x into convenient sets R_j and take $\widehat{Y}$ in that region to be the mean of the observed values of Y in that region. This is called a *bin smoother*. A variation on a bin smoother is a *running smoother*, which, at any given point x_0, uses weighted averages of observed values of Y corresponding to observed values of x near x_0. This is the idea behind *kernel smoothers*. A smoothing procedure based on local averaging directly uses the fact that the systematic component of the model is a conditional expectation of y, given x. The fitted systematic component then takes the form

$$\widehat{f}(x) = \sum K(g(x, x_i))y_i,$$

where $K(g(x, x_i))$ is a kernel and $g(x, x_i)$ is some function that increases in $\|x - x_i\|$. The kernel is often taken as a radially symmetric function of a scalar, and we can express the kernel smooth at x as

$$\widehat{f}_V(x) = \frac{\sum K_V(\|V^{-1}(x - x_i)\|y_i)}{\sum K_V(\|V^{-1}(x - x_i)\|)},$$

where V is a scaling matrix, perhaps one that also spheres the data (see page 391). The scaling matrix V controls the locality of influence of observations around x. As in the case of density estimation (see page 504), the scaling matrix, or the window width, has a major effect on the performance of the estimates. A wide window makes for a very smooth regression surface, whereas a very small window results in a highly variable surface.

Use of kernels is a common method in nonparametric regression. Another approach to local fitting is to use splines, as we discuss on page 404.

Projection Pursuit Regression

In Chapter 16, we discussed several methods for analyzing multivariate data with the objective of identifying structure in the data or perhaps reducing the dimension of the data. Two projection methods discussed were principal components and projection pursuit. In models for dependencies, we often apply these multivariate methods just to the independent variables. In linear regression, principal components can be used to reduce the effects of multicollinearity among the independent variables. The dependent variable is regressed on a smaller number of linear combinations of the original set of independent variables.

Projection pursuit is a method of finding interesting projections of data. In the discussion of projection pursuit beginning on page 564, it was applied to a multivariate dataset with the objective of identifying structure in the data.

In regression modeling, projection pursuit is often applied to an additive model of the form

$$Y = \beta_0 + \sum_j^m f_j(x_j, \beta) + E.$$

The idea is to seek lower-dimensional projections of the independent variables, that is, to fit the model

$$Y = \beta_0 + \sum_j^m f_j(\alpha_j^{\mathrm{T}} x) + E.$$

Projection pursuit regression involves the same kind of iterations described beginning on page 567 except that they are applied to the model residuals.

Even for the linear model, these computations can be extensive. The R or S-Plus function `ppreg` performs the computations for linear projection pursuit regression.

Fitting Models of Sequential Dependencies

On page 593 of Section 17.1, we briefly discussed models that have a sequential dependency structure, and on page 599 in Section 17.2 we described probability models that account for sequential dependencies. We now consider some of the issues in fitting such models to data.

Models for dependencies within a sequence such as a time series often are regression models in which an independent variable is the same as the dependent variable at a previous point in time. Data for fitting such models can be put in the same form as data for other regression models by simply adding variables that represent lagged values.

The model for sequential dependency often expresses a rate in a differential equation, such as equation (17.34) (in a slightly different form),

$$\frac{\mathrm{d}Y}{Y} = \mu\,\mathrm{d}t + \sigma\,\mathrm{d}B_t. \tag{17.61}$$

A rate cannot be directly observed; the data for fitting such a model are observations at discrete points in time, and derivatives are approximated by ratios of finite differences. In applications to financial data, for example, this model is fit by selecting a fixed time interval Δt, such as a day, and observing ΔY_t and Y_t at a set of points in time. (In this application, there are obvious problems because of time restrictions on the underlying process; stocks are not traded on weekends and market holidays.) The parameters in the model, μ and σ, are generally estimated by the method of moments. If B_t is a Brownian motion — that is, if the random variable has a normal distribution — the method of moments estimator is also the maximum likelihood estimator.

The geometric Brownian motion model leads to a lognormal distribution of prices with an expected value that decreases proportionally to the variance of the rate of return (see Exercise 17.4). The importance of this model is not because of the stock prices themselves but for applications in pricing derivatives.

A derivative is a financial instrument whose value depends on values of other financial instruments or on some measure of the state of the economy or of nature. The most common types of derivatives available to individuals are *call options* on stocks, which are rights to purchase the stock at a fixed price, and *put options* on stocks, which are rights to sell the stock at a fixed price. Both of these types of derivatives have expiration dates at which the rights terminate. There are many variations on calls and puts that are useful in academic analyses, but in the usual applications a call option conveys to the owner the right to buy a stated number of shares of the underlying stock at a fixed price anytime before the expiration date, and a put option conveys to the owner the right to sell a stated number of shares of the underlying stock at a fixed price anytime before the expiration date. (For no particular reason, such options are called "American options". Options that cannot be exercised prior to the expiration time are called "European options". Although European options are rare, their values are easier to analyze.) Stock options are *rights*, not obligations, so the value, and consequently the price, cannot be negative. Much study has been devoted to determining the appropriate price of a call or a put.

Based on use of equation (17.61) and the associated assumptions of normality with constant μ and σ as a pricing model, with certain assumptions about financial markets (no arbitrage and the existence of a riskless investment, for example) and additional assumptions about the underlying stock (no dividends, for example), a differential equation for the price of European options can be developed. This is called the Black-Scholes differential equation; it has a fairly simple solution (see Hull, 2008, for example).

The failure of any one of the assumptions can invalidate the Black-Scholes differential equation. In some cases, there is no simple differential equation that can take its place. In other cases, the resulting differential equation cannot be solved easily.

More realistic versions of equation (17.61), such as ones with jumps, or ones that are mean-reverting, can easily be simulated. Likewise, more realistic assumptions about exercise times, dividend payouts, and so on can easily be accommodated in a simulation. Our ability to simulate the process allows us to use Monte Carlo methods to study whether the assumptions about the process correspond to observable behavior.

If our assumptions about the process do indeed correspond to reality, we can use Monte Carlo simulation to determine various features of the process, including the appropriate prices of derivative assets. The way this is done is one of the standard techniques of the Monte Carlo method; that is, we express the quantity of interest as an expected value, simulate the process many times, each time computing the outcome, and then estimate the expected value as the mean of the outcomes.

17.4 Classification

In Section 16.1, we considered the general problem of clustering data based on similarities within groups of data. That process is sometimes called "unsupervised classification". In "supervised classification", usually just called "classification", we know to which groups all of our observed data, or at least a training set of data, belong. The objective is to develop a model for classification of new data.

Classification and Regression Trees

A rooted tree that is used to define a process based on a sequence of choices is called a *decision tree.* Decision trees that are used for classifying observations are called *classification trees.* They are similar to the cluster tree shown in Figure 16.5 on page 524 except that the nodes in the classification trees represent decisions based on values of the independent variables. Decision trees that are used for predicting values of a continuous response variable are called *regression trees.* The terminal nodes correspond to predicted values or intervals of predicted values of the response variable.

The objective in building a classification tree is to determine a sequence of simple decisions that would ultimately divide the set of observations into the given groups. Each decision is generally in the form of a comparison test based on the values of the variables that constitute an observation. Often, the comparison test involves only a single variable. If the variable takes on only a countable number of values, the comparison test based on the variable may be chosen to have as many possible outcomes as the values associated with the variable. If the variable has a continuous range of values, the comparison test is generally chosen to have a binary outcome corresponding to observations above or below some cutpoint. Trees with exactly two branches emanating from each nonterminal node are called *binary trees.* The terminal nodes of a classification tree represent groups or classes that are not to be subdivided further.

How well a test divides the observations is determined by the "impurity" of the resulting groups. There are several ways of measuring the impurity or how well a test divides the observations. Breiman et al. (1984) describe some methods, including a "twoing rule" for a binary test. For classifying the observations into k groups, at a node with n observations to classify, this rule assigns a value to a test based on the formula

$$n_L n_R \left(\sum_{i=1}^{k} \left| L_i n_L - R_i n_R \right| / n \right)^2 , \tag{17.62}$$

where n_L and n_R are the number of observations that the test assigns to the left and right child nodes, respectively, and L_i and R_i are the number of group i assigned to the left and right, respectively.

The classification tree can be built by a greedy divide-and-conquer recursive partitioning scheme, as given in Algorithm 17.2.

Algorithm 17.2 Recursive Partitioning for Classification Using a Training Set

1. Evaluate all tests that divide the given set into mutually exclusive sets.
2. Choose the test that scores highest, and divide the set based on this test.
3. For any subset that contains observations from more than one group, repeat beginning at step 1. ∎

The results of a binary partitioning may be presented as a tree, as shown in Figure 17.4, in which the rule for splitting at each node is shown. In this example, there are three groups and two numerical variables.

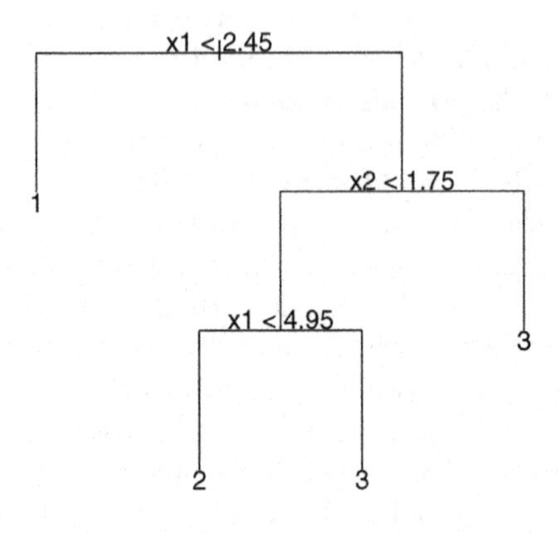

Fig. 17.4. A Classification Tree

The process of building a tree by recursive binary partitioning continues until either the data at the terminal nodes are sufficiently homogeneous or consist of a small number of observations. Homogeneity is measured (negatively) by *deviance*. For continuous response variables, deviance is the sum of squares; for factor variables, it is two times the log-likelihood of the full model (that is, all categories) minus the log-likelihood of the current model.

The R function `tree` in the `tree` package uses binary recursive partitioning to build a classification tree.

The rules in a decision tree can be used to define each of the classes by conjunctive combinations of the rules at each node. In the example of Figure 17.4, we can define the classes as in the following table, where "$\wedge$" represents "and" and "$\vee$" represents "or". Rules expressed in these forms are called conjunctive normal forms, or CNFs (if the major conjunctions are all $\wedge$) or disjunctive normal forms, or DNFs. (A formula in DNF is one written as a disjunction of terms, each of which may be a conjunction.)

Rules such as these are useful in describing the result of the classification, and they also aid in our understanding of the basis for the classification.

Table 17.1. Rules Defining the Clusters Shown in Figure 17.4

cluster	rule(s)
1	$x_1 < 2.45$
2	$(2.45 \leq x_1) \wedge (x_1 < 4.95) \wedge (x_2 < 1.75)$
3	$\big((2.45 \leq x_1) \wedge (1.75 \leq x_2)\big) \vee \big((4.95 \leq x_1) \wedge (x2 < 1.75)\big)$

A variety of other classification tree programs have been developed. Some classification tree programs perform multilevel splits rather than binary splits. A multilevel split can be represented as a series of binary splits, and because with multilevel splits predictor variables are used for splitting only once, the resulting classification trees may be unrealistically short. Another problem with multilevel splitting is the effect of the order in which the variables are used in splitting. The number of levels for splits of variables affects the interpretation of the classification tree. These effects are sometimes referred to in a nontechnical sense as "bias".

Instead of splitting on a single variable at each node, we can form splits based on combinations of values of several variables. The most obvious way of doing this is to use some linear combination of the variables, such as the principal components (see Section 16.3, page 548). Splits based on linear combinations of the variables are called *oblique linear splits*. Seeking good linear combinations of variables on which to build oblique partitions is a much more computationally intensive procedure than just using single variables.

Linear Classifiers

In the following we will assume that there are only two classes. While there are some apparently obvious ways of extending a binary classification method to more classes, the problem is not as simple as it may appear, and we will not pursue it here.

In a very simple instance of binary classification, a oblique linear split on the single root node of a classification tree yields two pure classes. This is illustrated in Figure 17.5, in which the points are classified based on whether they are above or below the line $w^{\mathrm{T}}x = w_0$.

This illustrates how the data in two dimensions, that is the 2-vector x, could be reduced to one dimension. That one dimension would correspond to points along a single line normal to the line $w^{\mathrm{T}}x = w_0$. The bivariate points projected onto that lower dimensional space would be perfectly classified in the sense that a single criterion would separate the points. It is obvious from the figure that the separating line (or in general, separating hyperplane) may not be unique.

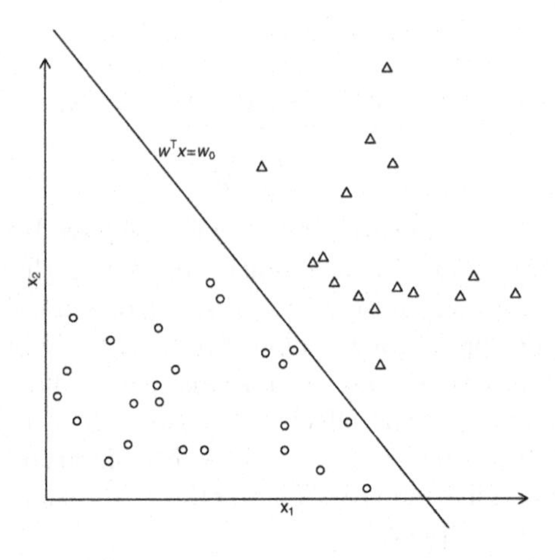

Fig. 17.5. Linear Classification

Real world data rarely can be perfectly separated by a linear classifier as in this simple example. Whenever the data cannot be perfectly separated, we might consider some kind of optimal separating hyperplane. A simple criterion of optimality is a separation such that the sum of squares from the means of the two classes is maximal with respect to the sums of squares from the means within the two classes. This is the kind of decomposition done in analysis of variance.

In the classification problem, however, we generally do not assume that the variance-covariance matrix is necessarily proportional to the identity, as we do in the usual analysis of variance. The appropriate sums of squares must be scaled by the variance-covariance matrix. Since the variance-covariance matrix is likely unknown, we use the sample variance-covariance matrix, equation (9.10), applied to the separate groups.

A second problem arises in this approach, and that is the question of whether the variance-covariance matrix is the same in the two groups. There are various alternative procedures to try to accommodate different variance-covariance matrices, but we will not pursue them here. (The Behrens-Fisher problem in a two-group t-test is one of the most familiar of this type of problem.)

If we assume that the variance-covariance matrices are equal, for two groups defined by the indices C_1 and C_2, the pooled within-class variance-covariance matrix

$$S_{\mathrm{W}} = \frac{1}{n_1 + n_2 - 2}\left(\sum_{i \in C_1}(x_i - \bar{x}_1)(x_i - \bar{x}_1)^{\mathrm{T}} + \sum_{i \in C_2}(x_i - \bar{x}_2)(x_i - \bar{x}_2)^{\mathrm{T}}\right), \tag{17.63}$$

where $\bar{x}_i$ is the mean of the n_i observations in group i, for $i = 1, 2$. The corresponding between-group variance-covariance matrix is simply

$$S_{\mathrm{B}} = (\bar{x}_2 - \bar{x}_1)(\bar{x}_2 - \bar{x}_1)^{\mathrm{T}}. \tag{17.64}$$

R. A. Fisher, following the ideas he had put forth in the analysis of variance, suggested that the optimal separating hyperplane w is one such that

$$J(w) = \frac{w^{\mathrm{T}} S_{\mathrm{B}} w}{w^{\mathrm{T}} S_{\mathrm{W}} w} \tag{17.65}$$

is maximized. Differentiating $J(w)$ and setting the derivative to zero, we have

$$(w^{\mathrm{T}} S_{\mathrm{B}} w) S_{\mathrm{W}} w = (w^{\mathrm{T}} S_{\mathrm{W}} w) S_{\mathrm{B}} w.$$

This yields the optimal direction for w as

$$w_* = S_{\mathrm{W}}^{-1}(\bar{x}_2 - \bar{x}_1). \tag{17.66}$$

This expression is called *Fisher's linear discriminant.*

If we project the data onto the space defined by w_*, we can determine an optimal separation in one dimension. If the data are linearly separable as in the example in Figure 17.5, all of the w_* coordinates of one class will be larger than all of the w_* coordinates of the other class.

If the data are not linearly separable, we have a simple one-dimensional classification problem in which we must optimally choose a point w_0 on the line determined by w_* as the point of separation. One obvious way of choosing this point is a data-based Bayes rule, by which we select the point such that the frequency with which observations are miss-classified into the two classes is minimized.

After the direction w_* is determined the projection matrix can be determined as described on page 379, possibly following a rotation of the full space as indicated on page 375.

Kernel Methods and Support Vector Machines

Identification of nonlinear structure in data is often preceded by a nonlinear mapping from the m-dimensional data space to a p-dimensional "feature space". The feature space may have more, even many more, dimensions than the data space. The mapping carries the basic data m-vector x to a p-vector $f(x)$. The feature space is a subspace of IR^p and thus is an inner-product space. (The original data space may be more general.) This is essentially the approach of *support vector machines* (see Mika et al. 2004). The basic idea is that while there may be no separating hyperplane in the data space as in Figure 17.5, there may be a separating hyperplane in the higher-dimensional feature space.

The function f may be quite general. It may treat the elements of its vector argument differently, but a common form for the argument $x = (x_1, \ldots, x_m)$, has the form

$$\left(x_1^{e_p}, \ldots, x_m^{e_p}, x_1^{e_{p-1}}, \ldots, x_m^{e_{p-1}}, \ldots, x_1^{e_1}, \ldots, x_m^{e_1}\right).$$

Another common form is homogeneous in the elements of the data space vector, for example, for the 2-vector $x = (x_1, x_2)$,

$$\tilde{x} = f(x) = \left(x_1^2, \sqrt{2}x_1x_2, x_2^2\right). \tag{17.67}$$

Figure 17.6 illustrates a situation in which there is no separating hyperplane (line) in the two-dimensional data space. After transforming the data space to a feature space, however, a hyperplane can be chosen to separate the data. In the case shown in the right side of Figure 17.6, the data were transformed into a three-dimensional feature space, a separating plane was determined, and then the data were projected onto a z_1-z_2 plane that is perpendicular to the separating plane.

A basic operation in the use of support vector machines in classification is the computation of inner products in the feature space; that is, the computation of $\tilde{x}^{\mathrm{T}}\tilde{y}$. The computations to transform a data vector to a feature vector $\tilde{x} = f(x)$, as well as the computation of the inner products in the higher-dimensional feature space, may be costly. The overall computations may be reduced by a simple expedient called the "kernel trick", which involves finding a kernel function (see page 23) in the data space that is equivalent to the inner product in the feature space. The form of the kernel obviously depends on f. For the function shown in equation (17.67), for example, we have

$$\begin{aligned}(f(x))^{\mathrm{T}} f(y) &= \left(x_1^2, \sqrt{2}x_1x_2, x_2^2\right)^{\mathrm{T}} \left(y_1^2, \sqrt{2}y_1y_2, y_2^2\right) \\ &= \left(x^{\mathrm{T}}y\right)^2 \\ &= K(x, y),\end{aligned} \tag{17.68}$$

where we define the kernel $K(x, y)$ as $\left(x^{\mathrm{T}}y\right)^2$.

Although support vector machines can be used for exploratory clustering, most of the applications are for classification with a given training set.

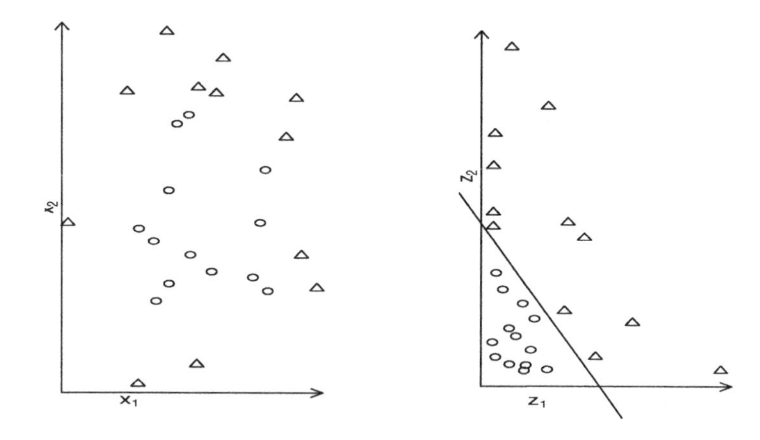

Fig. 17.6. Data Space Transformed to Feature Space and Linear Classification in the Feature Space

Combining Classifications

Classification methods tend to be unstable; that is, small perturbations in the training datasets may yield large differences in the classification rules. To overcome the instability, which manifests itself in large variability of the classes, various *ensemble methods* have been suggested. These methods make multiple passes over the training dataset and then average the results. In some cases, the multiple passes may use the full training dataset with different classification methods, different subsets of the classification variables, or with perturbation of the results. In other cases the multiple passes use subsets or resamples of the full training dataset.

The averaging is essentially a voting by the various classifications formed following resampling. The idea of averaging is an old one; in this context it is sometimes called "stacked generalization". A similar idea in linear regression in which subsamples of minimal size are used is called "elemental regression". (The minimal size in linear regression is a dataset that has the same number of observations as variables.) The coefficient estimates are averaged over the subsamples.

In some cases, the full training dataset is used on each pass, but different classification methods are used or else following each pass, the results are perturbed. One of the most effective ways of doing this is to reweight the variables following each classification pass. The variables that show the weakest discrimination ability in the most recent pass are given heavier weight in the next pass. This method is called "boosting". Similar methods are also called "arcing", for "adaptively resample and combine".

In another multiple-pass method that uses the full dataset on each pass, instead of modifying the weights or the method of classification in each pass, the results are randomly perturbed, that is, the classes to which the observations are assigned are randomly changed. The randomly perturbed classes are then averaged. This method, surprisingly, also sometimes improves the classification.

Other ensemble methods use either a resample or a subsample of the full dataset. A method of forming random training sets that uses bootstrapping (that is, it resamples) the given dataset is called "bagging", because of the bootstrap samples. For a given "bag" or bootstrap sample, the classification tree is constructed by randomly choosing subsets of the features to use in forming the branches from nodes. The process is continued until the full tree is formed (that is, the tree is not "pruned").

If random subsamples of the training dataset are used to form classification trees, the collection of random trees is called a *random forest.*

Within the broad range of ensemble classification methods there is a multitude of details that can be changed, resulting in an almost overwhelming number of classification methods. There does not appear to be a method that is consistently best. Some methods are better on some datasets than others, and there is no simple way of describing the optimality properties.

An easy way to explore the various methods is by use of the Weka program, which is a collection of many algorithms for clustering and classification. Weka can either be applied directly to a dataset or called from the user's Java code. It contains tools for data pre-processing, classification, regression, clustering, association rules, and visualization. It can also be used for developing new machine learning schemes. It is open source software available at
`http://www.cs.waikato.ac.nz/ml/weka/`

It is useful to compare classification methods using different datasets. An important collection of datasets is available at the Machine Learning Repository at the University of California at Irvine. The UCI Machine Learning Repository can be accessed at
`http://archive.ics.uci.edu/ml/`

17.5 Transformations

Most observable data can be measured in various ways without incurring any substantive change. If we make a one-to-one transformation on data, we can always recover the initial values. The method in which data are represented, however, can have an effect on how well any simple model fits the data.

Transformations to Make Data Fit Models

Often, a model is so tractable, both statistically and computationally, that even if it is not a good representation of observed reality, it may be worth using it as an approximation. Sometimes, the approximation can be made even better by transforming the data. The use of transformations to make the data fit simple models is an old idea.

The most common types of transformations are those that attempt to make the low-order moments of the transformed variable correspond to the assumptions in the model. In real data, often the second moment (the variance) increases as the first moment increases, but a common assumption is that the moments are constant; hence, we may seek a transformation that stabilizes the variance. Another common assumption in the model is that the distribution is symmetric. A transformation based on the third moment may make the data more symmetric.

Variance Stabilization

A transformation may be suggested by some pattern in the observations. For example, certain types of measurements are often observed to exhibit greater variability as the magnitude of the observations becomes larger. For a probability distribution such as the normal distribution, there are two distinct parameters for the mean and the variance, and we may wish to model populations with different means using normal distributions with the same variance. To use data pooled from the different populations, a *variance stabilizing* transformation may be useful. It may turn out, for example, that if the variance of y tends to be proportional to its magnitude, the variance of $y^{1/2}$ is relatively constant. In this case, the square root transformation is a variance-stabilizing transformation.

In general, if the variance of Y is some function V of the mean of Y, a transformation $h(Y)$ to stabilize the variance would have the property

$$\frac{\partial h(y)}{\partial y} \propto \frac{1}{\sqrt{V(y)}};$$

therefore, the appropriate transformation $h(y)$ is the integral of $1/\sqrt{V(y)}$.

In addition to assumptions in the model about the variance, there are generally assumptions about covariances, either among the variables or among the same variables in different observations. Often the assumption is that the correlations are zero. Transformations to reduce correlations would, therefore be of interest. The same kinds of ideas as those for controlling the variances would lead to transformations to reduce correlations, and, although the range of possibilities is greater (if for no other reason than the fact that two variables are involved), the problem of finding reasonable transformations to reduce correlations is more difficult.

Transformations of the Independent Variables

In addition to problems with the random component of the model not corresponding well with the data, the systematic component also may not fit the data. This may suggest transformations of either the independent or the dependent variables.

For a linear model such as equation (17.6), with the independent variables taking only positive values, a power transformation

$$w_j = \begin{cases} x_j^{\alpha_j} & \alpha_j \neq 0 \\ \log x_j & \alpha_j = 0 \end{cases} \tag{17.69}$$

may be useful. Values of the α_j's that minimize the residual sum of squares, can be determined by an iterative procedure. Transformations of the independent variables of this type are called *Box-Tidwell transformations.*

Transformations of the Box-Cox Type

Box and Cox (1964) study power transformations of the dependent variable (assumed to be positive) and suggest use of a maximum likelihood method to determine the power. The power transformation of equation (17.69) can be made continuous in the power by making a slight modification. This results in the *Box-Cox transformation*,

$$\tau(y;\lambda) = \begin{cases} (y^\lambda - 1)/\lambda & \lambda \neq 0 \\ \log y & \lambda = 0 \end{cases}. \tag{17.70}$$

If τ is the dependent variable in a linear model such as equation (17.22) and the elements of ϵ are from independent normals with mean 0 and variance σ^2, the log-likelihood function is

$$l(\lambda, \beta, \sigma; y) = -n\log\sigma - \frac{(\tau(y;\lambda) - X\beta)^{\mathrm{T}}(\tau(y;\lambda) - X\beta)}{2\sigma^2} - (\lambda - 1)\sum_i^n \log y_i. \tag{17.71}$$

(Recall that the y_i's are assumed to be positive.) For a fixed value of λ, the maximum of equation (17.71) with respect to β and σ yields the usual least squares estimates, with $\tau(y;\lambda)$ as the dependent variable. The function $\widehat{l}(\lambda; \widehat{\beta}, \widehat{\sigma}, y)$ is called the *profile likelihood.* In practice, the profile likelihood is often computed for a fixed set of values of λ, and the maximum with respect to λ may be chosen based on an inspection of the plot of the function. The choice of λ is often restricted to $\ldots, -2, -1\frac{1}{2}, -1, -\frac{1}{2}, 0, \frac{1}{2}, 1, 1\frac{1}{2}, 2, \ldots$. The maximum of $\widehat{l}$ with respect to λ is the same as the maximum of l with respect to λ, β, and σ. If one really believes the likelihood from which equation (17.71) is derived, confidence intervals on λ could be computed based on the asymptotic chi-squared distribution of the log-likelihood.

There are several obvious extensions to the transformations discussed above. The transformations can be applied whether the systematic component of the model is linear or of the more general form in equation (17.8). Instead of the power transformations in equations (17.69) and (17.70), more general transformations could be used, and the λ in $\tau(y;\lambda)$ could represent a much more general parameter. Both the Box-Tidwell transformations of the independent variables and the Box-Cox transformations of the dependent variables could be applied simultaneously; that is, transform both sides. Either the same transformation for both the dependent variable and the systematic components or different transformations could be used.

Alternating Conditional Expectation

In any model, we want to include independent variables that have strong relationships to the dependent variable. We may seek transformations of all of the variables to achieve stronger relationships.

Breiman and Friedman (1985a, 1985b) describe and study a method of fitting an additive model (17.7) that relates the independent variables to a transformation of the dependent variable:

$$\tau(Y) \approx \beta_0 + f_1(x_1) + \cdots f_m(x_m).$$

The f's are just transformations of the x's. The basic approach is to transform the variables iteratively using τ and the f's to maximize the sample correlation of the transformed variables.

The procedure is called *alternating conditional expectations*, or ACE. It attempts to maximize the sample correlation between $\tau(y)$ and $\sum_{j=1}^{m} f_j(x_j)$ or to minimize

$$e^2(\tau, f_1, \ldots, f_m) = \frac{M\left(\left(\tau(y) - \alpha - \sum_{j=1}^{m} f_j(x_j)\right)^2\right)}{S(\tau(y))},$$

where $M(\cdot)$ is the sample mean and $S(\cdot)$ is the sample variance. The method is shown in Algorithm 17.3, in which we write $e^2(\tau, f_1, \ldots, f_m)$ as $e^2(\tau, f)$.

Algorithm 17.3 Alternating Conditional Expectation (ACE)

0. Set $k = 0$. Set $\tau^{(k)}(y) = (y - M(y))/(S(y))^{1/2}$.
1. Set $k = k + 1$.
2. Fit the additive model with $\tau^{(k-1)}(y)$ (see page 604) to obtain $f_j^{(k)}$.

3. Set
$$\tau^{(k)}(y) = \frac{M\left(f^{(k)}(x) \mid y\right)}{\left(S\left(f^{(k)}(x) \mid y\right)\right)^{1/2}}.$$
4. If $e^2\left(\tau^{(k)}, f^{(k)}\right) < e^2\left(\tau^{(k-1)}, f^{(k-1)}\right)$, then go to step 1; otherwise terminate. ∎

If there is only one explanatory variable x, step 2 is just to compute

$$f^{(k)}(x) = M\left(\tau^{(k-1)}(y) \mid x\right).$$

When the algorithm terminates, the sample of y's and of the x vectors have been transformed so that the simple additive model provides a good fit.

Additivity and Variance Stabilization

Transformations to stabilize the variance of the residuals can be performed simultaneously with the transformations that achieve the strong additive relationship. Tibshirani (1988) introduced a technique called additivity and variance stabilization (AVAS) that attempts to do this.

Algorithm 17.4 Additivity and Variance Stabilization (AVAS)

0. Set $k = 0$. Set
$$\tau^{(k)}(y) = \frac{y - M(y)}{(S(y))^{1/2}}.$$
1. Set $k = k + 1$.
2. Fit the additive model with $\tau^{(k-1)}(y)$ (see page 604) to obtain $f_j^{(k)}$.
3. Determine the variance function
$$v(u) = S\left(\tau^{(k-1)}(y) \mid \sum f_j^{(k-1)}(x_j) = u\right);$$
compute the variance-stabilizing function (see page 629)
$$h(t) = \int_0^t v(u)^{-1/2}\,\mathrm{d}u;$$
set
$$\tilde{\tau}^{(k-1)}(t) = h(\tau^{(k-1)}(t));$$
set
$$\tau^{(k)}(t) = \frac{\tilde{\tau}^{(k-1)}(t) - M(\tilde{\tau}^{(k-1)}(y))}{S(\tilde{\tau}^{(k-1)}(y))}.$$
4. If
$$e^2\left(\tau^{(k)}, f^{(k)}\right) < e^2\left(\tau^{(k-1)}, f^{(k-1)}\right),$$
then go to step 1;
otherwise terminate. ∎

Assessing the Fit of a Model

We have described various approaches to fitting a model using data, such as maximum likelihood, based on an assumed probability distribution, fitting by minimizing the residuals, and fitting by matching moments. Each of these methods has its place. In some cases, a particular type of one of these general schemes is identical to a particular type of another; for example, maximum likelihood under a normal distribution is equivalent to minimizing the sum of the squares of the residuals. Whatever method is used in fitting the model, we are interested in the *goodness* of the fit. If the method of fit is to minimize some norm of the residuals, intuitively it would seem that the proportional reduction in the norm of the residuals by fitting a model gives some indication of the goodness of that model. In other words, we compare the norm of the "residuals" with no model (that is, just the responses themselves) with the norm of the residuals after fitting the model,

$$\frac{\|r\|}{\|y\|},$$

which should be small if the fit is good or, looking at it another way,

$$\frac{\|y\| - \|r\|}{\|y\|}, \tag{17.72}$$

which should be close to 1 if the fit is good. Expression (17.72) is the familiar R^2 statistic from least squares linear regression.

Although a particular method of fitting a model may yield a relatively small norm of the residual vector, certain observations may have very large residuals. In some cases of model fitting, deleting one or more observations may have a very large effect on the model fit. If one or more observations have very large residuals (the meaning of "very large" is not specified here), we may question whether the model is appropriate for these anomalous observations, even if the model is useful for the rest of the dataset. We may also question whether the method of fit is appropriate. The method of fit should come under special scrutiny if deleting one or more observations has a very large effect on the model fit.

Quantile plots are especially useful in assessing the validity of assumptions in a model. If the residuals are ranked from smallest to largest, the pattern should be similar to a pattern of ranked normal random variables. Given a sample of n residuals,

$$r_{(1)} \leq \cdots \leq r_{(n)},$$

we compare the observed values with the theoretical values corresponding to probabilities

$$p_{(1)} \leq \cdots \leq p_{(n)}.$$

Although the sample distribution of the residuals may give some indication of how well the data fit the model, because of the interactions of the residuals with the fitting process, the problem of assessing whether the model is appropriate is ill-posed. The most we can generally hope for is to assess the predictive ability of the model by use of partitioned data and cross validation, as discussed in Section 12.2.

Notes and Further Reading

Linear, General Linear, and Generalized Linear Models

Full rank linear statistical models of dependencies are called regression models, and linear classification models (not of full rank) are called ANOVA models or general linear models. Kutner, Nachtsheim, and Neter (2004) provide an extensive coverage of these linear statistical models. Nonlinear regression models are covered in some detail by Seber and Wild (2003). The generalized linear model is a special type of nonlinear model that was introduced in its general setting by Nelder and Wedderburn (1972). A thorough coverage of generalized linear models is available in McCullagh and Nelder (1990). The special logistic generalized linear models are the subject of Hosmer and Lemeshow (2000).

An extensive discussion about the modeling issues and the statistical interpretation of the errors-in-variables model is provided by Fuller (1987).

Algorithmic Models

The article by Breiman (2001), with discussion, emphasizes the role of algorithmic models when the objective is prediction instead of a simple description with the primary aim of aiding understanding.

A classic type of algorithmic model is a neural net. See Ripley (1993, 1994, 1996) for discussion of neural nets in modeling applications.

Transformations

Kutner, Nachtsheim, and Neter (2004) and Carroll and Ruppert (1988) have extensive discussions of various transformations for linear models. Velilla (1995) discusses multivariate Box-Cox transformations, as well as the robustness of the transformations to outliers and diagnostics for detecting the effect of outliers.

Recursive Partitioning

Many of the clustering methods discussed in Chapter 16 and the classification methods discussed in this chapter are based on recursively partitioning the dataset. Trees or other graphs are used to maintain the status during the recursion. General methods as well as specific applications of recursive partitions are summarized by Zhang (2004).

Conceptual Clustering

Hunt, Marin, and Stone (1966) called classification that is based on general concepts "concept learning", and developed concept-learning systems. (Their purpose was to study ways that humans learn, rather than to address a given classification problem.) Michalski (1980) and Michalski and Stepp (1983) described "conceptual clustering", which is a set of methods of classification that identifies sets of characteristics, or concepts.

Combining Classifications

Most of the best classifiers are based on ensemble methods in which classifiers are combined or improved by reweighting as in bagging and boosting. Bühlmann (2004) provides an overview and summary of ensemble methods for classification.

Some examples of applications in classification in which combined classifiers had markedly superior performance are described by Richeldi and Rossotto (1997) and by Westphal and Nakhaeizadeh (1997).

Regularized Fitting and Variable Selection

The use of ridge regression to deal with the problems of multicollinearity in regression analysis was described in a 1962 paper by A. E. Hoerl. Hoerl and Kennard (1970a,b) discussed various applications of ridge regression, described how it ameliorates the effects of multicollinearity, and gave several suggestions for the choice of the weighting factor λ. The regularization used in ridge regression is called Tikhonov regularization in the applied mathematical literature after A. N. Tikhonov who described it in a 1963 paper (in Russian).

The lasso was first described by Tibshirani (1996). The least angle regression (LAR) algorithm for variable selection, which happens to correspond to lasso with increasing weights on the L_1 penalty on the fitted coefficients, was described by Efron et al. (2004).

Many statistical issues relating to variable selection and regression model building are discussed in the book by Miller (2002).

In the linear regression model $y = X\beta$, instead of fitting the original set of x's, we may use a set of the principal components $\tilde{x}$'s from equation (16.15). This is called *principal components regression.* While principal components regression does not remove any of the original variables from the model, the model itself, $y = \widetilde{X}\beta$, is in a lower dimension. Jolliffe, Trendafilov, and Uddin (2003) describe a regularization similar to that in lasso some of the $\hat{w}_j$ in equation (16.14) to zero, so that the corresponding original variables are no longer in the model at all.

Stochastic Differential Equations

The extensive theory for stochastic differential equations is covered in a number of texts, such as Øksendal (2003). Kloeden and Platen (1999) dissuss the numerical solution of stochastic differential equations. Their book (Kloeden, Platen, and Schurz, 1994) that discusses simulation with stochastic differential equations is also interesting, although the computer programs are in Basic.

Local Fitting

Loader(2004) describes various approaches to smoothing of data based on local criteria. She also discusses the statistical properties of the fits.

Computational Issues of Fitting Models to Data

Many of the methods of fitting models to data discussed in Section 17.3 are computationally intensive.

Hawkins and Olive (1999) give an algorithm for the least trimmed absolute values case.

The least median of squares estimator (equation (17.44), page 607, with $\rho(t) = t^2$) was proposed by Rousseeuw (1984). Fitting the model by this criterion is obviously computationally intensive. Hawkins (1993b) gives a feasible set algorithm for the least median of squares estimator, and Xu and Shiue (1993) give a parallel algorithm.

Exercises

17.1. Consider the use of a simple model of the probability density for univariate data:

$$f(y; \theta_1, \theta_2) = \frac{\theta_1}{2} e^{-\theta_1 |y-\theta_2|} \quad \text{for } -\infty \leq y \leq \infty.$$

(This is the double exponential.) Choose

$$\theta_1 = 10,$$
$$\theta_2 = 100,$$

and generate a random sample of size 100.

a) Write out the likelihood function of θ_1 and θ_2, and use an optimization program to maximize it. Make appropriate transformations prior to optimizing a function.

b) For the moment, ignore θ_1, and estimate θ_2 by minimizing various functions of the residuals $y - \theta_2$. Use iteratively reweighted least squares to obtain the L_1, $L_{1.5}$, and L_2 estimates of θ_2. Now, in each case, obtain an estimate of θ_1. Using your estimate of θ_1 and rescaling, which of the L_p estimates is closest to the MLE above? Which one would you expect to be closest?

c) Obtain 95% confidence intervals for both θ_1 and θ_2 using your L_p estimates. (Assume asymptotic normality of your location estimator.)

17.2. Consider the simple linear model

$$y_i = \beta_0 + \beta_1 x_i + \epsilon_i.$$

a) Assume that the ϵ_i are independent realizations of a random variable that has a $N(0, \sigma^2)$ distribution. Taking

$$\beta_0 = 10,$$
$$\beta_1 = 10,$$

and x to be

$$0.0, 1.0, 2.0, \ldots, 100$$

(i.e., 101 observations equally spaced from 0 to 100):

i. Let $\sigma = 1$, generate the corresponding y's, estimate β_0 and β_1 by ordinary least squares, and plot the data and the fitted line.

ii. Repeat with $\sigma = 100$.

b) Assume that the ϵ_i are independent realizations of a random variable with a Cauchy distribution with scale parameter a. Now, for $a = 1$ and $a = 100$, repeat the previous part of this exercise.

c) Repeat the previous part of this exercise, except use the L_1 criterion for estimating β_0 and β_1.

17.3. Conduct a Monte Carlo study to determine the size of the test for

$$\beta_{q+1} = \beta_{q+2} = \cdots = \beta_p = 0$$

in the linear model

$$y_i = \beta_0 + \beta_1 x_{i1} + \cdots + \beta_{q+1} x_{i,q+1} + \cdots + \beta_p x_{ip} + e_i$$

when the parameters are fitted using Huber M estimation. Conduct the test at the nominal significance level of 0.05. (The "size of the test" is the actual significance level.) Use a normal distribution, a double exponential distribution, and a Cauchy distribution for the error term.

Summarize your findings in a clearly-written report.

17.4. The geometric Brownian motion model for changes in stock prices (17.61) on page 619 leads to a lognormal distribution for the prices themselves with an expected value that decreases proportionally to the variance of the rate of return. Study this decrease by Monte Carlo methods beginning with the model for changes in price. Give an intuitive explanation why it should decrease.

17.5. Choose any ten stocks traded on the New York Stock Exchange. Do the price histories of these stocks over the past 5 years support the validity of the geometric Brownian motion model?
Stock price data can be accessed online from a variety of sites. One of the most stable sites is
`http://finance.yahoo.com/`
Stock data for a particular issue can be obtained either by entering the symbol on the web page or directly (for IBM, for example) by
`http://finance.yahoo.com/q?s=ibm`
Historical price data can be downloaded into a csv file, and can be read in R by `read.csv`. The R function `read.csv` can go directly to a URL to get data. John Nolan wrote the following R statements:

```
url <- paste("http://ichart.finance.yahoo.com/table.csv?a=",
    start.date[1]-1,"&b=",start.date[2],"&c=",start.date[3],
    "&d=",stop.date[1]-1,"&e=",stop.date[2],"&f=",
    stop.date[3],"&s=",symbol,sep="")
x <- read.csv(url)
```

The `start.date` and `stop.date` are numeric arrays of length 3, containing the month number, the day, and the year in the four-digit format.

17.6. Use Monte Carlo to study and compare least-squares-based variable selection methods in the linear regression model

$$y_i = \beta_0 + \beta_1 x_{1i} + \cdots + \beta_m x_{mi} + \epsilon_i,$$

in which some $\beta_j = 0$. The problem in variable selection is to determine which of the β's are nonzero, and to determine "good" estimates of them. Fitting this regression model depends on two factors primarily: The magnitude of β_k and the variability of the associated x_{ki}, the correlations of the x_{*i}'s with each other, and the distribution of ϵ_i. (There are only three here; the magnitude of β_k and the variability of the associated x_{ki} make up only one factor.) Within a fixed distributional family, the variability of ϵ_i provides a scale for measuring the other factors.
This exercise is open-ended and encourages you to use your imagination in designing and conducting your study. As a simple start, however, let all x_{ki} be uncorrelated with each other and have the same variability. Generate them randomly with the same variance. Choose values for the β's, some as zero. Let ϵ_i be sampled independently from a $\mathrm{N}(0, 1)$ distribution. Use forward selection (`step` or `stepAIC` in R) and least angle regression (`lars` in R), and see if the nonzero β's enter the model first.
Summarize these initial findings in a clearly-written report.

Next, still with ϵ_i sampled independently from a N(0, 1) distribution, begin systematic explorations of the effect of different magnitudes of the component terms of the model and differences in the correlations of the x's. You can use the same variable-selection methods, or choose others. Summarize your findings in a clearly-written report.

17.7. Obtain the Weka program and read the introductory usage notes. Next obtain 5 datasets from the UCI Machine Learning Repository. (Almost any ones will suffice; just select the first 5 that you encounter.) See page 628 for the websites.

Now try three variations of classification tree methods and a support vector machine method from Weka on these 5 datasets. How do the methods that you selected compare?

Summarize your findings in a clearly-written report.

17.8. Suppose that you have data on x and y that follow a model of the form

$$y^{p_y} = \beta_0 + \beta_1 x^{p_x} + \epsilon.$$

(Of course you would not know that the data follow this model.) Choose some representative values of the parameters in the model and, for each combination, generate artificial data to study some of the data transformations we have discussed. (This is not a Monte Carlo study; it is just an exercise to yield a better understanding of the methods.) The following R or S-Plus code would generate a useful example:

```
nmod  <- 50
beta0 <- 3
beta1 <- 2
esd   <- 2
xdat  <- runif(nmod,0,2)
px    <- 2
py    <- 2
xmod  <- xdat^px
ymod  <- beta0 + beta1*xmod + esd*rnorm(nmod)
ydat  <- ymod^py
```

The data that would be observed are in `xdat` and `ydat`.

a) Plot the data.

b) Fit the linear model

$$y = \beta_0 + \beta_1 x + \epsilon.$$

Plot the residuals, and notice that a quadratic model seems to be in order.

c) Fit the model

$$y = \beta_0 + \beta_1 x + \beta_2 x^2 + \epsilon.$$

Plot the residuals, and notice that their variance seems to increase with the mean.

d) Fit the model

$$w = \beta_0 + \beta_1 x + \beta_2 x^2 + \epsilon,$$

where $w = y^{1/2}$. Comment on what you have observed.

e) Now, for a grid of values of λ, the model

$$w = \beta_0 + \beta_1 x + \beta_2 x^2 + \epsilon,$$

where $w = (y^\lambda - 1)/\lambda$. Compute a profile likelihood as a function of λ, and select a Box-Cox power transformation that seems appropriate.

f) Apply ACE to x and y (`ac <- ace(xdat,ydat)`). Use the fit to determine a predicted value of y for $x = 1$. (You may have to go through series of smoothing transformations to do this.)

g) Apply AVAS to x and y (`av <- avas(xdat,ydat)`). Use the fit to determine a predicted value of y for $x = 1$. (You may have to go through a series of smoothing transformations to do this.) Compare the AVAS results with the ACE results.

h) Using the original data, apply directly one step of the AVAS algorithm. (The interest here is in computing the variance-stabilizing transformation from the data.)

Summarize your findings in a clearly-written report.

Appendices

A

Monte Carlo Studies in Statistics

We have seen how Monte Carlo methods can be used in statistical inference, including Bayesian methods, Monte Carlo tests, and bootstrap methods. Simulation has also become one of the most important tools in the development of statistical theory and methods. If the properties of an estimator are very difficult to work out analytically, a Monte Carlo study may be conducted to estimate those properties.

Although high-speed computers have helped to expand the usage of Monte Carlo methods, there is a long history of such usage. Stigler (1991) describes Monte Carlo simulations by nineteenth-century scientists and suggests that "simulation, in the modern sense of that term, may be the oldest of the stochastic arts." One of the earliest documented Monte Carlo studies of a statistical procedure was by Erastus Lyman de Forest in the 1870s. Stigler (1978) describes how De Forest studied ways of smoothing a time series by simulating the data using cards drawn from a box.

Another early use of Monte Carlo was the sampling experiment (using biometric data recorded on pieces of cardboard) that led W. S. Gosset to the discovery of the distribution of the t-statistic and the correlation coefficient. (See Student, 1908a, 1908b. Of course, it was R. A. Fisher who later worked out the distributions.)

Often a Monte Carlo study is an informal investigation whose main purpose is to indicate promising research directions. If a "quick and dirty" Monte Carlo study indicates that some method of inference has good properties, it may be worth the time of the research worker in developing the method and perhaps doing the difficult analysis to confirm the results of the Monte Carlo study.

In addition to quick Monte Carlo studies that are mere precursors to analytic work, Monte Carlo studies often provide a significant amount of the available knowledge of the properties of statistical techniques, especially under various alternative models. A large proportion of the articles in the statistical literature include Monte Carlo studies. In recent issues of the *Journal of the*

American Statistical Association, for example, well over half of the articles report on Monte Carlo studies that supported the research.

A.1 Simulation as an Experiment

A simulation study that incorporates a random component is an experiment. The principles of statistical design and analysis apply just as much to a Monte Carlo study as they do to any other scientific experiment. The Monte Carlo study should adhere to the same high standards as any scientific experimentation:

- control;
- reproducibility;
- efficiency;
- careful and complete documentation.

In simulation, *control*, among other things, relates to the fidelity of a *non-random* process to a *random* process. The experimental units are only simulated. Questions about the computer model must be addressed (tests of the random number generators, and so on).

Likewise, *reproducibility* is predicated on good random number generators (or else on equally bad ones!). Portability of the random number generators enhances reproducibility and in fact can allow *strict* reproducibility. Reproducible research also requires preservation and documentation of the computer programs that produced the results.

The principles of good statistical design can improve the efficiency. Use of good designs (e.g., fractional factorials) can allow efficient simultaneous exploration of several factors. Also, there are often many opportunities to reduce the variance (improve the efficiency). Hammersley and Hanscomb (1964, page 8) note,

> ... statisticians were insistent that other experimentalists should design experiments to be as little subject to unwanted error as possible, and had indeed given important and useful help to the experimentalist in this way; but in their own experiments they were singularly inefficient, nay negligent in this respect.

Many properties of statistical methods of inference are analytically intractable. Asymptotic results, which are often easy to work out, may imply excellent performance, such as consistency with a good rate of convergence, but the finite sample properties are ultimately what must be considered. Monte Carlo studies are a common tool for investigating the properties of a statistical method, as noted above. In the literature, the Monte Carlo studies are sometimes called "numerical results". Some numerical results are illustrated by just one randomly generated dataset; others are studied by averaging over thousands of randomly generated sets.

In a Monte Carlo study, there are usually several different things ("treatments" or "factors") that we want to investigate. As in other kinds of experiments, a factorial design is usually more efficient. Each factor occurs at different "levels", and the set of all levels of all factors that are used in the study constitute the "design space". The measured responses are properties of the statistical methods, such as their sample means and variances.

The factors commonly studied in Monte Carlo experiments in statistics include the following.

- statistical method (estimator, test procedure, etc.);
- sample size;
- the problem for which the statistical method is being applied (that is, the "true" model, which may be different from the one for which the method was developed). Factors relating to the type of problem may be:
  - distribution of the random component in the model (normality?);
  - correlation among observations (independence?);
  - homogeneity of the observations (outliers?, mixtures?);
  - structure of associated variables (leverage?).

The factor whose effect is of primary interest is the statistical method. The other factors are generally just blocking factors. There is, however, usually an interaction between the statistical method and these other factors.

As in physical experimentation, observational units are selected for each point in the design space and measured. The measurements, or "responses" made at the same design point, are used to assess the amount of random variation, or variation that is not accounted for by the factors being studied. A comparison of the variation among observational units at the same levels of all factors with the variation among observational units at different levels is the basis for a decision as to whether there are real (or "significant") differences at the different levels of the factors. This comparison is called analysis of variance. The same basic rationale for identifying differences is used in simulation experiments.

A fundamental (and difficult) question in experimental design is how many experimental units to observe at the various design points. Because the experimental units in Monte Carlo studies are generated on the computer, they are usually rather inexpensive. The subsequent processing (the application of the factors, in the terminology of an experiment) may be very extensive, however, so there is a need to design an efficient experiment.

A.2 Reporting Simulation Experiments

The reporting of a simulation experiment should receive the same care and consideration that would be accorded the reporting of other scientific experiments. In addition to a careful general description of the experiment, the report should include mention of the random number generator used, any variance-reducing methods employed, and a justification of the simulation sample size.

Closely related to the choice of the sample size is the standard deviation of the estimates that result from the study. The sample standard deviations actually achieved should be included as part of the report. Standard deviations are often reported in parentheses beside the estimates with which they are associated. A formal analysis, of course, would use the sample variance of each estimate to assess the significance of the differences observed between points in the design space; that is, a formal analysis of the simulation experiment would be a standard analysis of variance.

The most common method of reporting the results is by means of tables, but a better understanding of the results can often be conveyed by graphs.

A.3 An Example

One area of statistics in which Monte Carlo studies have been used extensively is robust statistics. This is because the finite sampling distributions of many robust statistics are very difficult to work out, especially for the kinds of underlying distributions for which the statistics are to be studied.

As an example of a Monte Carlo study, we will now describe a simple experiment to assess the robustness of a statistical test in linear regression analysis. The purpose of this example is to illustrate some of the issues in designing a Monte Carlo experiment. The results of this small study are not of interest here. There are many important issues about the robustness of the procedures that we do not address in this example.

The Problem

Consider the simple linear regression model

$$Y = \beta_0 + \beta_1 x + E$$

where a response or "dependent variable", Y, is modeled as a linear function of a single regressor or "independent variable", x, plus a random variable, E, called the "error". Because E is a random variable, Y is also a random variable. The statistical problem is to make inferences about the unknown, constant parameters, β_0 and β_1, and about distributional parameters of the random variable, E. The inferences are made based on a sample of n pairs, (y_i, x_i), associated with unobservable realizations of the random error, ϵ_i, and assumed to have the relationship

$$y_i = \beta_0 + \beta_1 x_i + \epsilon_i. \tag{A.1}$$

We also generally assume that the realizations of the random error are independent and are unrelated to the value of x.

For this example, let us consider just the specific problem of testing the hypothesis

$$H_0:\ \beta_1 = 0 \tag{A.2}$$

versus the universal alternative. If the distribution of E is normal and we make the additional assumptions above about the sample, the optimal test for the hypothesis is based on a least squares procedure that yields the statistic

$$t = \frac{\widehat{\beta}_1 \sqrt{(n-2)\sum(x_i - \bar{x})^2}}{\sqrt{\sum r_i^2}}, \tag{A.3}$$

where $\bar{x}$ is the mean of the x's, $\widehat{\beta}_1$ together with $\widehat{\beta}_0$ minimizes the function

$$\begin{aligned} L_2(b_0, b_1) &= \sum_{i=1}^{n} (y_i - b_0 - b_1 x_i)^2 \\ &= \|r(b_0, b_1)\|_2^2, \end{aligned}$$

and

$$\begin{aligned} r_i &= r_i(\widehat{\beta}_0, \widehat{\beta}_1) \\ &= y_i - (\widehat{\beta}_0 + \widehat{\beta}_1 x_i) \end{aligned}$$

(in the notation of equation (17.37) on page 604).

If the null hypothesis is true, t is a realization of a Student's t distribution with $n-2$ degrees of freedom. The test is performed by comparing the p-value from the Student's t distribution with a preassigned significance level, α, or by comparing the observed value of t with a critical value. The test of the hypothesis depends on the estimates of β_0 and β_1 used in the test statistic t.

Another method of fitting the linear regression line that is robust to outliers in E is to minimize the absolute values of the deviations. The least absolute values procedure chooses estimates of β_0 and β_1 to minimize the function

$$\begin{aligned} L_1(b_0, b_1) &= \sum_{i=1}^{n} |y_i - b_0 - b_1 x_i| \\ &= \|r(b_0, b_1)\|_1. \end{aligned}$$

A test statistic analogous to the one in equation (A.3), but based on the least absolute values fit, is

$$t_1 = \frac{2\tilde{\beta}_1 \sqrt{\sum(x_i - \bar{x})^2}}{(e_{(k_2)} - e_{(k_1)})\sqrt{n-2}}, \tag{A.4}$$

where $\tilde{\beta}_1$ together with $\tilde{\beta}_0$ minimizes the function

$$L_1(b_0, b_1) = \sum_{i=1}^{n} |y_i - b_0 - b_1 x_i|,$$

$e_{(k)}$ is the kth order statistic from

$$e_i = y_i - (\tilde{\beta}_0 + \tilde{\beta}_1 x_i),$$

k_1 is the integer closest to $(n-1)/2 - \sqrt{n-2}$, and k_2 is the integer closest to $(n-1)/2 + \sqrt{n-2}$. This statistic has an approximate Student's t distribution with $n-2$ degrees of freedom (see Birkes and Dodge, 1993, for example).

If the distribution of the random error is normal, inference based on minimizing the sum of the absolute values is not nearly as efficient as inference based on least squares. This alternative to least squares should therefore be used with some discretion. Furthermore, there are other procedures that may warrant consideration. It is not our purpose here to explore these important issues in robust statistics, however.

The Design of the Experiment

At this point, we should have a clear picture of the problem: We wish to compare two ways of testing the hypothesis (A.2) under various scenarios. The data may have outliers, and there may be observations with large leverage. We expect that the optimal test procedure will depend on the presence of outliers, or more generally, on the distribution of the random error, and on the pattern of the values of the independent variable. The possibilities of interest for the distribution of the random error include:

- the family of the distribution (that is, normal, double exponential, Cauchy, and so on);
- whether the distribution is a mixture of more than one basic distribution and, if so, the proportions in the mixture;
- the values of the parameters of the distribution; that is, the variance, the skewness, or any other parameters that may affect the power of the test.

Our objective is to compare the relative power of two tests *within* different situations. If our objective were to be to compare the powers of the tests *across* those situations, we would need to choose the distributions of the random errors in such a way that they are comparable with respect to any reasonable test; that is, we would need to have models whose signal to noise ratios are essentially the same. In this Monte Carlo study, our interest includes the differences of the tests within different signal to noise ratios.

In textbooks on the design of experiments, a simple objective of an experiment is to perform a t test or an F test of whether different levels of response are associated with different treatments. Our objective in the Monte Carlo experiment that we are designing is to investigate and characterize the dependence of the performance of the hypothesis test on these factors. The principles of design are similar to those of other experiments, however.

It is possible that the optimal test of the hypothesis will depend on the sample size or on the true values of the coefficients in the regression model, so some additional issues that are relevant to the performance of a statistical test of this hypothesis are the sample size and the true values of β_0 and β_1.

In the terminology of statistical models, the factors in our Monte Carlo experiment are the estimation method and the associated test, the distribution of the random error, the pattern of the independent variable, the sample size, and the true value of β_0 and β_1. The estimation method together with the associated test is the "treatment" of interest. The "effect" of interest (that is, the measured response) is the proportion of times that the null hypothesis is rejected using the two treatments.

We now can see our objective more clearly: for each setting of the distribution, pattern, and size factors, we wish to measure the power of the two tests. These factors are similar to blocking factors, except that there is likely to be an interaction between the treatment and these factors. Of course, the power depends on the nominal level of the test, α. It may be the case that the nominal level of the test affects the relative powers of the two tests.

We can think of the problem in the context of a binary response model,

$$\mathrm{E}(P_{ijklqsr}) = f(\tau_i, \delta_j, \phi_k, \nu_l, \alpha_q, \beta_{1s}), \tag{A.5}$$

where the parameters represent levels of the factors listed above (β_{1s} is the sth level of β_1), and $P_{ijklqsr}$ is a binary variable representing whether the test rejects the null hypothesis on the rth trial at the $(ijklqs)^{\text{th}}$ setting of the design factors. It is useful to write down a model like this to remind ourselves of the issues in designing an experiment.

At this point, it is necessary to pay careful attention to our terminology. We are planning to use a statistical procedure (a Monte Carlo experiment) to evaluate a statistical procedure (a statistical test in a linear model). For the statistical procedure that we will use, we have written a model (A.5) for the observations that we will make. Those observations are indexed by r in that model. Let m be the sample size for each combination of factor settings. This is the Monte Carlo sample size. It is not to be confused with the data sample size, n, that is one of the factors in our study.

We now choose the levels of the factors in the Monte Carlo experiment.

- For the estimation method, we have decided on two methods: least squares and least absolute values. Its differential effect in the binary response model (A.5) is denoted by τ_i, for $i = 1, 2$.
- For the distribution of the random error, we choose three general ones:
  1. Normal $(0, 1)$;
  2. Normal $(0, 1)$ with $c\%$ outliers from normal $(0, d^2)$;
  3. Standard Cauchy.

We choose different values of c and d as appropriate. For this example, let us choose $c = 5$ and 20 and $d = 2$ and 5. Thus, in the binary response model (A.5), $j = 1, 2, 3, 4, 5, 6$.

For a given systematic component of the model, these choices result in different signal to noise ratios. If our objective were to be to compare tests across families of distributions, we might want to normalize the variances in some way. (With a Cauchy being one distribution in the study, this would not be possible, of course.)

- For the pattern of the independent variable, we choose three different arrangements:
  1. Uniform over the range;
  2. A group of extreme outliers;
  3. Two groups of outliers.

  In the binary response model (A.5), $k = 1, 2, 3$. We use fixed values of the independent variable.
- For the sample size, we choose three values: 20, 200, and 2,000. In the binary response model (A.5), $l = 1, 2, 3$.
- For the nominal level of the test, we choose two values: 0.01 and 0.05. In the binary response model (A.5), $q = 1, 2$.
- The true value of β_0 probably is not relevant, so we just choose $\beta_0 = 1$. We are interested in the power of the tests at different values of β_1. We expect the power function to be symmetric about $\beta_1 = 0$ and to approach 1 as $|\beta_1|$ increases.

The estimation method is the "treatment" of interest.

Restating our objective in terms of the notation introduced above, for each of two tests, we wish to estimate the power curve,

$$\Pr(\text{reject } H_0) = g(\beta_1 \mid \tau_i, \delta_j, \phi_k, \nu_l, \alpha_q),$$

for any combination $(\tau_i, \delta_j, \phi_k, \nu_l, \alpha_q)$. The minimum of the power curve should occur at $\beta_1 = 0$, and should be α. The curve should approach 1 symmetrically as $|\beta_1|$.

To estimate the curve, we use a discrete set of points, and because of symmetry, all values chosen for β_1 can be nonnegative. The first question is at what point the curve flattens out just below 1. We might arbitrarily define the region of interest to be that in which the power is less than 0.99 approximately. The abscissa of this point is the maximum β_1 of interest. This point, say β_1^*, varies, depending on all of the factors in the study. We could work this out in the least squares case for uncontaminated normal errors using the noncentral Student's t distribution, but for all other cases, it is analytically intractable. Hence, we compute some preliminary Monte Carlo estimates to determine the maximum β_1 for each factor combination in the study.

To do a careful job of fitting a curve using a relatively small number of points, we would choose points where the second derivative is changing rapidly and especially near points of inflection where the second derivative changes sign. Because the problem of determining these points for each combination of (i, j, k, l, q) is not analytically tractable (otherwise we would not be doing the study!), we may conveniently choose a set of points equally spaced between 0 and β_1^*. Let us decide on five such points, for this example. It is not important that the β_1^*'s be chosen with a great deal of care. The objective is that we be able to calculate two power curves between 0 and β_1^* that are meaningful for comparisons.

The Experiment

The observational units in the experiment are the values of the test statistics (A.3) and (A.4). The measurements are the binary variables corresponding to rejection of the hypothesis (A.2). At each point in the factor space, there will be m such observations. If z is the number of rejections observed, the estimate of the power is z/m, and the variance of the estimator is $\pi(1-\pi)/m$, where π is the true power at that point. (z is a realization of a binomial random variable with parameters m and π.) This leads us to a choice of the value of m. The coefficient of variation at any point is $\sqrt{(1-\pi)/(m\pi)}$, which increases as π decreases. At $\pi = 0.50$, a 5% coefficient of variation can be achieved with a sample of size 400. This yields a standard deviation of 0.025. There may be some motivation to choose a slightly larger value for m because we can assume that the minimum of π will be approximately the minimum of α. To achieve a 5% coefficient of variation at that point (i.e., at $\beta_1 = 0$) would require a sample of size approximately 160,000. That would correspond to a standard deviation of 0.0005, which is probably much smaller than we need. A sample size of 400 would yield a standard deviation of 0.005. Although that is large in a relative sense, it may be adequate for our purposes. Because this particular point (where $\beta_1 = 0$) corresponds to the null hypothesis, however, we may choose a larger sample size, say 4,000, at that special point. A reasonable choice therefore is a Monte Carlo sample size of 4,000 at the null hypothesis and 400 at all other points. We will, however, conduct the experiment in such a way that we can combine the results of this experiment with independent results from a subsequent experiment.

The experiment is conducted by running a computer program. The main computation in the program is to determine the values of the test statistics and to compare them with their critical values to decide on the hypothesis. These computations need to be performed at each setting of the factors and for any given realization of the random sample.

We design a program that allows us to loop through the settings of the factors and, at each factor setting, to use a random sample. The result is a nest of loops. The program may be stopped and restarted, so we need to be able to control the seeds (see Section 11.6, page 429).

Recalling that the purpose of our experiment is to obtain estimates, we may now consider any appropriate methods of reducing the variance of those estimates. There is not much opportunity to apply methods of variance reduction, but at least we might consider at what points to use common realizations of the pseudorandom variables. Because the things that we want to compare most directly are the powers of the tests, we perform the tests on the same pseudorandom datasets. Also, because we are interested in the shape of the power curves we may want to use the same pseudorandom datasets at each value of β_1 — that is, to use the same set of errors in the model (A.1). Finally, following similar reasoning, we may use the same pseudorandom datasets at each value of the pattern of the independent variable. This implies that our program of nested loops has the structure shown in Figure A.1.

Initialize a table of counts.
- Fix the data sample size. (Loop over the sample sizes $n = 20$, $n = 200$, and $n = 2,000$.)
  - Generate a set of residuals for the linear regression model (A.1). (This is the loop of m Monte Carlo replications.)
    - Fix the pattern of the independent variable. (Loop over patterns P_1, P_2, and P_3.)
      - Choose the distribution of the error term. (Loop over the distributions D_1, D_2, D_3, D_4, D_5, and D_6.)
        - For each value of β_1, generate a set of observations (the y values) for the linear regression model (A.1), and perform the tests using both procedures and at both levels of significance. Record results.
      - End distributions loop.
    - End patterns loop.
  - End Monte Carlo loop.
- End sample size loop.

Perform computations of summary statistics.

Fig. A.1. Program Structure for the Monte Carlo Experiment

After writing a computer program with this structure, the first thing is to test the program on a small set of problems and to determine appropriate values of β_1^*. We should compare the results with known values at a few points. (As mentioned earlier, the only points that we can work out correspond to the normal case with the ordinary t statistic. One of these points, at $\beta_1 = 0$, is easily checked.) We can also check the internal consistency of the results. For example, does the power curve increase? We must be careful, of course, in applying such consistency checks because we do not know the behavior of the tests in most cases.

Reporting the Results

The report of this Monte Carlo study should address as completely as possible the results of interest. The relative values of the power are the main points of interest. The estimated power at $\beta_1 = 0$ is of interest. This is the actual significance level of the test, and how it compares to the nominal level α is of particular interest.

The presentation should be in a form easily assimilated by the reader. This may mean several graphs. Two graphs, for the two test procedures, should be shown on the same set of axes. It is probably counterproductive to show a graph for each factor setting. (There are 108 combinations of factor settings.)

In addition to the graphs, tables may allow presentation of a large amount of information in a compact format.

The Monte Carlo study should be described so carefully that the study could be replicated exactly. This means specifying the factor settings, the loop nesting, the software and computer used, the seed used, and the Monte Carlo sample size. There should also be at least a simple statement explaining the choice of the Monte Carlo sample size.

As mentioned earlier, the statistical literature is replete with reports of Monte Carlo studies. Some of these reports (and likely the studies themselves) are woefully deficient. An example of a careful Monte Carlo study and a good report of the study are given by Kleijnen (1977). He designed, performed, and reported a Monte Carlo study to investigate the robustness of a multiple-ranking procedure. In addition to reporting on the study of the question at hand, another purpose of the paper was to illustrate the methods of a Monte Carlo study.

A.4 Computer Experiments

In many scientific investigations, we envision a relationship expressed by a model

$$y \approx f(x).$$

The quantity of interest y, usually called a "response" (although it may not be a response to any of the other entities), may be the growth of a crystal, the growth of a tumor, the growth of corn, the price of a stock one month hence, or some other quantity. The other variables, x, called "factors", "regressors", or just "input variables", may be temperature, pressure, amount of a drug, amount of a type of fertilizer, interest rates, or some other quantity. Both y and x may be vectors. An objective is to determine a suitable form of f and the nature of the approximation. The simplest type of approximation is one in which an additive deviation can be identified with a random variable:

$$Y = f(x) + E.$$

The most important objective, whatever the nature of the approximation, usually is to determine values of x that are associated with optimal realizations of Y. The association may or may not be one of causation.

One of the major contributions of the science of statistics to the scientific method is the experimental methods that efficiently help to determine f, the nature of an unexplainable deviation E, and the values of x that yield optimal values of y. Design and analysis of experiments is a fairly mature subdiscipline of statistics.

In computer experiments, the function f is a computer program, x is the input, and y is the output. The program implements known or supposed relationships among the phenomena of interest. In cases of practical interest, the function is very complicated, the number of input variables may be in the hundreds, and the output may consist of many elements. The objective is to find a tractable function, $\widehat{f}$, that approximates the true behavior, at least over ranges of interest, and to find the values of the input, say $\widehat{x}_0$, such that $\widehat{f}(\widehat{x}_0)$ is optimal. How useful $\widehat{x}_0$ is depends on how close $\widehat{f}(\widehat{x}_0)$ is to $f(x_0)$, where x_0 yields the optimal value of f.

What makes this an unusual statistical problem is that the relationships are deterministic. The statistical approach to computer experiments introduces randomness into the problem. The estimate $\widehat{f}(\widehat{x}_0)$ can then be described in terms of probabilities or variances.

In a Bayesian approach, randomness is introduced by considering the function f to be a realization of a random function, F. The prior on F may be specified only at certain points, say $F(x_0)$. A set of input vectors $x_1, \ldots, x_n$ is chosen, and the output $y_i = f(x_i)$ is used to estimate a posterior distribution for $F(x)$, or at least for $F(x_0)$. The Bayesian approach generally involves extensive computations for the posterior densities. In a frequentist approach, randomness is introduced by taking random values of the input, $x_1, \ldots, x_n$. This randomness in the input yields randomness in the output $y_i = f(x_i)$, which is used to obtain the estimates $\widehat{x}_0$ and $\widehat{f}(\widehat{x}_0)$ and estimates of the variances of the estimators.

Latin Hypercube Sampling

Principles for the design of experiments provide a powerful set of tools for reducing the variance in cases where several factors are to be investigated simultaneously. Such techniques as balanced or partially balanced fractional factorial designs allow the study of a large number of factors while keeping the total experiment size manageably small. Some processes are so complex that even with efficient statistical designs, experiments to study the process would involve a prohibitively large number of factors and levels. For some processes, it may not be possible to apply the treatments whose effects are to be studied, and data are available only from observational studies. The various processes determining weather are examples of phenomena that are not amenable to traditional experimental study. Such processes can often be modeled and studied by computer experiments.

There are some special concerns in experimentation using the computer, but the issues of statistical efficiency remain. Rather than a model involving ordinary experimental units, a computer experimental model receives a fixed input and produces a deterministic output. An objective in computer experimentation (just as in any experimentation) is to provide a set of inputs that effectively (or randomly) span a space of interest. McKay, Conover, and Beckman (1979) introduce a technique called Latin hypercube sampling (as a generalization of the ideas of a Latin square design) for providing input to a computer experiment.

If each of k factors in an experiment is associated with a random input that is initially a U$(0,1)$ variate, a sample of size n that efficiently covers the factor space can be formed by selecting the ith realization of the jth factor as

$$v_j = \frac{\pi_j(i) - 1 + u_j}{n},$$

where

- $\pi_1(\cdot), \pi_2(\cdot), \ldots, \pi_k(\cdot)$ are permutations of the integers $1, \ldots, n$, sampled randomly, independently, and with replacement from the set of $n!$ possible permutations; and $\pi_j(i)$ is the ith element of the jth permutation.
- The u_j are sampled independently from U$(0,1)$.

It is easy to see that v_j are independent U$(0,1)$. We can see heuristically that such numbers tend to be "spread out" over the space. Use of Latin hypercube sampling is particularly useful in higher dimensions.

Exercises

A.1. Write a computer program to implement the Monte Carlo experiment described in Section A.3. The R functions `lsfit` and `rq` in the `quantreg` package, or the IMSL Fortran subroutines `rline` and `rlav` can be used to calculate the fits. See Section 7.6 for discussions of other software you may use in the program.

A.2. Choose a recent issue of the *Journal of the American Statistical Association* and identify five articles that report on Monte Carlo studies of statistical methods. In each case, describe the Monte Carlo experiment.

  a) What are the factors in the experiment?
  b) What is the measured response?
  c) What is the design space — that is, the set of factor settings?
  d) What random number generators were used?
  e) Critique the report in each article. Did the author(s) justify the sample size? Did the author(s) report variances or confidence intervals? Did the author(s) attempt to reduce the experimental variance?

A.3. Select an article you identified in Exercise A.2 that concerns a statistical method that you understand and that interests you. Choose a design space that is not a subspace of that used in the article but that has a nonnull intersection with it, and perform a similar experiment. Compare your results with those reported in the article.

B

Some Important Probability Distributions

Development of stochastic models is facilitated by identifying a few probability distributions that seem to correspond to a variety of data-generating processes, and then studying the properties of these distributions. In the following tables, I list some of the more useful distributions, both discrete distributions and continuous ones. The names listed are the most common names, although some distributions go by different names, especially for specific values of the parameters. In the first column, following the name of the distribution, the parameter space is specified. Also, given in the first column is the root name of the computer routines in both R and IMSL that apply to the distribution. In the last column, the PDF (or probability mass function) and the mean and variance are given.

There are two very special continuous distributions, for which I use special symbols: the unit uniform, designated $\mathrm{U}(0,1)$, and the normal (or Gaussian), denoted by $\mathrm{N}(\mu, \sigma^2)$. Notice that the second parameter in the notation for the normal is the variance. Sometimes, such as in the functions in R, the second parameter of the normal distribution is the standard deviation instead of the variance. A normal distribution with $\mu = 0$ and $\sigma^2 = 1$ is called the standard normal. I also often use the notation $\phi(x)$ for the PDF of a standard normal and $\Phi(x)$ for the CDF of a standard normal, and these are generalized in the obvious way as $\phi(x|\mu, \sigma^2)$ and $\Phi(x|\mu, \sigma^2)$.

Except for the uniform and the normal, I designate distributions by a name followed by symbols for the parameters, for example, binomial(n, π) or gamma(α, β). Some families of distributions are subfamilies of larger families. For example, the usual gamma family of distributions is a the two-parameter subfamily of the three-parameter gamma.

There are other general families of probability distributions that are defined in terms of a differential equation or of a form for the CDF. These include the Pearson, Johnson, Burr, and Tukey's lambda distributions (see Section 14.2). Some families of distributions are particularly useful in Monte Carlo studies of the performance of statistical methods when the usual assumptions are violated. The t family may be useful in this connection, because as the degrees of freedom ranges from 1 to ∞, the tails of the distribution go from very heavy (that is, it is an outlier-generating distribution) to the standard normal. Other families that are useful in this same way are the skew-normal (see Gentle, 2003, page 170) and the stable family (*ibidem*, page 196). The "claw density", "smooth comb density", and "saw-tooth density", mentioned on page 508, are useful in Monte Carlo studies of the performance of nonparametric probability density estimators.

Most of the common distributions fall naturally into one of two classes. They have either a countable support with positive probability at each point in the support, or a continuous (dense, uncountable) support with zero probability for any subset with zero Lebesgue measure. The distributions listed in the following tables are divided into these two natural classes.

There are situations for which these two distinct classes are not appropriate. For most such situations, however, a mixture distribution provides an appropriate model. We can express a PDF of a mixture distribution as in equation (1.90) as

$$p_M(y) = \sum_{j=1}^{m} \omega_j p_j(y \mid \theta_j),$$

where the m distributions with PDFs p_j can be either discrete or continuous. A simple example is a probability model for the amount of rainfall in a given period, say a day. It is likely that a nonzero probability should be associated with zero rainfall, but with no other amount of rainfall. In the model above, m is 2, ω_1 is the probability of no rain, p_1 is a degenerate PDF with a value of 1 at 0, $\omega_2 = 1 - \omega_1$, and p_2 is some continuous PDF over $(0, \infty)$, possibly similar to an exponential distribution.

Another example of a mixture distribution is a binomial with constant parameter n, but with a nonconstant parameter π. In many applications, if an identical binomial distribution is assumed (that is, a constant π), it is often the case that "over-dispersion" will be observed; that is, the sample variance exceeds what would be expected given an estimate of some other parameter that determines the population variance. This situation can occur in a model, such as the binomial, in which a single parameter determines both the first and second moments. The mixture model above in which each p_j is a binomial PDF with parameters n and π_j may be a better model.

Of course, we can extend this kind of mixing even further. Instead of $\omega_j p_j(y \mid \theta_j)$ with $\omega_j \geq 0$ and $\sum_{j=1}^{m} \omega_j = 1$, we can take $\omega(\theta) p(y \mid \theta)$ with $\omega(\theta) \geq 0$ and $\int \omega(\theta)\, \mathrm{d}\theta = 1$, from which we recognize that $\omega(\theta)$ is a PDF and θ can be considered to be the realization of a random variable.

Extending the example of the mixture of binomial distributions, we may choose some reasonable PDF $\omega(\pi)$. An obvious choice is a beta PDF. This yields a standard distribution that is not included in the tables below, the *beta-binomial distribution*, with PDF

$$p_{X,\Pi}(x, \pi) = \binom{n}{x} \frac{\Gamma(\alpha+\beta)}{\Gamma(\alpha)\Gamma(\beta)} \pi^{x+\alpha-1}(1-\pi)^{n-x+\beta-1}.$$

This distribution may be useful in situations in which a binomial model is appropriate, but the probability parameter is changing more-or-less continuously.

We recognize a basic property of any mixture distribution: It is a joint distribution factored as a marginal (prior) for a random variable, which is often not observable, and a conditional distribution for another random variable, which is usually the observable variable of interest.

In Bayesian analyses, the first two assumptions (a prior distribution for the parameters and a conditional distribution for the observable) lead immediately to a mixture distribution. The beta-binomial above arises in a canonical example of Bayesian analysis.

Evans, Hastings, and Peacock (2000) give general descriptions of 40 probability distributions. Leemis and McQueston (2008) provide an interesting compact graph of the relationships among a large number of probability distributions.

Currently, the most readily accessible summary of common probability distributions is Wikipedia: `http://wikipedia.org/` Search under the name of the distribution.

Table B.1. Discrete Distributions (PDFs are w.r.t counting measure)

discrete uniform	PDF	$\frac{1}{m}$, $y = a_1, \ldots, a_m$
$a_1, \ldots, a_m \in \mathbb{R}$	mean	$\sum a_i/m$
R: `sample`; IMSL: `und`	variance	$\sum(a_i - \bar{a})^2/m$, where $\bar{a} = \sum a_i/m$
binomial	PDF	$\binom{n}{y}\pi^y(1-\pi)^{n-y}$, $y = 0, 1, \ldots, n$
$n = 1, 2, \ldots;$ $\pi \in (0, 1)$	mean	$n\pi$
R: `binom`; IMSL: `bin`	variance	$n\pi(1 - \pi)$
Bernoulli	PDF	$\pi^y(1-\pi)^{1-y}$, $y = 0, 1$
$\pi \in (0, 1)$	mean	π
(special binomial)	variance	$\pi(1 - \pi)$
Poisson	PDF	$\theta^y e^{-\theta}/y!$, $y = 0, 1, 2, \ldots$
$\theta > 0$	mean	θ
R: `pois`; IMSL: `poi`	variance	θ
hypergeometric	PDF	$\dfrac{\binom{M}{y}\binom{L-M}{N-y}}{\binom{L}{N}}$, $y = \max(0, N - L + M), \ldots, \min(N, M)$
$L = 1, 2, \ldots;$	mean	NM/L
$M = 1, 2, \ldots, L$; $N = 1, 2, \ldots, L$	variance	$((NM/L)(1 - M/L)(L - N))/(L - 1)$
R: `hyper`; IMSL: `hyp`		
negative binomial	PDF	$\binom{y+r-1}{r-1}\pi^r(1-\pi)^y$, y=0,1,2,...
$r > 0$; $\pi \in (0, 1)$	mean	$r(1 - \pi)/\pi$
R: `nbinom`; IMSL: `nbn`	variance	$r(1 - \pi)/\pi^2$
geometric	PDF	$\pi(1 - \pi)^y$, y=0,1,2,...
$\pi \in (0, 1)$	mean	$(1 - \pi)/\pi$
(special negative binomial)	variance	$(1 - \pi)/\pi^2$
logarithmic	PDF	$-\dfrac{\pi^y}{y\log(1-\pi)}$, y=1,2,3,...
$\pi \in (0, 1)$	mean	$-\pi/((1 - \pi)\log(1 - \pi))$
IMSL: `lgr`	variance	$-\pi(\pi + \log(1 - \pi))/((1 - \pi)^2(\log(1 - \pi))^2$
multinomial	PDF	$\dfrac{n!}{\prod \pi_i!}\prod \pi_i^{y_i}$, $y_i = 0, 1, \ldots, n$, $\sum y_i = n$
$n = 1, 2, \ldots$, $\pi_i \in (0, 1)$, $\sum \pi_i = 1$	means	$n\pi_i$
R: `multinom`; IMSL: `mtn`	variances	$n\pi_i(1 - \pi_i)$
	covariances	$-n\pi_i\pi_j$

Table B.2. Continuous Distributions (PDFs are w.r.t Lebesgue measure)

uniform	PDF	$\frac{1}{\theta_2 - \theta_1} \mathrm{I}_{(\theta_1,\theta_2)}(y)$
$\theta_1 < \theta_2 \in \mathbb{R}$	mean	$(\theta_2 + \theta_1)/2$
R: `unif`; IMSL: `unf`	variance	$(\theta_2^2 - 2\theta_1\theta_2 + \theta_1^2)/12$
normal	PDF	$\frac{1}{\sqrt{2\pi}\sigma} \mathrm{e}^{-(y-\mu)^2/2\sigma^2}$
$\mu \in \mathbb{R}$; $\sigma > 0 \in \mathbb{R}$	mean	μ
R: `norm`; IMSL: `nor`	variance	σ^2
multivariate normal	PDF	$\frac{1}{(2\pi)^{d/2}\lvert\Sigma\rvert^{1/2}} \mathrm{e}^{-(y-\mu)^{\mathrm{T}}\Sigma^{-1}(y-\mu)/2}$
$\mu \in \mathbb{R}^d$; $\Sigma \succ 0 \in \mathbb{R}^{d \times d}$	mean	μ
R: `mvrnorm`; IMSL: `mvn`	covariance	Σ
chi-squared	PDF	$\frac{1}{\Gamma(\nu/2)2^{\nu/2}} y^{\nu/2-1} \mathrm{e}^{-y/2} \mathrm{I}_{(0,\infty)}(y)$
$\nu > 0$	mean	ν
R: `chisq`; IMSL: `chi`	variance	2ν
t	PDF	$\frac{\Gamma((\nu+1)/2)}{\Gamma(\nu/2)\sqrt{\nu\pi}} (1 + y^2/\nu)^{-(\nu+1)/2}$
$\nu > 0$	mean	0
R: `t`; IMSL: `stt`	variance	$\nu/(\nu - 2)$, for $\nu > 2$
F	PDF	$\frac{\nu_1^{\nu_1/2}\nu_2^{\nu_2/2}\Gamma(\nu_1+\nu_2)y^{\nu_1/2-1}}{\Gamma(\nu_1/2)\Gamma(\nu_2/2)(\nu_2+\nu_1 y)^{(\nu_1+\nu_2)/2}} \mathrm{I}_{(0,\infty)}(y)$
$\nu_1 > 0$; $\nu_2 > 0$	mean	$\nu_2/(\nu_2 - 2)$, for $\nu_2 > 2$
R: `f`; IMSL: `f`	variance	$2\nu_2^2(\nu_1 + \nu_2 - 2)/(\nu_1(\nu_2 - 2)^2(\nu_2 - 4))$, for $\nu_2 > 4$
lognormal	PDF	$\frac{1}{\sqrt{2\pi}\sigma} y^{-1} \mathrm{e}^{-(\log(y)-\mu)^2/2\sigma^2} \mathrm{I}_{(0,\infty)}(y)$
$\mu \in \mathbb{R}$; $\sigma > 0 \in \mathbb{R}$	mean	$\mathrm{e}^{\mu+\sigma^2/2}$
R: `lnorm`; IMSL: `lnl`	variance	$\mathrm{e}^{2\mu+\sigma^2}(\mathrm{e}^{\sigma^2} - 1)$
gamma	PDF	$\frac{1}{\Gamma(\alpha)\beta^\alpha} y^{\alpha-1} \mathrm{e}^{-y/\beta} \mathrm{I}_{(0,\infty)}(y)$
$\alpha > 0$, $\beta > 0 \in \mathbb{R}$	mean	$\alpha\beta$
R: `gamma`; IMSL: `gam`	variance	$\alpha\beta^2$
exponential	PDF	$\lambda\mathrm{e}^{-\lambda y} \mathrm{I}_{(0,\infty)}(y)$
$\lambda > 0 \in \mathbb{R}$	mean	$1/\lambda$
R: `exp`; IMSL: `exp`	variance	$1/\lambda^2$
double exponential	PDF	$\frac{1}{2}\lambda\mathrm{e}^{-\lambda\lvert y-\mu\rvert}$
$\mu \in \mathbb{R}$; $\lambda > 0 \in \mathbb{R}$	mean	μ
(folded exponential)	variance	$2/\lambda^2$

Table B.2. Continuous Distributions (continued)

Weibull	PDF	$\frac{\alpha}{\beta} y^{\alpha-1} e^{-y^\alpha/\beta}\, \mathrm{I}_{(0,\infty)}(y)$
$\alpha > 0,\ \beta > 0 \in \mathrm{IR}$	mean	$\beta^{1/\alpha}\Gamma(\alpha^{-1}+1)$
R: `weibull`; IMSL: `wib`	variance	$\beta^{2/\alpha}\left(\Gamma(2\alpha^{-1}+1) - (\Gamma(\alpha^{-1}+1))^2\right)$
Cauchy	PDF	$\frac{1}{\pi\beta\left(1+\left(\frac{y-\gamma}{\beta}\right)^2\right)}$
$\gamma \in \mathrm{IR};\ \beta > 0 \in \mathrm{IR}$	mean	does not exist
R: `cauchy`; IMSL: `chy`	variance	does not exist
beta	PDF	$\frac{\Gamma(\alpha+\beta)}{\Gamma(\alpha)\Gamma(\beta)} y^{\alpha-1}(1-y)^{\beta-1}\, \mathrm{I}_{(0,1)}(y)$
$\alpha > 0,\ \beta > 0 \in \mathrm{IR}$	mean	$\alpha/(\alpha+\beta)$
R: `beta`; IMSL: `beta`	variance	$\alpha\beta/((\alpha+\beta)^2(\alpha+\beta+1))$
logistic	PDF	$\frac{e^{-(y-\mu)/\beta}}{\beta(1+e^{-(y-\mu)/\beta})^2}$
$\mu \in \mathrm{IR};\ \beta > 0 \in \mathrm{IR}$	mean	μ
R: `logis`	variance	$\beta^2\pi^2/3$
Pareto	PDF	$\frac{\alpha\gamma^\alpha}{y^{\alpha+1}}\, \mathrm{I}_{(\gamma,\infty)}(y)$
$\alpha > 0,\ \gamma > 0 \in \mathrm{IR}$	mean	$\alpha\gamma/(\alpha-1)$ for $\alpha > 1$
	variance	$\alpha\gamma^2/((\alpha-1)^2(\alpha-2))$ for $\alpha > 2$
von Mises	PDF	$\frac{1}{2\pi \mathrm{I}_0(\kappa)} e^{\kappa\cos(x-\mu)}\, \mathrm{I}_{(\mu-\pi,\mu+\pi)}(y)$
$\mu \in \mathrm{IR};\ \kappa > 0 \in \mathrm{IR}$	mean	μ
IMSL: `vms`	variance	$1-(\mathrm{I}_1(\kappa)/\mathrm{I}_0(\kappa))^2$

C

Notation and Definitions

All notation used in this work is "standard". I have opted for simple notation, which, of course, results in a one-to-many map of notation to object classes. Within a given context, however, the overloaded notation is generally unambiguous. I have endeavored to use notation consistently.

This appendix is not intended to be a comprehensive listing of definitions. The Subject Index, beginning on page 715, is a more reliable set of pointers to definitions, except for symbols that are not words.

C.1 General Notation

Uppercase italic Latin and Greek letters, A, B, E, Λ, and so on are generally used to represent either matrices or random variables. Random variables are usually denoted by letters nearer the end of the Latin alphabet, X, Y, Z, and by the Greek letter E. Parameters in models (that is, unobservables in the models), whether or not they are considered to be random variables, are generally represented by lowercase Greek letters. Uppercase Latin and Greek letters, especially P, in general, and Φ, for the normal distribution, are also used to represent cumulative distribution functions. Also, uppercase Latin letters are used to denote sets.

Lowercase Latin and Greek letters are used to represent ordinary scalar or vector variables and functions. **No distinction in the notation is made between scalars and vectors**; thus, β may represent a vector and β_i may represent the ith element of the vector β. In another context, however, β may represent a scalar. All vectors are considered to be column vectors, although we may write a vector as $x = (x_1, x_2, \ldots, x_n)$. Transposition of a vector or a matrix is denoted by a superscript ^{T}.

Uppercase calligraphic Latin letters, $\mathcal{F}$, $\mathcal{V}$, $\mathcal{W}$, and so on, are generally used to represent either vector spaces or transforms.

Subscripts generally represent indexes to a larger structure; for example, x_{ij} may represent the $(i, j)^{\text{th}}$ element of a matrix, X. A subscript in parentheses represents an order statistic. A superscript in parentheses represents an iteration: for example, $x_i^{(k)}$ may represent the value of x_i at the kth step of an iterative process.

x_i — The ith element of a structure (including a sample, which is a multiset).

$x_{(i)}$ — The ith order statistic.

$x^{(i)}$ — The value of x at the ith iteration.

Realizations of random variables and placeholders in functions associated with random variables are usually represented by lowercase letters corresponding to the uppercase letters; thus, ϵ may represent a realization of the random variable E.

A single symbol in an italic font is used to represent a single variable. A Roman font or a special font is often used to represent a standard operator or a standard mathematical structure. Sometimes, a string of symbols in a Roman font is used to represent an operator (or a standard function); for example, exp represents the exponential function, but a string of symbols in an italic font on the same baseline should be interpreted as representing a composition (probably by multiplication) of separate objects; for example, exp represents the product of e, x, and p.

A fixed-width font is used to represent computer input or output; for example,

```
a = bx + sin(c).
```

In computer text, a string of letters or numerals with no intervening spaces or other characters, such as `bx` above, represents a single object, and there is no distinction in the font to indicate the type of object.

Some important mathematical structures and other objects are:

$\mathbb{R}$ — The field of reals, or the set over which that field is defined.

$\mathbb{R}^d$ — The usual d-dimensional vector space over the reals, or the set of all d-tuples with elements in $\mathbb{R}$.

$\mathbb{R}^d_+$ — The usual d-dimensional vector space over the reals, or the set of all d-tuples with positive real elements.

$\mathbb{C}$ — The field of complex numbers, or the set over which that field is defined.

$\mathbb{Z}$ The ring of integers, or the set over which that ring is defined.

$C^0, C^1, C^2, \ldots$ The set of continuous functions, the set of functions with continuous first derivatives, and so forth.

i The imaginary unit, $\sqrt{-1}$.

C.2 Computer Number Systems

Computer number systems are used to simulate the more commonly used number systems. It is important to realize that they have different properties, however. Some notation for computer number systems follows.

$\mathbb{F}$ The set of floating-point numbers with a given precision, on a given computer system, or this set together with the four operators +, -, *, and /. ($\mathbb{F}$ is similar to $\mathbb{R}$ in some useful ways; see Chapter 2 and Table 2.1 on page 98.)

$\mathbb{I}$ The set of fixed-point numbers with a given length, on a given computer system, or this set together with the four operators +, -, *, and /. ($\mathbb{I}$ is similar to $\mathbb{Z}$ in some useful ways.)

e_{min} and e_{max} The minimum and maximum values of the exponent in the set of floating-point numbers with a given length.

ϵ_{min} and ϵ_{max} The minimum and maximum spacings around 1 in the set of floating-point numbers with a given length.

ϵ or ϵ_{mach} The machine epsilon, the same as ϵ_{min}.

$[\cdot]_c$ The computer version of the object $\cdot$.

NA Not available; a missing-value indicator.

NaN Not-a-number.

C.3 Notation Relating to Random Variables

A common function used with continuous random variables is a *density function*, and a common function used with discrete random variables is a *probability function.* The more fundamental function for either type of random variable is the *cumulative distribution function*, or CDF. The CDF of a random variable X, denoted by $P_X(x)$ or just by $P(x)$, is defined by

$$P(x) = \Pr(X \leq x),$$

where "Pr", or "probability", can be taken here as a primitive (it is defined in terms of a measure). For vectors (of the same length), "$X \leq x$" means that each element of X is less than or equal to the corresponding element of x. Both the CDF and the density or probability function for a d-dimensional random variable are defined over IR^d. (It is unfortunately necessary to state that "$P(x)$" means the "function P evaluated at x", and likewise "$P(y)$" means the *same* "function P evaluated at y", unless P has been redefined. Using a different expression as the argument *does not redefine* the function, despite the sloppy convention adopted by some statisticians.)

The density for a continuous random variable is just the derivative of the CDF (if it exists). The CDF is therefore the integral. To keep the notation simple, we likewise consider the probability function for a discrete random variable to be a type of derivative (a Radon-Nikodym derivative) of the CDF. Instead of expressing the CDF of a discrete random variable as a sum over a countable set, we often also express it as an integral. (In this case, however, the integral is over a set whose ordinary Lebesgue measure is 0.)

A useful analog of the CDF for a random sample is the *empirical cumulative distribution function*, or ECDF. For a sample of size n, the ECDF is

$$P_n(x) = \frac{1}{n}\sum_{i=1}^{n} \mathrm{I}_{(-\infty,x]}(x_i)$$

or, equivalently,

$$P_n(x) = \frac{1}{n}\sum_{i=1}^{n} \mathrm{I}_{[x_i,\infty)}(x).$$

See page 669 for definition of the indicator function I.

Functions and operators such as Cov and E that are commonly associated with Latin letters or groups of Latin letters are generally represented by that letter in a Roman font.

$\Pr(A)$	The probability of the event A.
$p_X(\cdot)$ or $P_X(\cdot)$	The probability density function (or probability function), or the cumulative probability function, of the random variable X.

$p_{XY}(\cdot)$ or $P_{XY}(\cdot)$	The joint probability density function (or probability function), or the joint cumulative probability function, of the random variables X and Y.
$p_{X\|Y}(\cdot)$ or $P_{X\|Y}(\cdot)$	The conditional probability density function (or probability function), or the conditional cumulative probability function, of the random variable X given the random variable Y (these functions are random variables).
$p_{X\|y}(\cdot)$ or $P_{X\|y}(\cdot)$	The conditional probability density function (or probability function), or the conditional cumulative probability function, of the random variable X given the realization y.
	Sometimes, the notation above is replaced by a similar notation in which the arguments indicate the nature of the distribution; for example, $p(x, y)$ or $p(x\|y)$.
$p_\theta(\cdot)$ or $P_\theta(\cdot)$	The probability density function (or probability function), or the cumulative probability function, of the distribution characterized by the parameter θ.
$Y \sim D_X(\theta)$	The random variable Y is distributed as $D_X(\theta)$, where X is the name of a random variable associated with the distribution, and θ is a parameter of the distribution. The subscript may take forms similar to those used in the density and distribution functions, such as $X\|y$, or it may be omitted. Alternatively, in place of D_X, a symbol denoting a specific distribution may be used. An example is $Z \sim \mathrm{N}(0, 1)$, which means that Z has a normal distribution with mean 0 and variance 1.
CDF	A cumulative distribution function.
ECDF	An empirical cumulative distribution function.
i.i.d.	Independent and identically distributed.
$X^{(i)} \xrightarrow{d} X$ or $X_i \xrightarrow{d} X$	The sequence of random variables $X^{(i)}$ or X_i converges in distribution to the random variable X. (The difference in the notation $X^{(i)}$ and X_i is generally unimportant. The former notation is often used to emphasize the iterative nature of a process.)

$\mathrm{E}(g(X))$ The expected value of the function g of the random variable X. The notation $\mathrm{E}_P(\cdot)$, where P is a cumulative distribution function or some other identifier of a probability distribution, is sometimes used to indicate explicitly the distribution with respect to which the expectation is evaluated.

$\mathrm{V}(g(X))$ The variance of the function g of the random variable X. The notation $\mathrm{V}_P(\cdot)$ is also often used.

$\mathrm{Cov}(X, Y)$ The covariance of the random variables X and Y. The notation $\mathrm{Cov}_P(\cdot, \cdot)$ is also often used.

$\mathrm{Cov}(X)$ The variance-covariance matrix of the vector random variable X.

$\mathrm{Cor}(X, Y)$ The correlation of the random variables X and Y. The notation $\mathrm{Cor}_P(\cdot, \cdot)$ is also often used.

$\mathrm{Cor}(X)$ The correlation matrix of the vector random variable X.

$\mathrm{Med}(X)$ The median of the scalar random variable X. It may be defined uniquely by means of equation (1.142) on page 62, with $\pi = 0.5$.

$\mathrm{Bias}(T, \theta)$ or $\mathrm{Bias}(T)$ The bias of the estimator T (as an estimator of θ); that is,

$$\mathrm{Bias}(T, \theta) = \mathrm{E}(T) - \theta.$$

$\mathrm{MSE}(T, \theta)$ or $\mathrm{MSE}(T)$ The mean squared error of the estimator T (as an estimator of θ); that is,

$$\mathrm{MSE}(T, \theta) = \big(\mathrm{Bias}(T, \theta)\big)^2 + \mathrm{V}(T).$$

$\mathrm{J}(T)$ The Jackknife estimator corresponding to the statistic T.

C.4 General Mathematical Functions and Operators

Functions such as sin, max, span, and so on that are commonly associated with groups of Latin letters are generally represented by those letters in a Roman font.

Generally, the argument of a function is enclosed in parentheses (for example, $\sin(x)$), but often, for the very common functions, the parentheses are omitted: $\sin x$. In expressions involving functions, parentheses are generally used for clarity, for example, $(\mathrm{E}(X))^2$ instead of $\mathrm{E}^2(X)$.

Operators such as d (the differential operator) that are commonly associated with a Latin letter are generally represented by that letter in a Roman font.

$|x|$ The modulus of the real or complex number x; if x is real, $|x|$ is the absolute value of x.

$\lceil x \rceil$ The ceiling function evaluated at the real number x: $\lceil x \rceil$ is the smallest integer greater than or equal to x.

$\lfloor x \rfloor$ The floor function evaluated at the real number x: $\lfloor x \rfloor$ is the largest integer less than or equal to x.

$\#S$ The cardinality of the set S.

$\mathrm{I}_S(\cdot)$ The indicator function:

$$\begin{aligned} \mathrm{I}_S(x) &= 1, \text{ if } x \in S, \\ &= 0, \text{ otherwise.} \end{aligned}$$

If x is a scalar, the set S is often taken as the interval $(-\infty, y]$, and in this case, the indicator function is the Heaviside function, H, evaluated at the difference of the argument and the upper bound on the interval:

$$\mathrm{I}_{(-\infty,y]}(x) = \mathrm{H}(y - x).$$

(An alternative definition of the Heaviside function is the same as this except that $\mathrm{H}(0) = \frac{1}{2}$.) It is interesting to note that

$$\mathrm{I}_{(-\infty,y]}(x) = \mathrm{I}_{[x,\infty)}(y).$$

In higher dimensions, the set S is often taken as the product set,

$$\begin{aligned} A^d &= (-\infty, y_1] \times (-\infty, y_2] \times \cdots \times (-\infty, y_d] \\ &= A_1 \times A_2 \times \cdots \times A_d, \end{aligned}$$

and in this case,

$$\mathrm{I}_{A^d}(x) = \mathrm{I}_{A_1}(x_1)\mathrm{I}_{A_2}(x_2) \cdots \mathrm{I}_{A_d}(x_d),$$

where $x = (x_1, x_2, \ldots, x_d)$. The derivative of the indicator function is the Dirac delta function, $\delta(\cdot)$.

$\delta(\cdot)$ — The Dirac delta "function", defined by

$$\delta(x) = 0, \quad \text{for } x \neq 0,$$

and

$$\int_{-\infty}^{\infty} \delta(t)\,\mathrm{d}t = 1.$$

The Dirac delta function is not a function in the usual sense. We do, however, refer to it as a function, and treat it in many ways as a function. For any continuous function f, we have the useful fact

$$\begin{aligned}\int_{-\infty}^{\infty} f(y)\,\mathrm{dI}_{(-\infty,y]}(x) &= \int_{-\infty}^{\infty} f(y)\,\delta(y-x)\,\mathrm{d}y \\ &= f(x).\end{aligned}$$

$\min f(\cdot)$ or $\min(S)$ — The minimum value of the real scalar-valued function f, or the smallest element in the countable set of real numbers S.

$\operatorname{argmin} f(\cdot)$ — The value of the argument of the real scalar-valued function f that yields its minimum value.

$\mathrm{O}(f(n))$ — Big O; $g(n) = \mathrm{O}(f(n))$ means that there exists a positive constant M such that $|g(n)| \leq M|f(n)|$ as $n \to \infty$. $g(n) = \mathrm{O}(1)$ means that $g(n)$ is bounded from above.

$\mathrm{o}(f(n))$ — Little o; $g(n) = \mathrm{o}(f(n))$ means that $g(n)/f(n) \to 0$ as $n \to \infty$. $g(n) = \mathrm{o}(1)$ means that $g(n) \to 0$ as $n \to \infty$.

$\mathrm{o_P}(f(n))$ — Convergent in probability; $X(n) = \mathrm{o_P}(f(n))$ means that, for any positive ϵ, $\Pr(|X(n)/f(n)| > \epsilon) \to 0$ as $n \to \infty$.

d — The differential operator. The derivative with respect to the variable x is denoted by $\frac{\mathrm{d}}{\mathrm{d}x}$.

$f', f'', \ldots, f^{k'}$ — For the scalar-valued function f of a scalar variable, differentiation (with respect to an implied variable) taken on the function once, twice, …, k times.

f^{T} — For the vector-valued function f, the transpose of f (a row vector).

∇f — For the scalar-valued function f of a vector variable, the gradient (that is, the vector of partial derivatives), also often denoted as g_f or D_f.

∇f — For the vector-valued function f of a vector variable, the transpose of the Jacobian, which is often denoted as J_f, so $\nabla f = \mathrm{J}_f^{\mathrm{T}}$ (see below).

J_f — For the vector-valued function f of a vector variable, the Jacobian; that is, the matrix whose $(i,j)^{\text{th}}$ element is

$$\frac{\partial f_i(x)}{\partial x_j}.$$

H_f or $\nabla\nabla f$ or $\nabla^2 f$ — For the scalar-valued function f of a vector variable, the Hessian. The Hessian is the transpose of the Jacobian of the gradient. Except in pathological cases, it is symmetric. The element in position (i,j) is

$$\frac{\partial^2 f(x)}{\partial x_i \partial x_j}.$$

The symbol $\nabla^2 f$ is sometimes also used to denote the diagonal of the Hessian, in which case it is called the Laplacian.

$f \star g$ — The convolution of the functions f and g,

$$(f \star g)(t) = \int f(x) g(t - x)\,\mathrm{d}x.$$

The convolution is a function.

$\mathrm{Cov}(f,g)$ — For the functions f and g whose integrals are zero, the covariance of f and g at lag t;

$$\mathrm{Cov}(f,g)(t) = \int f(x) g(t + x)\,\mathrm{d}x.$$

The covariance is a function; its argument is called the lag. $\mathrm{Cov}(f,f)(t)$ is called the autocovariance of f at lag t, and $\mathrm{Cov}(f,f)(0)$ is called the variance of f.

$\mathrm{Cor}(f,g)$ — For the functions f and g whose integrals are zero, the correlation of f and g at lag t;

$$\mathrm{Cor}(f,g)(t) = \frac{\int f(x) g(t + x)\,\mathrm{d}x}{\sqrt{\mathrm{Cov}(f,f)(0)\mathrm{Cov}(g,g)(0)}}.$$

The correlation is a function; its argument is called the lag. $\mathrm{Cov}(f,f)(t)$ is called the autocorrelation of f at lag t.

$f \otimes g$ — The tensor product of the functions f and g,

$$(f \otimes g)(w) = f(x)g(y) \quad \text{for} \quad w = (x, y).$$

The operator is also used for the tensor product of two function spaces and for the Kronecker product of two matrices.

$f^{\mathcal{T}}$ or $\mathcal{T} f$ — The transform of the function f by the functional $\mathcal{T}$.
$f^{\mathcal{F}}$ usually denotes the Fourier transform of f.
$f^{\mathcal{L}}$ usually denotes the Laplace transform of f.
$f^{\mathcal{W}}$ usually denotes a wavelet transform of f.

δ — A perturbation operator. δx represents a perturbation of x and not a multiplication of x by δ, even if x is a type of object for which a multiplication is defined.

$\Delta(\cdot, \cdot)$ — A real-valued difference function. $\Delta(x, y)$ is a measure of the difference of x and y. For simple objects, $\Delta(x, y) = |x - y|$; for more complicated objects, a subtraction operator may not be defined, and Δ is a generalized difference.

$\tilde{x}$ — A perturbation of the object x; $\Delta(x, \tilde{x}) = \delta x$.

$\text{Ave}(S)$ — An average (of some kind) of the elements in the set S.

$\langle f^r \rangle_p$ — The rth moment of the function f with respect to the density p.

$\bar{x}$ — The mean of a sample of objects generically denoted by x.

$\bar{x}$ — The complex conjugate of the object x; that is, if $x = r + \text{i}c$, then $\bar{x} = r - \text{i}c$.

$\log x$ — The natural logarithm evaluated at x.

$\sin x$ — The sine evaluated at x (in radians) and similarly for other trigonometric functions.

$x!$ — The factorial of x. If x is a positive integer, $x! = x(x-1) \cdots 2 \cdot 1$. For other values of x, except negative integers, $x!$ is often defined as

$$x! = \Gamma(x + 1).$$

$\Gamma(\alpha)$ The complete gamma function. For α not equal to a nonpositive integer,

$$\Gamma(\alpha) = \int_0^\infty t^{\alpha-1} \mathrm{e}^{-t} \, \mathrm{d}t.$$

We have the useful relationship, $\Gamma(\alpha) = (\alpha - 1)!$. An important argument is $\frac{1}{2}$, and $\Gamma(\frac{1}{2}) = \sqrt{\pi}$.

$\Gamma_x(\alpha)$ The incomplete gamma function:

$$\Gamma_x(\alpha) = \int_0^x t^{\alpha-1} \mathrm{e}^{-t} \, \mathrm{d}t.$$

$\mathrm{B}(\alpha, \beta)$ The complete beta function:

$$\mathrm{B}(\alpha, \beta) = \int_0^1 t^{\alpha-1} (1-t)^{\beta-1} \, \mathrm{d}t,$$

where $\alpha > 0$ and $\beta > 0$. A useful relationship is

$$\mathrm{B}(\alpha, \beta) = \frac{\Gamma(\alpha)\Gamma(\beta)}{\Gamma(\alpha+\beta)}.$$

$\mathrm{B}_x(\alpha, \beta)$ The incomplete beta function:

$$\mathrm{B}_x(\alpha, \beta) = \int_0^x t^{\alpha-1} (1-t)^{\beta-1} \, \mathrm{d}t.$$

A^{T} For the matrix A, its transpose (also used for a vector to represent the corresponding row vector).

A^{H} The conjugate transpose of the matrix A; $A^{\mathrm{H}} = \bar{A}^{\mathrm{T}}$.

A^{-1} The inverse of the square, nonsingular matrix A.

A^{+} The g_4 inverse, or the Moore-Penrose inverse, or the pseudoinverse, of the matrix A.

$A^{\frac{1}{2}}$ For the nonnegative definite matrix A, the Cholesky factor; that is,

$$(A^{\frac{1}{2}})^{\mathrm{T}} A^{\frac{1}{2}} = A.$$

sign(x) For the vector x, a vector of units corresponding to the signs

$$\begin{aligned} \text{sign}(x)_i &= 1 \quad \text{if } x_i > 0, \\ &= 0 \quad \text{if } x_i = 0, \\ &= -1 \quad \text{if } x_i < 0, \end{aligned}$$

with a similar meaning for a scalar. The sign function is also sometimes called the signum function and denoted sgn$(\cdot)$.

L_p For real $p \geq 1$, a norm formed by accumulating the pth powers of the moduli of individual elements in an object and then taking the $(1/p)^{\text{th}}$ power of the result.

$\|\cdot\|$ In general, the norm of the object $\cdot$. Often, however, specifically either the L_2 norm, or the norm defined by an inner product.

$\|\cdot\|_p$ In general, the L_p norm of the object $\cdot$.

$\|x\|_p$ For the vector x, the L_p norm:

$$\|x\|_p = \left(\sum |x_i|^p\right)^{\frac{1}{p}}.$$

$\|X\|_p$ For the matrix X, the L_p norm:

$$\|X\|_p = \max_{\|v\|_p=1} \|Xv\|_p.$$

$\|f\|_p$ For the function f, the L_p norm:

$$\|f\|_p = \left(\int |f(x)|^p \mathrm{d}x\right)^{\frac{1}{p}}.$$

$\|X\|_\mathrm{F}$ For the matrix X, the Frobenius norm:

$$\|X\|_\mathrm{F} = \sqrt{\sum_{i,j} x_{ij}^2}.$$

$\langle x, y\rangle$ The inner product of x and y.

$\kappa_p(A)$ The L_p condition number of the nonsingular square matrix A with respect to inversion.

diag(v) For the vector v, the diagonal matrix whose nonzero elements are those of v; that is, the square matrix, A, such that $A_{ii} = v_i$ and for $i \neq j$, $A_{ij} = 0$.

$\text{diag}(A_1, \ldots, A_k)$
The block diagonal matrix whose submatrices along the diagonal are $A_1, \ldots, A_k$.

a_{i*} The vector whose elements correspond to the elements in the ith row of the matrix A. a_{i*} is a column vector (as are all vectors in this book).

a_{*j} The vector whose elements correspond to the elements in the jth column of the matrix A.

$\text{trace}(A)$ The trace of the square matrix A; that is, the sum of the diagonal elements.

$\text{rank}(A)$ The rank of the matrix A, that is, the maximum number of independent rows (or columns) of A.

$\det(A)$ The determinant of the square matrix A, $\det(A) = |A|$.

$|A|$ The determinant of the square matrix A, $|A| = \det(A)$.

E_{jk} The elementary operator matrix that by premultiplication exchanges rows j and k of a matrix. E_{jk} is the identity matrix with rows j and k interchanged.

$E_{jk}(c)$ The elementary operator matrix that by premultiplication performs an axpy operation on rows j and k of a matrix. $E_{jk}(c)$ is the identity matrix with the 0 in position (j, k) replaced by c.

e_i The ith unit vector, that is, the vector with 0s in all positions except the ith position, which is 1.

C.5 Models and Data

A form of model used often in statistics and applied mathematics has three parts: a left-hand side representing an object of primary interest; a function of another variable and a parameter, each of which is likely to be a vector; and an adjustment term to make the right-hand side equal the left-hand side. The notation varies depending on the meaning of the terms. One of the most common models used in statistics, the linear regression model with normal errors, is written as

$$Y = \beta^{\mathrm{T}} x + E. \tag{C.1}$$

The adjustment term is a random variable, denoted by an uppercase epsilon. The term on the left-hand side is also a random variable. This model does not represent observations or data. A slightly more general form is

$$Y = f(x; \theta) + E. \tag{C.2}$$

A single observation or a single data item that corresponds to model (C.1) may be written as

$$y = \beta^{\mathrm{T}} x + \epsilon$$

or, if it is one of several,

$$y_i = \beta^{\mathrm{T}} x_i + \epsilon_i.$$

Similar expressions are used for a single data item that corresponds to model (C.2).

In these cases, rather than being a random variable, ϵ or ϵ_i may be a realization of a random variable, or it may just be an adjustment factor with no assumptions about its origin.

A set of n such observations is usually represented in an n-vector y, a matrix X with n rows, and an n-vector ϵ:

$$y = X\beta + \epsilon$$

or

$$y = f(X; \theta) + \epsilon.$$

The model is not symmetric in y and x. The error term is added to the systematic component that involes x. The way the error term is included in the model has implications in estimation and model fitting (see Chapter 17).

D

Solutions and Hints for Selected Exercises

Exercises Beginning on Page 75

1.5. $\|A\|_1 = 1.667$, $\|A\|_2 = 1.333$, and $\|A\|_\infty = 1.500$.

1.8. The distribution of $X = s^2/\sigma^2$ is a gamma with parameters $(n-1)/2$ and 2 (that is, a chi-squared with $n-1$ degrees of freedom). Using this fact, evaluate $\mathrm{E}(X^{1/2})$ and determine the scaling needed to form an unbiased estimator for σ.

1.16b.

$$\begin{aligned}\Psi(P_n) &= \int_{-\infty}^{\infty}\left(y - \int_{-\infty}^{\infty} u\,\mathrm{d}P_n(u)\right)^2 \mathrm{d}P_n(y) \\ &= \frac{1}{n}\sum_{i=1}^{n}(y_i - \bar{y})^2.\end{aligned}$$

1.17a. $\mathrm{E}(X)$, $\mathrm{E}(X^2)$.

1.17b. The median.

1.17c. No; it is a nonlinear combination of two linear functionals, $\mathrm{E}(X)$ and $\mathrm{E}(X^2)$.

1.20a. $\widehat{\alpha} = (n-2)\bar{y}^2/\sum(y_i - \bar{y})^2$;
$\widehat{\beta} = \sum(y_i - \bar{y})^2/((n-2)\bar{y})$.

1.20c. It does not have a closed-form solution.

Exercises Beginning on Page 102

2.8b. $g = 0.005$ under the assumption of rounding to an even digit.

2.9a. The first thing is to identify the computer numbers in the interval

$$[0,\ 1].$$

The computer numbers have the form of equation (2.2) on page 89 with $b = 10$, $0 \le d_i \le 9$, $p = 5$, $-9 \le e \le 9$. Therefore the numbers in this system that are in this interval are

A		B			J		K	
e	mantissa	e	mantissa		e	mantissa	e	mantissa
1	1 0 0 0 0							
0	9 9 9 9 9	-1	9 9 9 9 9		-9	9 9 9 9 9	-9	0 9 9 9 9
0	9 9 9 9 8	-1	9 9 9 9 8		-9	9 9 9 9 8	-9	0 9 9 9 8
	....		....			....		....
0	9 9 9 9 0	-1	9 9 9 9 0		-9	9 9 9 9 0	-9	0 9 9 9 0
0	9 9 9 8 9	-1	9 9 9 8 9		-9	9 9 9 8 9	-9	0 0 9 8 9
0	9 9 9 8 8	-1	9 9 9 8 8		-9	9 9 9 8 8	-9	0 0 9 8 8
	....		....	...		....		....
0	9 9 9 8 0	-1	9 9 9 8 0		-9	9 9 9 8 0	-9	0 0 9 8 0
0	9 9 9 7 9	-1	9 9 9 7 9		-9	9 9 9 7 9	-9	0 0 9 7 9
0	9 9 9 7 8	-1	9 9 9 7 8		-9	9 9 9 7 8	-9	0 0 9 7 8
	....		....			....		....
0	9 9 9 7 0	-1	9 9 9 7 0		-9	9 9 9 7 0	-9	0 0 9 7 0
	....		....			....		....
	....		....			....		....
0	1 0 0 0 0	-1	1 0 0 0 0		-9	1 0 0 0 0	-9	0 0 0 0 1
							0	0 0 0 0 0

Now all we do is count the points and distribute the uniform probability over them. There are 100,000 points in each column, except for columns A and K. Column A has one additional point, as shown. Column K is an interesting one. It contains the 9,999 nonnormalized numbers that allow "graceful underflow", plus the number 0. The probability content to distribute over the points in each column is the length of the interval represented by the numbers in the column, plus the proportionate amount from the interval above and from the interval below. For example, column A, which does not have an interval above receives 0.9 plus the proportionate amount (that is, half) of the probability in the interval between 0.1 (it smallest number) and 0.099999 (the largest number in the interval just below it, represented by column B); hence, the probability assigned to all of the numbers in column A is 0.900005, over those in column B is 0.0899955, over those in column C (not shown) is 0.00899955, and so on, through column J.

The discrete probability distribution can then be built up in the obvious way by assigning the proportional probability to each point that is not some integral power of 10. Those that are integral powers of 10, get a proportionate amount from the interval above and from the interval below. For instance,

$$\Pr(1.0000) = 0.000005$$

because that represents one half of the probability content in the interval $[.99999, 1]$. Notice that there is a certain asymmetry here; the probability assigned to 1 is the appropriate amount without consideration of a number outside of the interval. The probabilities associated with all of the other numbers in column A are equal except for the number 0.10000. By the same process, we get

$$\Pr(.00000) = 0.000000000000005.$$

Notice that both 0 and 1 have nonzero probabilities, but they are not equal. In the case of the U(0, 1) distribution, which is continuous, $\Pr(0) = \Pr(1) = 0$, and the support of the distribution can equivalently be considered to be $(0, 1)$ or $[0, 1]$. In the case of a distribution over $\mathbb{F}$, the distribution is discrete, and each point in the support matters.
Although the solution given above is the correct one as the exercise was stated, we consider a different type of problem in Chapter 7. In that problem we want to *simulate* the U(0, 1) distribution over the finite set $\mathbb{F}$, but because of other computations we may do, we want $\Pr(0) = \Pr(1) = 0$ in $\mathbb{F}$.

Exercises Beginning on Page 142

3.9. An elegant solution to this problem was given by Jay Kadane in 1984. (See Bentley, 2000.) Here's some R code that implements Kadane's algorithm:

```
### array is the array to be processed
### n is its lenght
subarray <-c(-Inf,0,0)
sum <- 0
start <-1
for (end in 1:n){
  sum <- sum +array[end]
  if (sum>subarray[1]) subarray <- c(sum,start,end)
  if (sum<0) {
    sum <- 0
    start <- end+1
  }
}
```

Notice the use of -Inf.

Exercises Beginning on Page 199

4.1. For $x \in D$, consider the change of variable $y = ax$ for $a \neq 0$. The Jacobian of the inverse transformation, J, is $1/a$. Let D_a represent the domain of y. For the L_1 norm,

$$\begin{aligned}\int_{D_a} |f_a(y)|\mathrm{d}y &= \int_{D_a} |f(y/a)||\mathrm{J}|\mathrm{d}y \\ &= \int_{D} |f(x)|\mathrm{d}x.\end{aligned}$$

For the L_2 norm (squared),

$$\begin{aligned}\int_{D_a}(f_a(y))^2\mathrm{d}y &= \int_{D_a}(f(y/a))^2|\mathrm{J}|\mathrm{d}y\\ &\neq \int_D (f(x))^2\mathrm{d}x.\end{aligned}$$

As a specific example, let $f(x)$ be the density $2x$ over the interval $(0,1)$, and let $a=2$.

4.2a. $3/(8\sqrt{\pi}\sigma^5)$.

Exercises Beginning on Page 237

5.1. $\kappa_1(A)=5001$, $\kappa_2(A)=3555.67$, and $\kappa_\infty(A)=5016$.

Exercises Beginning on Page 301

6.13b. The Hessian is

$$\frac{x_1}{(2+\theta)^2}+\frac{x_2+x_3}{(1-\theta)^2}+\frac{x_4}{\theta^2}.$$

Coding this in Matlab and beginning at 0.5, we get the first few iterates
0.6364
0.6270
0.6268
0.6268

6.13c. The expected value of the information is

$$\frac{n}{4}\left(\frac{1}{2+\theta}+\frac{2}{1-\theta}+\frac{1}{\theta}\right),$$

which we obtain by taking $\mathrm{E}(X_i)$ for each element of the multinomial random variable. Coding the method in Matlab and beginning at 0.5, we get the first few iterates
0.6332
0.6265
0.6268
0.6268

6.13d. To use the EM algorithm on this problem, we can think of a multinomial with five classes, which is formed from the original multinomial by splitting the first class into two with associated probabilities $1/2$ and $\theta/4$. The original variable x_1 is now the sum of x_{11} and x_{12}. Under this reformulation, we now have a maximum likelihood estimate of θ by considering $x_{12}+x_4$ (or x_2+x_3) to be a realization of a binomial with $n=x_{12}+x_4+x_2+x_3$ and $\pi=\theta$ (or $1-\theta$). However, we do not know x_{12} (or x_{11}). Proceeding as if we had a five-outcome multinomial observation with two missing elements, we have the log-likelihood for the complete data,

$$l_c(\theta) = (x_{12} + x_4)\log(\theta) + (x_2 + x_3)\log(1-\theta) + k,$$

and the maximum likelihood estimate for θ is

$$\frac{x_{12} + x_4}{x_{12} + x_2 + x_3 + x_4}.$$

The E-step of the iterative EM algorithm fills in the missing or unobservable value with its expected value given a current value of the parameter, $\theta^{(k)}$, and the observed data. Because $l_c(\theta)$ is linear in the data, we have

$$\mathrm{E}\left(l_c(\theta)\right) = \mathrm{E}(x_{12} + x_4)\log(\theta) + \mathrm{E}(x_2 + x_3)\log(1-\theta).$$

Under this setup, with $\theta = \theta^{(k)}$,

$$\begin{aligned}\mathrm{E}_{\theta^{(k)}}(x_{12}) &= \frac{1}{4}x_1\theta^{(k)}/(\frac{1}{2} + \frac{1}{4}\theta^{(k)}) \\ &= x_{12}^{(k)}.\end{aligned}$$

We now maximize $\mathrm{E}_{\theta^{(k)}}\left(l_c(\theta)\right)$. This maximum occurs at

$$\theta^{(k+1)} = (x_{12}^{(k)} + x_4)/(x_{12}^{(k)} + x_2 + x_3 + x_4).$$

The following Matlab statements execute a single iteration.

```
function [x12kp1,tkp1] = em(tk,x)
x12kp1 = x(1)*tk/(2+tk);
tkp1 = (x12kp1 + x(4))/(sum(x)-x(1)+x12kp1);
```

Exercises Beginning on Page 329

7.3b. Conditional on u, the distribution of T is geometric:

$$p_{T|u}(t) = \pi_u(1-\pi_u)^{t-1}, \quad \text{for } t = 1, 2, \ldots,$$

where

$$\pi_u = \int \mathrm{H}\big(g_Y(y) - u\big)\,\frac{p_X(y)}{c g_Y(y)}\,\mathrm{d}y,$$

and $\mathrm{H}(\cdot)$ is the Heaviside function. The marginal probability function for T is

$$p_T(t) = \int_0^1 \pi_u(1-\pi_u)^{t-1}\,\mathrm{d}u, \quad \text{for } t = 1, 2, \ldots.$$

Therefore, we have

$$\mathrm{E}(T) = \int_0^1 \frac{1}{\pi_u}\,\mathrm{d}u$$

and

$$\mathrm{V}(T) = \int_0^1 \frac{1}{\pi_u}\left(\frac{2}{\pi_u} - 1\right)\mathrm{d}u - (\mathrm{E}(T))^2.$$

For certain densities these moments may be infinite. Assume, for example, that for some y_0, $g_Y(y_0) = 1$, and that we can expand g_Y about y_0:

$$g_Y(y_0 + \delta) = 1 - a\delta^2 + \mathrm{O}(\delta^3).$$

For u close to 1, π_u is approximately proportional to $2a^{-1/2}(1-u)^{1/2}$, so $\mathrm{V}(T)$ is arbitrarily large (see Greenwood, 1976).

7.5a. First, we obtain the conditional copula for the Gumbel copula,

$$\begin{aligned} C_v(u) &= \frac{\partial}{\partial v} C(u,v)|_v \\ &= (-\log(u))^{\theta-1}\left((-\log(u))^\theta + (-\log(v))^\theta\right)^{(1-\theta)/\theta} \\ &\quad \exp\left(-\left((-\log(u))^\theta + (-\log(v))^\theta\right)^{1/\theta}\right)|_v . \end{aligned}$$

We notice that $C_v(u)$ must be inverted numerically, by methods discussed in Section 6.1. The steps now are

1. Generate indpendently w and v from $\mathrm{U}(0,1)$.
2. Set $u = C_v^{-1}(w)$. (This requires numerical root finding.)
3. Set $Y = -\alpha\log(1-u)$ and $Z = -\beta\log(1-v)$.

7.6. A small piece of R code for the main loop after the parameters have been initialized is

```
x1ind <- rnorm(n)
x2ind <- rnorm(n)
sdf1 <- sig1*sqrt(1-rho2)
mf1 <- rho*sig1
sdf2 <- sig2*sqrt(1-rho2)
mf2 <- rho*sig2
k <- 1
while (k <= n) {
   x1 <- sdf1*x1ind[k] + mu1+mf1*(x2-mu2)
   x2 <- sdf2*x2ind[k] + mu2+mf2*(x1-mu1)
   x[k,]=c(x1,x2)
   k <- k + 1
}
```

Exercises Beginning on Page 368

8.2. It is likely that the tails will be light because the median is smaller in absolute value than the mean.

Exercise 8.2$'$: Generate a sample of size 50 of maximum order statistics from samples of size 100 from a normal (0,1) distribution, and plot a histogram of it. Notice the skewed shape. Because much of our experience is with symmetric data, our expectations of the behavior of random samples often are not met when the data are skewed.

Exercises Beginning on Page 397

9.3. There are 63 planes in 4-D. They may be somewhat difficult to see because of the slicing and because of the relatively large number of planes.

9.13a. The rotation matrix Q rotates (x_{1i}, x_{2i}, x_3) into $(a_1x_{1i} + a_2x_{2i},\ 0,\ x_3)$. If

$$Q = \begin{bmatrix} c & -s & 0 \\ s & c & 0 \\ 0 & 0 & 1 \end{bmatrix},$$

then $a_2c = a_1s$ and $s = \sqrt{1-c^2}$. Thus, we have two equations in two unknowns for which we can solve in terms of a_1 and a_2. The projection matrix is just

$$P = \begin{bmatrix} 1 & 0 & 0 \\ 0 & 0 & 1 \end{bmatrix}.$$

9.13b. The first thing to consider here is whether to use the known model (that is, in the notation of the previous question, $a_1 = 5$ and $a_2 = 1$), or to use the data to determine coefficients that better fit the data. (In the latter case, we would first regress x_3 on x_1 and x_2, and then use the estimates $\hat{a}_1$ and $\hat{a}_2$.) At this point, the problem is almost like Exercise 9.13a. This depends on the signal-to-noise (that is, on whether the variation in ϵ dominates the variation in $5x_1 + x_2$). If the noise dominates, it is not likely that a good projection exists.

9.14. $m = 3$, and

$$X = \begin{bmatrix} 3.31 & 0.95 & -1.38 \\ -0.68 & 0.79 & 0.31 \\ -0.31 & 1.73 & 1.31 \\ -4.17 & -0.72 & -0.94 \\ 1.85 & -2.74 & 0.69 \end{bmatrix}.$$

Exercises Beginning on Page 414

10.2.

$$
\begin{aligned}
\mathrm{IMSE}\left(\widehat{f}\right) &= \int_D \mathrm{E}\left(\left(\widehat{f}(x) - f(x)\right)^2\right) \mathrm{d}x \\
&= \int_D \mathrm{E}\left(\left(\widehat{f}(x) - \mathrm{E}\left(\widehat{f}(x)\right) + \mathrm{E}\left(\widehat{f}(x)\right) - f(x)\right)^2\right) \mathrm{d}x \\
&= \int_D \mathrm{E}\left(\left(\widehat{f}(x) - \mathrm{E}\left(\widehat{f}(x)\right)\right)^2\right) \mathrm{d}x + \int_D \left(\mathrm{E}\left(\widehat{f}(x)\right) - f(x)\right)^2 \mathrm{d}x \\
&= \int_D \mathrm{E}\left(\left(\widehat{f}(x) - \mathrm{E}\left(\widehat{f}(x)\right)\right)^2\right) \mathrm{d}x \\
&\quad +2 \int_D \mathrm{E}\left(\left(\widehat{f}(x) - \mathrm{E}\left(\widehat{f}(x)\right)\right) \left(\mathrm{E}\left(\widehat{f}(x)\right) - f(x)\right)\right) \mathrm{d}x \\
&\quad + \int_D \left(\mathrm{E}\left(\widehat{f}(x)\right) - f(x)\right)^2 \mathrm{d}x \\
&= \mathrm{IV}\left(\widehat{f}\right) + \mathrm{ISB}\left(\widehat{f}\right).
\end{aligned}
$$

Exercises Beginning on Page 431

11.4. What we are to do is a goodness-of-fit test. These are hard (in the sense that there are too many alternatives). Just three $\mathrm{O}(n)$ statistics cannot result in a very powerful omnibus test. Most reasonable tests, such as a Kolmogorov-Smirnov test, use at least $\mathrm{O}(n \log n)$ test statistics. What we can do in this case are just some simple tests for specific aspects of the hypothesized distribution.
The unknown parameter θ complicates anything we might try; we cannot generate Monte Carlo samples without knowing this value. Therefore, we must consider two general approaches: one kind in which we use an estimate of θ, and another kind in which we address characteristics of the distribution that are independent of θ.
First of all, note that the distribution is symmetric; therefore, the mean is 0. Unfortunately, we cannot construct a test based on m unless we know θ or have an estimate of it.
Because $\mathrm{V}(X) = 2\theta^2$, where X is a random variable with the hypothesized distribution, we can estimate θ^2 with $m_2/2$. Using this to estimate θ, we can now construct various tests. We can, for example, do Monte Carlo tests on m, m_2, or m_4. These tests would in reality be on the population parameters corresponding to the expectations of the corresponding random variables (that is, the sample mean, the sample second central moment, and the sample fourth central moment). Each Monte Carlo test would be performed in the usual way by generating samples from a double exponential with the estimated value of θ, computing the appropriate statistic from each sample, and then comparing the single observed value of that statistic with the Monte Carlo set. To do a test at the α level, we have several possibilities. An easy way out is just to do one of the Monte Carlo tests described; the observed value of just one of the quantities m,

m_2, and m_4 would be compared with the $\lfloor n\alpha/2 \rfloor^{\text{th}}$ and the $\lfloor n(1-\alpha/2) \rfloor^{\text{th}}$ order statistics from the Monte Carlo sample of corresponding statistics. Another possibility would be to do three tests with appropriately adjusted significance levels. There are no (obvious) Bonferroni bounds, so the adjusted level would just be $\alpha/3$, and the rejection criterion would be rejection of any separate test.

Another approach would be to determine salient features of the hypothesized distribution that are independent of the unknown parameter. An example is the ratio of the fourth moment to the square of the second moment, which is 12. The expected value of m_4/m_2^2 is independent of θ. Therefore, we can choose an arbitrary value of θ and generate Monte Carlo samples. We would then do a Monte Carlo test using m_4/m_2^2. This test would compare the observed value of this ratio with the Monte Carlo sample of this ratio. (The test would not directly compare the observed value with the hypothesized population value of 12.)

Exercises Beginning on Page 449

12.4. Consider the numerator,

$$\begin{aligned}
\sum_{j=1}^{n}(T_j^* - T)^2 &= \sum_{j=1}^{n}\left((T_j^* - \overline{T}^*) - (T - \overline{T}^*)\right)^2 \\
&= \sum_{j=1}^{n}(T_j^* - \overline{T}^*)^2 - 2\sum_{j=1}^{n}(T_j^* - \overline{T}^*)(T - \overline{T}^*) + \sum_{j=1}^{n}(T - \overline{T}^*)^2 \\
&= \sum_{j=1}^{n}(T_j^* - \overline{T}^*)^2 - \sum_{j=1}^{n}(T - \overline{T}^*)^2 \\
&\leq \sum_{j=1}^{n}(T_j^* - \overline{T}^*)^2,
\end{aligned}$$

which is the numerator of V_{J} with $r = n$. The same arithmetic also holds for other values of r.

Exercises Beginning on Page 465

13.1a.

$$\begin{aligned}
\mathrm{E}_{\widehat{P}}(\bar{y}_b^*) &= \mathrm{E}_{\widehat{P}}\left(\frac{1}{n}\sum_i y_i^*\right) \\
&= \frac{1}{n}\sum_i \mathrm{E}_{\widehat{P}}(y_i^*) \\
&= \frac{1}{n}\sum_i \bar{y} \\
&= \bar{y}.
\end{aligned}$$

Note that the empirical distribution is a conditional distribution, given the sample. With the sample fixed, $\bar{y}$ is a "parameter", rather than a "statistic".

13.1d.

$$\begin{aligned} \mathrm{E}_P(V) &= \mathrm{E}_P(\mathrm{E}_{\widehat{P}}(V)) \\ &= \mathrm{E}_P\left(\frac{1}{n}\sum(y_i - \bar{y})^2/n\right) \\ &= \frac{1}{n}\frac{n-1}{n}\sigma_P^2. \end{aligned}$$

Exercises Beginning on Page 484

14.1a.

$$\int_0^{x_{(n)}} \left(\frac{1}{x_{(n)}} - \frac{1}{\theta}\right)^2 \mathrm{d}x + \int_{x_{(n)}}^{\theta} \left(0 - \frac{1}{\theta}\right)^2 \mathrm{d}x = \frac{1}{x_{(n)}} - \frac{1}{\theta}.$$

14.1b. $\frac{1}{n-1}$ (remember that the ith order statistic from a uniform distribution has a beta distribution with i and $n-i+1$).

14.1c. The minimum occurs at $c = 2^{1/(n-1)}$

Exercises Beginning on Page 510

15.6. For regular triangles, $c = \frac{1}{6\sqrt{3}}$; for squares, $c = \frac{1}{12}$; and for regular hexagons, $c = \frac{5}{36\sqrt{3}}$.

15.8a.

$$\begin{aligned} \int_{\mathrm{IR}} \widehat{p}_F(y)\mathrm{d}y &= \frac{1}{2}\frac{n_1}{nv_1}v_1 + \sum_{k=1}^{m-1}\frac{1}{2}\left(\frac{n_{k+1}}{nv_{k+1}}v_{k+1} + \frac{n_k}{nv_k}v_k\right) + \frac{1}{2}\frac{n_m}{nv_m}v_m \\ &= 1. \end{aligned}$$

15.15c. Take the density $f(x) = 1$ on $[0,1]$ and the estimator $\widehat{f}(x) = n$ on $[0, \frac{1}{n}]$.

15.16a. For the first moment,

$$
\begin{aligned}
\mathrm{E}_{\widehat{p}_H}(Y) &= \int_D y\widehat{p}_H(y)\mathrm{d}y \\
&= \int_D y\sum_{k=1}^{m} \frac{n_k}{nv_k}\,\mathrm{I}_{T_k}(y)\mathrm{d}y \\
&= \sum_{k=1}^{m} \int_{T_k} y\frac{n_k}{nv_k}\mathrm{d}y \\
&= \sum_{k=1}^{m} \frac{n_k\left(t_{k+1}^2 - t_k^2\right)}{2nv_k} \\
&= \sum_{k=1}^{m} \frac{n_k\left(t_{k+1} + t_k\right)}{2n} \\
&= \mu_H.
\end{aligned}
$$

This is just the weighted mean of the midpoint of the bins. The sample first moment, of course, is just the sample mean. The bins could be chosen to make the two quantities equal. The higher central moments have more complicated expressions. In general, they are

$$
\begin{aligned}
\mathrm{E}_{\widehat{p}_H}\left((Y-\mu_H)^r\right) &= \int_D (y-\mu_H)^r\widehat{p}_H(y)\mathrm{d}y \\
&= \int_D (y-\mu_H)^r\sum_{k=1}^{m} \frac{n_k}{nv_k}\,\mathrm{I}_{T_k}(y)\mathrm{d}y \\
&= \sum_{k=1}^{m} \int_{T_k} (y-\mu_H)^r\frac{n_k}{nv_k}\mathrm{d}y \\
&= \sum_{k=1}^{m} \frac{n_k\Big((t-\mu_H)_{k+1}^{r+1} - (t-\mu_H)_k^{r+1}\Big)}{(r+1)nv_k}.
\end{aligned}
$$

For the case of $r = 2$ (that is, for the variance), the expression above can be simplified by using $\mathrm{E}((Y-\mathrm{E}(Y))^2) = \mathrm{E}(Y^2) - (\mathrm{E}(Y))^2$. It is just

$$
\sum_{k=1}^{m} \frac{n_k\left(t_{k+1}^2 + t_{k+1}t_k + t_k^2\right)}{3n} - m\mu_H^2.
$$

15.16b. Let $K(t) = 1$ if $|t| < 1/2$ and $K(t) = 0$ otherwise.

$$
\begin{aligned}
\mathrm{E}_{\widehat{p}_K}(Y) &= \int_D y\widehat{p}_K(y)\mathrm{d}y \\
&= \frac{1}{nh}\sum_{i=1}^{n}\int_D yK\left(\frac{y-y_i}{h}\right)\mathrm{d}y \\
&= \frac{1}{2nh}\sum_{i=1}^{n}\int_{y_i-h/2}^{y_i+h/2} y\mathrm{d}y \\
&= \mu_K \\
&= \bar{y}.
\end{aligned}
$$

The higher central moments have more complicated expressions. In general, they are

$$
\begin{aligned}
\mathrm{E}_{\widehat{p}_H}\left((Y-\mu_K)^r\right) &= \int_D (y-\mu_K)^r\widehat{p}_K(y)\mathrm{d}y \\
&= \frac{1}{nh}\sum_{i=1}^{n}\int_D (y-\mu_K)^r K\left(\frac{y-y_i}{h}\right)\mathrm{d}y \\
&= \frac{1}{nh}\sum_{i=1}^{n}\int_{y_i-h/2}^{y_i+h/2} (y-\mu_K)^r\mathrm{d}y. \\
&= \frac{1}{(r+1)nh}\sum_{i=1}^{n}\left((y_i-\mu_K+h/2)^{r+1}-(y_i-\mu_K-h/2)^{r+1}\right).
\end{aligned}
$$

For the case $r=2$, as in Exercise 15.16a, this can be simplified considerably:

$$
\mathrm{E}_{\widehat{p}_H}\left((Y-\mu_K)^2\right) = \sum_{i=1}^{n} y_i^2 - n\bar{y}^2 + \frac{h^3}{12}.
$$

This is the same as the second central sample moment except for the term $\frac{h^3}{12}$.

Exercises Beginning on Page 580

16.1b. 1

16.1c. .9

16.1d. .8

16.5. This depends on the signal-to-noise. If the ratio is large, then there will be no more than two strong principal components. If x_1 and x_2 are independent, then there will be at least two strong principal components.

16.7b. The optimal projection for the data-generating process is clearly $(0,1)$. For a given dataset from this process, of course, it may be slightly different.

16.11. 10. This is a simple exercise in the application of the generalized Pythagorean Theorem.

E

Bibliography

The literature on computational statistics is diverse. Relevant articles are likely to appear in journals devoted to quite different disciplines, especially computer science, numerical analysis, and statistics.

There are at least ten journals and serials whose titles contain some variants of both "computing" and "statistics"; but there are far more journals in numerical analysis and in areas such as "computational physics", "computational biology", and so on that publish articles relevant to the fields of statistical computing and computational statistics. The journals in the mainstream of statistics also have a large proportion of articles in the fields of statistical computing and computational statistics because, as we suggested in the preface, recent developments in statistics and in the computational sciences have paralleled each other to a large extent.

There are two well-known learned societies whose primary focus is in statistical computing: the International Association for Statistical Computing (IASC), which is an affiliated society of the International Statistical Institute, and the Statistical Computing Section of the American Statistical Association (ASA). The Statistical Computing Section of the ASA has a regular newsletter carrying news and notices as well as articles on practicum. Also, the activities of the Society for Industrial and Applied Mathematics (SIAM) are often relevant to computational statistics.

There are two regular conferences in the area of computational statistics: COMPSTAT, held biennially in Europe and sponsored by the IASC, and the Interface Symposium, generally held annually in North America and sponsored by the Interface Foundation of North America with cooperation from the Statistical Computing Section of the ASA.

In addition to literature and learned societies in the traditional forms, an important source of communication and a repository of information are computer databases and forums. In some cases, the databases duplicate what is available in some other form, but often the material and the communications facilities provided by the computer are not available elsewhere.

E.1 Literature in Computational Statistics

In the Library of Congress classification scheme, most books on statistics, including statistical computing, are in the QA276 section, although some are classified under H, HA, and HG. Numerical analysis is generally in QA279, and computer science in QA76. Many of the books in the interface of these disciplines are classified in these or other places within QA.

Current Index to Statistics is an online index produced by the American Statistical Association and the Institute for Mathematical Statistics. It has a broad coverage of the statistical literature.

Mathematical Reviews, published by the American Mathematical Society (AMS), contains brief reviews of articles in all areas of mathematics. The areas of "Statistics", "Numerical Analysis", and "Computer Science" contain reviews of articles relevant to computational statistics. The papers reviewed in *Mathematical Reviews* are categorized according to a standard system that has slowly evolved over the years. In this taxonomy, called the AMS MR classification system, "Statistics" is 62Xyy; "Numerical Analysis", including random number generation, is 65Xyy; and "Computer Science" is 68Xyy. ("X" represents a letter and "yy" represents a two-digit number.) *Mathematical Reviews* is available to subscribers via the World Wide Web at MathSciNet:

`http://www.ams.org/mathscinet/`

There are various handbooks of mathematical functions and formulas that are useful in numerical computations. Three that should be mentioned are Abramowitz and Stegun (1964), Spanier and Oldham (1987), and Thompson (1997). Anyone doing serious scientific computations should have ready access to at least one of these volumes.

Almost all journals in statistics have occasional articles on computational statistics and statistical computing. The following is a list of journals, proceedings, and newsletters that emphasize this field.

ACM Transactions on Mathematical Software, published quarterly by the ACM (Association for Computing Machinery), includes algorithms in Fortran and C. Most of the algorithms are available through `netlib`. The ACM collection of algorithms is sometimes called *CALGO*.

ACM Transactions on Modeling and Computer Simulation, published quarterly by the ACM.

Applied Statistics, published quarterly by the Royal Statistical Society. (Until 1998, it included algorithms in Fortran. Some of these algorithms, with corrections, were collected by Griffiths and Hill, 1985. Most of the algorithms are available through `statlib` at Carnegie Mellon University.)

Communications in Statistics — Simulation and Computation, published quarterly by Marcel Dekker. (Until 1996, it included algorithms in Fortran. Until 1982, this journal was designated as *Series B*.)

Computational Statistics, published quarterly by Physica-Verlag (formerly called *Computational Statistics Quarterly*).

Computational Statistics. Proceedings of the xxth Symposium on Computational Statistics (COMPSTAT), published biennially by Physica-Verlag. (It is not refereed.)

Computational Statistics & Data Analysis, published by North Holland. Number of issues per year varies. (This is also the official journal of the International Association for Statistical Computing and as such incorporates the *Statistical Software Newsletter*.)

Computing Science and Statistics. This is an annual publication containing papers presented at the Interface Symposium. Until 1992, these proceedings were named *Computer Science and Statistics: Proceedings of the xxth Symposium on the Interface.* (The 24th symposium was held in 1992.) In 1997, Volume 29 was published in two issues: Number 1, which contains the papers of the regular Interface Symposium; and Number 2, which contains papers from another conference. The two numbers are not sequentially paginated. Since 1999, the proceedings have been published only in CD-ROM form, by the Interface Foundation of North America. (It is not refereed.)

Journal of Computational and Graphical Statistics, published quarterly by the American Statistical Association.

Journal of Statistical Computation and Simulation, published irregularly in four numbers per volume by Gordon and Breach.

Proceedings of the Statistical Computing Section, published annually by the American Statistical Association. (It is not refereed.)

SIAM Journal on Scientific Computing, published bimonthly by SIAM. This journal was formerly *SIAM Journal on Scientific and Statistical Computing*. (Is this a step backward?)

Statistical Computing & Graphics Newsletter, published quarterly by the Statistical Computing and the Statistical Graphics Sections of the American Statistical Association. (It is not refereed and it is not generally available in libraries.)

Statistics and Computing, published quarterly by Chapman & Hall.

A useful electronic journal for computational statistics is the *Journal of Statistical Software* at

`http://www.jstatsoft.org/`

Resources Available over the Internet

The best way of storing information is in a digital format that can be accessed by computers. In some cases, the best way for people to access information is by computers; in other cases, the best way is via hard copy, which means that the information stored on the computer must go through a printing process resulting in books, journals, or loose pages.

A huge amount of information and raw data are available online, much in publicly accessible sites. Some of the repositories give space to ongoing discussions to which anyone can contribute.

For statistics, one of the most useful sites on the Internet is the electronic repository `statlib`, maintained at Carnegie Mellon University, which contains programs, datasets, and other items of interest. The URL is

`http://lib.stat.cmu.edu`

The collection of algorithms published in *Applied Statistics* is available in `statlib`. These algorithms are sometimes called the *ApStat* algorithms.

The `statlib` facility can also be accessed by email or anonymous ftp.

Another very useful site for scientific computing is `netlib`, which was established by research workers at Bell Laboratories and national laboratories, primarily Oak Ridge National Laboratories. The URL is

`http://www.netlib.org`

The *Collected Algorithms of the ACM* (*CALGO*), which are the Fortran, C, and Algol programs published in *ACM Transactions on Mathematical Software* (or in *Communications of the ACM* prior to 1975), are available in `netlib`, under the TOMS link.

There is also an X Windows, socket-based system for accessing `netlib`, called `Xnetlib`; see Dongarra, Rowan, and Wade (1995).

The *Guide to Available Mathematical Software* (GAMS) can be accessed at

`http://gams.nist.gov`

A different interface, using Java, is available at

`http://math.nist.gov/HotGAMS/`

A good set of links for software are the Econometric Links of the *Econometrics Journal* (which are not just limited to econometrics):

`http://www.eur.nl/few/ei/links/software.html`

There are two major problems in using the WWW to gather information. One is the sheer quantity of information and the number of sites providing information. The other is the "kiosk problem"; anyone can put up material. Sadly, the average quality is affected by a very large denominator. The kiosk problem may be even worse than a random selection of material; the "fools in public places" syndrome is much in evidence.

It is not clear at this time what will be the media for the scientific literature within a few years. Many of the traditional journals will be converted to an electronic version of some kind. Journals will become Web sites. That is for certain; the details, however, are much less certain. Many bulletin boards and discussion groups have already evolved into electronic journals.

E.2 References for Software Packages

There is a wide range of software used in the computational sciences. Some of the software is produced by a single individual who is happy to share the software, sometimes for a fee, but who has no interest in maintaining the software. At the other extreme is software produced by large commercial companies whose continued existence depends on a process of production, distribution, and maintenance of the software. Information on much of the software can be obtained from GAMS. Some of the free software can be obtained from `statlib` or `netlib`.

The R software system and associated documentation is available at `http://www.r-project.org/`

The GNU Scientific Library (GSL) of C functions is available at `http://www.gnu.org/software/gsl/`

We refer to several software packages with names that are trademarked or registered. Our reference to these packages without mention of the registration in no way implies that the name carries a generic meaning.

E.3 References to the Literature

The following bibliography obviously covers a wide range of topics in statistical computing and computational statistics. Except for a few of the general references, all of these entries have been cited in the text.

The purpose of this bibliography is to help the reader get more information; hence I eschew "personal communications" and references to technical reports that may or may not exist. Those kinds of references are generally for the author rather than for the reader.

In some cases, important original papers have been reprinted in special collections, such as Samuel Kotz and Norman L. Johnson (Editors) (1997), *Breakthroughs in Statistics, Volume III*, Springer-Verlag, New York. In most such cases, because the special collection may be more readily available, I list both sources.

Abbott, Edwin A. (1884), *Flatland, A Romance of Many Dimensions*, Seeley & Co. Ltd., London. (Reprinted with an updated introductory note by Dover Publications, New York, 1992).

Abramowitz, Milton, and Irene A. Stegun (Editors) (1964), *Handbook of Mathematical Functions with Formulas, Graphs, and Mathematical Tables*, National Bureau of Standards (NIST), Washington. (Reprinted by Dover Publications, New York, 1974. Work on an updated version is occurring at NIST. This version is called the Digital Library of Mathematical Functions (DLMF). See http://dlmf.nist.gov/ for the current status.)

Albert, Jim (2007), *Bayesian Computation with R*, Springer, New York.

Albert, James; Mohan Delampady; and Wolfgang Polasek (1991), A class of distributions for robustness studies, *Journal of Statistical Planning and Inference* **28**, 291–304.

Allen, David M. (1971), Mean square error of prediction as a criterion for selecting variables (with discussion), *Technometrics* **13**, 469–475.

Allen, David M. (1974), The relationship between variable selection and data augmentation and a method of prediction, *Technometrics* **16**, 125–127.

Altman, Micah; Jeff Gill; and Michael P. McDonald (2004), *Numerical Issues in Statistical Computing for the Social Scientist*, John Wiley & Sons, New York.

Amit, Yali, and Donald Geman (1997), Shape quantization and recognition with randomized trees, *Neural Computation* **9** 1545–1588.

Ammann, Larry P. (1989), Robust principal components, *Communications in Statistics — Simulation and Computation* **18**, 857–874.

Ammann, Larry P. (1993), Robust singular value decompositions: A new approach to projection pursuit, *Journal of the American Statistical Association* **88**, 505–514.

Ammann, Larry, and John Van Ness (1988), A routine for converting regression algorithms into corresponding orthogonal regression algorithms, *ACM Transactions on Mathematical Software* **14**, 76–87.

Ammann, Larry, and John Van Ness (1989), Standard and robust orthogonal regression, *Communications in Statistics — Simulation and Computation* **18**, 145–162.

Amores, Jaume; Nicu Sebe; and Petia Radeva (2006), Boosting the distance estimation: Application to the K-nearest neighbor classifier, *Pattern Recognition Letters* **27**, 201–209.

Anderson, E. (1957), A semigraphical method for the analysis of complex problems, *Proceedings of the National Academy of Sciences* **13**, 923–927. (Reprinted in *Technometrics*, 1960, **2**, 387–391.)

Antoniadis, A.; G. Gregoire; and I. W. McKeague (1994), Wavelet methods for curve estimation, *Journal of the American Statistical Association* **89**, 1340–1353.

Antoniadis, Anestis, and Georges Oppenheim (Editors) (1995), *Wavelets and Statistics*, Springer, New York.

Ash, Peter; Ethan Bolker; Henry Crapo; and Walter Whiteley (1988), Convex polyhedra, Dirichlet tessellations, and spider webs, *Shaping Space. A Polyhedral Approach* (edited by Marjorie Senechal and George Fleck), Birkhäuser, Boston, 231–250.

Aurenhammer, Franz (1991), Voronoi diagrams — a survey of a fundamental geometric data structure, *ACM Computing Surveys* **23**, 345–405.

Banks, David L. (1989), Bayesian sieving, *Proceedings of the Statistical Computing Section, ASA*, 271–276.

Barnard, G. A. (1963), Discussion of Bartlett, "The spectral analysis of point processes", *Journal of the Royal Statistical Society, Series B* **25**, 264–296.

Barndorff-Nielsen, O., and D. R. Cox (1979), Edgeworth and saddle-point approximations with statistical applications, *Journal of the Royal Statistical Society, Series B* **41**, 279–312.

Barnett, V. (1976), The ordering of multivariate data (with discussion), *Journal of the Royal Statistical Society, Series A* **139**, 318–352.

Bentley, Jon (2000), *Programming Pearls*, second edition, Addison-Wesley Publishing Company, Reading, Massachusetts.

Bentley, Jon Louis, and Jerome H. Friedman (1978), Fast algorithms for constructing minimal spanning trees in coordinate systems, *IEEE Transactions on Computers* **27**, 97–105.

Bentley, Jon Louis, and Jerome H. Friedman (1979), Data structures for range searching, *ACM Computing Surveys* **11**, 397–409.

Besag, J. E. (1974), Spatial interaction and the statistical analysis of lattice systems (with discussion), *Journal of the Royal Statistical Society, Series B* **36**, 192–236.

Besag, J., and P. Clifford (1989), Generalized Monte Carlo significance tests, *Biometrika* **76**, 633–642.

Besag, J., and P. Clifford (1991), Sequential Monte Carlo p-values, *Biometrika* **78**, 301–304.

Bickel, Peter, and Kjell A. Doksum (2001), *Mathematical Statistics: Basic Ideas and Selected Topics, Volume I* second edition, Prentice Hall, Upper Saddle River, NJ.

Billingsley, Patrick (1995), *Probability and Measure*, third edition, John Wiley & Sons, New York.

Birkes, David, and Yadolah Dodge (1993), *Alternative Methods of Regression*, John Wiley & Sons, New York.

Bowyer, A. (1981), Computing Dirichlet tessellations, *The Computer Journal* **24**, 162–166.

Box, G. E. P., and D. R. Cox (1964), An analysis of transformations (with discussion), *Journal of the Royal Statistical Society, Series B* **26**, 211–252.

Box, G. E. P., and P. W. Tidwell (1962), Transformation of the independent variables, *Technometrics* **4**, 531–550.

Boyens, Claus; Oliver Günther, and Hans-J. Lenz (2004), Statistical databases, *Handbook of Computational Statistics: Concepts and Methods* (edited by James E. Gentle, Wolfgang Härdle, and Yuichi Mori), Springer, Berlin, 267–292.

Breiman, Leo (1998), Arcing classifiers (with discussion), *Annals of Statistics* **26**, 801–849.

Breiman, Leo (2000), Randomizing outputs to increase prediction accuracy, *Machine Learning* **40**, 229–242.

Breiman, Leo (2001), Statistical modeling: The two cultures (with discussion), *Statistical Science* **16**, 199–231.

Breiman, Leo, and Jerome H. Friedman (1985a), Estimating optimal transformations for multiple regression and correlation (with discussion), *Journal of the American Statistical Association* **80**, 580–619.

Breiman, Leo, and Jerome H. Friedman (1985b), Estimating optimal transformations for multiple regression, *Computer Science and Statistics: Proceedings of the Sixteenth Symposium on the Interface* (edited by Lynne Billard), North Holland, Amsterdam, 121–134.

Breiman, L.; J. H. Friedman; R. A. Olshen; and C. J. Stone (1984), *Classification and Regression Trees*, Wadsworth Publishing Co., Monterey, California.

Brigo, Damiano, and Jan Liinev (2005), On the distributional distance between the lognormal LIBOR and swap market models, *Quantitative Finance* **5**, 433–442.

Bühlmann, Peter (2004), Bagging, boosting and ensemble methods, *Handbook of Computational Statistics: Concepts and Methods* (edited by James E. Gentle, Wolfgang Härdle, and Yuichi Mori), Springer, Berlin, 877–907.

Burr, I. W. (1942), Cumulative frequency functions, *Annals of Mathematical Statistics* **13**, 215–232.

Burr, Irving W., and Peter J. Cislak (1968), On a general system of distributions. I. Its curve-shape characteristics. II. The sample median, *Journal of the American Statistical Association* **63**, 627–635.

Cabrera, Javier, and Dianne Cook (1992), Projection pursuit indices based on fractal dimension, *Computing Science and Statistics* **24**, 474–477.

Calinski, R. B., and J. Harabasz (1974), A dendrite method for cluster analysis, *Communications in Statistics* **3**, 1–27.

Caroni, C. (2000), Outlier detection by robust principal components analysis, *Communications in Statistics — Simulation and Computation* **29**, 139–151.

Carroll, R. J., and D. Ruppert (1988), *Transformation and Weighting in Regression*, Chapman & Hall, New York.

Casella, George, and Roger L. Berger (2002), *Statistical Inference* second edition, Thomson Learning, Pacific Grove, CA.

Chambers, John M. (2008), *Software for Data Analysis. Programming with R*, Springer, New York.

Cheng, R. C. H., and L. Traylor (1995), Non-regular maximum likelihood problems (with discussion), *Journal of the Royal Statistical Society, Series B* **57**, 3–44.

Chernick, Michael R. (2008), *Bootstrap Methods: A Practitioner's Guide*, second edition, John Wiley & Sons, New York.

Chernoff, Herman (1973), The use of faces to represent points in k-dimensional space graphically, *Journal of the American Statistical Association* **68**, 361–368.

Chib, Siddhartha (2004), Markov chain Monte Carlo technology, *Handbook of Computational Statistics: Concepts and Methods* (edited by James E. Gentle, Wolfgang Härdle, and Yuichi Mori), Springer, Berlin, 71–102.

Chou, Youn-Min; S. Turner; S. Henson; D. Meyer; and K. S. Chen (1994), On using percentiles to fit data by a Johnson distribution, *Communications in Statistics — Simulation and Computation* **23**, 341–354.

Chui, Charles K. (1988), *Multivariate Splines*, Society for Industrial and Applied Mathematics, Philadelphia.
Čížková, Lenka, and Pavel Čížek (2004), Numerical linear algebra, in *Handbook of Computational Statistics: Concepts and Methods* (edited by James E. Gentle, Wolfgang Härdle, and Yuichi Mori), Springer, Berlin, 103–136.
Clarkson, Douglas B., and James E. Gentle (1986), Methods for multidimensional scaling, *Computer Science and Statistics: The Interface* (edited by D. M. Allen), North-Holland Publishing Company, Amsterdam, 185–192.
Comon, Pierre (1994), Independent component analysis, a new concept? *Signal Processing* **36**, 287–314.
Conn, Andrew R.; Katya Scheinberg; and Luis N. Vicente (2009), *Introduction to Derivative-Free Optimization*, Society for Industrial and Applied Mathematics, Philadelphia.
Conway, J. H., and N. J. A. Sloane (1982), Fast quantizing and decoding algorithms for lattice quantizer and codes, *IEEE Transactions on Information Theory* **28**, 227–231.
Conway, J. H., and N. J. A. Sloane (1999), *Sphere Packings, Lattices and Groups*, third edition, Springer, New York.
Cook, Dianne; Andreas Buja; and Javier Cabrera (1993), Projection pursuit indexes based on orthogonal function expansions, *Journal of Computational and Graphical Statistics* **2**, 225–250.
Cook, Dianne, and Deborah F. Swayne (2008), *Interactive and Dynamic Graphics for Data Analysis with GGobi and R*, Springer, New York.
Cook, R. D.; D. M. Hawkins; and S. Weisberg (1993), Exact iterative computation of the robust multivariate minimum volume ellipsoid estimator, *Statistics and Probability Letters* **16**, 213–218.
Cormen, Thomas H.; Charles E. Leiserson; Ronald L. Rivest; and Clifford Stein (2001), *Introduction to Algorithms*, second edition, McGraw-Hill, New York.
Dale, M. B. (1985), On the comparison of conceptual clustering and numerical taxonomy, *IEEE Transactions on Pattern Analysis and Machine Intelligence* **7**, 241–244.
David, H. A., and H. N. Nagaraja (2004), *Order Statistics*, third edition, John Wiley & Sons, New York.
Davison, A. C., and D. V. Hinkley (1997), *Bootstrap Methods and Their Application*, Cambridge University Press, Cambridge, United Kingdom.
Davison, A. C., and S. Sardy (2000), The partial scatterplot matrix, *Journal of Computational and Graphical Statistics* **9**, 750–758.
Dawkins, Brian P. (1995), Investigating the geometry of a p-dimensional data set, *Journal of the American Statistical Association* **90**, 350–359.
De Boor, Carl (2002), *A Practical Guide to Splines*, revised edition, Springer, New York.
De Jong, Kenneth A. (2006), *Evolutionary Computation: A Unified Approach*, MIT Press, Cambridge.

Dempster, A. P.; N. M. Laird; and D. B. Rubin (1977), Maximum likelihood estimation from incomplete data via the EM algorithm (with discussion), *Journal of the Royal Statistical Society, Series B* **45**, 1–37.

Dempster, A. P.; Nan M. Laird; and D. B. Rubin (1980), Iteratively reweighted least squares for linear regression when errors are normal/independent distributed, *Multivariate Analysis V* (edited by P. R. Krishnaiah), Elsevier Science Publishers, Amsterdam, 35–57.

Devroye, Luc (1986), *Non-Uniform Random Variate Generation*, Springer, New York.

Diaconis, Persi, and David Freedman (1984), Asymptotics of graphical projection pursuit, *The Annals of Statistics* **12**, 793–815.

Dielman, Terry E.; Cynthia Lowry; and Roger Pfaffenberger (1994), A comparison of quantile estimators, *Communications in Statistics — Simulation and Computation* **23**, 355–371.

Dongarra, Jack; Tom Rowan; and Reed Wade (1995), Software distribution using `Xnetlib`, *ACM Transactions on Mathematical Software* **21**, 79–88.

Du, Qiang; Vance Faber; and Max Gunzburger (1999), Centroidal Voronoi tessellations: Applications and algorithms, *SIAM Review* **41**, 637–676.

Dufour, Jean-Marie (2006), Monte Carlo tests with nuisance parameters: a general approach to finite-sample inference and nonstandard asymptotics, *Journal of Econometrics* **133**, 443–477.

Dufour, Jean-Marie, and Lynda Khalaf (2002), Simulation based finite and large sample tests in multivariate regressions, *Journal of Econometrics* **111**, 303–322.

Dunkl, Charles, and Yuan Xu (2001), *Orthogonal Polynomials of Several Variables*, Cambridge University Press, Cambridge, United Kingdom.

Eckart, Carl, and Gale Young (1936), The approximation of one matrix by another of lower rank, *Psychometrika* **1**, 211–218.

Eddy, William F. (1985), Ordering of multivariate data, *Computer Science and Statistics: Proceedings of the Sixteenth Symposium on the Interface* (edited by Lynne Billard), North Holland, Amsterdam, 25–30.

Edgington, Eugene S., and Patrick Onghena (2007), *Randomization Tests*, fourth edition, Chapman & Hall/CRC, Boca Raton.

Efron, Bradley (1979), Bootstrap methods: Another look at the jackknife, *Annals of Statistics* **7**, 1–26.

Efron, Bradley (1982), *The Jackknife, the Bootstrap and Other Resampling Methods*, Society for Industrial and Applied Mathematics, Philadelphia.

Efron, Bradley (1987), Better bootstrap confidence intervals (with discussion), *Journal of the American Statistical Association* **82**, 171–200.

Efron, Bradley (1992), Jackknife-after-bootstrap standard errors and influence functions, *Journal of the Royal Statistical Society, Series B* **54**, 83–127.

Efron, Bradley, and Robert J. Tibshirani (1993), *An Introduction to the Bootstrap*, Chapman & Hall, New York.

Efron, Bradley; Trevor Hastie; Iain Johnstone; Robert Tibshirani (2004), Least angle regression, *Annals of Statistics*, **32**, 407–499.

Efromovich, Sam (1999), *Nonparametric Curve Estimation*, Springer, New York.

Einarsson, Bo (Editor) (2005), *Accuracy and Reliability in Scientific Computing*, Society for Industrial and Applied Mathematics, Philadelphia.

Epanechnikov, V. A. (1969), Non-parametric estimation of a multivariate probability density, Theory of Probability and its Applications **14**, 153–158.

Evans, Merran; Nicholas Hastings; and Brian Peacock (2000), *Statistical Distributions*, third edition, John Wiley & Sons, New York.

Evans, Michael, and Tim Schwartz (2000), *Approximating Integrals via Monte Carlo and Deterministic Methods*, Oxford University Press, Oxford, United Kingdom.

Everitt, B. S., and D. J. Hand (1981), *Finite Mixture Distributions*, Chapman & Hall, New York.

Filliben, James J. (1975), The probability plot correlation coefficient test for normality, *Technometrics* **17**, 111–117.

Filliben, James J. (1982), DATAPLOT — An interactive, high-level language for graphics, non-linear fitting, data analysis, and mathematics, *Proceedings of the Statistical Computing Section, ASA*, 268–273.

Fisher, R. A. (1935), *The Design of Experiments*, Oliver and Boyd, Edinburgh.

Fishman, George S. (1999), An analysis of the Swendsen-Wang and related methods, *Journal of the Royal Statistical Society, Series B* **61**, 623–641.

Fishman, George S., and Louis R. Moore, III (1982), A statistical evaluation of multiplicative random number generators with modulus $2^{31}-1$, *Journal of the American Statistical Association* **77**, 129–136.

Fishman, George S., and Louis R. Moore, III (1986), An exhaustive analysis of multiplicative congruential random number generators with modulus $2^{31}-1$, *SIAM Journal on Scientific and Statistical Computing* **7**, 24–45.

Flournoy, Nancy, and Robert K. Tsutakawa (1991). *Statistical Multiple Integration*, American Mathematical Society, Providence, Rhode Island.

Flury, Bernhard, and Alice Zoppè (2000), Exercises in EM, *The American Statistician* **54**, 207–209.

Forster, Jonathan J.; John W. McDonald; and Peter W. F. Smith (1996), Monte Carlo exact conditional tests for log-linear and logistic models, *Journal of the Royal Statistical Society, Series B* **55**, 3–24.

Freedman, D., and P. Diaconis (1981a), On the histogram as a density estimator: L_2 theory, *Zeitschrift für Wahrscheinlichkeitstheorie und Verwandte Gebiete* **57**, 453–476.

Freedman, D., and P. Diaconis (1981b), On the maximum deviation between the histogram and the underlying density, *Zeitschrift für Wahrscheinlichkeitstheorie und Verwandte Gebiete* **58**, 139–168.

Friedman, Jerome H. (1987), Exploratory projection pursuit, *Journal of the American Statistical Association* **82**, 249–266.

Friedman, Jerome H.; Jon Louis Bentley; and Raphael Ari Finkel (1977), An algorithm for finding best matches in logarithmic expected time, *ACM Transactions on Mathematical Software* **3**, 209–226.

Friedman, Jerome H., and Lawrence C. Rafsky (1979a), Multivariate generalizations of the Wald-Wolfowitz and Smirnov two-sample tests, *Annals of Statistics* **7**, 697–717.

Friedman, J. H., and L. C. Rafsky (1979b), Fast algorithms for multivariate lining and planing, *Computer Science and Statistics: Proceedings of the 12th Symposium on the Interface* (edited by Jane F. Gentleman), University of Waterloo, Waterloo, Ontario, 124–129.

Frigessi, Arnoldo; Fabio Martinelli; and Julian Stander (1997), Computational complexity of Markov chain Monte Carlo methods for finite Markov random fields, *Biometrika* **84**, 1–18.

Fuller, Wayne A. (1987), *Measurement Error Models*, John Wiley & Sons, New York.

Fushimi, Masanori (1990), Random number generation with the recursion $X_t = X_{t-3p} \oplus X_{t-3q}$, *Journal of Computational and Applied Mathematics* **31**, 105–118.

Garey, Michael R., and David S. Johnson (1979), *Computers and Intractability: A Guide to the Theory of NP-Completeness*, W.H. Freeman, San Francisco.

Gelfand, Alan E., and Adrian F. M. Smith (1990), Sampling-based approaches to calculating marginal densities, *Journal of the American Statistical Association* **85**, 398–409. (Reprinted in Samuel Kotz and Norman L. Johnson (Editors) (1997), *Breakthroughs in Statistics, Volume III*, Springer-Verlag, New York, 526–550.)

Gelfand, Alan E., and Sujit K. Sahu (1994), On Markov chain Monte Carlo acceleration, *Journal of Computational and Graphical Statistics* **3**, 261–276.

Gelman, Andrew; John B. Carlin; Hal S. Stern; and Donald B. Rubin (2004), *Bayesian Data Analysis*, second edition, Chapman & Hall/CRC, Boca Raton.

Geman, Stuart, and Donald Geman (1984), Stochastic relaxation, Gibbs distributions, and the Bayesian restoration of images, *IEEE Transactions on Pattern Analysis and Machine Intelligence* **6**, 721–741.

Gentle, James E. (2003), *Random Number Generation and Monte Carlo Methods*, second edition, Springer, New York.

Gentle, James E. (2007), *Matrix Algebra: Theory, Computations, and Applications*, Springer, New York.

Gentle, James E. (2009), *Optimization Methods for Applications in Statistics*, Springer, New York.

Gentle, James E.; Wolfgang Härdle; and Yuichi Mori (2004) *Handbook of Computational Statistics*, Springer, Berlin.

Gentleman, Robert (2009), *R Programming for Bioinformatics*, Chapman & Hall/CRC, Boca Raton.

Gentleman, Robert, and Ross Ihaka (1997), The R language, *Computing Science and Statistics* **28**, 326–330.

Gil, Amparo; Javier Segura; and Nico M. Temme (2007), *Numerical Methods for Special Functions*, Society for Industrial and Applied Mathematics, Philadelphia.

Gilks, W. R.; A. Thomas; and D. J. Spiegelhalter (1992), Software for the Gibbs sampler, *Computing Science and Statistics* **24**, 439–448.

Girard, Stéphane (2000), A nonlinear PCA based on manifold approximation, *Computational Statistics* **15**, 145–167.

Glaeser, Georg, and Hellmuth Stachel (1999), *Open Geometry: OpenGL + Advanced Geometry*, Springer, New York.

Glymour, Clark; David Madigan; Daryl Pregibon; and Padhraic Smyth (1996), Statistical inference and data mining, *Communications of the ACM* **39**, Number 11 (November), 35–41.

Golub, Gene, and James M. Ortega (1993), *Scientific Computing. An Introduction with Parallel Computing*, Academic Press, San Diego.

Golub, G. H., and C. F. Van Loan (1980), An analysis of the total least squares problem, *SIAM Journal of Numerical Analysis* **17**, 883–893.

Golub, Gene H., and Charles F. Van Loan (1996), *Matrix Computations*, third edition, The Johns Hopkins Press, Baltimore.

Gordon, A. D. (1999), *Classification*, second edition, Chapman & Hall/CRC, Boca Raton.

Goutis, Constantino, and George Casella (1999), Explaining the saddlepoint approximation, *The American Statistician* **53**, 216–224.

Gray, A. J. (1994), Simulating posterior Gibbs distributions: A comparison of the Swendsen-Wang and Gibbs sampler methods, *Statistics and Computing* **4**, 189–201.

Gray, H. L., and W. R. Schucany (1972), *The Generalized Jackknife Statistic*, Marcel Dekker, Inc., New York.

Greenwood, J. Arthur (1976), Moments of time to generate random variables by rejection, *Annals of the Institute for Statistical Mathematics* **28**, 399–401.

Grenander, Ulf (1981), *Abstract Inference*, John Wiley & Sons, New York.

Griffiths, P., and I. D. Hill (Editors) (1985), *Applied Statistics Algorithms*, Ellis Horwood Limited, Chichester, United Kingdom.

Griva, Igor; Stephen G. Nash; and Ariela Sofer (2008), *Linear and Nonlinear Optimization*, second edition, Society for Industrial and Applied Mathematics, Philadelphia.

Hall, Peter (1989), On polynomial-based projection indexes for exploratory projection pursuit, *The Annals of Statistics* **17**, 589–605.

Hall, Peter (1990), Performance of balanced bootstrap resampling in distribution function and quantile problems, *Probability Theory and Related Fields* **85**, 239–260.

Hall, Peter, and D. M. Titterington (1989), The effect of simulation order on level accuracy and power of Monte Carlo tests, *Journal of the Royal Statistical Society, Series B* **51**, 459–467.

Hammersley, J. M., and D. C. Handscomb (1964), *Monte Carlo Methods*, Methuen & Co., London.

Harrell, Frank E. (2001), *Regression Modeling Strategies with Applications to Linear Models, Logistic Regression, and Survival Analysis*, Springer, New York.

Harrell, Frank E., and C. E. Davis (1982), A new distribution-free quantile estimator, *Biometrika* **69**, 635–640.

Hartigan, J. A., and B. Kleiner (1981), Mosaics for contingency tables, *Computer Science and Statistics: Proceedings of the 13th Symposium on the Interface* (edited by William F. Eddy), Springer, New York, 268–273.

Hastie, Trevor, and Werner Stuetzle (1989), Principal curves, *Journal of the American Statistical Association* **84**, 502–516.

Hastie, Trevor; Robert Tibshirani; and Jerome Friedman (2009), *The Elements of Statistical Learning. Data Mining, Inference, and Prediction*, second edition, Springer, New York.

Hastings, W. K. (1970), Monte Carlo sampling methods using Markov chains and their applications. *Biometrika* **57**, 97–109. (Reprinted in Samuel Kotz and Norman L. Johnson (Editors) (1997), *Breakthroughs in Statistics, Volume III*, Springer-Verlag, New York, 240–256.)

Hathaway, R. J., and J. C. Bezdek (1988), Recent convergence results for the fuzzy c-means clustering algorithms, *Journal of Classification* **5**, 237–247.

Hausman, Robert E., Jr. (1982), Constrained multivariate analysis, *Optimization in Statistics* (edited by S. H. Zanakis and J. S. Rustagi), North-Holland Publishing Company, Amsterdam, 137–151.

Hawkins, Douglas M. (1993a), A feasible solution algorithm for minimum volume ellipsoid estimator in multivariate data, *Computational Statistics* **8**, 95–107.

Hawkins, Douglas M. (1993b), The feasible set algorithm for least median of squares regression, *Computational Statistics & Data Analysis* **16**, 81–101.

Hawkins, Douglas M., and David Olive (1999), Applications and algorithms for least trimmed sum of absolute deviations regression, *Computational Statistics & Data Analysis* **32**, 119–134.

Hesterberg, Timothy C., and Barry L. Nelson (1998), Control variates for probability and quantile estimation, *Management Science* **44**, 1295–1312.

Hewitt, Edwin, and Karl Stromberg (1965), *Real and Abstract Analysis*, Springer-Verlag, Berlin. (A second edition was published in 1969.)

Hoaglin, David C., and David F. Andrews (1975), The reporting of computation-based results in statistics, *The American Statistician* **29**, 122–126.

Hoerl, A. E. (1962), Application of ridge analysis to regression problems, *Chemical Engineering Progress* **58**, 54–59.

Hoerl, Arthur E., and Robert W. Kennard (1970a), Ridge regression: Biased estimation for nonorthogonal problems, *Technometrics* **12**, 55–67.

Hoerl, Arthur E., and Robert W. Kennard (1970b), Ridge regression: Applications to nonorthogonal problems, *Technometrics* **12**, 68–82.

Hogg, Robert V.; Joseph W. McKean; and Allen T. Craig (2004), *Introduction to Mathematical Statistics* sixth edition, Prentice-Hall, Upper Saddle River, NJ.

Hope, A. C. A. (1968), A simplified Monte Carlo significance test procedure, *Journal of the Royal Statistical Society, Series B* **30**, 582–598.

Horowitz, Ellis; Sartaj Sahni; and Sanguthevar Rajasekaran (1998), *Computer Algorithms*, W. H. Freeman and Company, New York.

Hosmer, David W., and Stanley Lemeshow (2000), *Applied Logistic Regression*, second edition, John Wiley & Sons, New York.

Hu, Chenyi; R. Baker Kearfott; Andre de Korvin; and Vladik Kreinovich (Editors) (2008), *Knowledge Processing with Interval and Soft Computing*, Springer, New York.

Huber, Peter J. (1985), Projection pursuit (with discussion), *The Annals of Statistics* **13**, 435–525.

Huber, Peter J. (1994), Huge data sets, *Compstat 1994: Proceedings in Computational Statistics* (edited by R. Dutter and W. Grossmann), Physica-Verlag, Heidelberg, 3–27.

Huber, Peter J. (1996), Massive Data Sets Workshop: The morning after, *Massive Data Sets*, Committee on Applied and Theoretical Statistics, National Research Council, National Academy Press, Washington, 169–184.

Hubert, Lawrence, and Phipps Arabie (1985), Comparing partitions, *Journal of Classification* **2**, 193–218.

Hull, John C. (2008), *Options, Futures, & Other Derivatives*, seventh edition, Prentice-Hall, Englewood Cliffs, New Jersey.

Hunt, Earl B.; Janet Marin; and Philip J. Stone (1966), *Experiments in Induction*, Academic Press, New York.

Hyvärinen, Aapo; Juha Karhunen; and Erkki Oja (2001), *Independent Component Analysis*, John Wiley & Sons, New York.

Ingrassia, Salvatore (1992), A comparison between the simulated annealing and the EM algorithms in normal mixture decompositions, *Statistics and Computing* **2**, 203–211.

Jensen, Jens Ledet (1995), *Saddlepoint Approximations*, Oxford University Press, Oxford, United Kingdom.

Jolliffe, Ian T.; Nickolay T. Trendafilov; and Mudassir Uddin (2003), A modified principal component technique based on the LASSO, *Journal of Computational and Graphical Statistics* **12**, 531–547.

Jolliffe, Ian T., and Mudassir Uddin (2000), The simplified component technique: An alternative to rotated principal components, *Journal of Computational and Graphical Statistics* **9**, 689–710.

Jones, M. C., and Robin Sibson (1987), What is projection pursuit (with discussion), *Journal of the Royal Statistical Society, Series A* **150**, 1–36.

Kaigh, W. D., and Peter A. Lachenbruch (1982), A generalized quantile estimator, *Communications in Statistics — Theory and Methods* **11**, 2217–2238.

Kaufman, Leonard, and Peter J. Rousseeuw (1990), *Finding Groups in Data: An Introduction to Cluster Analysis*, John Wiley & Sons, New York.

Kendall, M. G. (1961), *A Course in the Geometry of n Dimensions*, Charles Griffin & Company Limited, London.

Kennedy, W. J., and T. A. Bancroft (1971), Model building for prediction in regression based upon repeated significance tests, *Annals of Mathematical Statistics* **42**, 1273–1284.

Kennedy, William J., and James E. Gentle (1980), *Statistical Computing*, Marcel Dekker, Inc., New York.

Kieffer, J. C. (1983), Uniqueness of locally optimal quantizer for log-concave density and convex error function, *IEEE Transactions in Information Theory* **29**, 42–47.

Kim, Choongrak; Woochul Kim; Byeong U.Park; Changkon Hong; and Meeseon Jeong (1999), Smoothing techniques via the Bézier curve, *Communications in Statistics — Simulation and Computation* **28**, 1577–1598.

Kleijnen, Jack P. C. (1977), Robustness of a multiple ranking procedure: A Monte Carlo experiment illustrating design and analysis techniques, *Communications in Statistics — Simulation and Computation* **B6**, 235–262.

Klinke, Sigbert (2004), Statistical user interfaces, *Handbook of Computational Statistics: Concepts and Methods* (edited by James E. Gentle, Wolfgang Härdle, and Yuichi Mori), Springer, Berlin, 379–402.

Kloeden, Peter E., and Eckhard Platen (1999), *Numerical Solution of Stochastic Differential Equations*,

Kloeden, Peter E., Eckhard Platen, and Henri Schurz (1994), *Numerical Solution of SDE Through Computer Experiments*, Springer, Berlin. (Reprinted, with corrections, 1997.)

Knuth, Donald E. (1973), *The Art of Computer Programming, Volume 3, Sorting and Searching*, Addison-Wesley Publishing Company, Reading, Massachusetts.

Kruskal, Joseph B., Jr. (1956), On the shortest spanning subtree of a graph and the traveling salesman problem, *Proceedings of the American Mathematical Society* **7**, 48–50.

Kulisch, Ulrich (2008), *Computer Arithmetic and Validity*, Walter de Gruyter, Berlin.

Kutner, Michael; Christopher Nachtsheim; and John Neter (2004), *Applied Linear Statistical Models*, fifth edition, McGraw-Hill/Irwin, New York.

Lance, G. N., and W. T. Williams (1966), Computer programs for hierarchical polythetic classification ('similarity analyses'), *Computer Journal* **9**, 60–64.

Lance, G. N., and W. T. Williams (1967a), A general theory of classificatory sorting strategies, *Computer Journal* **9**, 373–380.

Lance, G. N., and W. T. Williams (1967b), Mixed-data classificatory programs. I. Agglomerative systems, *Australian Computer Journal* **1**, 15–20.

Lance, G. N., and W. T. Williams (1968), Mixed-data classificatory programs. II. Divisive systems, *Australian Computer Journal* **1**, 82–85.

Laviolette, Michael; John W. Seaman, Jr.; J. Douglas Barrett; and William H. Woodall (1995), A probabilistic and statistical view of fuzzy methods (with discussion), *Technometrics* **37**, 249–292.

L'Ecuyer, Pierre (2004), Random number generation, *Handbook of Computational Statistics: Concepts and Methods* (edited by James E. Gentle, Wolfgang Härdle, and Yuichi Mori), Springer, Berlin, 35–70.

L'Ecuyer, Pierre, and Richard Simard (2007), TestU01: A C library for empirical testing of random number generators, *ACM Transactions on Mathematical Software* **33**, 1–40. Available at
`http://www.iro.umontreal.ca/~simardr/testu01/tu01.html` .

Lee, D. T. (1999a), Computational Geometry I, *Algorithms and Theory of Computation Handbook* (edited by Mikhail J. Atallah), CRC Press, Boca Raton, 19-1–19-29.

Lee, D. T. (1999b), Computational Geometry II, *Algorithms and Theory of Computation Handbook* (edited by Mikhail J. Atallah), CRC Press, Boca Raton, 20-1–20-31.

Leemis, Lawrence M., and Jacquelyn T. McQueston (2008), Univariate distribution relationships, *The American Statistician* **62**, 45–53.

Lehmann, Erich L. (1975), *Nonparametrics: Statistical Methods Based on Ranks*, Holden-Day, Inc., San Francisco. (Reprinted by Springer, New York, 2006).

Lehmann, E. L., and George Casella (1998), *Theory of Point Estimation*, second edition, Springer, New York.

Lehmann, E. L., and Joseph P. Romano (2005), *Testing Statistical Hypotheses*, third edition, Springer, New York.

Lemmon, David R., and Joseph L. Schafer (2005), *Developing Statistical Software in Fortran 95*, Springer, New York.

LePage, Raoul, and Lynne Billard (Editors) (1992), *Exploring the Limits of Bootstrap*, John Wiley & Sons, New York.

Lewis, P. A. W.; A. S. Goodman; and J. M. Miller (1969), A pseudo-random number generator for the System/360, *IBM Systems Journal* **8**, 136–146.

Liem, C. B.; T. Lü; and T. M. Shih (1995), *The Splitting Extrapolation Method. A New Technique in Numerical Solution of Multidimensional Problems*, World Scientific, Singapore. (Also listed under Tao, Lü, 1995, with coauthors C. B. Liem and T. M. Shih.)

Liu, Jun S., and Chiara Sabatti (1999), Simulated sintering: Markov chain Monte Carlo with spaces of varying dimensions, *Bayesian Statistics 6* (edited by J. M. Bernardo, J. O. Berger, A. P. Dawid, and A. F. M. Smith), Oxford University Press, Oxford, United Kingdom, 389–413.

Liu, Regina Y. (1990), On a notion of data depth based on random simplices, *Annals of Statistics* **18**, 405–414.

Liu, Regina Y.; Jesse M. Parelius; and Kesar Singh (1999), Multivariate analysis by data depth: Descriptive statistics, graphics and inference (with discussion), *Annals of Statistics* **27**, 783–840.

Lloyd, S. (1982), Least square quantization in PCM, *IEEE Transactions in Information Theory* **28**, 129–137.

Loader, Catherine (2004), Smoothing: Local regression techniques, in *Handbook of Computational Statistics: Concepts and Methods* (edited by James E. Gentle, Wolfgang Härdle, and Yuichi Mori), Springer, Berlin, 539–563.

Manly, Bryan F.J. (2006), *Randomization, Bootstrap and Monte Carlo Methods in Biology*, third edition, Chapman & Hall/CRC, Boca Raton.

Marin, Jean-Michel, and Christian P. Robert (2007), *Bayesian Core. A Practical Guide to Computational Bayesian Statistics*, Springer, New York.

Maronna, Ricardo A., and Victor J. Yohai (1995), The behavior of the Stahel-Donoho robust multivariate estimator, *Journal of the American Statistical Association* **90**, 330–341.

Marriott, F. H. C. (1979), Barnard's Monte Carlo tests: How many simulations?, *Applied Statistics* **28**, 75–78.

Marron, J. S., and M. P. Wand (1992), Exact mean integrated squared error, *Annals of Statistics* **20**, 343–353.

Matsumoto, Makoto, and Takuji Nishimura (1998), Mersenne twister: A 623-dimensionally equidistributed uniform pseudo-random generator, *ACM Transactions on Modeling and Computer Simulation* **8**, 3–30.

McConnell, Steve (2004), *Code Complete: A Practical Handbook of Software Construction*, Microsoft Press, Redmond.

McCullagh, P., and J. A. Nelder (1990), *Generalized Linear Models*, second edition, Chapman & Hall/CRC, Boca Raton.

McCullough, B. D. (1999), Assessing the reliability of statistical software: Part II, *The American Statistician* **53**, 149–159.

McKay, Michael D.; William J. Conover; and Richard J. Beckman (1979), A comparison of three methods for selecting values of input variables in the analysis of output from a computer code, *Technometrics* **21**, 239–245.

McLachlan, Geoffrey J., and Thriyambakam Krishnan (1997), *The EM Algorithm and Extensions*, John Wiley & Sons, New York.

Mead, R. (1974), A test for spatial pattern at several scales using data from a trio of contiguous quadrats, *Biometrics* **30**, 295–307.

Metropolis, N.; A. W. Rosenbluth; M. N. Rosenbluth; A. H. Teller; and E. Teller (1953), Equations of state calculation by fast computing machines, *Journal of Chemical Physics* **21**, 1087–1092. (Reprinted in Samuel Kotz and Norman L. Johnson (Editors) (1997), *Breakthroughs in Statistics, Volume III*, Springer-Verlag, New York, 127–139.)

Michalski, R. S. (1980), Knowledge acquisition through conceptual clustering: A theoretical framework and an algorithm for partitioning data into conjunctive concepts, *Journal of Policy Analysis and Information Systems* **4**, 219–244.

Michalski, R. S., and R. E. Stepp (1983), Automated construction of classifications: Conceptual clustering versus numerical taxonomy, *IEEE Transactions on Pattern Analysis and Machine Intelligence* **5**, 396–409.

Mika, Sebastian; Christin Schäfer; Pavel Laskov; David Tax; and Klaus-Robert Müller (2004), Support vector machines, *Handbook of Computational Statistics: Concepts and Methods* (edited by James E. Gentle, Wolfgang Härdle, and Yuichi Mori), Springer, Berlin, 841–876.

Miller, Alan (2002), *Subset Selection in Regression*, second edition, Chapman and Hall/CRC, Boca Raton.

Mizuta, Masahiro (2004), Dimension reduction methods, in *Handbook of Computational Statistics: Concepts and Methods* (edited by James E. Gentle, Wolfgang Härdle, and Yuichi Mori), Springer, Berlin, 565–589.

Møller, Jesper, and Rasmus Plenge Wasgepetersen (2004), *Statistical Inference and Simulation for Spatial Point Processes*, Chapman & Hall/CRC, Boca Raton.

Moore, Ramon E., R. Baker Kearfott, and Michael J. Cloud (2009), *Introduction to Interval Analysis*, Society for Industrial and Applied Mathematics, Philadelphia.

Morton, Sally C. (1992), Interpretable projection pursuit, *Computer Science and Statistics: Proceedings of the Twenty-second Symposium on the Interface* (edited by Connie Page and Raoul LePage), Springer, New York, 470–474.

Murrell, Paul (2006), *R Graphics*, Chapman & Hall/CRC, Boca Raton.

Murtagh, F. (1984), A review of fast techniques for nearest neighbour searching, *Compstat 1984: Proceedings in Computational Statistics* (edited by T. Havránek, Z. Šidák, and M. Novák), Physica-Verlag, Vienna, 143–147.

Nakano, Junji (2004), Parallel computing techniques, *Handbook of Computational Statistics: Concepts and Methods* (edited by James E. Gentle, Wolfgang Härdle, and Yuichi Mori), Springer, Berlin, 237–266.

Nason, Guy P. (2001), Robust projection indices, *Journal of the Royal Statistical Society, Series B* **63**, 551–567.

Nelder, J. A., and R. Mead (1965), A simplex method for function minimization, *Computer Journal* **7**, 308–313.

Nelder, J. A., and R. W. M. Wedderburn (1972), Generalized linear models, *Journal of the Royal Statistical Society, Series A* **135**, 370–384.

Nelsen, Roger B. (2007), *An Introduction to Copulas*, second edition Springer, New York.

Newman, M. E. J., and G. T. Barkema (1999) *Monte Carlo Methods in Statistical Physics*, Oxford University Press, Oxford, United Kingdom.

Ng, Shu Kay; Thriyambakam Krishnan; and Geoffrey J. McLachlan (2004), The EM algorithm, *Handbook of Computational Statistics: Concepts and Methods* (edited by James E. Gentle, Wolfgang Härdle, and Yuichi Mori), Springer, Berlin, 137–168.

Nicole, Sandro (2000), Feedforward neural networks for principal components extraction, *Computational Statistics & Data Analysis* **33**, 425–437.

NIST (2000), *A Statistical Test Suite for Random and Pseudorandom Number Generators for Cryptographic Applications*, NIST Special Publication 800-22, National Institute for Standards and Technology, Gaithersburg, Maryland.

Nummelin, Esa (1984), *General Irreducible Markov Chains and Non-Negative Operators*, Cambridge University Press, Cambridge, United Kingdom.

Okabe, Atsuyuki; Barry Boots; Ksokichi Sugihara; and Sung Nok Chui (2000), *Spatial Tessellations: Concepts & Applications of Voronoi Diagrams*, second edition, John Wiley & Sons, New York.

Øksendal, Bernt (2003), *Stochastic Differential Equations. An Introduction with Applications*, sixth edition, Springer-Verlag, Berlin.

OpenGL Architecture Review Board (1992), *OpenGL Reference Manual*, Addison-Wesley Publishing Company, Reading, Massachusetts.

O'Rourke, Joseph (1998) *Computational Geometry in C*, second edition, Cambridge University Press, Cambridge, United Kingdom.

Overton, Michael L. (2001), *Numerical Computing with IEEE Floating Point Arithmetic*, Society for Industrial and Applied Mathematics, Philadelphia.

Pardo, Leandro (2005), *Statistical Inference Based on Divergence Measures*, Chapman & Hall/CRC, Boca Raton.

Park, Stephen K., and Keith W. Miller (1988), Random number generators: Good ones are hard to find, *Communications of the ACM* **31**, 1192–1201.

Picard, Richard R., and Kenneth N. Berk (1990), Data splitting, *The American Statistician* **44**, 140–147.

Politis, Dimitris N., and Joseph P. Romano (1992), A circular block-resampling procedure for stationary data *Exploring the Limits of the Bootstrap* (edited by Raoul LePage and Lynne Billard), John Wiley & Sons, New York, 263–270.

Politis, Dimitris N., and Joseph P. Romano (1994), The stationary bootstrap, *Journal of the American Statistical Association* **89**, 1303–1313.

Posse, C. (1990), An effective two-dimensional projection pursuit algorithm, *Communications in Statistics — Simulation and Computation* **19** 1143–1164.

Posse, Christian (1995a), Tools for two-dimensional exploratory projection pursuit, *Journal of Computational and Graphical Statistics* **4**, 83–100.

Posse, Christian (1995b), Projection pursuit exploratory data analysis, *Computational Statistics & Data Analysis* **20**, 669–687.

Quenouille, M. H. (1949), Approximate tests of correlation in time series, *Journal of the Royal Statistical Society, Series B* **11**, 18–84.

Quenouille, M. H. (1956), Notes on bias in estimation, *Biometrika* **43**, 353–360.

Ramberg, John S., and Bruce W. Schmeiser (1974), An approximate method for generating asymmetric random variables, *Communications of the ACM* **17**, 78–82.

Ramsay, J. O., and B. W. Silverman (2002), *Applied Functional Data Analysis: Methods and Case Studies*, Springer, New York.

Ramsay, J. O., and B. W. Silverman (2005), *Functional Data Analysis*, second edition Springer, New York.

Rand, William M. (1971), Objective criteria for the evaluation of clustering methods, *Journal of the American Statistical Association* **66**, 846–850.

Rank, Jörn, and Thomas Siegl (2002), Applications of copulas for calculation of value-at-risk, *Applied Quantitative Finance* (edited by W. Härdle, T. Kleinow, and G. Stahl), Springer, Berlin, 36–50.

Rao, J. N. K., and J. T. Webster (1966), On two methods of bias reduction in the estimation of ratios, *Biometrika* **53**, 571–577.

Renka, Robert J. (1997), Algorithm 772: STRIPACK: Delaunay triangulation and Voronoi diagram on the surface of a sphere, *ACM Transactions on Mathematical Software* **23**, 416–434.

Rice, John R. (1993), *Numerical Methods, Software, and Analysis*, second edition, McGraw-Hill Book Company, New York.

Richeldi, M., and M. Rossotto (1997), Combining statistical techniques and search heuristics to perform effective feature selection, *Machine Learning and Statistics: The Interface* (edited by G. Nakhaeizadeh and C. C. Taylor), John Wiley & Sons, New York.

Ripley, Brian D. (1987), *Stochastic Simulation*, John Wiley & Sons, New York.

Ripley, Brian D. (1993), Statistical aspects of neural networks, *Networks and Chaos — Statistical and Probabilistic Aspects* (edited by O. E. Barndorff-Nielsen, J. L. Jensen, and W. S. Kendall), Chapman & Hall, London, 40–123.

Ripley, Brian D. (1994), Neural networks and related methods for classification (with discussion), *Journal of the Royal Statistical Society, Series B* **56**, 409–456.

Ripley, B. D. (1996), *Pattern Recognition and Neural Networks*, Cambridge University Press, Cambridge, United Kingdom.

Rizzo, Maria L. (2007), *Statistical Computing with R*, Chapman & Hall/CRC, Boca Raton.

Robert, Christian P. (1998), A pathological MCMC algorithm and its use as a benchmark for convergence assessment techniques, *Computational Statistics* **13**, 169–184.

Roeder, Kathryn, and Larry Wasserman (1997), Practical Bayesian density estimation using mixtures of normals, *Journal of the American Statistical Association* **92**, 894–902.

Rosenblatt, M. (1956), Remarks on some nonparametric estimates of a density function, *Annals of Mathematical Statistics* **27**, 832–835.

Rousseeuw, P. J. (1984), Least median of squares regression, *Journal of the American Statistical Association* **79**, 871–880. (Reprinted in Samuel Kotz and Norman L. Johnson (Editors) (1997), *Breakthroughs in Statistics, Volume III*, Springer-Verlag, New York, 440–461.)

Rousseeuw, Peter J. (1995), Discussion: Fuzzy clustering at the intersection, *Technometrics* **37**, 283–286.

Rousseeuw, Peter J., and Mia Hubert (1999), Regression depth (with discussion), *Journal of the American Statistical Association* **94**, 388–433.

Rousseeuw, Peter J., and Ida Ruts (1996), Bivariate location depth, *Applied Statistics* **45**, 516–526.

Rousseeuw, Peter J., and Ida Ruts (1998), Constructing the bivariate Tukey median, *Statistica Sinica* **8**, 827–839.

Rubin, Donald B. (1987), *Multiple Imputation for Nonresponse in Surveys*, John Wiley & Sons, New York.

Rustagi, Jagdish S. (1994), *Optimization Techniques in Statistics*, Academic Press, Boston.

Ruts, Ida, and Peter J. Rousseeuw (1996), Computing depth contours of bivariate point clouds, *Computational Statistics and Data Analysis* **23**, 153–168.

Saad, Yousef (2003), *Iterative Methods for Sparse Linear Systems*, second edition Society for Industrial and Applied Mathematics, Philadelphia.

Sarkar, Deepayan (2008), *Lattice. Multivariate Data Visualization with R*, Springer, New York.

Sargent, Daniel J.; James S. Hodges; and Bradley P. Carlin (2000), Structured Markov chain Monte Carlo, *Journal of Computational and Graphical Statistics* **9**, 217–234.

Schafer, J. L. (1997), *Analysis of Incomplete Multivariate Data*, Chapman & Hall, London.

Schenker, Nathaniel (1985), Qualms about bootstrap confidence intervals, *Journal of the American Statistical Association* **80**, 360–361.

Schroeder, Will; Ken Martin; and Bill Lorensen (2004), *The Visualization Toolkit, An Object-Oriented Approach to 3D Graphics*, third edition, Kitware, Clifton Park, New York.

Schucany, W. R.; H. L. Gray; and D. B. Owen (1971), On bias reduction in estimation, *Journal of the American Statistical Association* **66**, 524–533.

Scott, David W. (2004), Multivariate density estimation and visualization, *Handbook of Computational Statistics: Concepts and Methods* (edited by James E. Gentle, Wolfgang Härdle, and Yuichi Mori), Springer, Berlin, 35–70.

Seaver, Bill; Konstantinos Triantis; and Chip Reeves (1999), The identification of influential subsets in regression using a fuzzy clustering strategy, *Technometrics* **41**, 340–351.

Sebe, Nicu; Michael S. Lew; and Dionysius P. Huijsmans (2000), Toward improving ranking metrics, *IEEE Transactions on Pattern Analysis and Machine Intelligence* **22**, 1132–1141.

Seber, G. A. F., and C. J. Wild (2003), *Nonlinear Regression*, John Wiley & Sons, New York.

Senchaudhuri, Pralay; Cyrus R. Mehta; and Nitin R. Patel (1995), Estimating exact p values by the method of control variates or Monte Carlo rescue, *Journal of the American Statistical Association* **90**, 640–648.

Shao, Jun, and Dongsheng Tu (1995), *The Jackknife and Bootstrap*, Springer, New York.

Sharot, Trevor (1976), The generalized jackknife: Finite samples and subsample sizes, *Journal of the American Statistical Association* **71**, 451–454.

Sibuya, M. (1961), Exponential and other variable generators, *Annals of the Institute for Statistical Mathematics* **13**, 231–237.

Silverman, B. W. (1982), Kernel density estimation using the fast Fourier transform, *Applied Statistics* **31**, 93–97.

Slifker, James F., and Samuel S. Shapiro (1980), The Johnson system: Selection and parameter estimation, *Technometrics* **22**, 239–246.

Solka, Jeffrey L.; Wendy L. Poston; and Edward J. Wegman (1995), A visualization technique for studying the iterative estimation of mixture densities, *Journal of Computational and Graphical Statistics* **4**, 180–198.

Solka, Jeffrey L.; Edward J. Wegman; Carey E. Priebe; Wendy L. Poston; and George W. Rogers (1998), Mixture structure analysis using the Akaike information criterion and the bootstrap, *Statistics and Computing* **8**, 177–188.

Spall, James C. (2004), Stochastic simulation, *Handbook of Computational Statistics: Concepts and Methods* (edited by James E. Gentle, Wolfgang Härdle, and Yuichi Mori), Springer, Berlin, 189–198.

Spanier, Jerome, and Keith B. Oldham (1987), *An Atlas of Functions*, Hemisphere Publishing Corporation, Washington (also Springer-Verlag, Berlin).

Spector, Phil (2008), *Data Manipulation with R*, Springer, New York.

Speed, T. P., and Bin Yu (1993), Model selection and prediction: Normal regression, *Annals of the Institute of Statistical Mathematics* **45**, 35–54.

Stevens, S. S. (1946), On the theory of scales of measurement, *Science* **103** 677–680.

Stewart, Ian (2001), *Flatterland: Like Flatland Only More So*, Perseus Books Group, Boulder, Colorado.

Stigler, Stephen M. (1978), Mathematical statistics in the early states, *Annals of Statistics* **6**, 239–265.

Stigler, Stephen M. (1991), Stochastic simulation in the nineteenth century, *Statistical Science* **6**, 89–97.

Stone, Charles J. (1984), An asymptotically optimal window selection rule for kernel density estimates, *Annals of Statistics* **12**, 1285–1297.

Struyf, Anja, and Peter J. Rousseeuw (2000), High-dimensional computation of the deepest location, *Computational Statistics & Data Analysis* **34**, 415–426.

Student (1908a), On the probable error of a mean, *Biometrika* **6**, 1–25.

Student (1908b), Probable error of a correlation coefficient, *Biometrika* **6**, 302–310.

Sullivan, Thomas J. (2002), *Classification Methods for Augmented Arc-Weighted Graphs*, unpublished Ph.D. dissertation, George Mason University, Fairfax, Virginia.

Sun, Jiayang (1992), Some computational aspects of projection pursuit, *Computer Science and Statistics: Proceedings of the 22nd Symposium on the Interface* (edited by Connie Page and Raoul LePage), Springer, New York, 539–543.

Sun, Jiayang (1993), Some practical aspects of exploratory projection pursuit, *SIAM Journal on Scientific and Statistical Computing* **14**, 68–80.

Swendsen, R. H., and J.-S. Wang (1987), Nonuniversal critical dynamics in Monte Carlo simulations, *Physical Review Letters* **58**, 86–88.

Tarter, Michael E.; William Freeman; and Alan Hopkins (1986), A Fortran implementation of univariate Fourier series density estimation, *Communications in Statistics — Simulation and Computation* **15**, 855–870.

Taylor, Malcolm S., and James R. Thompson (1986), Data based random number generation for a multivariate distribution via stochastic simulation, *Computational Statistics & Data Analysis* **4**, 93–101.

Thomas, William (1991), Influence on the cross-validated smoothing parameter in spline smoothing, *Computer Science and Statistics: Proceedings of the Twenty-third Symposium on the Interface* (edited by Elaine M. Keramidas), Interface Foundation of North America, Fairfax, Virginia, 192–195.

Thompson, William J. (1997), *Atlas for Computing Mathematical Functions: An Illustrated Guide for Practitioners with Programs in C and Mathematica*, John Wiley & Sons, New York.

Tibshirani, R. (1988), Estimating transformations for regression via additivity and variance stabilization, *Journal of the American Statistical Association* **83**, 395–405.

Tibshirani, R. (1996), Regression shrinkage and selection via the lasso, *Journal of the Royal Statistical Society, Series B* **58**, 267–288.

Tierney, Luke (1994), Markov chains for exploring posterior distributions (with discussion), *Annals of Statistics* **22**, 1701–1762.

Tierney, Luke (1996), Introduction to general state-space Markov chain theory, *Practical Markov Chain Monte Carlo* (edited by W. R. Gilks, S. Richardson, and D. J. Spiegelhalter), Chapman & Hall, London, 59–74.

Vapnik, Vladimir N. (1999a), *The Nature of Statistical Learning Theory*, second edition, Springer, New York.

Vapnik, Vladimir (1999b), Three remarks on the support vector method of function estimation, *Advances in Kernel Methods. Support Vector Learning* (edited by Bernhard Schölkopf, Christopher J. C. Burges, and Alexander J. Smola), The MIT Press, Cambridge, Massachusetts, 25–41.

Velilla, Santiago (1995), Diagnostics and robust estimation in multivariate data transformations, *Journal of the American Statistical Association* **90**, 945–951.

Vidakovic, Brani (2004), Transforms in statistics, *Handbook of Computational Statistics: Concepts and Methods* (edited by James E. Gentle, Wolfgang Härdle, and Yuichi Mori), Springer, Berlin, 200–236.

Vines, S. K. (2000), Simple principal components, *Applied Statistics* **49**, 441–451.

Virius, Miroslav (2004), Object oriented computing, *Handbook of Computational Statistics: Concepts and Methods* (edited by James E. Gentle, Wolfgang Härdle, and Yuichi Mori), Springer, Berlin, 403–434.

Walster, G. William (2005), The use and implementation of interval data types, *Accuracy and Reliability in Scientific Computing* (edited by Bo Einarsson), Society for Industrial and Applied Mathematics, Philadelphia, 173–194.

Walter, Gilbert G., and Jugal K. Ghorai (1992), Advantages and disadvantages of density estimation with wavelets, *Computing Science and Statistics* **24**, 234–243.

Ward, Joe H., Jr. (1963), Hierarchical grouping to optimize an objective function, *Journal of the American Statistical Association* **58**, 236–244.

Watson, D. F. (1981), Computing the n-dimensional Delaunay tessellation with application to Voronoi polytopes, *The Computer Journal* **24**, 167–171.

Wegman, Edward J. (1990), Hyperdimensional data analysis using parallel coordinates, *Journal of the American Statistical Association* **85**, 664–675.

Wegman, Edward J. (1995), Huge data sets and the frontiers of computational feasibility, *Journal of Computational and Graphical Statistics* **4**, 281–295.

Wegman, Edward J., and Ji Shen (1993), Three-dimensional Andrews plots and the grand tour, *Computing Science and Statistics* **25**, 284–288.

Westphal, M., and G. Nakhaeizadeh (1997), Combination of statistical and other learning methods to predict financial time series, *Machine Learning and Statistics: The Interface* (edited by G. Nakhaeizadeh and C. C. Taylor), John Wiley & Sons, New York.

Wilkinson, J. H. (1959), The evaluation of the zeros of ill-conditioned polynomials, *Numerische Mathematik* **1**, 150–180.

Wilkinson, Leland (2004), The grammar of graphics, *Handbook of Computational Statistics: Concepts and Methods* (edited by James E. Gentle, Wolfgang Härdle, and Yuichi Mori), Springer, Berlin, 337–378.

Woodruff, David L., and David M. Rocke (1993), Heuristic search algorithms for the minimum volume ellipsoid, *Journal of Computational and Graphical Statistics* **2**, 69–95.

Xu, Chong-Wei, and Wei-Kei Shiue (1993), Parallel algorithms for least median of squares regression, *Computational Statistics & Data Analysis* **16**, 349–362.

Zhang, Heping (2004), Recursive partitioning and tree-based methods, *Handbook of Computational Statistics: Concepts and Methods* (edited by James E. Gentle, Wolfgang Härdle, and Yuichi Mori), Springer, Berlin, 813–840.

Zhu, Lixing (2005), *Nonparametric Monte Carlo Tests and Their Applications*, Springer, New York.

Ziff, Robert M. (2006), Generalized cell–dual-cell transformation and exact thresholds for percolation, *Physical Review E* **73**, 016134–016136.

Zuo, Yijun, and Robert Serfling (2000a), General notions of statistical depth function, *Annals of Statistics* **28**, 461–482.

Zuo, Yijun, and Robert Serfling (2000b), Structural properties and convergence results for contours of sample statistical depth functions, *Annals of Statistics* **28**, 483–499.

Index